Understanding the Network

A Practical Guide to Internetworking

New Riders Professional Library

Inside Windows 2000 Server
William Boswell
ISBN: 1-56205-929-7

Planning for Windows 2000
Eric Cone, Jon Boggs, and Sergio Perez
ISBN: 0-7357-0048-6

Windows NT DNS
Michael Masterson, Herman Knief, Scott Vinick, and Eric Roul
ISBN: 1-56205-943-2

Windows NT Network Management: Reducing Total Cost of Ownership
Anil Desai
ISBN: 1-56205-946-7

Windows NT Performance: Monitoring, Benchmarking, and Tuning
Mark Edmead and Paul Hinsberg
ISBN: 1-56205-942-4

Windows NT Registry: A Settings Reference
Sandra Osborne
ISBN: 1-56205-941-6

Windows NT TCP/IP
Karanjit Siyan
ISBN: 1-56205-887-8

Understanding Directory Services
Beth Sheresh and Doug Sheresh
ISBN: 0-7357-0910-6

Windows NT Terminal Server and Citrix MetaFrame
Ted Harwood
ISBN: 1-56205-944-0

Windows NT Power Toolkit
Stu Sjouwerman and Ed Tittel
ISBN: 0-7357-0922-X

Cisco Router Configuration and Troubleshooting
Mark Tripod
ISBN: 0-7357-0024-9

Exchange System Administration
Janice Rice Howd
ISBN: 0-7357-0081-8

Implementing Exchange Server
Doug Hauger, Marywynne Leon, and William C. Wade III
ISBN: 1-56205-931-9

Network Intrusion Detection: An Analyst's Handbook
Stephen Northcutt
ISBN: 0-7357-0868-1

Understanding Data Communications, Sixth Edition
Gilbert Held
ISBN: 0-7357-0036-2

Internet Information Services Administration
Kelli Adam
ISBN: 0-7357-0022-2

Understanding the Network

A Practical Guide to Internetworking

New Riders
201 West 103rd Street
Indianapolis, IN 46290

Michael J. Martin

Understanding the Network

A Practical Guide to Internetworking

Michael J. Martin

International Standard Book Number: 0-7357-0977-7

Library of Congress Catalog Card Number: 99-63274

03 02 01 00 7 6 5 4 3 2 1

Interpretation of the printing code: The rightmost double-digit number is the year of the book's printing; the rightmost single-digit number is the number of the book's printing. For example, the printing code 00-1 shows that the first printing of the book occurred in 2000.

Composed in SABon and MCPdigital by New Riders Publishing

Printed in the United States of America

Trademarks

All terms mentioned in this book that are known to be trademarks or service marks have been appropriately capitalized. New Riders Publishing cannot attest to the accuracy of this information. Use of a term in this book should not be regarded as affecting the validity of any trademark or service mark.

Warning and Disclaimer

This book is designed to provide information about various internetworking technologies. Every effort has been made to make this book as complete and as accurate as possible, but no warranty or fitness is implied.

The information is provided on an as-is basis. The author and New Riders Publishing shall have neither liability nor responsibility to any person or entity with respect to any loss or damages arising from the information contained in this book or from the use of the discs or programs that may accompany it.

Publisher
David Dwyer

Associate Publisher
Brad Koch

Executive Editor
Al Valvano

Managing Editor
Gina Brown

Product Marketing Manager
Stephanie Layton

Acquisitions Editor
Karen Wachs

Development Editors
Ami Frank Sullivan
Lisa M. Thibault

Project Editors
Sheri Replin
Laura Loveall

Copy Editor
Lunaea Hougland

Indexer
Lisa Stumpf

Manufacturing Coordinator
Chris Moos

Book Designer
Louisa Klucznik

Cover Designer
Aren Howell

Composition
Scan Communications Group, Inc.
Amy Parker

Contents

Introduction

1 Understanding Networking Concepts 1

Computer Network Types 2

Transmission Media 2

Network Topologies 8

Network Transmission Protocols 10

Network Communication Protocols 16

Network Hardware 17

Data Communication Reference Models 23

Summary 34

2 The Networker's Guide to TCP/IP 35

The Origins of TCP/IP 36

TCP/IP and UNIX 37

Layer 3: IP Protocol 38

Layer 4: TCP and UDP 72

UDP 77

The Application Layer Protocols of the TCP/IP Suite 78

RFCs 82

Additional Resources 83

3 The Networker's Guide to AppleTalk, IPX, and NetBIOS 85

AppleTalk 86

IPX and NetBIOS 107

Summary 121

Related RFCs 121

Additional Resources 121

4 LAN Internetworking Technologies 123

IEEE 802 LAN Standards and Logical Link Control 124

LAN Physical Layer Signaling, Encoding, and Transmission Media 128

Ethernet 137

Token Ring 169

FDDI 183

Summary 190
Related RFCs 190
Additional Resources 190

5 WAN Internetworking Technologies 191
A Brief History of the PSTN 192
Digital Carrier Systems 198
ISDN 219
PSTN Packet- and Cell-Switched Networks 229
Data-Link Framing 245
Summary 250
Related RFCs 251
Additional Resources 251

6 Network Switches 253
The Need for Network Switching 253
Switching Fundamentals 254
The Types of Bridges 255
Bridging Functionality 257
Layer 2 LAN Switching Fundamentals 268
Layer 2 LAN Switching Services for Layer 3 282
ATM Switches 292
Summary 303
Related RFCs 304
Additional Resources 304

7 Introduction to Cisco Routers 305
Cisco Router Hardware 306
Memory on Cisco Routers 308
Talking to Your Cisco Router (Through the Console) 309
Cisco IOS 314
Configuring Your Router with <*copy*> and TFTP 326
Basic Cisco Router Configuration 331
Disaster Recovery 346
Setting the Bootstrap Behavior 346
Upgrading Your Router's IOS 350
Configuring the Router's Clock 356

IOS Message Logging 359
Setting Up Buffered Logging 360
Setting Up Trap Logging 361
IOS Authentication and Accounting 365
Summary 376
Related RFCs 377
Additional Resource 377

8 TCP/IP Dynamic Routing Protocols 379
An Introduction to General Routing Concepts and Terms 380
TCP/IP Static Routing 393
TCP/IP Interior Gateway Protocols 396
TCP/IP Exterior Gateway Protocols 414
Summary 419
Related RFCs 419
Additional Resources 419

9 Advanced Cisco Router Configuration 421
Access Control Lists 421
Policy Routing 439
Gateway Redundancy 441
Network Address Translation 445
Cisco Tunneling 450
Cisco Router Wide Area Interface Configuration 454
Summary 479
Related RFCs 480
Additional Resources 481

10 Configuring IP Routing Protocols on Cisco Routers 483
Choosing the Right Protocol 484
Route Selection 486
Displaying General Routing Information 486
Managing Static Routing 492
Configuring Dynamic IGP and EGP IP Routing Protocols 500
Route Control and Redistribution 547
Summary 556
Related RFCs 557
Additional Resources 557

11 Network Troubleshooting, Performance Tuning, and Management Fundamentals 559

Network Analysis and Performance Tuning 560

Developing Troubleshooting Skills 588

Network Management Fundamentals 591

SNMP 597

Summary 626

Related RFCs 627

Additional Resources 628

A Binary Conversion Table 629

About the Author

Michael J. Martin has been a UNIX/NT systems administrator for the last 10 years. Until recently, he was the manager of Desktop Systems and leader of the Tier 2 MIS support group at ANS Communications, Inc.—one of the original Internet service providers. During the last 10 years, he has designed and implemented multiprotocol data communications networks for financial, manufacturing, and biotech institutions. He is currently working for Sanford C. Bernstein & Company, Inc., as a network engineer and architect. He also provides network design and UNIX/NT services on a consulting basis for businesses and regional ISPs. He received a B.A. in philosophy from Manhattanville College in Purchase, New York. Michael enjoys spending time with his wife and children, camping, running, reading, and going to the movies.

About the Reviewers

These reviewers contributed their considerable hands-on expertise to this book's entire development process. As the book was being written, these dedicated professionals reviewed all the material for technical content, organization, and flow. Their feedback was critical to ensuring that this publication fits our readers' need for the highest quality technical information.

Allison S. MacFarlan is a UNIX systems administrator and Oracle DBA for UUnet Technologies. She specializes in financial systems and associated networking and security issues. She has a B.A. in philosophy.

Mark J. Newcomb is the IS operations supervisor for a major medical laboratory in the Pacific Northwest where he is directly responsible for both data and telephony communications for over 600 remote locations. He is a Cisco Certified Network Professional and a Microsoft Certified Systems Engineer with more than 18 years of experience in the microcomputer industry. His efforts are currently focused on Frame Relay, ISDN, and Microsoft networking systems. He also provides freelance consulting to numerous independent clients regarding network design and implementation. Mark can be contacted at `mnewcomb@paml.com`.

Ariel Silverstone has been involved in the computer industry for more than 15 years. He has consulted nationally for Fortune 1000 firms on the implementation of management information systems and networking systems (with emphasis on security). He has designed and set up hundreds of networks over the years, including using all versions of NetWare and Windows NT Server. For five years, he has been the chief technical officer for a computer systems integrator in Indiana. Although no longer a professional programmer, he is competent in a variety of computer languages, including both low- and high-level languages. Ariel is the author of two networking books about routers and connectivity, and has been technical reviewer on more than 25 books, including titles on Windows NT, NetWare, networking, Windows 2000, Cisco routers, and firewalls.

Dedication

This book is dedicated to my wife and best friend, Sarah. Not only has she given me the greatest gift in my life, my two sons, Zachary and Emmett, but without her love, support, encouragement, and understanding, this book would never have been possible.

Acknowledgments

First, I would like to thank everyone at New Riders Publishing, particularly Karen Wachs (who made it happen), Jennifer Garrett (who pulled it all together), David Gibson (who is responsible), Lisa Thibault, Ami Frank Sullivan, Lunaea Hougland, Gina Brown, Sheri Replin, and Lisa Stumpf (who made order out of it all).

I especially want to thank Allison MacFarlan, whose assistance with each of the chapters as an editor, writer, and friend was invaluable to me. I want to thank the technical reviewers, Mark Newcomb and Ariel Silverstone, for their creative suggestions and comments. All of you have made this book more accurate and readable—thanks again.

Like most projects of this sort, there is a group of people who functioned as a test audience for the early drafts. I want to thank those individuals who contributed their time and provided very helpful content feedback, in particular, James Davis, Ken Germann, Claude Brogle, Bill Friday, the folks at X2net, and my Tier 1 and Tier 2 support staffs at ANS.

Finally, I want to thank my wife, children, and friends for being so understanding about my absence from so many things during the last six months; my grandfather, Ernest A. Martin Sr., for my love of technology; my friend, John McMahon, for sharing the secret of writing; and my mother, Mary Martin, who has made everything possible.

This book was written on a Macintosh G3 running MachTen and MacOs 8.6, using OxTex, Alpha, and Canvas. It was later converted to Microsoft Word, because it does not matter if it's the best tool, it only matters that everyone has it.

Tell Us What You Think

As the reader of this book, *you* are our most important critic and commentator. We value your opinion and want to know what we're doing right, what we could do better, what areas you'd like to see us publish in, and any other words of wisdom you're willing to pass our way.

As the executive editor for the networking team at New Riders Publishing, I welcome your comments. You can fax, email, or write me directly to let me know what you did or did not like about this book—as well as what we can do to make our books stronger.

Please know that I cannot help you with technical problems related to the topic of this book, and that due to the high volume of mail I receive, I might not be able to reply to every message. When you write, please be sure to include this book's title, author, and ISBN, as well as your name and phone or fax number. I will carefully review your comments and share them with the author and editors who worked on the book.

Fax: 317-581-4663

Email: `nrfeedback@newriders.com`

Mail: Al Valvano
Executive Editor
New Riders Publishing
201 West 103rd Street
Indianapolis, IN 46290 USA

Introduction

If you are reading this, I am sure that you are already aware that this book is about networking. To be more specific, it describes the various technologies and techniques needed to design, build, and manage a multiprotocol local or wide area network (LAN/WAN).

Generally, there are two types of computer networking books: theory books and practical books. Theory books provide in-depth explanations of computer networking concepts. Often, theory books focus on a particular protocol or networking service. Practical books, on the other hand, teach a specific skill or skills related to a specific function. Books like *Windows NT Networking* or *Networking with Cisco Routers* provide the reader with what he or she needs to know about networking as it relates to a specific function or task. Both of these types of books are written under the assumption that the reader has an adequate background in networking.

There is another kind of book, you might recall from grade school, called a primer. The goal of a primer is to provide the reader with a foundational understanding of the subject matter, in most cases through practical examples. It was in the spirit of the grade school primer that *Understanding the Network: A Practical Guide to Internetworking* was conceived. The goal of this book is to provide both beginning and advanced computer professionals with a solid knowledge base that will enable them to design, build, and manage a multiprotocol computer network.

Who This Book Is For

This book is written for anyone who has a desire to understand the various aspects that make up the broad topic of computer networking. It is structured to serve as a teaching and reference sourcebook for both junior and senior NT and UNIX systems administrators who are involved with the design, management, and maintenance of large- and small-scale enterprise computer networks. Although the covered material should be accessible to anyone with a basic understanding of computers, the book's slant is toward system administrators. They are often responsible for building and maintaining the computer data networks that provide the connectivity between the systems they maintain.

What Is Covered in This Book

Networking is not a new field by any means. Historically, a network engineer's focus and training was oriented toward telecommunications, because data networking mostly dealt with interconnecting mainframe and mini computers, as well as maintaining local and remote serial data terminals. When LANs started to become common in the 1980s, many of them were designed and managed not by network engineers, but by computer support personnel. This condition occurred largely because LANs were initially developed to interconnect personal computers and workstations, which were perceived by many to be nothing more than toy computers.

Early LAN technologies were by no means simple to implement. For many, implementing these technologies involved a large learning curve. Today, LAN and even WAN technologies have become much easier to implement. Advances have shortened the learning curve required to get a network up and running. In many cases, however, the lack of a real understanding of data networking fundamentals and the various technologies involved becomes a significant issue when problems arise. With this fact in mind, this book follows a "ground-up" approach, with each chapter building on the one before. By the end of this book, you will know what you need to build and manage a multiprotocol LAN or WAN. The following is a breakdown of the coverage in each chapter. Keep in mind that the best way to read this book is from beginning to end, because the chapters build on each other.

Chapter 1, "Understanding Networking Concepts," introduces the core concepts of data networking. It defines the various types of computer networks. We discuss the basics surrounding computer data transmission such as transmission media, transmission techniques, network topologies, and components. The chapter concludes with a discussion of various data communication models used for reference and framing decisions about the various components of computer networks.

Chapter 2, "The Networker's Guide to TCP/IP," provides a functional overview of the TCP/IP networking protocol suite, perhaps the most widely used networking protocol in use today. TCP/IP is available on virtually all computer hardware platforms and provides networking connectivity for all the devices connected to the global Internet. This chapter examines major TCP/IP protocols (IP, TCP, UDP, and so on), their addressing, message format, and the services they provide.

Chapter 3, "The Networker's Guide to AppleTalk, IPX, and NetBIOS," covers the popular proprietary local area networking protocols used with desktop-class computers (Macintoshes and Intel-based PCs).

Chapter 4, "LAN Internetworking Technologies," examines 802.3 Ethernet, 802.5 Token Ring, and the Fiber Distributed Data Interface (FDDI) local area network data transmission protocols.

Chapter 5, "WAN Internetworking Technologies," covers general operation of the Public Switched Telephone Network (PSTN) and various analog and digital transmission protocols and services used to provide wide area networking transmission links. Topics include AT&T's T-carrier standard, SONET/SDH, and public packet- and cell-switched data transmission technologies (X.25, Frame Relay, and ATM).

Chapter 6, "Network Switches," examines the design and function of LAN and ATM switches. LAN network switches have become the basis for LANs and WANs today. We discuss the foundational technologies on which LAN network switches are based (PSTN switches and LAN bridges), their implementation, and the additional performance and network design capabilities they provide.

Chapter 7, "Introduction to Cisco Routers," provides the core skills required to configure and manage Cisco routers. Cisco routers make up the majority of installed routers throughout the world. If you do not already have one, chances are high that you will if you plan to implement any WAN or a large-scale LAN. Topics covered include:

- Basic router components
- Configuring IP, AppleTalk, and IPX
- Configuring terminal servers and remote access
- Configuring accounting, authentication, and logging
- Installing and upgrading the router's operating system

Chapter 8, "TCP/IP Dynamic Routing Protocols," examines the operational aspects of the various TCP/IP routing protocols RIP, OSPF, EIGRP, and BGP.

Chapter 9, "Advanced Cisco Router Configuration," picks up where Chapter 7 left off. Here we discuss how to configure many of the advanced Cisco router options. Topics include:

- Standard and extended access control list configuration
- Implementing router redundancy with Cisco's Hot Stand-By Router Protocol
- Virtual private network tunneling and WAN interface configuration

Chapter 10, "Configuring IP Routing Protocols on Cisco Routers," closes our discussion of Cisco routers by looking at how to implement IP routing and the various dynamic IP routing protocols that were covered in Chapter 8.

Chapter 11, "Network Troubleshooting, Performance Tuning, and Management Fundamentals," examines tools and techniques for monitoring and managing your network. Topics covered include:

- Network performance baselining
- Troubleshooting tips
- Network management basics
- A short introduction to SNMP

Appendix A, "Binary Conversion Table," provides the binary conversion table. This table is a number conversion chart for 8-bit numbers.

1

Understanding Networking Concepts

When personal computers became part of the workplace in the mid-1980s, computer networking was, for most companies, an additional task for the computer administrator or perhaps a group project for those who needed and used the network. The computer network was primarily a way for individual computers to share common resources, for example, a file server, printers, or maybe terminal access. In the 1990s, all this changed. The explosion of the Internet, mostly due to the World Wide Web, has transformed computers and their importance in the workplace. What were once spreadsheet stations and personal printing presses have now become powerful communication tools that utilize text, graphics, video, and sound-to-relay information. Now, computer networks are the central nervous systems of most companies and, to some extent, of our planet.

The goal of this chapter is to make you familiar with the concepts and processes associated with computer networking before you examine their specifics in the following chapters.

Computer Network Types

What is a computer network? Generically, a *network* is a collection of computers interconnected by a common method. Because this definition is rather abstract and provides little insight about how the interconnection is achieved, further definition is required. There are currently three classes of computer networks:

- Wide area networks (WANs)
- Local area networks (LANs)
- Metropolitan area networks (MANs)

LANs are limited to a single geographical area (usually a building or collection of buildings, such as a college or company campus complex). Most LANs utilize high-speed networking technologies and are limited by the network size constraints associated with those particular technologies.

A WAN is a collection of LANs in different geographical locations (even on different continents) that are interconnected using low-speed data communications links provided by the Public Switched Telephone Network (PSTN) or other means.

A MAN is a combination of LAN and WAN elements. MANs utilize transmission facilities provisioned from the PSTN, but they employ high-speed communication protocols that are commonly used in LANs.

Computer networks are a ground-up creation. Their complexity grows as the different elements required to send data are implemented. The following section provides an overview of the different elements involved in computer data network communication.

Transmission Media

Transmission media are used to carry computer network data. All transport systems operate over some kind of medium. The desired result of any transmission medium is to transport something with minimal loss. For example, interstate highways use concrete and blacktop for the roads that cars travel on. Water systems use iron, copper, and plastic piping to transport water. Computer networks use copper or fiber optic cable and wireless connection media to provide a conduit to transmit data signals.

Computer network transmission media have two elements: the transmission medium and the transmission signal.

Transmission Medium

The transmission medium for most computer networks is connected cable. Even if the transmission is connectionless or wireless, the telephone cable medium has been in use for some time in WANs. This wiring medium has been used only recently in LAN and MAN environments. When constructing a computer network, several factors should be considered when deciding on the type of data cabling to use:

- End use—What kinds of signaling technologies will be used totransmit data?
- Environment—What possible effects will the transmission medium be subjected to in the location? What security and reliability requirements are expected from the medium?
- Cost of ownership—Data communication cabling is expensive. Achieving a balance between performance and environmental factors is key.

When choosing a media type, make sure that it is suitable for use with different data signal transmission solutions and can accommodate future signaling innovations.

In theory, a computer network can operate over any transport medium that can carry transmission signals, but a desirable level of performance and reliability is required for data transmission. There are a variety of connected and connectionless data communication media. Connected media are used for dedicated links between two or more communication devices. Wireless media use transmitters and receivers to send modulated electromagnetic waves. Connected media offer reliability and security, but are dependent on the scope of the infrastructure. Wireless media are not constrained by physical access or distance, but they are susceptible to obstruction and noise and offer no media component security.

Connected Media

Coaxial cable, twisted-pair cable, and optical fiber are the most common media used for connected signal transmission (see Figure 1.1).

Coaxial cable looks essentially like community antenna television (CATV) cable. It varies from thin and flexible 50 ohm cable to thick, rigid, low-loss 70 ohm cable. Transmission signals are sent using direct current (DC) across the cable length. Coaxial cable has a single solid copper core and a braided or foil shield. The core and the shield are separated by a plastic insulator. Thin coaxial cable is used with connectors known as *British Naval Connectors (BNCs)*. Thick coaxial cable is used with coax or vampire taps. Installing taps requires that the cable have a small hole drilled into the casing and through the shielding. Then, the tap is inserted in the hole and tightened down so the tap pierces firmly into the copper core. Coaxial cable has very desirable transmission characteristics.

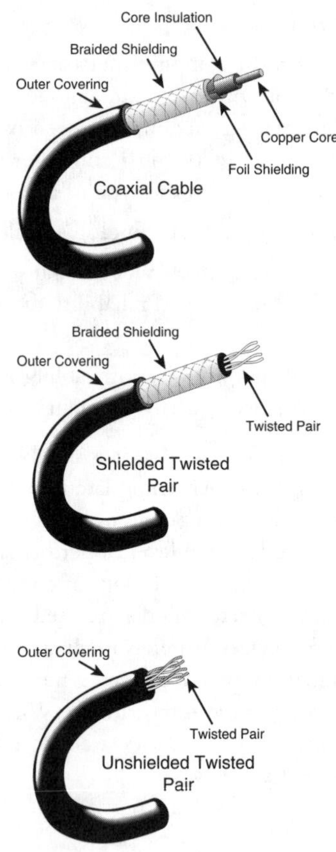

Figure 1.1 Basic cable types.

Twisted-pair cable is by far the most common data transmission medium. A twisted-pair cable is two solid copper wire strands of either 22 or 24 AWG, wound around each other in a double helix, similar to a DNA strand. As with coaxial cable transmission, signals are sent using DC voltage. Then, the twisted pairs are put inside a larger casing to finish the cable. Twisted-pair cables range in size from a single pair to 1,000 pairs. The amount of twists, known as the *pitch*, is measured in pitch per foot. The cable pitch reduces the amount of crosstalk (electrical interference) between signals sent across the cable. A cable's pitch, along with the medium's signal latency, resistance (its diameter), and physical strength, determines its grade. Under the cable grading system, there are service levels (transmission speeds) associated with different grades. Most enterprises use cable up to Category 5. Table 1.1 lists the common cable types and their application.

Note

AWG refers to the American Wire Gauge standard for thickness. The higher the AWG, the smaller the diameter of the wire.

Table 1.1 **Common Twisted-Pair Cable Types**

Category 1	Voice or low-speed data transfers up to 56Kbps.
Category 2	Data transfers up to 1Mbps.
Category 3	Data transfers up to 16Mbps.
Category 4	Data transfers up to 20Mbps.
Category 5	Data transfers up to 100Mbps.

Voice Grade Cable (VGC) is Category 3 and is used for telephone and low-speed data transmission. Data Grade Cable (DGC) is Category 5 and is used for high-speed data transmission. Twisted-pair cable has limited bandwidth, high attenuation and is susceptible to electronic interference. By using a signal repeater, the bandwidth and attenuation can be addressed.

Interference issues are minimized by using high quality shielded and unshielded cable casings. Although the shielding minimizes the effects of outside noise, it also reduces the signal strength of the twisted pairs, which use their crosstalk to help amplify the carrier signal. For this reason, not all signaling methods can operate on shielded twisted pair.

Optical fiber cable is the preferred choice for high bandwidth, high speed, and long distance signal transmission. Transmission signals are sent using light emitting diodes (LEDs) or laser diodes (LDs), and they are received using a pin field effect transistor (pinFET). Signals are sent across an insulated glass core using modulated (mixing a data signal and an electromagnetic waveform) lightwaves. These lightwaves travel across the cable using *pulse amplitude* (lightwaves of varying intensity) or *pulse frequency* (lightwaves sent at a controlled frequency) to represent data signals. The biggest advantage associated with using optical fiber is its immunity to external noise and its extremely low attenuation. Optical fibers in computer network transmission media come in two varities: multi-mode, which can carry multiple light signals across a single optical cable, and single-mode, which carries a single light signal. The relative characteristics of these two types will be discussed later in the book, in the context of their application in enterprise network.

Note

Advances in cable technology are constantly occurring. At the present time, work is being done on the IEEE standardization of Category 6 and Category 7 cable, but these are not widely deployed due to their expense. Cabling will be examined further in Chapter 4.

Note

A repeater is a device that is used between two cable lengths to amplify a transmission signal. The repeater receives a weak signal from the end of one cable length, reconstitutes it, and resends the signal across the other cable length.

Connectionless Media

Connectionless media are usually provided by an outside source, such as a telephone or satellite service company. Each medium is dependent upon a different form of electromagnetic radiation:

- Radio transmission—It is achieved by sending electromagnetic wave forms of a certain size (wavelength) and certain speed. The most familiar types of radio transmission are ultra-high frequency (UHF) and very-high frequency (VHF). UHF (100mm to 1m in length) wavelengths are used for satellites, mobile communications, and navigation systems. VHF (1m to 10m in length) wavelengths are used for television, FM radio, and citizen band (CB) radio. Microwaves (10mm to 100mm in length) are used for telephone and satellite communication systems.
- Lightwave transmission—It is used for line-of-sight transmission. It operates on the same principles as optic fiber cable without theconnection carrier.

Transmission Signals

To send data across any type of medium, some type of encoding method needs to be used. Data signals are transmitted across the transport media using varied electrical signal.

The data encoding methods used to convey data over copper-based media operate by sending electrical current to your home. Where direct current (DC) flows in a single direction across a conductor, alternating current (AC) flows in both directions across a conductor, alternating between positive and negative voltage.

Note

In case you forgot what you learned in physics class, here is a quick study on some basic electrical terms:

- Current (I)—The number of electrons that pass a given point. Current is expressed in units of amperes or amps.
- Voltage (V)—The electrical pressure or force of the DC flow. Voltage is expressed in units of volts.
- Resistance (R)—The ability of the medium to resist the flow of current. Resistance is expressed in units of Ohms.
- Power (P)—The amount of work performed by a current. Power is expressed in units of watts.
- Root-mean-square (RMS) voltage—The AC voltage equivalent to DC voltage that yields the same force. RMS is the transition point (0) between the positive (+) and negative (-) flow.
- Impedance (Z)—The opposition created by a conductor to the voltage change +/-of the AC flow. Like resistance, this is expressed in units of Ohms.
- Frequency (Hz)—The number of cycles that occur within one second. Frequency is expressed in hertz.
- Cycle—The shift in the electromagnetic wave from its peak positive amplitude to its peak negative amplitude. The completion of a single cycle is known as a period.

This alternating current flow is called *oscillation*. Common household current is 110 volts 60hz, which means the current alternates from a positive flow to a negative flow 60 times a second. Data is sent across a computer network using a similar method. Computers use binary ones (on) and zeros (off) to represent data. An electrical signal that has a smooth oscillation from positive to negative is known as a *sine wave*. To represent ones and zeros, a data signal or encoding scheme combines different sine waves and specific voltages to create square waves that are used to represent ones and zeros.

In order for the transmission signals to be understood correctly, they need to be sent across a transmission medium that will maintain its integrity. Loss of signal is attributed to attenuation and noise. *Attenuation* is the loss of structure in the signal as it travels across the medium. Attenuation is the result of the medium's resistance and impedance. If a signal suffers enough attenuation, it will become unrecognizable. *Noise* is the result of outside electrical interference. Copper cable is similar to a radio antenna in that it picks up outside electrical signals and electromagnetic radiation.

These elements can alter the shape of the transmission signals and make them unrecognizable. To ensure data integrity, network distribution schemes have maximum and minimum distance requirements for the length of a media segment and the distances between connected devices.

For each medium, there are ranges of frequencies that can be sent across the wire without significant loss within a certain range. The range specifies the amount of data that can be transmitted through the medium and is known as *bandwidth*. The amount of bandwidth available is dependent on the impedance and the signal-encoding scheme used.

Voltage Encoding Techniques

Baseband and *broadband* are voltage-encoding schemes used to send transmission signals across transmission media. You learned from the previous section that transmission signals send data digitally (using binary coding) across the medium using variable DC voltage. How the signal voltage is applied to the medium determines which encoding technique will be used.

- Baseband transmission applies the signal voltage directly to the transmission medium. As the voltage signal travels across the conductor, the signal is attenuated. To maintain signal integrity, a repeater is often needed. Twisted-pair cable is a baseband medium.
- Broadband transmission is used with voltage signals that have been modulated. By separating the medium's bandwidth into different frequency channels, multiple signals can be sent across the medium simultaneously. Coaxial and optical fiber cable can be used as both broadband and baseband media.

Note

The network cable can also act as a broadcast antenna instead of a receiving antenna. The network cable radiates the carrier signals as they move across the wire. These signals can be detected and decoded. Network wiring should be exposed as little as possible in environments where security is a major concern.

Network Topologies

Network devices connect to the transport medium through a transmitter/receiver called a *transceiver*. A transceiver can be an external device or can be built into the device's Network Interface Card (NIC). The pattern in which the network devices (nodes) are connected together is called the *network topology*. *Network cabling topology* is largely determined by the type of connection medium being used to connect the nodes. In addition, there is a *logical topology* that is defined by how the nodes interact with one another through the transmission media protocol. For example, it's possible for a collection of hosts to be physically connected in a star topology but to run as if they're connected as a logical ring. There are three topology types used in LANs, WANs, and MANs, as discussed in the following sections.

Star Topology

Star topology consists of a central node off which all other nodes are connected by a single path (see Figure 1.2). A pure star topology is almost never found in a modern computer network, but modified star or constellation topology is common.

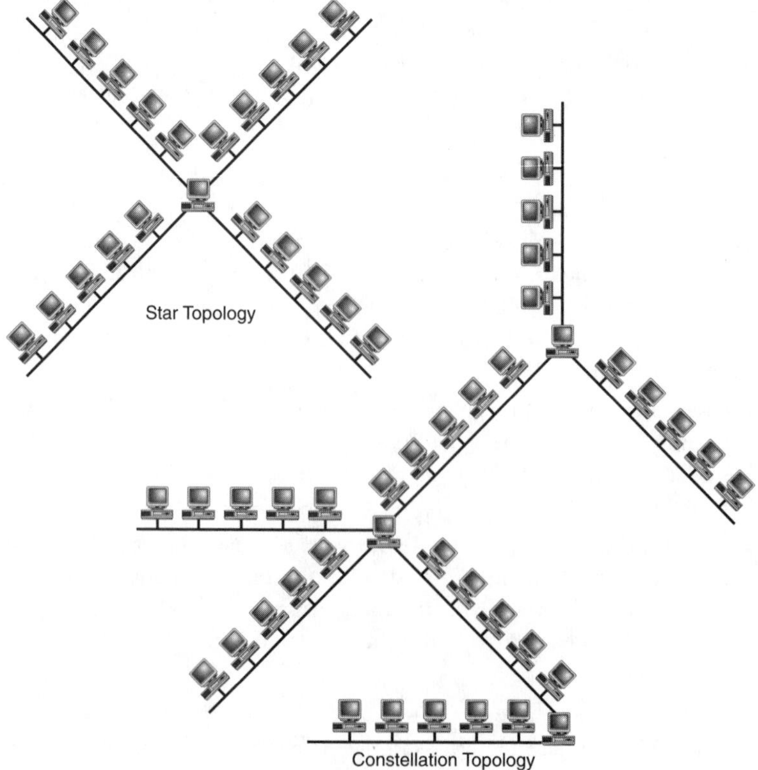

Figure 1.2 Star and constellation topology.

Constellation topology is a collection of star networks tied together with separate interconnection paths. The star/constellation topology is perhaps the most popular network topology in use today. This kind of architecture melds easily with the kind of structured wiring system used for telephone and data cabling in large buildings.

Bus Topology

Bus topology was the basis for most of the original LAN networks. Ideally suited for use with coaxial cable, the bus topology is a single length of transmission medium with nodes connected to it (see Figure 1.3). Most installations consist of multiple cable lengths connected by repeaters.

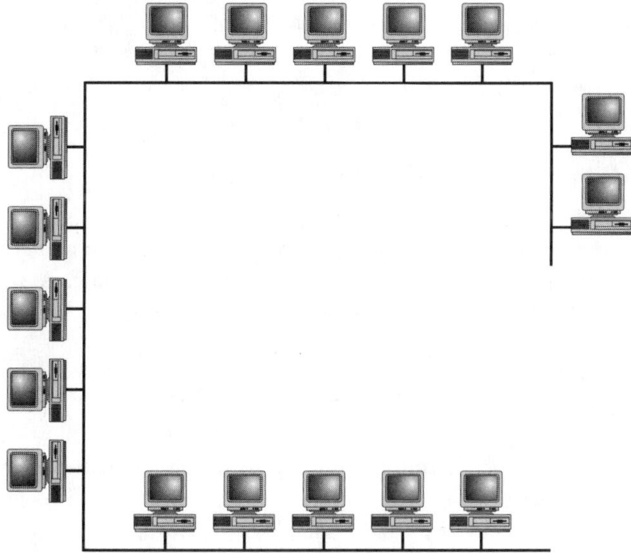

Figure 1.3 The bus topology.

Ring Topology

Ring topology uses lengths of transmission media to connect the nodes (see Figure 1.4). Each node is attached to its neighbor. The transmission signal moves around the ring in one direction and is repeated, instead of just passed, as it moves from node to node. When a station transmits a data message, the transmission is picked up by the next station on the ring, examined, then retransmitted to the downstream neighbor. This process is continued until the transmitted signal is returned to the host that started the transmission, which then removes the data from the network.

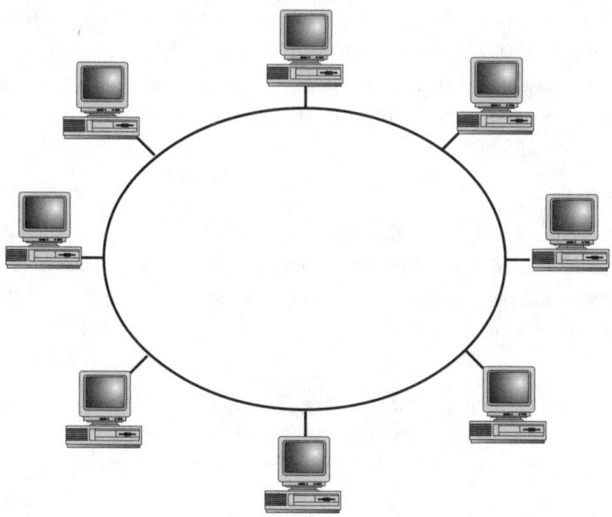

Figure 1.4 The ring topology.

It is not uncommon to mix different network topologies when designing cabling infrastructures. Because different topologies are associated with different transmission protocols, creating hybrid topologies can become a necessity.

Network Transmission Protocols

Computer network communication is primarily event driven. The most efficient way to accommodate event-driven communication is to use a technique known as *packet switching*. The idea behind packet switching is that network bandwidth will not be segmented and evenly dedicated to each network entity. Instead, each node will have the opportunity to use all of the available bandwidth to transmit data when it needs to.

Packet-switching computer network data is exchanged in *frames* (also called *packets*).

Note

Packets and frames are often used as mutually exclusive terms. For the purposes of this discussion, we'll use the term "packet" to indicate a transmission frame containing data, and "frame" to discuss the transmission format.

The frame contains the network data and node addresses of the sender and the receiver. Think of a network packet as an envelope. The letter (data) is inside. The recipient's address is on the outside, and so is the sender's address, in case the packet gets lost or cannot be delivered.

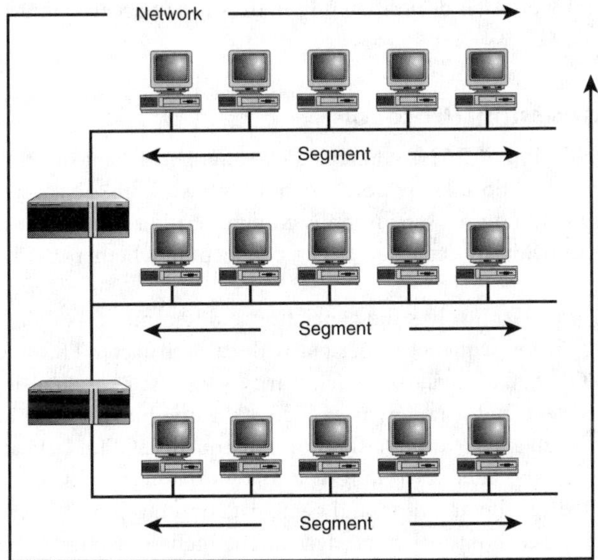

Figure 1.5 Network versus segment.

> **Note**
>
> Computer networks use two types of addresses to deliver network data. The transmission protocol address or node address is used to distinguish each of the different nodes connected to the transport medium segment. The transport protocol, known as Ethernet, uses a 48-bit address called the Media Access Control (MAC) address for this purpose. The MAC address is the machine address: the physical identifier for the host on a segment. It needs to be unique only on a given segment. Theoretically, two hosts could have the same transport layer address, provided they were not connected to the same network segment. There is a distinction between a transport address and a network address.
>
> The network protocol address or host/end-station address is used as a unique identifier throughout the entire network, which may consist of more than one transport medium segment. Unlike transport medium addresses, every host or node on the network must have a unique network protocol address. For example, the network protocol called Internet Protocol (IP) uses a statically assigned 32-bit address. Alternatively, AppleTalk dynamically assigns a network address to every host on its network. Network protocols are used to exchange data between different networks; they will be explained in the next section.
>
> Figure 1.5 illustrates the conceptual difference between network segments and the network as a whole.

Transmission protocols are responsible for two important activities required when network data communication is achieved. *Access control* relates to how the transport medium is accessed as transmission signals are sent across the medium. *Data framing* dictates how the signals are formatted and sent between nodes. Node addressing is also the responsibility of the transmission protocols. LAN transmission protocols and WAN transmission protocols each deal differently with issues of access control to the transport medium.

WAN Transmission Protocols

A WAN is a collection of LANs connected by some permanent or semi-permanent point-to-point connection. These point-to-point network links use transport protocols that employ asynchronous or synchronous communication to transmit data. In either case, the link bandwidth is dedicated to the connection, whether there is data to transmit or not.

Asynchronous transmission involves the generation and transmitting of characters that have a specific start and stop sequence associated with each character. This is known as *character framed data (CFD)*. CFD will sound familiar to you. It starts with a single bit (start bit). Data is then sent, and data flow is ended with a stop bit sequence. The receiving hardware must be able to determine the stop bit sequence. CFD data can be sent continuously or in varying intervals as long as it follows the start and stop sequence. Async data is transmitted one bit at a time, and async data transmission equipment looks at each bit that is transferred. The highest practical speed for async transmission is 1,800 bps. Data rates above this can be transferred with asynchronous equipment. This data, however, is really transmitted synchronously. The equipment converts the outgoing data to sync, and it is converted back to async on the receiving end.

Synchronous data transmission moves both character-oriented (async) and bit-oriented data, known as *binary stream data*. Where async data transmits data one bit at a time, synchronous data transmits data as messages, and so sync data is known as *message framed data (MFD)*. Sync data is buffered because the messages are transmitted one at time in a continuous stream. Sync data uses a separate TX (transmit) and RX (receive) path. Synchronous transmissions are commonly used for data transfers above 2,000 bps.

An easy way to distinguish between synchronous and asynchronous transmissions is that synchronous transmissions are dependent on a clock or timing source, whereas asynchronous transmissions only require that the ends agree upon a means of transmission flow (that is, start and stop bits, flow control, and so on).

There are many varieties of WAN transmission protocols available. These protocols operate over the AT&T/Bell Labs Digital Carrier System/Digital Transport System or Synchronous Optical Network (SONET). Examples of WAN transmission protocols include the following:

- HDLC
- ATM (Asynchronous Transfer Mode)
- X.25/Frame Relay
- PPP (Point-to-Point Protocol)

These protocols will be covered in Chapter 5, "WAN Internetworking Technologies."

LAN Transmission Protocols

In LANs where the medium bandwidth is shared, there are two mechanisms used to control access to the network:

- Contention-based mechanisms
- Non-contention-based mechanisms

Because all the nodes on the network are connected to one another across a common transmission medium, an access control mechanism is required. The *access control mechanism* makes it possible for each node to have access to the medium to transmit data signals. Contention-based mechanisms function on access demand, which results in the chattiest hosts dominating the segment. A non-contention-based mechanism provides an opportunity for each host to have equal access to the medium.

Contention-Based Access

Network contention is a difficult problem. If nodes cannot send or receive data, there is no point to having a computer network. With star and bus network topologies, the *Carrier Sense Multiple Access with Collision Detect (CSMA/CD)* method is used for access control. With CSMA/CD, protocol forces only one workstation to send data at any time (see Figure 1.6). Without this mechanism, all the hosts would send their data at the same time.

A network node continuously listens to the network cable segment. The cable segment shared by all of the nodes when CSMA/CD is employed is called a *collision domain*. When the node is ready to send data, it checks to see if there is any activity. If the cable is clear, it sends its data. If two stations do this at the same time, a "collision" occurs. When this happens, the participating stations begin the *Collision Consensus Enforcement Procedure (CCEP)*, under which a "jam" pattern is broadcast to all of the workstations on the network. This period lasts as long as it takes to reach the ends of the network. If the collision domain is quite large, it may take some time. After the CCEP is over, one of the affected nodes retransmits its frame. If it's successful and needs to transmit more data, the "listening" process starts over again.

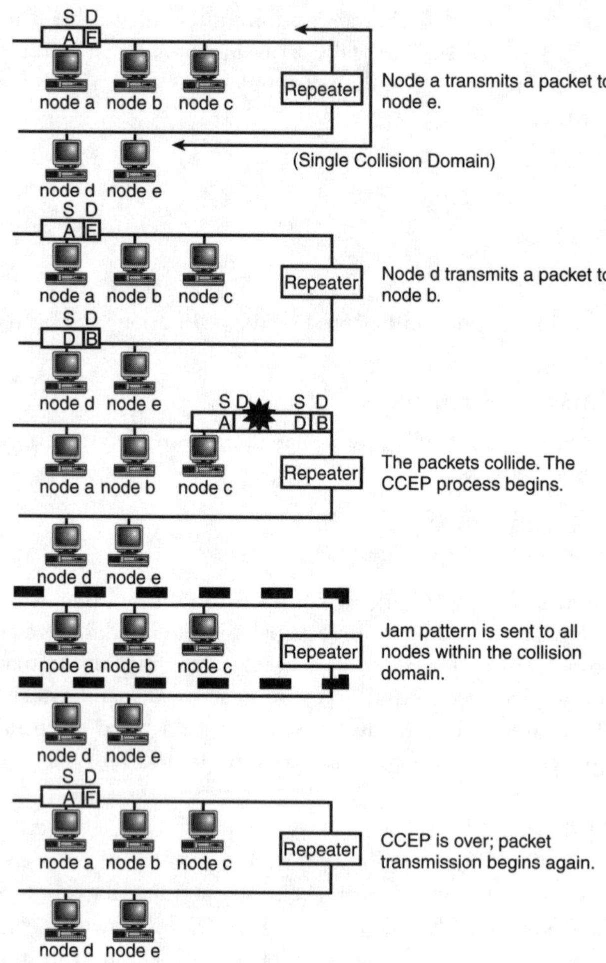

Figure 1.6 Collision handling under the CSMA/CD access method.

However, if a collection of nodes collide, they stop transmitting and try again after a retransmission delay. Each node computes its own delay time using the truncated binary exponential backoff algorithm. This algorithm basically takes the duration of the collision event, doubles its value, and uses it as a wait time. This time is usually longer then the collision domain's round trip delay, which should be no more than 51.2 microseconds and varies depending on the transmission protocol medium used. For example, according to the IEEE 10-BaseT specification, the round-trip delay should be no more than 5.76 microseconds. The theory is that if all the stations are transmitting at random intervals (with random delay periods after collisions), collisions will be minimized.

The major problem with CSMA/CD is that all transmissions are half-duplex (the node can either send or receive at one time), and it only allows one source-to-destination transmission at a time because the bandwidth is shared. That means as the domain adds more nodes or the collision domain becomes larger, the chance of a collision increases. As a result, overall performance decreases because more time is spent detecting collisions and performing collision recovery than actually transmitting data. To maximize the efficiency of a CSMA/CD collision domain, it is normal to engineer the domain to operate under a sustained load of no more then 30 percent of available capacity.

Non-Contention-Based Access

Token passing is the non-contention-based access method used with transmission protocols that utilize ring and bus topologies. Token passing works by having the connected nodes pass a "token" from one station to another. When a station needs to send data, it takes the token and transmits its data. Only the station that has possession of the token can send data. The station that has the token can only transmit one packet; then, it must release the token. The token handling process is different depending upon the transmission medium.

In a ring topology, a token is passed around the ring from node to node in one direction. When a node needs to transmit data, it removes the token from the network and transmits its packet. The packet, like the token, is passed from node to node, until it reaches its destination. At the destination node, the packet is copied, marked as received, and sent on to the next node around the ring. When the packet reaches the node that transmitted it, the node removes the packet, releases a new token, and passes it on to the next node.

When used in a bus topology, token passing operates in a manner similar to operating on a ring. The difference is that on a ring, since there is only one incoming and one outgoing connection, addressing information is not required. This is known as *implicit token*. On a bus, where all nodes are physically adjacent, explicit token is used. *Explicit token passing* uses a logical ring created by using the addresses of the connected nodes. The nodes are logically ordered from highest address to lowest. Each host passes the token to its next logical successor. This logical ring is created when the network is first activated. Each node at activation starts a countdown based on its node address. When a station's timer expires and there is no token on the network, it generates a token. This is the first station on the logical ring. After the "ring" is initialized by token generation, the other stations are invited to join the "ring" at periodic intervals.

With token passing, "fairness" is maintained by forcing each node to release the token after transmitting a single packet. The design ensures that each node on the ring gets a chance to send a packet. IEEE 802.x Ethernet and Apple's LocalTalk are CSMA/CD-based transmission protocols. IBM's Token Ring and ANSI's Fiber Distributed Data Interface (FDDI) are token passing based protocols. These transmission protocols will be covered in depth in Chapter 4, "LAN Internetworking Technologies."

Network Communication Protocols

Transmission protocols are used to send information across the transmission media. Network protocols are used to format and deliver information between computers or hosts.

Think of a transmission protocol as your voice; you expel air from your diaphragm and across the larynx to make sound. That sound is carried across a room as a sound wave. It reaches someone's ear, and the tympanic nerves and bones translate the sound waves into nervous impulses that are interpreted by your brain. This sequence is much like the way electrical signals are sent across a network cable.

Network protocols are similar to language. Language provides the rules and means of expression that allow us to transport our ideas to others. Also, like languages, there are varieties of network protocols, each of which has its own distinct operational "grammar," yet provides the same set of core services.

The most common service in the networking process is datagram delivery and addressing. *Datagrams* are the logical units that carry the actual data. Datagrams, like transmission frames, have information about the sender and receiver to include with the data (see Figure 1.7).

With network communication protocols, each node on the network is assigned a specified address within a unique address range. This address generally has a format that contains the network host identifier and the station identifier. When hosts communicate with one another, they use these network addresses to specify the sender and the destination of the data.

One of the primary characteristics of network addressing is the use of a broadcast domain, which allows stations to send and receive data which is not addressed to a particular station, but to a group of adjacent hosts that all share the same distinct network address.

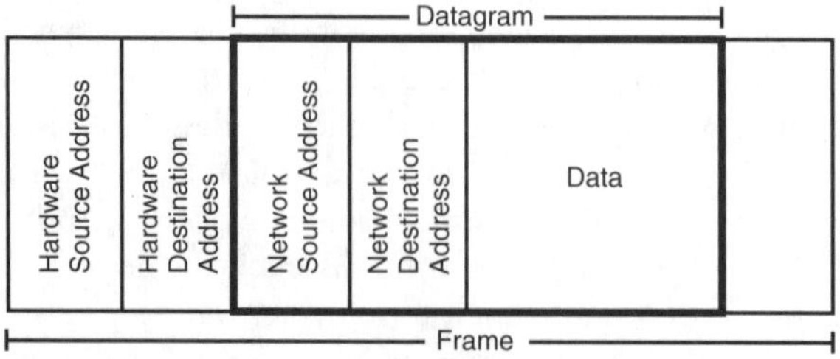

Figure 1.7 The transmission frame/network datagram relationship.

Network Hardware

The cable alone is not enough in large networks. Transmission media and the protocols that regulate them have finite limits on the number of nodes that can be supported. In order to expand the size and range of a computer network, devices that amplify, connect, and segment the transmission media need to be added. These hardware devices are called repeaters, hubs, bridges, and routers.

Repeaters

The repeater was introduced earlier in the chapter. This device is a signal-amplifier/transmitter-receiver whose purpose is to amplify and re-clock a transmission signal between two cable segments (see Figure 1.8).

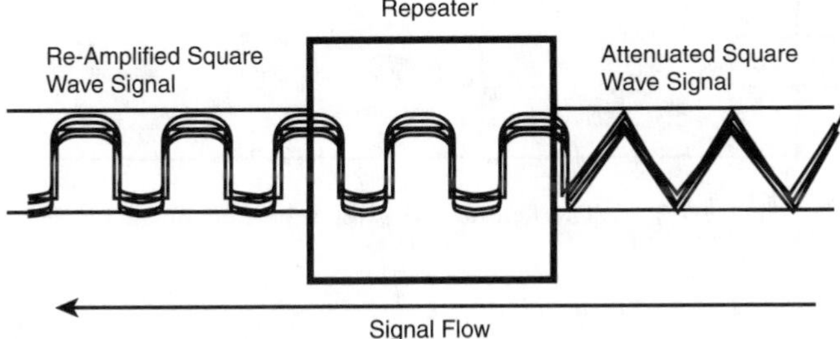

Figure 1.8 Transmission signal amplification using a repeater.

Technically, a repeater and a hub are identical. Repeaters are employed in bus network topologies to connect different network segments together. There are rules that govern the use of repeaters: If four repeaters are used, the network can be segmented into five logical segments. Only three of these segments can be used to terminate computer end stations; the other two must be used for inter-repeater links (this is known as the *3-4-5 rule*). Figure 1.9 illustrates the 3-4-5 rule. This rule applies only to Ethernet in a bus topology.

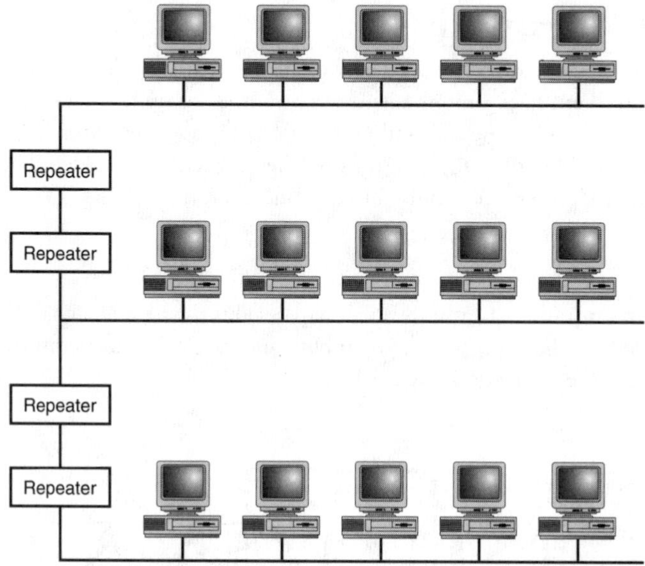

Figure 1.9 Using repeaters to extend the cable segment.

Hubs

A *hub* is a multiport repeater. In Ethernet technology, there are two types of hubs:

- Active hubs—Where each of the ports is a repeater that amplifies the signal for each connection (CMSA/CD)
- Passive hubs—Do not amplify the signal at all

In Token Ring technology, a hub is known as a *Multi-Station Access Unit (MAU)*, and it acts as the physical ring that all the hosts are connected to. For both Ethernet and Token Ring, there are strict specifications for how many hubs can be employed in a given context.

Although a repeater can be used to increase a network's size, it can, at the same, time reduce the network's performance. This reduction in performance is due largely to the expansion of the collision domain. Because the repeater just retransmits signals, the load on the network increases as more stations are added to the network. If your 50-node network needs to be extended because half of its users are being moved to another floor in the building, a repeater is the answer. If the network needs to be expanded so you can double the node count, a bridge is required (see Figure 1.10).

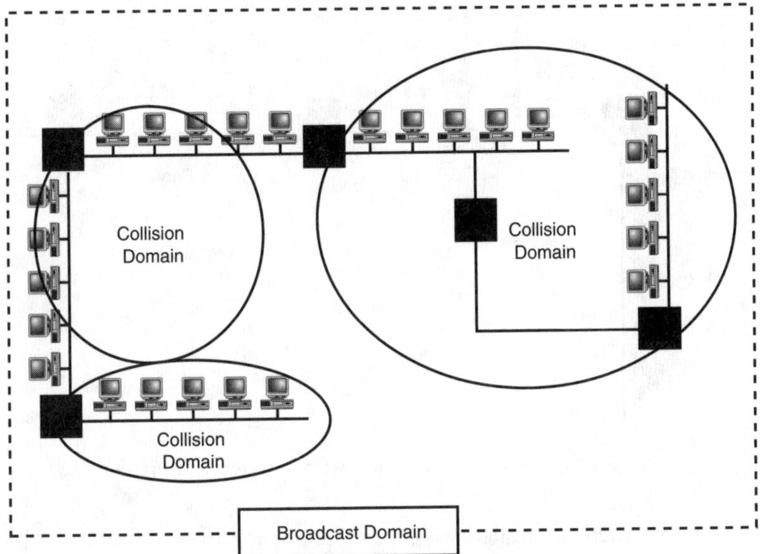

Figure 1.10 Extending the network versus increasing the network.

Bridges

A *bridge*, like a repeater, is used to connect similar and dissimilar transport media segments together without breaking the segments into separate networks. In other words, a bridge enables the separation of a local network into distinct segments while retaining the broadcast domain. The devices all believe they're all locally adjacent, even though they're physically segmented. A bridge is an "intelligent network device," meaning that the bridge is actively involved with how transmission signals move across the network. When a bridge is used to connect media segments, the network is affected in two ways:

- First, the network's transmission length is increased to the sum of the distance of both cable segments (just like a repeater).
- Second, both cable segments are now each a distinct collision domain, as shown in Figure 1.11.

The bridge acts as traffic-handler between the two cable segments. As a node sends a packet to another node, the bridge listens to the transmission. If the sending and receiving nodes are on the same cable segment, the bridge is passive. If the sending node is on one side of the bridge and the receiving node is on the other, the bridge passes the packet from one cable segment to the other. The bridge determines the destination segment by looking at the transmission protocol destination address that is included in the header of the packet.

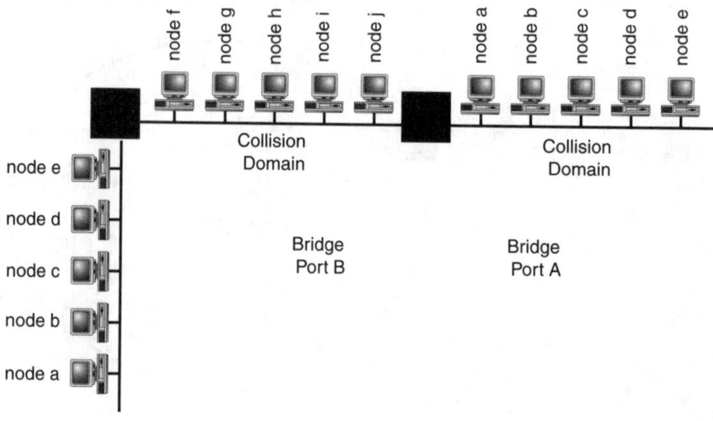

Figure 1.11 Network expansion with repeaters and bridges.

Modern network bridges are called *learning bridges* because they learn the transmission protocol addresses of the nodes they are adjacent to dynamically. The bridge looks at every packet transmitted on its connected segments. It takes the transmission protocol source addresses from the transmitted packets and builds a Bridge Address Table (BAT), as shown in Figure 1.12.

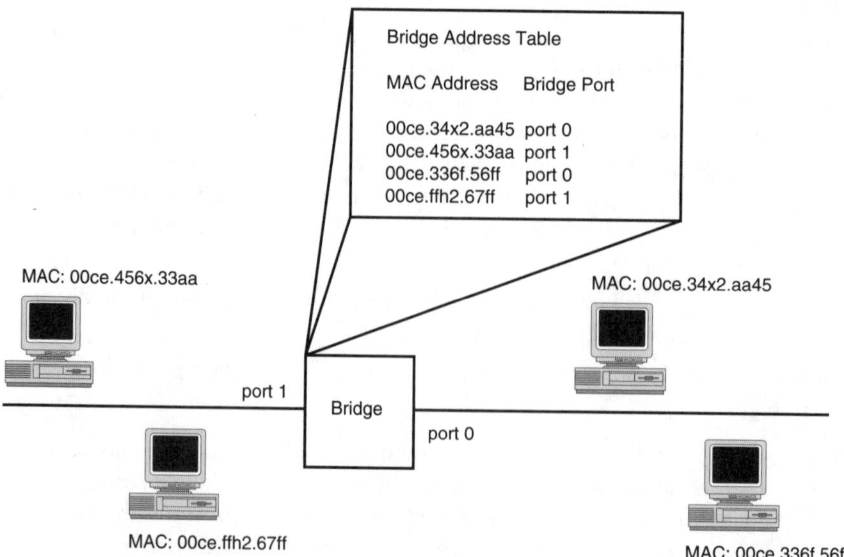

Figure 1.12 A collection of nodes attached to a bridged network and the BAT.

The BAT lists the transmission protocol address of each node on the network and associates it with the respective bridge port the node is adjacent to (refer to Figure 1.10). This table is constantly updated as nodes are added and removed from the network.

The bridge must relay all transmission protocol broadcast packets. Broadcast packets are destined for all nodes on the network (regardless of which side of the bridge they are on). Although a network can have multiple collision domains, it has only one broadcast domain. Broadcast packets are used to send information that all nodes on the network either require or must respond to (this situation arises more often than you might think). If the number of nodes on the network increases, broadcast packets can be a source of significant network performance loss. The only way to reduce the size of a broadcast domain is to reduce the number of nodes that participate in it. This is accomplished by creating a second network segment and moving nodes over to it. For the nodes on the separate networks to communicate, a network protocol and a device called a router must be used.

In the same way that a hub is a collection of repeaters, a device known as a *switch* functionally behaves as a collection of bridges. Switches have become more common in recent years due to reductions in price, and they represent a significant shift in the way data networks are designed. Bridging and switching will be addressed and explained thoroughly in Chapter 6, "Network Switches."

Routers

Bridges and repeaters provide connection services for segmenting and extending the network. *Routers* join multiple networks (that is, media segments) together and act as packet relays between them. Computer networks that consist of more than one logical network segment are called *internetworks*.

What distinguishes routers from bridges are the addresses they pay attention to in the data packet. A network bridge uses the transmission protocol node address to determine where the packet is going. The router uses the network protocol address to determine the packet's destination. The difference in the addresses is that a datagram's source and destination network addresses never change, whereas the transmission protocol packet addresses change each time the data packet transmits a router (because the datagram is placed into a new transmission frame).

Routers are used to perform three major functions:

- Remote network segmentation is made possible and efficient by using routers. Although you can use bridges to extend networks across WAN links, network performance suffers because the broadcast domain is not only extended, it is extended by a link that has substantially less bandwidth than the networks it is connecting (and wastes a good percentage sending broadcast packets). Routers only send traffic across the WAN link that's destined for the attached network, without broadcasts.

- Routers offer an easy way to use different types of transmission media. Because the router forwards the packet from network to network, the data needs to be extracted from one frame (received from the router's inbound network interface) and placed into another (the router's outbound network interface). The router has the ability to chop up the network data and place it in a different kind of frame if the transport media types are different.
- Local network segmentation is performed with routers when broadcast and collision domain reduction is needed. CSMA/CD collision domains can be reduced with bridges, but the broadcast domain cannot (it is used for information collection by both the transmission and network protocols). Excessive broadcasts can impact performance. By segmenting the network nodes onto separate networks, the collision and broadcast domains are reduced because fewer hosts are on more networks. Although there is a performance cost associated with the router processing the frames, this cost is substantially less than the bandwidth and operational loss that results from broadcast and collision storms.

Additional network segmentation can be accomplished with routers by restricting certain kinds of network traffic and (if need be) certain networks from reaching the greater internetwork. Because routers "process" the data packets, they can look at their destination, type, and content. Routers can look forspecific types of data, or packets from a particular source or destination, and discard them. Packet filtering is a popular form of network security.

Routers are a large element in LAN, WAN, and MAN network operations. Routing protocols (how routers exchange network reachability information), router configuration, and implementation will be dealt with at great length later in this book.

Note

Joining different media types together can also be accomplished with a special bridge called a translation bridge. Translation bridges repackage network data from one media frame type to another. However, although a translation bridge will create two collision domains, the broadcast domain is unaffected because the network's structure is at least logically a single transmission segment.

Note

Broadcast storms happen when the network is saturated with collisions or broadcast packets, and no data can be transmitted.

Data Communication Reference Models

Although each data communication protocol has its own operational reference model, all are contrasted to the Open Systems Interconnect Reference Model (OSI-RM) and the Institute of Electrical and Electronics Engineers (IEEE) model. The OSI-RM is the basis for discussing the various elements of the data communication process. The IEEE extension model defines the operational specifications for the transport layer protocols.

An understanding of these models and their relationship to the logical computer network elements, which was already covered, is crucial to forming an understanding of internetworking. Protocol reference models exist to serve as road maps to illustrate how the different protocols perform in relation to one another. Computer networking hardware and software vendors use the definitions contained in the reference models to ensure interoperability with other network elements. Most of the basic elements of network design and implementation can be accomplished without any detailed knowledge of the OSI-RM or protocols in general. Network troubleshooting and network security and management, however, are difficult without some understanding of the protocols in use and their interrelationships.

The Origins of the OSI-RM

The OSI-RM effort began in 1977 as a series of articles on the development of a standard reference model for networking. In 1978, the OSI-RM was defined as a standard by the International Standards Organization (ISO). The OSI-RM's purpose is to provide a framework for visualizing the data communication process. After it's described, it is then possible to segment this process into defined layers that represent natural separations in the communication process as a whole. Standards for how products and protocols should operate were specified so that the services at each layer could function with the adjoining layers. The lower three layers address host-to-host communications functions, and the upper four layers host-to-application communications functions.

The protocols associated with the OSI-RM are known as the *ISO protocol suite.* The ISO suite is entirely standards-based. These standards were developed by various working groups within the ISO and the International Telecommunication Union Telecommunications Standardization Sector (ITU-T), a UN treaty organization made up of the world's telecommunication providers.

> **Note**
>
> ITU-T was originally known as the CCITT, or the Consultative Committee for International Telegraph and Telephone.

It was widely believed that OSI-based protocol suites would replace the TCP/IP protocol suite (and others), but this never happened. One reason it never happened was the time it took for the standards to be developed. Although everyone who took part could agree on the need for standards, getting everyone to agree on a specific implementation was an entirely different story. The second reason was that they encountered cost and interoperability problems. There are only 12 laboratories accredited to certify products for OSI conformance, making for a slow and costly development cycle. This additional cost was passed on to the customer. In addition, although the products might have been tested and certified to meet the OSI standard, this did not mean that they would integrate with each other. These problems killed the OSI effort in the United States for the business sector.

However, for the U.S. government, OSI is not dead. All communications equipment purchased by the U.S. government must be GOSIP (Government OSI Profile) compliant (see Figure 1.13). In Europe, OSI product compliance was required early on in the OSI standards development process. Because the U.S. government and the European Union make up a rather large customer base, OSI has found its way into many of the systems you already use. For example, if you are a Novell 4.x or 5.x shop, you are using Novell's Directory Services, based on the X.500 standard. The X.25 network standard is also an ISO creation. X.25 is used to connect ATM machines, credit card validation systems, and some of the older networks run TCP/IP over it. X.25 has an error checking and correcting version. Frame Relay and X.25 are commonly used WAN protocols. All Cisco Systems products are GOSIP compliant.

There are several Network Communication Protocol (NCP) suites available to interconnect computers today. For example, DECnet was developed to connect end stations and Digital VAX systems. IBM has Systems Network Architecture (SNA) for mainframe and mini-communication. Apple Computer uses AppleTalk to provide print and file access. Novell's NetWare network operating system uses the IPX/SPX protocol suite based on Xerox's open standard XNS system.

Of course, there is the "mother" of all protocol suites: TCP/IP. Transmission Control Protocols, on the other hand, (before OSI) were largely considered separate from the complete data communication process. Data transport was seen as different from data communication. NCPs defined operational specifications and support for specific TCPs or operated over their own proprietary TCP.

The problem with having a large variety of network transmission and communication protocols is that they do not interoperate with one another from a protocol basis, and some are even closed proprietary systems with limitedsupport for proprietary computer systems. The OSI dream (created by telecommunications and business worlds drifting in the sea of incompatible protocols) was to fulfill a rather difficult task: to build data communication networks that work over a variety of infrastructures and allow communication between a variety of system types.

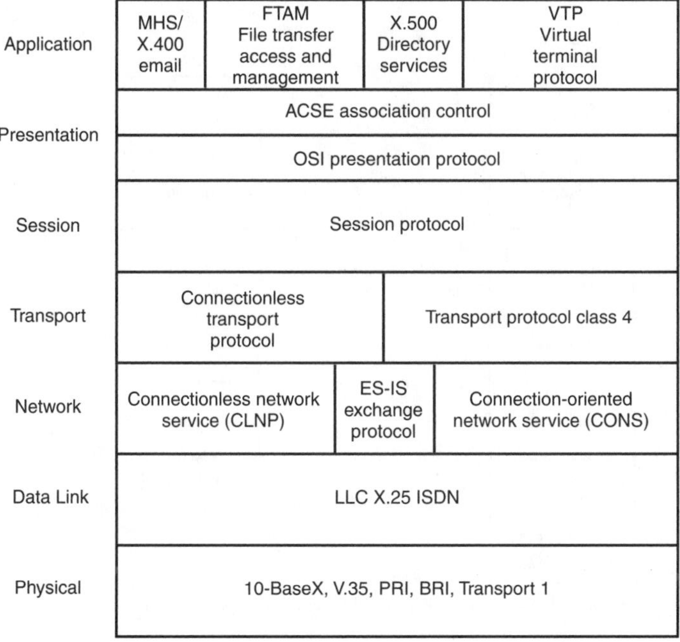

Figure 1.13 The U.S. government OSI profile.

Because the majority of NCPs were vendor-specific—that is, SNA, DECnet, AppleTalk, Novell—TCP/IP was the only contender. TCP/IP was a vendor-independent protocol suite developed in the government and research world. This meant that it was a public domain network communication protocol suite, free from royalty obligations. Until OSI came along, the standard networking reference model for many was the Internet-RM (based on the TCP/IP protocol suite), as shown in Figure 1.14.

Unfortunately, although TCP/IP was free and had a large support and user base, it was limited as a reference model for the whole data communication process. So, in 1984, the ISO fulfilled part of the dream and replaced the Internet-RM with OSI-RM (refer to Figure 1.13). What makes the OSI-RM so important is that, from the beginning, it was supposed to be the "networking" standard. It addresses each step of the communication process from the cable up. The OSI effort was international in scope and had the Herculean task of creating a model and a set of protocols to meet a wide community of users, quite unlike TCP/IP and the Internet-RM, which were created to address the needs of ARPANET. Over time, the OSI-RM became the standard context in which all of the previous network communication protocols are compared and discussed. Although the ISO protocol suite does not share the success that its reference model has, it is important to note that the majority of network communication protocols are still widely used today.

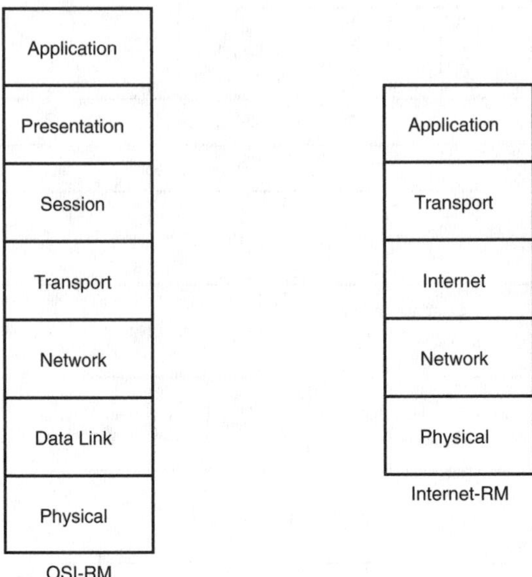

Figure 1.14 The OSI and Internet reference models.

The OSI Model

The OSI model has the following seven layers:

- Application
- Presentation
- Session
- Transport
- Network
- Data link
- Physical

Each layer defines the function(s) needed to support end-to-end networked communication. *End-to-end* means that it moves from the physical/hardware level to the application/user interface level.

The OSI functions are subdivided into two layer groups: upper and lower. The upper-layer services (7, 6, 5, and 4) relate to application services. Application services, such as Web browsers, network mail clients, and Graphic User Interface (GUI) client/server database applications, all use upper-layer services. The lower-layer services (3, 2, and 1) are where host-to-host communication takes place. Data created by the Upper-Layer Protocols (ULPs) is placed in datagrams. Datagrams, as you recall, are logical groupings of data of a specific size that are used to carry data created by the ULP. These datagrams are placed in transmission protocol (Ethernet, Token Ring, and FDDI) frames.

If a service processes user data, it is an upper-level OSI function. If a service transports data, it is a lower-level OSI function. Under the Internet-RM, the OSI upper layers fall into the application and transport layers and the lower layers fall into the Internet, network, and physical layers. Although the OSI-RM does provide a better model for examining the entire communications process, the Internet-RM still provides some value because it closely resembles a practical model of how the network protocol communication process works.

The Seven Layers of the OSI-RM

The OSI-RM is known as the *X.200 standard* in the ISO universe. Because the OSI-RM standard is the context for examining data communication protocols, this book provides some examples from various protocol suites to establish perspective for later chapters as to how the actual protocols function. The upper OSI reference model standards perform the following functions, as noted in the following sections.

Layer 7: Application

This layer provides the network service access to applications. All application programs are included at this layer (including ones that do not require any communication services), and each application must employ its own protocol to communicate with the Lower-Layer Protocols (LLPs). Basic file and print services also fall into the application layer.

Layer 7 services are the places where the application communicates with the user in terms of actual (human) meaningful input and output data. The following standards govern data between the user application and the network:

- ISO—X.500 Directory Services
  X.400 Message handling (email) services
  Virtual Terminal Protocol (VTP)
- TCP/IP—Telnet virtual terminal service
  Simple Mail Transfer Protocol (SMTP)
  Domain Name Service (DNS)
  Berkeley Remote Commands
  Sun's Network File System
  CMU's Andrew File System
- AppleTalk—AppleShare print and file services
- IPX—NetWare Core Protocol
  NetWare Shell (NetX)

Layer 6: Presentation

This layer addresses the problems of data representation as it appears to the user. Data syntax, character sets, and data formatting also fall under Layer 6. Layer 6 also provides the means for the various Layer 7 services to exchange information in an encoding scheme. Almost all systems use the ASCII encoding scheme to present data so it can be transmitted in a computer-independent form. This way, computers of various types can exchange information with one another. Overall, Layer 6 presents data in a common and universally acceptable form that can be transported and exchanged without concern for the terminal used for displaying the data.

Protocols associated with this layer are the following:

- ISO—Connection-oriented presentation protocol
- TCP/IP—Network Virtual Terminal (NVT)
- AppleTalk—AppleTalk Filing Protocol (AFP)
  Adobe PostScript
- IPX—NetWare File Service Protocol (NFSP)
- Server Message Block (SMB)

Layer 5: Session

Layer 5 manages the exchange of data between application processes. Session interposes communication, flow control, and error checking services. Most network operating systems (AppleTalk, Novell NetWare, and Microsoft Networking) provide service activities at this level. The variety of protocols used to provide session services are as follows:

- ISO—Connection-oriented session protocol
- TCP/IP—Berkeley socket service
  System V streams service
- AppleTalk—AppleTalk Data Stream Protocol (ADSP)
  AppleTalk Session Protocol (ASP)
  Printer Access Protocol (PAP)
  Zone Information Protocol (ZIP)
- IPX—NetBIOS

Layer 4: Transport

Layer 5 checks the data for integrity and keeps the application program apprised of the communications process, but Layer 4 is concernedwith end-to-end data transport. This transport can be managed as either connection-oriented or connectionless. Connection-oriented data transmission is needed for reliable end-to-end, sequenced data delivery. Because the data loss might occur because of LLP delivery problems, a variety of services are needed to address such a condition.

A connection-oriented transport must be able to perform the following data handling functions:

- Multiplexing—A connection-oriented transport service must be able to move the data in and out of the Layer 3 carrier.
- Segmenting—Data, in most cases, needs to be transmitted in several units. Segmenting is the process of breaking the data into segments and reassembling it at the remote end.
- Blocking—Some data segments are small enough to be moved in one data unit. Blocking is the process of putting multiple data segments into a single data unit and extracting them at the remote end.
- Concatenating—This is the process of putting multiple data units into a single Layer 3 carrier and extracting them at the remote end.
- Error detection and error recovery—The transport service must have a way of detecting if the data has become damaged during the Layer 3 carrying process and have the means to resend it.
- Flow control—The transport must be able to regulate itself as to the number of data units it passes to the adjacent layers.
- Expedite data transfer—The transport needs to be able to provide for special delivery service for certain data units and override normal flow control conditions.

Some connection-oriented transport protocols are the following:

- ISO—Transport Protocol class 4 (TP4)
- TCP/IP—Transmission Control Protocol (TCP)

A connectionless transport is also known as a *datagram transport*. Connectionless transport has no requirement for data sequencing, data integrity checking, or loss due to LLP delivery problems. Connectionless transport is used when fast delivery of unimportant data is required, for things like domain name service lookups or voice and video transport. The main requirement for this transport mechanism is consistent data delivery speed, but a slow, consistent stream is preferred over a fast, intermittent one.

Common connectionless transport protocols are the following:

- ISO—Transport Protocol class 0 (TP0)
- TCP/IP—User Datagram Protocol (UDP)
- AppleTalk—AppleTalk Transaction Protocol (ATP)
  Routing Table Maintenance Protocol (RTMP)
  AppleTalk Echo Protocol (AEP)
  Name Binding Protocol (NBP)
- IPX—Service Advertisement Protocol (SAP)

Regardless of the mode, it is the responsibility of Layer 4 to optimize the use of the network's resources.

The lower layers (3, 2, and 1) of the OSI model are where the "network" actually starts. Under OSI, the network is slightly more abstract than under the TCP/IP model because it was created from the perspective of telecommunications providers. Although this perspective is not apparent when you look at the standards themselves, it becomes clearer when you try to implement WAN technology.

Layer 3: Network

The network layer is where the actual delivery of the transport data units takes place. The network layer provides the delivery and addressing services needed to move the transport data units from host to host. This is accomplished by sequencing the data into data packets and adding a delivery information header with the source and destination addresses and any additional sequencing information. This packeted data is known as a datagram.

Layer 3 is also responsible for delivery of the datagrams, requiring some kind of routing service. Under the Internet-RM, this is handled at the Internet layer. Under OSI, the activity of datagram routing is seen as two processes: *routing* and *forwarding*. Routing is the process of seeking and finding network information. Forwarding is the actual process of using the routing information to move data from host to host. The OSI model sees routing as a three-tiered hierarchy:

- Tier 1 is route discovery.
- Tier 2 is intradomain routing.
- Tier 3 is interdomain routing.

In an OSI environment, there are two types of network layer transport services:

- Connectionless service handled by connectionless network (CLNP)
- Connection-oriented service handled by CONS X-25 (connection-oriented network service over X.25)

With connection-oriented transport, confirmation of delivery is required after each datagram is delivered. With connectionless transport, datagrams are sent by the best-known path; as the packet is moved from intermediate router to intermediate router, the delivery path might be different for each datagram, depending on network traffic conditions. The TCP/IP protocol suite only provides a connectionless network layer transport, because TCP/IP was built on the premise that it would be used in an internetworking environment where link reliability would be questionable and so left the responsibility for actual delivery to Layer 2, the data link layer. The IP protocol provides the connectionless network layer transport for the TCP/IP suite. AppleTalk and IPX are also connectionless delivery services. AppleTalk uses Datagram Delivery Service (DDS) and IPX uses Internetwork Packet Exchange (IPX).

Layer 2: Data Link

Data link is the facility that controls the transport of the upper layer protocol (ULP) data bits across the physical connection medium. ULP data is "enclosed" inside of a Layer 2 protocol "envelope," called a frame, which is then transmitted. Layer 2 has two data transport functions. The first (and most commonly associated) is Media Access Control (MAC), which defines the logical representation of the ULP data and access to transport medium. The second is link control (LC), which acts as the interface between the Layer 3 protocol(s) and the MAC. Depending on the Layer 2 protocol and its application (such as WAN or LAN use), the LC function is handled differently. WAN protocols define this process as part of their specification. The majorities of LAN protocols (and some variations on WAN protocols) utilize the Institute of Electrical and Electronics Engineers (IEEE) 802.2 LLC specification to perform this function. Figure 1.15 illustrates the IEEE "enhanced" OSI-RM. Advances in network speed, performance, and reliability for the most part, all occur at the data link layer (Layer 2).

All transport control protocols are considered Layer 2. Some of the more common protocols are the following:

- IEEE 802.X Ethernet, Fast Ethernet, Gigabit Ethernet. The most common CSMA/CD baseband LAN protocol.
- ANSI X3T9.5 FDDI, Fiber Distributed Data Interface. A LAN/MAN redundant transport technology that runs over fiber optic cable.
- ITU-T V.34 is the serial line standard used for modem transmissions up to 28.8Kbps.
- ITU-T V.90 is the serial line standard used for modem transmissions up to 53.3Kbps. This is the standard that replaced USR's X2 and Lucent/Rockwell's Kflex proprietary standards.

Note

Individuals who are just getting started with networking are often unfamiliar with Layer 2 technologies. This is the result of Layer 2 and Layer 3 technologies being commonly grouped together in discussions about topics (such as NT and UNIX System Administration) that depend on networking, but are not primarily concerned about its operation. The reality is that these two layers know nothing about what the other is doing (and do not care). Their separation means that the ULP can leave actual transport and data integrity to Layer 2. This partitioning of network delivery and network transport is a very important distinction. By separating these two functions, the ULP has no involvement in (or dependency on) the actual data transport. This makes it possible for data to be moved across various Layer 2 transmission types with little, if any, modifications to the ULP. This allows for a greater level of flexibility in what transport technologies are used, which means faster data delivery and low transmission delays.

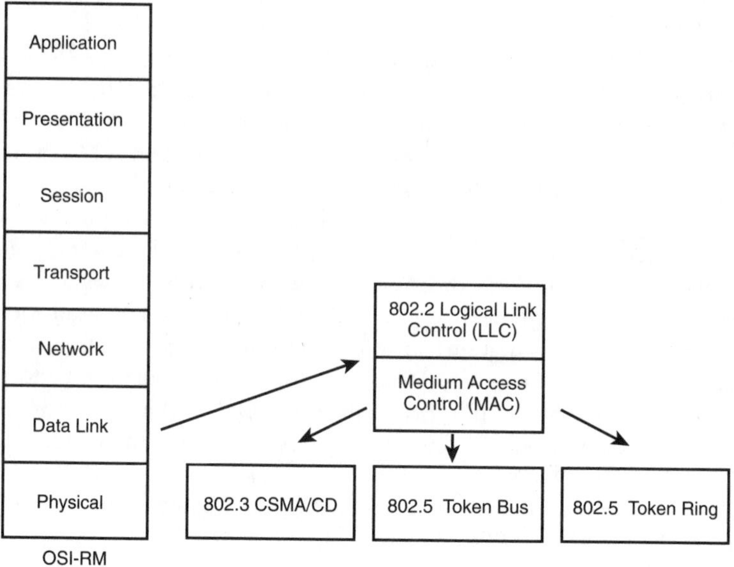

Figure 1.15 IEEE enhancement to the OSI-RM.

- ITU-T V.35 is the standard used for synchronous communications between routers and a public packet data network. The interface is usually a Data Service Unit/Channel Service Unit (DSU/CSU), a device used to provide data conversion so the data can be sent over a digital telephone loop.

As you can see from the previous examples, any means by which data is actually transported is addressed at the Layer 2 level.

Layer 1: Physical

The physical layer deals with specifications of the medium used to move bit data from point to point. All physical, electrical, and mechanical aspects of the transmission media are addressed at Layer 1. Layer 1 and Layer 2 are also commonly looked at together because the physical layer standards are usually taken for granted. Do not fall into the trap of grouping them. The physical layer of the network is one of the most complex and, next to configuration errors, the most common cause of problems found in networks. All physical media have corresponding standards. When working with any medium, you should review, or at least be aware of, the minimum operating specifications, such as connector type(s), maximum cable length, and any environmental installation requirements, that might interfere with the performance of the transport or affect the operation of other network/non-network equipment.

Common physical layer standards are the following:

- IEEE 10-BaseT—The cabling standard for using unshielded twisted-pair copper wire to transmit 802.3 Ethernet.
- IEEE 100-BaseT—The cabling standard for using unshielded twisted-pair copper wire to transmit 802.3 Fast Ethernet.
- EIA/TIA-232—The standard used for unbalanced (async) circuits at speeds up to 64Kbps. This is commonly known as the RS232 serialport standard. The actual RS232 standard was based on the ITU-T V.24 standard that is no longer used.

There are two key principles at the core of the OSI-RM. The first is the idea of open systems. The open systems principle is based on the idea of interoperability. Any computer system running any operating system can exchange data seamlessly by using communication protocols that employ OSI layer functions. This effort is fueled by advocating open communication standards that are royalty-free and are not dependent on any proprietary technology.

The other principle is a "bottom up" concept of peer-to-peer communication (see Figure 1.16). The idea is that the communication process is a layered one, so data passes from layer to layer unaltered and independent of any of the layers for anything except their functionality. Open systems can use different protocols, that is, improve upon or replace different layers without affecting the overall behavior of the data communication process.

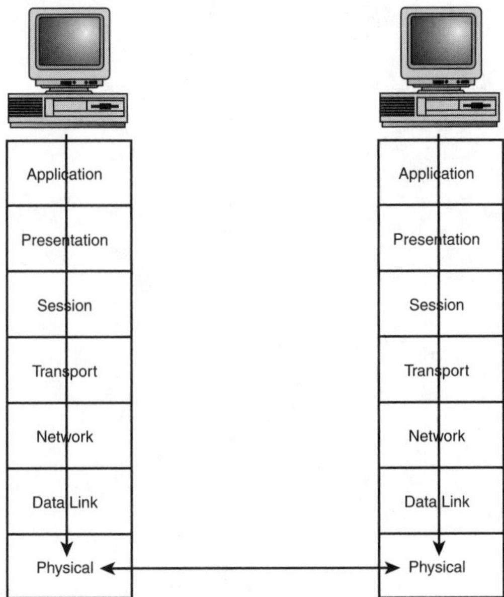

Figure 1.16 The "bottom up" communication process.

Summary

Up to this point, the logical aspects of computer networks have been focused on. Understanding the logical concepts will make grasping the practical applications of those concepts easier. Let's review what has been covered so far:

- The Open Systems Interconnect Reference Model—The internationally accepted standard for describing and categorizing the elements (and related network protocols) that make up the communication process.
- Transmission media and encoding techniques—Transmission media are the combination of the OSI Layer 1 transmission medium and the OSI Layer 2 transmission protocol.
- Network topologies—The star, bus, and ring transport medium (OSI Layer 1) configurations.
- Network transmission protocols (OSI Layer 2)—These are responsible for transmission signaling and framing used in transmitting signals across the transmission medium.
- Network communication protocols (OSI Layer 3)—These are responsible for packaging (into datagrams) and delivering data between hosts connected to the internetwork.
- Network connection hardware—The passive and active electromechanical elements used to relay OSI Layer 2 and OSI Layer 3 data packets.

2

The Networker's Guide to TCP/IP

TCP/IP IS THE PREDOMINANT NETWORK PROTOCOL IN USE TODAY. Although proprietary networking protocols, like AppleTalk and IPX, still enjoy widespread use on local area networks (LANs), the increased demand for desktop-level access to the Internet, which utilizes the TCP/IP protocol suite for host to host communication has lead many networking environments to implement TCP/IP and abandon their proprietary protocol implementations altogether or implement TCP/IP alongside their proprietary LAN protocols. TCP/IP is well suited for both local and wide area networking applications, and is easily scalable to support networks of any size. It is not only capable of coexisting with other Layer 3 protocols, IP can be used to transport proprietary LAN protocol datagrams over IP WANs. TCP/IP was developed as the second generation network protocol for ARPAnet (Advanced Research Projects Administration Network), which was an experimental network funded by the U.S. government to research packet switching networks. The ARPA network and the technologies developed for it formed the foundation of the modem Internet. Development on TCP/IP began in 1973; it became available as a production protocol suite in 1983. Figure 2.1 illustrates TCP/IP's developmental timeline.

Note

In the previous chapter, we examined the Internet-RM, OSI, and IEEE reference models. By becoming familiar with the different reference models, you gain a sense of the logical flow of the data communication process. Protocol theory, however, is often simpler than its physical deployment. Knowledge of the communication models is important, but it can be confusing. To minimize this confusion, the OSI-RM will be the default reference model in discussing the protocols and their respective layer of operation.

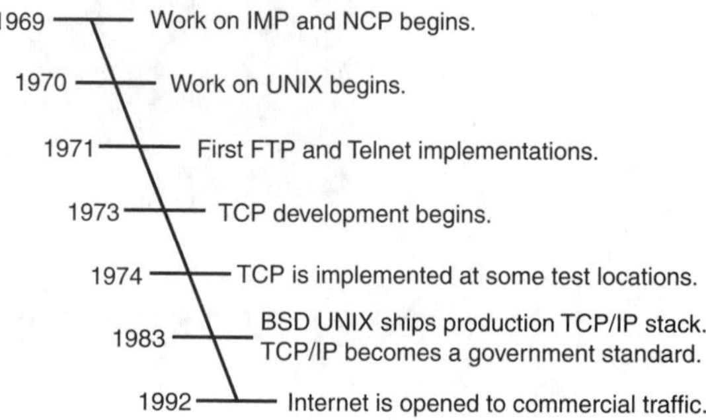

Figure 2.1 The TCP/IP development timeline.

The goal of this chapter is to describe the various protocols of the TCP/IP protocol suite. Using the OSI-RM model, the key elements of TCP/IP are covered: what each protocol does, its sequence in the suite, and services it needs to perform. This chapter also explores how the various protocols of the TCP/IP suite work together and how TCP/IP relates to other transport mechanisms (such as Ethernet and FDDI) that carry TCP/IP packets. An understanding of the material is crucial to anyone who is building or managing a TCP/IP based computer network.

The Origins of TCP/IP

TCP/IP gets its name from the most widely used protocols in the suite: *Transmission Control Protocol (TCP)*, the OSI-RM Level 4 protocol that provides connection-oriented data transport service, and *Internet Protocol (IP)*, the OSI-RM Level 3 protocol that provides datagram delivery and addressing facilities. Early in TCP/IP's development, its capability to provide network connectivity between different types of hosts was extolled as one of its greatest virtues.

This virtue has made TCP/IP the protocol on which most client/server computing is based. Most interactive application layer protocols use the client/server model to exchange data. *Client/server* is a transaction-based processing model. The client represents the user or device making a data request. This request is made using a common data exchange process or user interface to a server.

The server processes client requests by providing data that can be processed by the requesting client. Both the server and the client agree on the data exchange process before transactions are made. This is ensured because both the client and server are following a common application programming interface (API) that specifies how the transaction function calls will take place.

The capability of TCP/IP to function equally well across both WAN and LAN connections, coupled with its diverse computer platform support, makes TCP/IP the natural choice for connectivity in heterogeneous internetworking environments. Although TCP/IP is vendor-independent and in the public domain, it has required continuous development as computer and networking technologies have progressed. This development is overseen by the *Internet Engineering Task Force (IETF)*, which consists of more than 80 sub-committees that develop, review and revise the technologies standards used on the Internet. The IETF operates under the *Internet Society*, a nonprofit organization that coordinates the development of the Internet.

TCP/IP and UNIX

In the early days of computing, operating systems were something the user built. Early computer systems came with a limited set of tools, which provided the capability to load, compile program routines, and perform basic I/O tasks. The "operating system" was developed by the programmer/users because everything was written in the assembler language specific to that computer (there was no Plug and Play in the early days). In the late 1960s, two fellows at Bell Labs, Dennis Ritchie and Ken Thompson, started working on a different approach to computer operating systems, known as *UNIX*. The goal was to develop a core operating system (the *kernel*) to manage the various hardware elements of the computer (CPU processing, memory management, storage I/O, and so on), plus an extensive set of tools that were easily ported to any hardware platform.

Although UNIX was the property of Bell Labs, it was made available to universities and research facilities for noncommercial use with no support. As a result of its wide distribution, a lab project became an operating system that was improved by a large community of academic programmers who saw its potential. The notion that "the documentation is the code" and that "programs should be small, and do one thing and do it well" came out of this giant collaborative effort. In 1974, some of the people involved with the development of ARPAnet and TCP/IP saw a similar synergy between UNIX and TCP/IP. Along with the integration of TCP/IP protocols into ARPAnet, an effort was started at the University of California at Berkeley to integrate TCP/IP into the UNIX implementation. This beta (experimental) TCP/IP stack was introduced in 1981 as the Berkeley Software Distribution (BSD) 4.1a release, and the production stack was BSD 4.2. The result of this integration enabled the majority of ARPAnet sites to connect through TCP/IP. Along with its status as a government standard, TCP/IP's wide use in the UNIX operating system ensured its place as the de facto networking standard.

Layer 3: IP Protocol

The IP protocol is responsible for the delivery of datagrams. A *datagram* is a block of data, "packaged" by the upper layer protocols for transmission. This data is "packaged" by the Upper Layer Protocol (ULP). The contents of this data can be TCP or UDP transport data or the transport layer data of a different networking protocol, such as AppleTalk or IPX. IP is uninterested in the type of data it is providing delivery for and can accommodate any Layer 2 packet size limitation, which it handles by fragmenting the datagram into the supported Layer 2 frame size. The datagram fragments are then transmitted to the destination host where they are reassembled before being handed to the ULP. Additionally, IP does not guarantee datagram delivery. The Layer 2 transmission devices involved in the transport of datagrams can discard the Layer 2 frames containing IP datagrams if the adequate resources needed for transmission are unavailable. In the event that IP datagrams are lost, it is the responsibility of the ULP of the transmitting host to resend the lost data.

The following section explains the many components of the IP protocol and its operation, such as its interaction with the Layer 2 protocols that provide network address to transmission address translation, the structure of the IP datagram, and the various elements of network addressing, including IP datagram delivery.

IP Interactions with Layer 2 (ARP, Proxy ARP, and RARP)

All Layer 3 delivery is dependent upon Layer 2 for actual transport. This being the case, a mechanism needs to exist for translating Layer 3 network addresses to Layer 2 transport addresses. The ARP protocol was developed to perform these translations for IP.

When IP assembles a datagram, it takes the destination address provided by the ULP and performs a lookup in the host's route table to determine if the destination is local or remote to itself. If the destination is local, the datagram is delivered directly to the local host. If the datagram is not local, the datagram is sent to a network gateway, which takes responsibility to ensure that it is delivered correctly. In either case, the IP datagram needs to be placed inside a Layer 2 packet for delivery.

In order to determine the correct Layer 2 destination address, the sending host must have a mechanism for translating the Layer 2 to Layer 3 address map. It uses *Address Resolution Protocol (ARP),* which is a dynamic process, and *Reverse Address Resolution Protocol (RARP)*, which is an application service, to assist it in finding the locally connected hosts. ARP and RARP map the different Layer 2 protocol addresses (for example, Ethernet/and Token Ring) to the 32-bit addresses used by IP to identify the different hosts on the network. Technically, both the Layer 2 and Layer 3 IP addresses both provide the same basic function of host identification in terms of the data delivery process, but nevertheless, the addresses are incompatible. So, in order for data to be delivered between locally connected hosts, a remapping service is needed.

ARP runs as a local process on a host (computer or other network device) connected to the network. ARP accomplishes two tasks:

- First, it performs translations when a host needs to send data to an IP address for which it has no Ethernet address.
- Second, it keeps track of the hosts it discovers during its broadcasts.

Here's how it works, step by step:

1. The host makes an ARP request. The ARP request is a broadcast to all the locally connected stations, asking, "What is the Ethernet address for X.X.X.X?"
2. Destination hosts reply to the ARP request. The request is made using the first IP datagram destined for delivery. Because the datagram is busy asking, the data is lost during the ARP request process, so the datagram must be retransmitted. This retransmission is usually unseen by the user, but it can be demonstrated when a new host is connected to the network and a user pings another local host.
3. If the destination host is not in the user host's ARP cache, an ARP request is made, usually resulting in the loss of the first datagram. But, ARP also keeps a record of the Ethernet/IP address translations it discovers. These mappings are stored in RAM until the host is shut down.

In situations where the host knows little other than its Ethernet address at startup, RARP is used. RARP is an IP level client/server address provisioning service.

When the "dumb" host boots up, it sends a RARP broadcast across the local network segment. A RARP server that knows the IP address that corresponds to the Ethernet address of the requesting station responds to the request and provides the IP address to the host. This process only occurs at boot time.

Note

The lifetime of an ARP entry depends on a hardware platform's IP implementation. For example, Cisco routers flush the ARP table entries approximately every four hours.

Sometimes, these default purges are not fast enough. In a case where you need to change a bad Network Interface Card (NIC) on a host, when you first bring the host back online, you might not be able to exchange data with other hosts on the network. When this happens, the first thing you should do is flush the ARP cache on the hosts to which you are having trouble connecting. Corrupt and incorrect ARP entries are the usual suspects in situations where intermittent or sudden lapses in connectivity happen between locally connected hosts.

IP also supports a facility for hosts to masquerade and accept IP packets on behalf of other hosts. This is known as *Proxy ARP*. Proxy ARP is used in situations where sending hosts are unable to determine that IP routing is required. Proxy ARP maps one or more network prefixes to a physical Ethernet address. This permits the host running Proxy ARP to answer all ARP requests made for the IP network for which it acts as proxy. Proxy ARP is a limited type of routing and is rarely found in modern networks. Its original intent was to assist in situations where host IP implementations were limited and to provide the ARP facility for non-ARP-compatible transport links.

With local datagram delivery, the Layer 2 protocol handles most of the work. IP and its supporting protocols act as address translation mechanisms that assist Layer 2 in identifying the source and destination hosts for the delivery of the ULP data. The main function of IP (and all other Layer 3 protocols) is to provide a common transmission interface between the ULP and the Layer 2 transmission medium. This function is essential in order for hosts to have the capability to exchange data independent of the actual transmission protocol. The utilization of this common interface makes it possible for the exchange of data to occur over different transmission mediums utilizing different addressing, data encoding, and transmission schemes. In a LAN environment, this is often not a large concern because one transmission medium is used to connect the different hosts on the network. However, when data needs to be delivered to a host beyond the local network, a remote delivery model comes into play. Remote datagram delivery involves the possible use of different Layer 2 transmission protocols and the need for a mechanism for determining how to reach the remote host, known as routing.

IP Header

Now that you have an understanding of IP datagram delivery and its interaction with the Layer 2 protocol, let's look at how the IP header is structured. An IP header is appended to each actual upper layer protocol message that needs to be delivered across the network. In most cases, the IP header and its various parts have little effect on your environment beyond delivering your IP data. But, understanding the parts of the IP header gives you insight into how IP works in general. Figure 2.2 illustrates the different data fields that make up the IP header, followed by a brief description of each field's function.

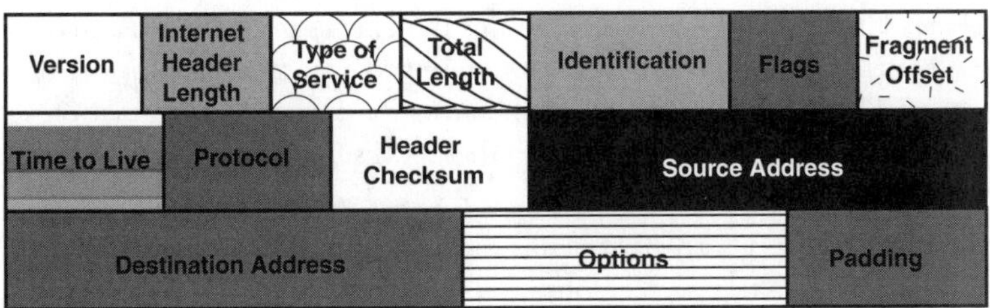

Figure 2.2 The IP header.

- Version—This field defines which IP version created the datagram so that all parties involved with handling the datagram know its proper formatting. IPv4 is the current standard. This field will become more and more significant as IPv6 begins to be implemented in the next few years. The field is 4 bits in size.
- IHL—The Internet Header Length (IHL) field describes the total length of the IP datagram header. The field has a maximum size of 4 bits.
- ToS—The Type of Service (ToS) field is an 8-bit field that provides the capability to define the preferred service treatment. The ToS field might not be used by every hardware or software platform.
- Total Length—This field describes the total size of the datagram, including both the IP header and the ULP data. The size of the field is 16 bits, which is expressed in octets (8 bits). There is no limitation on maximum or minimum datagram size. The size is limited by the specifications of the Layer 2 transport medium (Ethernet, for example).
- Identification—This value tells the destination host what part of the data fragment is contained in the datagram data payload. The ULP or IP can generate the value. The ID field is 16 bits long.
- Flags—This 3-bit field indicates whether fragmentation is permitted. If the "Don't Fragment" flag is on, fragmentation is not possible. If the "More Fragment" flag is on, fragmentation is used in conjunction with the packet ID to determine how to reassemble the datagram fragments.
- Fragment Offset—This field is used to determine what part of the fragmented data contained in the datagram is being carried relative to the original nonfragmented datagram. The first datagram fragment and a complete nonfragmented datagram have this set to off (zero). The maximum size for this field is 13 bits.
- TTL—The Time to Live (TTL) field is IP's way of destroying undeliverable data. A datagram has a maximum life span of 4.25 minutes, after which the packet is discarded. When the datagram is sent, the hop counter is set to 255. This number is decreased by 1 for each router that handles the datagram. This field is 8 bits long.
- Protocol—This field provides information about the data contained in the data portion of the datagram. The field is 16 bits long.

- Header Checksum—This field contains the checksum for the IP datagram header. It is recomputed at each route point to accommodate for the decrement to the TTL field. The recomputation allows the IP header information to be validated as correct. Datagrams that fail the checksum are discarded. The field is 16 bits long.
- Source Address—This field contains the address of the host that originated the datagram. The field is 32 bits long.
- Destination Address—This field contains the address of the host to which the datagram is to be delivered. The field is 32 bits long.
- Options—This field is optional and has no specified size. It is used for both defined special options and as a means for making modifications to the IP header for testing and development purposes.
- Padding—This field is used to add any additional bits that may be needed to make the header meet an adequate size for transmission.

IP Datagram Fragmentation

IP fragmentation provides the capability for IP datagrams to be sent across various types of networks that have different limitations on the size of their transport frames. Because the IP datagram is appended to an ULP message, the size of an IP datagram varies. Different transmission media have different message size limitations; in order to deliver an IP datagram over these different types, IP needs the ability to fragment a large ULP protocol message into smaller pieces that can be accommodated by the Layer 2 transmission medium. Layer 2 protocols transmit Layer 3 data in frames. Each Layer 2 protocol (Ethernet, FDDI. X.25, Frame Relay, and so on) has specifications for the size of the frames it can support. This value is known as the *Maximum Transmission Unit (MTU)*. The MTU is expressed in bytes, and, along with the Layer 3 data, must also contain the needed data required for the frame to be processed correctly. Because modern networks are made up of network segments of varying speeds, IP's capability to fragment itself into different MTUs is an important necessity.

In order to ensure this variability, IP provides mechanisms for managing fragmentation and reassembly. Fragmentation is performed by the gateway that needs to route the datagram over interfaces with incompatible MTUs. The gateway first fragments the original datagram into *datagram fragments*. These fragments are then delivered to the destination host where they are reassembled. Because IP provides no facility to guarantee datagram delivery or the order in which they are delivered, the destination host must cache the datagrams until all the fragments arrive, and then reassemble them.

To assist in the reassembly process, the Identification, Flags, and Fragment Offset IP header fields are used. When a fragmentation candidate is identified, the packet ID contained in the Identification field identifies all the IP datagrams that contain fragments of the original datagram. The first datagram containing a datagram fragment is sent with its Flags field set to "More Fragments." This tells the destination host that it should cache datagrams with the same packet identifier and that there are more datagram fragments coming with which to reassemble the original datagram. The additional datagrams containing datagram fragments are sent with the "More Fragments" flag and with a marker in the Fragment Offset field that describes the datagram fragment's relative place in the order of the original datagram. After all the original datagram has been sent, the final fragment is sent with its "More Fragments" flag off. When all the datagrams arrive at the destination host, they are reassembled and the data is delivered to the ULP.

The IP Address

Until now, the handling of IP datagrams has been described as a process, so you could gain an understanding of how IP handles ULP data delivery. But, addressing also plays an essential role in the delivery of datagrams. Here is a review of the practicalities of the IP address space.

Every device that exchanges information using the TCP/IP protocol suite needs a unique IP address. The IP address consists of two parts: the network address and the node address. IPv4 is the current protocol standard. This standard is provided by the IETF and is in the public domain. What distinguishes IPv4 from previous versions is

- Vendor independence—All IPv4 implementations can communicate with each other. Each implementation must follow the IETF standard, so all TCP/IP implementations can interoperate.
- 32-bit address space—All IPv4 addresses are 32 bits long and are expressed in four 8-bit octets. These octets are represented in Arabic (number), hexadecimal, and binary form.
- End-to-end communication—The sending host and receiving host can each acknowledge when datagrams are delivered.

Note

The reason "octets" are used to describe the IP address space instead of "bytes" is that some computers have byte sizes different than 8 bits. A 4-byte address implies that the address contains the number of bits that the computer platform defines as a byte.

The limitations of the IP address space were only realized later in IPv4's evolution. What this realization means to you as an IP networker is that you need to understand two methods of implementing IP addressing. The original method is known as *classful addressing,* and the revised method is known as *classless addressing*.

Classful IP Addressing

The earlier statement that an IP address has two parts is only half true. All IP addresses have two parts: the network and node address. There is, however, a second address known as the *subnet mask* that is required to make sense of the IP address.

TCP/IP is based on the idea that a collection of independent networks are connected together by gateways, and that any node on a network can connect to any other node by knowing only the node's IP address. To make this possible, routing is used to send IP datagrams between different networks. To make it easy to distinguish between the node address and the network address, IP's creators made different IP address classes. There are three common classes of IP address space (A, B, and C) and a defined number of networks available in each class. The *class type* determines how many bits can be used to define the network address and how many bits can be used for the node address. The way to determine the class of the network is to look at the placement of the first significant bit. To do this, you must look at the address in binary format.

If the first bit is 0, it is a Class A network, which can support up to 16 million hosts. If the first bit is 1 and the second bit is 0, it is a Class B network, which can support up to 64,000 hosts. If the first two bits are 1s and the third bit is 0, it is a Class C network, which can support 254 hosts.

There are also Class D and Class E address spaces. Class D is reserved for use by multicast services provided by IGMP, the Internet Group Management Protocol. Multicast broadcasts allow a single IP datagram message to be sent to specific groups of hosts that belong to the multicast group. Multicasts can be used across a single subnet, or they can be forwarded across IP network segments or subnets by setting up a router to perform multicast forwarding. Multiple multicast groups can exist on one subnet. The destination host addresses are specified in the address field of the multicast IP datagram. The Class E space is not used at all in networks. It was reserved for testing and development. Most networking equipment will reject addresses from the Class E space. See Figure 2.3 for a comparison of the networks.

> **Note**
>
> The IP subnet mask is the most important and misunderstood part of the IP address. This is mostly due to the confusion between classful and classless routing. According to RFC950, all hosts are required to support subnet mask addressing (some do it better than others). The subnet mask, like the IPv4 address, is 32-bits long, in binary format, and expressed as 4 octets. The masks are defined from left to right using ones or zeros.

> **Note**
>
> There is a Arabic/Hexadecimal//Binary conversion chart in Appendix A, "Binary Conversion Table."

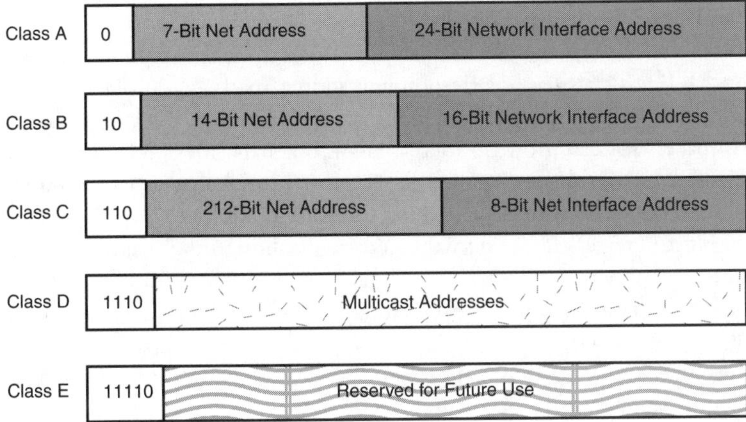

Figure 2.3 IP address classes compared, in terms of bit significance of the first octet.

With classful addressing, the natural mask is determined by the values of the first few bits contained in the first octet of the address. By examining the IP address in binary form, you'll see that the first three bits of a Class A network are 000, the first three in a Class B network are 100, and, in a Class C network, are 111.

The first eight bits in an address class natural network mask are

```
0XXXXXXX Class A 255.0.0.0
01XXXXXX Class B 255.255.0.0
011XXXXX Class C 255.255.255.0
```

The subnet mask's function is to mask the network address portion of an IP address. For example, within an IP subnet, the actual node for address 192.160.3.45 with its corresponding mask of 255.255.255.0 is 45.

> **Note**
>
> Imagine that a gateway is connected to three different IP networks. When the gateway receives an IP datagram, it looks at the network address portion of the datagram's destination address and then looks to see if it knows how to reach that network. It performs this lookup by comparing the datagram's IP address to a table of networks it knows how to reach. This table is known as a routing table.
>
> The routing table is created using the IP addresses and masks of its connected interfaces. The router uses the mask to determine the network address of its connected interfaces and then places the network address in the table. If the destination address matches one of the networks to which it is connected, the datagram is forwarded through the corresponding gateway interface. If the datagram is not destined for one of the networks to which the router is connected, the packet is forwarded to another gateway, or discarded.
>
> This same process is performed by the end-stations. When an end-station creates an IP datagram, it takes the destination address of the datagram, compares it to its own address, and then uses its subnet mask to see if the network addresses match. The datagram is either forwarded on to the destination or is sent to the gateway. The subnet mask is clearly important and will cause problems if not set correctly.

Classful Subnetting

Again, under the classful rules, all the usable IP address space is segmented into address classes. Each address class has a corresponding address mask. The problem is that the natural address masks do not work well in real-world applications. Networks made up of address spaces that can grow up to 16 million hosts are not efficient. To effectively use large address spaces, *subnetting* must be employed to break up the address space into usable subnets. Subnetting works by taking the natural mask of the address and adding additional bits to create a larger network address (resulting in a smaller number of usable hosts per subnet).

Figure 2.4 shows an example of subnetting a Class A network into smaller Class B and Class C networks.

When subnetting large address spaces into smaller subnets, setting the subnet mask is essential for datagrams to be passed correctly.

In Figure 2.5, network A's use of IP address space is totally inefficient, but it will work because the two network subnets are using different Class A address spaces. The host knows to forward the datagram to the IP gateway for delivery because the destination address has a network number different from its own. In network B, the host's subnet mask is incorrect, so it thinks that its network address is 90 when it's actually 90.4.0. Because the mask is wrong, any traffic that needs to be sent to a host with a starting network prefix of 90 will be lost. The host will try to deliver it locally (because the network address and subnet make it look to the sending host to verify the correct delivery path) and will not forward it to the IP gateway.

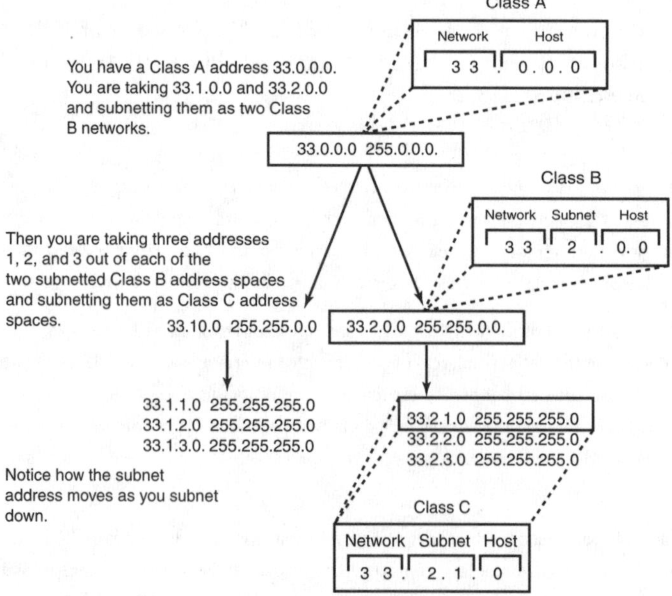

Figure 2.4 Address partitioning of a classful, subnetted address.

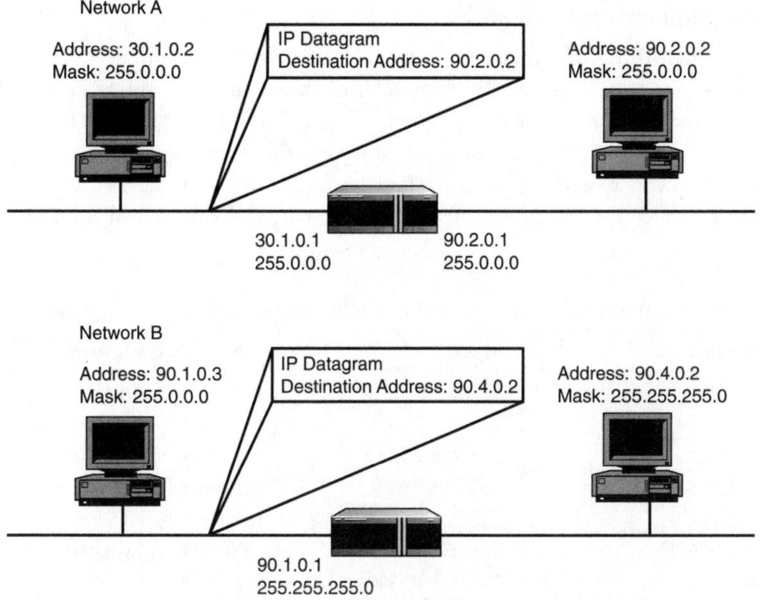

Figure 2.5 A subnetting example.

When using subnetting with large address spaces, follow these setup guidelines:

- Every node needs a unique IP address.
- Every IP network segment must also use a unique IP subnet address space.
- Every network must have a unique classful subnet address.
- Every node on the network using the same root IP address space must use the same subnet mask.
- Every node on the IP network segment must use the same subnet mask.

Note

When subnetting is used, the IP address is segmented into three parts: the network address, the subnet address (which is created by adding the additional masking bits to the natural mask), and the node address.

Classful Subnetting Examples

The important thing to remember about subnetting using classful addressing is that *all* the subnet masks must be the same for each IP subnet used on the same IP network. In other words, if you have a Class B address of 149.178.0.0 255.255.0.0 and a network of 10 separate Ethernet segments, and you want each segment to have a Class C address space, you would subnet the Class B address into 255 Class C networks (shown in Table 2.1) and use one of the subnetted address spaces on each of the Ethernet segments.

Table 2.1 **Subnet Network for 149.178.0.0 255.255.0.0 Address Space—Part I**

Network	Mask	Network Assignment
149.178.0.0	255.255.255.0	Ethernet Network 1
149.178.1.0	255.255.255.0	Ethernet Network 2
149.178.2.0	255.255.255.0	Ethernet Network 3
149.178.3.0	255.255.255.0	Ethernet Network 4
149.178.4.0	255.255.255.0	Ethernet Network 5
149.178.5.0	255.255.255.0	Ethernet Network 6
149.178.6.0	255.255.255.0	Ethernet Network 7
149.178.7.0	255.255.255.0	Ethernet Network 8
149.178.8.0	255.255.255.0	Ethernet Network 9
149.178.9.0	255.255.255.0	Ethernet Network 10

Say you want to add another IP segment to your network. The segment, however, is across town, so you need IP address space for the WAN link and three more Ethernet networks. Now, your subnet network table looks like Table 2.2.

Table 2.2 **Subnet Network for 149.178.0.0 255.255.0.0 Address Space—Part II**

Network	Mask	Network Assignment
149.178.0.0	255.255.255.0	Ethernet Network 1
149.178.1.0	255.255.255.0	Ethernet Network 2
149.178.2.0	255.255.255.0	Ethernet Network 3
149.178.3.0	255.255.255.0	Ethernet Network 4
149.178.4.0	255.255.255.0	Ethernet Network 5
149.178.5.0	255.255.255.0	Ethernet Network 6
149.178.6.0	255.255.255.0	Ethernet Network 7
149.178.7.0	255.255.255.0	Ethernet Network 8
149.178.8.0	255.255.255.0	Ethernet Network 9
149.178.9.0	255.255.255.0	Ethernet Network 10
149.178.10.0	255.255.255.0	Office WAN Network 1
149.178.11.0	255.255.255.0	Remote Ethernet Network 1
149.178.12.0	255.255.255.0	Remote Ethernet Network 2
149.178.13.0	255.255.255.0	Remote Ethernet Network 3

As you can see, every IP address subnet uses the same subnet mask. Because the same subnet mask must be used to subnet the entire space, each host on the network has a defined number of networks that it can accommodate with that mask. Table 2.3 shows the breakdown of possible hosts and networks in a Class A network space using classful subnetting. Remember, with classful subnetting, the entire space must be segmented using the same subnet mask.

Table 2.3 **Possible Class A Host/Network Classful Subnetting Breakdown**

Mask Size	Number of Hosts	Number of Networks	Decimal Mask in Bits
8	16,000,000	1	255.0.0.0
10	4,000,000	2	255.192.0.0
11	2,000,000	6	255.224.0.0
12	1,000,000	14	255.220.0.0
13	512,000	30	255.248.0.0
14	250,000	62	255.252.0.0
15	128,000	126	255.254.0.0
16	64,000	254	255.255.0.0
17	32,000	510	255.255.128.0
18	16,000	1,000	255.255.192.0
19	8,000	2,000	255.255.224.0
20	4,000	4,000	255.255.240.0
21	2,000	8,000	255.255.248.0
22	1,000	16,000	255.255.252.0
23	510	32,000	255.255.254.0
24	254	64,000	255.255.255.0
25	126	128,000	255.255.255.128
26	62	256,000	255.255.255.192
27	30	512,000	255.255.255.224
28	14	1,000,000	255.255.255.240
29	6	2,000,000	255.255.255.248
30	2	4,000,000	255.255.255.252

Table 2.4 shows the breakdown of possible hosts and networks in a Class B network address using classful subnetting.

Table 2.4 **Possible Class B Host/Network Classful Subnetting Breakdown**

Mask Size	Number of Hosts	Number of Networks	Decimal Mask
16	64,000	1	255.255.0.0
17	32,000	2	255.255.128.0
18	16,000	4	255.255.192.0

continues

Table 2.4 **Continued**

Mask Size	Number of Hosts	Number of Networks	Decimal Mask
19	8,000	6	255.255.224.0
20	4,000	14	255.255.240.0
21	2,000	30	255.255.248.0
22	1,000	62	255.255.252.0
23	510	126	255.255.254.0
24	254	254	255.255.255.0
25	126	510	255.255.255.128
26	62	1,000	255.255.255.192
27	30	2,000	255.255.255.224
28	14	4,000	255.255.255.240
29	6	8,000	255.255.255.248
30	2	16,000	255.255.255.252

Table 2.5 shows the breakdown of possible hosts and networks in a Class C network address using classful subnetting. Although it is possible, Class C address spaces are not subnetted.

Table 2.5 **Possible Class C Host/Network Classful Subnetting Breakdown**

Mask Size	Number of Hosts	Number of Networks	Decimal Mask
24	254	1	255.255.255.0
25	128	2	255.255.255.128
26	62	4	255.255.255.192
27	30	6	255.255.255.224
28	14	14	255.255.255.240
29	6	30	255.255.255.248
30	2	62	255.255.255.252

Here are some final notes on using classful addressing:

- Only one subnet mask can be used to subnet the entire network space. This rule applies to Class A, B, and C network address spaces.
- All interfaces connected to a subnet must have host addresses and subnet masks that reflect the address and subnet mask designated for the network segment.
- Subnets where the subnet address is all ones will not work, and subnet addresses of all zeros should be avoided (although some routers will permit a subnet address of all zeros).

Classless IP Addressing

Address space under IPv4 is tight, and with the Internet growing larger every day, the usable address space is getting smaller. As it turns out, the classful address scheme did not work out as well as its designers had thought it would back in 1974. So, they gave it another try, and came up with *supernet addressing* and *Classless Interdomain Routing (CIDR)*.

Supernetting came on the Internet scene in 1993. The idea was to give networks contiguous groups of Class C addresses that would reflect their actual need instead of a single large Class B address space that would be mostly wasted. Supernetting fixed the address shortage, but created routing problems because, although all the supernet addresses were sequential, they still needed a single classful route for each address space used. To fix the supernet routing problem, CIDR was developed. CIDR also arrived in 1993 (when Border Gateway Protocol version 4 was released). CIDR's fix for the Internet's routing issue was the abandonment of the classful-based address space model. Although classful routing would continue to be used at the local network end, the Internet backbones would use CIDR. Instead of building large tables of classful routes, CIDR builds single routes based on summaries of contiguous addresses.

Figure 2.6 shows an example where, under classful addressing, each of these Class C addresses would require a route for each address space.

The Classful Way	The Classless Way
192.160.0.0 255.255.255.0	192.160.0.0 /21 (255.255.248.0)
192.160.2.0 255.255.255.0	
192.160.3.0 255.255.255.0	
192.160.4.0 255.255.255.0	
192.160.1.0 255.255.255.0	
192.160.5.0 255.255.255.0	
192.160.6.0 255.255.255.0	
192.160.7.0 255.255.255.0	

Figure 2.6 A comparison of classful and classless network address expressions.

In classless mode, the address group is expressed as a single network. The network loopback address is 192.160.0.0, and the network broadcast address is 192.160.7.255. The goal with CIDR is to provide a single route to the gateway that leads to the network(s), as shown in Figure 2.7. How the actual address space is subnetted is of no concern to CIDR, which only provides the summarized route.

Because CIDR replaces classful addressing and masks, networks needed to be represented in a format that reflects classless addressing and masking. Under CIDR, an address space can be any number of addresses that is a power of 2, minus the network loopback and broadcast address. CIDR address summaries move from the lowest address space upward. In Figure 2.7, the eight summarized Class C addresses are represented as the classless network 192.160.0.0 /21 (255.255.248.0). Classless and classful addresses both use 32-bit address masks to distinguish the network and host portions of the address space. CIDR network address masks are expressed by indicating the number of binary ones contained in the address mask instead of the traditional series of four octets. This distinction is done mainly for semantic purposes, indicating that CIDR is being used for addressing. From a practical standpoint, the network mask is entered as four octets when configuring a device with an IP address. Because the CIDR mask representation is far more elegant than the four octet mask, in the last few years it has come into common use for describing IP address masks.

Table 2.6 details CIDR bit mask representations in comparison to the traditional classless four octet mask. Along with a number of possible nodes supported by each classless address space variation.

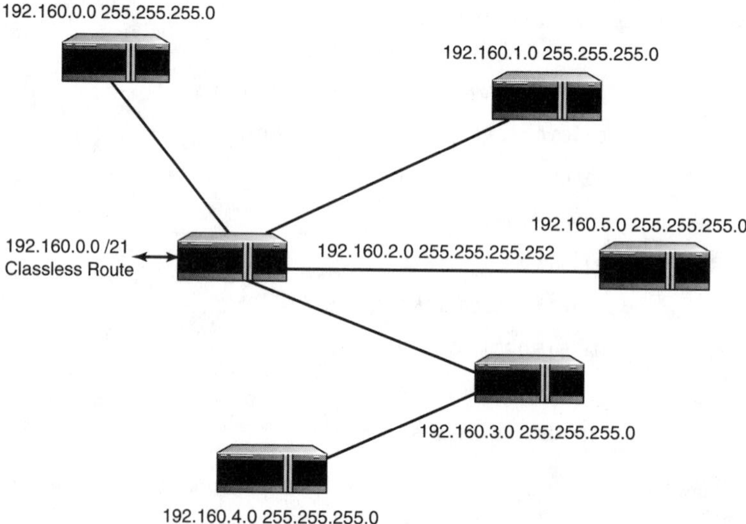

Figure 2.7 CIDR route summarization and classful subnetting.

Table 2.6 **Classless Network Address Variations**

Classless	Subnet Mask	Number of Hosts
/0	0.0.0.0	0
/1	128.0.0.0	2.14B
/2	192.0.0.0	1.07B
/3	224.0.0.0	536M
/4	240.0.0.0	268M
/5	248.0.0.0	134M
/6	252.0.0.0	67M
/7	254.0.0.0	33M
/8	255.0.0.0	16M
/9	255.128.0.0	8M
/10	255.192.0.0	4M
/11	255.224.0.0	2M
/12	255.240.0.0	1M
/13	255.248.0.0	524,288
/14	255.252.0.0	262,144
/15	255.254.0.0	131,072
/16	255.255.0.0	65,536
/17	255.255.128.0	32,678
/18	255.255.192.0	16,384
/19	255.255.224.0	8,192
/20	255.255.240.0	4,096
/21	255.255.248.0	2,048
/22	255.255.252.0	1,024
/23	255.255.254.0	512
/24	255.255.255.0	256
/25	255.255.255.128	128
/26	255.255.255.192	64
/27	255.255.255.224	32
/28	255.255.255.240	16
/29	255.255.255.248	8
/30	255.255.255.252	4
/32	255.255.255.255	1

B = Billion

M = Million

To get a better feel for CIDR, examine how you can subnet a single Class C space. CIDR permits all the variations shown in Table 2.7.

Table 2.7 **Possible Classless Variations of a Class C Address Space**

Classless	Mask	Number of Networks	Number of Usable Hosts
124	255.255.255.0	1	254*
125	255.255.255.128	2	126*
126	255.255.255.192	4	62*
127	255.255.255.224	8	30*
129	255.255.255.248	32	6*
130	255.255.255.252	64	2*

* *Two hosts are lost per network for use as network loopback and broadcast.*

Classless addressing is understood by routers—not by hosts. The hosts use their subnet mask to determine if the destination host is local or not. When you create supernets out of IP address spaces, use classful address subnets as beginning and end points.

VLSM

Under classful addressing, subnetting is permitted with any classful space provided the same mask is used to subnet the entire address space. With Variable-Length Subnet Masks (VLSMs), this limitation is removed.

VLSM permits networks of the same class to be variably subnetted and used on the same network. The important thing to understand about VLSM is that it is a function of routing rather than addressing. VLSM enables routers to announce classless variations of the same classful root address, allowing the address space to be used more efficiently. All the subnetting options available in the classful subnetting tables can be used with VSLM. The practical application of VLSM is discussed later in this chapter. Calculating VLSM addresses follows the same process used to calculate CIDR address spaces. What makes VLSM special is that it is possible to break up a network address space into several different usable networks. Table 2.8 illustrates such a partitioning.

Note

Because CIDR is used for address provisioning by all ISPs, you will find that addressing commonly will be provisioned in blocks of two or multiples of two.

Table 2.8 **A Possible VLSM Subnetting Variation for a Class C Network**

Subnet	Available Addresses
255.255.255.252	x.x.x.5 through x.x.x.6
255.255.255.252	x.x.x.9 through x.x.x.10
255.255.255.252	x.x.x.13 through x.x.x.14
255.255.255.252	x.x.x.17 through x.x.x.18
255.255.255.252	x.x.x.21 through x.x.x.22
255.255.255.252	x.x.x.25 through x.x.x.26
255.255.255.252	x.x.x.29 through x.x.x.30
255.255.255.252	x.x.x.241 through x.x.x.242
255.255.255.252	x.x.x.245 through x.x.x.246
255.255.255.240	x.x.x.33 through x.x.x.46
255.255.255.240	x.x.x.49 through x.x.x.62
255.255.255.240	x.x.x.225 through x.x.x.238
255.255.255.224	x.x.x.193 through x.x.x.222
255.255.255.192	x.x.x.65 through x.x.x.126
255.255.255.192	x.x.x.129 through x.x.x.190
255.255.255.252	x.x.x.249 through x.x.x.250

VLSM is supported by only some routing protocols, so when you decide on a routing protocol, check and see ifVLSM is supported by the protocol you want to use.

The IP Address Space

Table 2.9 shows all the usable address ranges by class, along with the reserved classes and addresses.

Table 2.9 **Usable IPv4 Address Spaces, Expressed Classfully**

Class	Range	Number of Nets	Number of Hosts per Net
A	1.0.0.0 to 126.254.254.254	126	16,000,000
B	128.0.0.0 to 191.254.254.254	16,000	64,000
C	192.0.0.0 to 223.254.254.254	2,000,000	256
D	224.0.0.0 to 239.0.0.0	NA	multicast
E	240.0.0.0 to 255.0.0.0	NA	test
	127.0.0.0 to 127.254.254.254	NA	loopback
	255.255.255.255	NA	broadcast

The *loopback address* (127.0.0.x) is used for IP loopback testing. The host interface sends datagrams from itself (using its assigned IP address) to itself (using the loopback address). The loopback address is also used in some applications as a way to internally access the network interface through IP. The first address of the IP subnet address space—192.160.1.0, or the "zero" address—is also known as the *network loopback* or *network address*. This address cannot be used as a host address.

The *broadcast address* (255.255.255.255) is the address used by IP to send broadcast messages to all stations. The broadcast of all ones or 255.255.255.255 is a broadcast only to the local subnet. Routers do not forward IP broadcasts. The last address of the IP subnet address space, 192.160.1.255, is known as the *network broadcast address*. Like the network loopback address, this address cannot be used as a host address.

The *multicast address* is used for IGMP messages and other multicast type broadcasts.

Public and Unregistered IP Address Space

Because the IP address space is used to address all the hosts on the Internet, and each host must have its own distinct address, networks that are going to connect to the Internet need to use publicly registered IP addresses. These addresses were assigned by the *Internet Assigned Numbers Authority (IANA)* until 1999, when it was replaced by a Commerce Department unit called the *Internet Corporation for Assigned Names and Numbers (ICANN)*. To be on the Internet, you must apply to one of the ICANN-designated entities for address space or get address space from your Internet service provider (ISP). If your network is not on the Internet, you can use any address space you want. However, because you never know whether you will connect in the future, it is best to use the recommended unregistered network address spaces:

Class A: 10.0.0.0 255.0.0.0 or /8

Class B: 172.16.0.0 to 172.31.0.0 255.255.0.0 or /16

Class C: 192.160.0.0 to 192.168.255.0 255.255.255.0 or /24

These spaces are set aside for unregistered network use by IANA/ICANN. If you later decide to connect to the Internet, you can continue to use the unregistered space and use a Network Address Translation (NAT) gateway to map your unregistered host's IP datagrams to registered IP addresses.

Note

Remember, if your IP addressing is not set up correctly, no datagrams will be able to be delivered. When using classful IP addressing, follow these rules:

- Each host interface must have its own unique IP address.
- Each IP subnet must have its own unique IP address space and corresponding subnet mask.
- On each physical network segment, all the hosts must use a common unique IP address space.
- A network address of all ones (network loopback) and all zeros (network broadcast) cannot be used for host addresses (except under certain conditions when classless addressing is used).

IP Datagram Delivery Process

Two processes, routing and forwarding, accomplish IP datagram delivery. The routing process is how the path to the destination host is discovered and route selection accomplished. *Routing* is a Layer 3 activity concerned with the delivery (in most cases) of datagrams between two unique physical identities. *Forwarding* is the activity of placing the IP datagram into physical transport (Layer 2) frames (such as Ethernet, FDDI, Frame Relay) to carry the IP datagram across the physical network segment to the destination host or intermediate router.

The IP datagram delivery process is based on a concept known as *hop-to-hop forwarding*. Because IP is a connectionless protocol, IP datagrams are forwarded on next-hop basis. The essence of this idea is that the forwarding host may not know exactly how to reach the destination, but it does know a host that claims to know the destination address, so it will send the IP datagram on to that host. This forwarding process is performed over and over until the datagram is delivered to an intermediate that can actually forward the IP datagram to its destination.

IP Datagram Delivery Models

When any host/end-station sends an IP datagram, the routing process first decides whether the destination host is on the same local IP network and then determines how the IP datagram will be forwarded. There are two forwarding models used (as shown in Figure 2.8):

- Local Delivery
- Remote Delivery

Local delivery occurs when host-to-host datagram forwarding occurs on the same physical or logical IP subnet. *Remote delivery* occurs when the destination and source hosts are not local to one another and the datagram must be delivered by way of an intermediate router.

Figure 2.8 illustrates this process rather simply. If host 192.168.7.66 wants to deliver a datagram to host 192.168.7.23, the local delivery model is usedbecause the hosts are local to one another. If host 192.168.7.23 wants to deliver a datagram to host 192.168.8.3, the remote delivery model is used.

Note

Routers are also called intermediate systems (ISs) because a router acts as a forwarding intermediary. Layer 2 protocols and functions will be covered in Chapter 4, "LAN Internetworking Technologies," and Chapter 5, "WAN Internetworking Technologies."

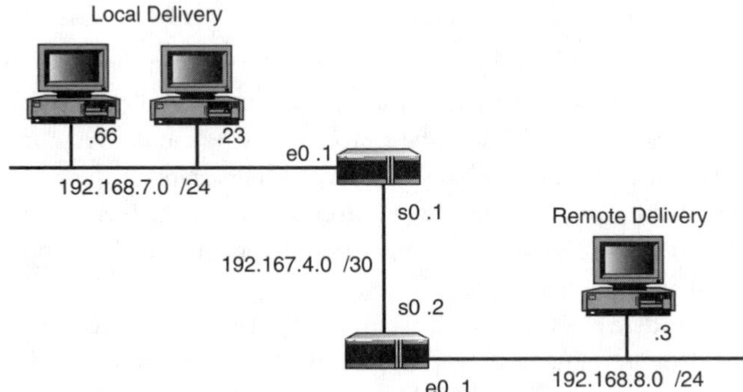

Figure 2.8 Local and remote delivery models.

Here, the destination host is not local to the source, so the datagram is forwarded to an intermediate router. The intermediate router then determines where to forward the IP datagram next. If the host is local to the intermediate router, it forwards the datagram on to the destination host (as in this example). If it's not local, the router forwards the IP datagram on to a router that it believes can deliver the datagram. Routing table lookups make these determinations.

IP Routing and IP Datagram Forwarding

All IP datagram delivery starts with a router table lookup. Every host that is part of a TCP/IP network has a routing table. The *routing table* is created either by static or dynamic means. There are four basic values that make up an IP routing table entry:

- Interface—The interface with which the route is associated
- Network address—The network portion of IP address for the destination network
- Subnet mask—The bitmask used to distinguish the network address from the host address
- Gateway—The next hop to which the interface forwards the IP datagram for delivery

All IP-based hosts create a local routing table at startup. Almost all end-station routing tables are created statically. Every device that uses TCP/IP for host-to-host communication has at least one static route: the default gateway.

Table 2.10 provides the routing table for the network illustrated in Figure 2.9. The routing table has entries for all its connected interfaces and the default gateway, which is defined with a network address of 0.0.0.0 and a network mask of 0.0.0.0.

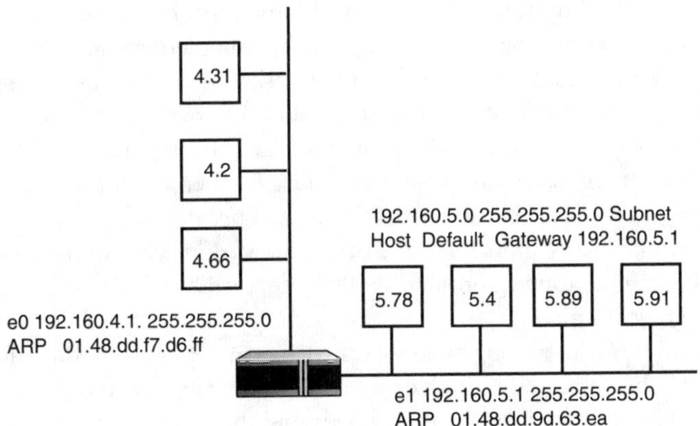

Figure 2.9 An example IP network.

Table 2.10 **A Simple Network**

Interface	Network	Mask	Gateway
Ethernet port 0	192.160.4.0	255.255.255.0	Directly connected
Ethernet port 1	192.160.5.0	255.255.255.0	Directly connected
Serial port 0	12.127.247.108	255.255.255.0	Directly connected
0.0.0.0	0.0.0.0		Via 12.127.247.109

Additional IP network routes can be added to end-station or routers by hand or by configuration files that are read when the end-station boots up. When routes are added by either of these methods, the routing table is considered to be created statically because the routes can only be changed by active intervention. Small IP networks are well-suited for using static routing. In a hop-to-hop delivery model, the router only needs to know where the next hop is. In small IP networks where there is only one gateway between IP subnets, static tables are perfectly adequate to create the routing table needed for IP delivery.

Note

On end-stations, the default gateway is set as part of the IP configuration. On routers, it is set as part of the configuration process. Although it is common to many configurations, the default gateway is not required for IP datagram delivery to work. As long as the source host has a route entry for a destination host, the datagram will be sent. The purpose of the default gateway is to designate an intermediate for all IP datagrams for which the local host does not have a route entry in its local routing table.

Dynamically generated routing tables are usually found on ISs/routers rather than end-stations. Dynamic routing tables are created by using a dynamic routing protocol, and are required when multiple routers and network exit points exist in the network. A *dynamic routing protocol* enables the router to build the routing table on its own by communicating with other routers designated to share routing information with each other. The routers then have conversations with each other about the state of their links, their type and speed, how long they have been up, and so on. Each router can use the information it has about the state of the network to choose the best route path. TCP/IP dynamic routing protocols will be covered in Chapter 8, "TCP/IP Dynamic Routing Protocols."

Regardless of how the routing table is created, it is imperative that the information in the routing table be correct. If a host's or router's routing table has incorrect information, it will not be able to forward datagrams properly and they will not be delivered.

The actual route table lookup is handled by a process known as a *match lookup*. The match lookup compares the datagram's network destination address to those listed in the routing table. This lookup is performed by comparing the binary network address to each network address that is in the routing table. The destination network address is compared, reading left to right, until the longest match is found. Look at the two following examples.

In Example 1, the router has two interfaces: e1:192.168.3.21 /24 and s1:147.28.0.1 /30. The destination address of the datagram is 12.14.180.220. The binary routing table looks like this:

Decimal Network	**Binary Network Address**
	ADDRESS
e1: 192.168.3.0	11000000.10101000.00000011.00000000
s1: 147.28.0.0	10010011.00011100.00000000.00000000

The binary address of the destination address looks like this:

Decimal Network	**Binary Network Address**
	ADDRESS
12.14.180.220	00001100.00001110.10110100.11011100

Because there is no match, the IP datagram is undeliverable. Here, you can add a default route. Now the routing table looks like this:

Decimal Network	**Binary Network Address**
	ADDRESS
e1: 192.168.3.0	11000000.10101000.00000011.00000000
s1: 147.28.0.0	10010011.00011100.00000000.00000000
dg: 0.0.0.0	00000000.00000000.00000000.00000000

Now, if we send a datagram to the same host, the datagram will be forwarded to the default gateway.

In Example 2, the router has four interfaces, each of which is addressed using a /24 subnet of the Class B network 172.100.0.0. The sending host is on one of the interfaces, and the destination host is on another. Which interface should the packet be forwarded to? The destination address is as follows:

10101100.01100100.11111100.11101100

The gateway addresses are

e1: 10101100.01100100.00000001.00000000

e2: 10101100.01100100.00001000.00000000

e3: 10101100.01100100.10100101.00000000

e4: 10101100.01100100.11111100.00000000

dg: 00000000.00000000.00000000.00000000

The answer is interface e4.

The goal of each match lookup is to find an exact or explicit match. The router will search its entire routing table until all entries are examined. If no exact match is found, the packet is sent to the default gateway.

Forwarding

While routing finds the way, *forwarding* does the work. After the routing table lookup is completed and the next hop is determined, the IP datagram is forwarded according to the local or remote delivery model. If the destination network is on the host's locally connected network:

1. The Layer 3 to Layer 2 address mapping lookup is complete.
2. The IP datagram is placed in the Layer 2 packet.
3. The datagram is forwarded directly to the destination host.

If the datagram's destination is remotely connected:

1. The Layer 3 to Layer 2 address mapping lookup is complete.
2. The IP datagram is placed in the Layer 2 packet.
3. The datagram is sent directly to default gateway.

Note

Every time an IP datagram is forwarded, the host/router must perform a route lookup to determine the next hop. This is done so the best route will be used to deliver the IP datagram. Although this might seem like a waste of time on an Ethernet network where each subnet is connected by only one Ethernet interface, it is not a waste of time where there is a choice between sending the datagram over a 56K ISDN dial-up link or a TI dedicated link. Because IP has no knowledge of the Layer 2 transport, it handles all datagrams the same way.

It is important to keep in mind that IP datagram delivery is a Layer 2 and Layer 3 activity. The IP protocol is uninterested in the actual physical layer transport. This gives IP the flexibility to send datagrams across any physical transport medium, provided there is a means to provide physical transport (OSI Layer 2) to network (OSI Layer 3) address mapping. The address resolution protocol is the most common example of this service and is used with most Layer 2 LAN protocols (Ethernet, FDDI, and so on). ARP, as you recall from the earlier discussion, provides the mapping service used to map the Layer 3 addresses to Layer 2 addresses.

Mapping is needed because the Layer 2 address is only valid for the physical transport segment to which the source host (and, in the case of local delivery, the destination host) is attached. In other words, the Layer 2 address associated with an IP datagram changes each time the IP datagram is moved between intermediate points. The Layer 3 addresses, however, are always valid. They never change, except in the case where a proxy host or NAT service is used.

The dependency on both Layer 2 and Layer 3 addressing for IP datagram delivery can sometimes be confusing. It's important to be clear about the IP delivery models and the roles that the Layer 2 and Layer 3 addressing play.

- In instances where local delivery is used, the source and destination hosts' Layer 2 and Layer 3 addresses correspond to one another because the delivery is local.
- Under the remote delivery model, the source and destination Layer 2 addresses changes each time the IP datagram is forwarded through an intermediate router. The Layer 3 source always reflects the originating host, and the Layer 3 destination address is always that of the actual destination host.

Before moving on, here is a review of the IP datagram delivery. In Figure 2.10, a private WAN uses a VLSM subnetted Class B 172.168.0.0 network and two natural Class C networks: 192.168.20.0 and 192.168.21.0. There is also a host on the 172.168.0.0 /24 subnet that sends a datagram to a host on the 192.168.21.0 /24 subnet.

Use this example to review the various processes needed to deliver the datagram:

1. Starting at the sending host—The host performs a route lookup. No match is found, so the network is not local. The host then performs an ARP lookup to find the gateway's Layer 2 address. The IP datagram (source: 172.168.0.14; destination: 192.168.21.23) is placed into an Ethernet frame (source: a3; destination: 25f) and forwarded to the default gateway. This begins the remote delivery process.

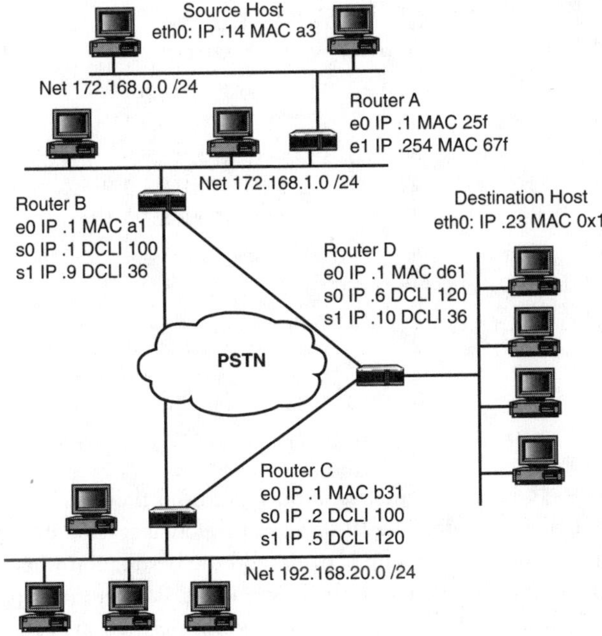

Figure 2.10 Another example network.

2. Router A—Router A receives the packet on interface e0. It strips out the IP datagram, discards the Ethernet frame and performs a route table lookup. The next hop is 172.168.1.1 /24 via interface e1. The datagram is placed into a new Ethernet frame (source: 67f; destination: a1) and forwarded to the next hop. Because all datagram delivery is based on the "next hop," the router does not know or care if the actual destination is the next hop or 10 hops away. It is only concerned with building a table that (to its knowledge) lists forwarding points to the networks that are part of its internetwork. After the datagram is forwarded, it is the next hop's responsibility to deliver the datagram.
3. Router B—Router B receives the packet on interface e0. It extracts the IP datagram and performs route lookup. Table 2.11 shows two gateways that advertise routes to the destination network.

Table 2.11 **Routing Table for Router B**

Network Address	Netmask	Gateway	Interface
172.168.1.0	/24	172.168.0.1	direct connect e0
172.168.3.0	/30	172.168.3.1	direct connect s0
172.168.3.8	/30	172.168.3.9	direct connect s1
172.168.0.0	/24	172.168.1.254	e0
172.168.3.4	/30	172.168.3.2	via s0
172.168.3.10	via s1		
192.168.20.0	/24	172.168.3.2	via s0
172.168.3.10	via s1		
192.168.21.0	/24	172.168.3.2	via s0
172.168.3.10	via s1		
0.0.0.0	/0	172.168.3.2	s0

4. Router D—Router D is the transition point from the remote delivery model to the local delivery model. The IP datagram is now extracted from the Frame Relay frame, the route lookup is made, and the destination is local to the router. Now, the router is interested in the host address. Routers are only interested in a datagram's host address when the destination is on the router's own network. The router must perform an ARP lookup to find out the destination host's Layer 2 address. After the ARP lookup is complete, the datagram is placed in an Ethernet frame (source: d61; destination: 0x1) and forwarded directly to the destination host, where the Layer 4 data is extracted from the Layer 2 and Layer 3 packets.

Note

All things being equal, the router uses both interfaces to deliver traffic to a network. Determining just how equal these two paths are is left to the routing protocol. In this case, the route via the 172.168.3.2 address adds an additional hop, so the route via the 172.168.3.10 address will be used. The IP datagram is placed inside a Frame Relay packet and forwarded to the next hop. This part of the journey provides an example of a router being used as a Layer 2 media gateway.

You should also keep in mind that the Layer 3 IP source and destination addresses have not changed; only the Layer 2 addresses changed as the IP datagram moved from network to network.

This example walks you through the Layer 2/Layer 3 forwarding process. The same process is performed by every router that forwards IP datagrams. Also in this example, you see the use of redundant paths to deliver datagrams. If the link between Router A and Router D went down, the datagram could have been delivered over the Router A to Router C, Router C to Router D link.

The Internet Routing Process

Most people's experience with IP routing started with their involvement with the Internet. The Internet is the common name of the global internetwork of tens of thousands of distinct inter/intranetworks. The Internet was the driving force behind the development of various dynamic IP routing protocols. Although the intricacies of these protocols are covered in Chapter 8, an introduction on the use of dynamic routing protocols is needed in the context of a chapter on TCP/IP.

When large enterprise and regional networks use centralized routing protocols to manage their routers, an *intradomain* routing policy is followed. In instances where multiple routing protocols are being used to cross-pollinate routing information, an *interdomain* routing policy is followed because more than one routing domain is being used to manage the route discovery across the entire network.

The distinctions between intradomain and interdomain make more sense when you examine how large and extremely large IP networks are structured. Intradomain routing is generally used in intranetworks. Most company-wide networks can be considered intranetworks. An *intranetwork* can be a single LAN or a collection of LANs connected by private data links. What defines an intranetwork is its usage. Intranets have an exclusive set of users, and access is restricted only to that user community. Most intranets pursue an intradomain routing policy. The term "intranetwork" is relatively new. The distinction arose as more and more office LANs and company-wide private data networks began installing access points to the global Internet.

Intranetworks use *Regional Internet Service Providers (RISPs)* and/or *National Internet Service Providers (NISPs)* to gain access to the Internet. RISPs and NISPs operate *internetworks*, which are networks made up of multiple distinct networks that operate as separate routing domains. These networks connect to a common internetwork backbone that exists as a routing domain apart from the distinct connected networks. Figure 2.11 illustrates an internetwork connection to the backbone.

> **Note**
>
> Dynamic routing protocols are used by a designated collection of routers (known as a routing domain) to exchange information on the availability and status of the networks to which the individual routers are connected. This information is exchanged so the members of the routing domain can have accurate route entries in their routing tables in order to forward IP datagrams using the most efficient and direct path.

In the Internet world, there are special internetworks known as *autonomous systems (ASs)*. When operating as an AS, the ISP is responsible for maintaining the intradomain routing scheme for all its connected networks. It does this by running two types of routing protocols: *Interior Gateway Protocol (IGP)* and *Exterior Gateway Protocol (EGP)*.

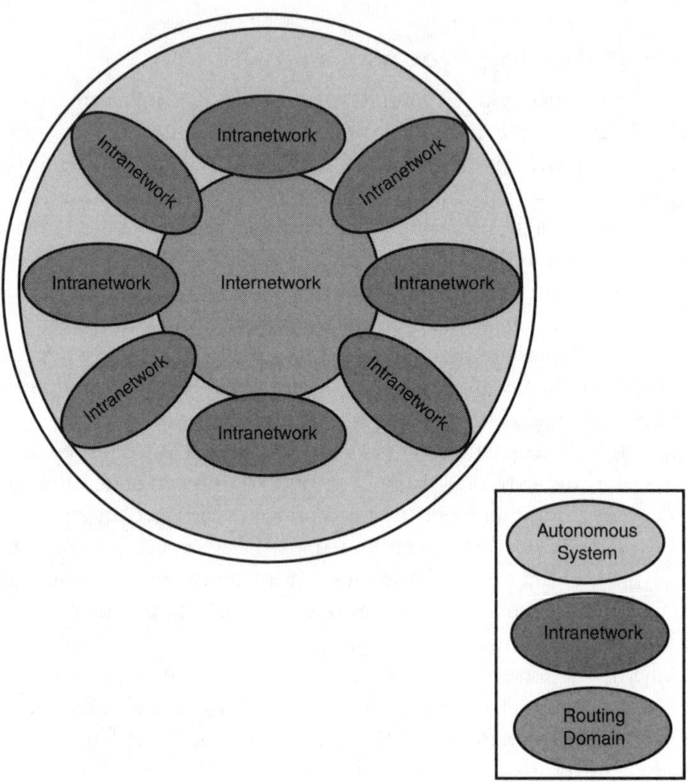

Figure 2.11 The Internet network hierarchy.

Note

IGP is also used to exchange routing information in intranetworks. EGP is used for exchanging routing information between ASs.

The ISP runs an IGP on its network backbone routers to provide its supporting networks with inbound/outbound routing information. The connected intranetworks do not have to exchange routing information with the ISP; they only need to have an interexchange point to which the inbound/outbound datagrams can route. This concept is illustrated in Figure 2.12.

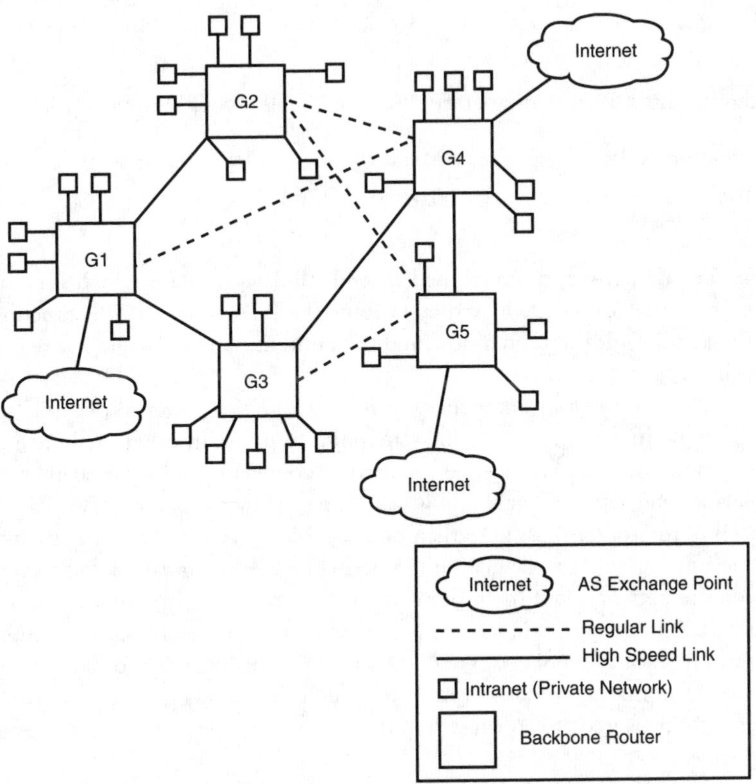

Figure 2.12 A simple regional (ISP) internetwork.

In the regional internetwork example, there are two types of interexchange points. The first is the *intranetwork/backbone router interexchange point*. This is the routing demarcation between the ISP and the intranet. Both the ISP and the intranet run their own IGP processes on the interexchange gateway router. In the case of the ISP, they need to have an entry in their routing table for all the IP address spaces that are used by the intranet.

As the IANA administering party, they are responsible for advertising the routes for all the networks for which they provide Internet connectivity. Using our earlier LAN example, the routing table entry for the ISP would look like this:

Destination Network	Mask	Gateway
192.160.4.0	255.255.255.0	12.127.247.108
192.160.5.0	255.255.255.0	12.127.247.108

The routing entry for the intranet side (same example) would look like this:

Destination Network	Mask	Gateway
0.0.0.0	0.0.0.0	12.127.247.109

As far as the intranetwork is concerned, it sends all datagrams for which it has no local route to the interexchange, where they become the ISP's problem. This functions the same way as a default route on a host, and the entire intranet's datagrams are sent to the backbone router.

The *backbone-to-backbone interexchange point* is where the power of dynamic routing comes through. In most networks, links are point to point. In others, redundancy is important, so multiple network links are used to connect some of the segments. In our example, the internetwork backbone is using a partial mesh topology. With a *partial mesh*, each of the routers has more than one possible route path to any network point. A full mesh is where every device on the network has a redundant path to every other device on the network. Full mesh networks are extremely cost prohibitive; in most cases, a partial mesh approach can adequately meet most networking requirements. The problem from a routing point of view is managing a routing table for a network that is comprised of so many alternative routes. Although diversity is good to ensure that the datagrams get delivered, it can cause havoc with routing. With diversity comes a large routing table.

To get an idea of how large a regional network's table could get, look at the possible backbone paths available for intranets homed on GW1. To get to the intranets homed on the other backbone routers, the paths available to route packets to the Internet exchange points are as follows:

GW1 to GW2

Path	Number of Hops
gw1, gw2	2
gw1, gw3, gw5, gw2	4
gw1, gw3, gw4, gw2	4
gw1, gw4, gw2	3

Path	Number of Hops
gw1, gw3, gw4, gw5, gw2	5
gw1, gw3, gw5, gw2	3
gw1, gw3, gw5, gw4, gw2	5

GW1 to Internet GW	Number of Hops
gw1	1
gw1, gw3, gw5	3
gw1, gw4	2
gw1, gw2, gw4	3
gw1, gw3, gw4	3
gw1, gw2, gw5	3

As you can see, there are 13 possible routes just for routing traffic within the routing domain for gw1. GW1 would also have to build tables to route traffic to each of its connected intranets. Although the intranets can use default routes to forward their traffic onto the backbone, the backbone network cannot. Because the whole ISP internetwork operates as an AS, it needs to advertise routes for all its connected networks. In other words, it needs to pursue an interdomain routing policy.

With an interdomain routing policy in place, the backbone routers run a single IGP routing domain to manage routes between all the backbone routers. An EGP-like BGP is used to exchange routing information between other ASs. The backbone routers also run separate IGP domains for all the intranet exchange points and then distribute the routing information between the intranet routing domains and the backbone domain. This process looks something like what's illustrated in Figure 2.13.

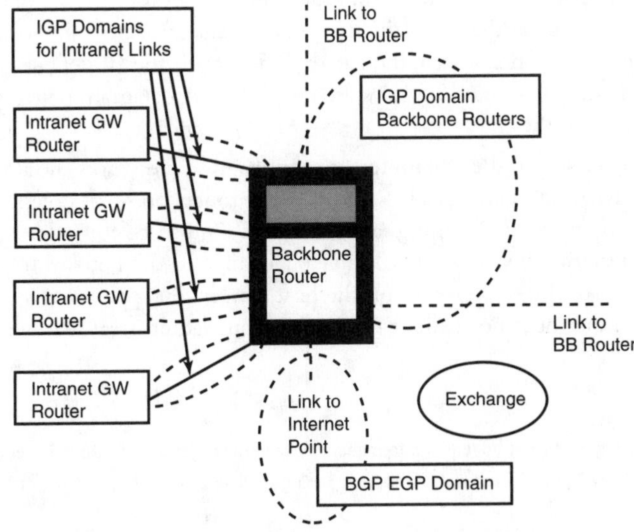

Figure 2.13 The interdomain routing process model.

Under this model, the backbone router builds a route table based on the various routing domains it is connected to. It then distributes the information it has about its connected domains to the other members of the backbone IGP domain. This model is followed by all the other backbone IGP domain members. Each backbone router uses the information it exchanges with its domain partners to create dynamic routing tables to determine the best possible routes for datagrams based on the status of the overall network.

Some Closing Notes on Routing

IP routing configurations for most small local networks do not change very often because they have single traffic exit points, and new devices are not being added to the network all the time. In larger environments where multiple gateways are available between segments, and devices are constantly changing, configurations are dynamic. The difficulty with routing is the initial setup of the network. Deciding how to handle route discovery (static versus dynamic), and determining how to subnet and segment the network into usable and manageable partitions, are decisions that need to be made at the early stages of development. Most networks do not require the operation of both IGPs and EGPs. The variety of IGPs available meet most routing needs and, if they're set up correctly at the start, they require little involvement from a configuration management perspective.

When deciding on a routing strategy, keep some of the IP routing basics in mind:

- Remember how routing works. IP datagram delivery is based on source and destination network address and mask comparison. The comparison is relative to the source address and mask. Two calculations are performed. First, the source and masks are calculated together (using Boolean logic); the subnet masks and the IP address yield the network address. Then, the same operation is performed on the destination IP address using the subnet address of the sending host. If the network address is the same, the datagram is sent out across the local network. If the addresses do not match, the datagram is sent to the IP default gateway. If network address subnetting rules are not followed, datagram delivery does not work. Follow the rules.
- Routing tables on the ISs/routers in multigateway networks should be minimal and as dynamic as possible. Do not set static routes on hosts unless it is absolutely necessary. Routing tables on hosts tend to be forgotten and impose specific routing behaviors. If you make a change to the network that affects the usefulness of the static routes on the host, you might spend hours searching for a problem that does not exist at the gateway, but rather is caused by a bad static route.

Note

Dynamic routing might not be required to manage your network. However, a solid understanding of the "hows" and "whys" prove beneficial if you are planning to have any involvement with the Internet.

- Document your network before you set it up, and after it is up and running, document any changes or special conditions (such as static routes on hosts) that could cause problems if changes to the network occur.
- Keep your subnetting rules straight. With classful addressing, subnetting can be used as long as the same mask is used for the entire address space. Disconnecting classful subnets with other address spaces should be avoided, but it is allowed. You can use VLSM to subnet classful subnets if your router supports it. VLSM only affects routing; hosts do not require any modification to operate on networks that use VLSM.
- Document your IP address subnetting. If you use VSLM to subnet address spaces, be sure to map out the address tables before you use them. A poorly thought-out VLSM strategy can create problems that will cause your network to perform badly at the very least, and not at all at the very worst. Make sure you use a dynamic routing protocol that supports VSLM if you decide to use VLSM to partition your network address space.

ICMP

Internet Control Message Protocol (ICMP) is the IP error reporting messaging service. ICMP is used to send error reports to the datagram sender from the receiver or transit point (router) to indicate some kind of problem. One problem could be that the destination host or service is unavailable or that the network is down. If you have ever tried to Telnet or FTP to a host and were told that the "service was unavailable" or "host not found," you have received an ICMP message.

Because IP is a connectionless transport, the sender is unaware if the destination host and service are even available when it starts to send IP datagrams. Many things can fail along the datagram's transit path, so ICMP is needed to report failures when they arise.

ICMP's most visible manifestation is in the testing of host availability with *Packet Internet Groper Application (PING)*. PING enables IP connected hosts to send IP datagrams of various sizes to destination hosts to test their reachability. Depending on the implementation, PING can tell you simply if the host is just reachable, or it can tell you how many packets were delivered and lost and what the traveling time was for each packet from source to destination. Although all TCP/IP implementations are required to provide support for ICMP, not all implementations provide the tools. ICMP, though a part of the IP protocol (Layer 3), operates as a Layer 4 service. ICMP messages are wrapped and transported as IP datagrams in the same way that TCP and UDP are transported.

Layer 4: TCP and UDP

Transmission Control Protocol (TCP) and *User Datagram Protocol (UDP)* are the transport protocols (OSI Layer 4) for the TCP/IP protocol suite. The *transport layer*, as you might recall, provides the network-independent process connection services between host client/server applications. These processes can be connection-oriented (similar to making a phone call), an immediate point-to-point communication, or connectionless, like sending a letter through the mail. After it is sent, it might or might not arrive within a reasonable time frame, and if it gets lost, you can always send another.

Connection-oriented processes are point-to-point communication streams. These connections are thought of as *virtual circuits (VCs)*. Each VC connection is distinct and used only for host-to-host communication for a single process. In other words, a host with multiple TCP communication sessions would have a separate VC for each process. Connection-oriented transports are used for interapplication communication processes that have the following requirements:

- Dedicated connection setup and maintenance
- Delivery acknowledgment (by the destination host)
- Guaranteed byte-stream sequencing

Connectionless transport processes use standalone message units or datagrams to transport interapplication data. Connectionless transport is faster than connection-oriented transport because it does not have the flow control and data integrity mechanisms. Connectionless transport is used when the application is looking for fast data transport and needs to control the transmission process, such as with remote file and print services and fast query response transactions. The hallmarks of connectionless transport are as follows:

- Response-based, event-oriented messages
- No guaranteed delivery
- No data sequencing

The transport layer as a whole represents the gateway of the host level communication process. The transport bridges the internal data communication process and the network data communication process. This is evident in the fact that the process of encapsulating the application layer data into and out of wrappers occurs at this level. Figure 2.14 shows the relationship between the layers and the wrapping of the data.

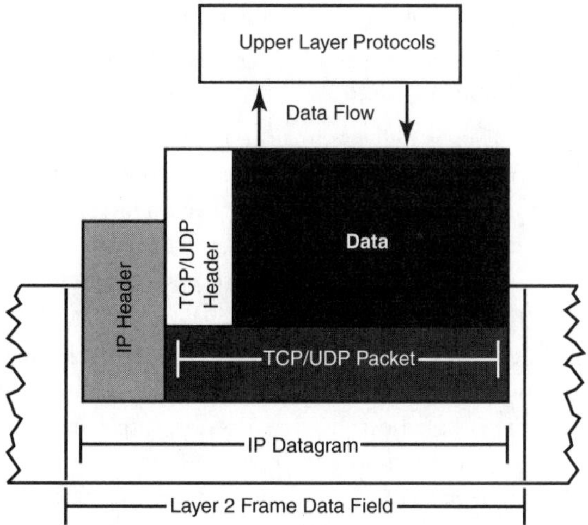

Figure 2.14 ULP data handling.

This wrapping of ULP data is an outbound/inbound process. The ULP data follows this passthrough process on leaving the source host and then follows the reverse on its arrival at the destination host:

1. Data stream is handed to the transport layer.
2. ULP data is wrapped in the transport packet, which consists of a header and the data.
3. The transport packet is handed to the network layer.
4. At the network layer, the transport packet is wrapped in the network packet, which consists of the network header and the transport packet.
5. The network packet is handed to the data link layer.
6. At the data link layer, the network packet is wrapped in a header and footer that are appropriate to the data-link protocol.

After the destination host receives the network packet, the process is reversed and the data is handed back to the ULP. Remember that the transport protocol is independent from the ULP and the LLP (Lower Layer Protocols). Its job is only to provide data transport. The actual delivery is in part a joint effort between the ULP and the LLP.

TCP

TCP provides connection-oriented data transport over dedicated process owned by a VC. It works over almost any type of network and is designed to provide reliable data transport between nodes over unreliable network delivery services. For TCP, *reliable transport* means in-sequence, adjustable-rate, and error-free transmission of ULP data.

TCP provides the following services to facilitate these transport goals:

- Full duplex—TCP connections are bidirectional. Continuous data flow is provided to communicating with the ULPs.
- Data sequencing—Data is delivered to the ULP in the byte order in which it was sent. TCP provides for transport level and below data fragmentation. TCP assigns sequence numbers for each ULP data segment. These sequence numbers are used at the peer end of the TCP session to reassemble the ULP data back into its original form. If a sequence is lost, the peer requests it to be resent, and the data is not sent to the ULP until all the data is reassembled.
- Flow control—TCP uses a sliding window strategy to transmit data. TCP establishes the speed of the entire path at the start of the session. It then creates a window whose size is based on the speed of the path between the source and destination. TCP uses the window to frame the amount of data it will send at one time and uses the sequencing numbers to keep track of each byte it transmits. TCP sends a burst of data and waits for confirmation from the peer that all the data has been received. If no response is received, TCP will retransmit the data until it gets an acknowledgment that the peer has received all the data. TCP uses *timers* (to measure between when the data was sent and the acknowledgment was received) to adjust the size of the window to reflect changes in the path speed. TCP will always send data at the rate of the slowest link.
- Error correction—TCP uses checksums to ensure that the data is error free.
- Timely service notification—If conditions prevent the maintenance of the required service, TCP will notify the ULP of the condition and follow the ULP response.

TCP Connection Setup

TCP uses the abstraction of *ports* (also referred to as sockets) to describe its transport scheme. These ports are used by the hosts to establish TCP VCs over which they can exchange ULP data. After the data exchange session is complete, the TCP VC is torn down, and the ports are free to be used to establish a TCP VC with another host.

There are no rules for which ports are used to establish connections, but there are a set of "known" reserved ports that are used for services providing known application layer communication services. All ports under 1024 are reserved for server use. For example, the SMTP listens on TCP port 25, the Telnet service listens on port 23, and HTTP, the protocol used for sending WWW data, listens on port 80. TCP establishes these client/server interapplication connections using a three-way handshake connection scheme. This process allows both sides to synchronize and establish the process endpoints. Figure 2.15 illustrates this connection process.

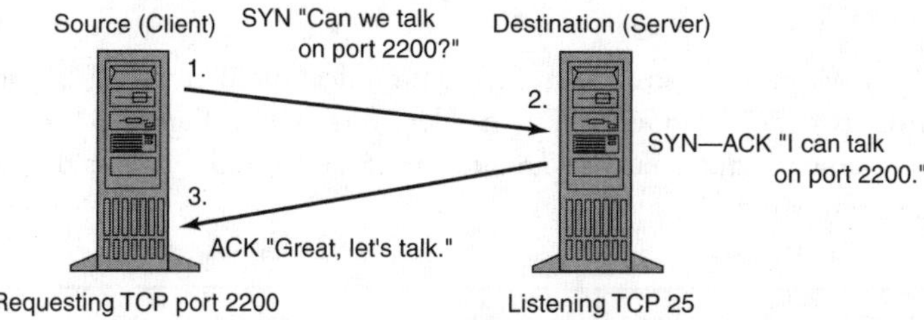

Figure 2.15 The TCP connection process.

To set up a TCP connection, the client host sends a SYN (synchronization) packet to the application service port. The server host then sends a SYN and an ACK (acknowledgment) to the client's originating TCP port confirming that the connection is established. The client then sends an ACK back to the server. Now the dedicated VC is established and full duplex data exchange can take place.

TCP ports can support many simultaneous connections or processes. TCP keeps these processes organized by using the process endpoints to track the connections. The endpoint address is the process port number plus the IP address of the host that started the connection. In this example, the process endpoint on the server would be 192.160.33.20.2200 and the process endpoint on the client would be 90.16.44.8.25. The most common TCP service ports are listed in Table 2.12.

Table 2.12 **Common TCP Service Ports**

Port Number	Service
1	TCPMUX
21	FTP
20	FTP-DATA
22	SSH (Secure Shell)
23	Telnet
25	SMTP
53	DNS (Domain Name Service)
80	HTTP (WWW)
139	WINS
119	NNTP (Network News Transport Protocol)
110	POP3 (Post Office Protocol)
543	Klogin (Kerberos login)
544	Kshell (Kerberos shell)
751	Kpasswd (Kerberos password)
750	Kerberos server
512	Berkeley rcommands
513	login
443	HTTPS secure WWW server
2105	eklogin (encrypted Kerberos login)
2049	NFS (Network File System)

The TCP Header

The *TCP header* provides the means for communication about the TCP process between the two TCP endpoints. The header (as shown in Figure 2.16) provides the data sequencing and acknowledgment information, and serves as the facility for connection setup and disconnection.

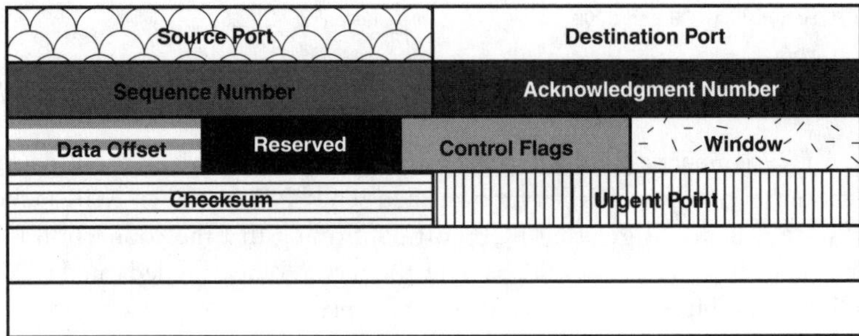

Figure 2.16 The TCP packet header.

The TCP packet can be up to 65KB in size. Because TCP is network independent, packet size limitation is not a priority. IP fragments the TCP packets into sizes it can transport efficiently. The TCP header is 40 bytes in size, which is rather large for a network packet header. The header is so large because it needs to carry all the data control information.

The header dedicates memory segments or fields to provide the information needed to deliver and reassemble the ULP data into an unaltered state. These fields are as follows:

- Source Port—This is the TCP port the process is coming from.
- Destination Port—This is the TCP port the process is sending data to.
- Sequence Number—This is the sequence number for data contained in the packet.
- Acknowledgment Number—This field contains the sequence number the sender expects to receive from the destination.
- Data Offset—This field indicates how large the TCP header is.
- Control flags indicate the status of the TCP connection:
    - SYN sets up TCP connections.
    - ACK indicates if the information in the Acknowledgment field is relevant.
    - RST resets TCP connections.
    - PSH tells the destination that the DATA should be delivered to the ULP upon delivery.
    - FIN ends the TCP connection.
- Window—This field is used to provide flow control information. The value is the amount of data the sender can accept.

UDP

The connectionless transport service of the TCP/IP protocol suite is *user datagram protocol (UDP)*. UDP uses the same port process abstraction that TCP uses. Its function is to provide low-overhead, fast-transport service for ULP data transport. UDP has none of the flow control and data synchronization mechanisms that TCP offers. It processes one packet at a time and offers best-effort delivery service.

The UDP header (see Figure 2.17) is 32 bits long and contains information on the source and destination ports, the packet size, and an optional checksum for applications that require checksum support (BOOTP and DHCP).

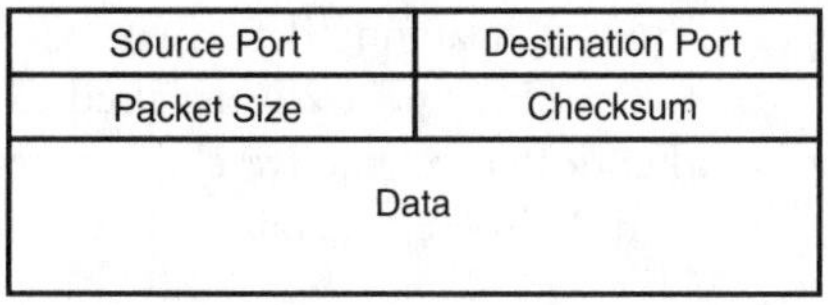

Source Port	Destination Port
Packet Size	Checksum
Data	

Figure 2.17 The UDP header.

Like TCP, UDP has a set of reserved ports used for different application server data exchange points. The most commonly used UDP ports are shown in Table 2.13.

Table 2.13 **Commonly Used UDP Ports**

Port Number	Service
49	TACACS authentication server
53	DNS (Domain Name Service)
67	BOOTP server
68	BOOTP client
69	TFTP
137	NetBIOS name service
138	NetBIOS datagram service
123	NTP (Network Time Protocol)
161	SNMP (Simple Network Management Protocol)
1645	RADIUS authentication server
1646	RADIUS accounting server
2049	NFS (Network File System)

These ports are used by application layer servers to listen for UDP datagrams from client applications on network hosts. Although the port abstraction is similar to that of TCP, there is no VC "continuous datastream" connection with UDP. In the case of client/server applications, the server accepts datagrams on the known UDP port on a packet-by-packet basis. Server replies to client queries are returned using the same packet-by-packet, best-effort delivery process.

The Application Layer Protocols of the TCP/IP Suite

Now that you have an understanding of how the network and transport layers of the TCP/IP suite function, take a look at the application layer protocols and services that make the network useful.

The following sections outline the major application protocols.

FTP

File Transfer Protocol (FTP) is used to exchange data between two undefined hosts. FTP is uninterested in the types of hosts exchanging data. Two TCP connections, ftp-control (TCP port 21) and ftp-data (TCP port 20), are used to perform the data exchange. FTP uses a host-authenticated interactive client/server model. The client connects to the known ports, and the server establishes the session(s) on the client's requesting port address.

TFTP

Trivial File Transfer Protocol (TFTP) is a simple data exchange program. It is well suited for applications that require basic file transfer service. TFTP uses a simple unauthenticated client/server model. Because it is designed to operate under the simplest conditions, it uses UDP. The server listens on UDP port 69. TFTP is primarily used for bootstrap applications such as booting diskless workstations and networking equipment. TFTP is also used for uploading and downloading configuration and OS files to networking equipment that use flash and NVRAM memory for storage.

Telnet

Telnet is used to provide a dedicated host-to-host communication session. It uses a connection-oriented 8-bit byte session to exchange character data. Its primary function is to provide remote terminal service. To maintain host/terminal independence, the Telnet protocol is assisted by *network virtual terminal service (NVT)*, as shown in Figure 2.18.

An NVT is a pseudo-device that provides a standard terminal interface used by both client and server. The NVT acts as a translation bridge between the client and server by translating the local terminal commands into commands that work on the remote system's terminal. Like FTP, it uses a host-authenticated interactive client/server model. The server operates on TCP port 23. Like FTP, the client requests the service at the known port and the server responds to the client's requesting port.

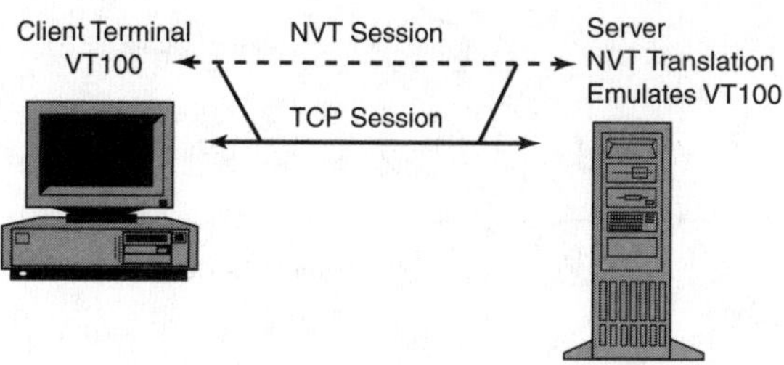

Figure 2.18 NVT over TCP.

SNMP

Simple Network Management Protocol (SNMP) is the facility for making queries about the status of a network device. (Using SNMP is anything but simple. In fact, it is complex and is dealt with in detail in Chapter 11, "Network Troubleshooting, Performance Tuning, and Management Fundamentals.") The server operates on UDP port 161.

SMTP

Simple Mail Transfer Protocol (SMTP) is the process for exchanging electronic mail between hosts. SMTP defines how messages are passed from host to host. It is generally a noninteractive, nonauthenticating client/server event and operates on TCP port 25. SMTP is only involved with the activity of passing electronic mail, specifically the address translation and formatting. How the mail is composed, sent, and stored is managed by other protocols. The most common associated mail protocol is Post Office Protocol (POP).

DNS

Domain Name Service (DNS) is one of the most complex application layer protocols of the TCP/IP suite. DNS provides the function of translating the IP addresses of a host connected to the Internet into meaningful human understandable names. DNS uses a tree-like naming structure to provide a naming hierarchy. The upper level deals with the universal hierarchy, such as root:country:state:organization (see Figure 2.19).

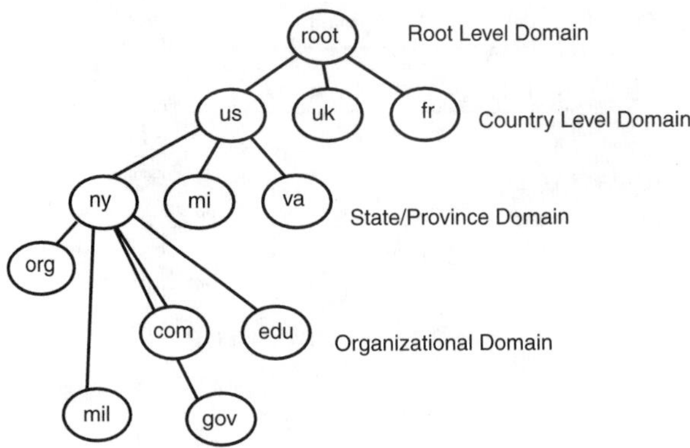

Figure 2.19 The universal domain tree.

At the organizational level, the domains are segmented into locally administered domains and subdomains. These subdomains provide name translation for themselves and the branches below them (see Figure 2.20).

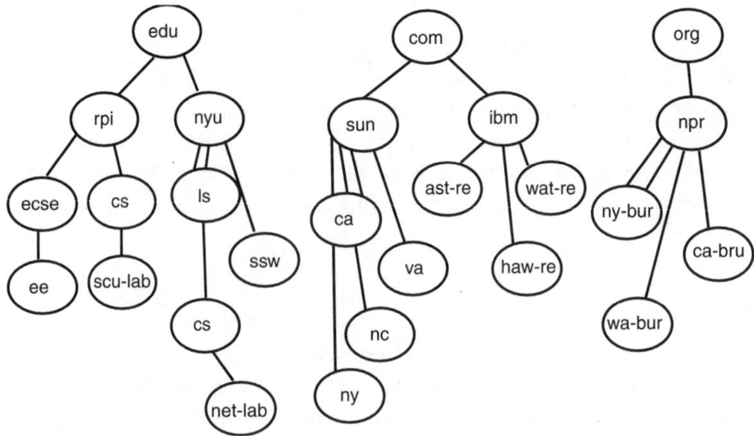

Figure 2.20 A view below the organizational branch.

When a domain name request is made for a host outside of the locally administered domain, the request is relayed to a caching DNS server or to the server that is responsible for providing hostnames for the requested domain. The DNS service uses both TCP and UDP for service delivery. TCP port 53 is used for server-to-server data transfers. The data consists of the forward (hostname to IP address) and reverse (IP address to hostname) translation files. Name requests are performed using UDP.

RARP, BOOTP, and DHCP

In order for a host to exchange data with other hosts, each host must have its own distinct IP address. *Reverse Address Resolution Protocol (RARP)* is the simplest way to provide IP addresses to hosts. The server listens to broadcasts sent across the Ethernet segment. If it receives a broadcast from a host with an Ethernet address that it knows, it responds to the broadcast with the IP address of the requesting host. This is great if all that's needed is an IP address, and if the host and server are on the same Ethernet segment. RARP's functional shortcomings leave it little place in today's networks.

BOOTP, on the other hand, picks up where RARP leaves off. BOOTP uses IP to provide IP configuration and boot service, hence the name BOOTP, for *bootstrapping protocol*. The requesting host sends a BOOTP request, which is either answered by a BOOTP server that has a configuration for the requesting station, or the request is relayed to a BOOTP server on another Ethernet segment that contains the host configuration information.

The BOOTP server runs as a UDP service on port 68. Although there is no interactive authentication, hosts must be configured in the BOOTP server database and have corresponding BOOTP images in the TFTP server for the BOOTP service to work. After the configuration information has been transferred to the host, the BOOTP server then directs the host to the TFTP server where the actual boot file resides.

This boot file is then downloaded through TFTP and used to boot the host. BOOTP provides more services than RARP, so the two cannot be compared directly. BOOTP's capability to provide all the network configuration information (default gateway, DNS servers, and so on), its capability to boot the host from a remote server, and its capability to have service requests relayed across separate Ethernet cable segments make BOOTP more versatile than RARP. In terms of providing IP addresses to hosts, BOOTP is a better choice because it uses IP and does not have the direct hardware access requirement RARP has. BOOTP does not work well in environments that are constantly changing (in other words, lots of laptops) and where configuration information needs to be changed on a constant basis.

Dynamic Host Configuration Protocol (DHCP) provides IP host configuration services dynamically with no host configuration requirements. To use DHCP, the host's IP implementation must support configuration by DHCP. As with BOOTP, the host makes a DHCP broadcast request and the DHCP server replies with configuration information. DHCP is different from BOOTP in that the configuration information provided to the host is only temporary. DHCP provides addresses to hosts in the form of *leases*. These leases have limited life spans that are set by the server administrator. This way, IP addresses are recycled as systems come on and leave the network. The DHCP server operates on UDP port 68 and, like BOOTP, operates on an unauthenticated client/server process.

Note

BOOTP needs to be supported by the host device. The IP address can be entered in the host's boot flash, either obtained through RARP or by using an IP broadcast (255.255.255.255) and the Ethernet address (like RARP) to map the address.

RFCs

Requests for Comments (RFCs) are the building blocks of the Internet. All the protocols, applications, and standard operating procedures (SOPs) start as RFCs. These comment documents are then taken by various standards bodies and individuals and reviewed, implemented, and eventually published as standards. The RFC process is the cornerstone of the open systems approach. The following is a list of RFCs relevant to the topics covered in this chapter. A good online resource for RFCs is `http://www.it.kth.se/docs/rfc/`.

RFC 7	Internet Protocol
RFC 768	User Datagram Protocol
RFC 791	Internet Protocol
RFC 792	Internet Control Message Protocol
RFC 793	Transmission Control Protocol
RFC 821	Simple Mail Transfer Protocol
RFC 826	Subnet Access Protocol ARP for 802.x networks
RFC 903	Reverse Address Resolution Protocol (RARP)
RFC 917	Internet subnets
RFC 919	Broadcasting Internet datagrams
RFC 919	Internet Protocol
RFC 922	Broadcasting Internet datagrams in the presence of subnets
RFC 922	Internet Protocol
RFC 932	Subnetwork addressing scheme
RFC 947	Multinetwork broadcasting within the Internet
RFC 950	Internet Protocol
RFC 950	Internet standard subnetting procedure
RFC 974	Mail routing and the domain system
RFC 1034	Domain names—concepts and facilities
RFC 1035	Domain names—implementation and specification
RFC 1093	NSFNET routing architecture
RFC 1157	Simple Network Management Protocol
RFC 1219	On the assignment of subnet numbers
RFC 1375	Suggestion for New Classes of IP Addresses
RFC 1467	Status of CIDR Deployment in the Internet
RFC 1541	Dynamic Host Configuration Protocol

RFC 1550	IP: Next Generation (IPng) White Paper
RFC 1700	Assigned TCP/IP port numbers
RFC 1878	Variable Length Subnet Table for IPv4
RFC 1881	IPv6 Address Allocation Management
RFC 1883	Internet Protocol, Version 6 (IPv6)
RFC 1884	IP Version 6 Addressing Architecture
RFC 1918	Address Allocation for Private Internets
RFC 1958	Architectural Principles of the Internet
RFC 1985	Nonroutable IP addresses

Additional Resources

Comer, Douglas E. *Internetworking with TCP/IP, Volume I: Principles, Protocols and Architecture.* Third Edition. Upper Saddle River, NJ: Prentice Hall, 1995.

Comer, Douglas E., and David L. Stevens. *Internetworking with TCP/IP, Volume II: Design, Implementation, and Internals.* Second Edition. Upper Saddle River, NJ: Prentice Hall, 1994.

Stevens, W. Richard. *TCP/IP Illustrated, Volume I: The Protocols.* Reading, MA: Addison-Wesley, 1994.

Thomas, Stephen A. *IPng and the TCP/IP Protocols: Implementing the Next Generation Internet.* New York: John Wiley and Sons, 1996.

3 The Networker's Guide to AppleTalk, IPX, and NetBIOS

UNTIL THE EARLY 1990S, TCP/IP WAS REALLY ONLY PREVALENT in large government and research facilities where UNIX and other supercomputing operating systems used it as a common network communications protocol. When PCs came into the picture, they were not networked. Rather, they were used either as front-ends to big micro or mainframe systems (IBM was a big fan of this approach) or as standalone systems. In the early 1980s, as PCs grew in number and in performance, three strategies emerged to provide PCs with networking services: AppleTalk, Novell NetWare, and IBM's NetBIOS.

The goal of this chapter is to give you an understanding of the various protocols that make up the protocol suites and the roles they perform. It is not intended to explain how to design, set up, and manage a network. Chapter 7, "Introduction to Cisco Routers," and Chapter 10, "Configuring IP Routing Protocols on Cisco Routers," discuss configuration issues for these protocols. Because NetBIOS is a session layer protocol rather than a protocol suite, it will be described in the context of its operational behaviors at the end of this chapter.

AppleTalk

AppleTalk was an outgrowth of the Apple Macintosh computing platform. First introduced in 1984 and updated in 1989, it was designed to provide the Macintosh with a cohesive distributed client/server networking environment. AppleTalk, like the Macintosh, is a "user friendly" network operating system (NOS). All the actual communication operations are masked from the user. To facilitate this, AppleTalk incorporates a dual network identity structure, both operationally and contextually. The operational structure uses binary network addressing to represent network segments, end-nodes, and network services (such as file transfer, printing, and so on). The contextual structure uses names and logical groupings, called *zones*, as the user interface for addressing network visible entities (NVEs). The contextual structure functions like a mask, covering the actual physical network infrastructure. This provides the network administrator with the ability to group users and network services into logical groupings instead of physically segmenting the network to achieve the same effect. Figure 3.1 illustrates this concept.

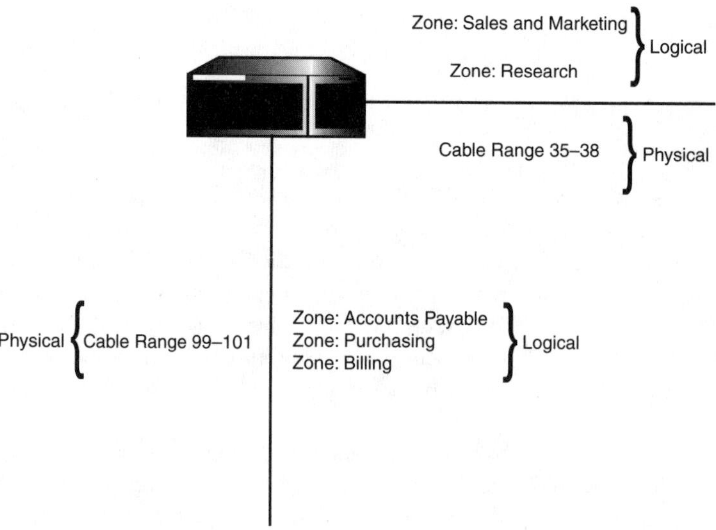

Figure 3.1 *Physical versus logical AppleTalk network segmentation.*

All AppleTalk network service interactions are based on a client/server model. *Clients* are the end-nodes requesting the service; *servers* are the end-nodes providing the service. The protocols that provide the services for the operational identity structure are provided on OSI-RM Layers 1, 2, 3, and 4. Contextual services are provided on Layers 4 and 5.

AppleTalk Phase 1 and Phase 2

There are two flavors of AppleTalk network: AppleTalk Phase 1 and AppleTalk Phase 2. The Phase 1 network approach is oriented toward interconnecting workgroups. Phase 1 supports a limited network diameter of a single network segment containing no more than 127 clients and 127 servers. Phase 1 networks use a single network number (0) for the entire physical network.

Phase 2 networks support multiple logical networks over the same cable segment. Each logical network supports up to 253 clients or servers. To maintain compatibility with Phase 1 networks and provide support for multinetwork cable segments, Phase 2 supports two different network configurations: nonextended and extended. With Phase 2, an AppleTalk logical network is defined by its cable range. The *cable range* is the network number or numbers used by the end-nodes connected to the transmission media. Each AppleTalk cable range supports 253 hosts. The size of the cable range determines the number of hosts that can be connected on the media simultaneously.

The cable range is a number range or contiguous sequence of numbers from 1 to 64,000, expressed in a start-end format. The size of the cable range determines if the network is a nonextended or extended type. A nonextended Phase 2 network uses a single cable range (to maintain compatibility with the Phase 1 network structure) and can support 253 connected users; 60001-60001 is an example of a nonextended Phase 2 network. The start and end range are the same number. An example of an extended cable range would be 60001-60011. With this range, 253 hosts can be supported on each range, so theoretically, 2,530 end-stations could be connected to this media segment. As you can see, the main advantage of extended over nonextended is the amount of hosts that can be supported over a single cable segment.

There are some compatibility issues between Phase 1 and Phase 2 networks, so it is best to use Phase 2, if possible. The major incompatibilities are with Phase 1 and Phase 2 EtherTalk (AppleTalk's Ethernet implementation), and with using Phase 1 and Phase 2 extended networks together. EtherTalk Phase 1 and Phase 2 use different frame formats and are not compatible.

It is possible to run Phase 1 and Phase 2 over the same Ethernet cable, but they cannot exchange data with each other without a router. Phase 1 networks and Phase 2 extended networks also cannot interoperate because Phase 1 cannot understand extended cable ranges. If you need to use Phase 1 and Phase 2 together, use nonextended Phase 2 networks.

AppleTalk operates over all the IEEE and ANSI Layer 2 protocols and WAN (both point-to-point and dial-on-demand configurations) transports. Apple Computer has also defined its own transport media specification known as LocalTalk. *LocalTalk* is a proprietary network architecture, left open to development by any vendor, as long as interpretability and standards compliance is assured.

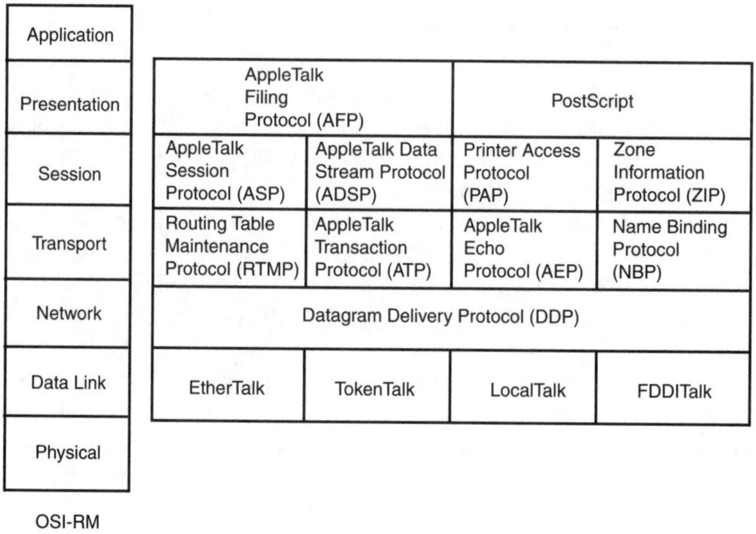

OSI-RM				
Application				
Presentation	AppleTalk Filing Protocol (AFP)		PostScript	
Session	AppleTalk Session Protocol (ASP)	AppleTalk Data Stream Protocol (ADSP)	Printer Access Protocol (PAP)	Zone Information Protocol (ZIP)
Transport	Routing Table Maintenance Protocol (RTMP)	AppleTalk Transaction Protocol (ATP)	AppleTalk Echo Protocol (AEP)	Name Binding Protocol (NBP)
Network	Datagram Delivery Protocol (DDP)			
Data Link	EtherTalk	TokenTalk	LocalTalk	FDDITalk
Physical				

Figure 3.2 The AppleTalk protocol suite compared with the OSI-RM.

AppleTalk Layers 1 (Physical) and Layer 2 (Data Link)

AppleTalk supports four LAN media access implementations: LocalTalk, EtherTalk, TokenTalk, and FDDITalk. These implementations are supported over most WAN point-to-point access protocols. AppleTalk also uses *AppleTalk Address Resolution Protocol (AARP)* to manage Layer 3 AppleTalk network address to network hardware controller address translation.

AppleTalk Node Addressing

All AppleTalk clients and servers require a unique AppleTalk address to participate on the network (see Figure 3.3). The network address is 24-bits long and consists of a 16-bit network address and an 8-bit node address. Unlike most network protocols, however, AppleTalk does not require that nodes have a preconfigured address. Instead, the node acquires an address when it first accesses the network.

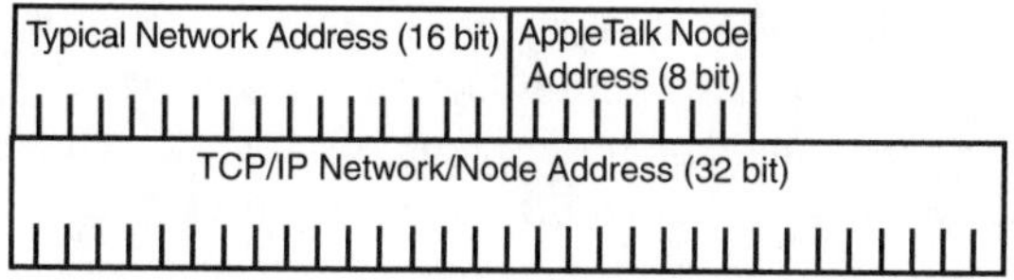

Figure 3.3 The AppleTalk address structure.

The network and node addresses are assigned dynamically when the node joins the network.

When a node first joins the network, it acquires a provisional address. The node address portion is randomly selected from the number range 1 to 254 (0 and 255 are reserved). The network portion is assigned by the Layer 2 protocol (ELAP, LLAP, TTAP, or FDAP) from a reserved class of network addresses spanning from 65,280 (FFF0 hexidecimal) to 65,534 (FFFE hexidecimal). These network addresses are recognized by all AppleTalk nodes as provisional addresses. After the provisional address is acquired, a permanent address is needed.

The node then sends a `GetNetInfo` request using the Zone Information Protocol (ZIP). If a router responds, the available network number(s) is returned with the ZIP reply. The node then uses that network number, generates another random node ID, and broadcasts this number across the segment. If no nodes respond claiming rights to that address, the node uses the address. If a node responds claiming ownership to that ID, the node must repeat the node generation and validation process until it selects a node address that is not in use.

If no router is available or the network type is Phase 1, the network address is set to 0 and the node sends a broadcast to see if its address conflicts with another node. If it doesn't, it becomes the node's permanent address. After a node address is acquired, it is stored by the end-station for later use. If the node leaves the network and returns at a later point, it attempts to validate its previous address. If there are no conflicts, it continues to use the address.

AppleTalk Address Resolution Protocol

AppleTalk, like IP, is unable to understand Layer 2 hardware addresses. *AppleTalk Address Resolution Protocol (AARP)* is used to resolve AppleTalk to Layer 2 Media Access Control (MAC) addresses. AARP has two roles, the primary one being to build an address mapping table (AMT) which contains the AppleTalk to Layer 2 hardware addresses translations. Each node builds its own AARP AMT. Each time a node resolves a network-to-hardware address, it is entered into the AMT with an associated timer. After a period of time, the entry expires and another AARP request is needed to validate the entry. AARP requests are made using broadcast packets, in much the same way as an IP ARP request is made.

It is also possible to update the AMT by reading the hardware and network addresses on incoming data packets. This is known as *address gleaning*. This process has an associated packet processing cost, however, so it is not widely used in end-stations, but rather on routers where it is incorporated into the packet handling process. Address gleaning is helpful in terms of network performance because it reduces AARP requests.

AARP packets use the packet header type appropriate to the link making the request, such as ELAP, TLAP, and so on. There are three types of AARP messages:

- Request
- Response
- Probe

Request and probe messages are sent as hardware level address broadcasts, and are processed by every node on the network segment. AARP response messages are sent as unicast messages to the originator of the probe or request message. Figure 3.4 shows the AARP message formats for each of the AARP message types.

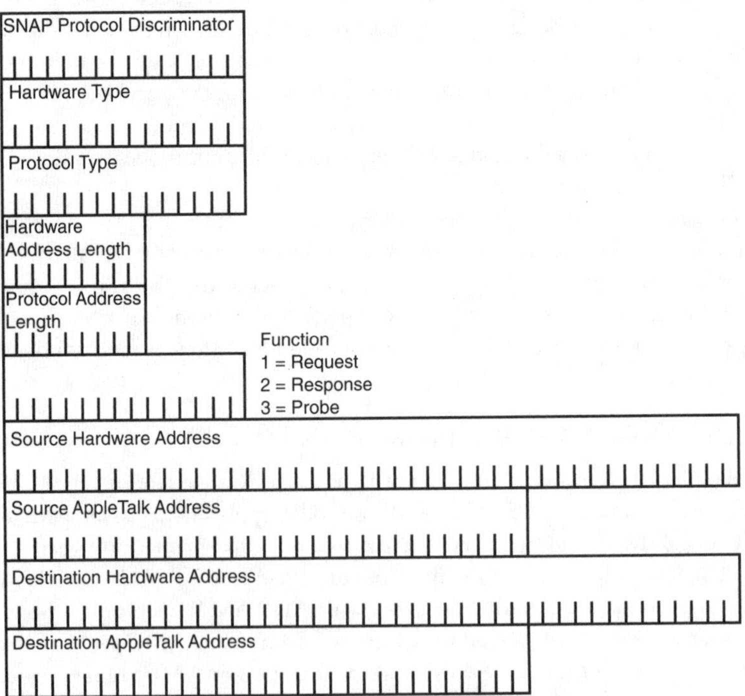

Figure 3.4 AARP message formats.

The second role of AARPs is to assist in the address acquisition process. When the node asks Layer 2 for a network address, a random provisional address is chosen and then checked against entries in the AMT. If the address is not in its AMT, the host AARPs for it. If the address is not in use, the provisional address is used and the address acquisition process continues.

LocalTalk

LocalTalk is the Apple Computer plug-and-play network cabling infrastructure. In its original form, LocalTalk operated over shielded twisted-pair cable using a bus topology. The protocol was later implemented over two-pair voice-grade twisted-pair, with a star topology known as *PhoneNet* developed by the Farallon Corporation. LocalTalk operates at 230Kbps using Carrier Sense Multiple Access with Collision Avoidance (CSMA/CA) as an access control method. With CSMA/CA, the network must be clear for an interdialog gap (IDG) of 400 microseconds, in addition to a random wait, based on the current network traffic level and collision history, before a node starts its data transmission. The transmissions themselves use handshaking between nodes to establish the connection and effectively own the transmission medium until the exchange is completed. Each transmitted packet can have no more than a gap of 200 microseconds between frames. See Figure 3.5 for a LocalTalk message frame format.

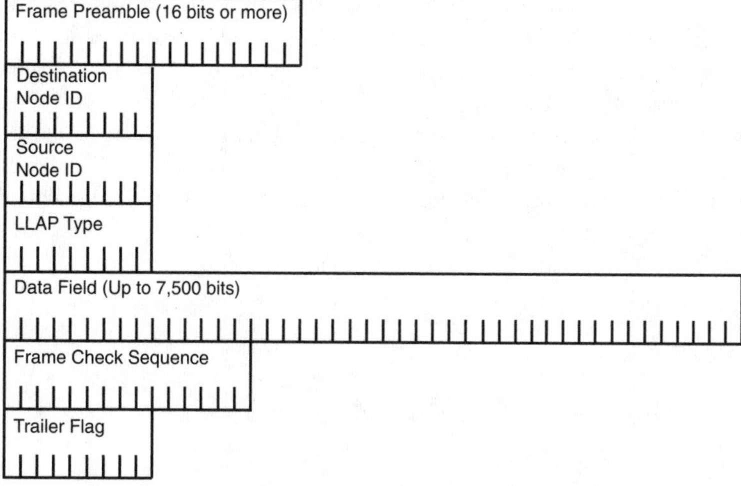

Figure 3.5 LocalTalk message frame format.

Layer 2 transport is handled by the LocalTalk Link Access Protocol (LLAP). LLAP provides "best effort" transport service and uses the node and network numbers for source and destination addressing so no hardware-to-network address resolution service is required. Addressing is assigned dynamically. When a host joins a network, it generates a random network number that it broadcasts to the network for validation. If the address is not in use, the node uses it. If the address is in use, the node generates another address and validates again until it finds an address it can use. LocalTalk's bus implementation network diameter is limited to a 300-meter total cable distance with no more than 32 active nodes. PhoneNet supports longer span distances, but is still limited to 32 active nodes a segment. LocalTalk has no extended Phase 2 support; if it's used with other media (EtherTalk, TokenTalk, and so on), a router or translation bridge is required.

EtherTalk

EtherTalk provides collision-based access control (using CSMA/CD) over 10Mbps and 100Mbps Ethernet with EtherTalk Link Access Protocol (ELAP). ELAP handles all the AppleTalk Upper Layer Protocols (ULPs) interaction with the transmission medium. The version of AppleTalk used (Phase 1 or Phase 2) determines how the EtherTalk frame is formatted. AppleTalk version 1 uses the Ethernet-II frame specification. AppleTalk version 2 uses the IEEE 802.3 SNAP (Subnetwork Access Protocol) frame specification. AppleTalk protocols do not understand Layer 2 hardware addresses. ELAP uses AARP for determining proper frame source and destination addressing.

It is possible to operate clients and servers on the same media segment using both Phase 1 and Phase 2 packets. However, types 1 and 2 frame types are only recognized by similar clients, so a translation router must be installed if the networks need to exchange data with one another. ELAP transmits data bytaking the client destination address from the DDP datagram, performing an AARP address mapping table lookup, then constructing the Ethernet frame appropriate to the network: Ethernet 2 for AppleTalk Phase 1 or 802.3 SNAP for AppleTalk Phase 2.

> **Note**
>
> Ethernet, Token Ring, and FDDI Layer 2 protocols are all covered in detail in Chapter 4, "LAN Internetworking Technologies." The discussion that follows does not require an extensive knowledge of these protocols, but you might want to skip ahead if you have questions.

All AppleTalk link access protocols for standards-based media (Ethernet, Token Ring, and so on) use the IEEE 802.2-type logical link control standard for MAC. The 802.2 standard provides the capability for different network protocols running on the same computer to discern which incoming frames belong to them. It accomplishes this by using *service access points (SAPs)* to identify which protocol the packet is destined for. The 802.2 header consists of a destination and source SAP value; the value used to indicate a non-IEEE standards-based protocol is `$AA`. AppleTalk uses this value.

Along with the SAP, there is a 5-byte Subnetwork Access Protocol (SNAP) discriminator (see the SNAP header portion of Figure 3.6). This is used to identify the protocol family in which the packet belongs. Two SNAP protocol discriminators used to define AppleTalk packets exist:

- `$080007809B` defines that the frame contains an AppleTalk data packet.
- `$00000080F3` defines that the frame is an AARP packet.

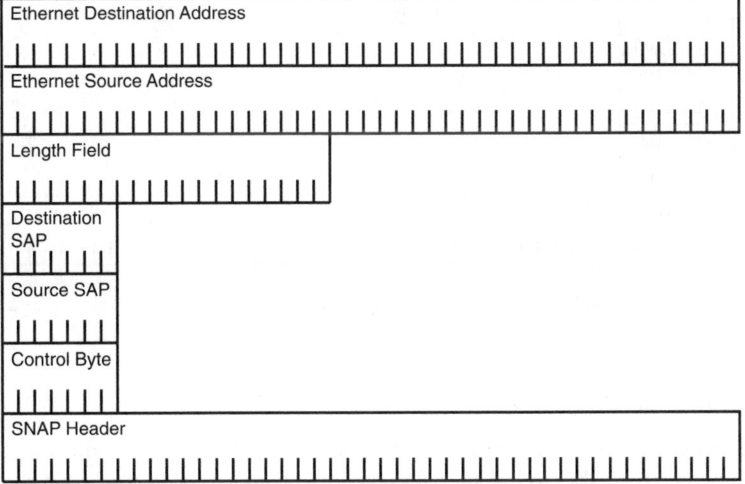

Figure 3.6 EtherTalk message frame formats.

For all protocols (Ethernet, TokenTalk, and so on), the 80F3 always identifies the AARP packet. Data packets vary.

TokenTalk

TokenTalk provides non-collision-based media access control (token passing) over 4, and 16Mbps IEEE 802.5 Token Ring (see Figure 3.7). TokenTalk is only supported in AppleTalk Phase 2 networks. Like EtherTalk, TokenTalk has a transmission-media-specific access control protocol called *TokenTalk Link Access Protocol (TLAP)* that manages all the UPL interaction with the transport and uses AARP for hardware-to-AppleTalk address translation. TLAP constructs packets by first extracting the destination address out of the DDP datagram, then checking it against the AARP AMT to retrieve the hardware destination address. After the destination address has been confirmed, TLAP assembles the frame consisting of the DDP datagram plus the SNAP, LLC, and 802.5 Token Ring message headers.

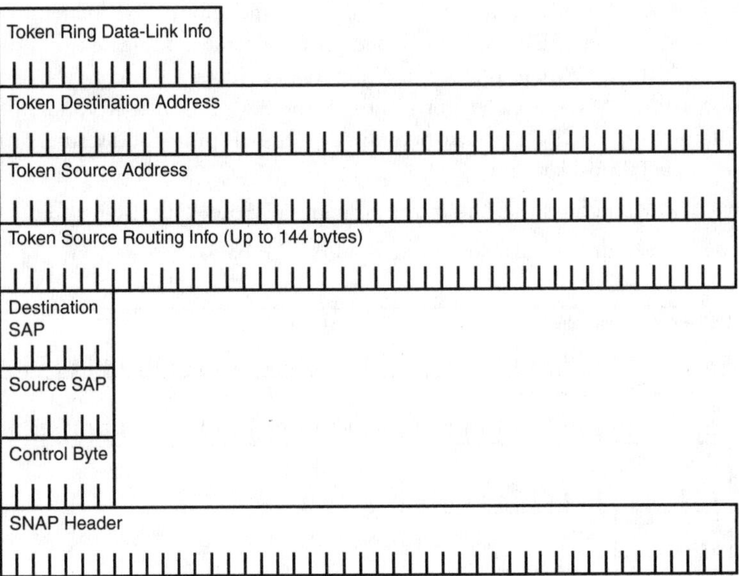

Figure 3.7 TokenTalk message frame formats.

FDDITalk

FDDITalk Link Access Protocol (FLAP) provides access control for 100Mbps single and dual attach ANSI X3T9.5 Fiber Distributed Data Interfaces (FDDI). FLAP uses AARP for hardware-to-network address resolution. FDDITalk frames are constructed by attaching the DDP to the SNAP, LLC, and FDDI headers.

AppleTalk Layer 3 (Network) Protocol

AppleTalk's Layer 3 protocol, *Datagram Delivery Protocol (DDP),* is responsible for addressing and transport of ULP data between clients and servers. The DDP formats the ULP data into datagrams for delivery to destination nodes across the network. DDP datagram delivery is a best-effort delivery service and has no provision for error recovery, leaving these services for Layer 2 to provide. The DDP datagram has a maximum size of 599 bytes (a 13-byte header and 586 bites of payload) including the datagram header and the checksum, which is used for error checking.

AppleTalk Sockets

All AppleTalk network functions and NVE services are provided using the socket interface. Sockets function along the same lines as a post office box. A letter is mailed to a building address (the end-node), and is then routed to a specific P.O. box (the service) to reach a specific person. Socket services are provided using *socket listeners*, which listen on a specific socket address for a service request.

Different socket addresses are used for different services. Socket addresses are 8-bit numbers that originate from specific number ranges to reflect whether the socket assignments are of a static or dynamic type. *Static assigned sockets (SAS)* range from 1 to 127; these numbers are reserved for use with known AppleTalk services. Numbers 1 through 63 are used for AppleTalk maintenance services such as SAS 1 (RTMP), SAS 2 (Names Information Socket), SAS 4 (Apple Echo Protocol), and SAS 6 (ZIP). Socket numbers 64 to 127 are reserved for experimental use. *Dynamically assigned sockets (DAS)* use port numbers 128 to 254. These sockets are randomly assigned by the node—for example, DAS socket 253 is a possible DAS Apple Echo Protocol's ping service.

Socket services' context identities are discovered and available to the user through the *Name Binding Protocol (NBP)*. Each node generates a socket table to maintain a list of open socket listeners, describing the services, and their port address, if available. DDP is used to transport datagrams between locally and remotely accessible (client/server) end-node sockets. To provide this service, DDP has two different datagram formats, as described in the following sections.

DDP Datagram Headers

The full source and destination addresses used to exchange data between end-nodes are 32 bits in size. The network address uses 16 bits, the node uses 8 bits, and the socket uses 8 bits. In the case of a Phase 1 network, the network address is zero, so only the node and socket address are relevant for delivery. In Phase 2 networks, the network address is anything but zero, so it is needed for datagram delivery for hosts outside the local segment.

Because addresses needed for proper DDP datagram delivery are varied, there are two DDP packet headers used for datagram addressing. The DDP short header is used on AppleTalk Phase 1 networks and for local datagram delivery on nonextended Phase 2 networks, as only the node and socket address are needed to successfully deliver the packet. The long header was developed for remote delivery of DDP datagrams on AppleTalk Phase 2 extended networks. It is used for internetwork datagram delivery, where network, node, and socket addresses are needed for delivery. Long headers were originally intended for internetwork delivery. Although long headers are not as efficient as shortheaders, they can be used for local delivery (if specified by the application).

The Phase 1 and Phase 2 headers are illustrated in Figure 3.8.

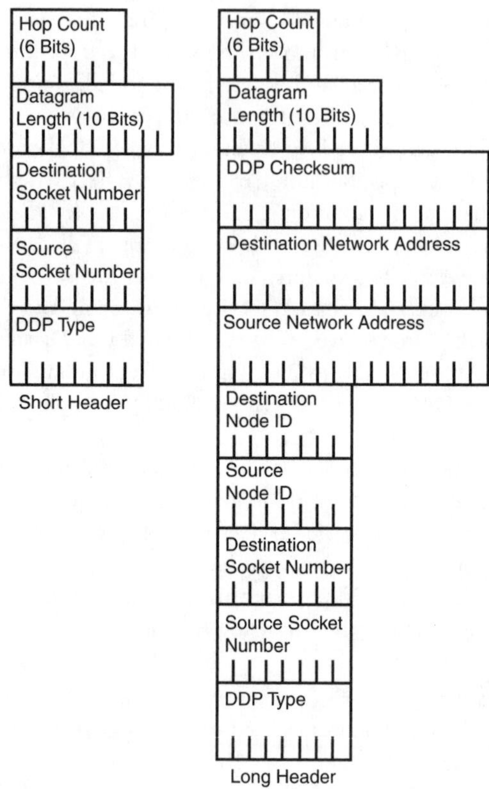

Figure 3.8 Phase 1 (short) and Phase 2 (long) DDP headers.

The function of each header field is listed here:

- Hop count (long header only)—This 6-bit field is used for tracking packet life. The counter starts at zero and is incremented by each time it traverses a router.
- Datagram length (short and long header)—This 10-bit field describes the entire size of the datagram and the header; anything larger than 599 bytes is discarded.
- DDP checksum (long header only)—This 16-bit field is used for error detection resulting from router-to-router transmissions. The checksum, together with the datagram at the source, is used by the router at arrival to verify data integrity.
- Destination network address (long header only)—This is a 16-bit network address.

- Source network address (long header only)—This is a 16-bit network address.
- Destination node ID (long header only)—This is an 8-bit node address.
- Source node ID (long header only)—This is an 8-bit node address.
- Destination socket number (short and long header)—This is an 8-bit socket address.
- Source socket number (short and long header)—This is an 8-bit socket address.
- DDP type (short and long header)—This is an 8-bit field used to indicate the transport layer protocol.

DDP Broadcasts

Three types of DDP broadcasts exist:

- Network-specific broadcast
- Network-wide broadcast
- Zone-specific broadcast

The way the broadcast is interpreted by the node is determined by the node address field in the packet's destination network address. If the network number is any value other than zero, a datagram with a destination node ID of `$FF` (255) is examined by all nodes. However, the datagram is only accepted by nodes with the corresponding network number.

If the network number is zero (`$00000000`), the broadcast is intended for the local network segment. If the packet is a network-wide packet, the network address is zero and the node address is `$FF`, and all nodes on the segment accept the packet. If the packet is intended to be a zone-specific broadcast, it has the same addressing as a network-wide packet. However, it is up to the ULP to determine if the packet is relevant to the node. Because the packet is addressed as a network-wide broadcast, the node accepts the packet. After it's accepted, the zone information is checked, and anything not intended for the node isdiscarded. Zone-specific broadcasts with DDP are dependent on ZIP, zone multicasting function and addressing, for correct handling.

DDP Datagram Assembly and Delivery

DDP datagram delivery uses a local and remote delivery model. ULP sends data to DDP for delivery. DDP determines which delivery model to use based on the network number of the destination address. If the destination address is within the range of the local network, a DDP short header is encapsulated along with the data and sent to Layer 2 for local delivery. If the destination is out of range, the datagram is encapsulated with a long header, handed to Layer 2, and sent to the router for delivery. AppleTalk networks are limited to 15 hops. Extended DDP headers have a hop count field, which is incremented by 1 each time the datagram passes through a router. When the counter reaches 15, the packet is discarded.

AppleTalk Layer 4 (Transport) Protocols

AppleTalk's Layer 4 protocols all contribute to providing the following end-to-end transport services for ULP data between end-nodes:

- Routing table creation and maintenance
- AppleTalk internetwork transport services over TCP/IP
- End-node ratability
- Binary network addressing (physical addressing) to network-named entity (contextual addressing) translation services
- Connection-oriented socket data transport

Routing Table Maintenance Protocol

AppleTalk routing is a dynamic process. Although network addresses are statically set, node addresses are usually assigned dynamically, so static addressing has a very limited value. End-nodes can determine if a datagram is to be delivered locally or remotely. If the destination is remote, the router takes over. The router's main job is to maintain information about different network segments that are reachable within the internetwork. This information includes the following:

- Network (cable) range
- Distance to network in hops
- Router interface used to reach the destination network
- Network address of the next hop node

There are three types of routers used in AppleTalk internetworks:

- *Local routers* are used to connect locally adjacent AppleTalk network segments. A local router is used to segment a large physical network into different network segments.
- *Half routers* are used for point-to-point WAN connections. One half is connected to a local AppleTalk segment, and the other half is connected to the WAN link. The nature of transport used by the link can be a modem, a public data network, and so on.
- *Backbone routers* are used to transport AppleTalk traffic across another non-AppleTalk network. The backbone transit network encapsulates AppleTalk data in its transport format.

In the case of local and half routers, AppleTalk protocols are used throughout the interconnect path. Backbone routers use AppleTalk in conjunction with another protocol (usually TCP/IP) to provide data transport. Regardless of router type, only AppleTalk reliability information is contained in routing tables.

AppleTalk routing table creation all starts with a single router known as the *seed router*. The job of the seed router is to provide non-seed routers with network address information. The seed router has the network range statically set on its ports. A non-seed router does not. For a one-router or multiple router network, one seed router is needed.

When an AppleTalk router starts up, it creates a table of all the connected network segments. This is known as a *routing seed* (not to be confused with a seed router). Each defined network (with a nonzero network number) is entered as a local network with a hop distance of zero. A seed router builds a table network range associated with each router interface. A non-seed router builds a table with all the interfaces using a network address of zero. After the routing table is created, the router sends out a routing update containing all the networks it can reach out of each of its connected interfaces. The seed router sends updates to routers with all of the correct network address information. The non-seed routers then use this network address information to update their tables.

Routing Table Maintenance Protocol (RTMP) is similar to the Routing Information Protocol (RIP) covered in Chapter 8, "TCP/IP Dynamic Routing Protocols," except one value is used to determine which route is the best route. This value is called a *routing metric*. AppleTalk uses a routing metric known as the *hop count*. Hop count is determined based on the number of routers that must be traversed in order to reach the destination network. If a network is directly connected to a network, the hop count to reach the network is zero. An RTMP uses a technique called split horizon to prevent routing loops (discussed in Chapter 8).

RTMP's goal is to have a routing table with the best single route to each given network. The job of the seed router is to provide network number information to routers as they join the network. The following provides a simple example illustrating an RTMP table and corresponding network. The first zone listed for each entry is its default (primary) zone.

```
R Net 20-20 [1/G] via 900.82, 9 sec, FastEthernet1/0, zone Phase II SunLAN
R Net 51-51 [1/G] via 900.82, 9 sec, FastEthernet1/0, zone VaxLAN
R Net 55-55 [1/G] via 900.82, 9 sec, FastEthernet1/0, zone SunLAN-D
R Net 57-57 [1/G] via 900.82, 9 sec, FastEthernet1/0, zone UtilLAN
R Net 64-64 [1/G] via 900.82, 9 sec, FastEthernet1/0, zone PcLAN
R Net 68-68 [1/G] via 900.82, 0 sec, FastEthernet1/0, zone MediaLAN
A Net 789-789 [1/G] via 0.0, 1330 sec, Tunnel1, zone GraceLan
                     Additional zones: 'FatherLan','OutLAN'
C Net 900-900 directly connected, FastEthernet1/0, zone TestLAN
```

Note

Split horizon dictates that routing information learned from an interface cannot be sent back out in its routing update. Only network information from the router's other interfaces are sent in routing updates.

There are four types of RTMP messages:

- Request
- Data
- Response
- Route data request

These messages are carried in DDP datagrams, and the format varies depending on the network type (extended or nonextended). Data messages are used by routers to exchange routing information. Network information is sent as tuples of network distance.

Along with the tuple, the sending router's network address and the interface that sent the update (node ID) are sent. RTMP updates occur every 10 seconds. All route entries have a validity time associated with them. If a route is not verified after 10 seconds, it is marked as suspect. After 20 seconds, it is marked as bad and removed. The request, response, and route data messages are used for communication between nodes and the router for node address assignment (request and response) and routing table acquisition from a specific router (route data request). See Figure 3.9 for the RTMP message format.

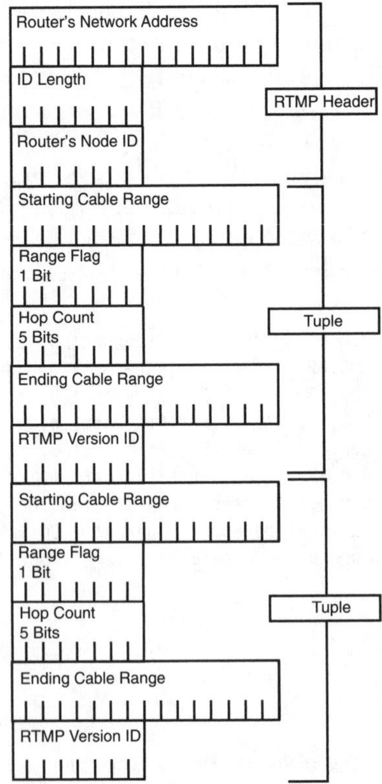

Figure 3.9 RTMP data message.

AppleTalk Update Based Routing Protocol

AppleTalk Update Based Routing Protocol (AURP) is not a routing protocol. Rather, it provides a means for connecting AppleTalk internetworks across TCP/IP networks. The AURP transport mechanism operates as a tunnel through which all AppleTalk protocol data is encapsulated using TCP/IP User Datagram Protocol (UDP) packets as the transport and IP for delivery. AURP implementations have two parts: *exterior routers* and the *AURP tunnel*.

Exterior routers (backbone routers) are the bridge routers between the AppleTalk internetwork and the AURP tunnel. The AURP tunnel is the logical conduit built on top of the local and remote IP interfaces of the exterior routers involved with the AURP exchange. Two types of AURP tunnels exist:

- A single point-to-point tunnel is where all AppleTalk data is exchanged between the two internetworks.
- A multipoint tunnel is where three or more AppleTalk internetworks are connected. Multipoint tunnels can be fully connected or partially connected where only some internetworks are available.

AURP routing updates are adjustable and triggerable when changes in one of the connecting internetworks occur. Initially, routing and zone information tables are exchanged across the tunnel between the connected internetworks when the AURP tunnel is initialized. Updates use TCP or AURP-Tr which both provide reliable transport. This makes AURP a more efficient alternative for connecting AppleTalk internetworks as compared to traditional AppleTalk point-to-point connections.

AURP also introduces the concept of *AppleTalk domains*. A domain identifier is associated with each internetwork connected to the tunnel. Domain identifiers can be statically or dynamically set.

```
64-bit domain-id/ 16-bit network address/ 8-bit node address/ 8-bit socket address
```

AURP also provides facilities for hiding networks, internetwork address conflicts, route path redundancy, and hop count reduction. AURP configuration will be covered in Chapter 10.

AppleTalk Echo Protocol

AppleTalk Echo Protocol (AEP) is used for packet generation to test node reachability and network performance. AEP uses static socket 4 as the echoer socket (receiver) and a dynamically assigned socket as the sender. AEP supports two functions: request and reply.

> **Note**
>
> Remember, AppleTalk internetworks have a distance limitation of 15 hops.
>
> Packets with a distance of 16 are discarded.

Name Binding Protocol

Name Binding Protocol (NBP) is the basis of AppleTalk's contextual addressing scheme. Its purpose is to identify each network service available on a given end-node with a symbolic name. After a service has an entity name associated with it, it becomes an NVE. Name registration is a process similar to dynamic address assignment. The network service registers its name with the end-host. The end-host then checks its name table to ensure that there is no conflict. If no conflict is found on the local name table, the service name is broadcast to all the nodes on the zone/cable range. If no conflict is found, the name is used. If a conflict arises during the local and network verification stage, the registration process is halted. After the NVE is available, its entity name has three parts:

```
object:type@zone
```

- `object` is the service's symbolic name (Montana, Moe's printer, and so on). This can be any name up to 32 characters in length.
- `type` is the service classification. This could be a mail server, printer, file server, and so on.
- `@zone` is the logical contextual network group where the printer resides.

This approach works well from a user interface perspective. Because AppleTalk uses dynamic addressing, statically named entities are easy for users to relate to as compared to a changing 32-bit network/node/socket address.

NBP name table creation occurs on each node on the network. NBP tables are initially created when the node joins the network and are updated as interaction with entities occurs. The NBP name table entries are sent and stored as tuples, which contain translations of network, node, and socket numbers to object, type, and zone for each available service. Lookups and updates are performed using local broadcasts (local name enquiries) and zone broadcasts (for remote name enquiries that are redirected by AppleTalk routers). Every host within the directed segment responds to a lookup request, checking its local name table and sending the result to the requester.

There are four services used for name table maintenance and lookups:

- *Name registration* is the process of services registering their entity names with the local end-node and network (cable range/zone) segment.
- *Name deletion* occurs when an NVE is removed from the network.
- *Name lookup* is performed whenever a node wishes to access an NVE. Requests are queried as either specific or global searches. The type of query dictates the request type (local broadcast or zone specific broadcast/multicast).
- *Name confirmation* is used to verify aged name table entries. Confirmations are performed prior to session establishment with an NVE. The process is different (and more efficient) than a name lookup, as the inquiry is sent directly to the NVE's hosting end-node.

NBP is the end-node-oriented protocol used for providing AppleTalk's contextual network naming scheme. ZIP is the session layer element used for contextual network segmentation. These two protocols are used in conjunction with RTMP to establish network data flow and user interaction. Each element can be used to modify network behavior and performance. ZIP is covered in the Layer 5 session protocols section.

AppleTalk Transaction Protocol

AppleTalk Transaction Protocol (ATP) provides acknowledged transmission service between sockets. Each network transaction consists of two actions: request and response. In most cases, the transmission is a client end-station interacting with a file server or printer. Each ATP request and response must be acknowledged with a transaction request and transaction response to report the outcome of the action. This approach is used by ATP to provide data acknowledgment, packet sequencing, data segmentation, and reassembly, which is needed to handle data loss due to network transmission errors. There are three types of ATP transactions: request, response, and release. The release transaction is used to end an ATP session.

ATP uses two types of transaction services to handle error correction:

- At-Least-Once (ALO)
- Exactly-Once (EO)

ALO transaction services are used by applications that return the same outcome if the transaction is executed more than once. For example, if a host performs a name lookup, the response is the same regardless of which transaction is successful. EO transaction services are used if duplicate requests would affect the success of the transaction. With EO, a transactions list is maintained, so duplicate transactions are performed only once if a data loss condition exists. All ATP transmissions are timed, and the duration varies depending on the type of ATP transaction.

AppleTalk Layer 5 (Session), Layer 6 (Presentation), and Layer 7 (Application) Protocols

Six protocols make up AppleTalk's upper layer protocol suite: four session layer protocols (Layer 5) and two presentation layer protocols (Layer 6). The session layer protocols are used for session negotiation and communication between the lower layer network protocols and end-node application data, which is provided by the presentation and application layers. AppleTalk has no protocol defined application protocol suite; rather, it uses the AppleTalk Filing Protocol (AFP) and PostScript to provide presentation services and application interface hooks.

Zone Information Protocol

Zone Information Protocol (ZIP) is used to create and maintain Zone Information Tables (ZITs) on AppleTalk routers. AppleTalk Phase 1 network supports a direct network address-to-zone association. Phase 2 networks can support up to 255 zone names per extended cable range. *Zones* are used for creating logical contextual network groups to provide user level network segmentation. The idea is that you can group clients, servers, and printers within the same logical group, making user resources more easily accessible.

Like RTMP and NBP, ZIT entries are stored as tuples: network number and zone name. AppleTalk Phase 1 network tuples are a single network number to a single zone name. AppleTalk Phase 2 network tuples are cable range(s) (extended and nonextended) to zone names. The RTMP and ZIP are used in conjunction on the AppleTalk router to direct NBP packets to the correct router interface.

Client/server interaction with ZIP is limited in use to selecting a zone. The client zone name setting is stored in the system's boot PRAM ZIP; when it boots, it verifies the zone name (and its corresponding network address). If no setting is available, the router provides the default zone to the client. ZIP uses five different message requests for table maintenance and zone verification.

- *ZIP query messages* are used to request a router's zone list.
- *ZIP response messages* are used to return the zone list.
- *ZIP extended reply messages* are used to fragment the list into multiple packets if the ZIT cannot fit into a single packet.
- *ZIP GetNetInfo* is used by clients to verify its zone name at boot time.
- *ZIP GetNetInfoReply* is used to respond to client zone verification requests and provide the zone's multicast address.

An illustration of a zone table appears in Table 3.1.

Table 3.1 **A Sample Zone Table**

Zone Name	Cable Range
MediaLAN	68-68
Phase II SunLAN	20-20
SunLAN-D	55-55
VaxLAN	51-51
OutLAN	789-789
InLANHappyLAN	
PcLAN	64-64
UtilLAN	57-57
FatherLan	789-789
TestLAN	900-900HOMElan
GraceLan	789-789

Each message type has its own message format and is sent using DDP datagrams, illustrated in Figure 3.10.

Each zone also has an associated binary multicast address. The multicast address provides a way to send broadcast messages between nodes belonging to the zone. The address is generated by processing the zone name through the DDP checksum algorithm and dividing the result by 255. The address is provided to the host as part of its initial zone registration/verification process.

AppleTalk Data-Stream Protocol

AppleTalk Data-Stream Protocol (ADSP) is used to provide reliable full-duplex data transmission for client/server socket data delivery. ADSP is directly encapsulated into DDP datagrams, and provides facilities for flow control and packet sequencing.

ADSP data exchange requires a socket-to-socket connection stream to be established before data can be exchanged. If either node drops or is unable to establish the stream connection, the session is dropped. To establish communication sessions, ADSP uses *control packets*, which are used for connection-related processes such as opening or closing connections, retransmission requests, or connection acknowledgment. Data is sent in ADSP data packets. Out of data flow messaging is also available and is accomplished with ADSP *message packets*. Each packet uses a specific ADSP header, and all are transported inside of DDP datagrams.

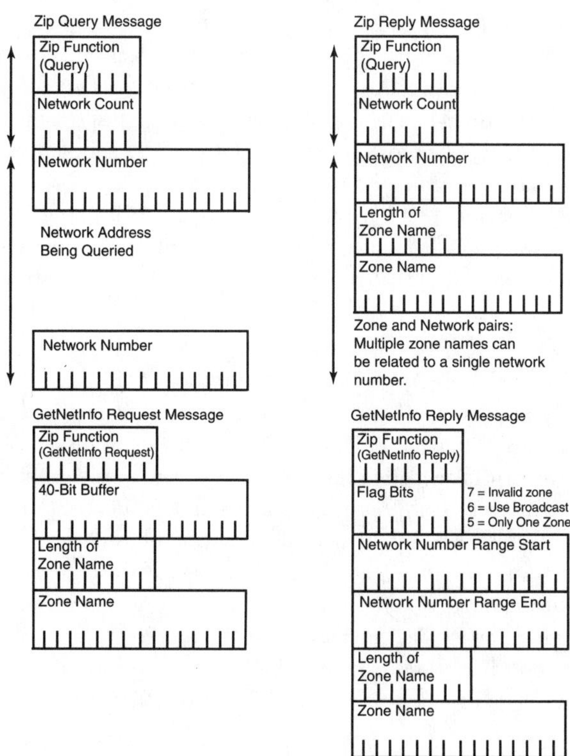

Figure 3.10 ZIP message formats.

To ensure proper packet data sequencing, ADSP uses a 32-bit sequence number along with a packet identifier for each packet. If the sequence number and packet identifier do not coincide, the packet is dropped. Flow control is achieved by the destination sending periodic updates to the sender on the amount of available buffer space. This value is known as the *reception window size*.

AppleTalk Session Protocol

AppleTalk Session Protocol (ASP) provides a connection-oriented facility to exchange multiplexed client/server session communications. ASP is used between clients and servers to exchange session commands. Clients' commands are delivered in sequential order to servers who use ASP to return command results. ASP, however, provides no means to ensure that the server executes them consecutively. ASP operates as a multiplexed process, providing servers with the means to maintain multiple client sessions at the same time. The server has no means for sending commands to the client. Only an attention mechanism is available to the server to notify the client if any action is required on its behalf. ASP uses ATP for transport and NBP for service socket identification.

Printer Access Protocol

Printer Access Protocol (PAP) is used for client/server to printer communication. It performs connection setup and tear down, as well as data transfer. ATP is used as the transport protocol and NBP (like ASP) is used for service (socket) addressing.

PAP transactions are time-based, as ATP is used as the transport mechanism. PAP will maintain half-open connections for the duration of the timeout. After a connection expires, the session is terminated.

Because AppleTalk printing is a device direct activity, PAP provides for a keepalive facility. *Tickle packets* are sent periodically from the clients with open sessions to maintain the connection and ensure that the printer is online and processing requests.

Presentation Layer Protocols

AppleTalk uses two protocols, AppleTalk Filing Protocol (AFP) and PostScript, for translating data responses (lower layer protocol) and requests (application layer) into a common data encoding language. AFP is used for client remote file access. *AFP* is a command translator that takes native file system calls and translates them into AFP calls that the server understands. *PostScript* is a stack-based page-description language used by printers and applications to mathematically describe objects. Apple QuickDraw is the native page-description language used to display Macintosh characters and graphics and is also used for printing to low resolution printers. QuickDraw acts as an operating system level translator for data images to PostScript. PostScript is used to communicate with the printer hardware to render the image for printing. Most common printing errors are related to corrupted QuickDraw-to-PostScript translations.

IPX and NetBIOS

Once upon a time, there was no such thing as Windows NT (and the world was a nice place for UNIX system administrators). Novell NetWare and Internetworking Exchange Protocol (IPX) ran on 60 to 70 percent of all networked Intel/DOS-based computers. Novell is a proprietary 100 percent DOS-compatible network operating system (NOS). Its basic design goal was to provide shared file system and printer access transparently to desktop PCs through the I/O interfaces provided by DOS. Networked file systems were available to users as drive letters (such as E:\), and networked printers were available through virtual (LPR) printer ports. Novell NOS runs on almost any Layer 2 protocol and is available for almost every major computer platform, keeping in mind, however, its first love (and primary orientation) is to DOS. Novell uses its own proprietary and closed architecture, based on Xerox's open standard, Xerox Network Systems (XNS).

The IPX is the original Novell NOS network layer protocol used for all network layer communication. The Novell NOS versions 4.0 and later also operate over TCP/IP.

Novell also supports both proprietary and standards-based session protocols, which under the Novell model acts as the bridge between user applications and network transport. Novell's session protocols are as follows:

- NetWare Core Protocol (NCP) (Novell-specific)
- NetWare Shell (NWS) (Novell-specific)
- NetWare Remote Procedure Call (NRPC) (Novell-specific)
- NetBIOS (open standard)

Novell's support of NetBIOS was driven by a need for NetWare systems to interoperate with NetBIOS-based NOS, like IBM's LAN Manager, which was the foundation of Microsoft's Windows for Workgroups (Windows.9x) and Windows NT networking environments. Microsoft's implementation of IPX is called *NWLink*. NWLink is the Microsoft version of the IPX protocol suite, and it is fully compatible and operationally identical to Novell's IPX/SPX protocols. It provides Windows-based systems native protocol access to both Novell NetWare and Microsoft networking services.

The IPX/NWLink protocols are implemented on Intel-based PCs, using either network device interface specification (NDIS) or open data-link interface (ODI) network driver interfaces. NDIS is a standard for interfacing between media access control (MAC) sublayer and network protocols. NDIS acts as a protocol multiplexer or traffic director between Layer 3 (network protocol) and Layer 1 (hardware network adapter), so multiple network protocols, such as TCP/IP and IPX, can be used on the same computer. ODI is the Novell proprietary specification for providing the same facility.

NetWare (IPX) Architecture: OSI Layer 1 and Layer 2

The NetWare architecture model uses a five-layer model in contrast to OSI's seven communication layers, as shown in Figure 3.11.

- Layer 0, the transmission media layer, is responsible for data exchange between the end-node and the transmission media.
- Layer 1, the Internet layer, provides a data exchange facility between end-nodes connected on different networks.
- Layer 2, the transport layer, handles end-to-end communication between end-nodes.
- Layer 3, the control layer, provides session control and data presentation services.
- Layer 4, the application layer, manages data semantics between client and server interactions such as login, file, and print services.

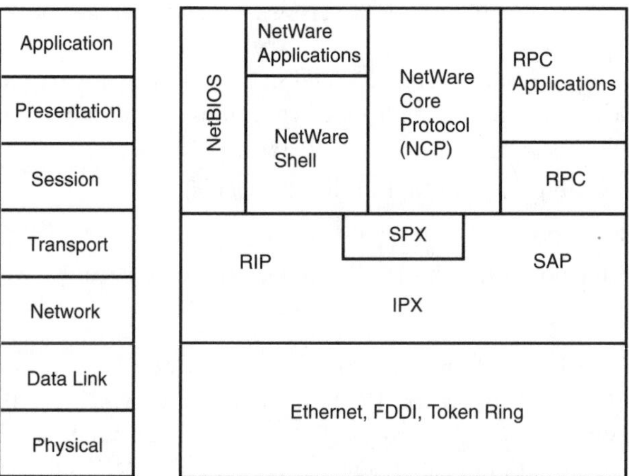

Figure 3.11 Novell (IPX) protocol suite.

Where TCP/IP and AppleTalk are unaware of Layer 2 (OSI-RM), IPX operates in conjunction with it. The most obvious example of this symbiosis is the IPX's end-node number. The *IPX end-node number* is the NIC's unique hardware address. The other, slightly more complex dualism is IPX's use of hardware encapsulation.

IPX operates over several LAN and WAN transmission media formats, including Ethernet, Token Ring, FDDI, and Point-to-Point Protocol (PPP). NetWare, in its original form, supported a single proprietary encapsulation format. However, as Layer 2 technologies evolved (just as with AppleTalk), IPX was adjusted to operate with the new encapsulation formats, of which IPX supports several (see Table 3.2).

Table 3.2 **IPX Encapsulation Schemes**

Media Type	Encapsulation Scheme	Frame Type
Ethernet	Novell	802.3 RAW
	ARPA	DEC Ethernet v2
	802.3	IEEE 802.3 Standard
	SNAP	Ethernet SNAP
Token Ring	IEEE 802.5 (802.2 LLC)	IEEE 802.5 Standard
	SNAP	IEEE 802.5 SNAP
FDDI	SNAP	FDDI SNAP
	SAP	FDDI 802.2 LLC
	Novell	ANSI FDDI RAW
Serial	IPXWAN	PPP
	IPXWAN	HDLC

Characteristics of different IPX encapsulation schemes include the following:

- Novell RAW uses the standard protocol frame without 802.2 logical link control.
- Ethernet version 2 is the standard pre-IEEE 3COM/DEC Ethernet version 2 standard frame specification.
- SNAP uses the 802.2 LLC frame format and the protocol type field.
- IEEE 802.* uses the IEEE 802.x standard frame format.
- IPXWAN is a WAN-specific protocol used for IPX routing (and transporting) communication between routers connected over dedicated serial lines.

For compatibility and, as it turns out, increased flexibility, IPX can support multiple network segments using different encapsulation schemes over the same physical (Layer 1) medium. If no router is in place, none of the networks can exchange data between one another. This is because each network is using a specific frame type. Only end-nodes that are configured to process the same frame type can exchange information.

Note

This same result occurs when AppleTalk Phase 1 Ethernet and Phase 2 Ethernet are used on the same media. Because the frame types are different, they can only be understood by like end-nodes (Phase 1 or Phase 2) but are unable to cross-communicate.

The other nodes on the network that are configured to use other frame types discard the frames, believing them to be malformed. The encapsulation type must be set correctly on all routers, servers, and clients that need to interact locally. Incorrect encapsulation is a common IPX network problem, so when in doubt, check the settings.

NetWare (IPX) Architecture: OSI Layer 3

IPX, like TCP/IP's Internet Protocol (IP) and AppleTalk's Datagram Delivery Protocol (DDP), is the sole network layer delivery protocol for NetWare (and NWLink-based LAN manager implementations). IPX is a routable, connectionless datagram delivery protocol. Its original implementation operated around the Routing Information Protocol (RIP) as a routing protocol that is part of the IPX process and operates automatically whenever IPX is used (similar to AppleTalk's RMTP). Today, IPX can utilize NetWare Link State Protocol (NLSP) and Cisco's Enhanced Interior Gateway Routing Protocol (EIGRP) routing protocols to exchange route information.

IPX (NWLink) Addressing

IPX datagram delivery provides facilities for local network and (remote) internetwork data exchanges. IPX datagram delivery points are known as *ports*. IPX ports are just like AppleTalk sockets. While a ULP is responsible for the actual data transport, the source/destination port is part of the IPX packet address. Figure 3.12 illustrates the IPX network address format.

Figure 3.12 IPX address format.

The IPX address has three components:

- Network
- Node
- Port

The network address is 32 bits in length, generally expressed as a single string of hexadecimal digits. The node address is 48 bits in length, expressed as three dotted triplets of a pair of hexidecimal numbers or six dotted pairs of hexidecimal numbers. The port address is 16 bits in length expressed as a single four-digit hexidecimal number.

The IPX address is therefore a total of 96 bits in size, which is large for a network address. Datagram sizes vary depending on the encapsulation type and network media type being used for transport.

The IPX network address needs to be set by the network administrator. Like AppleTalk, the network address can be a random number, but each must be unique.

IPX uses the end-station's NIC hardware or MAC address for the node address. This makes IPX stations, in a sense, self-configuring. This approach also eliminates the need for a Layer 2 to Layer 3 address resolution protocol. This, in turn, reduces packet delivery complexity and network traffic. However, there is an associated disadvantage of having to replicate the Layer 2 address twice in the data frame. This reduces the amount of actual data that can be transported in each frame.

IPX's port communication exchange process is quite simple. Known services use known port numbers, and dynamic data exchanges (file transfers, for example) use dynamic port numbers (see Table 3.3).

Table 3.3 **Port Address Assignments for IPX**

Port Service Assignment	Number
Wild Card (all sockets)	0
NetWare Core Protocol	451
Service Advertisement Protocol	452
IPX RIP	453
NetBIOS	455
Novell Diagnostic Packet	456
Novell Serialization Packet	457
Dynamic Sockets	4000-6000

IPX Message Format

The IPX datagram header is, basically, all the addressing information needed for IPX's simple original orientation toward LAN-based datagram exchange. Figure 3.13 describes the message format.

- Checksum is a 16-bit field. Checksumming is not enabled by default in IPX, so the field is often unused and set to a default (FFFF). IPX relies on Layer 2 for error checking.
- Packet length is a 16-bit descriptor expressing the size of the entire IPX packet.
- Transport control is an 8-bit value that describes the number of hopsan IP packet has traversed. It is decremented by 1 each time it passes through a router. When 16 is reached, this is the maximum hop count for an IPX network, and the packet is dropped.

- Packet type is used to indicate the kind of data contained in the datagram. 0 = Unknown, 1 = RIP, 4 = SAP, 5 = SPX, 17 = NCP, 20 = NetBIOS.
- Destination network is a 32-bit field. If the sender is local, this value is 0.
- Destination node is a 48-bit field. Unicast messages use the MAC/IPX address of the destination end-node. Broadcast messages use all zeros.
- Destination port is a 16-bit field that indicates the ULP service port destination address.
- Source network is a 32-bit field. A 0 here indicates that the datagram is either unknown or a network broadcast.
- Source node is a 48-bit field, indicating the sender's address.
- Source port a 16-bit field that describes the sender's originating port (should be the same as the destination port number).

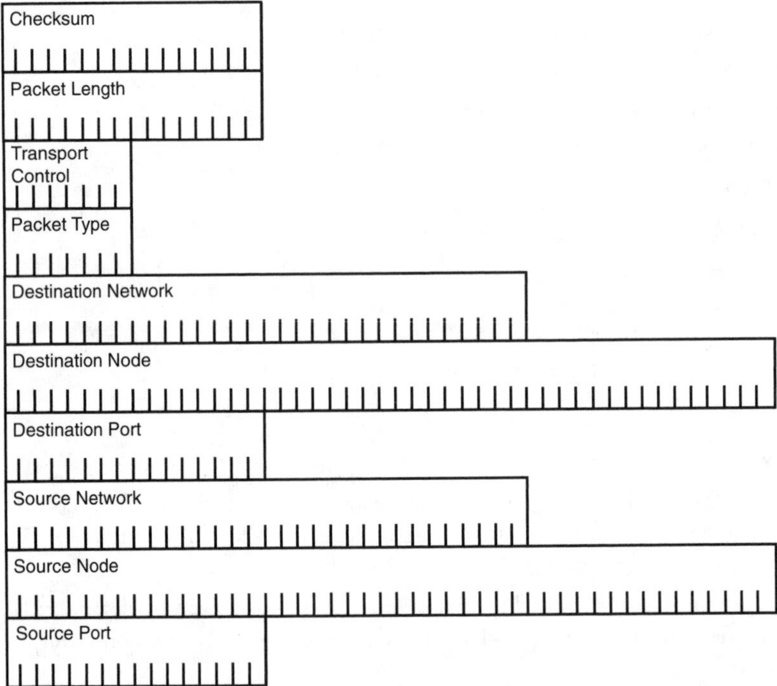

Figure 3.13 IPX datagram format.

The data field will contain the ULP header and data. IPX is used to transport all network messages, including SAP and IPX messages, which are technically part of IPX.

IPX Datagram Delivery

Like IP and AppleTalk, IPX uses a local and remote delivery model. IPX is used for datagram delivery, SAP is used to announce the services on the network, and RIP is used to determine how to reach these services. Both SAP and RIP will be examined in detail in the following sections.

When a server or router joins the network, it constructs a table listing all the services it provides (server) or all the servers and services it knows about (router). Both routers and servers announce this information periodically to the network. When a client joins the network, it needs to find out what its network address is and which server to attach to. It accomplishes these functions by listening for an IPX RIP message or by sending a `GetLocalTarget` broadcast request. A `GiveLocalTarget` response is sent to the client in response to the broadcast message. The client, in turn, learns its network address from the source network address of the update or response packet. The client then sends a `GetNearestServer` broadcast to learn the address of the nearest server. This is responded to by all available servers (and routers) with a `GiveNearestServer` message. These responses are stored locally in the client's SAP table and used to determine the best server to connect to.

IPX datagram delivery is determined by first determining the destination's network address. The end-station's address is compared to the server's address (in the SAP table), and if they are local to one another a connection is established. If the client and server are not local, a RIP request is made for the shortest and fastest path. The client then determines which path is the best path based on the information provided by the router(s). This is important to note because the client determines the network path that will be taken instead of just forwarding the datagram on to a router that makes that determination.

Service Advertising Protocol

Service Advertising Protocol (SAP) is an IPX support protocol through which service-providing end-nodes (servers) advertise their specific services (see Figure 3.14). SAP information needs to be available across the entire internetwork in order for IPX hosts who have knowledge of their available services to share this with other hosts. Routers collect local SAP updates and broadcast a single SAP update based on the cumulative information gleaned from other router updates. This allows clients to become aware of servers that are local. Updates contain the server name, network address, and SAP service identifier (indicating the type of service available).

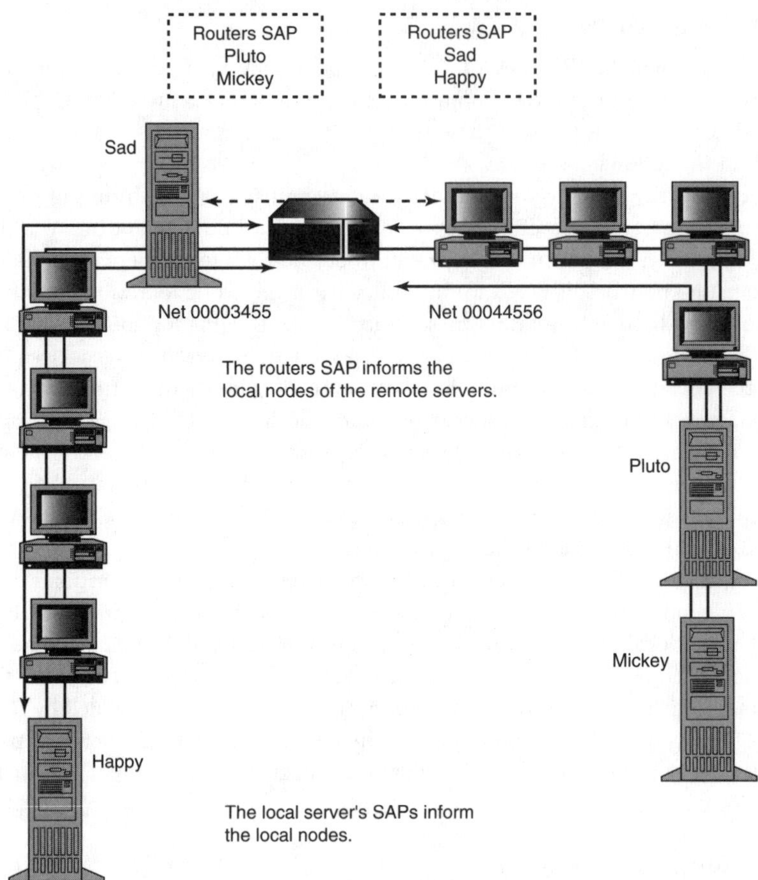

Figure 3.14 SAP message flows between servers and routers.

SAP is used by both clients and servers for requesting and responding to information about available network resources. SAP messages support four distinct operation types and contain information about eight nodes per message (see Figure 3.15).

- Operation is a 16-bit field that describes the packet's informational purpose.
  - *Request* is a general informational request about all the servers on the network.
  - *Response* is a reply to a SAP request or a general SAP announcement sent by servers and routers every 60 seconds (default).
- `GetNearestServer` is a specific request sent by a host to locate the nearest server.
- `GiveNearestServer` is a response sent by a router or server with the SAP information about the server closest to the requesting end-station.

- Service Type is a 16-bit field that indicates the kind of service being provided.
- Server Name can be up to 384 bits in size. It describes the unique name of the server. SAP provides the naming service used by NetWare. Each server on the internetwork requires its own unique name. SAP is used exclusively for name service in all versions of NetWare up to version 4.0. With NetWare 4.0, NetWare Directory Services (based on the ISO x.500 standard) is used for name service. SAP, however, is still needed for server locating and printing.
- Network Address is a 32-bit value.
- Node Address is a 48-bit value.
- Socket Address is a 16-bit value.
- Hop Count is the number of routers that must be traversed to reach the server.

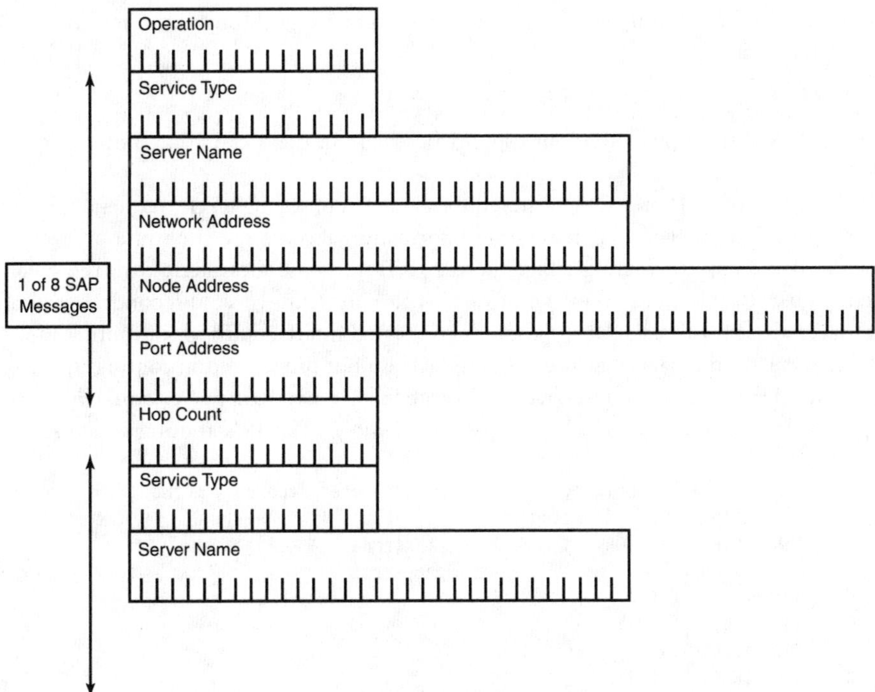

Figure 3.15 SAP message format.

Updates are sent out every 60 seconds. In large IPX environments, SAPs can consume a noticeable amount of bandwidth. One way to alleviate this is by having routers broadcast "combined" SAP updates. It is also possible to adjust the intervals during which routers send out SAP updates. This is especially useful over WAN links where bandwidth is limited. To calculate the SAP bandwidth network load, add 32 (SAP message header) plus 64 (SAP entry) per SAP device. The total will tell you the total bandwidth per minutes used by SAP.

IPX Routing Information Protocol

IPX RIP is a variation of TCP/IP RIP version 1. The RIP version 1 routing protocol is fully explained in Chapter 9. Only the IPX specifics are covered in this section. IPX RIP uses a "best path" route metric value to determine the best path to reach a network. The best path value is based on two factors:

- Hop count—Number of routers a packet must travel through to reach its destination
- Ticks—Used to represent network delay

Not all IPX RIP implementations support ticks, and in these cases only hop count is used.

Two types of RIP messages exist: requests and responses. IPX RIP requests are sent by clients (or another router) requesting information about a specific route or a router's entire current routing table. The IPX RIP response message returns the specified request. RIP response message broadcasts also are sent every 60 seconds. Broadcast messages are sent out each of the router's connected interfaces. To avoid routing loops, routes learned on a particular routing interface are not propagated in routing updates announced from the same interface. This practice is known as split horizon, which is explained in further detail in Chapter 10. An example of an IPX routing table follows:

```
Codes:
       C - Connected primary network, c - Connected secondary network
       S - Static, F - Floating static, L - Local (internal), W - IPXWAN
       R - RIP, E - EIGRP, N - NLSP, X - External, A - Aggregate
       s - seconds, u - uses, U - Per-user static
C    5 (SNAP)      Et0/0
C    45 (SNAP)      Et0/1
C    556 (NOVELL-ETHER)     Et0/1.1
C    899 (SNAP)     Fa1/0
R    88 [02/01] via     5.00e0.b06a.460e     36s     Et0/0
```

IPX RIP routers have a specific startup and shutdown procedure:

1. On startup, the router creates a routing table using only its connected interfaces.
2. It then sends this table as a network broadcast RIP response message.
3. After sending a RIP response, it sends a RIP request message to learn about the rest of the available network segments.

4. The response messages are used to create a complete routing table.
5. The routing table is sent out every 60 seconds.
6. When shutting down, the router broadcasts a shutdown notice over each of its connected interfaces. If the router fails and the shutdown message is not sent, the other routers will time out the route entries attributed to the failed router. This timeout is four minutes long, or four routing updates.

Only one router needs to reach the timeout limit for the route(s) to be deleted. After a single router deletes the routes, the change will be automatically announced by all the routers as soon as it gets the information. The same is true for adding a new network. After a change is detected and passed on, an update will be distributed in the updated routers' updates, in a ripple effect. After the failed routes have been timed out, a new route path will be calculated, if possible.

What makes IPX RIP different from its IP counterpart is its implementation as a client/server protocol. IPX RIP responses and requests are used to make dynamic routing decisions instead of having the end-station use its localrouting table to determine a datagram's route path, as with IP. End-station routing tables are usually created statically.

Sequenced Packet Exchange

Sequenced Packet Exchange (SPX) extends IPX's connectionless datagram service by providing a facility for reliable connection-oriented delivery. Based on Xerox's Sequenced Packet Protocol, SPX supports a virtual circuit connection approach similar to TCP. The source and destination ports are defined between the sender and receiver. SPX datagrams use the IPX format with additional fields for packet sequence identifier and acknowledgement number. IPX/SPX does not support data fragmentation, so each packet can be processed upon receipt. SPX provides facilities for determining the toleration level of the connection, by setting a limit on the amount of unacknowledged packets. After the limit is reached, the connection is dropped. SPX is used for transporting NCP and NetWare shell transactions.

Upper Layer Protocols

IPX is used to transport NetWare and non-NetWare-specific transactions. For NetWare-to-NetWare interactions, the NetWare Shell, NCP over SPX, and NetWare RPC are used. For NetBIOS exchanges, IPX/NWLink is used.

NetWare Shell, NCP, and NetWare RPC

The *NetWare shell* is the client-side command execution front-end. Its task is to monitor application input/output and act as an I/O traffic director. Local requests are passed to the local I/O systems. Network requests are redirected to the appropriate network service. The NCP protocol provides connection control and service request client/server function routines called by the NetWare shell.

NetWare RPC allows clients to remotely execute commands on the NetWare server as an alternative to using the NetWare shell and NCP.

NCP is a proprietary Novell protocol that operates as a simpler version of TCP. NCP performs message sequencing instead of TCP's byte-level sequencing. NCP messages use a sequence number that is employed by the client and server to track responses. If the server sends a packet with the sequence number 8, the server will reply to the request with the sequence number 8. NCP data exchanges flow in one direction: client (using NCP request packets) to server (using NCP response packets). The messages are handled one sequence number at a time. When the client sends a request message, it waits for a reply message with the correct sequence number. If no reply is received, a timeout is reached and the transaction starts again using a sequence number increased by 1. If a server receives a packet with the same sequence number, it retransmits its response. If a client receives a message with a sequence number it has already received, the packet is dropped.

NCP function routines include the following:

- Remote file system access
- System accounting
- Name service
- Printing access

NetBIOS

In 1984, Sytec, an IBM subcontractor, created the *Network Basic Input/Output System (NetBIOS)*. It was designed to provide OSI-RM Layer 4 full-duplex transmission service and OSI-RM Layer 5 session services. NetBIOS was originally published in the *IBM PC Network Technical Reference Manual* and has evolved as a de facto standard. The standard defines a collection of functions to be used with NetBIOS's message control block (MCB) scheme. The MCB scheme takes a block of data, formats it as an MCB, and delivers it, utilizing the defined NetBIOS functions.

Note

Remember, NCP and the NetWare shell are used only to provide NetWare specific functions; they are not used for processing NetBIOS interactions.

NetBIOS has been implemented on a variety of networking platforms. The most common NetBIOS platform in use today is Microsoft's *Windows Networking*, an environment based on a NetBIOS derived protocol called *Server Message Block (SMB)*. With respect to transport, NetBIOS provides little except for some route (delivery) handling guidelines. Both NetWare and Microsoft Networking provide routed network support for NetBIOS over IPX/SPX (NWLink) and TCP/IP.

NetBIOS defines four functions it can perform:

- Status and control—General command and reporting functions used for all NetBIOS session and interface calls:

`Msg.Reset`	Resets the NetBIOS interface
`Msg.Cancel`	Kills a NetBIOS command
`Msg.Status`	Provides status on the interface
`Msg.Trace`	Allows a trace of all the commands issued to the interface

- Name service—Each NetBIOS interface or entity has a name. NetBIOS entities can be a user, application, computer, printer, and so on. NetBIOS uses a single, flat namespace where all active participants are represented. Names are limited to 15 characters in size. Both users and end-nodes are represented by names. Names are not permanently bound to any element. They can be moved and reassociated with different elements if the previous relation has been deleted from the namespace.
- Naming structure—NetBIOS's naming structure allows users to log in on different workstations and still have the same privileges. Groups are also supported, so a collection of names can be associated to a single name and have privileges assigned to the group name, which are then passed on to the group members. The only hard requirement is that all names must be unique across the entire namespace. There are four NetBIOS function calls associated with the name service:

`Msg.Add.Name`	Adds a unique name to the namespace table
`Msg.Add.Group.Name`	Adds a group name to the namespace table
`Msg.Delete.Name`	Removes a name from the namespace table
`Msg.Find.Name`	Finds the associated information related to a name

- Session service—Provides NetBIOS's full duplex, sequenced data transfer between two NetBIOS-named entities. Entities can have more than one session. In such cases, the accessed entity is shared by the connecting names. The accessing sessions are identified by a session ID assigned to each of the connections. Data transfers use sequence numbers and acknowledgments; out of sequence packets trigger retransmission requests. Flow control is managed by the establishment of an adjustable buffer window at the beginning of the session. The window determines the number of messages that can be outstanding at any time. Session messages can be up to 64KB in size. There are eight NetBIOS session calls:

`Msg.Call`	Calls a NetBIOS entity to open a session
`Msg.Listen`	Opens a session with a named entity
`Msg.Hang.Up`	Closes a session with a named entity
`Msg.Send`	Sends a message across the session; failed acknowledgment closes the session
`Msg.Chain.Send`	Sends a stream of messages across the session
`Msg.Receive`	Receives a message from a specific named entity session; failed acknowledgment closes the session
`Msg.Receive.Any`	Receives a message from any named entity session; failed acknowledgment closes the session
`Msg.Session.Status`	Retrieves information on the status of one or all the active sessions

- Datagram service—Used to send messages to a named entity, without prior session establishment. Datagram service provides unreliable, best-effort, connectionless delivery for standalone messages used for data exchange scenarios where data retransmission does not affect operation. Datagram messages can be sent to single and group entities or as namespace broadcasts. Datagram messages have a maximum size of 512 bytes. There are four datagram service calls:

`Msg.Send.Datagram`	Sends a NetBIOS message as a datagram
`Msg.Send.Broadcast.Datagram`	Sends a NetBIOS message as a broadcast datagram to the namespace
`Msg.Receive.Datagram`	Receives a datagram message designated to the entity
`Msg.Receive.Broadcast.Datagram`	Receives a broadcast datagram

NetBIOS message delivery uses source routing for message delivery. This requires that the sending station knows and provides the specific route path used for delivering messages outside of the local network. The route path information is obtained using the `Msg.Find.Name` command. The route path is stored as part of the NetBIOS message and is referred to by the router as the message is processed. Up to eight network entries are stored in the NetBIOS message, forcing a network diameter of eight hops for any NetBIOS implementation. It is the source routing requirement that makes the NetBIOS name service so important to packet delivery in NetBIOS-based enterprise networks.

Summary

The overall focus of this chapter was to provide you with an understanding of the protocol mechanics of AppleTalk, IPX, and NetBIOS. Despite TCP/IP's increased usage in PC LAN environments, there are a large number of legacy installations in place today. Rather than replacing existing LAN protocolnetworks with TCP/IP, it is more common to find multiprotocol LANs being implemented. This is due largely to the improved stability of ODI and NDIS drivers that are being provided with NetWare and Microsoft Networking.

It is important as a network administrator and planner that you understand the operational processes that occur at each layer of protocol implementation, so you can troubleshoot effectively. In this chapter, we have reviewed the following:

- AppleTalk Phase 1 and Phase 2 protocol suite
- IPX and NWLink network protocol suite
- NetWare proprietary network protocols
- NetBIOS operational specification (the basis of Windows NT/95 networking)

In the next chapter, the various LAN and WAN OSI-RM Layer 2 protocols are reviewed. Chapter 4 will cover LAN protocols, such as Ethernet and FDDI. Chapter 5, "WAN Internetworking Technologies" will provide you with an understanding of the AT&T digital circuit "T" standard, second-generation digital transport technologies, such as ISDN and SONET, and the data link protocols that operate over them, such as Frame Relay and ATM.

Related RFCs

RFC 1001	Protocol Standard for a NetBIOS Service on a TCP/UDP Transport: Concepts and Methods
RFC 1002	Protocol Standard for a NetBIOS Service on a TCP/UDP Transport: Detailed Specifications
RFC 1088	Standard for the Transmission of IP Datagrams over NetBIOS Networks
RFC 1634	Novell IPX over Various WAN Media (IPXWAN)

Additional Resources

Apple Communications Library. *AppleTalk Network System Overview.* Addison-Wesley, 1989.

Sidhu, Gurshuran, Richard F. Andrews, and Alan B. Oppenheimer. *Inside AppleTalk,* Second Edition. Addison-Wesley, 1990.

4

LAN Internetworking Technologies

IN THIS CHAPTER, WE LOOK AT THE THREE most prevalent LAN protocols: Ethernet, Token Ring, and Fiber Distributed Data Interface (FDDI). Of the three, Ethernet easily has the largest installation base, which continues to expand into the foreseeable future. Token Ring and FDDI, although technically more efficient, resilient, and complex than Ethernet, will more than likely continue to lose market share. The total dominance of Ethernet in the LAN, however, is a relatively recent phenomenon. Until the development of a 100Mbps Ethernet or *Fast Ethernet*, it was accepted wisdom that Token Ring and FDDI installations would one day outnumber Ethernet installs.

But, Fast and Gigabit Ethernet have relegated FDDI and Token Ring to near "legacy" network status. Although their current institution base will remain viable for at least the next few years, their long-term usefulness is questionable. For this reason, this chapter is weighted toward Ethernet.

Each of these protocols began as proprietary standards that were later standardized by standards bodies. When these protocols were introduced into the marketplace, it was common for both proprietary and standards-based versions to be for sale and to be implemented. Today, only standards-based versions of these products are available. In the interest of completeness, however, both the proprietary and standards-based versions are discussed here when relevant.

IEEE 802 LAN Standards and Logical Link Control

In 1980, the Institute of Electrical and Electronic Engineers (IEEE) began work on the development of LAN/MAN standards: IEEE project 802. Its charter was to develop a set of OSI-RM Layer 1 and Layer 2 protocol standards for LANs and MANs. These standards specified the operational definitions for a collection of logical interfaces that make up the transmission interface that resides on OSI-RM Layers 1 and 2.

Logical interfaces are needed so that different transmission protocols can interact with one another in an efficient manner. Each logical interface defines a "handoff" point that must be able to provide a specific service. Establishment of such a point enables different transmission protocols to hand off upper layer protocol (ULP) data between them without corruption or additional interaction with the ULP. As mentioned in our discussion of the OSI-RM, this point also provides a basis for service enhancement or replacement without affecting the communication process as a whole. To this end, the IEEE's standards development focuses on distinct functional categories. These categories are as follows:

- General Architecture and Internetworking—This category covers bridging, switching, Layer 2 addressing, and management. These standards are developed by the 802.1 subcommittee. The majority of these standards define how OSI-RM Layer 2 transmission protocols will function in an operational and implementation way.
- Logical Link Control (LLC)—LLC provides the software interface between the Layer 2 transmission protocol and the Layer 3 networking protocol. The LLC standard was developed by the 802.2 subcommittee. LLC is used by all the IEEE standards-based protocols and several non-IEEE protocols. It is classified as part of OSI Layer 2 functionality. Nevertheless, it functions as a distinct service interface to manage the actual delivery of data between end-stations. Data link (the term used to describe this function with WAN protocols) or link control is provided differently, depending on the transmission protocol, but the service's function is the same regardless.
- Media Access Control (MAC)—MAC is responsible for the data transmission frame format, data integrity checking, and managing station access to the physical transmission medium (for example, collision detection in the case of Ethernet). For each MAC protocol there is a separate 802 subcommittee. Currently eight different IEEE subcommittees develop MAC standards for physical media and wireless communication. In terms of the LAN, the 802.3 subcommittee develops standards for Ethernet, and 802.5 develops standards for Token Ring. The MAC subcommittee standards encompass both OSI-RM Layers 1 and 2 functionality. The MAC service interface resides in Layer 2, and the PHY service interface—which defines the signaling and encoding used to

transmit the MAC interface frames—resides at Layer 1. The MAC protocol is often used to generalize the Layer 2 function because it represents for the most part the common denominator between the PHY and LLC service layers of the data transmission function.

- Security—All security issues and standards dealing with the 802.x standards are developed by the 802.10 subcommittee.

The development and implementation of standards-based technologies assures compatibility between networking products manufactured by different product vendors. The standards development process is a collaborative effort between the IEEE, networking product vendors, and other interested parties. Actually, quite often, vendors develop proprietary implementations of technologies and then offer them as a standard or as an addition to the existing standard. When a new technology is developed, a new IEEE subcommittee is formed to codify a standard. In the event that an innovation is made to an existing IEEE standard, it is handled as a subcommittee's supplemental project. Each supplement is assigned a one- or two-letter designator. For example, the IEEE standard for 100Mbps Fast Ethernet is 802.3u. The IEEE standardization process usually takes a few years, so quite often, there will be competing "proprietary" versions of the same base technology in the marketplace. Although these technologies might provide the same functionality, they are usually incompatible. For more information about the IEEE 802 committee, you can check out their Web site at `http://grouper.ieee.org/groups/802/index.html`.

802.2 Logical Link Control

In the discussion of the OSI-RM in Chapter 1, "Understanding Networking Concepts," the IEEE variation of the OSI-RM Layer 2 was also introduced. Under the IEEE/OSI variation, Layer 2 is split in half so that the lower half is the media access layer. The media access layer controls Layer 2 source and destination addressing, transmission signaling and encoding (framing), and error and data flow control. The upper half of the IEEE/OSI Layer 2 variation is the Logical Link Control (LLC) interface. The IEEE subcommittee 802.2 is responsible for the development of the LLC specification. The LLC layer is used by the MACs (and some other protocols, such as FDDI) to provide an interface between Layer 3 protocols and the Layer 2 MAC protocols.

The LLC standard has two interfaces (see Figure 4.1). The first is theNetwork/LLC interface, which defines the interface between the LLC and the Layer 3 protocol. The second is the LLC/MAC interface, which handles communication between the LLC and the MAC sublayer. These interfaces allow the LLC to interpret and generate commands between the Layer 3 network protocol and Layer 2 transmission protocol. This communication includes data flow control, error detection and recovery, and the exchange of Layer 2 and Layer 3 address information. Although the MAC is responsi-

ble for the transmission of the Layer 3 data between end-stations,
it is the responsibility of the LLC to ensure that data is delivered to the "service" or application for which it is intended. It accomplishes this through the use of Service Access Points.

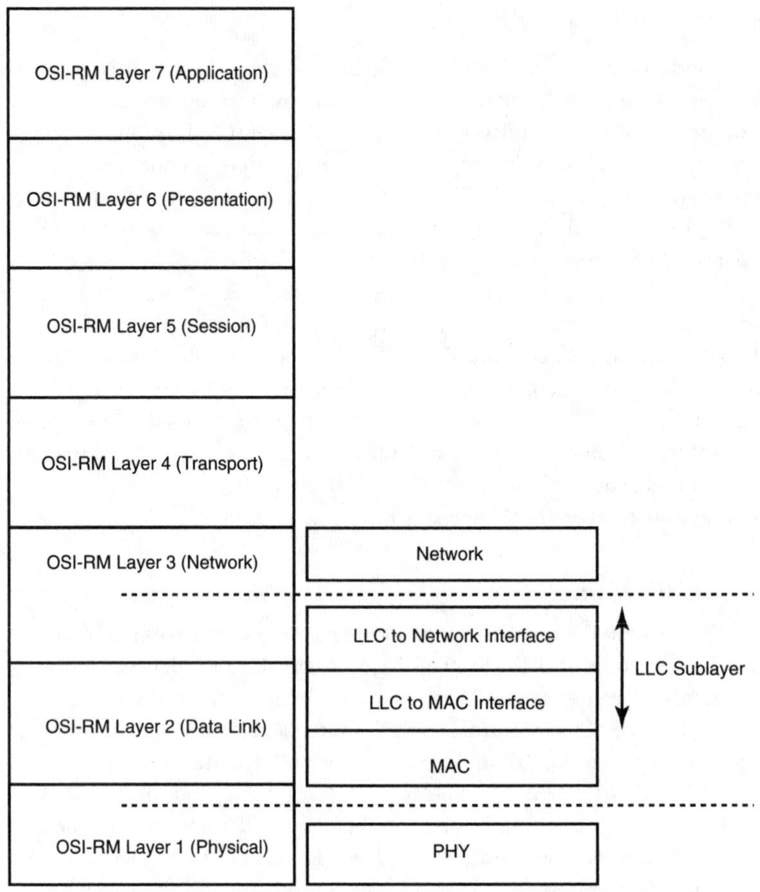

Figure 4.1 The LLC interfaces in relation to the OSI-RM Layers 1, 2, and 3.

The LLC Data Unit

The LLC data unit (LLC PDU) is generated by the LLC Layer 3 interface and appended with the Layer 3 data, and both are encapsulated in the data portion of the MAC frame. The LLC protocol data header consists of two parts. The first part is SAP information. The SAP is used by the Layer 3 protocol to identify what service the data is associated with. Two fields in the LLC header are used for this purpose:

Destination Service Access Point (DSAP) and Source Service Access Point (SSAP). The DSAP indicates the service protocol to which the data in the LLC PDU belongs. The SSAP indicates the service protocol that sent the data. These are the same value, as the source and destination protocols must be the same in order to actually exchange data.

The other part of the LLC header is the LLC control field, which indicates the LLC type and service class that should be used to deliver the Layer 3 data. The LLC control messages are used to send messages about the transmission of data between the source and destination end-stations. The LLC PDU is actually the "data" contained in the data field of the transmission packet, because the LLC PDU is used for both control messaging and the actual delivery of the Layer 3 datagrams. Three types of LLC data delivery services exist. Figure 4.2 illustrates the LLC PDU format.

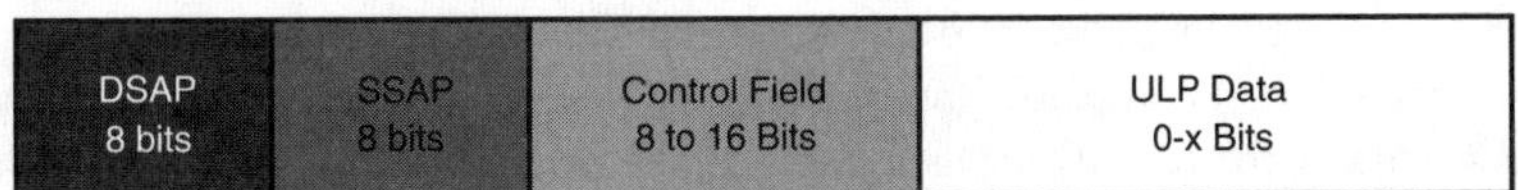

Figure 4.2 The LLC PDU.

LLC Service Types and Classes

The first LLC service, Type 1, provides unacknowledged connectionless data delivery. Type 1 is a best-effort delivery service. It offers no facilities for packet acknowledgement, flow control, or error recovery. No logical connection is established between the source and destination devices; rather, the source device sends the packet over a shared channel, where all the connected devices see the packet, but only the destination device accepts it. LLC Type 1 is the most commonly used LLC service, because most protocol suites have a reliable Layer 4 (transport) protocol. Type 1 LLC service is the fastest of all the LLC delivery services.

LLC service Type 2 is a connection-oriented data transmission. LLC 2 requires that a logical connection be established between the source and destination devices. Data delivery still occurs over a shared channel, however. The connection is established by the source host when the first LLC PDU is sent. When it receives the LLC PDU, the destination host responds with the control message "LLC PDU," no data, just a connection acknowledgement. When connection is established, data can be sent until the connection is terminated. LLC command and LLC response LLC PDUs are exchanged between the source and destination during the transmission to acknowledge the delivery of data, establish flow control, and perform error recovery if needed. The transmissions of certain Layer 3 protocols require the use of LLC Type 2.

The LLC Type 3 provides acknowledged connectionless data delivery. LLC Type 3 provides facilities for the establishment and termination of data delivery. However, it does not provide for error recovery and uses a more primitive flow-control mechanism than LLC Type 2.

The LLC station service class is determined by its capability to support different service types. All LLC stations can support class 1 service, which means they can support LLC Type 1 delivery. LLC class 2 service supports Type 1 and Type 2 delivery. Class 3 supports LLC Type 1 and Type 3 delivery. Class 4 stations can support all LLC service types.

LAN Physical Layer Signaling, Encoding, and Transmission Media

The purpose of signaling and encoding schemes is twofold. First, the schemes provide methods to allocate the available transmission medium's bandwidth. Second, they provide a signal formatting language of sorts so that signals transmitted across the medium will have some actual meaning and can be interpreted logically. Two types of signaling exist: baseband and broadband. LAN transmission protocols commonly utilize baseband signaling schemes, whereas WAN transport protocols tend toward broadband signaling.

With broadband signaling, the bandwidth of the transmission medium is partitioned into different frequency ranges. Each frequency range represents a different channel, each of which can support an independent data transmission. Broadband signaling is more complex than baseband signaling because the transmission and receiving equipment needs to distinguish between different frequency ranges. Broadband transmission occurs by having an end-station convert its digital data into a specific frequency modulated signal, which is demodulated back into digital form by the listening end-stations.

Baseband signaling is much simpler. All the transmission medium's bandwidth is used as one channel to transmit data. The advantage of baseband is that data can be transmitted in digital form, whereas with broadband, digital signals need to be modulated and demodulated. To transmit digital signals across the medium, DC current or light pulses are used to represent ones and zeros. In its most basic form, baseband signaling uses the presence of voltage or light to represent a one and the absence of voltage or light to represent a zero.

Note

In the case of Ethernet, there are PHY implementations that utilize both signaling schemes. It is for this reason that Ethernet PHY implementations contain either Base and Broad in their IEEE standard names. This identifies the type of transmission signaling used by the PHY implementation.

There are several baseband line signaling techniques. The most common is Non-Return-to-Zero (NRZ). NRZ uses the presence and absence of voltage to represent ones and zeros. Non-Return-to-Zero Invert on Ones (NRZI) is another popular baseband line signaling technique. NRZI uses the shift in line level to indicate a one and the absence of a line level shift to represent a zero. In other words, when there is no line signal, it's a zero, and when there is a signal, it's a one. For NRZ, a signal on the line represents a one, but the line starts off with no signal, whereas with NRZI there is always a line signal. Hence, the transition from an active state to passive state can indicate a zero. However, neither of these methods is easily implemented for LAN use, because a timing element is needed to distinguish the bit periods. Without timing, the sending and receiving stations will not be able to discern how many bits are actually being transmitted when a stream of consecutive ones or zeros are sent. To use these signaling systems, the protocol must specify a timing mechanism.

Manchester and MTL-3 Line Encoding

Because LAN transmissions are intermittent in nature and cannot be used to provide a constant source of clocking, a variation on NRZ was developed, Manchester encoding. Manchester encoding is used by 10Mbps Ethernet and Token Ring PHY implementations. Manchester encoding transmits bit data in a bit-by-bit serial stream across the transmission path. Positive and negative voltage is used to represent binary ones and zeros. Timing is derived through the voltage transitions in the middle of each bit. High to low transitions represent a zero bit and low to high represent a one bit. Manchester encoding works well, but is not very efficient because it requires a lot of signal bandwidth to accommodate the positive and negative voltage transitions. NRZ, by comparison, only requires half as much, as it utilizes just two voltage states (none and positive).

Manchester encoding also utilizes fairly high voltage states (+/- 3VDC). This high signal load becomes problematic at high frequencies (required for high-speed transmission) because it becomes susceptible to electromagnetic interference (EMI) and radio frequency interference (RFI). For 10Mbps Ethernet, Manchester's serial-bit stream and self-timing attributes make it attractive and its limitations go unnoticed.

For high-speed line encoding over copper transmission media, a different encoding scheme is used: Multilevel 3 (MLT-3). MLT-3 is used by 100Mbps Ethernet over UTP and Copper Distributed Data Interface (CDDI), which is an implementation of FDDI over copper cabling. MLT-3, like Manchester, is a bitstream "line" encoding mechanism, but it uses a low bandwidth circular ternary signaling system where bit changes are indicated with voltage transitions 0 to +1, to 0 to -1. MLT-3's efficient and smooth pattern flow enables MTL-3 to operate over bandwidth less than that of the original rate, without the excessive susceptibility to EMI and RFI that comes with Manchester.

Block Encoding Schemes

For high-speed operation, however, Manchester's serial bitstream transmission mechanism inefficiencies and its EMI and RFI vulnerabilities were unacceptable. So, a different encoding method was used: block encoding. Block encoding works by taking small bites or nibbles of bit data of a defined size and using a bit code that encodes in accordance with a set of defined symbols. The advantage of using block encoding is that it provides a way to send both data and control data, such as clock or error correction information. There are three popular block encoding methods used for high-speed data transmission: 4B/5B, 8B/6T, and 8B/10B.

4B/5B uses a 5-byte code to represent 4 bytes of data. There are 16 5-bit codes used to represent data and 5 codes used to send control information. 8B/6T uses 6-bit ternary codes (each bit can be VDC +1, -1, or 0) to represent the 256 possible 8-bit words. 8B/10B encoding works by using the same approach as 4B/5B, except that an 8-bit word is encoded using a 10-bit symbol. The actual encoding is accomplished with two different block code sets, 5B/6B and 3B/4B, in combination.

The use of symbols to represent data has some advantages over variable voltage line signaling. First, the symbols can be formed to minimize signal transitions, which makes them more efficient and less susceptible to misinterpretation. Second, the use of the additional bits provides a mechanism to send control and timing information independent of the transmission method.

The use of block encoding methods such as 4B/5B and 8B/10B also has the advantage of being independent of DC voltage transitions for symbol representation that results from symbol construction. This makes them suitable for use with fiber optic media, which can only transmit signals using an on or off signal state.

Physical Cabling Basics

In Chapter 1, we briefly discussed the types of transmission media used to transmit data signals, such as coaxial, shielded and unshielded twisted-pair, and fiber optic cabling. During that discussion, the idea of cable specifications or categories was also introduced. These categories and their operational specifications are defined in North America by the Telecommunications Industry Association (TIA) and Electronics Industries Association (EIA) in their 568 standards document. The TIA/EIA 568 standard describes the operational characteristics of telephone and data communication cabling system components. These components include the following:

- Horizontal and backbone copper or fiber optic transmission cable interface connectors for copper or fiber optic cabling
- Data center and wiring closet path panels and end-node wall jacks
- Path panel and end-node copper or fiber optic patch cables

Horizontal cabling, as defined by the TIA/EIA, is the cable that connects the end-node to the telecommunications/data equipment distribution closet. Backbone cabling is cable used to connect telecommunications/data equipment installed in distribution closets. Both the horizontal and backbone cabling standards use the same media types. The specific differences have to do with the permissible cable lengths. Table 4.1 contains TIA/EIA standards for copper horizontal and backbone cabling.

Table 4.1 **TIA/EIA Standards for Copper Horizontal and Backbone Cabling**

Medium	Use	Supportable Bandwidth	Distance	Connector
50-ohm thick coaxial	10Base-5 backbone cabling	500Mhz	500M	N-Series
Single mode fiber 10/125 micron*	Long distance data links up to 1,000Mbps and higher	27GHz	40km to 5000km	ST/SC/FDDI-MIC
Multimode fiber 62.5/125 micron	Short distance data links up to 1,000Mbps	160 to 500MHz	2,000M	SC/ST/FDDI-MIC
Category 1 22/24 AWG UTP	Voice and data up to 56KBps	.25MHz	Non-specified	RJ-11
Category 2 22/24 AWG UTP	Voice and data up to 1Mbps	1MHz	Non-specified	RJ-11
Category 3 24 AWG 100-ohm UTP	Voice and data up to 10Mbps	16MHz	Horizontal length 90M, 10m for patch cabling, backbone length 800m	RJ-45
Category 4 AWG 100-ohm UTP	Voice and data up to 20Mbps	20MHz	Horizontal length 90M, 10m for patch cabling, backbone length 800m	RJ-45

continues

Table 4.1 **Continued**

Medium	Use	Supportable Bandwidth	Distance	Connector
Category 5/5E 22/24 AWG 100-ohm UTP	Voice and data up to 100Mbps	100MHz	Horizontal length 90M, 10m for patch cabling, backbone length 800m	RJ-45
Category 6** 22/24 AWG 100-ohm UTP	Voice and data up to 1,000Mbps	200MHz	Non-specified Enhanced	RJ-45
Category 7** 22/24 AWG 100-ohm STP	Voice and data beyond 1,000Mbps	600MHz	Non-specified	Non-specified

* *One millionth of a meter*

** *Still in standards development*

Note

The transmission media is one of the most important components in the network when it comes to error-free data delivery. Cable length specifications (both minimum and maximum) exist to ensure that the transmission signal can be accurately interpreted. With this said, it is important that each node and device interconnect be cabled within specifications. Cable length violations can result in network timing errors (one of the more common Ethernet cable-related errors) and corrupted data transmissions due to signal loss as a result of cable attenuation (copper cable) or inadequate luminosity (fiber optic cable).

Fiber Optic Transmission Media

Fiber optic cable has been in use in the telecommunications industry for more than 20 years, and it provides many advantages over traditional copper-based cabling. The most significant advantage is its large bandwidth capability in terms of the amount of bits it can transport in a second. Fiber optic cabling is capable of transporting data at multi-gigabit rates (up to 160Gbps with current technology) and beyond. This has made fiber optic cabling the most attractive means for transporting high-speed/high-bandwidth transmission signals. In addition to its impressive bandwidth proprieties, fiber optic cabling has several additional properties that make it far superior to copper wire:

- Photons versus electrons—Fiber optic technology uses gallium-arsenide-based LEDs and lasers to send photons through the fiber to relay signaling information. Copper media use alternating electrical signals. This advantage can be summarized as the use of light over electricity, which makes fiber optic cable immune to electromagnetic and radio frequency interference (EMI/RFI) and excessively hot or cold environmental conditions.
- Low loss and higher data integrity—Electrical data signals lose their signal strength as they pass through copper cabling, because of the inherent resistance of copper (this is known as signal attenuation). To counteract signal loss, signal amplifiers (repeaters) are used along the data path to reconstitute the signal. In fiber optics, light is the signaling element, so there is little or no inherent resistance and therefore low signal attenuation. The lack of attenuation is a primary factor in fiber's high bandwidth qualities. This permits the use of fiber in long distance transmission scenarios without the need for signal regenerators, and results in a higher data integrity rating (low-bit error rate) than copper, because little or no signal regeneration is needed. Regeneration is the source of most cases of transmission errors.
- Security—Because light is used to relay data signals, fiber optic cabling does not emit any residual signaling. To tap into the signal channel, the cable must be severed. This results in a total signal loss until the path is refused together, making any unauthorized modification to the medium immediately detectable.
- Installation and cost—Historically, fiber optic cabling has had a reputation for being more costly to install and more prone to physical damage than copper cable. Although fiber optic cabling does require some additional expertise to install, the actual cost differential between copper and fiber cabling is negligible today. In addition, fiber optic cable is lighter, more durable, and can be installed in many environments where copper-based medium cannot be used at all (such as areas where high levels of EMI/RFI exist).

Fiber Optic Transmission Basics

Fiber optic transmission has three basic components:

- Light-generating source—A light-emitting diode (LED), but more recently a laser
- Fiber transport medium—Consists of a glass (or plastic) fiber core, cladding (an outer layer of glass that covers the core and reflects back any light that is lost due to refraction), an outer covering, and a connector
- Light-receiving photoelectric receiver

When light is sent from one substance to another, two things occur: the light is reflected and is absorbed. The light that is absorbed is skewed due to the absorption process. This is known as refraction. Data transmission is possible through the combination of light refraction and reflection, using materials with differing refraction indices. The refraction index determines how the wavelength of light is bent when it passes through a substance and also the rate of photon passage through the substance. Though copper and fiber cabling both transport data at the same rate (0.69 percent of the speed of light), the difference between the two is the amount of useable bandwidth, due to fiber optic's low attenuation rating. Fiber optic cable suffers minimal signal loss in the form of actual resistance, because photons instead of electrons are used. This results in a more effective use of the transmission medium. The actual amount of bandwidth available for transmission is determined by the width of the light source and the rise and fall rate of the light. The narrower the light wavelength and the faster the rise and fall rate, the greater the bandwidth of the fiber transport.

Three types of light wavelengths are used to transmit signals: short, long, and very long. Short wavelength light has a wavelength size between 770 and 910 nanometers. Short wavelength light is usually generated by a gallium-aluminum-arsenide infrared laser or LED with nominal yield of 850 nanometers. Long wavelength light ranges between 1,270 and 1,355 nanometers. Very long wavelength light ranges between 1,355 and 1,500 nanometers. Both long and very long wavelength light sources are generated with gallium-indium arsenide-phosphate infrared lasers. Short and long wavelength light is used for LAN transmission applications. Very long wavelength light, 1,500 nanometers, is used exclusively for telecommunications signaling applications such as Synchronous Optical Network (SONET).

As it passes through the fiber strand, the light slowly spreads out, and this phenomenon is known as dispersion. Three forms of dispersion are present in fiber transport: modal, chromatic, and material. The dispersion factor determines how far the data can travel across the fiber path before the signal will be unusable (the signal can be repeated or amplified if the cable length exceeds the dispersion factor). Fiber optic cable comes in multimode and single mode. Two differences between multimode and single-mode fiber are core size and how the light travels across the core.

The TIA/EIA describes cable using two measurements: the core size and the cladding size. The TIA/EIA defines two types of multimode fiber: 62.5/125 micron and 50/125 micron. 62.5/125 micron is the standard used by FDDI, 10Base-FL, 100Base-FX, and 1000Base-SX/LX. 50/125 micron is used (and preferred over 62.5/125 micron) only with 1000Base-SX and 1000Base-LX. Single-mode fiber has a much smaller core size. The TIA/EIA defines the single-mode fiber standard as 10/125 microns. With multimode fiber, the light bounces off the cladding in a rise and fall pattern across the fiber core. With single-mode fiber, the light travels directly through the core, as a single beam.

Multimode's larger size permits multiple pulses to be carried across the strand. With this larger size, it is more forgiving of transport errors, which commonly occur at the media interconnects. The "forgiveness" factor of multimode makes it better suited for LAN applications, where multiple path interconnections are the norm. The downside of multimode is that it suffers from modal dispersion. As different light pulses reflect at different angles off the fiber core, there is a possibility that pulses sent at the same time will reach the end at different times due to the number of times the pulse has been reflected. This delay can result in signal misinterpretation. Single-mode, with its smaller core diameter, can only carry one signal, so it does not suffer from modal dispersion and can therefore transfer data farther without any regeneration assistance. Figure 4.3 illustrates the differences between multimode and single-mode.

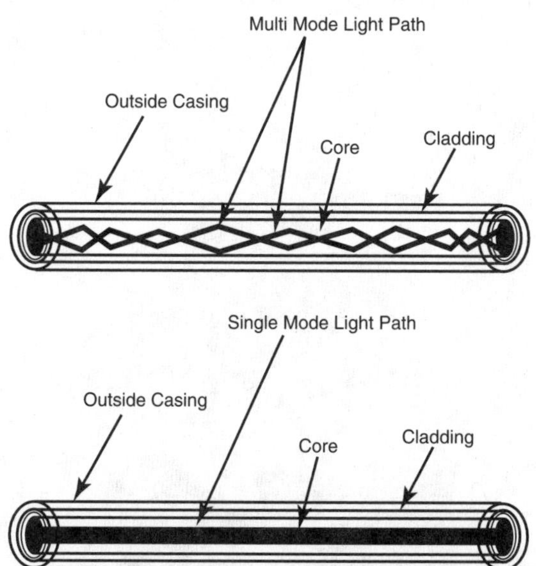

Figure 4.3 Light paths in multimode and single-mode fiber.

> **Note**
> Cladding is the glass "shell" that surrounds the fiber inside the plastic casing. Its reflective properties affect the performance of the fiber.

However, both single-mode and multimode still fall victim to chromatic and material dispersion. Chromatic and material dispersion result from imperfections in the transmission source. Ideally, the laser wants to send each pulse at the optimum wavelength specified for the media. This is not possible, however, so there is some variation on the wavelengths, which results in reduced signal strength. This is known as chromatic dispersion. Material dispersion is a result of imperfections in core and cladding.

Three types of connectors are used with fiber optic cabling:

- FDDI-MIC—The FDDI-MIC type of connector was specified as part of the FDDI PHY standard; it is only used for this purpose. The MIC connector uses two fiber cable strands on a special slotted connector. When used properly, the MIC connector only permits the connector to be inserted one way.
- ST—ST-type connectors, also known as "bayonet style" connectors, are used with 10Base-FL. ST connectors are the most common type of fiber optic connectors. Today, they are used almost exclusively for fiber optic patch panel applications.
- SC—SC-type or "box" connectors are used for 100Base-FX and 1000Base-X applications. They are easier to work with than the ST-type connectors. Quite often, ST-type to SC-type patch cables are needed to patch fiber optic equipment together or to build wide fiber cabling infrastructures.

The FDDI-MIC, ST-type, and SC-type connectors are depicted in Figure 4.4.

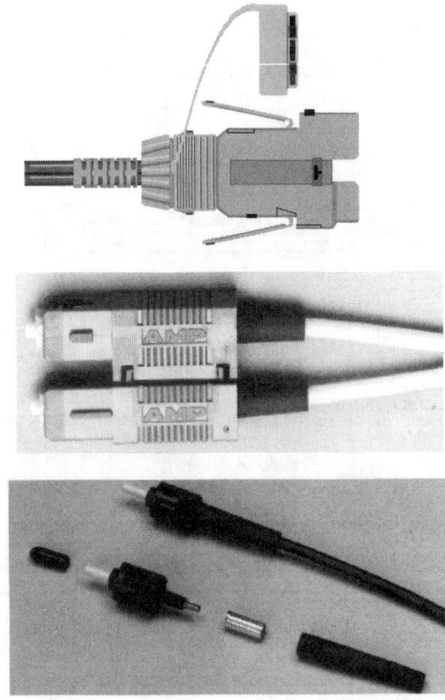

Figure 4.4 The FDDI-MIC, ST-type, and SC-type connectors.

Ethernet

The original Ethernet protocol was developed by Dr. Robert Metcalfe and his development team at Xerox PARC (Palo Alto Research Center) in Palo Alto, California, in 1972. Ethernet's media access mechanism, Carrier Sense Multiple Access Collision (CSMA/CD), was based on work done by Dr. Norman Abramson at the University of Hawaii on the radio transmission WAN known as the ALOHA system. Metcalfe's experimental Ethernet network, called ALOHA ALTO, supported 100 nodes on 1km length of 50-ohm coaxial cable and operated at a rate of 2.94Mbps. The nodes were connected to the cable with transmitter-receivers (transceiver) and a cable tap. The combination of the cable tap and transceiver is known as an MAU (media access unit).

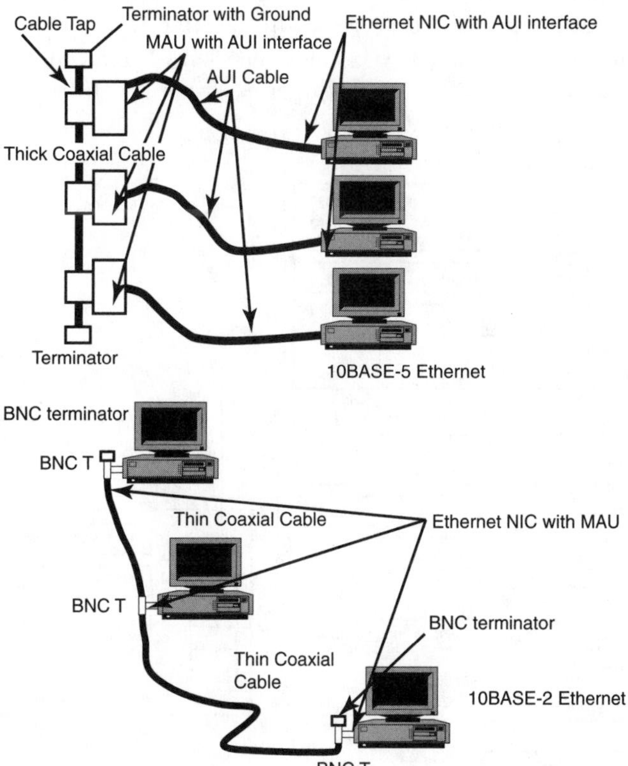

Figure 4.5 10Base-5 and 10Base-2 interconnection models.

In 1979, the Ethernet DIX consortium was formed between Digital Equipment Corporation, Intel, and Xerox. The following year (1980), the DIX "Blue Book" Ethernet V 1.0 specification was published (based largely on the experimental ALTO Ethernet). Then in 1982, the DIX specification was updated to Ethernet Version 2.0 (commonly referred to a the Ethernet II specification). Then in 1981, the IEEE subcommittee 802.3 began the process of developing a draft standard based on the DIX Ethernet specifications. In 1983, the IEEE finalized the first standards-based Ethernet version, 10Base-5, which was based on the DIX V2 or Ethernet II specification. The following year 3Com (Computers, Communication, and Compatibility) and Hewlett-Packard submitted the "Thin-Net" standard to the IEEE 802.3. Their standard, 10Base-2, operated over a thinner grade of 50-ohm coaxial cable, and the transceiver was integrated into the Network Interface Card (NIC). Instead of using the cable tap and AUI drop cable approach (between the NIC and the transceiver) used by 10Base-5, 10Base-2 used BNC (British Navel Connectors) and T-barrel connectors to connect the end-stations to the bus cable. The IEEE formally approved 10Base-2 in 1989.

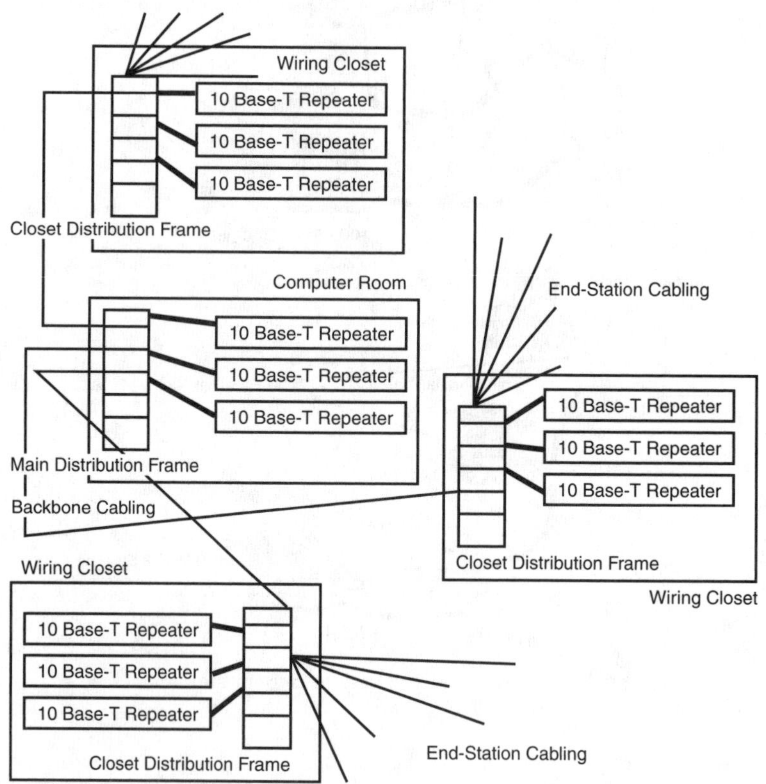

Figure 4.6 10Base-T multiport repeaters in a hub configuration.

The event that really solidified Ethernet as the dominant LAN protocol came in 1987, when a company Called SynOptics started shipping a version of Ethernet that used multiport repeater and Unshielded Twisted-Pair (UTP) in star topology instead of the standard Ethernet BUS topology. The product was called LATTISNET. Around the same time, the IEEE began meeting to develop a standard for the operation of Ethernet over UTP. Three years later, in 1990, the IEEE issued an Ethernet-over-UTP standard known as the 802.3i 10Base-T standard. The development of Ethernet over UTP had significant implications for the development of Ethernet-based LANs. In 1985, IBM began shipping its Token Ring LAN product. Though Token Ring was slower than Ethernet, it utilized star topology, and it offered a structured wiring model, which made large office installations much easier. StarLAN was the first star topology-based Ethernet implementation that could run over existing UTP telephone cables, but it only operated at 1Mbps. 10Base-T provided "bus-based" Ethernet operating rates, and it could operate over existing UTP telephone cabling infrastructures that were installed in most large buildings (refer to Figure 4.4). An example of the need for this integration between Ethernet and existing telephone cable infrastructures is in the RJ-45 pins selected to transmit and receive singles. This pin setup sets 1 and 2 for transmit and 3 and 6 for receive, leaving pins 4 and 5 available for the tip and ring pair (used for standard telephone service), so both signals could be run over the same UTP cable. In addition to easy integration into existing cabling plants, the use of multiport repeaters and bridges (and now switches) gave administrators the capability to effectively "grow" their Ethernet LANs to support very high end-station densities.

The 1990s were the evolutionary turning point for Ethernet. After large-scale Ethernet installations were possible, the increasing performance demands from faster computers soon outgrew the limitations of shared Ethernet. In response to these limitations, a flurry of performance-enhancing developments to Ethernet were introduced. The first was the development of Ethernet multiport bridges (commonly called switches) and full-duplex Ethernet. With the advent of 10Base-T Ethernet switching (by a company called Kalpana), the overall operating throughput of Ethernet was significantly enhanced and Ethernet's network size limitations (with 500m the maximum distance allowed between 10Base-T end-stations) were effectively removed (see Chapter 6, "Network Switches," for further information on Ethernet Switching).

In 1992, a company called Grand Junction Networks started work on Fast Ethernet, based in part on the signaling and physical media implementations specified in FDDI, PHY, and PMD sublayer standards. This work would eventually become the basis of the IEEE 802.3u, the 100Base-X Fast Ethernet standard, which was formally approved in 1995. The development of a 100Mbps implementation of Ethernet further increased Ethernet's already large installation base and brought high-speed data networking to the desktop.

100Mbps technology prior to Fast Ethernet was perceived largely as a backbone technology. Although FDDI can support large end-station densities on a single ring, its dependency on fiber optic cabling (and expensive NICs) made it too costly for use as a workstation technology. ATM was not much better, even though it provided a 25Mbps and 155Mbps solution that operated on UTP (622Mbps over fiber). ATM switches and NICs are also expensive, and ATM has the added bonus of not being compatible with any traditional Layer 2 LAN protocols. Therefore, for ATM to be implemented in the LAN, a LAN emulation (LANE) service needs to provide compatibility services so ATM-attached stations can interact with LAN-attached stations. Even as backbone technologies, FDDI and ATM support different frame and packet sizes, so Ethernet packets need to be processed in order to be carried over the high-speed backbone. This additional processing results in a performance cost that can only really be adjusted when transporting a slower protocol.

When Fast Ethernet first became a standard, it was seen as a backbone technology due to its limited network growth capability and short collision domain distance. However, once Fast Ethernet multiport bridges became available, Fast Ethernet product development took off. When its distance limitations were removed and its cost came down relative to traditional 10Mbps Ethernet, Fast Ethernet was installed to meet the growing bandwidth requirements of Internet, multimedia, and relational database applications. The utilization of Fast Ethernet as a desktop technology had been foreseen, but not at the rate it actually occurred. The traditional 100+ Mbps backbones based on FDDI (100Mbps), ATM (25 and 155Mbps), and switched Fast Ethernet (100Mbps full-duplex) could no longer provide enough adequate bandwidth for efficient operation. In many cases, network administrators with no real backbone technology alternative to increase their backbone capacity, enforced 10Mbps-only operating rates for the desktop end-stations on their LANs.

Note

Shared medium transmission protocols traditionally operate in half-duplex mode, where stations can either send or receive data at one time. With the advent of multiport bridges, it became possible to have stations connect to the multiport bridge and operate in full-duplex mode, which enables the simultaneous transmission and receipt of data. Full duplex is possible because each port on the multiport bridge (in combination with another nodes) represents a collision domain. If only one host is attached to the bridge port, the need for collision detection is removed, because only the bridge and the host are communicating. Communication with other hosts is managed across the multiport bridge's internal bus, so the bridge's port essentially acts as a transmission proxy for the remote host who is connected to another port on the bridge.

Although the IEEE could not have predicted all the effects Fast Ethernet has wrought, they did realize that a need for an even faster Ethernet transport existed. In late 1995, an IEEE study group was formed, and by the middle of 1998, a Gigabit Ethernet standard 802.3z was completed. Recognizing the need for a faster backbone technology, the IEEE specifically prioritized the development of a fiber optic based standard over traditional copper media. They did this for two reasons. The first was technical: The highest rated copper media did not offer adequate bandwidth, but fiber optic media did provide more than adequate bandwidth. This also permitted the utilization of existing fiber optic based, high-speed data transmission standards as a foundation for the Gigabit standard development (which was done with the development of the Fast Ethernet standard). The second reason was the perceived initial demand for the application of Gigabit Ethernet: to replace existing fiber-based LAN backbones that were already using fiber optic cable for the transmission of FDDI, ATM, and Fast Ethernet.

Since the introduction of the IEEE Ethernet standard in 1983, it has evolved from a shared medium, passive bus protocol operating at a rate of 10Mbps, with a maximum cable distance of 2500 meters and node count of 500 end-stations, to a protocol capable of operating at 1,000Mbps. In a point-to-point implementation, the standard can provide a distance of up to 80km between two endpoints and support more than 1,000 end-stations on a logical segment. Ethernet is an old technology by computer industry standards. By reinventing itself while maintaining backward compatibility, it continues to be the dominant LAN networking protocol, with an ever-increasing installation base. This section looks at the Ethernet MAC (Layer 2) and various PHY (Layer 1) implementations, as well as Gigabit Ethernet.

Ethernet/IEEE 802.3 MAC

Although Ethernet and 802.3 Ethernet are usually considered to be the same, technically, Ethernet and 802.3 are two different technologies. Ethernet is the CSMA/CD shared media protocol developed by the DIX consortium. DIX Ethernet has two versions: V1 (a 10Mbps version of the original Xerox experiment and quite obsolete), and V2 (Ethernet II), which is the basis of the IEEE 802.3 standard. IEEE 802.3 is the CSMA/CD standard and protocol based on the DIX Ethernet II standard. Although there are only slight differences in the frame formats of the Ethernet II and IEEE 802.3 standards, they are incompatible with each other. In addition to differences in frame format, Ethernet's PHY implementation is limited to passive bus thick and thin 50-ohm coaxial cable with a maximum operating rate of 10Mbps. The 802.3 standard has PHY implementations on coaxial, UTC, and fiber optic cable, and operating rates up to 1,000Mbps.

> **Note**
>
> In most respects, Ethernet II and 802.3 are almost identical in terms of implementation and operation. There is a rather important difference to be aware of, however, in terms of their collision detection implementations. Ethernet II utilizes a Signal Quality Error (SQE) test signal to verify the proper operation of the transceiver's collision detection circuitry. If enabled, this SQE signal causes conflicts with 802.3 transmission equipment. Because of this potential conflict, SQE is normally disabled on external Thin and Thick coaxial transceivers that are 803.3/Ethernet II compliant. If you are experiencing problems on an 802.3 network using Ethernet II/802.3 compatible external transceivers, check the SQE switch setting and make sure it is disabled.

All Ethernet MAC time-based functions are measured in bit-periods, which is the amount of time it takes for 1 bit of data to be transmitted. The bit-period, in terms of actual time, is based on the transmission rate of the Ethernet MAC protocol. As the operating rate of the MAC increases, the bit-period is reduced in actual time. For example, with standard 10Mbps Ethernet, it takes 1MHz or 1,000 nanoseconds to move 1 bit of data across the transmission medium. Therefore, for standard Ethernet, which operates using a 10MHz bandwidth, a bit-period is 0.1 µsec or 100 nanoseconds. The concept of time and its proper interpretation is vitally important for all data transmission protocols, in particular, those that use collision detection for transmission medium access. In order for the Ethernet collision detection mechanism to work correctly, each 802.3 MAC implementation has defined three "time periods" for managing access to the transmission medium. These time periods are slot-time, InterFrame Gap (IFG), and jam-period.

Slot-time represents the minimum period of time that a transmitting station needs access to the transmission medium to send the smallest legal frame size. Under the Ethernet CSMA/CD rules, all end-stations connected to the medium segment "listen" to all the frame transmissions on the wire, but only the destination accepts the packet. After the destination receives the packet, it checks it for errors, and if there are none, the data is handed to the ULP for further processing. If the frame is corrupt, it is discarded. When an end-station wants access to the media to transmit a frame, it listens to the wire for a period of one slot-time for carrier signal or active frame transmission. If no carrier signal is heard after this period, the wire is considered clear and the station can begin its transmission. The original IEEE 802.3 standard states that all Ethernet MACs have a 512 bit-period slot-time. When Gigabit Ethernet was introduced, the slot-time was extended to 4,096 bit-periods to accommodate for the increased transmission rate. MAC implementation's slot-time is also the basis for determining the physical size of the collision domain. Under Ethernet's CSMA/CD implementation, all stations on the segments must be able to "hear" the frame transmission before it is completed, in order to avoid a collision.

To ensure that the transmitted frames are distinguishable from one another, Ethernet's CSMA/CD implementation enforces a mandatory "dead time" between each frame called the *IFG*. The IFG is a 96 bit-period "space" of inactivity on the wire between each transmitted frame. Depending on the Ethernet NIC implementation, after the end-station has detected a "clear" wire, the end-station needs to wait an additional 96 bit-periods to ensure the wire is clear, effectively increasing the slot-time an additional 96 bit-times. The IFG period is also the basis for "defer time" clock used by the NIC when a station, after the slot-time has elapsed, still detects a carrier signal on the wire. The IFG, although not technically part of the Ethernet frame, needs to be accounted for as a percentage of the utilized bandwidth on the transmission segment.

When determining the segment's effective utilization, each IFG represents a finite period of transmission inactivity on the medium, which detracts from the total available bandwidth used to transmit Ethernet frames. For this reason, when looking at segment Frames Per Second (FPS) rates, the IFG is added to each frame transmitted.

In the event that two Ethernet end-stations send frames at the same time, a collision occurs. Although it would seem that collisions are undesirable, they are a natural part of Ethernet's operation. Collisions provide the mechanism for regulating access to the transmission medium so that a single station cannot control the transmission medium. Collisions are also used as a primitive flow-control mechanism when an end-station cannot handle the rate of data being transmitted; it can transmit a frame and halt the data flow. Any attempt of two or more stations to transmit frames on the segment results in a collision. When a collision occurs, it lasts 1 slot-time (including the IFG, for a total period of 576 bit-periods), during which time, all transmission activity on the medium stops. The collision event itself results in the following:

- All the end-stations involved in the collision (if functioning properly) stop transmitting data frames and then transmit a 32 bit-time (in some cases, a 48 bit-time) jam signal.
- Each end-station attached to the transmission medium needs to hear the jam signal. Upon hearing the jam signal, any transmission attempts in process on the segment are stopped.
- With the wire now quiet, the end-stations involved in the collision event use the truncated binary exponential backoff algorithm to determine a random listening period to wait before attempting to retransmit their frames. At the end of this wait period, if the wire is clear of traffic, the station can begin to retransmit its data. If a collision occurs again, the wait period is doubled. This process can continue for up to 16 times, after which the packet is discarded. The end-stations not involved with the collision event can initiate transmission processes using the standard CSMA/CD algorithm.

Table 4.2 lists the different time operating parameters for all the IEEE 802.3 Ethernet MAC implementations.

Table 4.2 **802.3 Ethernet Implementation's Operating Parameters**

Ethernet MAC	**10Base**	**100Base**	**1000Base**
Slot-Time†	512 bit-periods	512 bit-periods	4,096 bit-periods
Bit-Period*	100ns	10ns	1ns
Round Trip Delay	51.2µ	5.12µ	512µ
Interframe Gap	9.6µ	0.96µ	0.96µ
Minimum Frame Size**	64 bytes	64 bytes	64 bytes††
Maximum*** Frame Size	1,518/1,522 bytes	1,518/1,522 bytes	1,518/1,522 bytes

† *To accommodate for 1,000 Base Ethernet's increased transmission rate, an increased slot-time is required.*

†† *Under half-duplex operation, actual minimum frame size for Gigabit Ethernet is actually 512K. To maintain backwards compatibility with the other Ethernet MAC implementations, Gigabit uses a bit-stuffing process called Carrier Extension to pad frames so they meet the minimum frame requirements for the enlarged Gigabit MAC slot-time.*

**One bit time equals the cycle rate for the transmission path bandwidth. For example, 10Mbps Ethernet operates at 10Mhz so the bit time is 100ns.*

*** The minimum 64 byte Ethernet frame consists of the frame header, data PDU, and a CRC checksum for the frame. This value does not include the additional 8-byte Ethernet frame preamble that is discarded when the packet is actually processed. For the practical purposes of measuring the actual time it takes to transmit an Ethernet frame and measure the FPS transmission rate of the segment, however, the preamble must be accounted for in the packet size calculation. Therefore, by including the Preamble/Start of Frame Delimiter minimum, Ethernet packet size is actually 72 bytes.*

****The addition of the 8-byte preamble must also be done when calculating transmission times for the maximum Ethernet frame size. Adding the preamble increases the maximum frame size to 1,526/1,530 bytes.*

Ethernet/IEEE 802.3 MAC Framing

There are four Ethernet/802.3 frame types: the original DIX/Ethernet II and three variations of the 802.3 frame. The Ethernet II and 802.3 frame formats are illustrated in Figure 4.7. All the variations on the 802.3 frame relate to the format of the data field (also known as the protocol data unit). These variations are discussed in the next section.

> **Note**
>
> The 802.3 frame type used in Figure 4.5 is the latest iteration of the 802.3 frame, which is specified in the 802.3ac standard. This frame type adds a 4-byte field for the 802.1q VLAN standard, increasing the maximum frame size from 1,518 bytes to 1,522 bytes. VLAN is covered in Chapter 6.

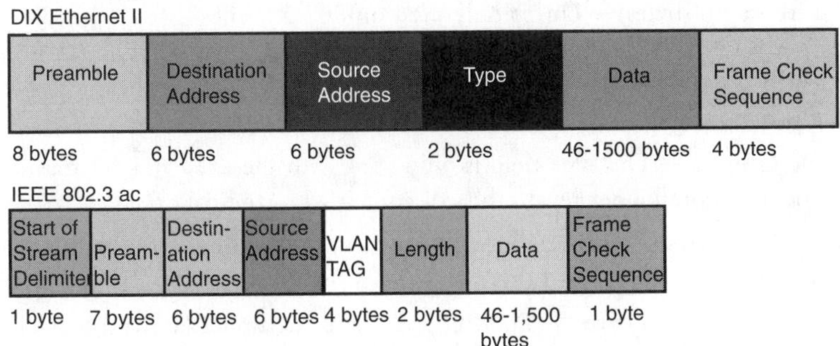

Figure 4.7 Ethernet II and IEEE 802.3ac frames.

Now let's look at the fields that make up the Ethernet/802.3 frames.

- Preamble (7 bytes)—Every device on the Ethernet segment looks at all the traffic transmitted. The preamble consists of 7 (1010101) or 8 bytes (10101010). The preamble tells all the devices on the network to synchronize themselves to the incoming frame.
- Start of Frame Delimiter (SFD) (1 byte)—This is a continuation of the preamble. The first 6 bits are in the binary order (101010), the last 2 bits are (11) so the SFD has a binary signature of (10101011). This breaks the sync and alerts the devices that frame data is following. The preamble and the start of frame delimiter are dropped by the Ethernet controller, and when transmitting, it adds the correct frames for the Ethernet format in use. The preamble and SFD are not included in the calculation of the minimum and maximum frame size.
- Destination address (6 bytes)—This tells the network where the data is being sent. There are slight differences between the Ethernet and IEEE 802.3 frame. The destination field also indicates if data is destined for a single station (unicast), a group of stations (multicast), or all the stations on the network (broadcast). The address itself can be either locally or universally assigned.
- Source address (6 bytes)—This address tells the network where the data was sent. Like the destination address, it can be locally or universally assigned. Locally administered addresses are 2 bytes long. The addresses are loaded onto flash memory on the Ethernet card. Locally administered addresses are generally found in mainframe environments. In most workstation environments, universal addresses are used. Universal addresses are "burned" into the card. Because each Ethernet card must have its own unique address, organizational unique identifiers (OUI) are assigned to every NIC manufacturer. The NIC identifier is the first 3 bytes of the address. The last 3 bytes of the address are uniquely assigned by the manufacturer.

- VLAN tag (4 bytes)—This is only used on 802.3 frames that are handled by 802.1q VLAN-enabled network switches (multiport bridges). The tag is used to identify a packet with a specific VLAN. When the packet leaves the switch, the tag is removed.
- Type (2 bytes)—The type field is only used with the `Ethernet II` frame type. It identifies the higher level protocol contained in the data field.
- Length (2 bytes)—This provides information on the size of the frame. The minimum size of an Ethernet frame is 64 bytes.
- Data (46 to 1,500 bytes)—The PDU contains the upper layer protocol information and the actual data. The minimum size is 46 bytes, which consists of a LLC PDU header of 4 bytes + 42 bytes of data or padding, and the maximum is 1,500 bytes. The PDU construction is the responsibility of the LLC interface. 802.2 (covered at the beginning of the chapter) is the standard LLC interface used to construct the PDU.
- Frame check sequence (4 bytes)—This provides the means to check errors. Each transmitter performs a Cyclic Redundancy Check (CRC) on the address and data fields. When the receiver gets the packet delivery, it performs the same check before accepting and moving the data along.
- Start and end stream delimiters (1 byte)—The SSD and the ESD are used only with Fast and Gigabit Ethernet frames. With these, PHY implementations frames are considered "streams," hence the name. These bytes are used to mark the beginning and end of the stream. The SSD is used to align a received frame for decoding; it does not add any additional length to the frame, rather, it uses the first byte of the preamble. The ESD indicates that the transmission has terminated and the stream was properly formed. The ESD technically does not add any length to the 802.3 frame because it falls into the .96ms gap on 1,522-byte packets.

The only real difference between the Ethernet II and 803.2 frames is the protocol type field (this field is used to indicate the packet size with 802.3) used by the Ethernet II frame. An easy way to tell the difference is by looking at the section of the packet that would contain the length/type field. If the value here is greater than 05DC (hex; 1500 decimal), we know that this is an Ethernet II frame, because the maximum size of an 802.3 Ethernet PDU is 1500. This service is now provided by the 802.2 Subnetwork Access Protocol (SNAP) frame. In fact, the Ethernet II frame has long since been replaced by the 802.3 frame as the basis for Ethernet. Therefore, it is more than likely you will never see it in use. Now, let's look at the three variations on the 802.3 frames.

The 802.3 PDU and Logical Link Control

Each of these variations uses the 802.3 MAC frame for transmission, only the structure of the PDU is different. The first variation is the IEEE "standard" frame, known as the 802.2 frame. The 802.2 frame uses the 802.2 LLC PDU. The LLC PDU contains a 1-byte DSAP, 1-byte SSAP, and a 2-byte control field, which is appended to the ULP data.

The second variation is the Ethernet SNAP frame, a vendor-extensible variation of the IEEE 802.3 that builds on the 802.2 LLC. The SNAP frame was the product of the 802.1, created to provide Ethernet II and IEEE 802.3 computability by adding a 2-byte type field and a 3-byte OUI field. It is ideal when an end-station is operating more than one ULP interface. The SNAP frame can be used to transport multiple protocols such as AppleTalk Phase II, NetWare, and TCP/IP. The type field is used to indicate which protocol is the intended recipient of the ULP payload contained in the SNAP PDU. When SNAP is in use, the DSAP, SSAP, and control fields are set to specific settings ("AA" in the SSAP and DSAP fields) to indicate that the packet is a SNAP frame. Figure 4.8 illustrates the SNAP and 802.3 (Novell RAW) frame variations.

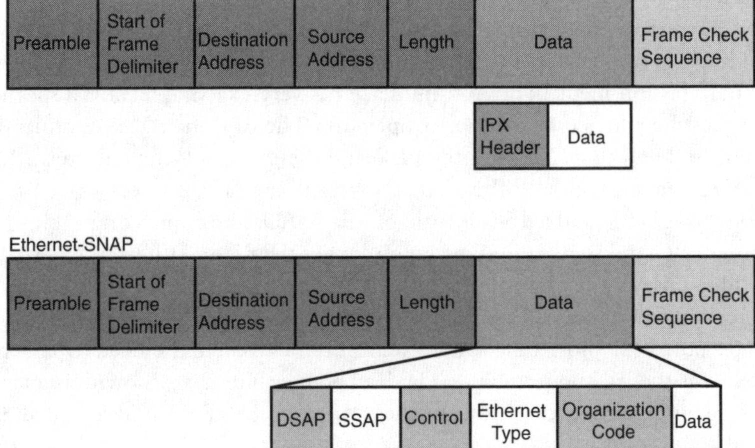

Figure 4.8 The SNAP and 802.3 (Novell RAW) frame formats.

The 802.3 frame is a proprietary variation on the IEEE 802.3 packet. It is used primarily by end-stations that are running Novell's IPX protocol (it can also be used with NWlink). The Novell RAW frame uses the 802.3 MAC without the 802.2 LLC PDU. Instead, it drops the IPX header and ULP data raw into the 802.3 data field. Because the frame is proprietary, it is not compliant with other 802.3 implementations. In terms of hardware support, 3COM and other NIC vendors market cards that can support both frame formats. However, NIC's cards are only IEEE 802.3 compliant and will not support Novell RAW 802.3 frames.

As stated earlier, the significant difference between DIX Ethernet and 802.3 is the supported physical transmission medium and operating rates. To facilitate changes to the operational characteristics and provide support for different transport technologies, the IEEE 803.3 standard provides for separate interfaces between the MAC and the PHY. Additionally, the PHY is segmented into different service-logical interfaces, which differ as the transmission rate increases. One of the main functions of the PHY is to provide data transmission signaling.

802.3 10Mbps (Classic Ethernet) Layer 1 (PHY) Implementations

For 10Mbps operation, there are four logical PHY interfaces:

- Physical layer signaling (PLS)
- Attachment unit interface (AUI)
- Physical medium attachment (PMA)
- Medium-dependent interface (MDI)

The PLS-to-MAC interface is responsible for transmitting and receiving bit data, and performing the carrier sense and collision detection duties. The AUI interface passes signals between the PLS and the PMA interface.

The AUI interface can be either integrated into the NIC or exist as an external physical interface. Its purpose is to give end-stations the independence to connect to different transmission media. The AUI interface has very exacting technical specifications that must be adhered to for proper operation. The AUI interface transmits data in a bitstream, 1 bit at a time, between the NIC and the media attachment unit (MAU) at a 10MHz operating rate (10Mbps). As an external interface, it uses a 15-pin female interface (on the NIC) and a 15-pin male on the MAU. The connecting cable between the NIC and the MAU must be no longer than 50m. When an external AUI is used with an MAU in a bus topology, the MAU represents the transmission medium interconnect. In multiport repeater or star topology, the cabling between the NIC and the repeater port is the horizontal cable connection. The cabling the MAU attaches to is referred to as the backbone cabling. The backbone cabling has its own distance and connector requirements, which are dependent on the type of transmission medium being used.

The MAU consists of the PLA and PMA interfaces. Substantial differences exist between the MAUs used for baseband and broadband. Baseband Ethernet is what is commonly used for LAN connectivity, whereas broadband implementations are used almost exclusively for consumer networking over public cable systems. Broadband Ethernet implementations have gained popularity in the last few years because they allow cable systems to provide Internet access over a cable TV infrastructure. Both broadband and baseband MAUs perform the same function, they just do it differently.

With a baseband MAU, the PMA takes signals from the AUI, converts them appropriately, and sends them on to the medium through the MDI, performing in reverse when transmissions are being received. The PMA is also responsible for detecting collisions, and it relays notification about them to the PLS over the AUI. In addition, the PMA acts as a transmission regulator, suppressing transmission activities when signals are being received and "listening" to the wire for collisions and incoming transmissions. The PMA also enforces the 20-millisecond maximum transmission period (a function known as jabber protection) rule, when long datastreams need to be transmitted. The PMA functions remain the same regardless of the MDI.

A broadband MAU operates in many respects like a modem. The MAU is responsible for modulating the digital signals sent through the AUI interface and demodulating the analog signals before transmitting them across the AUI. Its similarity with a modem comes from how it handles collision detection, data transmission, and reception. When a station transmits data, it receives a copy of the transmitted data like all the other stations on the medium; this is also the case with modem operation. This process is used for collision detection. When a packet is transmitted, a copy is saved by the MAU. The saved copy and the returned copy are then compared. If no errors are found, it is assumed that no collision occurred.

Today, many of the available NICs integrate the AUI and MAU into the interface card, or provide the choice between using the integrated AUI/MAU or the traditional AUI. The integration of the AUI/MAU has provided greater cost savings and reduced possible points of failure. Now let's look at the different 10Mbps 802.3 PHY implementations. The 10Mbps MAC and PHY model is illustrated in Figure 4.9.

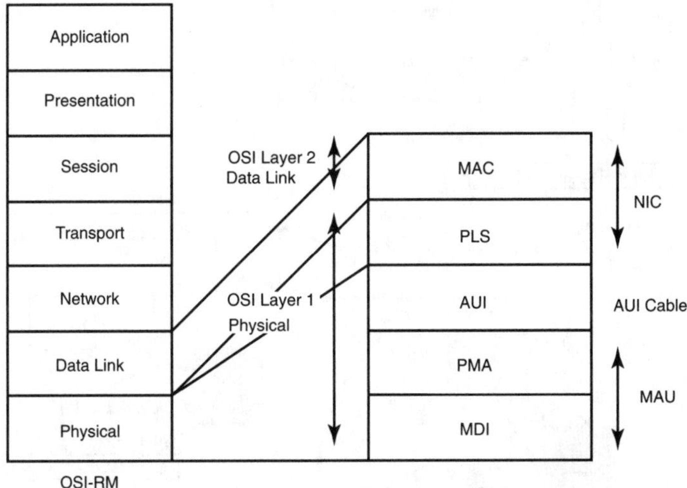

Figure 4.9 10Mbps Ethernet, data link (MAC), and physical (PHY) layer interfaces.

> **Note**
> As mentioned earlier, the terms "base" and "broad" describe the PHY transmission signaling used. The full IEEE naming convention includes the transmission speed, which will be 10, 100, or 1,000. The transmission signaling for almost every case will be base, for baseband. The name will end with either the maximum allowable cable segment length (for coaxial standards) or the transmission medium type (for everything else).

- 10Base-5—Also known as "Thicknet," 10Base-5 is virtually identical to the DIX Ethernet II standard, except for the framing type, which is IEEE 802.3. 10Base-5 uses thick 50-ohm coaxial cable (sometimes called Xerox wire) for the transmission medium. Its data transmission rate is 10Mbps, with a maximum of 100 users per segment. Up to four repeaters can be used to extend the backbone cabling length up to 500m, with a maximum total backbone length of 2500m. Only three of the cable segments can be used to attach end-stations. The unusable lengths are called inter-repeater backbone links. This topology requirement is known as the 5-4-3 rule: 5 cable segments, 4 repeaters, 3 usable cable segments. Alternatively, network expansion can be implemented with the use of bridges between separate cable segments. Figure 4.10 illustrates a 10Base-5 topology utilizing repeaters and bridges.

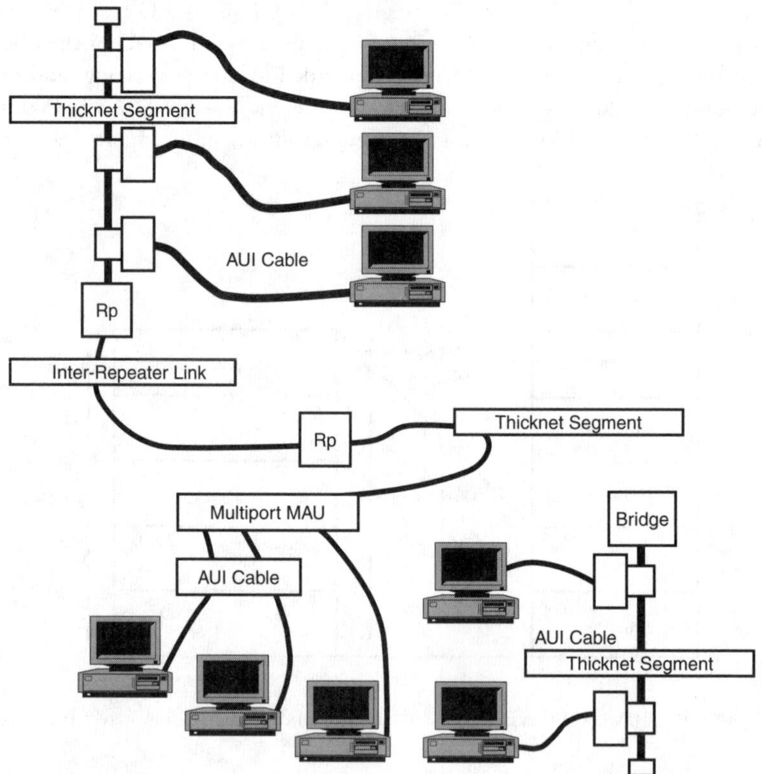

Figure 4.10 10Base-5 network expansion with repeaters and bridges.

10Base-5 uses a separate network controller and MAU, which connects through an AUI interface cable, and which can be no longer than 50 meters. MAUs connected to the backbone cable segments must be spaced at least 2.5 meters apart. The MAU attaches to the cable through a cable tap (vampire tap) interface. Alternatively, multiport MAUs can be used which electronically simulate the distance and signaling functions of the backbone cable. Multiport MAUs are governed by the same segment density rules as traditional MAUs (no more than 100 stations per segment).

- 10Base-2—10Base-2 uses thin (0.2 inch) 50-ohm coaxial cable as a transmission medium and supports data transmission at 10Mbps. 10Base-2 can support up to 30 end-stations per cable segment. The same 5-4-3 rule applies to 10Base-2. Each cable segment is limited in length to 185m, with a maximum backbone cable length of 925m. 10Base-2 stations connect to the transmission medium using BNC (British Naval Connector) "T" connectors. The connector can be attached to a NIC with an integrated MAU or to an external MAU directly attached to the NIC's AUI interface. Thin coaxial cabling is also supported by DIX Ethernet. Both 10Base-5 and 10Base-2 require grounding and termination. Grounding is only required at one end of the backbone cable—this is commonly done with the use of a cable terminator with a ground wire. The other end of the backbone cable only needs to be terminated. The terminator is essentially a resistor that absorbs the current flow at the ends of the cable. Without proper termination, the network will not function properly. Figure 4.11 depicts a 10Base-2 MAU with a BNC "T" connector and terminator.

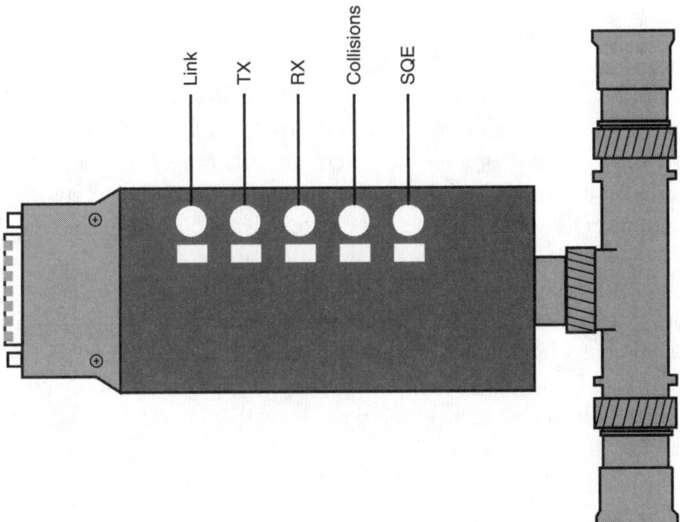

Figure 4.11 A 10Base-2 MAU.

- 10Broad-36—10BROAD-36 uses 75-ohm coaxial cable as a transmission medium and, like its baseband counterparts, supports a 10Mbps transmission rate. 10BROAD-36 uses specific frequency ranges to transmit and receive 803.3 packets. As mentioned earlier, 10BROAD-36 networks use special MAUs, which are essentially Ethernet modems, to send and receive packets. Along with the job of modulating and demodulating signals, the "modem" also modifies the frame slightly so timing can be maintained. 10BROAD-36 uses NRZ encoding (instead of Manchester) and two different frequencies to send and receive packets. Its operation over 75-ohm cable gives it a maximum cable segment distance of 3,600 meters that can support up to 300 users.

10Base-T

10Base-T is the most commonly used Ethernet transport used today. It operates over Category 3, 4, or 5 UTP and uses a multiport repeater (commonly referred to as a hub). The UTP cable has two or four wire pairs, but only two pairs are needed. One pair is used for transmitting and the other for receiving. The use of separate "wires" for transmission and reception is significant because it enables the station to operate in full-duplex mode: the end-station can transmit and receive data transmission at the same time. Full duplex is not supported on 10Base-T hubs, only on multiport bridges. Full-duplex operation is also not easily implemented with coaxial-based PHYs because the cable has only one wire pair. This limits the end-station to be either transmitting or receiving.

Stations can connect to the hub through a NIC that has a RJ-45 female MDI interface and are limited to a horizontal cable segment of 100m. Alternatively, a station equipped with NICs that have an AUI interface can use a directly connected 10Base-T MAU to connect to the hub. Only one station can be connected to a 10Base-T port. 10Base-T networks are commonly implemented with a structured wiring system, whereas 10Base-T hubs are installed in a central location and horizontal cable segments are installed at end-station locations. In these scenarios, the horizontal cable distance should not exceed 90m, because each time the cable is segmented, the signal between the end-station and the hub is considered to be broken. Because most structured wiring models use three cable segments to actually provide an interconnection between the end-station and the hub, the signal loss and distances of the interconnect cable segments must be accounted for in the total cable distance. The signal loss resulting from interconnections varies depending on the category grade of cable and the type of connectors used, so when you plan cable installations, signal loss due to connectors needs to be accounted for. Figure 4.12 provides a simple example of a 10Base-T end-station to hub connection path.

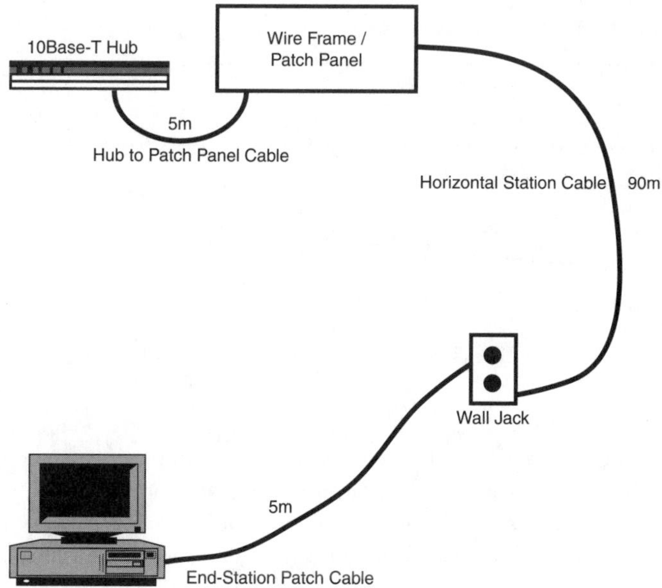

Figure 4.12 A basic 3-cable 10Base-T horizontal cable path.

The 10Base-T hub behaves differently than a multiport MAU, which simply emulates the behavior of the coaxial segment. The 10Base-T hub must perform specific functions: it operates like a standard two-port repeater, and it receives signals from end-devices connected to its ports, then retimes, regenerates, and retransmits them to all its active ports. The hub also monitors the collision rates and packet integrity of all its ports. On some hubs, in the event a station generates an excessive amount of consecutive collisions, the hub will disable the port. This prevents the network segment from being effectively halted by a broadcast or collision storm that results from a defective NIC or circumstances when end-stations are connected to hubs over cable segments that exceed the distance specification. The hub will not regenerate packets that exceed the maximum permitted packet length of 1,518 bytes. These packets are commonly the result of collisions or the byproduct of the PMA interface enforcing jabber control. 10Base-T hubs also use a special signal generated by end-stations, known as a link pulse, to verify port operation. If the link between the end-station and the hub is lost, the hub disables the port until a link signal returns.

Because the hub is effectively the backbone, there is no backbone cabling specification. Instead, hubs can be linked together using a special crossover cable. The pinout is illustrated in Figure 4.13.

> **Note**
>
> As an alternative to using a crossover cable, certain hub manufactures provide an MDI/MDI-X port. This port is wired with a switch that can change the pin signals so the port can behave as a crossover or straight-through port. This is a nice feature as it eliminates the need for the crossover cable. It is not available on multiport bridges (or switches), however, which follow the same device interconnection scheme.

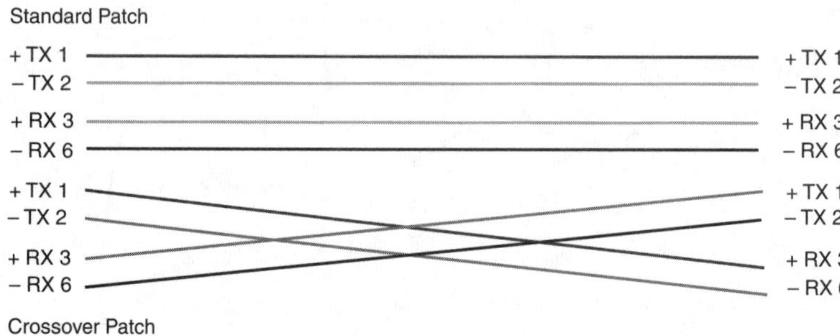

Figure 4.13 Ethernet 10Base-T station-to-hub interconnect and hub-to-hub crossover cable.

A 10Base-T hub-to-hub crossover link has the same maximum distance (100m) as a station-to-hub interconnection. In addition to the maximum distance between hubs, no end-station can be separated by more than four hubs connected tougher by five 10Base-T cable segments for a total cable distance of 500m. It is possible to have more than one hub connected to an end-station (see Figure 4.14).

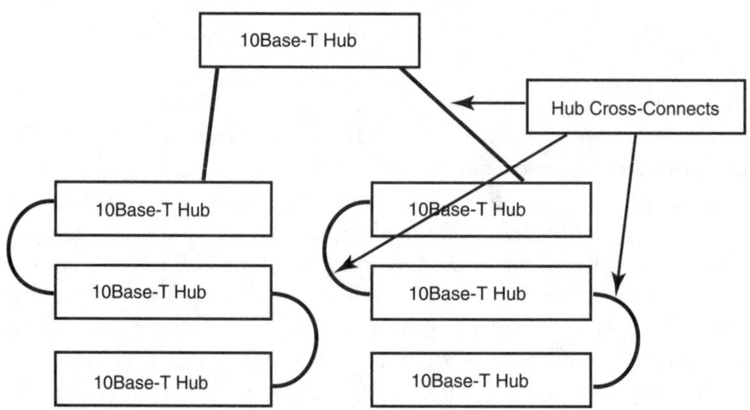

Figure 4.14 10Base-T hub-to-hub interconnection using 10Base-T cabling.

To provide for greater network distance, it is common to use a 10Base-5 for backbone cabling. This allows for longer distances between hubs and eliminates the hub cascading limitation by forcing the 10Base-5 backbone to provide the interconnection link between 10Base-T hubs. Most manufacturers provide an AUI or BNC interface on 10Base-T hubs just for this purpose. For longer distances beyond 500M, 10Base-FL (which can support link distances up to 2,000m) can be used. Figure 4.15 illustrates a 10Base-5 backbone with 10Base-T hubs.

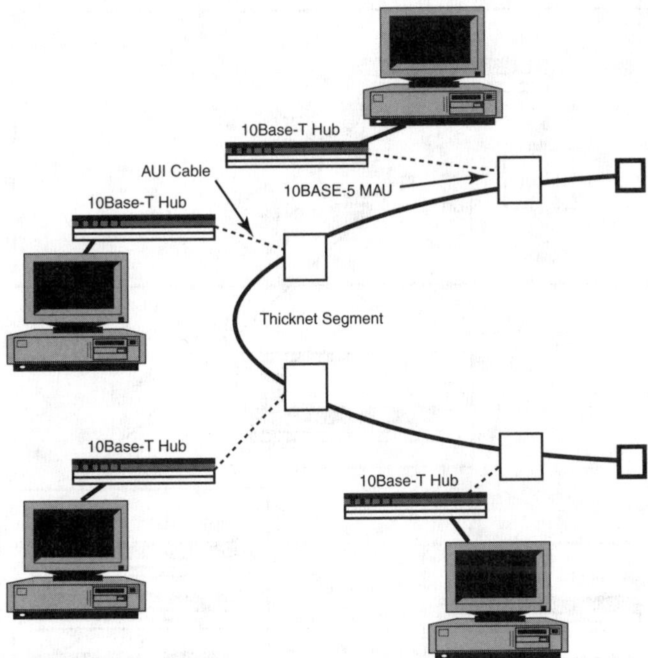

Figure 4.15 A 10Base-5/10Base-T network topology.

An alternative to using a 10Base-5 backbone is to employ a bridge to connect 10Base-T repeater segments. By using bridges, the hub interconnection limitation is essentially removed, because the addition of the bridge restarts the hub interconnection count. Bridges can also be used in conjunction with 10Base-FL to provide wire distribution closet links. This approach not only addresses the 500m segment limitation, it also segments each wiring closet as its own collision domain, which can improve performance between the nodes locally attached within the same closet. Using this approach, however, also requires all the traffic between stations connected in different closets to traverse the wiring closet links. If the majority of the network traffic is inter-closet instead of intra-closet, a performance loss might be incurred. This loss is due to the added latency contributed by the bridges regenerating the transmission signals. Figure 4.16 illustrates this kind of topology.

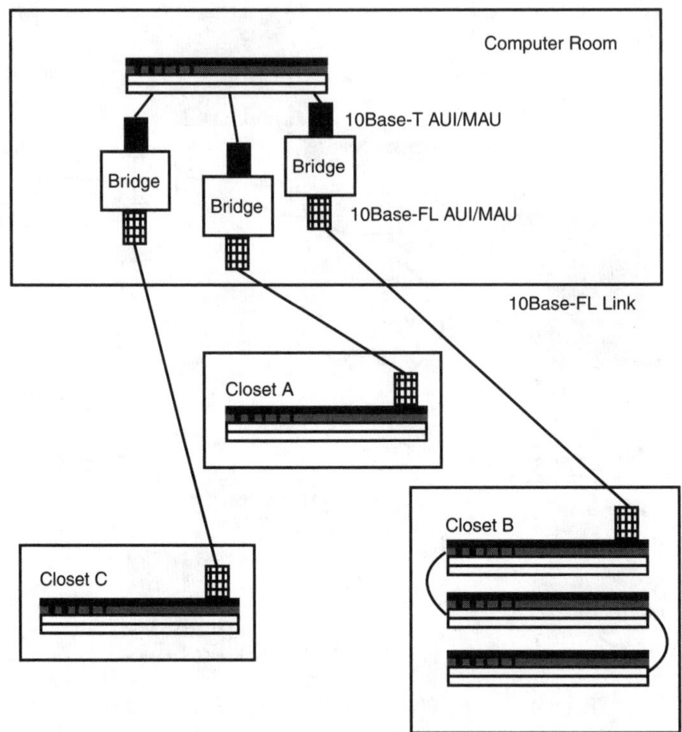

Figure 4.16 10Base-T in a consultation topology, utilizing 10Base-FL backbone cabling.

There are primarily two types of bridges used with 10Base-T: two-port transparent and multiport transparent bridges. Transparent bridges connect two or more segments of the same kind of transmission medium together. In most cases, two-port transparent bridges will have two AUI interfaces instead of two 10Base-T RJ-45 interfaces, for instance. Multiport bridges, alternatively, will have a set of PHY media-specific interfaces. There is also a third type of bridge called a translation bridge. Translation bridges use different PHY media-specific interfaces to translate data PDUs from one Layer 2 protocol to another. One of the most common types of translation bridges is an FDDI-to-10Base-T encapsulation bridge that enables FDDI to be used as a high-speed backbone transport between 10Base-T hubs. As mentioned previously, multiport bridges (or switches) will be covered in Chapter 6.

10Base-FL

10Base-FL is the IEEE standard for running Ethernet over multimode or single-mode fiber optic cable. Its prestandard name was Fiber Optic Inter-Repeater Link (FOIRL). It was created to provide repeater and hub interconnections for distances beyond 500m. 100Base-FL is implemented as a point-to-point link between devices using external fiber MAUs with the standard AUI interface or in a star topology using multiport 10Base-FL hubs. Up to four 10Base-FL repeaters can be used to construct a backbone link with a maximum distance of 2,000m between repeaters. 10Base-FL is quite common in large enterprise network installations as the basis for interconnection links between wiring closets and building-to-building links, not only for its capability to transmit over long distances, but also for its complete immunity to radio and electrostatic emissions. Its transmission mechanism is an LED laser sending light pulses through a glass fiber to a photodetector, instead of sending DCV current across a wire. The 10Base-FL uses 850-nanometer short wavelength light on 62.5/125 micron fiber optic cable with ST-type connectors.

803.3 100Mbps (Fast Ethernet) Layer 1 (PHY) Implementations

100Mbps operation of 802.3 Ethernet required that adjustments be made to the MAC (OSI Layer 2) and PHY (OSI Layer 1). Although the MAC (frame format) layer itself and its operational proprieties (slot-period, and so on) are identical to the 10Mbps standard, the 10Base PLS, AUI, and PMA interfaces were inadequate for the new higher transmission rate. To replace them, the developers of Fast Ethernet developed a new PHY interface for the 100Base-T4 implementation and utilized existing FDDI PHY interface standards for the 100Base-TX and 100Base-FX implementation. From an operational perspective, the 802.3u standard defines the following operational characteristics for 100Base-T:

- Backward compatibility with existing 10Mbps implementations—The CSMA/CD MAC utilized by 10Base-T implementations is the same used by the 100Base-T standard. The slot-period (512 bit-periods), maximum (1518-bit) and minimum (64-bit) frame size, and interframe gap (96 bit-periods) are identical. However, the actual bit-period for 100Base-T is decreased by a factor of 10 due to 100Base-T's increased operating rate. Consequently, the total segment length (in other words, the collision domain size) for 100Base-T is limited in comparison to 10Base implementations.
- 10/100 interpretability and autonegotiation—100 Base Ethernet repeaters and NICs support end-station transmissions at either 10Mbps or 100Mbps. Additionally, 10/100 operation and auto-negotiation between the NIC and repeater/multiport bridge to the highest operating rate.

- Support for full-duplex operation—When multiport bridges were introduced for 10Mbps Ethernet, one of the derivative features was full-duplex operation for 10Base-T devices. This enhancement, however, was added out of the 802.3 specification. The 802.3u standard defines a mechanism for full-duplex operation, recognizing its valuable potential in terms of actual performance and possible use in supporting larger 100Mbps segment topologies.
- Exclusive use of an active star topology—Only a point-to-point "active hub" (end-station to repeater) transmission media model is supported. There is no PHY specification for "passive bus" transmission media such as 10Base-5. Passive bus transmission media cannot easily support the simultaneous dual-speed operation and transmission and receive functionality required for full-duplex operation.

The 802.3u standard defines two UTP implementations: 100Base-T4, which operates over four pairs of Category 3 UTP, and 100Base-TX, which utilizes two pairs of Category 5 UTP. The 100Base-FX standard operates over a single pair of 62.5/125 or 50/100 micron multimode fiber. 100Base-T2, a third UTP implementation of 100Mbps operation using only two pairs of Category 3 UTP, was added a year or so after the 100Base-T standard was finalized. No commercial products have ever shipped using this standard.

10Mbps Ethernet employed Manchester encoding for both the MAC-to-PDM communication between the end-station controller and its MAU and for end-station-to-end-station transmissions across the network medium. As mentioned earlier, this approach is not suitable for high-speed data transmission. Therefore, a new set of PHY interfaces was developed based on the FDDI Physical Media Dependent (PMD) sub-layer mechanisms. There are six 100Mbps PHY interfaces or sublayers (listed from the MAC down).

The Media Independent Interface was developed to provide interpretability between the MAC layer and the new PHY implementations. This interface defined a Physical Coding Sublayer (PCS), which performs data encoding, data transmission and receive functions, and CSMA/CD duties. The PCS in essence performs the same duties as the 10Mbps PLS interface. The Physical Medium Attachment (PMA) interface handles the mapping of the PCS to and from the transmission medium. The PMD interface defines the connection interface and the usage designations for the transmission medium elements—for example, which wire pairs are used for what function. The Auto-Negotiation (AutoNeg) interface exists between the NIC and the repeater as a management interface to determine the link's operating rate. Auto-negotiation is not supported on all 100Mbps PHY implementations. It is accomplished by using the normal link pulse and fast link pulse tests employed by repeaters to check link integrity when an end-station comes online and establishes its connection with the repeater. The repeater uses the link pulses to determine the optimal operating rate.

This interface is only specified for UTP PHY implementations and was later revamped in the Gigabit Ethernet standard as a common interface for all PHY implementations. The Media Dependent Interface (MDI) is the actual electromechanical interface used between the end-device and the transmission medium. The 100Mbps MAC and PHY interface model is illustrated in Figure 4.17.

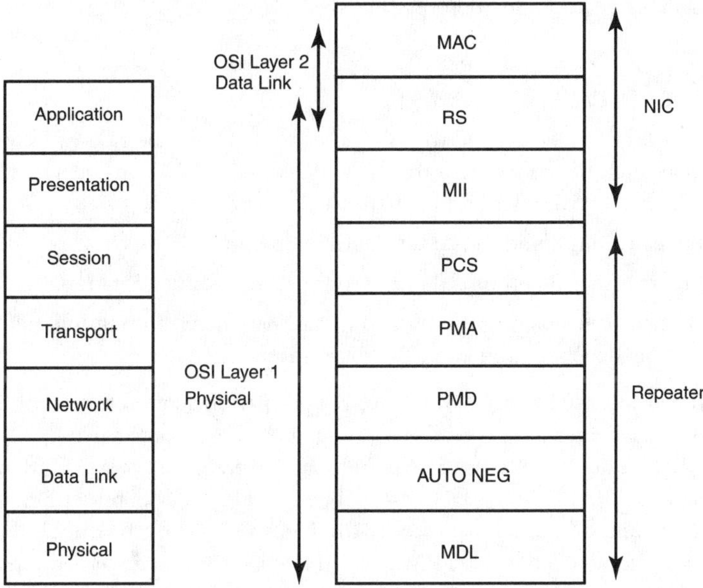

Figure 4.17 100Mbps Ethernet, data link (MAC), and physical (PHY) layer interfaces.

The MII Interface

Logically, the MII performs the same function as the AUI interface. With 10Mbps Ethernet, the AUI was the interface that carried the MAC frame in a bit-by-bit serial stream to and from the Manchester encoder/decoder. With Fast Ethernet and the MII, a new sublayer called the Reconciliation Sublayer (RS) operates between the MAC and the MII. Its purpose is to act as a translation interface between the original MAC PLS interface, which transmitted data a bit at a time, and the MII interface, which transmits data in 4-byte words called "nibbles" using NRZ encoding.

Although the different 100Mbps PHY implementations essentially operate the same way from a topology and transmission rate perspective, the actual PHY interface mechanisms, in terms of encoding and media access, are different between PHY implementations. The MII provides a common data interface that can used by the different PHY implementations to encode transmitted data in accordance with PCS (and lower) interfaces defined for use with a specific transmission media.

The AUI uses a 6-pin (2 transmit, 2 receive, 2 control) electromechanical interface, with each 10MHz serial bitstream operating at a rate of up to 10Mbps over a maximum distance of 50 feet. The MII uses an 18-pin (4 transmit, 4 receive, 10 control) interface, with 4 transmit and 4 receive serial bitstream's operating at either 2.5Mbps (for 10Mbps operation) or 25Mbps (for 100Mbps operation) over a maximum distance of 0.5m. The two different operating rates are needed for the support of both 10Mbps and 100Mbps operation. In addition, the RS sublayer and MII interface are full-duplex capable.

100Base-T4 PHY

100Base-T4 PHY was designed to utilize the large installation base of Category 3 UTP on which the majority of the early 10Base-T installations operated. The 100Base-T4 PMD uses all four wire pairs:

- One pair (pins 1 and 2) is designated for transmission.
- One pair (pins 3 and 6) is designated for receive-only operation.
- Two pairs (pins 4 and 5, and 7 and 8) alternate between transmit and receive function, depending on what the station is doing.

The odd pair not used to send or receive data is used for collision detection and resolution—again, depending on the data operation being performed.

The 100Base-T4 PHY uses the MII interface to transmit the MAC frame into 4-byte nibbles. The PCS then takes two nibbles and encodes them with 8B/6T. Station input and output are handled by the PMA using the 6T symbols. These symbols are sent across the three PMD 25MHz channels at a 33.3Mbps operating rate. 100Base-T4 does not support full-duplex operation because all four pairs are used for transmission. End-station-to-repeater lengths cannot exceed 100m. A maximum distance of 205m is permitted between end-stations.

The 100Base-TX and 100Base-FX PHY

The 100Base-TX/FX PHY standards are based on the FDDI PHY and FDDI PHY sublayer specifications.

The 100Base-TX PMD operates over two pairs of a Category 5 UTP cable (pins 1 and 2 and 3 and 6 are the same as 10Base-T) or Type-1 or Type-2 STP using RJ-45 connectors. The 100Base-FX PMD operates over a pair of multimode (850-nanometer short wavelength light on 62.5/125 micron) fiber optic cables using SC type connectors.

The MII interface hands off 4-bit data blocks called "nibbles" to the PCS, which encodes them using 4B/5B encoding. Depending on the PMD, the PMA transmits the 4B/5B symbols serially over 125MHz data paths. Each path is capable of 100Mbps operation. One is used for data transmission and collision resolution, and the other is used for data reception and collision detection. With the 100Base-FX PMD, the 4B/5B symbols are sent by the PMA across the medium using NRZI signaling.

Category 5 UTP, however, can only provide 100MHz of bandwidth, so additional signaling must be used to reduce the bandwidth requirement. The data is processed further by the PMA and encoded with MLT-3 line encoding, which operates at 100 Mbps on only 32MHz of bandwidth. The MTL-3 signal is then handed to the PMD for transmission over the medium.

100Base-TX supports dual 10/100 operation and auto-negotiation. This standard, like 100Base-T4, can support hub-to-end-station cable segment distances up to 100m. Also like 100Base-T4, the maximum cable segment length permitted between end-stations is 205m. 100Base-FX can be implemented in a point-to-point topology between end-stations utilizing segment lengths up to 400M. A 100Base-FX hub can also be utilized which reduces the hub-to-end-station segment length to 300m. 100Base-FX supports full-duplex operation, but not auto-negotiation. In full-duplex mode, the point-to-point segment distance increases substantially, to 2,000m.

100Base-T Repeaters

Whereas 10Base-T and 100Base-T differ in their repeater implementations, they operate using the same collision detection mechanisms and the same slot-periods (512 bit-periods). With 10Base-T, the network size is limited, so a collision can be detected within one slot-period. This reduced size enables a station sending the smallest legal packet size to acknowledge a collision if one occurred with a station located at the farthest possible distance from itself. However, 100Base-T's transmission rate is 10 times faster than 10Base-T's. This has the effect of reducing the actual bit-period by a factor of 10. Therefore, in order for the original CSMA/CD collision mechanism to work, the collision domain needs to be reduced so a collision can still be detected within one slot-period.

To accommodate this reduced collision domain of 205m (compared to 10Base-5's 2,500m and 10Base-T's 500m domain), 100Base-TX can only use, at the most, two repeaters per segment. There are two types of repeaters used for Fast Ethernet: Class 1 (the most common) and Class 2. Class 1 repeaters have a certain advantage over Class 2 repeaters, because Class 1 repeaters provide translation services between different PHY encoding implementations. This enables manufacturers to support all the 100Mbps media types (TX, FX, and 4T) on a single repeater. Using one repeater does add an additional level of delay on what is already a time-sensitive media platform, so only a single Class 1 repeater can exist within a segment (collision domain). Some Class 1 repeater implementations provide an interconnect bus to connect repeaters together so they appear as single repeaters with a high port density.

Class 2 repeaters only support a single PHY implementation. This enables the repeater to simply repeat the incoming signals to its ports, without the added delay of translating different PHY encoding schemes. This limitation to one standard also enables Class 2 repeaters to use the traditional 10Base-T hub-to-hub interconnection model (which is, in part, why they were specified). The interconnection cable is limited to a length of no more than 5m. Class 2 repeater networks can only run two repeaters deep, which keeps them within the maximum 205m end-station-to-end-station segment distance.

100Base-T's small segment diameter makes the large-scale implementation of shared 100Base-T impracticable. With the advent of 100Base-TX multiport bridges, however, 100Base-TX became an extremely viable network technology. Fast Ethernet switching coupled with 100Base-T's full-duplex operation eliminated the segment distance limitations required for proper 100Mbps CSMA/CD functionality. Today, most 100Base-T implementations are based on multiport bridge topologies. This approach has effectively laid waste to the concept of using FDDI or ATM for LAN backbone infrastructures.

100VG-AnyLAN

During the Fast Ethernet standards development process, there were two technical models assessed. The first (by Grand Junction Networks) proposed using the existing CSMA/CD access protocol with the FDDI PMD sublayer. The second (by AT&T and Hewlett-Packard) proposed discarding the CSMA/CD access protocol and replacing it with a new protocol called demand-priority. Demand-priority was optimized to support the transmission of real-time multimedia applications. It did this by using network access demand prioritization levels that are assigned to different application types. The debate between the CSMA/CD-compatible and demand-priority MAC architectures escalated to a holy war of sorts during the standardization process. In the end, Demand-Priority did not make it, largely because it was not compatible with existing Ethernet. It didn't fade away—it became a new access method and a new IEEE working group. 802.12 was started and developed as a separate standard, 100VG-AnyLAN. In addition to the replacement of the CSMA/CD access protocol, the 802.12/100VG-AnyLAN standard can support any MAC frame type (Token Ring, FDDI, or Ethernet).

100VG-AnyLAN operates over Category 3, 4, and 5 UTP. It uses a star hub topology, like 10Base-T's, but the hubs can cascade three hubs deep. 100VG-AnyLAN hubs are deployed in a parent/child hierarchical topology, with one hub (the parent) from which all other hubs cascade down. Each child hub must have one uplink port that can reach the parent; the other hub ports can be used as downlink ports or for end-station connections. Hubs and end-stations can be no more than 100m (Category 3 UTP) or 150m (Category 5 UTP) apart. The maximum cable distance for a Category 3 UTP network is 600m, or 900m if Category 5 UTP is used.

The reason for the parent/child topology is that, under the demand-priority MAC, the parent hub controls end-station access to the network. The effect is that all the connected hubs act as a single large repeater. When a station wants to transmit data, it queries the hub. If the network is idle, the hub signals the requesting station that it's clear to transmit. If more than one end-station needs to transmit data, a round-robin mechanism alternates transmission access to the hub between the requesting stations. The demand-priority MAC also provides a priority access mechanism for special data types (such as real-time audio and video). When a station makes a request to transmit one of the prioritized data types, the normal access queue mechanism is overridden and they are permitted access to the medium. Transmission access requests for these special priority types override the processing of other kinds of traffic.

1000Base-X PHY Implementations

1000Base Ethernet (Gigabit Ethernet), for all intents and purposes, is 100Base Ethernet (Fast Ethernet) on steroids. Like 10Base and 100Base Ethernet, at the data link layer (OSI-RM Layer 2), the 1000Base implementation utilizes the "standard" IEEE 802.3 MAC frame format, making it completely backwards-compatible with pre-existing Ethernet implementations. To achieve this "compatibility," however, certain adjustments need to be made to the 1000Base Ethernet CSMA/CD mechanism in order to support segment lengths that could be useful, in light of the greatly reduced slot-time.

In terms of the physical layer (OSI-RM Layer 1), architecturally, 100Base and 1000Base Ethernet are similar. With the most obvious being the retention of the RS sublayer and the development of an updated variation of the Fast Ethernet MII, called the Gigabit Media Independent Interface (GMII) to interact with the new PHY data rate requirements. 1000Base Ethernet also maintains 100Base Ethernet's Auto-Negotiation capabilities in an updated form to accommodate operation over for fiber optic transmission mediums, Gigabit Ethernet's primary transmission medium.

Where the difference between Gigabit and its predecessors is readily apparent is in the 1,000Mbps transmission speed and in the actual PHY implementation. This (like 100Base Ethernet) is based on the pre-existing, high-speed serial interconnect PHY standard: Fibre Channel.

Fibre Channel is a high-speed, point-to-point data transmission standard. (It was designed to establish bidirectional high-speed serial datastreams between supercomputers, mass storage systems, and dedicated multimedia and imaging applications. Connections are made between devices using a circuit-switching device that provides full bandwidth rate connections between all the connected devices. These rates can be anywhere from 100Mbps to 800Mbps. Fibre Channel is an ANSI standard approved in 1994. It uses a five-layer protocol model as follows:

- FC-0—Defines the electromechanical characteristics of the transmission media (optical fiber), transmitters, receivers, and operating rates
- FC-1—Defines the data encoding schemes, 8B/10B
- FC-2—Defines data signaling and framing, circuit management, data sequencing, identification, and classes of service, of which there are four:
    - Class 1—Dedicated point-to-point circuit
    - Class 2—Point-to-point, guaranteed connectionless delivery with conformation
    - Class 3—One-to-many connectionless service without conformation
    - Class 4—Connection-based point-to-point, with guaranteed bandwidth

- FC-3—Defines common services for bandwidth allocation and connection allocation services
- FC-4—Defines a mapping (reconciliation) interface for integration with legacy communications protocols (FDDI, Ethernet, Token Ring, SCSI, and so on)

Fibre Channel is not a LAN/WAN data solution. It is designed for high-volume data transmissions between like devices. Because of its limited application, Fibre Channel has not enjoyed the household recognition that other high-speed communication protocols have, such as ATM. Industry, however, has found a quite suitable application for Fibre Channel: Storage Area Networks (SANs). SANs are touted as the next big thing in data storage systems. Instead of using dedicated tape and disk storage on large-scale computer systems, large disk arrays and high capacity tape storage systems can be shared among different computers through Fibre Channel.

The GMII Interface

Functionally, the MII and GMII are the same. The reconciliation sublayer translates the incoming and outgoing MAC data between the GMII and the original MAC/PLS interface. The role of the GMII is to provide a media-independent interface between the MAC and different PHY implementations. The GMII, unlike the MII (and AUI), is not a physical interface. Rather, an electrical IC-level interface is part of the Gigabit controller card. The reason for the GMII is the need for a wider data path between the RSand the Gigabit physical coding sublayer (PCS). The GMII uses two 8-wire data paths (one TX (transmit) and one RX (receive) 8-wire path) for data transmission. On the TX side, MAC data is transmitted in octets (8-bit groups) from the RS to the PCS. Over the 8-wire GMII TX path, at a rate of 1 octet/cycle with a clock rate of 125MHz, providing a total data transmission rate of 1,000Mbps. After the octet reaches the PCS end of the GMII path, the octet is encoded into a 8B/10B symbol. On the RX side, PCS decodes the octet and transmits it to the RS sublayer at the equivalent TX transmission rate. The GMII can also operate at 2.5 and 25MHz to be backward compatible with 10Mbps and 100Mbps operation. The GMII is also responsible for relaying control information (such as collision detection) to the MAC. Figure 4.18 illustrates the Gigabit Ethernet MAC and PHY interfaces.

Note

Where Gigabit Ethernet differs from its predecessors is its preference for full-duplex operation, and its operational orientation (in terms of operational media preference—that is, fiber optic cable) toward network backbone and high-speed server applications.

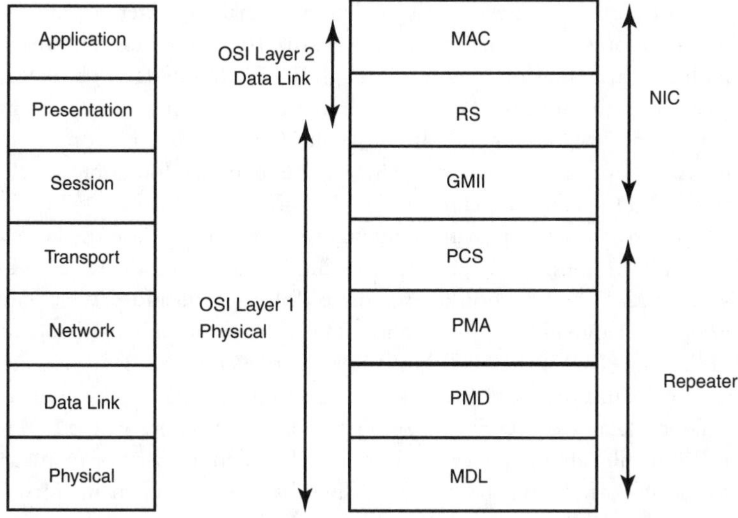

Figure 4.18 1,000Mbps Ethernet, data link (MAC), and physical layer (PHY) interfaces.

The Gigabit PHY

The Gigabit PHY consists of four interfaces: PCS, PMA, PMD, and MDL. This model closely resembles the Fast Ethernet PHY, with the difference being the integration of the auto-negotiation function into the PCS interface. The model is also an updated form of the 802.3u specification that now supports auto-negotiation for fiber optic media. The PCS interface is based on a variation of the Fibre Channel FC-1 layer. The FC-1 encoding mechanism is based on IBM's 8B/10B block encoding mechanism. Under the FC-1 specification, a maximum operation rate of 1.065 GBaud or 850Mbps is defined. To achieve 1,000Mbps transmission rate, it was increased to 1.25 GBaud.

The PCL performs the same essential functions performed by the 802.3u PCS interface (such as collision detection monitoring and so on). Its primary function is to encode/decode the 8-bit binary groups to and from the GMII into 10B symbols. These symbols are then handed to the PMA interface. The PCS also manages the auto-negotiation function. With 100Base-TX, the repeater issues a link pulse test to verify link quality. This mechanism is replaced with 8B/10B code words to establish the proper operating rate. By default, all standards-based gigabit implementations should perform auto-negotiation by default, with full-duplex operation being the preferred operating mode. The GMII/MII management interface is used to set full-duplex operation on the GMII/MII.

The PMA interface is, in essence, a parallel to a serial converter. It provides the 10B symbol serialization service required to convert the 10-bit symbols for transmission over the physical transmission medium through the PMD interface. The PMA uses two 10-bit data paths from the PCL encoder and decoder. These data paths in turn interface with the PMA input and output parallel/serial datastream converters. The 10-bit interfaces operate at a rate of 125MHz. The incoming/outgoing serial data paths to the PMD operate at a clock rate of 1,250MHz.

The PMD is the electromechanical interface to the transmission medium. It provides the transmission mechanism for sending and receiving the serialized 8B/10 symbols. The 802.3z specification defines three PMD specifications, which are derived from the Fibre Channel FC-0 specification. The most commonly implemented Gigabit Ethernet PMD is the 1000Base-SX standard. With 1000Base-SX, the "S" stands for short haul. 1000Base-SX was developed for use as an upgrade or to be installed alongside an existing fiber-based network infrastructure where FDDI or 100Base-FX are already implemented. It is suitable for short distance (comparatively to LX) end-station-to-hub/switch and hub-to-hub backbone implementations. The 1000Base-SX standard specifies an 850-nanometer short wavelength light source with four multimode fiber variations. It employs 62.5/125 micron fiber (commonly used for FDDI and 100Base-FX), and a maximum (full-duplex) distance of 220m to 270m is achievable between end-points. With 50/125 micron fiber, maximum distances between 500m and 550m are supported. All four multimode implementations use SC-type connectors. 1000Base-SX implementations are the most common because they can be run on existing cabling infrastructures. One byte 1000Base-LX was developed for use primarily in long backbone transport scenarios, such as building-to-building or floor-to-floor connectivity. 1000Base-LX utilizes a long wavelength (1300nanometer) light source over multimode and single-mode fiber. With 62.5/125 and 50/125 micron fiber, multimode fiber distances up to 550m are supported. With 10/125 micron single-mode fiber, distances up to 5,000m (and beyond) are supportable. The last PMD specification uses 150-ohm twinax cabling with either the Fibre Channel Type-2 balanced connector or D-subminiature connector.1 byteThe purpose of 1000Base-CX is to provide a more cost-effective method for short haul (up to 25M) interconnections between repeaters and switches. It operates by connecting the PMA interfaces of the two connected devices, essentially extending the PMA over separate cable paths. Due to the extremely high frequency operating rate (1,250MHz), the connecting cables and connectors must be shielded and balanced.

1000Base-T

The 1000Base-T standard IEEE 802.3ab is expected to be ratified by the end of 1999. The proposed standard uses four pairs of Category 5/5E cabling and the standard RJ-45 connector. All four wire pairs are used for data transmission reception using a technique known as dual duplex. The 1000Base-T standard specifies a new PHY layer from the GMII down. The PCS interface uses a variation of the Pulse Amplitude

Modulation 5 (PAM5) line encoding scheme. This is a five-state encoding system. PAM5 uses five voltage variations to represent 2 bits of data, with each wire pair operating at a clock rate of 125MHz. This new scheme is called 4D-PAM5. To actually interpret the simultaneous transmission and reception signals, a digital signal processor is part of the transceiver circuitry. Although the 1000Base-T standard has been anxiously awaited by some, its added technological complexity will undoubtedly come at a cost greater than the current 1000Base-X standard-based products that are already shipping. 1000Base-T also has a smaller maximum segment distance (100M) than the currently shipping products, making it only practical for end-station, not backbone, connectivity use.

1000Base-X CSMA/CD

Although the PHY implementation of Gigabit Ethernet is quite similar to Fast Ethernet, there are some real differences as far as implementation topology is concerned. Gigabit Ethernet uses the same CSMA/CD mechanism and 802.3 frame format used by 10Mbps and 100Mbps Ethernet implementation. Minimum frame size is 64 bytes; the maximum is 1,518 bytes.

CSMA/CD is sensitive to operating rate and network size, because a station must be able to detect a collision before the transmission is complete. If a collision is not detected during the transmission, it's assumed that the transmission is successful. It is CSMA/CD's sensitivity to network distance and the round-trip packet delay that resulted in Fast Ethernet's relatively short maximum segment distance between end-stations. This same problem confronted the engineers who developed the Gigabit Ethernet specification. Retaining all Ethernet MAC's original properties would result in a maximum distance between stations of about 20M, which is a rather impractical distance for LAN networking. To overcome this problem, the slot-period for the Gigabit MAC has been increased to 512 bytes (4,096 bit-periods), as opposed to the 512 bit-periods specified in the original Ethernet MAC. This alone does not correct the problem of maximum distance, because the slot-period and minimum packet time are directly related. With CSMA/CD, the assumption is that the smallest packet size will take one slot-period to transmit. If the destination station receives a packet smaller than the minimum slot-period provides for, it is assumed to be a collision fragment and is discarded. Therefore, for the extended slot-period to work, the minimum packet size needed to be increased to 512 bytes.

However, a change in the minimum frame size would result in incompatibilities with existing Ethernet implementations. So, the original Ethernet MAC packet size was retained and a technique called carrier extension was used to add padding to packets smaller than 512 bytes in size. With carrier extension, the actual packet is not altered; instead, the PHY transmits the carrier extension padding after the FCS field of the packet. The transmission of the actual frame and the carrier padding is seen by the transmitting host as a single slot-period. In the event that a collision is detected during the transmission of the frame or the carrier extension, the transmission is aborted and a jam notification message is generated and transmitted.

When the extended frame is received, only the real packet is copied to the buffer and the carrier bits are discarded. If the slot-period has passed and if the actual packet is a valid size (at least 64 bytes), it is handed to the MAC for delivery. If a collision occurs during the slot period (even if the host already has the actual packet in its buffer) or if the frame size is not valid, the packet is discarded. With packets, sizes equal to 512 bytes and higher are handled normally. They are transmitted in unaltered form and, once received, passed to the MAC for delivery to the ULPs.

The downside to carrier extension is that it adds transmission time to any packet smaller than 512 bytes. Because packets smaller than 512 bytes make up at least half the packets transmitted on the LAN, this reduction in performance is addressed by a new technique called frame bursting. Frame bursting enables a transmitting station to send a consecutive stream of data, so it can send a string of smaller packets without adding carrier extensions to each packet. Instead, carrier extension bits are used to fill the interframe gap between the frames, so the burst stream appears to other stations as a single frame. Frame bursting utilizes a timer, called the burst timer. When a transmitting station sends a packet to a destination, the burst timer starts. At the end of the transmission (assuming it was successful), the station checks to see if it has another packet to send to the destination and if it still has time on the burst timer. If both are true, the transmitting station sends 96 bits of carrier extension symbols and then starts sending the next packet without carrier extension bits. This state exists until the burst timer expires or the transmitter has no more data to transmit.

Gigabit Repeaters

The need for carrier extension and frame bursting for half-duplex operation makes full-duplex operation a much more attractive transmission mechanism, because under full-duplex operation, CSMA/CD is disabled. In fact, during the standards process, it was argued that a half-duplex operation was unnecessary considering the application scope for which Gigabit Ethernet was being designed—network backbone and server applications that can utilize full-duplex transmission effectively. Both half- and full-duplex operation are part of the standard, however, mainly for the sake of backward compatibility with the previous Ethernet standards.

Although a half-duplex transmission mechanism exists, full-duplex is the common implementation for Gigabit Ethernet. Gigabit full-duplex operation is based on the 802.3x flow control mechanism (covered in Chapter 6). To achieve flow control, manufacturers have developed both full-duplex Gigabit repeaters and Gigabit multiport switches. Gigabit switches operate like any other Ethernet switch, in that each port is its own collision domain, and the switch provides the capability to sustain simultaneous connections between devices and switch ports. The maximum cable distance between the device and the switch port is the same as what is defined for full- and half-duplex repeater operation.

Full-duplex repeaters (FDR) operate differently than traditional half-duplex repeaters (HDR). FDRs permit one station to transmit and then replicate the transmission to all the repeater's ports. They use memory buffers on each of the repeater's ports to store frames as they are transmitted. The FDR then uses a round-robin repeater bus access method to share the bus with each port. This works by having the repeater constantly check the buffer of each port to see if it contains any packets for transmission. The round-robin check works on a circular basis from one port to the next, giving each port a chance to transmit. If a buffer contains no data, it is passed over and the next port gets a chance until the next buffer check. Where FDR differs from a switch is that when the port gets access to the repeater bus, the packet is transmitted to each repeater port. In a switch, the switch performs a port address lookup to see which switch port is connected to the destination host. When the switch knows the destination port, it forwards the packet only to the designated port.

The main advantage of using FDR over HDR is the elimination of Gigabit Ethernet's CSMA/CD implementation. Like switches, FDRs also use 802.3x flow control to regulate the traffic flow rate between the stations and the repeater. This way, the repeater will not drop packets in the event that its buffers overflow. FDRs also check packet integrity (also like switches), though this function is not performed by traditional half-duplex switches. FDR's advantage over switches is cost. FDRs are not as complex as switches, and they do not require the additional programming and processing to perform MAC-to-port address lookups. They also do not require the same amount of backplane capacity.

Token Ring

Token Ring was originally developed by IBM in the 1970s to provide a transaction-oriented communications protocol for its Systems Network Architecture (SNA) protocols, which were used to provide data communication services for its mainframe computers. When IBM introduced its PC product in 1981, it provided no networking solution, because IBM had envisioned the PC primarily as consumer product or as a front-end device for its larger computing systems, which would communicate serially with its mainframe systems. In 1984, in response to Ethernet, IBM released a 4Mbps version of Token Ring to provide LAN transmission services to its PCs that would be compatible with its existing networking architecture technologies. 4Mbps Token Ring uses STP and UTP cabling in a logical star topology that uses active hubs (known as Multistation Access Units [MAUs]) to interconnect end-stations. IBM did not endorse Ethernet as a networking technology when it was released in 1981, mainly because it was not compatible with IBM's existing technology. In 1985, the IEEE developed a Token Ring standard, IEEE 802.5, which is largely based on IBM's Token Ring architecture. IBM Token Ring and IEEE 802.5 are essentially compatible; the major difference is that IBM's support for source route bridging which is accommodated by the data field in the 802.5 packet and additional network management functionality. In 1989, IBM released a 16Mbps version of Token Ring.

In 1992, IBM licensed its proprietary Token Ring implementation to other network equipment vendors. This move resulted in some substantial improvements to Token Ring, particularly, Token Ring multiport bridges (switches), full-duplex Token Ring operation, and most recently, the development of high-speed Token Ring (which is currently in the IEEE standards process, IEEE 802.5t). High-speed Token Ring uses the standard Token Ring MAC and operates on a variation of the Fast Ethernet PHY standard.

Token Ring makes up about 10 percent of the installed LAN networking base. It still enjoys a small popular following, particularly in shops where IBM is the major vendor. The main drawbacks to Token Ring have been and will continue to be cost and limited vendor support. Token Ring adapter cards and hubs (MAUs) traditionally cost anywhere from two to six times more than their Ethernet equivalents because there is really only one manufacturer: IBM. The Token Ring controller chip set is produced by IBM. Currently, only two major vendors produce new Token Ring products: IBM and Madge. Most of the large national resellers sell refurbished versions of the IBM Token Ring MAU's bridges, and NICs.

IEEE 802.5/Token Ring MAC

Token Ring uses a token-passing mechanism for a media access protocol. When an end-station needs to transmit data, the station seizes what is called a free token. A free token is a 3-byte frame (packet) consisting of three fields: start delimiter, access control, and end delimiter. When the station has the free token, it appends the data it wants to transmit to the token and changes the token's configuration to what is known as a dirty token. The frame/token is then passed on to the next station on the ring. The token is passed on from station to station until it reaches its destination. When it reaches its destination, the frame is copied by the end-station and sent back to the ring again. Before retransmitting the frame, however, the receiving station sets the control bits in the frame's *Frame Status Field* to indicate that the frame reached its destination and how the data was handled after it was received. This provides the capability for the transmitting station to become aware of any errors that might have occurred. After the frame has reached its destination, it continues around the ring until it reaches the station that originally transmitted the token. When the sending station receives the frame back, it removes it from the network, regenerates a new token, and transmits a new token onto the ring where it is grabbed by the next station on the ring. If the station has no data to transmit, it just passes on the token to the next station. Each station can only hold the token for a specific amount of time, after which it must release the token, to allow another station to transmit data.

The ring's performance is largely determined by the ring's average token rotation time (TRT). The TRT averages the mean average measurement of the amount of time it takes a token to go around the ring. The TRT is a product of the ring's operating rate and station density. To improve the TRT and overall performance of the ring, the 16Mbps implementation of Token Ring provides support for early token release.

With early token release, when the frame has reached its destination, the receiving station copies the frame, marks it as received, and sends it and a new token on to the ring. This is possible because the ring traffic is synchronously clocked and only flows in a single direction. The received frame is removed from the ring by the station that transmitted it, and the token is accepted by the next station, in accordance to ring's order or the traffic priority mechanism.

Token Ring also provides an optional traffic priority mechanism that enables certain types of traffic to be prioritized over others. When in use, stations prioritize their traffic. When they receive a free token, they compare the priority of the traffic they have to transmit against the priority of the free token. This is set in the access control field. If the token's control field priority is less than or equal to their traffic's priority, they can transmit. If the token's priority is greater than the station's traffic priority, they pass the token on without transmitting. The token priority is set using the token priority reservation mechanism. If any station has high priority traffic to send, when it gets the token it can set its frame's priority using the reservation bits in the access control field, provided the reservation bits are not already set with a priority higher than the traffic the station wants to send. When the reservation bits are set and the currently transmitting station receives the transmitted frame back, it reads the reservation bits, sets the new token's priority value to the reserved value, and sends the token on to the ring.

Ring Management and Operation

The IBM and IEEE 802.5 Token Ring architectures both define a set of management agents that operate on the ring station to provide facilities that maintain the ring's proper operation. These ring management facilities include token generation and traffic monitoring, ring error detection, ring error data collection, and ring recovery functions. Each of these defined agents or "server" functions is performed (in most cases) in addition to the station's primary function, which is to transmit and receive tokens containing ULP application data. Any Token Ring-defined management facility can be performed by any properly functioning ring station connected to the ring. In this respect, like access to the transmission medium, each station is equal. Each management server function is essentially a finite state machine, which is to say that any station can change its operational "state" dynamically in response to changes in the ring's larger operational state.

This dynamic state-based approach to implementing the ring management functions reflects the underlying operational basis of Token Ring's design. The design provides a fault-tolerant, self-recovering data transmission environment that provides contentionless access to the transmission medium for all stations. Each of these qualities is essential to provide data transmission services for the transaction-oriented processing applications that operated on IBM's mainframe computers (for which Token Ring was originally developed). With this background in mind, it is easier to understand why the IEEE and IBM Token Ring specifications differ in some areas, particularly when it comes to the defined ring management functions.

In the case of the IEEE 802.5 specification, its focus is on defining a transmission protocol and the basic management services needed for proper ring operation to take place. Therefore, the IEEE specification focuses on the management server functions of the active monitor and the standby monitor. The IBM Token Ring architecture, on the other hand, is focused on providing a data transmission protocol for transaction processing, which requires not only ring operation management, but also ring station configuration and monitoring services, ring error data collection, and centralized network management. Therefore, in addition to the active and standby monitor functions, the IBM Token Ring architecture defines additional management servers to provide the needed additional functionality. These additional Token Ring management servers work primarily in conjunction with IBM's LAN Network Manager product.

The primary ring station management role is known as the active monitor (AM). The AM is responsible for maintaining the balanced flow of management and end-station data over the ring. Each ring has one station that serves as the AM. Any station on the ring can assigned the role of the AM, which is determined through the token claiming process. The token claiming process is initiated when the first station joins the ring and whenever it is perceived that no active monitor exists on the ring, by any of the attached ring stations. The token claiming process (for example, the AM "election"), is won by the station with the highest active MAC address on the ring. The following list details the AM's ring management functions:

- Neighbor notification—The AM is responsible for initiating and monitoring the neighbor notification process. The ring's data flow rotation travels only in one direction. In order for the ring to manage and identify error conditions, each ring station needs to know the MAC address for its nearest active upstream neighbor (NAUN). Because data on the ring flows in only one direction, the downstream neighbor detects any errors transmitted by the upstream neighbor. In the event of an error condition, the section of the ring between the NAUN and the station detecting the error is considered the fault domain (in most cases). Depending on the error, the stations within the fault domain can enter an error condition known as *beaconing*, which alerts the other stations on the ring that a ring fault error has occurred. Stations on the ring learn the address of their NAUN through the neighbor notification process. The AM transmits an active monitor present (AMP) frame to each station at least once every 15 seconds. As each station receives the AMP, it responds with a standby monitor present (SMP) message to notify its downstream neighbor of its presence and MAC address.
- Ring recovery—During some types of ring error conditions, the ring needs to be "reset" to its normal operational state. This is accomplished through the ring purge procedure. To purge the ring, AM transmits a ring purge token onto the ring. As each station receives the token, it resets itself and initiates the neighbor notification process.

- Token monitoring—The Token Ring specification dictates that it should take no more then 10msec for a token to traverse the ring. If the AM fails to see a good token within this period, the AM initiates ring recovery. Additionally, the AM interrogates each token that is transmitted on the ring. The AM does this by checking the frame status bit. If the frame status bit is incorrect, the AM interprets the token as errored and initiates a ring purge.
- Ring timing maintenance—The AM provides two ring time-related functions. The first is to maintain the ring's master reference clock for timing data transmission and management functions. On a 4Mbps ring, the transmission clock rate is 8MHz. On a 16Mbps ring, the transmission clock is 32MHz. The token, which is originated by the AM, is used by each station to maintain ring timing and its local transmitter clock. The AM is responsible for adjusting the ring's reference clock to accommodate for jitter, which phase shifts in the ring's timing due to environmental characteristics that exist in the ring. Many of the Token Ring operations are managed through the use of different protocol timers to measure essential ring communication functions on the ring. So, accurate timing is
  essential for the ring's proper operation.
- In addition to maintaining the ring's reference clock, the AM is also responsible for maintaining the ring's transmission delay. In order to ensure that the entire transmitted token reaches the destination station before being returned to the sending station, a transmission delay is needed. The AM inserts a 24-bit delay pattern into the ring to provide enough delay to ensure proper token processing.

After the ring has an active monitor, the remaining stations on the ring function as standby monitors (SMs). As each station completes the ring insertion process, verifying the existence of an active monitor, it assumes the role of an SM. The SM's primary function is to monitor the ring for errors and the presence of an active monitor. If an SM fails to detect an active monitor at any time, the SM can initiate the token claiming process.

The Ring Error Monitor (REM) is the one of the more important management server definitions in the IBM Token Ring architecture. The REM collects ring error reports from stations that have detected ring error conditions. Its purpose is to provide a central point for collecting troubleshooting information on types of errors that occur on the ring. The REM is typically not a normal ring station; rather, the REM is a dedicate station running management software, like IBM's LAN Manager or a dedicated network monitoring probe or protocol analyzer.

The Ring Parameter Server (RPS), like the REM, is defined as part of the IBM Token Ring architecture specification. The RPS's function is to provide ring initialization parameters to each new station that is inserted into the ring. One of the RPS most important functions is the monitoring of the ring's station density. In the event that a new station inserted into the ring exceeds the maximum numbers of stations permitted on the ring, the RPS notifies the station and the station removes itself from the ring. Additionally, the RPS maintains a list of all the active stations according to their station address, mircocode level, and NAUN address.

The Configuration Report Server (CRS) is essentially the IBM LAN Manager console. Its role is to collect information on the overall operation of the ring. This information includes ring station insertions and removals, token claiming process initiations, NAUN changes, and throughput and utilization statistics.

In a multi-ring Token Ring environment, bridges are use to interconnect rings. Token Ring supports two bridging methodologies. IEEE 802.5 transparent bridging (the same method defined for use with Ethernet) is the defined bridging methodology. In IBM Token Ring environments, an additional bridging method known as source route bridging can also be used. With transparent bridging, only one bridge link is permitted between two ring segments. With source route bridging, multiple bridge interconnections are permitted between ring segments. Source route bridging has two advantages over transparent bridging in large multi-ring Token Ring LANs. First, because source route bridging enables multiple paths to exist between different segments, this provides for a higher level of fault tolerance, then transparent bridging. In case of a bridge failure, the two rings can still communicate over a different bridge link. Second, because multiple paths exist between rings, it is possible to set a "preferred" delivery path for the tokens, which ensures that the token will travel the shortest path to its final destination. The LAN Bridge Server (LBS) monitors the traffic passing through both successful and dropped tokens on the bridge ports in the LAN, and keeps track of the available route path options. The LBS sends its bridge status data to the LAN management console.

Now that we have looked at the various ring station management roles that have been defined by the IEEE 802.5 and IBM Token Ring specifications, let's look at how the ring's operation is established and maintained by examining the different phases of the *ring insertion* process. Before a station can become an active member of the ring, each station must go through the ring insertion process. The ring insertion process is a series of test and management functions that ensure the station's NIC is configured for the proper ring speed (each station on the ring must operate at the same speed) and is functioning properly. The ring insertion process also, if need be, initiates any additional ring management functions that might be needed to stabilize the ring in response to the addition of a new ring member. Additionally, certain phases of the ring insertion process are used in the ring recovery process in response to certain types of Token Ring errors. The following details each phase of the ring insertion process:

- Phase 0: Lobe test—The interconnect cable between the station and the Token Ring MAU is known as the *lobe cable*. When a station join's the ring, it generates a lobe test signal that verifies the integrity of the connection and verifies the ring's operating speed. Each MAU is a collection of looped electrical gates that form a logical ring. When nothing is attached to an MAU port, the port's gate is looped so the port is effectively bypassed from the active ring. When a station is initially attached to the MAU, it transmits a lobe test signal, which is looped back to the station because the MAU port is looped. The station uses this signal to verify the port's operating rate and to check the lobe cable. If the lobe test signal is returned without error, the station transmits a ring attach signal, which opens the MAU port's gate. With the MAU's gate open, the connected station is now part of the active ring.

- Phase 1: Active monitor check—The insertion of a new station interrupts the ring's transmission signal. This causes an error in the transaction occurring during the new ring station's insertion. This error forces the active monitor to purge the ring, notify the connected stations of the AM's existence, and start the neighbor notification process. The newly attached station waits to see the management messages, which indicate that the ring is functioning correctly. If the new station fails to see these messages within an appropriate amount of time, the new station initiates the token claiming process. This action is needed because the failure to receive these messages indicates that either the active monitor for the ring has failed or it is the first station on the ring and no active monitor exists.
- Phase 2: Duplicate address check—After the active monitor is detected (or elected), the new station generates a duplicate address message. This message contains the same source and destination address of the new ring station. If the new station receives the message back without the frame status bit being set as received, it assumes that no other station has the same destination address as itself, and removes the message from the ring. If the message is returned with the frame status bit set as received, the station removes itself from the ring. This test is done mainly to accommodate for Token Ring implementations that use locally assigned addresses instead of universal addressing.
- Phase 3: Neighbor notification—This process is normally initiated by the AM. So, each station periodically announces itself to its downstream neighbor. When a new station joins the ring, it announces itself to its downstream neighbor and waits for a notification announcement from its upstream neighbor.
- Phase 4: Ring initialization request—Each station has its own default operational configuration. The station's defaults might not be adequate by themselves, depending on the special operational needs of the ring. The last step of the new station's insertion process is to send a ring initialization request message to the ring parameter server, containing the new stations MAC address, its microcode version and its NAUN. The ring parameter server replies with basic information about the ring and any additional configuration information needed in order for the requesting station to function properly. If no RPS exists or no additional configuration information exists, the requesting station uses its default settings.

Token Ring MAC Frame Format and MAC Frame Types

There are three Token Ring frame formats: the token, which is 16 bits in size, the IEEE 802.5 data frame, and the IBM Token Ring data frame. The 4Mbps data frame has a minimum frame size of 22 bytes and maximum size of 4,521 bytes. The 16Mbps frame has a minimum size of 22 bytes and a maximum size of 18,021 bytes. Figure 4.19 illustrates the three frame formats.

IBM Token Ring Data Frame Format

Start Delimiter 8 bits	Access Control 8 bits	Destination Address 16 to 48 bits	Source Address 16 to 48 bits	Routing Info 32 bits	Data	FCS 32 bits	End Delimiter 8 bits	Frame Status 8 bits

802.5 Data Frame Format

Start Delimiter 8 bits	Access Control 8 bits	Destination Address 16 to 48 bits	Source Address 16 to 48 bits	Data 0-X Bytes	FCS 32 bits	End Delimiter 8 bits	Frame Status 8 bits

IBM/802.5 Token Format

Start Delimiter 8 bits	Access Control 8 bits	End Delimiter 8 bits

Figure 4.19 The IEEE and IBM Token Ring frame formats.

Now let's look at the fields that make up the 802.5/IBM Token Ring frames:

- Starting Delimiter (8 bits)—The starting delimiter contains non-data encoding values. This is used by the stations as an indicator that a data or control frame has arrived.
- Access Control (8 bits)—This field identifies if the frame is a data frame or a token. It specifies the token's priority, sets the reservation for the next generated token, and contains the monitor bit.
- Frame Control (8 bits)—Indicates whether the frame contains control information or data.
- Destination Address (48 bits)—Like Ethernet addresses, these can either be locally assigned or use a combination of an Organizationally Unique Identifier (OUI) and a manufactured assigned address. Ethernet MAC and Token Ring MAC addresses are not compatible. The Token Ring address has five subfields. The first bit identifies the destination address as belonging to a single end-station or as a group address. The second bit indicates if the address is locally administered or if it is derived from the NIC. The next subfield contains six hex digits (a null address is used if the address is locally administered), making up the

manufacturer's OUI. The next subfield indicates which function, if any, is associated with the destination (such as gateway, active monitor, or bridge). The remaining bits are used for the station's unique ID, group address, broadcast address, and so on. In addition to the formatting incompatibilities, it should also be noted that Token Ring addresses are expressed in big-endian (most significant byte first) form, and Ethernet addresses are expressed in little-endian (least significant byte first) form.

- Source Address (48 bits)—This address has three subfields. The first bit is known as the Routing Information bit (RI). The RI indicates if routing information is contained in the data portion of the frame. The second bit indicates if the address is locally or universally administered. The remaining bits contain the OUI (or NULL address) and the sending station's unique address.
- Data (1 byte to 4,500 bytes or 18,000 bytes)—The data field contains either control information, which is made up of vectors and subvectors which relay the Token Ring MAC control information, or a standard 802.2 LLC-PDU. In addition, the LLC-PDU frame can also contain source routing information. Source routing is an IBM transparent bridge variation that permits the use of more than one bridge to connect a ring to other rings within the greater network. Source routing works by having end-stations send out network learning packets that send information back to the originator about the available paths to reach destinations on the network. Up to 16 bytes of routing information can be sent with a packet. If source route bridging is enabled and a packet contains no routing information, the packet cannot be delivered to destinations outside the local ring. Source routing is not part of the IEEE standard.
- Frame Check Sequence (32 bits)—The FCS contains the CRC value of the frame. When the destination host receives the frame, it performs a CRC check on the frame and compares the results to the CRC in the FCS field. If they match, the frame is processed further if it fails, the frame is discarded.
- End Delimiter (8 bits)—This field, like the Start Delimiter, uses non-data symbols to indicate the end of the frame. This frame also contains identification bits that indicate if the frame is part of a multiframe transmission and if a frame error has been detected by another host.
- Frame Status (8 bits)—This field is used by the destination station to indicate the status of the frame and whether it was successfully received by the destination.

To provide the inter-station messaging needed to manage the ring management processes, the IBM Token Ring architecture defines 25 different MAC frame types. The IEEE 802.5 standard defines 6 MAC frame types. The MAC frame types' identification (called a vector) is contained in the first 32 bits of the frame's information field. The first 16 bits is the vector message's length. The following 16 bits contain the

MAC's vector code (that is, MAC type). The remainder of the information field contains the actual management data (known as sub-vectors). One or more management messages can be contained in a single Token Ring frame. The IEEE 802.5-defined MAC frame types are as follows:

- Beacon (BCN)—Two classifications for Token Ring erroblrs exist: soft errors and hard errors. Soft errors affect the transmission of data, but do not represent a condition that alters the configuration or disables the ring. Depending on the severity of the soft error, different ring recovery procedures are initiated so that the ring can return to a normal state as quickly as possible. A Token Ring hard error represents failure that halts the ability for the ring to transmit data. When hard errors occur, the ring attempts to reconfigure itself by removing the faulty elements from the ring. This process, known as beaconing, is initiated by having the station that has detected the hard error transmit beacon frames. When a station enters a beaconing state, the beacon frames notify the other stations of the ring error and the status of the recovery process.
- Claim Token (CTK)—When a standby monitor has reason to believe the active monitor has failed or has been removed from the ring, it generates a Claim Token in order to initiate the token claiming process.
- Ring Purge (PRG)—When the active monitor needs to reset the ring due to an error, it transmits a ring purge frame. The "ring purge" resets the ring back to the ring's normal operational state.
- Active Monitor Present (AMP)—The AMP frame is transmitted by the active monitor to all the standby monitors as part of the active monitor's neighbor notification monitoring process. If a standby monitor fails to get an AMP once every 15 seconds, the standby monitor generates an error and a claim token, initiated the token claiming process.
- Standby Monitor Present (SMP)—The SMP is sent as part of the neighbor notification process when the station joins the ring. It is used to notify its nearest active downstream neighbor (NADN) of its MAC address.
- Duplicate Address Test (DAT)—The DAT is used by a ring station to verify that no other station on the ring has the same address. In addition to the ring insertion process, DAT frames are also transmitted when a ring station is involved in a hard error condition.

Of the 25 IBM MAC frame types, the majority of them are used to notify the configuration report server of changes (and, in some cases, failures) in the ring's operation. Here are descriptions of the IBM MAC frames that pertain to error reporting in conjunction with IBM's LAN Manager product or equivalent product that can function as an REM and the Ring Parameter server:

- Report Soft Error (RSE)—When a station detects a soft error, its first action is to collect data on the nature of the error. The error collection process is a timed event. When the soft error timer has expired, it generates a report and sends it to the REM (provided one exists).
- Report Neighbor Notification Incomplete (RNNI)—The neighbor notification process is a timed event. If the notifying station fails to receive a response from its NAUN in the allotted time, an RNNI is generated and sent to the REM.
- Report Active Monitor Error (RAME)—In the event that the AM detects an error with any of its functions (and during the token claiming process). It sends a notification message to the REM.
- Ring Station Initialization (RSI)—This message type is sent to the RPS containing its configuration information announcing that it needs any ring configuration parameters that might exist.

Token Ring Service Interfaces

The Token Ring specification defines a single PHY interface for sending and receiving serialized, Manchester encoded data symbols using baseband signaling. Four symbols are used for data encoding: binary one and zero and two non-data symbols for the start and end frame delimiters. There are four service interfaces, defined as LLC-to-MAC, MAC-to-PHY, MAC-to-NMT, and PHY-to-NMT. The LLC-to-MAC interface is responsible for data exchange between the ULPs and the MAC interface. The MAC-to-PHY interface performs the encoding and decoding of symbol data. The MAC-to-NMT interface enables the NMT to perform the monitor and control functions needed for the MAC interface operations. These functions include token hold timing, monitor election, beacon acknowledgment, and frame error acknowledgment and reporting. The PHY-to-NMT interface provides the control and monitoring functions that enable the NMT to control the operation of the PHY. These functions include the insertion and removal of the local station from the ring, and notification of the NMT from the PHY to the existence of an error or status change in the ring. Figure 4.20 illustrates the Token Ring service interfaces and their relation to the OSI-RM.

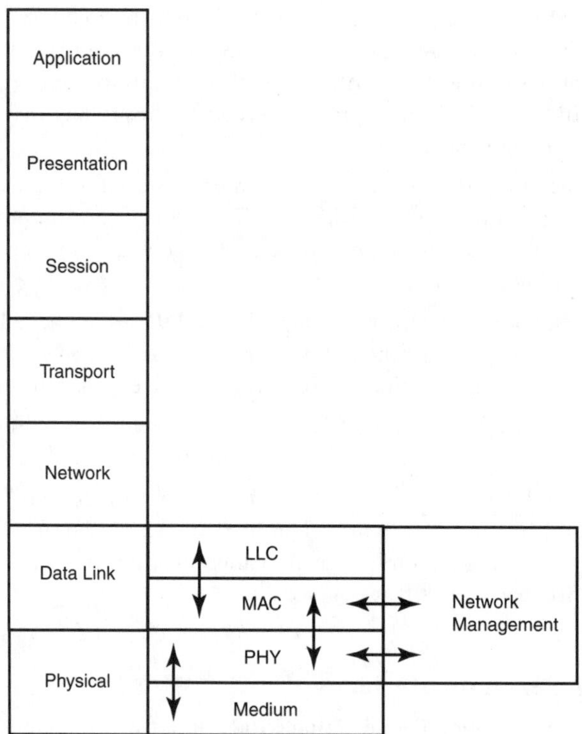

Figure 4.20 4/16 Token Ring data link (MAC), and physical (PHY) layer interfaces.

Token Ring PHY Implementations

The PHY-to-MAC service interface performs the encoding/decoding of the MAC frame into Manchester encoded symbols. The encoded/decoded serial bitstream operates at a rate of 4MHz or 16MHz depending on the ring operating rate. Signal timing for all stations is derived from the active monitor. The ring functions as a timing loop, which is set by the AM. Stations defer to their upstream neighbor when they first join the ring or in the event that they lose phase lock with the ring. The active monitor is also responsible for setting the minimum latency for the ring, which is based on the size of the active token. The minimum latency is 24 bits. The IBM and IEEE electro-mechanical interface specifications for the PHY differ somewhat. IBM defines the use of three different cable and connector specifications with an active hub or MAU. The IEEE specification provides a description for a two-wire shielded 150-ohm cable, known as a Medium Interface Cable (MIC). This provides the interconnection between the end-station network adapter and the Trunk Coupling Unit (TCU). This model is in essence the IBM Type 1 cabling specification.

According to IBM, a Token Ring network consists of three elements:

- Token Ring Network Adapter Card (TR-NAC)—The TR-NAC provides the logic and signaling functions to implement the MAC and PHY layers of the network. It provides the physical interface between the connected end-station and the other adjacent end-stations attached to the physical network media.
- Token Ring Network Multistation Access Unit (MAU)—The IBM MAU (IEEE TCU) using Type 1 cabling is a ten-port active hub that can connect up to eight end-stations. The remaining two ports are used for interconnecting with additional MAUs, which enables multiple MAUs to be connected together to form a logical ring. One port is the "ring in," the other is the "ring out" (see Figure 4.21). Connections between MAUs are made with patch cables, and connections between MAU and end-stations are made with lobe cables (IEEE MIC). Today, Token Ring MAUs are almost identical to Ethernet repeaters, with port densities ranging between 12 and 24 ports, using RJ-45 connectors instead of the type 1 MIC connector (see Figure 4.22).

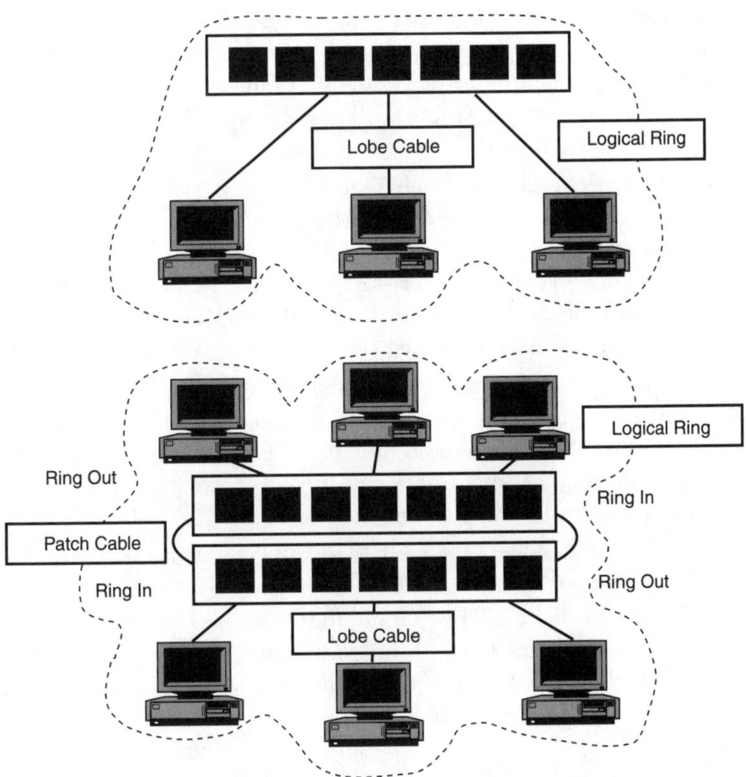

Figure 4.21 Single and multiple MAU Token Ring topologies.

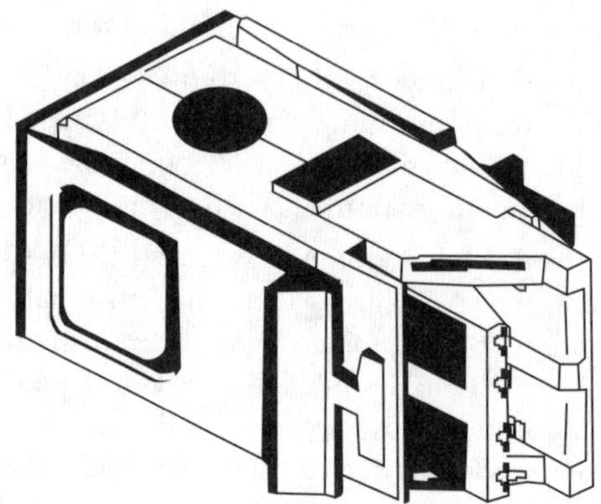

Figure 4.22 The IBM type 1 MIC connector.

- Cabling system—The IBM Token Ring specification specifies three cable types for baseband transmission. The IEEE specification only specifies the NIC-to-TCU interconnection cable. Other than the medium attachment interface, no formal specification for transmission medium or topology is included in the standard. This is done to provide the option of implementing the IEEE token-passing access protocol on various media types and topologies. The IBM cabling specifications are as follows:
  - IBM Type 1 cabling—When Type 1 cabling is used, each end-station can be up to 101m from the MAU. Type 1 cabling consists of two shielded pairs of 22 AWG with 150-ohm impedance. The two shielded pairs are placed in a shielded outer casing. Type 1 cabling uses a Type A or IEEE Medium Interface Connector (MIC). Both 4Mbps and 16Mbps operation are supported. A Type 1 based ring can support up to 260 end-stations.
  - IBM Type 2 cabling, shielded twisted-pair (STP)—STP cable can consist of two or four wire pairs. It can support end-stations to MAU lengths up to 100m. STP uses a Type A or MIC connector; only two pairs are used for transmission. Both 4Mbps and 16Mbps operation are supported. STP rings support the same density as Type 1 rings, 260 end-stations.
  - IBM Type 3 cabling, data-grade unshielded twisted-pair (UTP)—Token Ring operates at 4Mbps on Category 3 and 4, and at 16Mbps on Category 4 and 5 UTP. UTP uses RJ-11 and RJ-45 connectors. The IBM specification states a UTP ring can support a density of no more than 72 end-stations. The maximum cable distance between the end-station and the MAU cannot exceed 45m.

A ring can consist of up to 33 MAUs using Type 1 and Type 2 cabling. This count is reduced to 9 MAUs when you deploy eight ports per MAU for a maximum of 72 stations per UTP ring. Connections between MAUs can be no more than 45m. These distances can be increased through the use of repeaters.

With Type 3 cabling, which is by far the most common today, there are some limits with regard to the maximum functional cable length between end-stations and MAUs, and with MAU-to-MAU interconnections. The IBM specification for station-to-MAU connections is 45m. There are, however, manufacturers that support site distances up to three times that amount. It is best to check with your equipment's manufacturer for the maximum supported cable lengths.

Token Ring Bridges and Dedicated Token Ring

It was mentioned previously that IBM Token Ring supports a type of bridging called source route bridging (SRB). SRB is part of the IBM Token Ring specification; it is not part of the IEEE 802.5 specification. 802.5, like all the other 802 specifications. uses 802.1 spanning tree bridging. Both of these bridging methods will be covered in Chapter 6. There are two types of bridges used with Token Ring: traditional two-port bridges and multiport bridges.

The 802.5r Token Ring standard defines the operation of full-duplex or Dedicated Token Ring (DTR). With DTR, the end-station can transmit and receive tokens simultaneously, which, when utilized, gives the end-station an effective data transmission throughput rate of 32Mbps. DTR requires the use of a Token Ring multiport bridge and a DTR adapter. Token Ring multiport bridges operate the same as Ethernet multiport bridges. Each bridge port is its own network segment—in this case, a ring. The bridge enables all its active ports to establish connections between each other simultaneously. The DTR adapter is different than a traditional Token Ring adapter. With traditional Token Ring adapters, a repeater loop between the transmit and receive path facilitates the ring operation. DTR adapters do not have this loop; the transmit and receive paths are not connected. When the DTR interface is connected to a Token Ring multiport bridge, the Transmit Immediate Protocol (TXI) is used to facilitate simultaneous data transmission and reception between the end-station and the bridge port, without acquiring a token before transmission. If a non-DTR end-station is directly connected to a Token Ring multiport bridge port, the station must wait for a token before it can transmit data.

FDDI

Work on the Fiber Distributed Data Interface (FDDI) began in 1982 by both standards bodies (ANSI) and vendors (Digital, Control Data, Sperry Rand, and others) out of the need for a higher-speed LAN protocol. In 1990, FDDI became an ANSI X3 standard. It specifies a 100Mbps token-passing (similar in operation to Token Ring) shared transmission medium LAN.

The original FDDI standard specified a passive, counter-rotating dual-ring topology over fiber optic cable. In a dual-ring topology, each end-station is a Dual-Attachment Station (DAS). One ring is the primary, on which data flows in one direction, and it is used for data transmission. The secondary ring is backup. The same data that is transmitted on the primary ring is transmitted on the secondary ring, but it flows in the opposite direction. The secondary ring remains idle during normal operation. In the event of a station failure or break in the ring, the primary and secondary rings wrap back on themselves, creating a single ring. In the event that more than one break occurs, the ring will segment itself into separate rings until the damaged state is resolved. In addition to the dual-attachment topology, FDDI can utilize a dual-attachment concentrator, which is part of the dual ring, but provides the capability for FDDI Single Attachment Stations (SAS) to connect to the primary ring (they move to the secondary ring in the event of a failure). A dual-attachment implementation of the FDDI protocol utilizes copper cabling, called Copper Distributed Data Interface (CDDI), developed by Crescendo Communications (now part of Cisco Systems). The CDDI specification was added to the ANSI FDDI PHY specification.

FDDI has enjoyed a lot of success as a LAN backbone and high-speed server technology. Typically, a FDDI backbone is constructed and bridges are used to interface low-speed (Ethernet and so on) LAN hubs and concentrators. Then, through the use of DAS NIC and DAS concentrators, high-speed servers are attached to the network backbone. Figure 4.23 illustrates a FDDI backbone topology utilizing this approach.

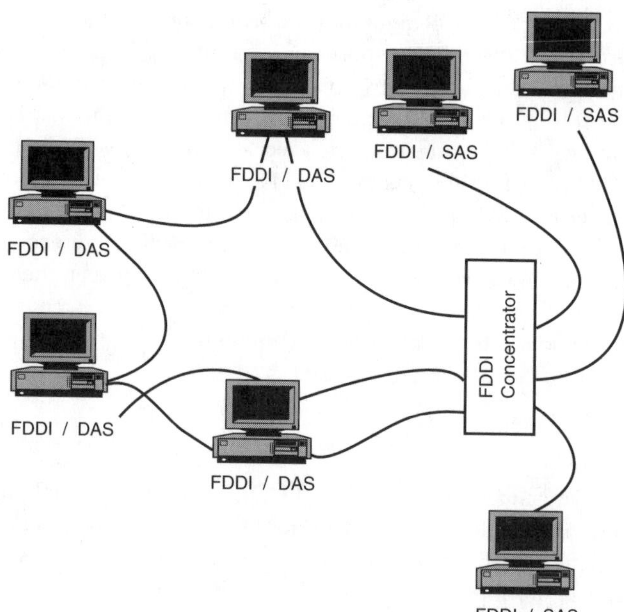

Figure 4.23 FDDI backbone utilizing DAS and SAS with a FDDI concentrator.

FDDI has historically had wide vendor support, some of whom believed that FDDI would survive even in the face of Fast Ethernet. Although today that seems like poor judgment, when looking at FDDI's strengths it does not seem so farfetched. FDDI's features include the following:

- High fault tolerance with built-in redundancy.
- Efficient media access protocol, which uses token-passing and has an efficiency rate of 85 to 90 percent of the medium's bandwidth. Ethernet with CSMA/CD has around a 65 percent efficiency.
- Large segment diameter and end-station density; a single ring can support 1,000 end-stations and has a maximum distance of 200km.
- Fiber optic cable as its transmission medium, which is superior to a copper-based medium in terms of bandwidth, reliability, resistance to environmental conditions, and security.

However, all this comes at a cost, which for most is too high. FDDI has never been seen as a desktop technology, because of its expense and its use of fiber as a transmission medium. When CDDI became available, its costs were still too high. So, when the demand for high-speed desktop network access arose, FDDI was never really considered.

The FDDI Specification

The FDDI specification is not a single specification. Rather, it is a set of four standards, each providing a specific function. These functions interoperate to provide high-speed transmission capabilities. The original FDDI specification's primary goals were large station support and unattended, fault-tolerant, decentralized operation. This makes FDDI quite well-suited for enterprise LAN backbone and MAN networking applications. It was with these applications in mind that the use of a fiber optic transmission medium seemed ideal. The four standards that make up the FDDI architectural model are the following:

- Medium Access Control (MAC)
- Physical Layer Protocol (PHY)
- Physical Layer Medium (PMD)
- Station Management (SMT)

Figure 4.24 illustrates their relation to one another and placement within the OSI-RM.

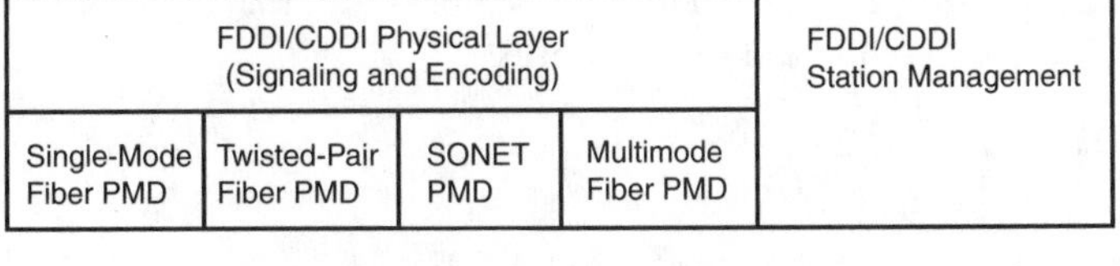

<table>
<tr><td colspan="4">FDDI/CDDI Media Access Control</td><td rowspan="3">FDDI/CDDI
Station Management</td></tr>
<tr><td colspan="4">FDDI/CDDI Physical Layer
(Signaling and Encoding)</td></tr>
<tr><td>Single-Mode
Fiber PMD</td><td>Twisted-Pair
Fiber PMD</td><td>SONET
PMD</td><td>Multimode
Fiber PMD</td></tr>
</table>

Figure 4.24 The FDDI architecture model.

The Media Access Control Standard

The Media Access Control (MAC) layer is responsible for describing, transmitting, and receiving MAC frames and media access. FDDI uses timed token-passing for media access control. Data is transmitted from station to station in a serial bitstream. Each station receives the stream, then regenerates and retransmits it to the next station. When a station has data to transmit, it captures the token and transmits its data. The receiving station copies the transmitted frames and then retransmits them back on to the ring. The amount of time a station has to transmit data is governed by the token holding rules. When its transmission is complete or when the holding timer has expired, the sending station transmits a new token. It is the responsibility of the transmitting station to strip its transmitted frames off the ring when they return from the destination station.

The stations attached to the ring determine the token holding rules through the use of the Token Rotation Protocol (TRP). The TRP defines three classes of data traffic: two synchronous classes and one asynchronous.

- Synchronous (Class 1) supports guaranteed bandwidth and response times
- Synchronous (Class 2) is restricted token traffic that supports point-to-point sessions between stations.
- Asynchronous supports multiple traffic priorities

The synchronous classes allocate a dedicated amount of bandwidth; the remaining bandwidth is used for asynchronous transmissions. This async bandwidth is shared by all the stations through the use of timers. The TRP is tunable, and some of these parameters are Target Token Rotation Timer (TTRT), Token Rotation Timer (TRT), and the Token Holding Timer (THT). These and other tuning options allow network performance to be tuned to best meet higher level application access needs.

The token holding rules are negotiated by the station management interface. The negotiation process occurs whenever a new station joins the ring. This process is in essence a bidding war between the stations to decide who will initialize the ring and to agree upon a guaranteed service time. The bidding is done by having each station announce how often it needs access to the token, based on its synchronous service requirement. The station with the lowest bit wins. The winning bid is stored as the TTRT, and the guaranteed service time is two times the TTRT. The winning station sends the first token, and on the second pass other synchronous devices might start to transmit. When the synchronous "class" stations have transmitted their data, the asynchronous stations can have access to the token.

The FDDI frame has a maximum size of 4,500 bytes. The token and frame are two different frame entities. They are not a single frame appended with delivery information and data, as with Token Ring. The FDDI MAC frames carry SMT control information and use 802.2 LLC-PDUs to transport ULP data. Figure 4.25 illustrates the FDDI MAC data frame and token formats.

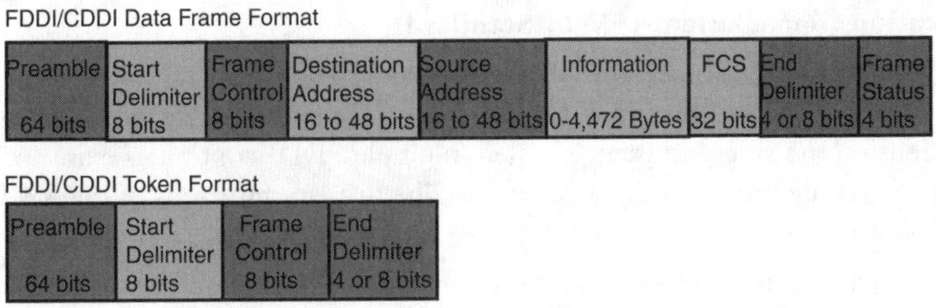

FDDI/CDDI Data Frame Format

Preamble	Start Delimiter	Frame Control	Destination Address	Source Address	Information	FCS	End Delimiter	Frame Status
64 bits	8 bits	8 bits	16 to 48 bits	16 to 48 bits	0-4,472 Bytes	32 bits	4 or 8 bits	4 bits

FDDI/CDDI Token Format

Preamble	Start Delimiter	Frame Control	End Delimiter
64 bits	8 bits	8 bits	4 or 8 bits

Figure 4.25 FDDI MAC data frame and token formats.

Now let's take a look at these formats:

- Preamble (64 bits or more)—Contains idle non-data symbols; it functions as an interframe gap between data frames and tokens.
- Starting Delimiter (8 bits)—Indicates the beginning of the data frame or token and uses a unique data symbol, distinct from normal data symbols.
- Frame Control (8 bits)—Indicates the frame priority class, if it is a beacon or claim frame. It is also used to partition the data contained in the incoming frame and to indicate the type of data contained in the information field (LLC-PDU or SMT).
- Destination Address (16 or 48 bits)—FDDI uses the same address format as Token Ring. The first 2 bits indicate if the address is a station or group and if local or universal addressing is used. The remaining bits are used for the station or group address.
- Source Address (16 or 48 bits)—Indicates the station that sent the frame.
- Information (0 to 4,472 bytes)—802.2 LLC-PDU or SMT data.
- Frame Check Sequence (32 bits)—Contains the CRC value for the frame. The destination host recalculates the CRC value for the frame and compares it to the one contained in the frame. If the FCS sequences match, the frame is processed further.
- End Delimiter (32 bits)—Indicates the end of the data frame or token. It uses a unique data symbol, distinct from normal data symbols.
- Frame Status (4 or 8 bits)—Used by the destination host to indicate that the frame has been received.

The Station Management (SMT) Standard

The SMT acts in unison with the MAC, PHY, and PMD layers. Its role is to ensure the proper operation of the ring and to manage interpretability between all the stations on the ring. In instances where multiple FDDI controllers exist on a given station, only one SMT interface exists. The primary functions of the SMT are as follows:

- Connection management—Connection establishment, link testing, and error monitoring between adjacent PHYs. MAC-to-PHY configuration, bypass control, and medium availability acknowledgment are part of this function.
- Ring management—Beacon detection and generation, transmission problem resolution, address management, duplicate address detection, and MAC availability monitoring.
- Control frame generation—Generation and processing of SMT control frames, which come in the following flavors:
  - Neighborhood information frames—These frames determine duplicate addresses and the development of a local ring address map and ring path.
  - Status information frames—Two types of status information frames, configuration and operation, are used to glean additional status information about the stations connected on the ring.
  - Parameter management frames—These frames are used in conjunction with the station's management information base attributes to facilitate management information collection and station configuration.
  - Status report frames—These frames are used to send station and ring status announcements to other stations on the ring.

The Physical Layer Protocol

The Physical Layer Protocol (PHY) is independent of the PMD. This enables it to interface with a variety of PMD interfaces. The PHY is responsible for encoding and decoding data, using the 4B/5B encoding scheme. The MAC-to-PHY interface is a 4-bit data path operating at 25MHz. The PHY converts the MAC datastream into 4-bit symbols and then, using a 4B/5B encoder, converts them into a 5-bit symbol code. The 5-bit code is then serialized and transmitted through a NRZI converter at a line rate of 125MHz. The NRZI encoded stream is then handed to the PMD for transmission across the medium, also transmitted at a clock rate of 125MHz, which yields a 100Mbps transmission rate. FDDI gets its speed from the encodfting method. The PHY is also responsible for managing timing, and with FDDI, this is a distributed process. Each station has its own clock and gets its clocking signal from the incoming datastream. The data is buffered and transmitted back out to the ring. The internal station buffer expands or contracts the inter-packet gap to make up for timing errors. This avoids line jitter and affords a much more efficient data flow, because the inter-packet gap is optimized to a minimum.

The Physical Layer Medium Standard

The Physical Layer Medium (PMD) interface describes the electromechanical interfaces, media and transmission specifications, and station attachment topologies needed for the transmission of bitstream data between stations. The PHY layer hands the 4B/5B serial bitstream to the PMD for transmission. The PMD converts these data bitstreams into optical or electrical (depending on the PMD) signals that can be understood by the transmitting and receiving PMD stations. The following are the most common FDDI PMD implementations:

- Multimode PMD (MM-PDM)—This is the PMD original definition. It specifies the use of 62.5/125 micron multimode fiber with a long wavelength 1300-nanometer light, a power budget of 11 dB, and a maximum signal loss of 1.5 dB per km. Multimode DAT might be separated by as much as 2km.
- Single-mode PMD (SM-PMD)—This calls for the use of 10/125 micron single-mode fiber with a long wavelength 1300-nanometer light. A power budget of 10 dB to 32 dB is permissible. Single-mode DAS can operate between 40 to 60km apart, depending on the types of transmitters used.
- Twisted-pair PMD (TP-PMD)—This specifies the use of shielded IBM Type-1 or Category 5 UTP. A cable distance of up to 100m is supported between stations. The NRZI encoding scheme is not used with the TP-PMD; MLT-3 is used instead. The MM-PMD and SM-PMD use keyed media interface connectors (see Figure 4.26). The keys are used to ensure optical alignment and the correct ring connections between the connection and the FDDI NIC. The TP-PMD uses RJ-45 connectors. SC connectors are also used by some manufacturers of MM-PMD NICs. These cards should be used with some caution, because they provide no ring keying facility.

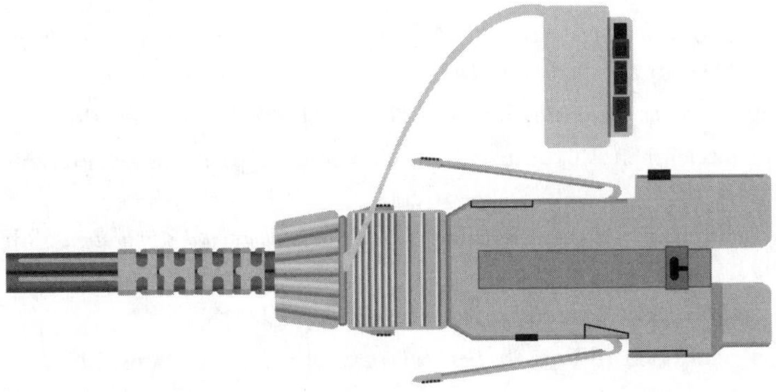

Figure 4.26 A FDDI media interface connector.

The PMD standard also defines the operation of the Dual-Attachment Concentrator (DAC), Single-Attachment Concentrator (SAC), and the optical bypass switch. DAC and SAC FDDI are used to provide SAT connections to the FDDI ring. The optical bypass switch is a device that provides the capability for continuous operation of the dual ring in case of a station failure or shutdown. It does this through the use of mirrors that redirect the light away from the DAS. The disabled station is bypassed, and the ring maintains its primary and secondary ring functionality.

Summary

At this point, you should have a solid understanding of the operation and implementation of Ethernet, Token Ring, and FDDI. You should also have a basic understanding of data signaling, encoding, fiber optic and copper-based transmission media, and LAN cabling components. Now, let's move on to WAN data link and transport protocols.

Related RFCs

RFC 894	Standard for the transmission of IP datagrams over Ethernet networks
RFC 1042	Standard for the transmission of IP datagrams over IEEE 802 networks
RFC 1390	Transmission of IP and ARP over FDDI networks

Additional Resources

Breyer, Robert, and Sean Riley. *Switched, Fast, and Gigabit Ethernet*, Third Edition. Macmillan Technical Publishing, 1999.

Codenoll Technology Corporation, *The Fiber Optic LAN Handbook*, Fourth Edition. Codenoll Technology Corporation, 1990.

Held, Gilbert. *Ethernet Networks*, Second Edition. John Wiley & Sons, 1996.

Kadambi, Jayant, Mohan Kalkunte and Ian Crawford. *Gigabit Ethernet*. Prentice Hall, 1998.

Mirchandani, Sonu, and Raman Khanna. *FDDI: Technology and Applications*. John Wiley & Sons, 1993.

Partridge, Craig. *Gigabit Networking*. Addison-Wesley, 1993.

Quinn, Liam, and Richard Russell. *Fast Ethernet*. John Wiley & Sons, 1997.

5 WAN Internetworking Technologies

IN THIS CHAPTER, WE LOOK AT WAN INTERNETWORKING TECHNOLOGIES. When we examine this subject, the first thing we notice is the clear separation between physical transport (OSI-RM Layer 1) and data transmission (OSI-RM Layer 2) services. This separation contrasts sharply with LAN technologies, where the services are handled by both layers under the same protocol specification. This delineation might seem odd to administrators who have most of their experience with LAN protocols, but it is characteristic of WAN networking, and reflects the utilitarian nature of the Public Switched Telephone Network (PSTN) that provide transport service.

PSTN was originally developed for voice transport, which required that each transmission be distinguishable from every other, but that the continuity of each separate transmission be maintained. To facilitate this kind of service, each voice transmission was given its own dedicated channel. When the PSTN was later used for transmitting data, this same model was followed. Each data path has its own channel for transmitting signals across the PSTN, and for the duration of the "call," the available bandwidth of that channel can be used to transmit any signal. In order for data to be transmitted over the PSTN channel, some kind of transmission scheme (framing) needs to be employed to format the data so it can be interpreted correctly when it reaches the other end of the channel. The PSTN is uninterested in what is being carried over the channel. Its only function is to ensure that the data is in a useable form when it reaches its destination.

As the transmission of data became a larger and larger percentage of the traffic being carried over the PSTN, WAN packet-switching and cell-switching technologies were developed. Fundamentally, these technologies utilize the dedicated physical layer transport mechanisms to send data signals over the PSTN. They also provide intelligent signaling mechanisms that enable the available transport bandwidth to be used to support multiple, discrete connections between different endpoints. These technologies afford a more dynamic, efficient, and cost-effective utilization of the channel bandwidth and expand the functional usefulness of the PSTN beyond that of a large channel-switching system.

To start, we look at the PSTN's history and basic structure. Then, we review three transport technologies: AT&T/Belcore's Transport-1 (T1) digital carrier system, SONET (Synchronous Optical Network), and SDH (Synchronous Digital Hierarchy), which are ANSI and ITU-T international standards. We then take a brief look at the major PSTN packet- and cell-switching technologies: ISDN, Frame Relay, and ATM. We close the chapter by examining the popular "data-link" protocols: HDLC (High-Level Data Link Control) and PPP (Point-to-Point Protocol). These two protocols are used to provide data-link transmission services over dedicated and dial-on-demand synchronous and asynchronous point-to-point PSTN transport links. The overall goal of this chapter is to provide you with a basic functional understanding of the available WAN transports and data-link protocols.

A Brief History of the PSTN

Almost all WAN communications—dedicated point-to-point (D-PPP), dial-on-demand routing (DDR), and remote access dial (RA)—links are dependent on the PSTN for data transport. PSTN access is provided on a tariff regulated subscriber basis, depending on the country; the provider might be a private/publicly owned telephone exchange carrier or government-owned utility. PSTN access pricing is dependent on the type of PSTN connection, but it is generally either bandwidth/usage-based or facility-based. Bandwidth/usage-based pricing is billed on a time usage rate based on bandwidth utilization. The basic home telephone uses this pricing. Users are billed on a per minute or per call basis for the use of a 64K PSTN channel. What is transmitted over that channel, voice or data, is irrelevant; the usage cost is the same. Facility-based pricing is based on the allocation of a specific bandwidth path over a finite distance. This pricing is used for dedicated PSTN connections. With dedicated connections, the link's bandwidth is yours to use. It is always available and the link is always up, or "off hook" in telephone lingo, whether or not you have actual data to send.

In the United States, PSTN and its operating companies are regulated by the Federal Communications Commission (FCC). PSTN access is considered a utility service and basic service is guaranteed by law. Basic service is a standard single pair 64Kbps (with an actual throughput limitation of 53Kbps), plain old telephone service (POTS) line that provides local network access and supports essential services such as

911, long distance (interexchange service), touch-tone dialing, and basic operator assistance. The Universal Service system, updated most recently in the Telecommunications Act of 1996, defines the scope of basic service and the subsidy programs used to pay for low-income consumer access.

The original PSTN system (the Bell System) in the United States was constructed in the 1930s by the American Telephone and Telegraph Company (AT&T). It was a private/public utility. This is to say that AT&T and its operating companies, known as Bell Operating Companies (BOCs), owned the physical infrastructure from the telephone and wires in your home to the national cabling and switching infrastructure. Subscribers could use the telephone system, but they couldn't attach anything directly to it. Telephone equipment, telephones, switches, modems, and so on were built by Western Electric (WECO). The system itself was a network of electromechanical switches and wire, and signals were carried using analog transmissions (sine wave). In the 1960s, AT&T began to develop digital switch transmission technologies and to convert the PSTN transmission facilities from analog to digital. We look at these transmission methods a little later.

Placing a Call

The functional model of the PSTN is simple, and although the technologies have changed over the years, the telephone system behaves the same way it did when it was initially comprised of electromechanical switches in the 1930s. In the simplest of terms, a telephone call is a completed circuit. The telephone system is a large collection of looped circuits between subscriber locations (homes, offices, and so on) and a local central switching office. These looped circuits are called local loops.

The PSTN is based on DC voltage. Each local loop has a test tone of .001 watts (1 milliwatt) running through it. The maximum line voltage permitted across the PSTN is 2.2 +/- VDC (Volt's Direct Current); this is the baseline for measuring signal gain and loss across the network. Gain and loss is measured in decibels (dB).

> **Note**
>
> The formula for calculating gain or loss is dB = 10 log 10 P0/P1, where P0 = output (transmit power) and P1 = input (received power). 10 log 10 is also known as the common logarithm (base 10). The base 10 logarithm indicates how many times 10 must be multiplied by itself to yield a given number, for example, 10 log 10 1000 = 10 x 10 x 10 = 3.

A call is placed by taking the telephone "off hook," which breaks the test tone circuit and places the line in "short" state. This short is acknowledged by the central office (CO) switch, which listens for a series of pulses (variations in the applied voltage) that tell the telephone switch the subscriber number you want to connect to, and the connection is made. The original PSTN used a rotary dial, which sent a series of voltage pulses down the line to represent numbers. Today, tones are used to represent different numbers. In the U.S., subscriber numbers are assigned using the North American Numbering Plan (NANP). The NANP defines a 10-digit subscriber number for establishing calls within the PSTN in North America. The first three digits identify the numbering plan area (NPA), which is the geographical area where the call is being placed. The second three digits identify the local dial exchange, and the last four digits identify the subscriber ID.

When the switch has the subscriber ID, it establishes a dedicated circuit between the two parties, who then have use of the circuit's transmission bandwidth for the duration of the call. When the call is completed, the bandwidth will be reallocated for another call.

The PSTN Hierarchy

The U.S. PSTN is a collection of switching offices or call centers all over the country. The original PSTN used a hierarchy of six different types of call centers (listed in order from the customer's premise):

- Class 5—This is the local switching office or local CO between the PSTN and its subscribers. The subscribers are connected to the local CO over 24 or 26 AWG copper twisted-pair cable. Most residential homes have between two and four twisted pairs, whereas a standard office building can have anywhere from 25 pairs and up. Only a single twisted-pair is needed to provide a basic 56/64Kbps (4KHz bandwidth) PSTN circuit. Depending on the geographical area being served by the Class 5 CO, it can support one or more local dial exchanges. A local dial exchange is limited to 9,999 subscribers. The Class 5 office also has trunk lines that lead to other local offices or toll offices.
- Class 4—There are two types of Class 4 offices: tandem and toll. A tandem office acts as an interconnect point between Class 5 offices. It provides no local loop service, only interconnections between Class 5 offices and, in some cases, a Class 4 toll office. A Class 4 toll office provides interconnections between long distance or interexchange carriers and different local area exchanges.
- Class 3—This primary toll center terminates Class 4 (and sometimes Class 5) centers. It can also offer spillover termination services for overcrowded Class 4 and 5 centers. The center's main function is to perform call routing between different local area exchanges and long distance exchanges through Class 1 and 2 call centers.

- Class 2—This sectional toll center can terminate Class 5, 4, and 3 call centers. The center routes toll level service for a BOC's area of operation, and performs gateway services for Class 1 centers.
- Class 1—This regional toll center performs the same function as a Class 2 center except that it also connects to international gateway offices. These centers are operated by long distance carriers.
- International gateway—These offices interconnect with other international gateway offices in other countries and Class 1 call centers. They perform international call exchange and operator and support assistance.

In 1974, the U.S. government filed an antitrust suit against AT&T that resulted in the breakup of AT&T and the BOCs. Under the divestiture agreement, AT&T retained its long distance and equipment manufacturing businesses (the manufacturing business is now Lucent Technologies). The BOCs were spilt up into seven Regional Bell Operating Companies (RBOCs). The divestiture agreement segmented the country into different local area transport areas (LATAs). The RBOCs and other independent telephone companies (ITC) were given jurisdiction over different LATAs to provide local exchange service, but they were limited to providing only intraLATA traffic exchanges. InterLATA exchanges were provided by AT&T and the other long distance phone companies, which were now referred to as Interexchange Carriers (IXCs). This reorganization effectively dismantled the old call center hierarchy and created a new one with three office types: LEC-CO, LEC-Tandem, and IXC-Point of Presence (IXC-POP).

Note

Of course, the breakup of AT&T and the BOCs was not enough for the U.S. government. They needed to make the U.S. telecommunications system even more dysfunctional (in my opinion) by passing the Telecommunications Act of 1996 (TCA-1996). The TCA-1996 opened what was believed to be the doors of free trade to the once heavily regulated telephone communication industry. It was perceived by some that the separation between IXCs and LECs was too constrictive and did not foster competition. So the TCA-1996 removed the separation between IXCs and LECs by allowing them to compete in each other's markets. Now, an LEC could provide long distance and an IXC could provide LATA service. This resulted in a long series of mergers between the RBOCs, ITAs, and IXCs, and created a new type of LATA carrier, called the Competitive Local Exchange Carrier (CLEC). Because the original divestiture agreement had handed all the Class 4 and lower offices to AT&T, and the Class 5 offices to the RBOCs, neither the LECs nor the IXCs had the infrastructure to provide service in the other's market. So both had to develop "wholesale" pricing to sell network access service to each other to be able to provide service. Wholesale pricing also allows the CLECs to purchase network access at wholesale rates from the LATA's LEC, or what is called the Incumbent Local Exchange Carrier (ILEC), and use their own CO switch to provide telecommunications service.

There is a new generation of IXCs constructing new high-speed networks to compete in the CLEC/IXC marketplace. What the long-term results of these different competing carriers will be is still too difficult to predict. What you need to know as a network administrator is that your carrier might or might not own the physical transport your circuits are being provided on, and you need to determine if this situation is acceptable for your company from a service and support perspective.

The following is a list of the "new" office classification types:

- LEC-CO—This is the local loop office maintained by the ILEC. It maintains the local loop circuits between PSTN termination blocks that are installed at subscriber sites and the local CO. Like a Class 5 office, the wire pair counts between the subscriber and the CO can be four or more twisted pairs. The LEC-CO has interconnection trunk lines to LEC-Tandem offices and IXC-POPs. Many people still refer to the LEC-CO as the Class 5 office.
- LEC-Tandem—The LEC-Tandem offices provide intraLATA (CO trunk line and CLEC POP trunk line) connections. The interLATA trunk line connections are used for routing inter-exchange call traffic between different ILEC-COs. The CLEC POP trunk lines provide CLEC POP's access to the ILEC network. No subscriber terminations are made in an LEC-Tandem office, they exist only to provide intermediate distribution points between different COs.
- IXC-POP—These offices are the interexchange points between IXCs and ILEC and CLEC. These facilities can be maintained by one or many different IXCs. IXC-POPs are also used by network service providers (NSPs) for providing customer attachment points to their data networks.

PSTN Basic Line Transport

The local loop transport circuit is the basis of all PSTN transport provisioning. The minimum PSTN transport is a 4000Hz bandwidth data path, commonly called plain old telephone service (POTS). POTS service is used for voice transmission between the 300Hz and 3400Hz frequency ranges, which are the optimal ranges for transmitting the human voice. It is also capable of data transmissions up to 64Kbps. The actual transmission rate is limited to 52Kbps by the telephone company (TelCo) for operational reasons. The transmission of digital data over POTS lines is provided through the use of a codec (coder/decoder), which functions much like a modem. A codec's job is to convert one signal format into another. A modem (modulator/demodulator) takes the binary data path sent by the computer and converts it into analog signals for transmission over the PSTN. The modem on the receiving end converts the analog signals back into digital format.

All POTS circuits operate in full-duplex mode because they need to transmit and receive on the same wire pair. Consequently, when your voice is being transmitted over the line, you and the other party both hear your voice in the receiver. This echo normally does not present a real problem for voice transmission, and because most people are not speaking at the same time, they never even notice. With modem transmissions, however, the echo from this full-duplex operation does present a small problem. To overcome the echo, modems can spilt the circuit's bandwidth in half, so that one modem transmits on the upper half of the passband and the other transmits on the lower half. This approach can severely limit the amount of data that can be transmitted at one time. So, for any modem that operates faster than 1200bps, a symbol-encoding scheme is used to represent data, and a technique called Echo Cancellation (EC) is employed. Modems that employ EC save a copy of their transmitted data and then use the copy to filter the data when it is sent back over the line.

The PSTN was originally designed for voice transport, which was easily achievable using a single full-duplex transmission pair. For the actual transmission to take place, however, it must have a dedicated transmission path for the duration of the call. It is possible to have multiple parties share the same transmission path, but it becomes difficult to keep straight who you are talking to, and it is not very private. This need for each transmission path to have an actual physical transport channel is not a very efficient use of bandwidth. It also presents significant scaling problems for communications systems because it is just not practical for every connection to have its own discrete wire path to transmit.

FDM

One way of addressing this problem is to send multiple signals over the same wire pair using a technique called Frequency Division Multiplexing (FDM). At one time, FDM was used to transmit multiple analog signals over CO interconnection lines or trunk lines. FDM works like this: A single copper trunk loop has a bandwidth 96KHz. A basic voice transmission only requires 3.1KHz of bandwidth for transmission. FDM separates the available bandwidth into 4KHz "channels" (3.1KHz for transmission and 900Hz channel separation). This permits the transmission of 24 channels over one trunk loop. FDM transmission uses modulators to merge the different local loop transmissions into specific passbands (the frequency ranges of a specific channel on the medium), and then all the channels are transmitted as a single signal. Demodulators are then used on the remote CO's switch to separate the different channels back into their discrete forms for local loop transmission.

Although FDM works well for voice transport, it is not suited to data, mainly because voice transmission can be "dirty," but data transmission must be "clean." To transmit any analog signal, it needs to be amplified and sent across a transmission medium. When an analog signal is transported over a copper wire, it adds noise. When the analog signal is passed through the transport segment from amplifier to amplifier, any noise it picks up along the way is added to the signal and amplified. So the longer the transmission distance, and the more times the signal is passed between amplifiers, the worse the signal will be when it reaches its destination. With audio transmission systems, it is possible to filter noise, but with data transmission systems, you cannot filter without degrading the signal, which results in corrupted data.

Digital Carrier Systems

Although FDM makes more efficient use of the transport medium's available bandwidth, it has some practical limitations in terms of signal quality and distance. Therefore, in major metropolitan areas, interCO trunks were originally deployed using discrete wire pairs. By the 1960s, a real need arose for a CO-to-CO transmission system that could reduce cable densities and provide reliable transport over longer distances. To meet this requirement, AT&T developed the T-carrier digital transmission system.

Digital transmission is not affected by the line transmission noise and the amplification problems associated with analog transmission, because digital information is transmitted as Binary Coded Decimal (BCD) words instead of actual sine-wave signals. When a sine-wave signal collects noise during transmission, it is interpreted to the receiver as part of the signal. When a transmitted pulse code signal gathers noise, it is not significant enough to alter the "code" from being properly interpreted at the receiving end. Of course, over a long enough distance, a pulse code signal can become distorted enough so it is no longer recognizable. To overcome this problem, digital transport systems use signal regenerators at specific distances along the transmission path. The regenerator takes the incoming signal, regenerates a clean signal, and then transmits it out.

Pulse Code Modulation

Pulse code modulation (PCM) is the process used to convert analog signals into digital signals. There are two PCM sampling methods: linear and non-linear. PCM works by segmenting the available analog frequency range into steps, where each step is a certain frequency level and has a specific voltage level associated with it. The sampling quality is dictated by the size of the steps and the number of available voltage increments that can be used to represent those steps. With linear PCM, the analog signal is segmented into equal voltage increments. The problem with linear PCM is that in limited bandwidth scenarios the steps are too large because there are not enough voltage steps to accurately represent the signal. Consequently, certain linear PCM samples come out distorted and full of noise. To improve the sample quality and still accommodate the limited available bandwidth, non-linear PCM was developed. Non-linear PCM algorithms use variably sized steps. Broad steps are used for frequencies that are not important to the accurate reproduction of the signal source, and ones that are finer are used for the important ones. There are two non-linear PCM algorithms used for digital signal transmission: μ-Law and A-Law. In the United States, μ-Law is used for PCM encoding. In Europe, A-Law is used for PCM encoding. The actual analog-to-digital conversion is accomplished through a three-step process using an analog-to-digital (A-to-D) codec. The three steps are described here:

- Sampling—The first step in the conversion is to take the analog signal and sample it. This results in a Pulse Amplitude Modulated (PAM) signal. The sampling rate is determined using Nyquist's Theorem, which states that a sampling rate needs to be at least two times the highest frequency in the analog signal. For analog voice transmission, which is based on FDM, the sampling rate is 8,000 times per second. For digital audio, the standard is 47KHz.
- Quantizing—The next step is to round off each PAM signal to the closest defined voltage value. Each defined value equals a specific analog frequency. Although this rounding does add some noise to the signal, the symbol steps are close enough that this "quantizing noise" is essentially inaudible.
- Encoding—Each defined voltage signal has a corresponding 8-bit word. The final step is to convert quantized analog values into their 8-bit binary words and transmit them.

After the analog signals have been converted into 8-bit words, they can be grouped together and transmitted using a multiplexing technique. There are two types of multiplexing techniques used for digital voice and data transmission: time division multiplexing (TDM) and statistical time division multiplexing (STDM).

TDM and STDM

TDM is used for the majority of time-sensitive voice and data transmission applications. With FDM, the transmission medium's available bandwidth is divided into separate passbands, and each discrete circuit is assigned its own passband and time-slot to carry the transmission signal. The circuits are then combined together to form a single channel. Instead of each channel getting its own transmission passband, however, each discrete circuit gets a chance to send a PCM word, and then the next channel gets a turn. With TDM, each channel gets a turn to transmit one PCM word during a transmission cycle. If the channel has no actual data to transmit, bit-stuffing is used to fill the slot, because the transport frame is made up of data from each of the discrete channels. Therefore, although TDM is great for transporting data in real time (like voice and video), if there is no data to transmit, the bandwidth is wasted.

This is where STDM comes into play. With STDM, each discrete circuit is not assigned its own channel. Instead, STDM looks at which discrete channels actually have data to transmit and then allocates the available time-slots to those active channels to transport their data. The major advantage to using STDM over TDM is that you can "over-subscribe" the transport circuit. Over-subscribing is the practice of assigning more devices to a transport circuit than the circuit can transmit at one time. When using TDM, each device is assigned its own channel. If the device has no data to transmit, stuffing bits are sent in place of data, and this wastes useable bandwidth. With STDM, a collection of devices can share the bandwidth, the idea being that, statistically speaking, not all the devices need to transmit data at the same time. The practice of over-subscription is commonly used in voice applications.

Multiplexers

TDM and STDM data transport is accomplished through the use of a multiplexing device. A multiplexer has three basic components: discrete channel inputs (DCI), a framer, and a carrier interface.

- DCI—DCI is the input/output interface of the multiplexer that connects to the data terminal device, which generates the data signal that is to be multiplexed and transmitted over the carrier transport.
- Framer—The framer is a scanning and distributing device that samples the various DCI signals and encodes them, assigns them a time-slot, frames them, and transmits them over the carrier transport. Depending on the type of multiplexing scheme used (TDM/STDM), buffering might be used to accommodate the data flow over the transport.
- Carrier interface—Multiplexers can support one or more carrier transport lines. The carrier interface is where the PSTN digital carrier line terminates. Depending on the type of multiplexer, the carrier circuit might terminate directly or through an external Channel Service Unit (CSU). We will look at the functions performed by the CSU a little later in this section. Figure 5.1 illustrates the logical structure of a multiplexer.

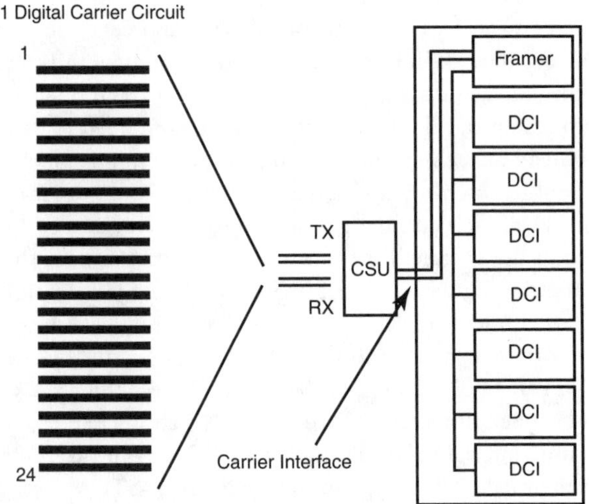

Figure 5.1 A multiplexer from a logical component perspective.

There are several different types of multiplexers: carrier class channel banks, which are used by PSTN carriers for POTS circuit provisioning; M-Class multiplexers, which are used by PSTN carriers to multiplex groups of digital transport circuits into faster digital transport circuits; and DS-X multiplexers (DMUX), inverse multiplexers (IMUX), and analog channel banks, which are CPE (Customer Premise Equipment) type multiplexers used to allocate the channels of digital carrier circuits for different voice and data applications.

A channel bank is an analog-to-digital multiplexer. It channelizes a group of discrete channels into a high-speed carrier transport interface. Channel banks are used to provision standard analog local loop service. The local loop circuits are terminated into the channel bank. PCM is then used to encode the analog signals for transport over digital carrier lines. A channel bank typically supports 48 discrete channels. Although PCM is used by telephone carriers, there are other types of CPE channel banks that employ more advanced time domain coding schemes that require less transport bandwidth. One such coding scheme is called Adaptive Differential PCM (ADPCM), which can transmit two discrete channels in the same amount of bandwidth required to send one PCM-coded channel. ADPCM channel banks are often employed when using PSTN carrier lines between Private Branch Exchanges (PBX).

A DMUX is a CPE device that provides different types of DCIs to support the transport of data from different sources. These sources can be LAN interfaces, standard analog POTS circuits, video systems, and so on. The job of the DMUX is to channelize these different datastreams and transport them across the carrier interface. DMUXs were developed to enable PSTN T-carrier customers to maximize the available bandwidth of their dedicated carrier circuits. There are DMUXs that support both TDM and STDM. These devices can allocate the carrier circuit's bandwidth in either a dedicated or dynamic mode. In dedicated mode, the available carrier transport channels are allocated to a task. For example, half the carrier circuit's bandwidth can be allocated for voice and the other half for data. In a dynamic mode, the DMUX allocates the carrier transport channels according to the usage requirements of the connected DCIs. Therefore, if analog telephones, a router, and a videoconference system are all connected to the multiplexer, it can share the available circuit bandwidth depending on which application needs to transport data.

An Inverse Multiplexer (IMUX) is similar in function to an M-Class multiplexer. It is used for aggregating a group of data channels (that is, time-slots) on a single digital transport circuit or group of digital transport circuits to create a single high-speed data transport channel. IMUX transport solutions are popular for high-speed/high-availability solutions because it is possible to take digital circuits that come from different COs and IMUX them together. IMUXing creates a fast data pipe with some fault tolerance. In the event of circuit failure in one feeding CO, the other CO's circuits will still be in place and passing traffic.

Digital Signal Hierarchy

The Digital Signal Hierarchy (DSH) describes the digital transmission rates supported by digital transport systems. In the United States, the DSH is defined by the American Standards Institute (ANSI T1 Committee). In the rest of the world, it is defined by the International Telecommunications Union Telecommunications Sector (ITU-T). The foundation of the DSH is based on Digital Signal Zero (DS0), which is the digital transmission equivalent to the standard 4KHz single pair POTS line. DS0 supports a transmission rate of 64Kbps. Under the ANSI DSH, the first increment is Digital Signal One (DS1). The DS1 signal supports the transmission of 24 DS0 signals. The ITU-T DSH defines the DS1 signal at an equivalent of 30 DS0s. DS1 is the typical carrier circuit used by customers for voice and data transmission over the PSTN. Table 5.1 lists the American and international DSH standards.

Table 5.1 **The American (DS) and International (I-DS) DSH**

DSH Signal Level	Transmission Rate	Number of DS0	Carrier System	Transmission Medium Channels
DS0	64Kbps	1	Analog	24/26 AWG Copper Wire
DS1	1.544Mbps	24	T1	24/26 AWG Copper Wire
I-DS1	2.048Mbps	30	E1	24/26 AWG Copper Wire
DS1C	3.152Mbps	48	T1C	24/26 AWG Copper Wire
I-DS2	8.448Mbps	130	E2	24/26 AWG Copper Wire
DS2	6.312Mbps	96	T2	24/26 AWG Copper Wire
I-DS3	34.368Mbps	480	E3	Coaxial/MM Fiber/ Microwave
DS3	44.736Mbps	672	T3	Coaxial/MM Fiber/ Microwave
I-DS4	44.736Mbps	672	E4	Coaxial/MM Fiber/ Microwave

DSH Signal Level	Transmission Rate	Number of DS0	Carrier System	Transmission Medium Channels
DS4	274.176Mbps	4032	T4	Coaxial/MM Fiber/ Microwave
I-DS5	565.148Mbps	7680	E5	Coaxial/MM Fiber/ Microwave

T1/DS1

The T1 is the basic transport facility of the T-carrier system. Through the use of multiplexers, faster T-carrier transmission rates are achieved to meet the DS-X transport requirements. T1 transmission is facilitated with two conditioned AWG 24/26 copper twisted pairs. A T1 circuit operates as a synchronous transport, using a common timing source that the transmission equipment on both ends of the circuit use to derive timing. The first use of T-carrier transport was in 1962 to provide interoffice trunk links in major metropolitan areas. Until 1983, T1s were only used by the telephone system and the U.S. government. This was due to pre-divestiture pricing that based cost on the bandwidth of the circuit. After divestiture, T1 service pricing was tariffed and made available to the consumer market.

It is rumored that T1 got its name from the fact that one mile (the distance between manholes in large cites) was the most reliable distance that could be supported between signal regenerators. Actually, the specification calls for 6,000 feet between repeaters, with the first leg and last leg of circuit being no more than 3,000 feet between the repeater and the termination point of the CO.

Note

The "DS" abbreviation is used by both the ANSI and ITU-T DSH definitions. The "I-DS" abbreviation is only used in Table 5.1 to make distinctions between the two standards.

T1 Bandwidth Provisioning

T1 circuits are provided to customers as Type III (4-wire interface with PSTN signaling) and Type IV (4-wire interface without signaling). Pricing is based on distance, not usage. This makes it advantageous to the user to make the most efficient use of the available bandwidth (if applicable) through the use of a DMUX. Due to the cost of T1s, for many years the LECs offered dedicated fractional T1 service, which was bandwidth based on 64K increments. Due to the large drop in T1 pricing and the availability ISDN and Frame Relay, fractional T1 service in many areas is being phased out by LECs.

T1 facilities are used for all types of primary rate digital transmission services (ISDN, Frame Relay, and so on). The T-carrier transport system operates on OSI-RM Layer 1. The T1 circuit's bandwidth is partitioned into 24 DS0 channels, or 24 × 64Kbps channels, for a user data transfer rate of 1,536 Kbps. To provide in-band control and management data, an additional 8Kbps were added to provide transport for signaling, making the total aggregate DS1 data rate 1.544Mbps. In the case of ISDN and Frame Relay service, which are transport services provided by telephone carriers, there are OSI Layer 2 and 3 operations involved. These services are implemented by using different signaling and multiplexing facilities. The digital transport system, however, is still T-carrier.

T1 Framing

T-carrier uses PCM and TDM to transport digital signals. The T1 is made up of 24 (DS0) 64Kbps channels or time-slots and transmits at 1.544 clock intervals per second. Each time-slot/channel gets a turn to transmit its data (650ns per interval). The incoming analog signal on a time-slot/channel is sampled 8,000 times per second. The sample is then converted into an 8-bit word (pulse code). This creates a digital stream of 8 bits × 8,000 samples/second = 64Kbps, the DS0 data rate.

In order for the remote end to use the data sent across the time-slots, it must know how the bits are organized. Framing provides the organization needed for the bit-stream to be understood in a meaningful way. T1 framing information is provided through the use of the 8,000 additional signaling bits added to the T1 datastream. The M24/D4 frame standard is the basis for all T1 framing. It contains an 8bps PCM word from each of the 24 time-slots and one framing bit, for a total frame length of 193 bits. The D4 frame format is illustrated in Figure 5.2.

The original T-carrier standard used the D4 superframe format. The D4 superframe contains 12 D4 frames and 12 framing bits. The framing bit serves two purposes: first, to provide transitions between each D4 frame, and second, to provide synchronization for the identification of the in-band signaling information (typically carried in frames 6 and 12). For this purpose, the 12 framing bits are in the same pattern for every D4 superframe (100011011100). Figure 5.3 depicts the D4 superframe format.

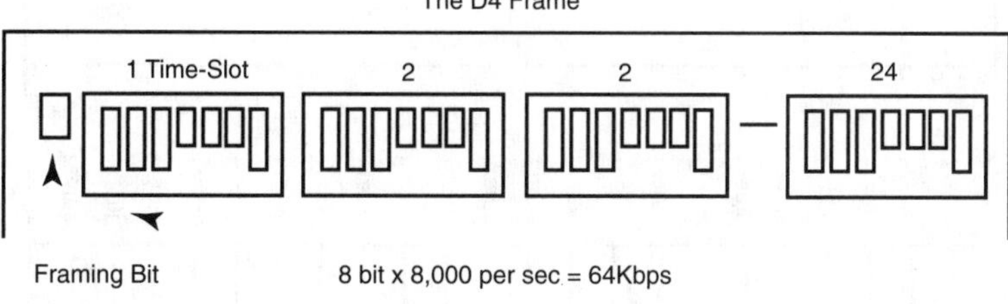

Figure 5.2 The D4 frame.

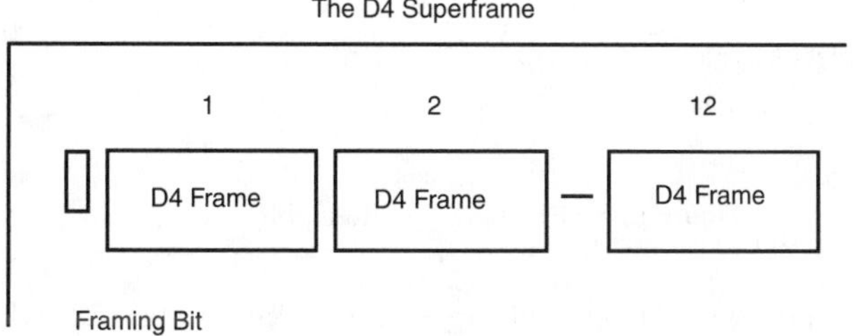

Figure 5.3 The D4 superframe.

To provide additional in-service error testing, reporting, and diagnostics, the T-carrier standard was enhanced with the development of the Extended Superframe Format (ESF). In addition to enhanced testing and reporting, ESF provides up to 16 different signaling states (D4 only supports four) and adds a 4Kbps Facilities Data-Link (FDL) outside the 24 data channels for signaling information. ESF supports a 98.4 percent error detection rate through the use of CRC-6, which permits the entire circuit to be segmented for testing without any interruption to service.

ESF accomplishes all this by increasing the frame size to 24 M24/D4 frames and reallocating the use of the available framing bits, only 6 of which are used for actual frame alignment. The framing bits are in frames 4, 8, 12, 16, 20, and 24. The CRC error checking facility uses 6 bits, which appear in frames 2, 6, 10, 14, 18, and 22. The FDC utilizes the last 12 bits, which are used to carry management information and alarm information. Figure 5.4 illustrates the assignments of each of the framing bits in the ESF frame format.

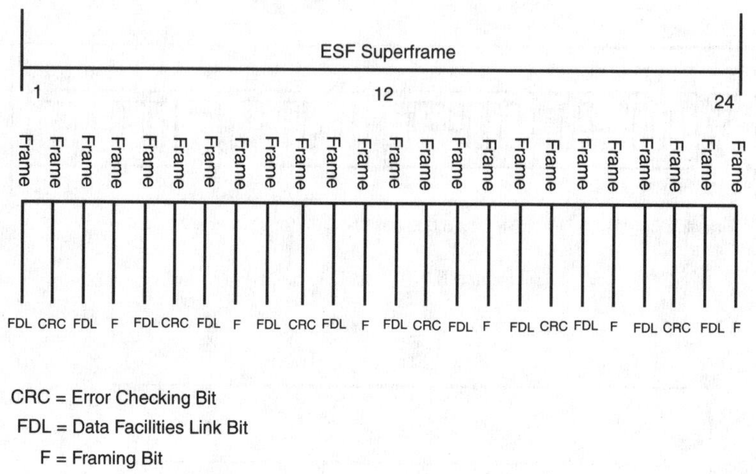

Figure 5.4 ESF framing bit assignments.

ESF management and data collection information is stored on the CSU, where it can be accessed by the carrier when non-obtrusive circuit testing is enabled. Because remote testing is one of the big advantages of using ESF, some kind of testing system must be employed by the carrier. One of the most commonly used is AT&T's Automated Bit Access Test System (ABATS).

T1 Line Coding

Formatting the bitstream is only half of the data transport process. The other half is the line-coding scheme, which determines how the actual bits are interpreted. Digital data transmission is a basically simple process, because there are only two data types: ones and zeros. In a basic transmission scheme, a binary one is interpreted as the presence of a signal, and zero is interpreted as the lack of signal. Transmission would be easy if all your data was like this: 10101010. Digital data contains consecutive ones and zeros, however, so timing is needed to determine how long a signal lasts, so that consecutive ones and zeros can be counted correctly. The T-carrier system transports data synchronously, and a timing scheme is essential for proper data delivery.

To maintain proper timing, the CSU counts the one bits in the datastream. It uses a bipolar signaling mechanism that uses a 3.3 VDC current stream. Binary ones are represented by either +/- 3.3 volts, and a binary zero is an absence of voltage flux. There are two bipolar coding schemes used with T1 circuits: Alternate Mark Inversion (AMI) and Bipolar 8 Zero Substitution (B8ZS).

AMI coding represents a binary one (called a mark) with alternate voltage pulses between zeros (called spaces) which are represented by an absence of a pulse. This is to say that if the first binary 1 is +3 volts, the next one will be –3 volts, regardless of how many zeros are in between, hence the name alternate mark inversion. If the ones are not represented by the correct polarity, a condition known as a bipolar violation results.

As mentioned earlier, the T1 circuit timing is maintained by counting the ones in the datastream. If the datastream does not have enough binary ones, the timing circuit in the CSU cannot derive reliable timing. To avoid this problem, the AT&T Compatibility Bulletin 119 specifies that a certain percentage of the bitstream traffic must be binary ones. This is commonly referred to as "ones density." The ones density specification requires that "no more than 15 contiguous zeros will present in a transport frame" and "there shall be at least 'N' ones in every 8 (N+1) bits for an average ones density of 12.5%."

To maintain compliance with the ones density rule, the CSU checks the datastream to ensure that the bitstream is in compliance. If it is not, the CSU inserts "extra" binary ones into the bitstream. Needless to say, this does affect how the data is interpreted at the receiving end. Because of this, AMI encoding is not well-suited for data transmissions, which cannot tolerate changes to the bitstream.

B8ZS

To transmit data, the capability to send any data pattern, including an indefinite period of all zeros, is an essential requirement. Because of the highly variable nature of data, a "clear channel" T-carrier circuit is required. To provide clear channel T-carrier service, B8ZS encoding is used. B8ZS uses alternating polarities to represent marks, the same way AMI does. When the MUX recognizes eight consecutive zeros in the bitstream, however, it inserts a bipolar violation in bit positions 4 and 7. This violation is recognized by the receiving MUX as a substitution code, and the zeros are interpreted properly. The only trick with using B8ZS is that all the equipment in the data path must recognize the B8ZS violation.

Ordering Your T1

Today, the common provisioning for T-carrier circuits is ESF/B8ZS. However, it is always best to specify ESF/B8ZS with your carrier when you are ordering your circuit. You also need to specify what your timing source is. Depending on the carrier, they might or might not provide a line timing source, but if they do, have them provide timing. As far as D4/AMI provisioned circuits are concerned, there are some channel banks that prefer D4/AMI to ESF/B8ZS. So, for channel bank applications, it's best to check with your equipment manufacturer before ordering.

CPE T-Carrier Network Equipment

When your T1 is installed on your premises, it will be terminated on an network interface unit (NIU) or what is commonly referred to as the "smart jack." NIU is the demarcation between the CPE and the carrier network. Everything beyond the NIU is considered CPE. The NIU provides the DC power termination, loopback capabilities, and a bypass access for Bit Error Rate (BER) testing. The NIU provides an RJ-45 interface for interfacing with the CSU.

In many cases, the LEC's D-mark will not be close to where your equipment is located. Consequently, you might be required to install an extension between the NIC and your equipment location. Some LECs will install the in-house extension wiring to a specified D-mark point, such as an office or computer room. Having the LEC install the extension has an advantage over doing it yourself. When the LEC installs the extension, it becomes responsible for the extension cabling along with the NIU. This can save you aggravation later if a problem arises. If you have installed the extension, the LEC will claim it is only responsible for problems up to the NIU. If the LEC installs the extension, they are responsible. Whenever possible, it is always better to have the smart jack installed in the same place as your equipment. The NIC can provide valuable information about the status of the circuit in the form of display indicators, which you will need to see in some cases when troubleshooting circuit problems over the telephone with the LEC.

Channel Service Unit

The channel service unit (CSU) is the CPE termination point for the T1 carrier circuit. Before the AT&T divestiture, the CSU was the telephone company demarcation. The CSU provides the following functions:

- Circuit termination—The CSU terminates the local T1 circuit and provides isolation between the CPE network and the PSTN.
- Line loopback/line build-out (LBO) facilities—Although the CSU is CPE, it is also used by the telephone company for line testing and signal equalization. The CSU can generate a test signal and loop itself to appear as if the circuit is a functional state to the remote end Data Terminal Equipment (DTE) for line and bit error rate (BER) testing and for signal loss measurement. It also provides testing for proper transmission framing format, per channel BER, and keepalive facilities. Actual loop testing is performed using automated testing tools or the DSX-1 digital patch panel at the CO. Loop testing can either be performed in-band with the actual data flow or out-of-band using the FDL.
- Signal regeneration—The CSU is the last single regenerator on the circuit path before the DTE.

- Transmission timing and ones density enforcement—T-carrier data transmissions are synchronous and very sensitive to timing problems. The circuit clock is maintained (and sometimes generated) by the CSU. Transmission timing is maintained by counting the binary ones that are contained in the datastream. In order for proper timing to be maintained, a certain amount of ones must be transmitted in the datastream. This is known as ones density (see the AT&T formula discussed previously). Ones density enforcement is handled by the CSU. If the datastream is not in compliance, the CSU inserts additional ones into the datastream.
- Basic diagnostics and performance data collection—The CSU provides circuit status and an alarm indicator. It can also provide ESF statistics, calculate and collect circuit error statistics, and store them locally for retrieval by the carrier provider or the user.

A CSU is required for any device that connects to the T-carrier network. For channel banks and PBXs, this is often a standalone unit. For data transmission applications, it is incorporated with a data service unit.

The Data Service Unit

The data service unit (DSU) performs two functions: interfacing and diagnostics. It provides a synchronous or asynchronous (V.35/TIA/EIA RS232) interface to the CPE data terminal equipment. This interface performs the DTE-signal-to- bipolar-signal conversion needed to transport digital data signals over the PSTN. This conversion is accomplished by an asynchronous-to-synchronous converter that is part of DSU's circuitry. For digital data service (DDS) applications that operate more than 56/64Kbps, the DSU has an integrated CSU, and it converts the unipolar DTE bitstream from the router or gateway device. The DSU/CSU defines the framing (AMI/B8ZS) used on the circuit and which T1 channels will be used to transport data when fractional T1 or frame-based transport is being used for DDS. The DSU also provides testing and diagnostic facilities such as BER testing, loopback facilities, and DTE status indicators. Most integrated DSU/CSU provide a serial configuration and testing interface and support SNMP management.

High-Speed Digital Transport: T-Carrier (DS-3/DS-4) and SONET/SDH (OC1/OC3)

To meet the growing demand for voice and data communications services, telephone carriers require interoffice trunk lines that can provide transport rates and DS0 destinations beyond that of the primary rate (DS1) trunk carrier circuit. To provide these transport services, one of two digital transport solutions is employed: T-carrier or SONET.

DS3 Service over T3

To provide DS3 service over T-carrier, an asynchronous T3 transport circuit is used. It is asynchronous transport because timing is derived from the two output circuits independently from one another. T3 transmission circuits operate over microwave radio, multimode fiber, satellite, and coaxial cable (for short distances only). The formation of a T3 utilizes 28 T1 circuits, multiplexed together using a two-step multiplexing process called M13 multiplexing.

The first step multiplexes the 28 T1s into seven T2 circuits. Each T2 channel is constructed out of four T1s and consists of the 6.176Mbps from the four T1s, plus 136KBs of framing and stuff bits to accommodate the timing differences between the four T1s. This is called plesiochronous multiplexing, where two or more bitstreams operating at the same bit rate are controlled by different timing sources. The actual T2 frame is formed by interleaving the T1 bits together. It consists of four subframes, each consisting of six blocks of 49 bits each. Figure 5.5 illustrates this process.

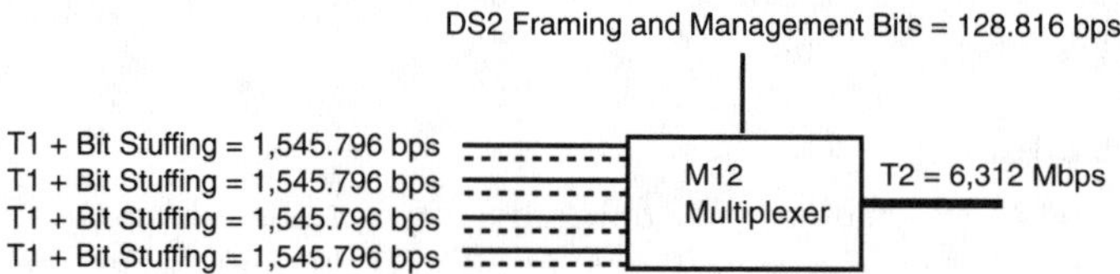

Figure 5.5 T1-to-T2 multiplexing.

The second step takes the seven T2 circuits and multiplexes them together to form the T3 circuit. The combination of the seven T2 circuits is also a plesiochronous process. The T3 multiplexer adds stuffing bits (552Kbps worth) on the output interface and strips them out on the input interface of the multiplexer on the far end of the circuit. This behavior is different from the T2 multiplexing step, which adds the stuff bits on the T1 input interface to the T2 multiplexer. The reason for the two different stuffing methods is to provide the option to have the T2 interfaces come from different locations. After the T3 circuit is complete, it might be multiplexed with other T3 circuits to form a T4 interface (six T3 circuits), which is the common interoffice trunk transport link (transmitted over fiber). This can be used as a standalone interoffice trunk or terminated at a customer presence for clear channel use. Figure 5.6 illustrates the T2-to-T3 multiplexing step.

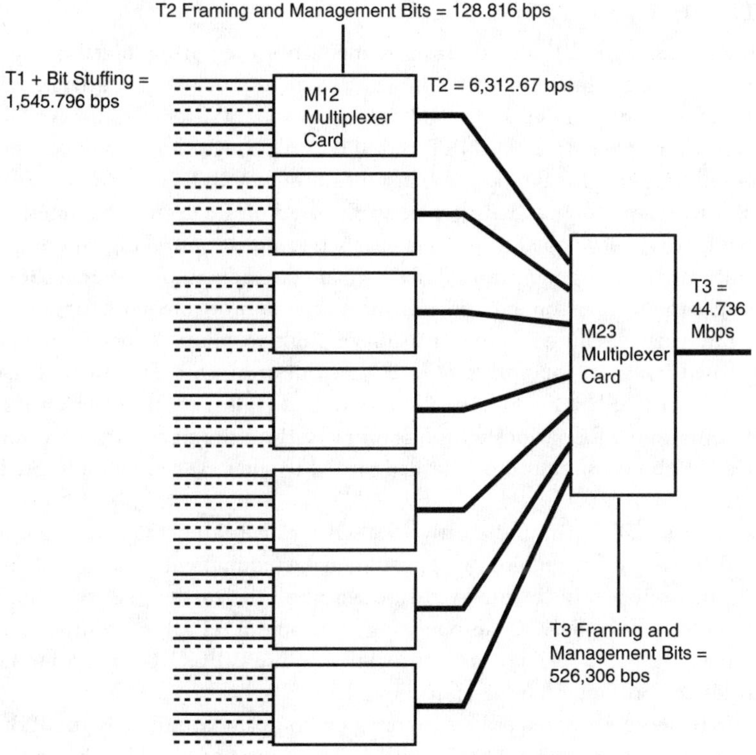

Figure 5.6 T2-to-T3 multiplexing.

As a carrier trunking circuit, when the T3/T4 has reached the CO, a two or three step demultiplexing process must be used to provision the transport at the T1/DS1 level. As a clear channel data transport, the T3 terminates into a T3 DSU/CSU equipped with a High-Speed Serial Interface (HSSI). HSSI was originally developed by Cisco Systems and T3plus Network to provide a serial interface that could handle data rates up to 52Mbps for T3/DS3 WAN applications. HSSI is now an ANSI and ITU-T standard, and it can be used for STS-1/SONET transport applications as well. Cisco also provides an integrated DSU/CSU T3 interface card for its 7x00 routers.

SYNTRAN

Synchronous Transmission (SYNTRAN) is an ANSI standard that defines a reworking of the T3 transport implementation (SDS3). SYNTRAN provides direct accessibility to the DS1 and DS0 channels without the complex multiplexing and demultiplexing involved with standard T3 implementations. To do this, SYNTRAN uses bit-interleaved multiplexing and a different transport framing scheme, with two different operational modes. Direct DS0 access is provided in byte-synchronous mode. When only DS1 access is needed, bit-synchronous mode is available. SYNTRAN was implemented by some of the RBOCs, but never enjoyed large-scale use in the PSTN.

SONET/SDH Transport Basics

Today, in the U.S., SONET is used for most applications requiring digital transport at rates higher than DS3. In Europe and elsewhere, Synchronous Digital Hierarchy (SDH) as defined by the ITU-T is used. SONET is an ANSI standard for digital data transmission over optical fiber. SONET and SDH utilize a common hyper-accurate network-timing source for timing the transmission of SONET/SDH frames. SONET is a second-generation digital transport network based on single-step, bit-interleaved multiplexing. It was developed to provide an open standard optical digital transport that permits carrier equipment from different manufacturers to interoperate. SONET utilizes single-mode, fiber-optic cable as its primary transport medium, supporting several "self-healing" switching schemes that utilize point-to-point, point-to-multipoint, and ring-based transport topologies. SONET provides real-time fault analysis, diagnostics, and sophisticated bandwidth allocation, as well as fault recovery mechanisms and channelized transport for various transmission rates. Both the transport as a whole and the embedded transmission channels (called virtual tributaries) can be monitored distinctly.

Work on SONET began in the early 1980s. After divestiture, the need arose for a high-speed "mid-span" technology that would enable equipment made by different vendors to interoperate. In the predivestiture world, all the carrier-class switching equipment was built by WECO, an operating company of AT&T. After divestiture, IXCs, such as MCI and Sprint, were not inclined to use WECO T-carrier, and no standard existed for digital transmission above T3.

In 1985, Bellcore (now Telcordia), the research company founded by the RBOCs after divestiture (it is similar to AT&T Bell Labs), recommended a fiber-based transport network standard and a signal transmission rate hierarchy to the ANSI T1 committee. The committee provided transmission rates based on fixed rate increments. The original SONET standard proposal defined a base operating rate of 50.688Mbps, which was called Synchronous Transport Signal number one (STS-1). The ANSI proposal was then handed to the ITU-T for review. The ITU-T felt the STS-1 operating rate was inadequate and instead selected a base operating rate of 155.520Mbps. This letter standard was called Synchronous Transport Module number one (STM-1) for their Synchronous Digital Hierarchy (SDH) transmission scheme. To enable compatibility between SONET and SDH, the SONET standard and operating rate were adjusted to 51.84Mbps. This adjustment made STS-1 compatible with STM-1, by multiplexing STS-1 by a factor of three. Table 5.2 lists the ANSI/Bellcore SONET signaling and ITU-T SDH signaling operating rates' hierarchy.

Note

T4 implementations are proprietary and differ between carrier equipment types.

Table 5.2 **The SONET and SDH Transmission Rates**

STS/STM Level	OC Level	Line Rate
STS-1	OC-1	51.840Mbps
STS-3/STM-1	OC-3/STM-1O	155.520Mbps
STS-9	OC-9	466.560Mbps
STS-12/STM-4	OC-12/STM-4O	622.080Mbps
STS-18	OC-18	933.120Mbps
STS-24	OC-24	1244.160Mbps
STS-36	OC-36	1866.230Mbps
STS-48/STM-16	OC-48/STM-16O	2488.32Mbps
STS-96	OC-96	4876.64Mbps
STS-192/STM-64	OC-192/STM-64O	9953.280Mbps

OC = Optical Carrier, which specifies the fiber optic transport for SONET.

STM-XO = Synchronous Transport Module (Digital Signal Level) Optical, which specifies the fiber optic transport for SDH.

Although STS-1 is the basic rate transport for SONET, STS-3 is the common implementation used for trunk and customer provisioning.

SONET/SDH Operation

Although SONET and SDH are strictly OSI-RM Layer 1 activities, (like T-carrier) there is a definite separation of the transport functions. The separation of functions into distinct operational areas is common in open standards. Open standards allow users of the standard to make enhancements to a particular function or combine functions in their particular implementation. As long as all the functions are met in accordance to the standard, compatibility will be maintained. SONET and SDH protocol use a four-layer architecture model (illustrated in Figure 5.7). The four layers are as follows:

- Photonic layer
- Section layer
- Line layer
- Path layer

The photonic layer defines the electromechanical properties that relate to the transmission of the binary data across the physical transport. This is accomplished by converting the STS/STM electrical signals into optical OC/STM-x0 signals. These optical signals are typically transmitted using 1300 and 1500 nanometer wavelength signals, generated with gallium-indium arsenide-phosphate infrared lasers.

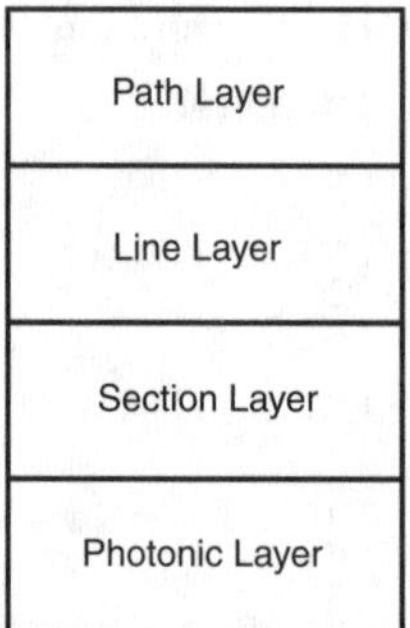

Figure 5.7 The SONET/SDH architecture model.

The section layer deals with regeneration (about every 10Km) and transport of the optical signal. This includes the framing, management information contained in the transport overhead portion of the frame, signal scrambling, and error monitoring. The STS-1 frame is 9 rows by 90 columns; each column is 1 octet (8 bits). It is comprised of 27 octets of transport overhead and a 783 octet payload. The STM-1 frame is 9 rows by 270 columns, with each column 1 octet. There are 81 octets used for transport overhead and 2,349 octets are used for the STM-1 payload. STS-3 or three multiplexed STS-1 frames are the equivalent to the STM-1 frame. Regardless of frame size, a frame is transmitted once every 125 microseconds. The SONET/SDH transmission rate is increased by incrementing the frame size by 3 times the STS/STM base rate (9×90/9×270, respectively). Figure 5.8 illustrates the STS and STM base rate frame formats.

The line layer refers to the maintenance operations of the SONET/SDH span, which is the link between two SONET/SDH devices. The actual span can run across several different SONET links. These operations include synchronization, span monitoring, and circuit recovery (which will move the span to another SONET/SDH circuit in response to a failure or drop in quality). The line layer also defines provisioning, error monitoring, and general link management.

The path layer provides the OSI-RM Layer 1 protocol mapping to the SONET/SDH payload. This can be T-carrier signals (T3, T2, and T1), FDDI, broadband ISDN, or ATM. These services are provided by termination equipment (that is, multiplexer or dedicated SONET/SDH interface). It is the service device's responsibility to provide the transport services between the transport protocol and SONET/SDH's lower layers on both ends of the path. SONET deals with these different signal formats by placing them in virtual tributaries (VT). These tributaries are subframe payload containers that include their own framing and management information. SONET/SDH define VT containers for most of the smaller transport signaling rates, such as T1/E1, FDDI, and others. There is no defined VT container for T3/DS3 transport. Rather, it is directly mapped into the STS-1 frame.

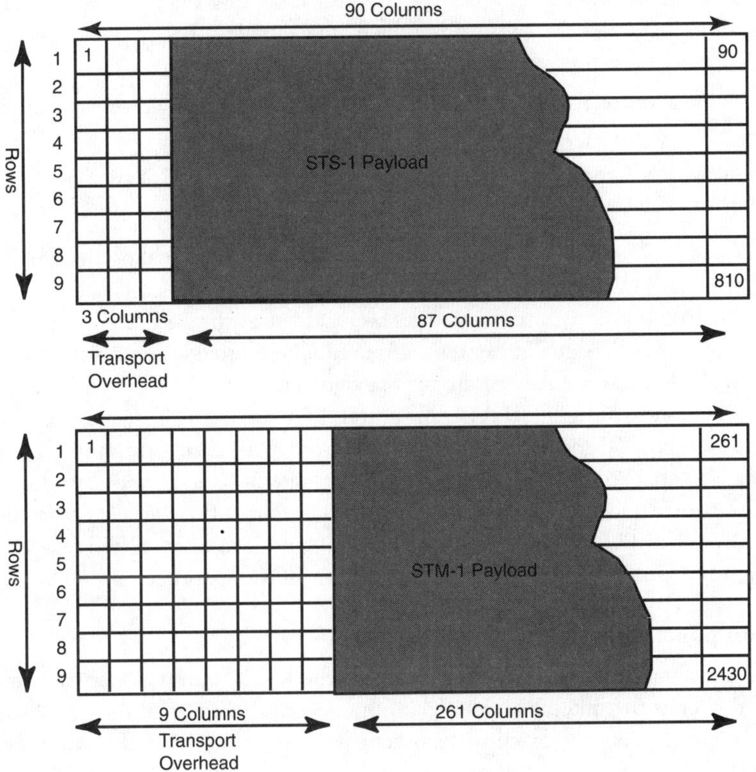

Figure 5.8 STS and SDH base rate frame formats.

Common Channeling Signaling System 7

To set up any type of call on the PSTN, there must be a signaling system. In its most basic form, the signaling system conveys the subscriber information (source and destination) to the PSTN equipment so the call can be established. When the call is established, the signaling system is used to verify that the call is still established and to relay "busy" messages to subscribers trying to establish calls with the subscribers engaged in the call. After the call is complete, the signaling system is used to reclaim the resources used by the call, so they can be reallocated. There are five classes of signals used to relay circuit operation and management instructions between subscribers' terminals and CO switching equipment:

- Address signals—These signals are used to relay call or switched circuit destination information.
- Call progress tones—These are informational signals relayed to the user on the status of a call (dial tone, busy signal, or ringing).

- Alerting signals—These signals are used for incoming call notification. The telephone ringer and the call-waiting beep are examples of alert signals.
- Control signals—These are supplemental or enhanced call service signals used to provide additional subscriber information services such as automatic number identification (ANI), which is part of caller ID.
- Supervisory signals—These signals are sent between the CO switches to perform call setup and teardown and other call maintenance functions.

Until the early 1970s, PSTN used in-band signaling to relay signaling information. This system was limited in the number and types of signals that could be sent, and it became increasingly slow as more features were added to the PSTN.

Today, out-of-band signaling systems are used. These provide the capability to accommodate the transmission of circuit management, the subscriber and billing database information used by interLATA and intraLATA carriers to manage circuit connections. The out-of-band signaling system used today is called Common Channeling Signaling System 7 (SS7). SS7 operates over a separate 56K dedicated digital network outside the voice and data carrier lines used to transport subscriber data over the PSTN. SS7 is an ITU-T standard (issued in 1980) and is used by all the telephone carriers in the United States and throughout most of the world.

SS7 Components

The SS7 signaling system is constructed outside the PSTN transport network through dedicated circuits that interconnect telecommunications switches of IXCs, ILECs, and CLECs. The use of an out-of-network system provides some significant advantages. Among the more significant is the capability to enhance the signaling system without impacting the performance of the transport network. Faster call processing is one result, because the signaling is out-of-band and is not impacted by transport network congestion or contention. The out-of-band signaling system also removes direct customer access, which prevents fraud and hacking attacks. It also provides facilities needed for implementing the Advanced Intelligent Network (AIN), which is a uniform maintenance, support, and service system for managing the PSTN. The major purpose of the AIN is to provide the capability for IXCs and LECs to be able to implement enhancements and new services to the PSTN in an efficient manner that ensures continuity. The AIN also provides the "intelligence" for enhanced subscriber services such as Centrex, Network-wide Automatic Call Distribution, Enhanced 800 service, Network-wide ISDN switched digital service support, and universal subscriber IDs.

There are two types of messages sent by SS7: circuit-control and database-access messages. Circuit-control messages are messages relating to circuit operation and management. Database-access messages query call completion databases to retrieve network resource availability information to complete call setups. There are five components to the SS7 network:

- PSTN/ISUP trunks—These are PSTN CO trunk lines (generally DS4/OC-3 links) that will transport ISDN and non-ISDN voice and data calls. These lines are connected to the service switching points (SSPs), which use them to complete calls that originate on their subscriber local loop lines (DS0).

- Service Control Point (SCP)—SCP is a database of call completion information, network resource and configuration information, and AIN information. SCP databases are typically installed as a pair to provide redundancy and load-sharing capabilities.
- Signaling Transfer Point (STP)—STP is a switching device that routes signaling information requests from CO switches to SCP databases. STPs do not typically generate their own signaling messages, except for SS7 network management messages. STPs, like SCPs, are commonly installed in pairs.
- Service Switching Point (SSP)—The SSP is any CO-class switch (LEC CO, LEC-Tandem, or IXC-POP) that is connected to the SS7 network by way of a CCSL. The SSP relays signaling and database queries to the STPs, which generate the actual data requests to the SCPs.
- Common Channel Signaling Links (CCSL)—These are the 56Kbps links used to carry SS7 signaling messages.

SS7 signaling components are connected using a redundant mesh of CCSLs, of which there are six types (A through F). Figure 5.9 illustrates the relation between these components.

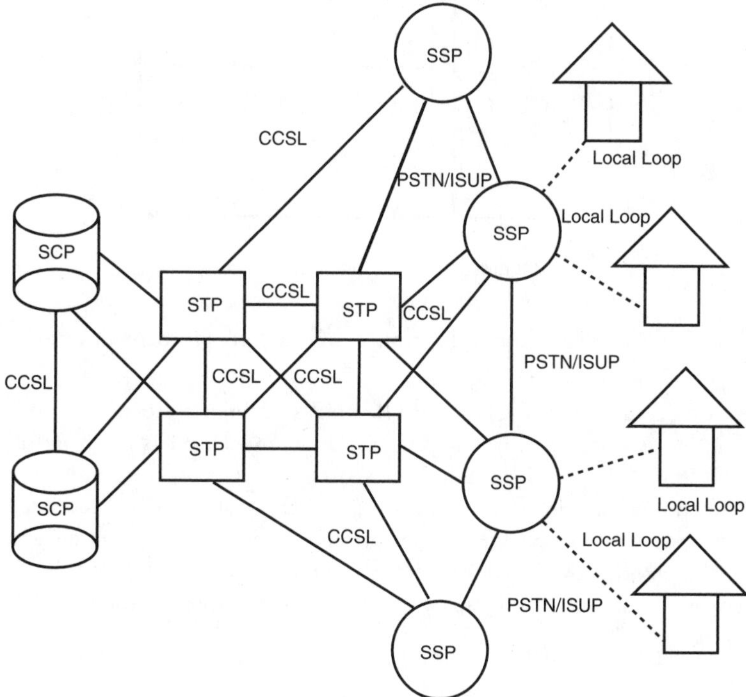

Figure 5.9 SS7 network components.

SS7 Protocols

The protocols that make up the SS7 are categorized into abstract reference model consisting of four functions areas:

1. Application
2. Transaction
3. Network
4. Data Link

Each of the various protocols that provide SS7's functionality falls into one or more of these functional areas. Figure 5.10 depicts the SS7 reference model and the placement of various SS7 protocols within the SS7 reference framework.

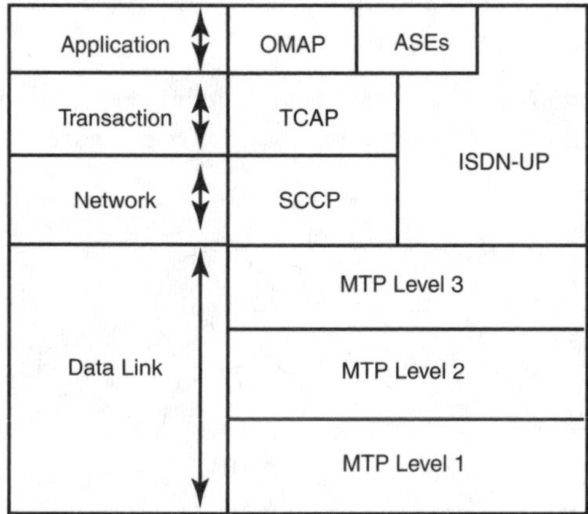

Figure 5.10 SS7 reference model.

The following list provides a brief description of each SS7 protocol and its relation to the OSI-RM:

- Message Transfer Part (MTP)—MTP consists of three protocols (MTP Levels 1, 2, and 3) that provide physical, data link, and connectionless network functions. MTP level 1 defines the physical transport and interfaces used with SS7. Defined interfaces include T1, E1, DSO (64Kbps), and 56Kbps dedicated. MTP Level 2 defines the SS7 message format and provides end-to-end message delivery and error recovery. MTP level 3 provides message routing and (internal) network management functions. The MTP levels operate at the OSI-RM Layers 1 and 2.

- Signaling Connection Control Part (SCCP)—This element provides message delivery services for SS7 ULPs, except for ISDN-UP. SCCP provides four classes of network delivery service:
  - Class 0, connectionless datagram service
  - Class 1, datagram service with sequencing
  - Class 2, temporary and permanent connection service; supports message segmentation and reassembly
  - Class 3, flow control connection-oriented service, along with OSI addressing capabilities

  SCCP operates as an OSI-RM Layer 3.
- ISDN User Part (ISDN-UP)—ISDN-UP messages are used to control the claim and release of CO trunk circuits that are employed for interLATA and intraLATA transport of voice and data. Calls that are within the same CO switch do not use ISDN-UP. The ISDN-UP is not related to the Integrated Services Digital Network (ISDN) switched digital service. However, ISDN and SS7 do use different messaging protocols (Q.913 and Q.700, respectively) and these message types are not compatible. So, for an ISDN call to be established with another ISDN terminal attached to a different switch, the ISDN signaling information must be translated into the SS7 equivalent. This function is performed by the SSP. ISDN-UP operates at OSI-RM Layers 3 through 7.
- Transaction Capabilities Applications Part (TACP)—Part of the fourth layer, TACP is used to exchange AIN, non-circuit-related SS7 message exchanges. These messages are exchanged using SCCP Class 0 delivery. TACP operates at OSI-RM Layer 7.
- Operations, Maintenance, Administration, and Provisioning (OMAP)—Also part of the fourth layer, OMAP messages are used to perform management functions and send operational and maintenance messages pertaining to common overall PSTN operations. OAMP operates at OSI-RM Layer 7.

ISDN

ISDN started to become available in 1980, around the same period that SONET was beginning to be deployed in major population centers in the United States. It was believed that ISDN would meet the increasing demand for reasonably affordable small office/home office (SOHO) digital telephony, enhanced voice and data services, and flexible digital transport for dedicated data and video services. However, the dream of ISDN, as it turned out, was long in coming. In the United States, starting in the 1970s, an effort had been underway by AT&T and the BOCs to convert the original electromechanical switch-based FDM telephone system to a TDM-based infrastructure. The TDM system provided better transport facilities for voice and accommodated data transport requirements. TDM-based systems were also cheaper to build, maintain, and operate. Consumers in the 1970s and 1980s saw this effort in the form of the conversion from rotary/pulse to touch-tone dialing. One of the downsides to TDM systems was that they require two wire pairs to provide digital transport. Therefore, implementation of digital services over the last mile between the consumer and the CO, which uses a single wire pair, presented some problems. ISDN's local loop implementation, known as the Basic Rate Interface (BRI), enabled digital transport over a single wire pair. When ISDN came on the scene, it provided the technology needed to effectively "digitize" the BOCs' local loop circuit. This spawned the possibility of a digital transmission path, end to end.

The problem with ISDN was its actual implementation. ISDN is a collection of international standards developed by the ISO/ITU-T (formerly known as CCITT). The original assumption was that by using an international standard, global communications systems could be easily integrated. The ISDN standard, however, like most "standards," offered no insight on real implementation, but, rather, was an abundance of conceptual elements (such as services, access, and interfaces) that were left to be interpreted and implemented by the various manufacturers and service providers. The standards approach worked well in Europe and Japan, where the telephone systems were owned by the government. In the U.S., however, ISDN had some major implementation problems. Although the Bell System was perceived as a single system, the reality was quite different. Each of the BOCs—and after divestiture, the RBOCs—all used different carrier switch implementations. So, although ISDN was a standard, and provided a standard set of interfaces and services, the actual implementation of ISDN varied depending on the carrier switch installed in the CO. This resulted in incompatibilities between CPE ISDN devices at the CO and the CPE ISDN applications that needed transport outside of the RBOCs' service area. In addition to ISDN's implementation incompatibilities, ISDN service has a finite service distance. A maximum cable distance of 18,000 feet (5.5Km) is permitted between the CO and the CPE. Cable distances beyond this cannot support ISDN service.

It was not until the mid 1990s that ISDN became a useable technology. This was largely as a result of the actual need for ISDN in the consumer market for increased Internet access performance. Once there was a customer demand for ISDN, there was a motivation by the carriers and switch manufacturers to develop compatibility solutions to make ISDN an applicable technology. ISDN services are tariff-priced based on the kind of usage. If voice services are being utilized over the circuit, pricing is based on (local/toll) distance rates. If data is being transmitted over the line, pricing is based on (local/toll) distance rates, plus a per minute charge based on bandwidth. This makes ISDN very lucrative for the providers and quite expensive for users. So, even with the technology issues resolved, ISDN has limited application because its costs are high for general consumer use of ISDN over a BRI.

What is even more ironic is that when ISDN became stable enough for consumer use around 1995-96, telephone carriers began to look at implementing Asymmetric Digital Subscriber Line (ADSL), which provides T1 transport rates (and higher) over a single wire pair and is tariffed with flat rate pricing. There are some limitations to ADSL, however, as it can only provide a high-speed downstream transport path. ADSL functions by connecting an ADSL modem on each end of the circuit. The line is partitioned into three data channels: a high-speed downstream channel, a full-duplex channel, and a 4Khz channel for POTS service. The full-duplex channel is used to send upstream data. The downstream channel is for receiving data. ADSL is currently being implemented as a high-speed modem technology for Internet access in major U.S. markets. Its limited upload bandwidth makes it unsuitable for WAN point-to-point links. It can be quite effective for providing access to the Internet and download-intensive applications such as streaming video and audio.

Due to ISDN's cost and limited bandwidth provisioning, its long-term future in the consumer market for high-speed data services is questionable. Its T1 equivalent, the Primary Rate Interface (PRI), is becoming the preferred means of deploying voice transport for PBXs. PRIs are also used to provide digital modem service.

ISDN Application Services

ISDN circuits are used for a variety of application-specific, variable-bandwidth applications. The most common of these applications are videoconferencing, online imaging, dial-on-demand network access, leased line backup, and digital telephone service. In these limited service applications, ISDN is a cost-effective solution over a dedicated digital transport solution. To provide these services, ISDN provides a core feature set which can be categorized into three service categories:

- Bearer services—These are traditional (analog/digital) data transport services that can be provided through traditional TDM transport methods.
- Telecommunication services—These are data services that can be transported over bearer services. These services can be provided by the ISDN service provider or by the user.
- Supplemental services—These are enhancements to bearer or telecommunication services. These services are not specific to ISDN per se, but they do require the use of an out-of-band signaling mechanism to implement the additional service processing.

All the available services are listed in Table 5.3.

Table 5.3 **ISDN Service**

Description Type of Service	Bearer/Telco/Supplemental
Digital Telephony	Bearer
Circuit-Switched Data Services (at 64Kbps)	Bearer
Packet-Switched (X.25)	Bearer
Frame Relay Data Services	Bearer
Email (Digital Network)	Telco
Videotex (Digital Video)	Telco
Teletex	Telco
Facsimile (Group IV)	Telco
Speed Dialing	Supplemental
Caller ID Services	Supplemental
Call Waiting	Supplemental
Call Conference	Supplemental
Call Forwarding	Supplemental

ISDN Provisioning

ISDN is a circuit-switched technology. It operates by dividing up the available transport bandwidth into transport and signaling channels. The transport channels are known as bearer channels or B channels. B channels are used to transport user data. The signaling channels are known as data or D channels. D channels are used to transport control information about the user data being carried in the B channels.

ISDN comes in two implementations. The first is the BRI, which provides two 64Kbps bearer (B) channels and one 16Kbps data (D) channel. The second is the PRI, which provides 23 64Kbps B channels and one 64Kbps D channel (in Europe, a PRI is 30 64Kbps B channels and one 64Kbps D channel). The PRI can also support what are known as H channels. There are two types of H channels: H0, which operates at 384Kbps, and H1, which operates at 1.536Mbps. The broadband ISDN (B-ISDN) standard defines operating rates above 1.544Mbps. ATM is based on the B-ISDN protocol reference model.

ISDN is compatible with TDM digital transmission systems. It accomplishes this by using a 64Kbps transport channel size, the same size channel used for TDM transmission. ISDN operates over the existing digital transmission infrastructure utilized by TDM systems. ISDN services and circuit provisioning are implemented in software on the CO switch and used for local loop transport. When the data reaches the CO, it is multiplexed with other data and transmitted over trunk lines to the endpoint's CO.

In order to provide the various ISDN transmission services and ISDN's out-of-band management, a number of different protocols are used. These protocols fall under one of two conceptual planes: user or control. Figure 5.11 illustrates the ITU-T ISDN protocol reference model.

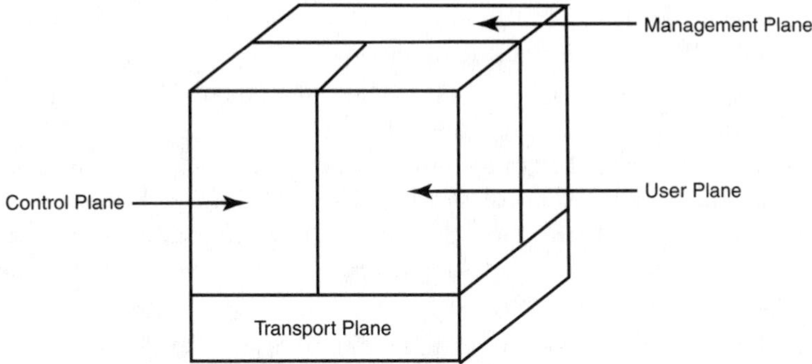

Figure 5.11 The ISDN protocol reference model.

Control plane protocols provide management services that control user data services, such as call establishment and termination, transport channel provisioning, and supplementary service management. User plane protocols transport user data. The management plane provides traffic management services on the link. Both the B and D channel information is transported over the same physical transport medium. Although ISDN is a "transport" mechanism, which traditionally resides at OSI-RM Layer 1, only the B channels operate exclusively at OSI-RM Layer 1. The ISDN D channel operates at OSI-RM Layers 1, 2, and 3.

ISDN Access and Reference Points

With a TDM network, there are defined demarcations between the CPE and LEC/CLEC equipment. Because ISDN is an international standard and the CPE/telephone company demarcation changes depending on the PSTN system, ISDN defines four demarcation reference points. These are sometimes referred to as user premise interfaces (UPI). Along with these demarcation reference points, ISDN also defines different types of CPE equipment. The type of CPE equipment generally specifies the type of physical ISDN interface and functions the equipment will support. First, let's look at the different types of ISDN CPE interfaces and their functions:

- Terminal Equipment type 2 (TE2)—This is the Data Terminal Equipment (DTE) that has no direct support for ISDN.
- Terminal Adapter (TA)—This provides ISDN access for TE2 equipment types. This can either be an external adapter connected by a common interface, such as an RS232 or V.35 interface port, or an internal BUS card.

- Terminal Equipment type 1 (TE1)—This is an ISDN-compatible device that can directly support ISDN protocols, interfaces, and management functions.
- Network Termination type 2 (NT2)—This is an ISDN-compatible switch or multiplexing device, or interface card on an ISDN-capable switch. It can provide support for ISDN Layer 1, 2, and 3 functions along with maintenance and circuit monitoring. The most common example of an NT2 device is a PBX. This is also the PRI circuit termination point.
- Network Termination type 1 (NT1)—This is the ISDN equipment responsible for the BRI circuit termination. The NT1 in the United States is a CPE device. In Europe and Japan, the NT1 is provided by the telephone company. The NT1 provides the same function as a smart jack in a TDM network configuration. It provides the circuit termination, the two-pair-to-single-pair conversion needed for PSTN transmission, and a testing interface used by the telephone company for troubleshooting and monitoring.
- Line Termination (LT)—LT is the telephone company CO ISDN circuit termination point. This is the port on the CO switch provisioned for your ISDN circuit termination. It is through this interface that the ISDN adapter negotiates its Terminal Endpoint Identifier (TEI). The TEI is a unique ID used between the CO switch and the ISDN CPE for identification.
- Exchange Termination (ET)—This is also the CO switch. This interface represents the switch trunk interface responsible for providing the interconnection services between LT and the rest of the ISDN (PSTN) network.

Now, let's look at the four ISDN demarcation reference points. Figure 5.12 illustrates these demarcation points in relation to the wire pair path and their physical CPE device interfaces.

- The R interface acts as a point between TE2 and the TA interface to the ISDN network. It comes in many forms, the most common being RS232 or Centronix Parallel. The R interface is only defined by the CCITT as an interconnect point, so the method is left up to the device manufacturer.
- The S interface is between the TA or TE1 and the NT2 device.
- The T interface is between the NT1 and the NT2. This is a logical interface for the most part. Its purpose is to provide a communications distinction between the NT2 and the NT1, and the NT1 and the telephone company.
- The U interface is the demarcation between the NT1 and the telephone company's LT interface.

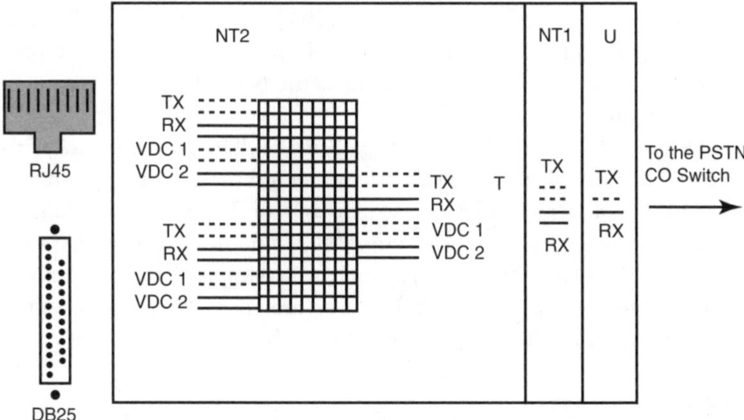

Figure 5.12 ISDN reference points and their physical interface implementations.

The U interface is only relevant in the United States because the NT1 device is considered CPE. Most ISDN adapters sold in the United States will come with a U type interface. ISDN equipment sold for use internationally will only come with an S/T interface. A separate NT1 will need to be purchased to terminate the ISDN BRI circuit. An ISDN PRI will utilize a CSU, or the NT2 will provide an integrated CSU.

The design principle behind all the different ISDN interface types is the physical transport path. Although the BRI and PRI function the same way, conceptually they each use different interface termination interfaces, framing, signaling, and data paths.

ISDN Layer 1

The BRI uses two different frame formats (ITU-T I.430) to send user data: one for network to terminal, and another from terminal to network. The frames are 48 bits in length (32 bits for user data and 16 bits for control). The BRI data path from the terminal adapter to the NT1 is a two-wire pair data path. The data path transports frames at a rate of 4,000 frames per second for a total operating rate of 192Kbps. The BRI data path up (from the TE to the U) uses a line encoding scheme called Pseudo-Ternary Signaling (PT-S). P-TS is similar to bipolar signaling, which is used with the TDM digital transport. P-TS uses a 750 millivolt (mv) signal that alternates from + to -, to represent a zero bit. A one bit is represented by an absence of line voltage.

In the United States, the NT1 represents the conversion from a two-pair data path to a single-pair data path. The single-pair local loop data path operates as a full-duplex link (ANSI T1.601). The NT1/U and the LT both employ an echo cancellation scheme, which inserts a negative image of the data transmission into the receiving circuit, which canceling out the full-duplex feedback of the transmitter's signal. Here, the user data is converted from P-TS to 2B1Q two binary-one quaternary signaling scheme. This line-encoding scheme uses four line voltages to represent 4-bit symbols (called quats) that are 2 bits of actual data. The 2BIQ transport frame is 240 bits in length (216 bits of B and D channel data and 24 bits for control data), and operates at a rate of 666 frames per second, which maintains the 64bps B channel rate and the 16bps D channel rate.

The PRI essentially uses the TDM data path model. The PRI is a two-pair full-duplex transmission path. It uses the same signaling and framing scheme employed by the standard T1/E1 carrier specification, which is 24 or 30 64-bit time-slots. Each frame contains 193 bits with a sampling rate of 8,000 frames per second, for a total operating rate of 1.544Mbps. When H channels are configured, the remaining 64-bit channels are used for B and D channels. PRI also provides the means for using a single D channel to transport control data for multiple PRIs.

ISDN Layer 2

To provide reliable communications for ISDN signaling and control messages, the ITU-T has developed the Link Access Procedure on the D channel (LAPD). LAPD is defined in ITU-T Q.920 AND 1.921. The LAPD frame format is illustrated in Figure 5.13.

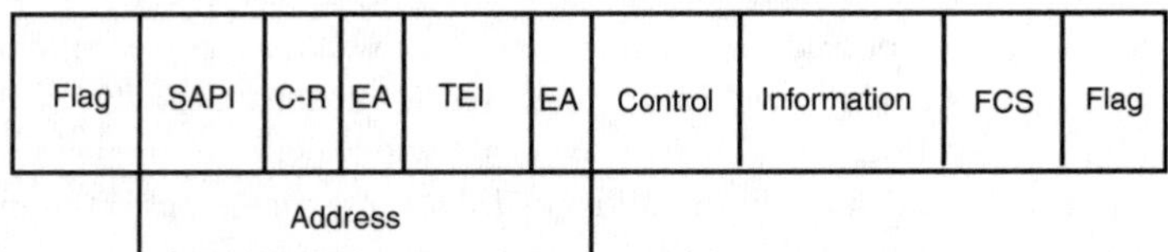

Figure 5.13 LAPD frame format.

> **Note**
>
> It is not uncommon to get T1 and PRI circuits confused, as they both support the same data transmission rate (1.544Kbps). PRIs are used almost exclusively for providing voice and digital modem services. They are not used for point-to-point data transport. What makes a PRI distinct from a T1 is the ISDN signal provisioning used between the LEC and CPE device, which is used to allocate the transport circuit's bandwidth. On a standard T1, this additional provisioning does not exist. The circuit is clear, and no additional management information is exchanged between the CPE and the LEC. The datastream is just sent across the transport, with the actual bandwidth allocation being left to the CPE devices attached on either end of the circuit.

The LAPD frame contains the following fields:

- Flag—This is an 8-bit (01111110) field that indicates the start and end of the frame.
- Address—This is a 16-bit field that contains the Service Access Point Identifier (SAPI). This is a 6-bit subfield, which indicates which Layer 3 service is associated with the message data. The Command and Response bit (C-R) indicates if the message data is a command or a response. The Terminal Endpoint Identifier (variable) specifies the unique identifier used between the ISDN device and the CO switch. One or more TEIs can be assigned to an ISDN device. TEI addresses can either be dynamically assigned when the interface first comes up or set manually. Dynamic TEI assignment is the common practice. The Extended Addressing (EA) bits are used for bit stuffing.
- Control—This is an 8-bit field that indicates the type of frame being transmitted. There are three types:
  - Information frames relay signaling or user data in a sequenced order.
  - Supervisory frames relay flow control, acknowledgements, retransmission requests, and so on. Like information frames, these messages are sequenced.
  - Unnumbered frames are used for transmitting data-link messages, sending unsequenced data, and so on.
- Information—This is a variable length field that contains the ISDN Layer 3 signaling information.
- FCS—This field contains a 16-bit CRC of the messages used for data integrity checking.

ISDN Layer 3

ISDN Layer 3 services provide the user-network signaling communications for other functions that establish, monitor, and terminate ISDN connections. These functions include voice call setup and teardown, virtual-circuit establishment and termination for data transmissions, and the use of supplemental ISDN services. The "user" does not necessarily mean an individual person, but rather an "interface" that is establishing a connection over the ISDN network.

ISDN uses its own D channel Layer 3 protocol, which is defined in ITU-T recommendations Q.930 (General Signaling, Q.931 (Call Control), Q.932 (Supplemental Services), and Q.933 (Frame-Mode Services). These documents describe the framing and signaling messages used for various ISDN services. The D channel protocol is used by the ISDN CPE device and the CO switch to exchange signaling information. Signaling communications outside of the local loop (that is, LE to LE) are done using SS7.

ISDN Addressing

ISDN PRI and BRI circuits have two numbered addresses associated with them: the ISDN address (or service profile identifier) and a regional-compliant NPA subscriber address. In most cases, with BRI circuits, the SPID and the NPA subscriber address are the same, except for a two- or four-digit subaddress, which is at the end of the ISDN address. This subaddress is used by the ISDN device and the LD for channel identification. To establish a voice/data call with an ISDN device, the NPA subscriber number is used. Depending on the BRI provisioning, the ISDN circuit might have one or two (one for each B channel) ISDN/NPA subscriber addresses. The subscriber address can be used to access only one of the B channels or both using a "rollover" configuration. In ISDN BRI configurations with two NPA/SPID assignments, each B channel is assigned its own primary NPA and SPID. Like the single NPA/SPID configuration, the subscriber address can be used to access only one of the B channels or both. This same approach is followed with ISDN PRI circuits, using a slightly different NPA/ISDN address format. Figure 5.14 illustrates the NPA/ISDN addressing schemes.

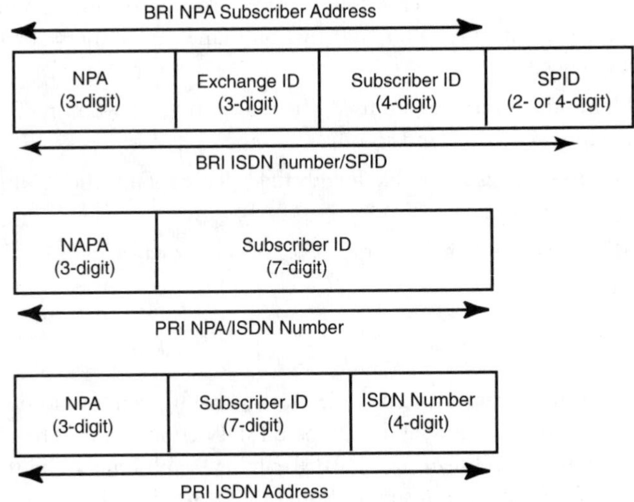

Figure 5.14 BRI/PRI address formats.

PRIs, which are commonly used with PBXs or digital modem banks, are assigned a single NPA address/ISDN number. This number is used to identify the ISDN switch device (for example, a PBX). The ISDN device is the root subscriber ID for the ISDN terminals attached to the ISDN switch device. When a call is placed to the ISDN number, the call is routed to the first available B channel. Along the same lines, if a PBX has 100 internal switch ports and 23 external PRI channels, when an internal ISDN terminal places a call outside the PBX, the first available B channel is used to establish the call. To establish calls with ISDN terminals directly or indirectly attached to the PRI, the ISDN address is used. An ISDN address is a distinct NPA subscriber address assigned to an ISDN B channel or ISDN terminal. When this address is assigned to a B channel, it is basically a telephone number, so when a call is placed using an ISDN address, the call is routed to a specifically assigned B channel. When the address is assigned to an ISDN terminal, the CO switch has a list of assigned ISDN addresses associated with a PRI/ISDN number. Call requests using one of the assigned ISDN addresses are routed to the PRI(s) connected to the ISDN switch device hosting the ISDN terminals assigned these addresses. When a call is placed using an ISDN address, it is routed to the CO switch. If a B channel is available to the ISDN switch, the call is assigned a channel and the ISDN switch (PBX) then routes the call to the correct ISDN switch port connected to the ISDN terminal that has the assigned ISDN address.

These two different addressing models reflect the ISDN signaling capability of the CPE. A CPE device using a BRI circuit depends on the LT to handle call routing and so on, whereas a PRI terminates at the NT2 interface and is capable of performing call routing and other ISDN signaling functions.

PSTN Packet- and Cell-Switched Networks

The operating concept of Packet-Switched Networks (PSNs) should be familiar to you. It is the basis for the majority of LAN networking technologies. A group of end-stations connected to the same medium share the available bandwidth to exchange information in variable-length packets (or frames). With LANs, after the hardware and cabling expense, the cost of operation is minimal. With WAN networking, however, along with the cost of hardware there is a monthly PSTN access cost. PSNs provide a usage-based data transport facility as an alternative to dedicated bandwidth transport provided with T-carrier point-to-point links.

The PSTN, however, was not originally designed for shared bandwidth operation. Its design is focused on providing dedicated, connection-oriented data exchanges between logical connections. Packet-switched networks were developed to provide shared bandwidth data transport services in line with the PSTN's operational model. With PSNs, each network endpoint is provided a dedicated PSTN connection and its own unique PSN address. This connection is provided over traditional PSTN transport facilities (T-carrier/POTS), and the circuits are provisioned to provide a dedicated network bandwidth rate.

The endpoint, which can be a router, modem, or any properly equipped DTE, is connected to a buffered PSN packet switch, which functions as the station's gateway to the PSN. When two endpoints exchange information over a PSN, the do so over a Virtual-Circuit (VC) connection. VCs are established between PSN endpoints (much like a phone call, in some cases), using their PSN addresses to establish and maintain the "call" over the shared bandwidth infrastructure provided by the network of interconnected buffered switches. A PSN endpoint can support single or multiple VCs between different PSN endpoints. When the PSN endpoints have established a VC, they can exchange data. The actual data exchange is handled by formatting the data into a frame format appropriate to the PSN type. After they are formatted, the frames that comprise the data are handed to the PSN switch (indicating in the frame which VC should be used to deliver the data) which then provides the transport over PSN.

The advantage of PSNs is efficiency. PSNs use STDM, and share the total available network bandwidth between all the connected endpoints, so the bandwidth is always being used to transmit data. The available bandwidth of the PSN is determined by the number of subscribers; the trick for the carrier is to provide enough bandwidth to facilitate adequate performance for all the connected subscribers. The downside is that PSNs operate with a certain amount of inherent latency. This is because each data request is first sent to a PSN switch that buffers the request and then transmits it when there is enough bandwidth available to complete the request. The bandwidth rate of the PSN circuit defines how much bandwidth the PSN endpoint can sustain through its interface. The PSN carrier, however, does not guarantee dedicated bandwidth, only that the bandwidth rate will be supported. For many data applications, error-free communication is what is important, not speed, so the additional latency does not present a problem. How the actual bandwidth of the PSN is managed depends on which PSN technology is used to provide the transport service.

X.25

The first large-scale PSN implementation was X.25. This is an ITU-T networking standard that operates at OSI-RM Layers 1, 2, and 3. X.25 networks first began to appear in the 1970s, and they provided only moderate data transmission rates. Nevertheless, their lack of speed was offset by network-level error-checking and recovery services, making X.25 a very reliable connection-oriented data transport option. X.25 provides delivery acknowledgement for each packet sent, if required. After a packet is transmitted, an acknowledgement must be received before any more are sent. This approach, needless to say, adds significant overhead to the data transmission process. These services were needed mainly because the original PSTN transport network for X.25 was based on electromechanical switches and analog transmission facilities, which were considered unreliable for data transmission.

X.25 networks are comprised of four components: Data Terminal Equipment (DTE), Packet Assembler/Disassembler (PAD), Data Circuit-terminating Equipment (DCE), and Packet-Switching Exchanges (PSEs). The DTE is the device that sends and receives the network data over the packet-switched network. This can be a computer, point of sale/credit card processing system, or even a data router. This device sends data to the PAD (or might incorporate this service). The PAD operates on OSI-RM Layers 2 and 3, and provides the buffering and X.25 packet assembly/disassembly of the DTE data. At Layer 3, this process includes creation and removal of the X.25

Packet Layer Protocol (PLP) header. X.25 PLP provides packet delivery services over the X.25. Layer 2 services needed for formatting the data are provided by an HDLC derivative protocol called Link-Access Procedure, Balanced (LAPB), which is also similar to IEEE 802.2. From the PAD, the X.25 packets are handed to the DCE. The DCE provides the OSI-RM Layer 1 facilities, which are provisioned using a TIA/EIA RS232 interface on a modem or TIA/EIA V.35 on a DSU/CSU. This provides the PSTN circuit termination and bitstream conversion. The type of DCE termination device depends largely on the type of VC used for transporting the DTE data.

The PSEs (also OSI-RM Layer 1) are what make up the actual PSN. These devices are located at the carrier facilities. Different PSEs are interconnected over standard PSTN trunk facilities. Figure 5.15 illustrates the relation between the different X.25 network elements.

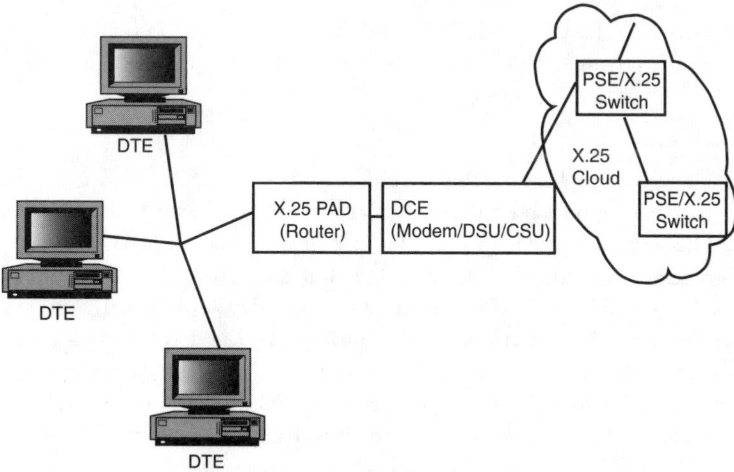

Figure 5.15 X.25 network elements.

> **Note**
>
> PSN networks are commonly referred to as "clouds." This reference alludes to the shared approach used to allocate network bandwidth and the open nature of PSN networks.

X.25 provides for two types of full-duplex VC: switched and permanent. The VCs are in essence just like telephone calls. They have several operational modes: call setup, data-exchange, idle, call teardown, and reset. A switched VC (SVC) is a temporary connection established between two X.25 connected endpoints. The process is as follows:

A DTE has data to send, and an SVC call-setup request is made.

The X.25 endpoint establishes the SVC.

Data transfer occurs between the two DTEs.

An idle time is reached on the SVC.

The SVC is torn down.

With a permanent VC (PVC) the call is established once and remains active all the time, so DTEs can send data at any time. X.25 uses the X.121 address format to establish SVCs. The X.121 address is an International Data Number (IDN) which consists of a data network identifier code indicating the PSN the X.25 endpoint is attached to, and a National Terminal Number (NTN) which identifies the X.25 endpoint.

Frame Relay

Frame Relay is fast packet-switching technology based on STDM switching service. Like X.25, it can support multiple logical VC connections over the same physical connection. The service is provided mostly over a primary rate (T1/E1/PRI) PSTN interface, and some carriers offer Frame Relay at data rates above DS1 (but not many). Frame Relay is a ITU-T/CCITT standard that was developed during the ISDN standards development process. It is strictly a connection-oriented data-link protocol for packet-switched networks. Frame Relay was originally developed as a high-speed transport alternative to X.25. It supports both SVC and PVC operation, but does not have the error-control and recovery capabilities that X.25 supports. Because Frame Relay was designed to operate over digital transmission facilities, these capabilities are not required, as digital facilities have a significantly lower BER than the analog electromechanical facilities X.25 was designed to use.

Frame Relay, like the ISDN effort in general, faced standardization and implementation problems due to carrier equipment incompatibility conflicts inherent in the PSTN. In the 1990s, the Frame Relay Consortium (FRC) was formed to develop a common interface standard to overcome these incompatibility problems; this interface is called the Local Management Interface (LMI). There are three versions of the LMI: the FRC version (commonly called the Cisco version), the ANSI version, and the ITU-T/CCITT version. In the United States and Europe, the ANSI/ITU-T, respectively, are used. The FRC version is used mostly in private network Frame Relay implementations.

Frame Relay is a packet-switched network, and it is based on the same functional model as X.25. Endpoints (DTEs) are connected to DCE over PSTN T-carrier local loop, four-wire links. These links are provisioned at the sustainable data rate contracted for with the Frame Relay provider. The Frame Relay circuit's Committed Frame Rate (CFR) can be anywhere from 128Kbps to 1.544Mbps. Frame Relay is billed based on usage, so it is possible to have both a usage bandwidth rate and a peak bandwidth rate. The peak rate can represent a different rate cost when the circuit throughput exceeds the normal usage rate. This can be a handy option: when you need more bandwidth, it will be there, and yet you do not have to pay for a higher usage rate you will not normally use.

Frame Relay Implementations

Frame Relay supports both SVC and PVC operation, but the most common implementation of Frame Relay is PVC. Frame Relay supports two types of Frame Relay PVC topologies: point-to-point and point-to-multipoint. A Frame Relay point-to-point topology operates almost identically to a dedicated T-carrier circuit. The connection is always up, and data can be sent at any time. Frame Relay PVCs between the DTE and the DCE use Data-Link Connection Identifiers (DCLIs). The DCLI is significant because it allows the DCE to distinguish between the different locally connected DTEs. In addition, if the circuit supports certain LMI extensions, the DCLI can have significance outside the local loop. In these cases, the DCLI serves as the DTE's Frame Relay network address.

In Frame Relay point-to-multipoint topologies, LMI extensions are used. In a multipoint configuration, a collection of Frame Relay endpoints function as if they are all connected together over a common transmission medium. This topology provides support for broadcast and multicast OSI-RM Layer 3 transmissions. Frame Relay multipoint configurations are not supported by all Frame Relay providers and are typically limited to only 64 endpoints.

Another Frame Relay topology variation is to have a single DTE support multiple point-to-point PVCs between different DTEs. In this topology scenario, the multipoint DTE acts as a connectivity interexchange point between the other DTEs. This topology is limited to the capabilities provided by the OSI-RM Layer 3 protocol used to provide delivery, and it is commonly used as a cost savings dodge. With this configuration, you only need to have the multipoint host contract for a large bandwidth Frame Relay connection, and the single PVC sites can contract for just a percentage of the supportable bandwidth of the multipoint site.

Frame Relay Congestion Control

This latter approach works well because Frame Relay supports congestion control mechanisms. These mechanisms are known as forward and backward explicit congestion notification (FECN and BECN). To enable FECN or BECN, a notification bit is set in the Frame Relay frame address field. BECN is used by DTEs to indicate congestion problems. If a BECN bit is enabled on a received frame, the receiving DTE reduces CFR for the interface. If consecutive BECN messages are received, the CFR is reduced further. When the host begins to receive, BECN "clears" messages; the DTE can begin to increase its CRF. FECN messages are implemented by the DCEs to indicate to the destination DTE that the frame encountered congestion over its delivery path through the Frame Relay network.

If the BECN and FECN mechanisms are not adequate to accommodate network congestion in the Frame Relay cloud, the DCE will implement further congestion recovery by discarding frames based on the Discard Eligibility (DE). A DE setting of 1 indicates the frame can be discarded in times of congestion. A setting of 0 indicates that the frame should be spared unless there is no alternative. The DE can be set by the DTE based on the type of traffic contained in the frame or depending on the DTE's CFR. If the DTE is in excess of its standard usage CFR, the frames will be tagged with a DE of 1.

The Frame Relay Frame Format

The standard Frame Relay frame has five fields, described in the following list. In addition to the standard frame, an LMI Frame Relay frame is used when the LMI extensions are enabled. The standard Frame Relay frame is illustrated in Figure 5.16 and described in the following bulleted list.

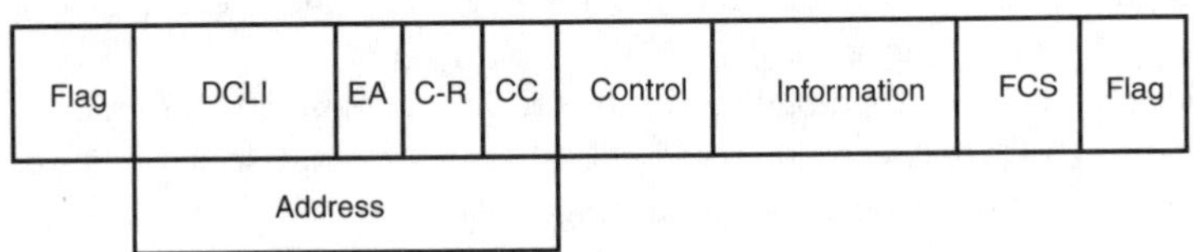

Figure 5.16 The standard Frame Relay frame format.

- Flag—This functions as a frame delimiter. It's an 8-bit field contacting the binary number 01111110 at the beginning and end of the frame.
- Address—The address frame is a 16-bit field made up of the following information: The first 10 bits are the DCLI address. This is followed by an EA bit that indicates where the DCLI address ends. Then comes the C/R bit, which is unused. The C/R is followed by congestion control (CC) subfield, which is 3 bits in size. The first bit is the FECN, the second is the BECN, and the last bit is the DE.

- Data—This field contains the ULP data being carried in the Frame Relay frame. It can be up to 16,000 octets in size.
- Frame Check Sequence (FCS)—This is a checksum of the frame generated by the source DTE. The destination DTE performs the same checksum test of the frame upon delivery and compares the two values.

Caveat Emptor (Buyer Beware)

Frame Relay is commonly used by ISPs to provide fractional primary rate and primary rate dedicated Internet access connections. This kind of connection is quite cost effective for them because it places the burden of infrastructure cost on the Frame Relay providers and does not require ISPs to maintain large numbers of dedicated circuits. It is also popular for large enterprise WAN transport networks, because it is cheaper than using dedicated T-carrier circuits. However, before you run out and contract some Frame Relay circuits, you should consider two things. First, Frame Relay is a packet-switched technology. It shares a finite amount of bandwidth with a group of customers. Frame Relay networks have real resource limitations from time to time. In times of heavy congestion, performance problems will be directly perceived and your data will be dropped. There have also been some rather significant mishaps in recent years by the major Frame Relay providers (AT&T and MCI WorldCom) which have illustrated what exactly happens when a Frame Relay cloud collapses. The answer is nothing, because your network is completely down until the frame cloud is restored.

ATM

Asynchronous Transfer Mode is a WAN and LAN cell relay transport technology developed for broadband network applications. The term "broadband" here is not a reference to the signaling mechanism, but rather defines any WAN data network application that utilizes a transport speed beyond that of the primary rate (T1/E1/DS1). ATM is the transport mode for the ITU-T B-ISDN standard, which provides standards defining the signaling, transport, and management aspects of digital transmission services beyond the primary rate interface.

ATM uses a connection-oriented, virtual circuit (VC) transport model that transports data in 53-byte cells between ATM endpoints. This small cell size makes ATM well suited to transmit voice, video, and data. ATM supports point-to-point bidirectional SVCs, unidirectional point-to-multipoint SVCs, and bidirectional point-to-point PVCs to establish connections between DTE endpoints. These VC connections, unlike traditional packet-switched WAN technologies, support Quality of Service (QoS) features that can provide guaranteed bit rate delivery services.

As you are already familiar with the general operation concepts behind packet-switched networks (VCs, congestion control, and so on), we will look at how ATM implements these operational functions, using the ATM architecture model as a reference.

ATM Architecture

During the ISDN development process, the ITU-T realized that a need existed for developing standards to provide digital transmission services beyond that of the ISDN primary rate. The B-ISDN reference model, illustrated in Figure 5.17, uses the same basic control/user/management plane model used by the ISDN reference model. The B-ISDN layers and planes define the following functions:

- Control plane—Manages the signaling and addressing functions needed to establish VCs (call setup and teardown)
- User plane—Addresses the processes pertaining to the user data delivery functions
- Management plane—Addresses system-wide management functions between the control and user layers
- Physical layer—Defines the transmission of the ATM cells between the various DTE and DCE interfaces
- ATM layer—Addresses the construction of the ATM cells (header generation, flow control, and VC addressing issues)
- ATM adaptation layer—Provides the "processing" for different types of user data (voice, video, binary) between the ULPs and the ATM layer

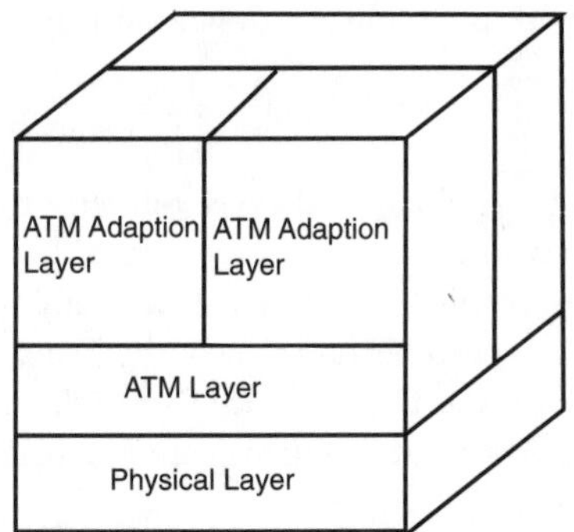

Figure 5.17 The B-ISDN reference model.

In 1990, the ITU-T revised the B-ISDN standards. In 1991, the ATM Forum was created (made up of Cisco, Stratcom, Newbridge, and NET) to develop a set of interoperability standards to facilitate the development of ATM-based products (ATM network adapters and switches). Without the ATM Forum, ATM would probably still be just a set of standards documents and no more. What the ATM Forum provided was a set of specifications to implement ATM utilizing the ITU-T standards. The ATM Forum also defined implementation extensions to ATM that specifically dealt with utilizing ATM as a data networking technology. Figure 5.18 illustrates the ATM architecture model.

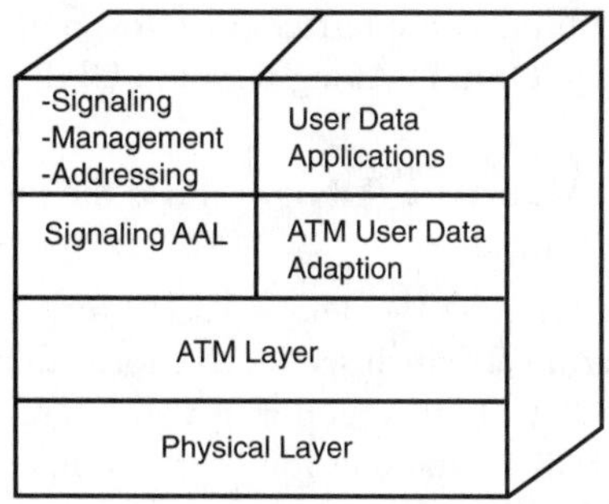

Figure 5.18 The ATM architecture model.

The Physical Layer

The physical layer is divided into two sublayers: transmission convergence (TC) and physical medium dependent (PMD). The TC sublayer handles the packing and extracting of ATM cells in and out of the SONET transport frames. There are four functions associated with this process:

- Cell delineation—Defines the cell boundaries so they can be identified in the bitstream
- Header Error Control (HEC)—Generates a checksum of the cell header and inserts it into the cell for comparison, to verify the validity of the header data
- Cell-rate decoupling—Maintains the cell flow to meet the supported rate of the PHY interface
- Transmission-frame adaptation—Performs the actual cell packing into the transmission medium's frame format

The PMD sublayer defines physical transport and timing mechanisms used to physically transmit the ATM cells. ATM cell transmission in both LAN and WAN environments is implemented using ATM switches, utilizing in most cases SONET (STS-3) for transport and management over fiber optic interfaces.

There are two functional ATM User-to-Network Interfaces: the public-UNI and the private-UNI. The UNI defines the connection between the DTE (user) and the ATM switch (network). The difference mostly has to do with the type of physical interface and its application. Private UNIs are defined for LAN applications that have public UNI support for transmission interfaces that are compatible with PSTN transports (such as T-carrier and SONET). Along with the UNI, there are also three other ATM interface types:

- Network-to-Network Interface (NNI)—This is the interconnection point between two ATM switches.
- Data Exchange Interface (DXI)—This is a serial (or equivalent) interface between a router/terminal, which sends DXI-formatted information over a DSU/CSU. The DSU/CSU then performs the data segmentation and formats the data into ATM cells for transmission over a compatible transport interface connected to an ATM network.
- Frame User Network Interface (FUNI)—This provides the capability for DTEs to transmit FUNI frames directly to an ATM switch, which does the frame-to-cell conversion necessary for data transmission over the ATM network.

Most common ATM implementations utilize single- or multimode fiber optic cable and SONET/SDH STS-3 transport for public/private UNI and NNI connections. Table 5.4 lists all the defined ATM public and private UNIs.

Table 5.4 **Common ATM Transport Implementations**

Transport Medium Application	Operating Rate	NNI or UNI
ATM over DS1 (Category 3)	1.544	Public
ATM over UTP (Category 3)	26.6	Private
ATM over DS3 (Category 3)	44.736	Public
ATM over STP (Category 3)	155.52	Private
ATM over SONET (SMF/MMF)	155.52	Public/Private
ATM over SONET UTP (Category 3)	155.52	Private
ATM over SONET UTP (Category 5)	155.52	Private
ATM over SONET (SMF)	622.08	Public/Private

SMF = Single Mode Fiber

MMF = Multimode Fiber

STP = Shielded Twisted Pair

UTP = Unshielded Twisted Pair

The ATM Layer

The ATM layer is concerned with the construction of the ATM 53-byte cells, the transmission and routing of the ATM cells over the different VCs, and traffic and congestion control.

Cell Formatting

There are two ATM cell formats: the UNI cell and the NNI cell. Both use 5 bytes of the 53 byte payload, leaving 48 bytes for data. Figure 5.19 illustrates the UNI and NNI cell formats.

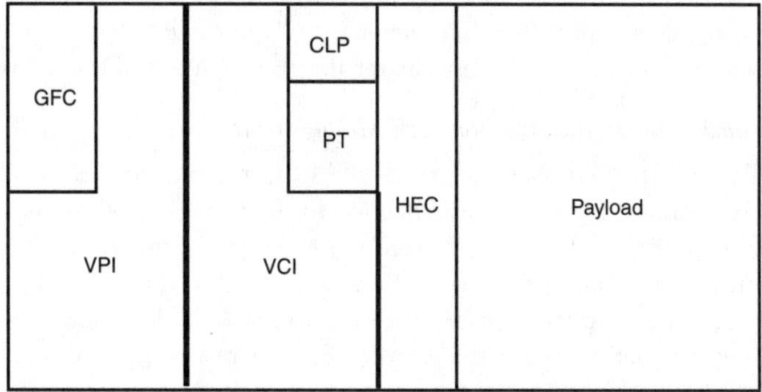

Figure 5.19 ATM cell format.

The ATM cell contains the following data fields:

- Generic Flow Control (GFC)—This field provides CPE equipment to regulate traffic flow between different stations sharing a common ATM interface. This field is not used in most cases and is not present on the NNI cell.
- Virtual Path Identifier/Virtual Channel Identifier (VPI/VCI)—ATM uses VC to establish data transport channels between ATM endpoints. This is called a Virtual Channel Connection (VCC), and there are switched and permanent VCCs. A VCC can support both bidirectional and unidirectional ATM traffic flows and its own QoS contract. VCCs operate through ATM virtual paths, which exist between each ATM switch in the transport path between the two endpoints. A Virtual Path Link (VPL) exists between each ATM switch, and each VPL has its own Virtual Path Identifier (VPI). Contained in the VPL are the virtual channels set up between the ATM endpoints. Each ATM VC has its own Virtual Channel Identifier (VCI). As the ATM cells are passed from switch to switch, each ATM switch constructs its own VPI/VCI translation table to perform remapping for each cell that passes through. Depending on the ATM switch type, the switch might use the VPI alone or both the VPI and VCI to determine what cells are the next destination point in the path.
- Payload Type (PT)—This field indicates if the cell contains control information or user data. The PT field is 3 bits in size. The first bit indicates the payload, the second bit indicates the congestion state of the network, and the third bit indicates management information relating to the cell's data.

- Cell Loss Priority (CLP)—This field indicates if the cell can be discarded in the event of congestion on the network.
- Header Error Check (HEC)—This check is performed by the physical layer interface to see if there are errors in the ATM cell header. The checksum of the header is placed in this field.

As user data flows down from the ATM adaptation layer, the ATM layer constricts the cell header with the appropriate VPI/VCIs for the first segment of the delivery.

VPI/VCI Translation, Traffic, and Network Management

After the cells are formatted, they are prioritized in accordance to the VC's QoS contract, multiplexed into a cell stream, and handed to the PHY layer for transmission over the physical transport. When they reach the next hop in the VC path, the cell stream is demultiplexed, and new VPI/VCIs are appended to the cells. Then they are multiplexed again into a stream and handed back to the PHY layer for transport. At each junction of the delivery path, the ATM layer must maintain the QoS requirement for the VC.

The ATM layer is responsible for monitoring the network cell flow and ensuring that adequate resources exist to meet the QoS requirements of the various VCs. The QoS contract is established when the VC is set up. This contract defines a set of traffic parameters that must be maintained to meet the QoS requirements of the VC:

- Peak Cell Rate (PCR)—Defines the fastest rate that cells for the specified VC can be introduced into the PHY bitstream
- Sustainable Cell Rate (SCR)—Defines the average rate that cells for the specified VC can be introduced into the PHY bitstream
- Maximum Burst Size (MBR)—Defines the duration of time that a specified VC can accept cells at the PCR
- Minimum Cell Rate (MCR)—Defines the minimum cell rate that must be guaranteed for a VC

To enforce the QoS contracts and implement congestion control mechanisms, ATM employs two traffic management functions: traffic policing and traffic shaping. Traffic shaping occurs during the cell multiplexing phase, when various VC cell streams are partitioned into buffers, which in turn adjust the cell flow into the PHY layer according to the QoS contract. During periods when resources are limited, cells associated with VCs that provide no service guarantee will be discarded.

To guard against cell loss, traffic policing is performed by the UNI interface, which relays network congestion information to the ATM layer. Based on the traffic shaping management information, if the network does not have the resources to meet a VC's QoS requirements, the UNI will reject the establishment of the VC until adequate network resources are available.

When the VC is established, a QoS contract is established. There are four QoS categories supported by ATM for data delivery over VCs:

- Constant Bit Rate (CBR)—This provides a fixed rate bandwidth link and is required for real-time video and audio applications. A CBR connection establishes a PCR at call setup, and the resources to meet this requirement are reserved at the same time. The MBS rate for a CBR connection is unlimited, and the bandwidth is reserved for the duration of the VC.
- Variable Bit Rate (VBR)—VBR VCs utilize a PCR, SCR, and MBR to regulate cell flow. There are two variations on the VBR contract. VBR-rt VCs can tolerate some controlled network delay and generally operate at a consistent bit rate (that might vary depending on traffic flow). VBR-nrt VCs have the same operating rate requirements as VBR-rt, but are not affected by network delays.
- Unspecified Bit Rate (UBR)—UBR VCs are used for network communications that require no QoS contract. These applications can tolerate both data loss and network delays and have no specific bandwidth availability requirements. UBR is commonly used for ATM LAN applications where the ULPs can provide recovery services for inconsistencies in the network transport.
- Available Bit Rate (ABR)—ABR VCs make use of the bandwidth available on the network based on network congestion notification. Congestion information is relayed by resource management cells.

ATM Adaptation Layer (AAL)

The adaptation layer has a control element and a user element. The control adaptation layer element deals with transporting signaling information between two ATM endpoints. The user adaptation layer deals with the flow of user data information into the ATM layer.

To accommodate different user data applications, ATM supports five different data adaptation layers. Each AAL service has two functions: segmentation and reassemble (SAR) and common part convergence sublayer (CPCS). The SAR segments the ULP user data into the 48-byte cells needed for transmission and extracts the cell data to reconstruct the UPL data as the cells are delivered. The CPCS "preps" the ULP data by inserting additional buffer and segment tags so the ULP data can be segmented and reassembled cleanly by the SAR.

AAL1

AAL1 is a CBR connection-oriented adaptation service. It is used for applications that have (PVC) connections between the source and destination endpoints that require traditional dedicated circuit transport, such as voice and video applications. AAL1 requires that synchronization timing be shared between the two ATM endpoints. It uses a 47-Byte Protocol Data Unit (PDU) and a 1-Byte header that contains a 4-bit

Sequence Number (SN), which is used to provide timing and cell order information, and a 4-bit Sequence Number Protection (SNP), which is a CRC of the SN. The SNP is used to perform integrity checks on the SN field.

AAL2

AAL2 is a VBR connection-oriented adaptation service. AAL2 has never been fully developed by the ITU-T standard and is not supported in current ATM implementations.

AAL3/4

AAL3/4 is a VBR connection-oriented and connectionless service. AAL3/4 does not require a (PVC) connection between source and destination endpoints. AAL3/4 uses a 44-byte PDU combined with a 2-byte SAR header and a 2-byte SAR footer. The SAR header contains three fields:

- Segment type—Indicates what part of the ULP message is contained in the PDU-payload (beginning, middle, or end)
- Sequence number—Indicates the PDU-payload's placement in the ULP message reassembly
- Multiplexing identification number—Indicates the source of the PDU-payload so the data can be discerned from other ULP messages that are contained in the multiplexed VC stream

The SAR footer contains two fields:

- Length indication—Displays the size of the actual data content in the PDU-payload
- CRC—A CRC value for the entire PDU for error checking

AAL5

AAL5 is also a VBR connection-oriented adaptation service. AAL5 does not have the SAR overhead that AAL3/4 has. The AAL5 PDU contains a 40-Byte PDU-payload and an 8-Byte CPCS footer. The CPCS footer contains user and cell boundary information, the size of the actual data portion (versus buffer) contained in the PDU-payload, and a CRC for integrity checking. The lack of SAR processing and processing information makes AAL5 attractive compared to AAL3/4 because it provides a payload savings and reduces the processing overhead and requirements on the ATM switch. AAL5 is used for most ATM LAN transport applications, such as LAN emulation and classical IP over ATM.

ATM Signaling

Signaling AAL (SALL) provides transport of SVC call setup and teardown messages. Due to the importance of signaling messages, SALL provides a robust transport mechanism, and uses the AAL5 CPCS and SAR functions. The SALL PDU consists of a 40-Byte PDU-payload with an 8-Byte CPCS footer. SALL uses two sublayers to process ULP signaling information: the Service-Specific Coordination Function (SSCF) and the Service-Specific Connection-Oriented Protocol (SSCOP). The SSCF provides the mapping service from the ULP application to the SSCOP. The SSCOP provides the connection-oriented reliable transport mechanism for the signaling messages. In addition to the error detection capabilities provided by AAL5, SSCOP provides frame sequencing, frame recovery, and error correction capabilities.

ATM VCs are established manually or dynamically. Manual VCs are employed with ATM PVC-type circuits. Dynamic VCs are used for SVC applications. ATM connection setup uses a hop-to-hop signaling method. This approach sets up calls by cascading the call setup request from switch to switch in the call path. This establishes the VCC path between each switch along the way to the destination. When the destination endpoint gets the request, it then accepts or denies the call.

The SVC call dialog for establishing a bidirectional point-to-point connection between two ATM endpoints goes like this:

1. The ATM end-device issues a `setup` request (containing the destination address and QoS contract) to its directly connected ATM switch. This then propagates the request to the network.
2. When the `setup` request is received by the destination, a `call proceeding` message is sent in response.
3. If the QoS contract is acceptable to the destination, it sends a `connect` message to the source, accepting the connection request.
4. After the connection acceptance has been received, the source sends a `connection ack`.

To tear down the call, a `release` message is sent by one of the endpoints, and a `release complete` message is sent in response. The `release` message indicates that the VCC is terminated and the VCI associated with the VCC should be discarded. A restart function is provided to reset a VCC. ATM also supports unidirectional point-to-multipoint SVC connections. To add and remove endpoints from the SVC, `add party` and `drop party` and corresponding `acknowledgements` are used. All the SVC call signaling messages are sent on a predefined VCI/VPI (0, 5).

ATM Addressing

ATM uses addresses to distinguish different endpoints on the network. All PVC or SVC connections are established using an ATM address. Like any PSTN network function, some kind of subscriber identification is required to establish any type of circuit-based call. To provide addressing to ATM endpoints on public ATM networks, the ITU-T standard specifies the E.164 address, which is similar to the NPA subscriber ID. For private ATM networks, the ATM Forum has provided three private network address formats. These formats are illustrated in Figure 5.20.

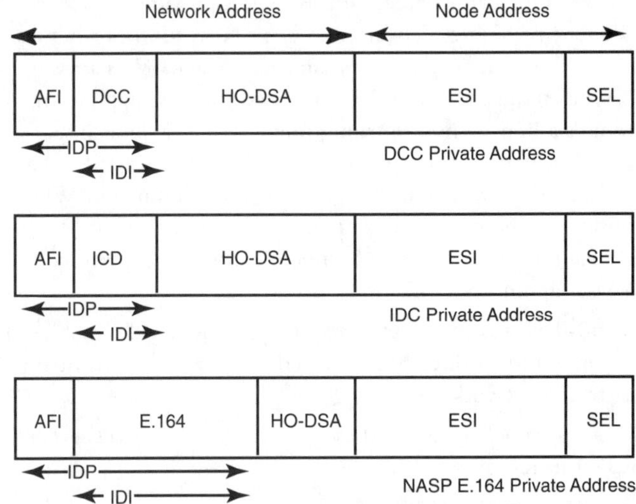

Figure 5.20 ATM Forum-defined private ATM address formats.

Each address format consists of two parts: the Initial Domain Part (IDP) and the Domain-Specific Part (DSP). The IDP consists of an Authority and Format Identifier (AFI) and an Initial Domain Identifier (IDI). The AFI indicates the type of ATM address format being used. The IDI will be Data Country Code (DCC), International Code Designator (IDC), or a public E.164 address. The remainder of the address is DSP, which consists of two parts. The first is the network address, which is indicated by the High-Order Domain Specific Address (HO-DSA). The HO-DSA, a multipart address that permits a structured addressing scheme within the ATM network. The second part is the node address, which consists of the End System Identifier (ESI) which is an endpoint specific address and a Selector address (SEL).

Address registration between endpoints and switches is facilitated through the use of the Interim Local Management Interface (ILMI). This protocol is used to exchange the node information on the endpoint with the network information on the ATM switch. This information exchange is performed between the switch and all its connected endpoints. These endpoint addresses are then exchanged between connected switches, so they can construct address databases, which are needed to establish SVCs between endpoints attached on different switches within the ATM network.

Data-Link Framing

Point-to-point dedicated DS0 and higher PSTN-based communications links require a data-link protocol to format ULP data for transmission. The data-link protocol is responsible for the encapsulation of ULP data into transmission frames, as well as their transmission and delivery over the synchronous or asynchronous transport link.

For serial data transmissions using Universal Synchronous Asynchronous Receiver/Transmitter (USART) and Universal Asynchronous Receiver/Transmitter (UART) based transmission devices, there are two commonly used data-link framing protocols: the ISO's HDLC and the open standard, PPP. Both are derivatives of IBM's Synchronous Data-Link Control (SDLC), which was developed in the 1970s to provide serial data transport between IBM mainframes and terminal controllers.

It is perhaps more accurate to say that PPP is a variation of the IP serial transport protocol, Serial Line Internet Protocol, which is built on a subset of the HDLC protocol. SLIP at one time was the basis for all serial-based IP transport (and still is among IP diehards). Today, however, it has been largely replaced by PPP, which supports multiprotocol transport.

HDLC

HDLC provides multiprotocol framing over synchronous full-duplex transport links. HDLC is the default framing mechanism employed by most router serial interfaces, including Cisco products. It operates using one of three primary/secondary control/transfer schemes:

- Normal Response Mode (NRM)—This is the SDLC operational model, where the primary station must grant permission to the secondary stations before they can transmit data to the primary station.
- Asynchronous Response Mode (ARM)—This mode supports unrestricted data transmission between the primary and secondary nodes.
- Asynchronous Balanced Mode (ABM)—With ABM, nodes are hybrids, which can act either as the primary or the secondary. Hybrid nodes also operate in an unrestricted transmission mode.

The primary station is responsible for establishing, managing, and terminating the transmission link. HDLC supports point-to-point (one primary and one secondary) and multipoint (one primary and multiple secondary) configurations.

HDLC transmission is a two-phase process. The first phase is the assembly of the frame, and the second is the transmission of the medium-dependent frame. The HDLC medium-dependent frame consists of an 8- or 16-bit address field, an 8- or 16-bit control field, and a variable length information field, which contains the ULP data and a 16- or 32-bit frame check sequence field. You should note that HDLC employs a Least Significant Bit (LSB) transmission scheme. This has the odd effect of making all the frame octets appear is if they are backward. The HDLC basic frame format is illustrated in Figure 5.21.

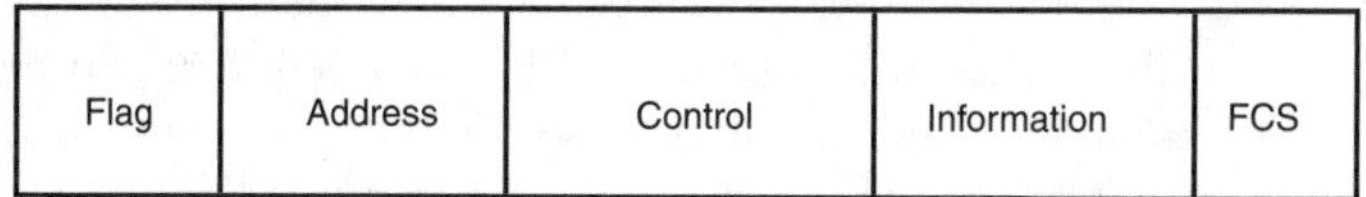

Figure 5.21 HDLC frame format.

The HDLC frame contains the following fields:

- Address—This field is used for multipoint configurations. The address can be a host or "all-stations" (broadcast) address. The host addresses can be a decimal number between 1 and 126. The address 127 is used to indicate an "all-station" address.
- Control—This field is used to indicate the type of message being sent. Three types of message frame formats are used. The information frame is used with connection-oriented data transmissions. This field contains send and receive sequence numbers which are used for message tracking. The supervisory frame is used to send control information. The unnumbered frame is used to establish sessions with secondary nodes and send connectionless data messages.
- Information—This field contains the ULP data being transmitted. There is no generic size definition for HDLC frames. PPP, which uses the HDLC frame format, specifies an information field of 1502 bytes.
- Frame Check Sequence (FCS)—This is a 32-bit CRC check for the address, control, and ULP information fields. This value is matched with the CRC generated by the receiver, and a mismatch indicates an error frame.

Although it is possible to use HDLC for dial-on-demand multiprotocol data-link transport over PSTN connected serial devices, it is not recommended, because HDLC provides no security or protocol level authentication. HDLC is fine for dedicated point-to-point transport over leased PSTN lines. However, PPP is generally becoming the preferable solution for dedicated link transport, as it is an open standard, and permits easy interoperation between serial devices made by different manufacturers. HDLC implementations, though standards-based, do vary between vendor implementations.

PPP

PPP can provide multiprotocol data-link transmissions over any full-duplex (and some half-duplex) asynchronous or synchronous link. Along with standard PSTN DSO (and higher) and SONET transports, PPP can also operate over ISDN, frame relay, and ATM. PPP is largely a creation of the Internet, much like TCP/IP, and is modeled on SLIP and the ISO's HDLC. The protocol itself is described in more than 40 RFCs that define PPP's general operation, LCP operation, authentication mechanisms, NCP implementations, data compression, and encryption schemes. Everything you want or need to know about PPP is in the RFCs. The following is intended merely as an overview to PPP's operation.

In addition to providing data-link framing, PPP supports two protocol-level authentication schemes: Challenge Handshake Authentication Protocol (CHAP) and Password Authentication Protocol (PAP). CHAP is a three-way authentication method that utilizes a shared secret between two hosts. Throughout the duration of the connection, the two hosts send periodic CHAP authentication requests that require the responding host to send its local identifier and shared password used by the two communicating hosts. PAP uses a standard username and password authentication scheme for authentication when the connection is first established. Along with transmission framing and authentication services, PPP also provides the following:

- Link quality monitoring and error detection—PPP employs a verbose Link Control Protocol (LCP) that establishes and tests the link configuration, provides link quality monitoring, data compression negotiation, error detection, and link state reporting.
- Network layer address negotiation—PPP provides a network control protocol to manage the various facilities needed to operate the Layer 3 protocols it supports (IP, AppleTalk, IPX, DecNet, and so on). Among the supported services are network address configuration, multiprotocol traffic multiplexing, and TCP header compression.
- Multilink operation—This is a mode of operation that uses a collection of ISDN B-channels or DS1 circuits as a single -high-speed logical transmission path. Packets are bit-striped across the channels, ensuring that the bandwidth is evenly utilized. After the connection is established, channels can be added or removed depending on bandwidth requirements. Dynamic allocation is not a function of PPP. It is handled through the use of the Bandwidth Allocation Control Protocol (BACP).

PPP link negotiation is a rather complex process. It uses a primary/secondary link establishment model similar to the one used by HDLC. One station negotiates the link establishment. There are three phases of PPP link establishment:

- LCP link connection and negotiation
- PPP authentication and link quality management
- Network control protocol negotiation

Link Establishment

Before any PPP link can be established, the physical layer transport link must be established. When used with dedicated links, the process is invoked when the interface is enabled. With a dial-on-demand link configuration, the link is either established manually or by a ULP request. When a physical link state has been established and verified, PPP sends LCP Configure-Request messages. The link source and destination peers then negotiate the link configuration parameters (Is the link single or multipoint? What kind of compression, if any, will be used with this link? and so on). After all the configuration parameters have been determined, each peer sends a Configuration-Acknowledgment LCP message.

PPP Authentication

After the link has been established, the LCP notifies the authentication layer that PPP link has been established. If authentication is enabled, the authentication process is started. PPP authentication will make several attempts to authenticate the link. If these fail, the LCP is notified and the link is torn down. If authentication is successful or if no authentication option is configured, the NCP layer is notified. PPP also supports link quality verification. This is an optional and additional step before NCP negotiation is invoked. Its purpose is to verify that the link is of a suitable transport quality to sustain network layer protocol traffic. If the link quality tests fail, the PPP link is terminated and re-established.

NCP Negotiation

When authentication and link quality testing are complete, NCP negotiation can begin. A separate NCP is used for each Layer 3 protocol being transported across the link. NCPs can be brought up and taken down at any time during the links duration.

When the link is brought down, either by an adverse transport event or through a user-initiated termination, the LCP will notify each of the NCPs in operation, so they can close open events and notify dependent applications of the connectivity loss.

PPP Framing

There are three HDLC frame variations used by PPP to transmit ULP data. The type of transmission frame employed is determined by kind of transport connection being used for transmission. These variations are defined in RFC 1662.

For asynchronous transport using modems, AHDLC framing is used. AHDLC framing is implemented in software as part of the PPP interface driver. It makes use of a defined frame delimiter character for marking the beginning and end of AHDLC frames. It also employs an escape character to indicate to the receiver how to interpret the next octet value. In order for the escape characters to be interpreted properly with certain octet values, the link transport must support 8-bit values, otherwise the data will be misinterpreted. These character combinations are never used to represent user data. AHDLC also utilizes some type of in-band or out-of-band flow control mechanism, but that is left up to the actual hardware and software implementation. AHDLC's processing of the datastream on a per octet basis does add some additional overhead to the data handling process.

For synchronous transport over dedicated DS-X leased lines or ISDN, bit-synchronous HDLC (BHDLC) framing is used. BHDLC is ideal for transport implementations where framing is accomplished by the hardware interface. Where AHDLC uses special escape characters for structuring the user data, BHDLC instead uses bit stuffing techniques to present the user data in interpretable form. Also like AHDLC, an end-of-frame marker is employed to indicate the end and beginning of a new frame (01111110). To ensure that the end-of-frame marker and user data are not mistaken, bit stuffing is used after any sequence of five consecutive ones. If after five consecutive ones the next bit is a zero, the zero is a stuffing bit and should be ignored. If a one is the next bit, the octet is complete. By using bit stuffing, the octet boundaries are no longer valid, so the entire bitstream must be interpreted to transport the data correctly. Hence the name, bit-synchronous. BHDLC's bit-synchronous transmission mechanism is also more efficient than AHDLC because it allows the datastream to be processed on a per packet basis instead of a per octet basis.

The last variation is octet-synchronous HDLC. This variation is rarely used except for SONET and SDH applications. Octet-synchronous uses the same start/end delimiter and escape character that AHDLC uses, the difference being in the handling of the escape characters. Octet-synchronous HDLC was originally developed for ISDN transport applications. It was never implemented, though, as BHDLC is more efficient for ISDN transport applications.

Regardless of the framing variation, the PPP frame format is essentially the same for each transmission variation, as illustrated in Figure 5.22.

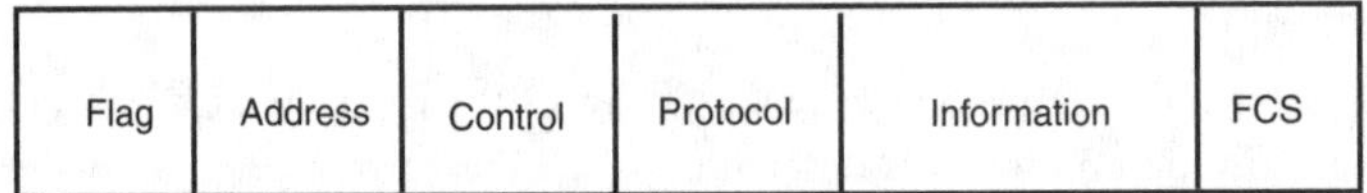

Figure 5.22 PPP frame format.

The PPP frame contains the following fields:

- Start/End Flag—This flag indicates the start/end of a new frame. The binary sequence of 01111110 is always used to mark the start of a new frame.
- Address—This field contains the binary sequence of eight ones or the hexadecimal FF, which indicates all stations or a broadcast. PPP does not support station addressing.
- Control—All PPP frames use the HDLC unsequenced frame type. This is indicated by a hexadecimal 03 or the binary equivalent 00000011.
- Protocol—The protocol field is technically inside the HDLC information field. This 2-byte field identifies the protocol associated with the data contained in the data field.
- Data—This field can contain a maximum of 1,500 bytes. The total data field length, with the protocol field, is 1502 bytes.
- Frame CRC—This field contains the 16- or 32-bit CRC information for the frame. The default is 16 bits, however, a 32-bit CRC can be negotiated as part of the link establishment process.

In addition to data, the PPP frame is also used to transmit LCP messages. There are three types of LCP message frames: link-establishment, link-maintenance, and link-termination. All the PPP message frames use the control tag for the HDLC unsequenced frame type.

Summary

This chapter discussed the various technologies and protocols associated with WAN networking. WAN networking is typically seen as a distinct function from LAN networking because WANs are almost exclusively centered around telephony, which is commonly disassociated from computing services in large information technology environments. Due to the ever-increasing demands to be "wired" all the time, however, this distinction is quickly disappearing. At the end of this chapter, you should have an understanding of the following concepts and terms:

- The PSTN, T-carrier, SONET, and SDH digital transport systems
- Packet-switched PSTN network technologies such as X.25 and Frame Relay
- Asynchronous Transfer Mode and Broadband-ISDN
- The HDLC and PPP data-link protocols

In the next chapter, we will look at network switches. In Chapter 4, "LAN Internetworking Technologies," and in this chapter, we mentioned briefly what switches (sometimes referred to as multiport bridges) are and their basic capabilities. In the next chapter, we look at switches from the ground up, examining their structure and functional elements, the capabilities they provide, and their important role in constructing large-scale enterprise networks.

Related RFCs

RFC 1332	The PPP Internet Protocol Control Protocol
RFC 1483	Multiprotocol Encapsulation over ATM Adaptation Layer 5
RFC 1570	PPP LCP Extensions
RFC 1577	Classical IP and ARP over ATM
RFC 1598	PPP in X.25
RFC 1618	PPP over ISDN
RFC 1619	PPP over SONET/SDH
RFC 1626	Default IP MTU for Use over ATM AAL5
RFC 1661	The Point-to-Point Protocol (PPP)
RFC 1662	PPP in HDLC-like Framing
RFC 1989	PPP Link Quality Monitoring
RFC 1990	The PPP Multilink Protocol (MP)
RFC 2125	The PPP Bandwidth Allocation Protocol (BAP) The PPP Bandwidth Allocation Control Protocol (BACP)
RFC 2153	PPP Vendor Extensions

Additional Resources

Ali, Syed. *Digital Switching Systems, System Reliability, and Analysis*. McGraw-Hill, 1997.

Black, Uyless. *ATM* Volume I, *Foundation for Broadband Networks*. Prentice Hall, 1997.

———. *ATM* Volume II, *Signaling in Broadband Networks*. Prentice Hall, 1998.

Black, Uyless, and Sharleen Walters. *SONET and T1 Architectures for Digital Transport Networks*. Prentice Hall, 1997.

Goralski, Walter. *SONET: A Guide to Synchronous Optical Network*. McGraw-Hill, 1997.

Kessler, Gary, and Peter Southwick. *ISDN Concepts, Facilities, and Services,* Third Edition. McGraw-Hill, 1996.

6

Network Switches

IN THIS CHAPTER, WE LOOK AT A TECHNOLOGY THAT HAS BECOME increasingly important to the implementation and development of enterprise networks: network switches, specifically, LAN switches (also known as multiport bridges), and ATM switches. We first take a brief look at the requirements for and operational fundamentals of network switches. Then, we examine Layer 2 bridging, which supports LAN technology and, to some extent, the functionality of ATM switches. After we have an understanding of how Layer 2 bridging works, we then compare Layer 2 bridging and Layer 3 routing from an implementation and appropriate use perspective. Next, we examine Layer 2 network switches, and look at their basic architecture, operation, and implementation. We close the chapter by examining ATM switch specifics and the integration of ATM with traditional LAN networks.

The Need for Network Switching

Enterprise networks traditionally have been built with shared transmission medium Layer 2 LAN protocols and point-to-point dedicated medium Layer 2 WAN protocols. Multiport repeaters, MAUs, and bridges were used to construct LANs, so that two connected end-stations would have momentary access to all the available LAN segment bandwidth to exchange data. These LANs were designed and operated using the 80/20 rule: 80 percent of the LAN segment traffic was local (on one side of a router), the other 20 percent was remote (on the other side of a router). On the WAN side, Transport-1 based PSTN links using synchronous serial interfaces were used to tie LANs together. In fact, much of the Internet backbone in the early 1990s operated only at T1 speeds.

Then, the network age was born. The idea of the "network as computer" took hold. In this period, thick and thin client/server computing became popular, and this, together with FTP, Gopher, Usenet, and eventually Web browsers such as Mosaic, helped turn computers into network communications tools. Browser or "Web-based" applications became bandwidth-intensive as they matured. Their integration of text, graphics, and streaming audio and video helped fuel the need for the development of high-speed data networking technologies.

The once-static LAN/WAN of shared LANs and dedicated WAN links just could not provide the bandwidth needed to meet the new demands of the network age.

LANs now use dedicated, high-speed point-to-point links to connect workstations and servers to the WAN, and WANs in turn use groups of inverse MUX (multiplexed) T1s, Frame Relay, and ATM switching technologies to operate multiple virtual circuits over a single PSTN access link. All these enhancements are made possible in part, or in whole, through the use of network switches.

Switching Fundamentals

ATM, Ethernet, Token Ring, Frame Relay, and FDDI switches are not based on a new technology, but rather on the same basic design principles of standard Central Office (CO) PSTN switches. This design assumes that there is a finite amount of bandwidth, which limits the number of point-to-point connections that can be established simultaneously. These connections are established across a switching matrix (also known as the switch backplane), which is essentially a connecting grid that provides the capability for all a switch's ports to establish connections with each other. The most important aspect of the switch matrix is that it needs to be large enough to provide a clear path for all these interconnections. If the switch matrix is not large enough to allow all the switch's ports to simultaneously connect with each other, the ports must wait until bandwidth is available to establish connections (a condition known as *blocking*). Figure 6.1 illustrates the behavior of an 8-port switch with adequate and inadequate matrices.

Packet- and cell-switching technologies such as ATM and Frame Relay were natural, evolutionary results of PSTN's point-to-point operating design. The evolution of the LAN followed its natural progression as well, which was to develop faster shared-media transmission protocols. Although faster shared technologies did contribute to increasing the performance of the LAN, it was the coupling of Layer 2 bridging with PSTN switching concepts that resulted in the major performance gains that are common in today's LAN and ATM switches. Now that you have a basic understanding of the principles of switching, let's take a closer look at how Layer 2 bridging works.

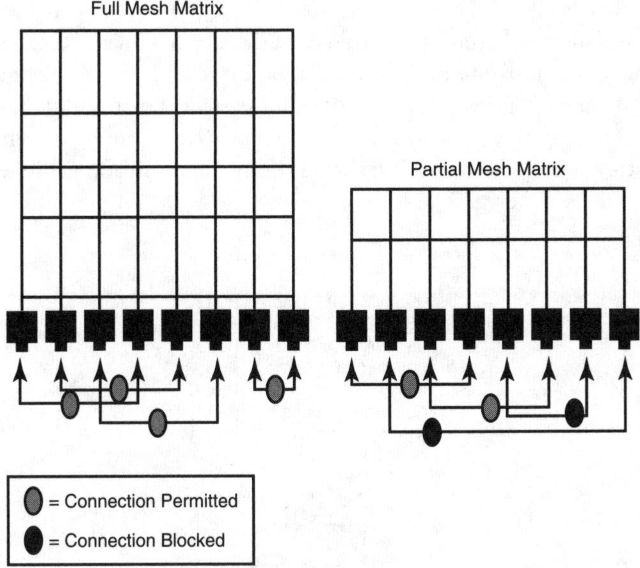

Figure 6.1 Partial and full mesh matrix switches.

The Types of Bridges

For many network engineers, bridging was avoided unless a situation absolutely required it for two reasons; First, it is believed by many that bridging is a Layer 2 answer to what is really a Layer 3 problem. Second, it is widely held that bridging negatively impacts Layer 2 network performance. In essence, both of these beliefs are true. Bridges do forward traffic between network segments, and packet forwarding is a function best performed as part of Layer 3 routing process. In addition, the insertion of a bridge in an efficient shared-medium LAN will result in slower network performance. However, when they are properly applied, network bridges can be used to improve traffic flow on a congested network. Switches can attach shared-medium network segments to high-speed backbones or interconnect remote networks running non-routable Layer 3 network protocols. There are three basic types of network bridges used in LAN and WAN networking:

- Transparent bridges—These bridges are used to join network segments that are based on the same Layer 2 protocol. Figure 6.2 illustrates the most common transparent bridge, an Ethernet-to-Ethernet bridge.

- Translation bridges—These bridges are used to join network segments that are based on different Layer 2 protocols. An Ethernet-to-Token Ring bridge is an example of a translating bridge. A variation on the translation bridge is the encapsulating bridge, which is used to join networks that use the same Layer 2 protocol across a network that uses a different Layer 2 protocol. Figure 6.3 illustrates the application of FDDI-to-Ethernet encapsulation bridges to interconnect three Ethernet LANs.

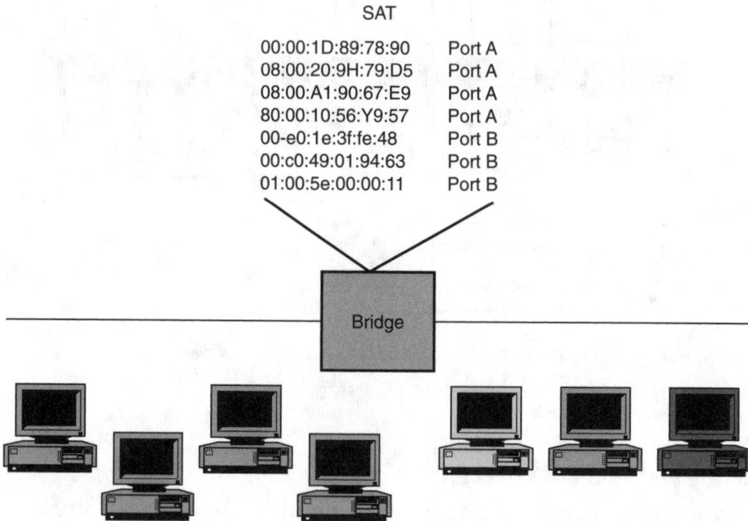

Figure 6.2 A Ethernet-to-Ethernet bridge with source address table (SAT).

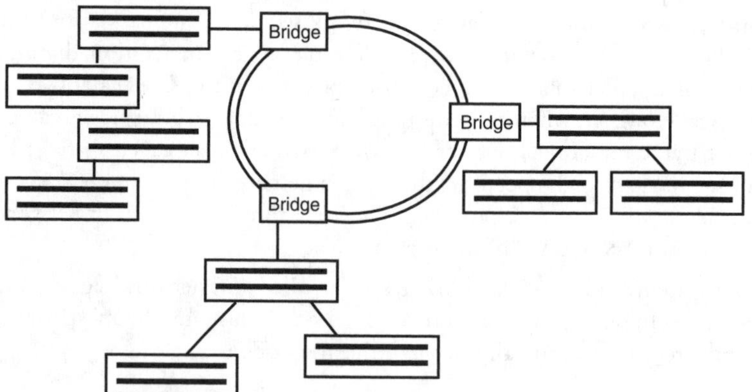

Figure 6.3 FDDI-to-Ethernet encapsulation bridge.

Translation bridges also come in a low bandwidth WAN variety known as a half-bridge. Half-bridges use an HDLC or PPP synchronous serial link to connect to same type Layer 2 LAN segments.

- Source route bridges—These bridges (in most cases) are used to join network segments that are based on the same Layer 2 protocol, however, they rely on the packet's source to provide the path the packet will take through the Layer 2 network. Source route bridges then use this information to forward the packet from segment to segment.

Although different types of bridges exist, there is one condition that is common to all bridges. Bridges, like all other nodes connected to a shared wire, operate under the protocol transmission rules governing the networks they connect. For example, an Ethernet bridge must wait until the wire is clear before it can forward packets to its destination, and token-based bridges can only transmit data when they have the token. This dependency on network availability, coupled with the processing required for bridges to forward traffic between network bridges, is why bridges can impact network performance.

Bridging Functionality

The first LAN network bridges to be developed were Ethernet transparent bridges. They were created by the Digital Equipment Corporation, one of the members of the original DIX Ethernet standards consortium. The transparent bridge standard was later co-opted by the IEEE as part of the 802.1 standard, which describes Layer 2 addressing and management concerns. Source route bridges were developed by IBM. Originally, source route bridging vied with transparent bridging to be the 802.1-bridging standard. It lost out, but later emerged as part of the 802.5 Token Ring standard. Source route bridging is dependent on the utilization of the Routing Information Field (RIF), which is an optional field in the Token Ring frame (see Chapter 4, "LAN Internetworking Technologies," for review). Translating and encapsulation bridges, although they represent a special type of bridge functionally, can operate as transparent, source route, or as a hybrid (part transparent and part source route) bridge. This functionality is largely determined by the types of Layer 2 transmission media the switch is connecting together. It should also be noted that hybrid-based translating bridges represent a significant level of complexity because they interconnect two bridging protocols that are basically opposed to each other.

Transparent Bridges

The standard transparent bridge consists of two or more LAN media ports. IEEE 802.1 standard, hardware-based transparent bridges are plug-and-play and require no configuration. You should recall from Chapter 1, "Understanding Networking Concepts," that after the bridge is connected to the network, it "learns" the location of the end-stations connected to its own segment in respect to the stations' adjacency to the bridge's ports. The interfaces of the bridge operate in what is known as "promiscuous mode" this is to say, they examine each packet transmitted across the wire. During this process, the bridge reads the Layer 2 source (MAC) address of the packets as they enter the bridge's interfaces. These addresses are then stored in a list, along with the port from which the packet was received. This table is known as the Bridge or Source Address Table (BAT or SAT) depending on who you are talking to. The bridge " learning" process is known as the Backward Learning Algorithm (BLA).

Note

All hosts can see the traffic that passes across a shared media segment. Normally, they only look at the packets addressed to them. It is possible to have your computer operate in promiscuous mode by using a packet sniffer application. A packet sniffer allows you to examine and capture network traffic that passes across the transmission medium. If you have a shared-media network, a packet sniffer is an indispensable tool for diagnosing network problems. It can also be used illegally to capture information that travels across the network, such as passwords and credit card numbers.

A variety of sniffer applications are available for the Windows platform. One of the more popular applications is a package originally called NetXRay. It is currently sold by Network Associates under the name Sniffer Pro. On the Macintosh platform, your best bet is EtherPeek, by the AG Group.

A number of hardware-based dedicated packet sniffers exist on the market. The Distributed and Expert sniffer product line by Network General (now also Network Associates) is perhaps the most popular among the dedicated packet sniffer products. Other hardware-based manufacturers include Fluke and Hewlett Packard.

Although packet sniffers are very useful, and in some cases essential, they can be complicated to operate and are expensive. So before running out and spending a large sum of money on a dedicated sniffer, first check out the PC-based software tools. This way, you can familiarize yourself with what you can actually do with a packet sniffer before making a large commitment of time and money. These tools have nice GUI front ends and come with some good packet collection and analysis tools out of the box. In most cases, software-based sniffers will meet your needs.

One of the essential requirements of a transparent bridge is to deliver packets between the connected segments completely unaltered. In other words, this type of bridge is passive, and it operates undetected by the other connected end-stations on both segments. Because the bridge, like all the other stations on the network segment, has to wait until the wire is available for it to forward any packet, it has a memory buffer that stores the packets it needs to forward. To ensure that no packet data is lost, the bridge actually copies every packet sent on the wire into its storage buffer. This process is known as filtering. After the packet is in the buffer, its destination address is checked against the bridge's SAT. If the packet's destination address is on the same bridge port from which the packet was received, the packet is discarded. If the destination is on a different bridge port, the packet is stored until the bridge can transmit the packet.

Each learned SAT entry has an associated aging time (usually around 300 seconds). If no traffic is received from a particular source address within the timeout window, its entry in the SAT is removed. There is no standard SAT timeout. If the SAT age is too long, packets might be misdirected or lost if an end-station is moved. Alternatively, an SAT age that is too short is inefficient. It places additional processing load on the bridge because the SAT table needs to be constantly updated. In certain situations, it might be desirable to shorten SAT entries. For example, in scenarios where end-stations move between segments rather frequently, a shorter SAT timeout might be desirable. With applications that tend to move data in one direction for extended periods (like network backups), a longer SAT timeout will reduce incidences of port flooding.

Port flooding is the mechanism used by bridges to deliver packets that have a destination address other than what's in the bridge's SAT table. When a bridge receives a packet with an unknown destination, it floods the packet through all of its interface ports. If the bridge only has two ports, it will simply forward the packet on to the other network. This approach in certain situations can be a great source of problems for transparent bridges.

Spanning Tree Algorithm

The problem with transparent bridges is that their "line of sight" to the network is limited to the contents of their SAT. If more than one bridge connects the same two network segments together, packet forwarding loops and excessive traffic (due to packet re-creation) results.

> **Note**
>
> Why would you want more than one bridge to connect LAN segments together? Because bridges represent a single point of failure in a network, and with enterprise networks, single points of failure can be catastrophic.

To get a better understanding of this problem, let's look at the example illustrated in Figure 6.4.

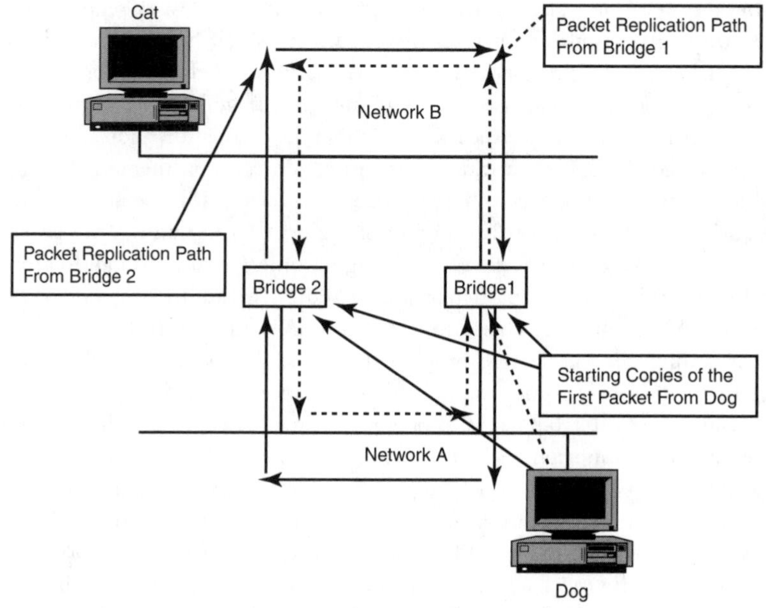

Figure 6.4 A transparent bridge packet loop.

The hosts dog and cat have just come up on network A and network B. The host dog sends a packet to cat. This packet is copied by both bridges, 1 and 2. Because the hosts dog and cat are new to the LAN, neither of them is entered in either of the bridge's SATs. So, both bridges add the host dog to their SAT tables when they see its packet hit the port. However, because the host cat has yet to send anything, its destination is unknown, both bridges blindly send or flood the packet on to network B.

Because both bridges have to wait until the wire is clear before they can actually forward the packet, one gets to forward the packet first. In this case, it is bridge 1, so its packet reaches network B, where it is immediately copied by bridge 2. It is at this point that the trouble begins.

Bridge 2 only has an SAT entry for the host dog, and still no entry for the host cat. The problem starts because bridge 2 believes (correctly, mind you) that dog is attached to network A. However, sensing that it has just received a packet from the host dog on network B (just forwarded by bridge 1), it assumes that its (bridge 2) SAT is wrong, and updates its SAT to reflect that the location of host dog is on network B. To make matters worse, because it still has no SAT entry for the host cat, bridge 2 then forwards the packet to network A again.

While all this is going on with bridge 2, the original packet that bridge 2 copied and forwarded has now reached bridge 1. The boomerang of the original packet causes a repeat performance of the same incorrect process thathas just occurred on bridge 2.

Now both packets are back on network A, where the same mistake is made all over again. With each loop, the packets proliferate, and the packet looks the same each time it is retransmitted.

To prevent loops from occurring, the spanning tree algorithm is used. According to its creator Radia Perlman (from her book *Interconnections: Bridges and Routers*), its purpose "is to have bridges dynamically discover a subset of the topology that is loop-free (a tree) and yet has enough connectivity so that where physically possible, there is a path between every pair of LANs (the tree is spanning)."

The spanning tree algorithm became IEEE standard 802.1d in 1990. It was specifically designed to address the looping and packet replication issues that result from multiple transparent bridge paths.

The spanning tree is created when all the bridges on the LAN elect a root bridge. After the root bridge is selected, all the other bridges connected to the LAN calculate the shortest possible distance to the root bridge. If a bridge has multiple paths, the other paths are placed in a "blocking state" until the primary path fails; then, the next shortest path will become active. A spanning tree bridge map is illustrated in Figure 6.5.

The root bridge election process and all other interbridge communication are accomplished by having the bridges exchange special messages known as configuration Bridge Protocol Data Units (BPDUs). These messages are sent in a Layer 2 packet with the BPDU information contained in the data portion of the packet. The destination address of the message is set to a Layer 2 multicast address used by all bridges. BDPUs are exchanged periodically between all the connected bridges to ensure that the spanning tree remains stable. So, when spanning tree changes occur, the bridges can update their path calculations accordingly or reelect a root bridge if needed.

The root election process is won based on the bridge with the lowest serial number when normal plug-and-play bridge configuration is used. However, because the root bridge can set certain behavioral properties for all the bridges in the network, it is often desirable (though not always possible) to configure in advance the bridge that will become the root bridge. On bridges that do support configuration, it is possible to configure most of the following variables:

- Bridge priority—Adds or subtracts value to the bridge's ID.
- Port priority—Allows one port to be favored over another in determining which port will be selected as the active port in the tree. In situations where a bridge (or a switch) has ports with different bandwidth connected to the same bridge, the higher bandwidth port will be selected.
- Hello time—Adjusts the period between the generation of BDPU messages. When adjusting the hello time on a bridge, you must adjust the hello time on the other bridges on the LAN segments to the same value.

- Max age—Determines when the bridge's stored BDPU messages will expire.
- Forward delay—Sets the amount of time the bridge should wait before forwarding packets on a port. This allows all the other bridges to run spanning trees and adjust to the change.
- Path cost—A value that is set on each port. It provides the capability to add additional cost to the root path cost for the port.

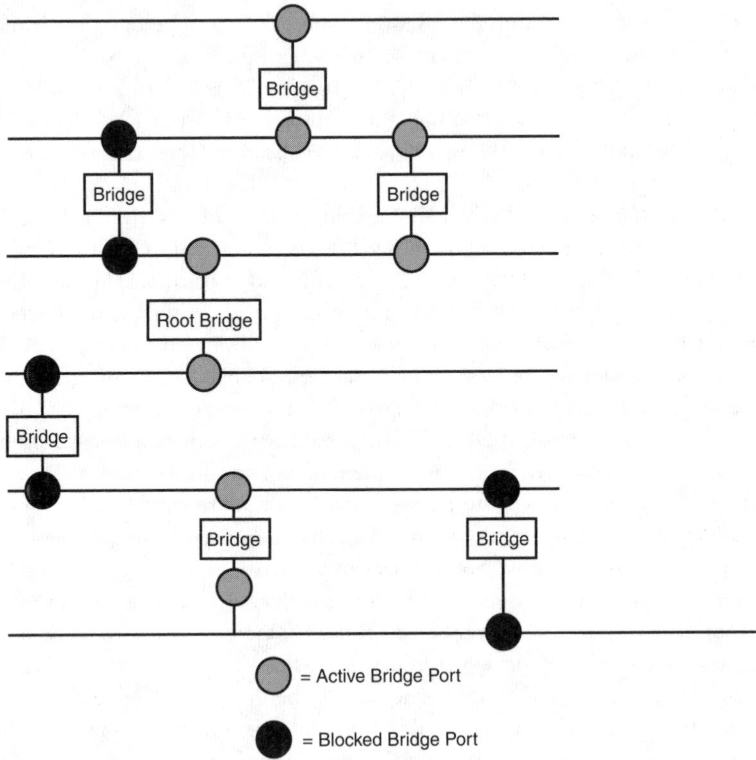

Figure 6.5 A bridged LAN with spanning tree enabled. Only one path can exist between each LAN; all other others are in a blocking state.

Of all these variables, the bridge and port priority are the most significant in terms of controlling how the spanning tree will control itself. If you plan to implement path redundancy, it is wise to take the time to configure all the bridges to function in such a manner that ensures a predictable fallover in the case of failure. Alternatively, if you are not planning to use link redundancy, it is also possible on most bridges to disable spanning tree, to minimize the amount of unneeded traffic on your network. This option should be considered carefully before implementation.

Source Route Bridges

Source route bridging is not subject to the looping and packet duplication characteristic of transparent bridging, because the packet's source, not the LAN, determines the packet delivery path. This method comes at a price, mainly in the form of additional complexity, because the Layer 2 transmission protocol needs to support a path discovery method. Source routing is only supported as an option on 802.5 Token Ring. The source routing information is included in the Routing Information Field (RIF). If source routing is implemented, the bridge checks the source address field of the packet to see if the RI bit is enabled. If the bit is enabled, it reads the routing information and forwards the packet on. If source routing is enabled on the bridge and the RI bit is off or no routing information is in the RIF, the packet will not leave the local ring.

Path discovery is accomplished by the source packet's end-station sending out a discovery packet. These packets are sent across all possible paths, and they send copies of themselves back to the source when they reach the destination host. The copies contain the path they took to reach the destination. The source then saves all the possible path information to generate source route information for the packets it sends to the destination. Each path, which consists of a segment or LAN ID and a bridge ID, is 2 bytes in size. Up to 15 routes are permitted, which means that any source-routed Token Ring network is limited in size to 15 bridges between any two rings. Figure 6.6 illustrates a Token Ring source route topology.

Along with its complexity and limited network length, each bridge and each LAN must be configured with a unique bridge ID and the LAN numbers to which it connects (as you can see from Figure 6.6). An unconfigured or misconfigured bridge would result in problems resembling those experienced by transparent bridges when multiple active bridge links exist between LANs.

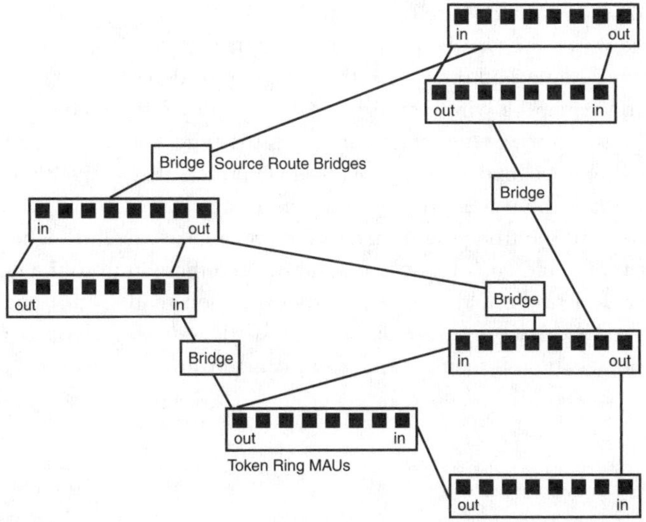

Figure 6.6 A Token Ring network interconnected with source route bridges.

Bridging Versus Routing

Until switching appeared in the early 1990s, network administrators were limited to using shared network transmission media to provide data transport. The size of the LAN network was determined largely by the Layer 2 technology used to provide transport. With 10Mbps Ethernet, depending on which Layer 1 medium was used, the network could be as small as 90 hosts or as large as 1,024 hosts. Token Ring network diameter sizes ranged from 72 to 260 hosts. The problem is that as shared-medium networks increase in size, their bandwidth utilization decreases. So, after the LAN grows beyond its operational or performance limitations, network administrators had two choices: extend the Layer 2 network segment with a bridge, or partition (or add) an additional network segment with a segmentation router. Both of these solutions created as many problems as they solved.

In the solution column, routers and bridges provide similar functions by two different means: Bridges operate at Layer 2, and routers operate at Layer 3. Functionally, however, they both can provide translation service to different Layer 2 transport media. They both can be used to reduce CSMA/CD collision domains, which are extended by the use of Ethernet repeaters. And, of course, they both enable the network to expand to its theoretical operational and/or performance limitation again.

The major downside to growing the network occurs irrespective of whether a router or bridge is used: increased network latency. *Latency* is the amount of time it takes for the first bit of the packet to reach the destination interface and the last bit of the packet to leave the source interface. Because both bridges and routers store or process the packets for a period of time, this increases the overall packet delivery time between hosts. Latency is unavoidable in large networks, because they are dependent on bridges and routers to relay data between end-stations. However, the amount of latency between a router and a bridge can be significant, and it is a direct result of the communication layer they operate on.

Routing requires the Layer 3 datagram to be extracted from the Layer 2 packet, stimulate a route lookup, and then place the datagram into another Layer 2 packet to be forwarded on when the wire is available. A bridge only copies the packet, checks its SAT, filters it (discards it), or forwards it on when the wire is available. Both, of course, are subject to wire availability, so they have storage buffers that enable them to process packets/datagrams while waiting to transmit. The router, by comparison, requires more processing time than the bridge, and therefore represents a source of higher network latency. Routers also add complexity and additional configuration to the network. Because routers function at Layer 3, their implementation requires that end-stations on different parts of the network utilize different network addressing. Router redundancy is also harder to accomplish than Layer 2 redundancy from an end-station perspective.

In terms of LAN implementation, routers and bridges are affected by similar factors, mainly proper placement. When installing a router or a bridge, the most important performance factor is placement. Ideally, common user end-stations and network resources (servers, printers, and so on) should be localized as groups on each side of the router or bridge. This way, the available bandwidth is shared between localized end-stations, and the bridge or router is only used for segment/network data exchanges. Although this seems like a relatively straightforward idea, it is the most common network design mistake made. When expanding or changing the topology of the network, it is vital that you identify how the end-stations on the network interact with each other.

If you have 300 users sending print jobs over a router because all the printers are on a separate network, or have six different workgroups of users who, for the most part, exchange files among themselves on their own local common Ethernet segment, the network is not designed to meet the data flow of the users connected to it. A more sensible approach is to have printers distributed across different networks so the traffic is localized. Segmenting Ethernet users into smaller localized collision domains will minimize bandwidth contention, resulting in better available bandwidth utilization. This also is true for Token Ring users, even though token-based technologies allocate bandwidth more efficiently. The larger the node count on the ring, the longer the distance between transmission times. The network's traffic flow dictates how well the network will perform. If routers and bridges are placed badly, the network performance can suffer substantially, and might be even worse in some cases than before they were installed.

In most LAN situations, bridging is a more efficient (from a performance and cost perspective) means for addressing Layer 2 performance and expansion problems than introducing a new router. This is particularly true when you are integrating media types of similar or higher speeds. One popular and common example is the use of translation (or encapsulation) bridges to connect Ethernet and Token Ring network segments to FDDI backbones. This can provide reliable redundant transport at 5 to 25 percent performance improvements for Token Ring and 10 percent for Ethernet. Another very common application of the translation bridge is the variation known as the WAN or remote bridge. Remote bridges use dedicated WAN point-to-point links to connect remote LAN segments together. This approach is common on networks that use Layer 3 protocols that do not support routing well, such as AppleTalk phase 1 or DEC's Local Area Transport (LAT), which doesn't route at all.

It is appropriate to use routers to address Layer 3 networking issues. Most often, these are issues like network security or access. Other common Layer 3 issues include network application performance, network data encryption, WAN reachability, and Layer 3 broadcast domain reduction. Enterprise-scale LANs often use routed network segmentation instead of bridged segmentation. However, Layer 3 segmentation can create problems when network expansion is required and adequate network address space is not available. Readdressing can be painful to outright destructive if not carried out appropriately. As the network administrator, you will need to assess your network's potential growth needs. The overall goal is to have a network that is efficient and scalable. Both routers and bridges have roles in achieving this goal.

The Evolution from Bridging to Switching

Realizing the performance shortcomings of standard bridges, network hardware manufacturers developed intelligent bridges. These machines were groups of two port bridges connected together by a common backplane. These bridges enabled the bridge to process more than one transaction at a time because the bridge itself acted like a LAN segment with only two nodes. These early intelligent bridge attempts eventually evolved into the modern LAN switch. Remember, functionally all that makes a switch different from a bridge is the switch's capability to perform simultaneous data exchanges between its ports. Otherwise, switches and bridges function identically.

There are no "switching" standards per se because switching is mainly an enhanced form of bridging. The only thing reasonably close to an operational standard (for LAN switches) is the IEEE 802.1d spanning tree/transparent bridge standard. As a result, most of the "standards" that are associated with switching are really enhancements to the actual (IETF, IEEE, and ANSI) Layer 2 protocol standards governing the operation of the Layer 2 interfaces the switch uses to connect to the network and end-stations. There are also many proprietary switching options that modify the behavior (and, in some cases, the size and format) of the various Layer 2 protocols. This is particularly true of some of the network switches that came to the marketplace in network switching's early years. Cabletron Systems' SecureFast networking and Kalpana/Cisco's Fast EtherChannel are two examples of proprietary switching technologies. At this point, most of these early proprietary designs have evolved into standards. The most significant of these options have been link aggregation, VLANs, Class of Service (CoS) traffic prioritization, and full-duplex transmission, discussed in detail in the following list.

- Link aggregation or fat pipe technology—This is the aggregation of multiple switch ports operating as a single interface. This technology was originally developed by Kalpana (later acquired by Cisco Systems under the name Fast EtherChannel). Link aggregation for Ethernet is now IEEE standard 802.3ad.
- Virtual LANs (VLANs)—VLANs are manufactured Layer 2 broadcast domains. Using VLANs, it is possible to partition the switch's ports into one or more logical, Layer 2 broadcast domains. The practical result is the capability of the switch to support multiple Layer 3 network transports (for example, Layer 3 broadcast domains) on the same switch (and if need be, on the same port). In other words, Layer 3 broadcast and unicast are only sent to ports designated to receive them. VLANs are created by either static or dynamic means, defining certain ports on the switch to be members of particular VLANs (Layer 2 broadcast domains). Cabletron's SecureFast, Cisco Systems' ISL VLAN protocol, and the ATM Forum's LANE are all variations of proprietary VLAN schemes. The IEEE 802.1q standard is now the supported VLAN standard for most switch manufacturers.
- Traffic prioritization—In addition to logical Layer 2 broadcast domain partitioning, certain VLAN implementations provide CoS-based forwarding. In contrast to the standard switch queuing method which is known as first in, first out (FIFO), with CoS forwarding, VLANs and physical ports can be set with a CoS

priority level. The CoS priority level definition, in conjunction with multiple transmission queues (one for each CoS level, defined on the switch), is then used to provide prioritized or "weighted" packet delivery. The IEEE standard 802.1p defines the Layer 2 traffic prioritization scheme used with 802.1q VLANs. Along with 802.1q, there are several proprietary prioritization schemes; Intel's High Priority Queue (HPQ) and 3COM's Dynamic Access are two examples of proprietary-based queuing.

- Full-duplex transmission—Full-duplex data transmission is not a new concept to data networking. Until switching arrived, however, it was not possible to achieve with LAN transmission protocols. Full-duplex operation provides the capability to simultaneously send and receive packets over the transmission medium. Full-duplex operation is available with FDDI (as a proprietary option only), 10BaseT, 100BaseTX and FX, Gigabit Ethernet (covered under the IEEE 802.3x standard), and Token Ring (which is defined as part of the IEEE 802.5r standard).

The Impact of Switching on Traditional Layer 2 LAN Protocols

The development of LAN switches has, to a certain extent, reinvented networking—particularly Ethernet networking. Before LAN switching, FDDI, CDDI, and ATM were considered the backbone technologies. Each offered 100Mbps or higher dedicated transfer rates, and by using routers and bridges, Ethernet and Token Rings could be scaled to efficient sizes and interconnected through the backbone. It was believed by many that Fast Ethernet would be a viable contender against these protocols, particularly because Fast Ethernet uses the identical MTU size of standard Ethernet, and that protocol accounts for the majority of installed LAN infrastructures. However, the different protocol variations, shared bandwidth format, and limited distance and expansion options deterred its usage initially as a backbone technology.

Token Technologies RIP

Although there is a substantial installation base of Token Ring in production, its long-term viability is questionable. Token Ring's cost, complexity, additional configuration requirements, and lack of built-in support on any computer platform have, in a sense, made it a legacy protocol. This is also becoming the case with FDDI, where cost and waning hardware support over time will relegate it to legacy status.

Oddly enough, their demise was brought about largely from the evolution of bridging to switching, even though both protocols were commonly used with bridging. Although they, like Ethernet, benefited from the performance gains made possible by switching, subsequent switching technology enhancements did not translate into the FDDI and Token Ring protocols as well as they did for Ethernet. For example, full duplex support, which was invented for the efficient use of Fast Ethernet and Gigabit Ethernet for backbone applications, was only available with certain FDDI implementations and provided little performance enhancements for Token Ring.

This is not to say that they will disappear tomorrow. IBM still touts Token Ring as a cutting edge networking solution. FDDI backbones are also still common, particularly in MAN scenarios where redundancy and security are still often required performance assumptions. Keep in mind, however, that these solutions do have a limited life span and you should plan accordingly.

With the evolution of bridging to switching, Ethernet has been transformed from a shared-media network transmission technology to a dynamic multi-application encapsulation protocol. In addition to providing dedicated bandwidth connections between end-stations, Ethernet is now used for high-bandwidth point-to-point and QoS/CoS virtual packet-switching applications (through the use of 802.1q/p VLANs) a capability, mind you, that historically had been associated with ATM, and is still heralded as one of ATM's greatest strengths. None of these services would have ever been possible without the advent of LAN switching.

Layer 2 LAN Switching Fundamentals

Layer 2 switches come in two basic formats: workgroup and enterprise. Workgroup switches have a fixed port count, usually between 12 to 24 ports, and one or two port expansion modules which are provided either for adding some of the same media or for high-speed uplink ports. Enterprise switches are essentially multislot chassis with power supplies. The chassis provides one or more interconnect buses that connect the switch modules. Some enterprise chassis vendors require a chassis control module, which controls the switch module's access to the chassis backplane. This, of course, creates a central point of failure and typically uses up one of the chassis slots. Other vendors build backplane management into the switch blade, so if a switch fails, only the connected end-stations are affected.

The enterprise switching modules themselves are quite often similar in port density and high-speed port expansion options to the vendor's workgroup line. This allows the vendor to save development costs by only having to develop one core workgroup/switch module architecture, which it can modify to use in both formats.

Regardless of format and Layer 2 protocol, all switches have the same basic architecture, consisting of the following:

- Layer 2 media interface—The Layer 1 and Layer 2 switch ports that support one or multiple end-stations.
- Switch backplane—Also referred to as a switch matrix. This is the interconnect used by the switch ports to establish connection paths and exchange packets.
- Switching logic—Contains four elements:
    - A CPU manages the operation of the switch and packet forwarding.
    - The switch logic module is generally an Application-Specific Integrated Circuit (ASIC) that performs Layer 2 address lookups, SAT table creation, and sets up and tears down port connections across the backplane as needed for packet forwarding. ASIC-based switches are the predominant type of switch on the market, mainly because they are less expensive and perform better than using RISC or CISC processors and software. Because the use of ASIC technology enables the switch, you can perform all the switch operations in hardware. The downside is that the switch vendor

needs to update or re-spin the ASIC every time a new technology arrives that cannot be handled with software. This can become an annoyance because often you need to buy a new switch for any significant technology or feature change.

- DRAM is used for packet buffer and SAT storage. Two different approaches are used to implement port buffering: dedicated and shared. With dedicated buffering, each port has its own dedicated DRAM buffer. With shared buffering, all the ports use a common memory pool. Dedicated buffering is faster, but is prone to overflows (which essentially halts packet processing) because the buffers are often not very large (32KB is the average size). The shared approach is slower but more efficient because the pool is shared to allow the ports that need the buffer to get it. With the dedicated model, the buffer goes unused. Packet buffer size is very important to the efficient operation of the switch the more the better. Inadequate port buffering results in packet loss, excessive collisions, and overall poor switch performance.
- Flash memory is used to hold the switch's operating system and switch configuration information. Older switches can use EPROMs to hold the switch's OS and NVRAM to hold configuration information. This, however, is quite a drag when the time comes for an OS upgrade because the PROM usually needs to be replaced.

Switch Ports and Related Functions

Each interface port (like a bridge) has its own distinct MAC address. Attached to each interface is a DRAM memory buffer, which is used to store the packets as they enter the switch port. Switching, like routing and bridging, is a packet-by-packet process. The packet comes in to the switch and is stored in its port buffer. A destination lookup is made using the SAT. If a destination is found, the packet is forwarded on to the correct port, where it is forwarded to its destination. If a destination cannot be found, the packet is flooded to each switch port. The flooding process is also used for SAT generation. When the switch first comes online, it knows nothing about the network. As it begins to process packets, it learns addresses the same way a bridge does, by examining the source addresses of the packets on their way in. The switching process does result in some latency, for the same reasons that latency occurs with routing and traditional bridges. The amount of latency depends on a variety of factors, such as the following:

- The speed of the ports involved
- The size of the packet being processed
- The forwarding scheme used by the switch
- The switch's backplane capacity and forwarding rate

We look at latency in more detail when we discuss the different switch forwarding methodologies.

Port Configuration Options

Most switches manufactured today autonegotiate between 10 and 100Mbps, half or full duplex. Earlier, we mentioned that, with switching, Ethernet has evolved into an encapsulation protocol for point-to-point dedicated connections. With Ethernet (as with FDDI and Token Ring), this capability requires the port to operate at full duplex. Full-duplex operation is possible because each port on the switch is effectively its own collision domain. When a single end-station (or switch port) is connected to a switch port, full-duplex operation is possible, because there are only two devices in the collision domain. CSMA/CD media access is not required because the stations are communicating with each other simultaneously. In half-duplex mode, the port operates under CSMA/CD and collisions occur. Even with only one device, collisions still occur between the switch and the connected end-station.

Switch Port Flow Control

In full-duplex mode, collisions do not occur, which presents some problems when flow control is needed. Flow control in some form is required by all communications devices to slow the flow of data to a rate where a device can process it efficiently. Without flow control, the device's buffers overflow, resulting in data loss. With half-duplex Ethernet networks, collisions are a form of flow control. If a receiving device's buffer begins to overflow, it can start a transmission and halt the flow of data with a collision. Additionally, as the Ethernet segment's bandwidth becomes increasingly saturated, actual transmission throughput is reduced, so by design Ethernet has built-in flow control. As for token-based technologies, only a single host can transmit at any one time, and for only a specified period of time, so again, flow control is built into the protocol. In full-duplex mode, however, collisions are lost and flow control is really needed. Generally, full-duplex links are used for network backbone and server links where the simultaneous transmission of data is actually possible.

To address this problem with Ethernet, two flow control schemes were developed:

- The backpressure method was the pre-standard answer to flow control. With backpressure, the switch port monitors its buffer utilization, and when it reaches a particular level, the port generates a jam pattern, as if it has detected a collision. This jam pattern is then sent to the device (or devices) connected to its port, effectively halting all transmission until the jam timeout (57.2 microseconds) is over. During the timeout evoked by the jam, the switch has the opportunity to forward the contents of its buffer and then resumes normal operation.

- The 802.3x standard requires both the switch and the end-station NIC to be capable of 802.3x full-duplex operation. Under 802.3x, the switch port or NIC monitors its buffer size, in a fashion similar to that used by the backpressure method. After the buffer reaches its limit, however, it sends a pause message to the remote end as a multicast packet. The switch or NIC then slows or halts the flow of data until the pause stops. At this point, data flow resumes at a normal rate.

Port Trunking (Link Aggregation)

Ethernet bridges operate in accordance with the 802.1d spanning tree/transparent bridge standard. The standard allows parallel links to exist between switches, but only one link is permitted to be active. The others are placed in a "blocking state" and only become active when the active link goes down. This limits the link bandwidth between switches to a single interface operating in full-duplex mode.

To address this problem, port trunking can be used. Figure 6.7 illustrates a single and trunked port switch configuration. Port trunking provides the capability to group switch ports to form a single link to another switch or computer equipped with a multilink interface. Port trunking will work with Fast, Gigabit, or standard Ethernet half- or full-duplex switch ports. Trunks can be constructed out of any combination of port types (provided the port configurations match on each end) and can consist of up to eight ports. Operationally speaking, a trunked interface behaves like a standalone port. In each trunk, one port in the group will function as the "lead" port. It is from the lead port that the other members of the trunk derive their VLAN and spanning-tree information.

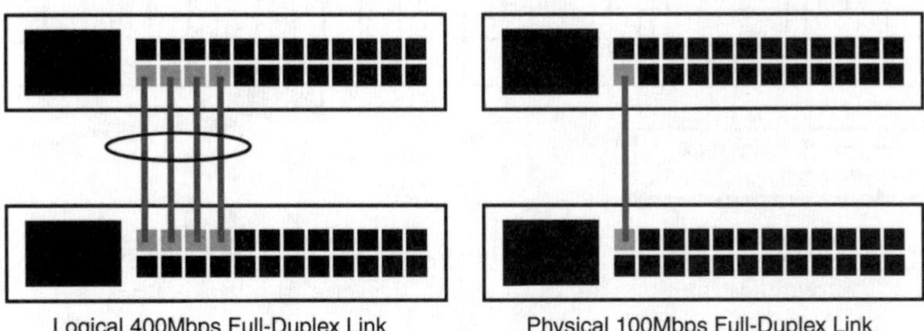

Figure 6.7 A comparison of parallel links between switches with and without port trunking.

With spanning tree enabled, it is possible to have two trunk interfaces connecting to the same switch. One of the interfaces will be placed in a blocking state. The spanning tree cost value is calculated in the total aggregate bandwidth of the trunk link. The link with the highest aggregate bandwidth will be the active link. In a scenario where both trunks are configured identically, the trunk with the lowest port number will be the primary link.

In the event that the primary trunk loses one of its ports or fails completely, the second trunk becomes active. Figure 6.8 illustrates this configuration combined with a physically diverse infrastructure. This approach (with and without the use of port trunking) is commonly used to provide link redundancy between the main distribution frame and intermediate distribution frames installed in wire distribution closets.

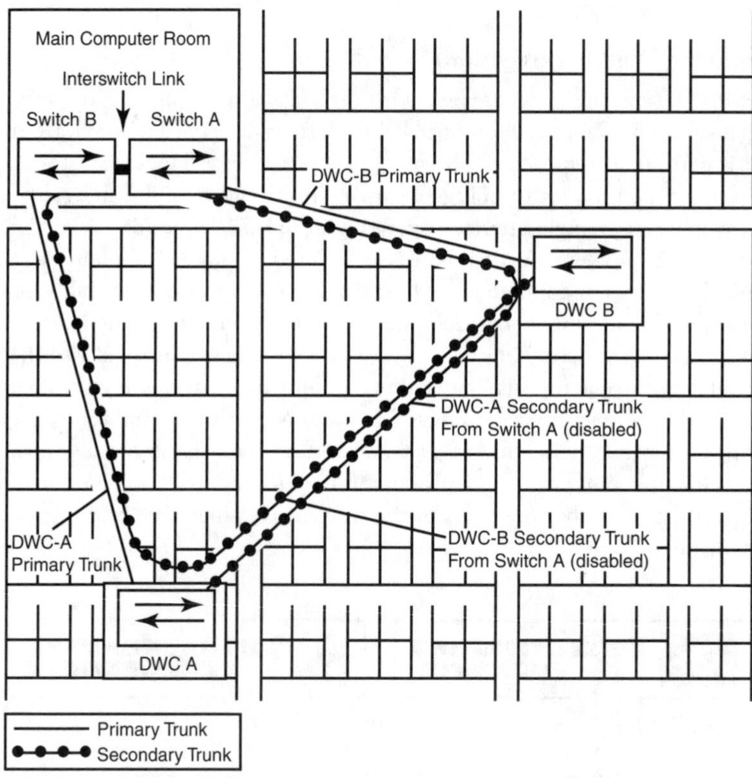

Figure 6.8 A trunked redundant link connection from computer room to wire distribution closet that utilizes a diverse path backbone.

Port trunk interfaces balance the traffic load across all the trunk's member links. If one of the member links fails, the switch redirects the traffic load across the remaining active links. This failure also results in a recalculation of the interface's port cost. After completion, the spanning tree is also recalculated and, if the blocked interface has a better cost, it becomes primary. However, port trunks can readjust the traffic flow to accommodate for lost member links. The need for redundancy needs to be evaluated against the cost (in terms of usefulness) of the idle ports.

Until Gigabit Ethernet became available, port trunking was the only way to provide backbone network links beyond 200Mbps (100Mbps full-duplex Ethernet). At the time of this writing, Gigabit Ethernet is only available over single or multimode fiber optic cable. This allows it to be easily implemented over existing 100BaseFX cabling infrastructures. However, many early nonstandard Gigabit Ethernet implementations only delivered around 400Mbps, and many of the standards-based Gigabit implementations suffer packet loss when utilization reaches around 80 percent. Alternatively, most workgroup-class switches with today's Application-Specific Integrated Circuits (ASICs) can provide "wire rate" full-duplex transmission without any problem. So, right now, port trunking will probably perform better than the current Gigabit offerings. Port trunking also has the added benefit of redundancy, due to the load balancing and recovery capabilities inherent in its functional design.

Port Mirroring

In the beginning of the chapter, a brief overview of packet sniffers was given. Packet sniffers are hardware and/or software tools used to monitor Layer 2 packets as they are transmitted across the shared transmission medium. The operative word here is "shared." With network switches, each port is a collision domain, which makes each port on the switch effectively its own shared network segment (see Figure 6.9).

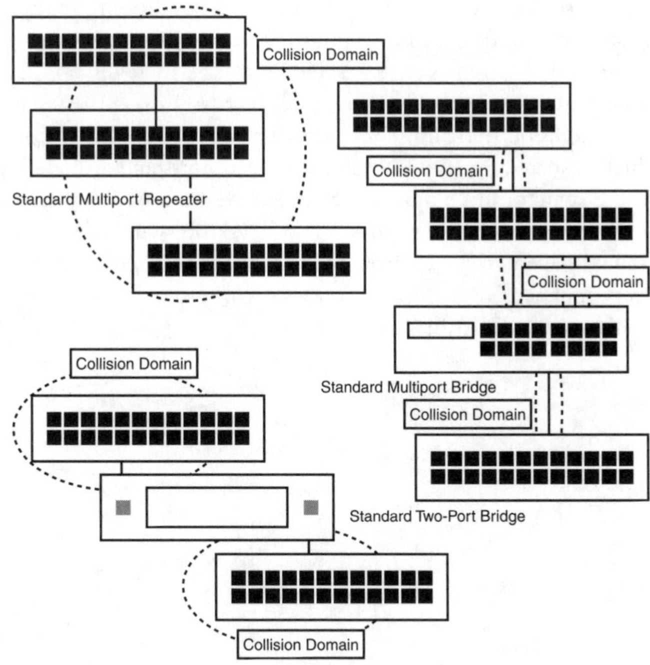

Figure 6.9 Layer 2 collision domains on repeaters, bridges, and switches.

This fact severely limits the usefulness of packet sniffers where looking at the traffic on the network as a whole is concerned. In fact, it is not possible at all. When a packet sniffer is connected to a port on a switch, it will only see the broadcast traffic. If the sniffer is plugged into a multiport repeater that is connected to a switch, its visibility will be limited to the unicast traffic generated by hosts connected to the repeater and network broadcast traffic. This relative immunity to sniffers is a very popular benefit of using switching for locations at which security is a concern. It is also a large headache for network administrators who are used to having the diagnostic capability of packet sniffers.

Realizing this last point, most switch manufacturers provide the capability to configure port mirroring or conversation steering. Port mirroring does just what its name says. It provides the capability to watch the packet exchange between two ports on the switch. When configured, the switch sends copies of all the traffic generated on both of the defined ports. After mirroring is set up, a packet sniffer can be attached to the mirror port, and the packet exchanges between the hosts connected to the mirrored ports can be examined.

Port End-Station Density

The number of end-stations a switch port can support is known as its End-Station Density (ESD) or Port Address Support (PAS). There are two methods of PAS: single port addressing and segment port addressing. Single port addressing switches have limited SAT table size and are designed to support one end-station per port. This is a method common to desktop-class switches. Desktop-class switches are designed to provide the same function as a multiport repeater or MAU except that each port is switched (its own collision domain). These switches often have limited backplane bandwidth which is shared by a several standard speed ports and a single high-speed uplink port so the switch can be uplinked to another switch (see Figure 6.10).

The major advantage to these switches is cost. Desktop switches have no network or configuration management or features like VLAN support. Their appeal is the fact that they are very inexpensive in comparison to a workgroup switch or enterprise switch blade with a similar port density.

Note

When using port mirroring, be sure that your packet sniffer is capable of supporting the speed and configuration (half- or full-duplex) of the port you are monitoring.

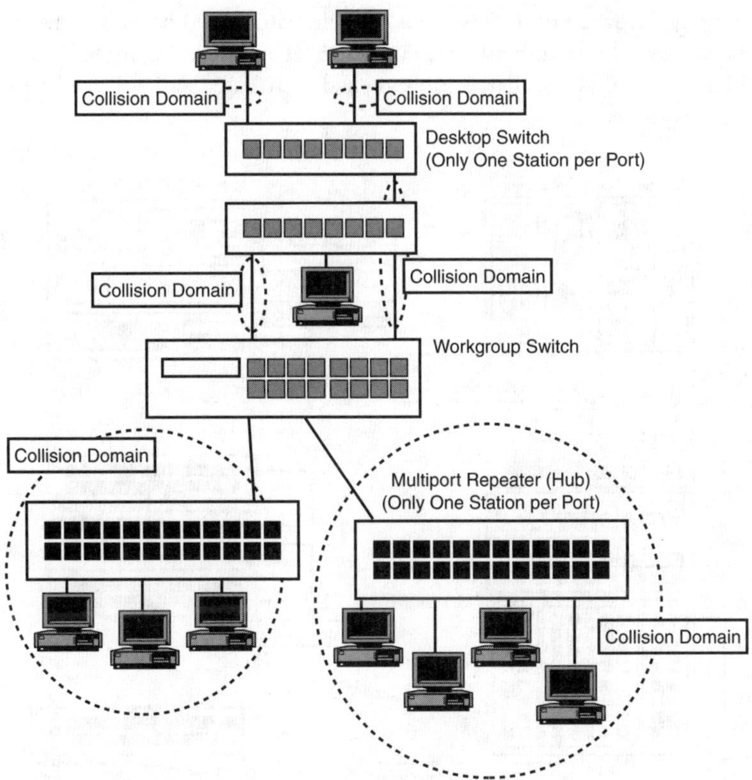

Figure 6.10 Multiport repeaters and desktop switches using a workgroup switch as a network backbone.

Segment port addressing was the PAS methodology originally used with bridges. Therefore, it makes sense that it is the primary PAS method employed by almost all workgroup and enterprise-class switches. Segmented PAS support provides the capability for these switches to associate multiple MAC addresses with a single switch port. Accordingly, although desktop-class switches have limited SAT entry capacity, workgroup and enterprise-class switches can support SAT tables ranging in size between 15,000 to 150,000 entries.

The capability to associate multiple MAC entries to a single switch port enables the switch to perform segment-based switching. This provides a great deal of networking flexibility, because you can attach multiple multiport repeaters or MAUs to a single switch. Segment-based switching is the one of the core functions that make switching so attractive for use in collapsed and distributed network backbone topologies. The backbone can function as a traffic multiplexer, enabling shared media segments to simultaneously exchange data. This is not possible with shared backbones because the backbone can only transmit between two stations at any time. Figure 6.11 illustrates the traffic flow dynamics of shared and switched, distributed and collapsed network backbone topologies.

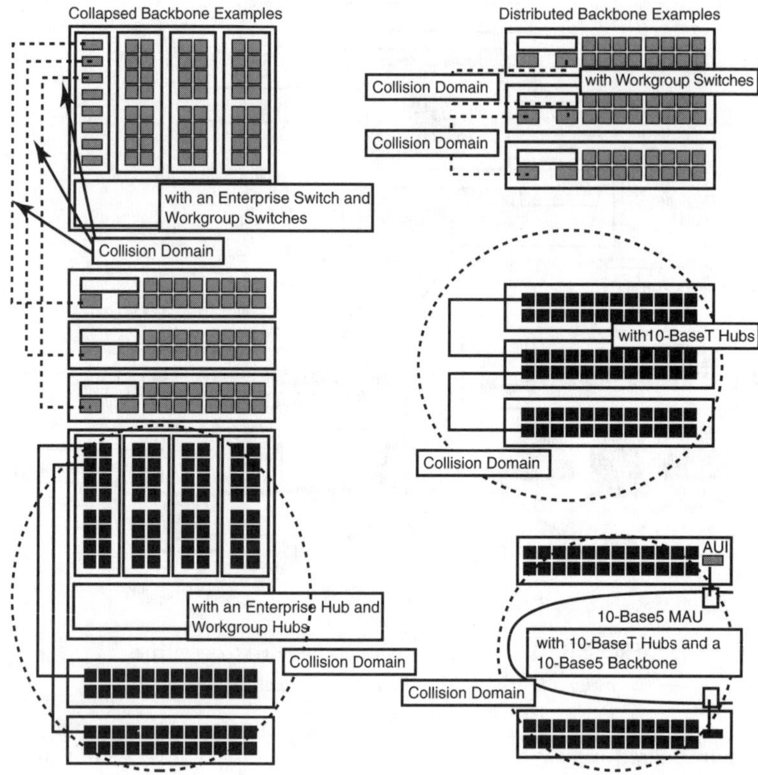

Figure 6.11 Distributed and collapsed network backbone topologies using shared and switched media.

Understanding Switch Backplane Capacity

The switch backplane is the logical fabric or matrix that acts as the interconnect for all the switch's Layer 2 interface ports. Its size or bandwidth capacity determines the throughput, or the amount of frame exchanges that can occur between the interconnected switch ports. Because switches can perform simultaneous frame exchanges (ideally, at the wire rate of the ports) between all the switch's interface ports, the required rate of throughput varies depending on the configuration and type of interfaces used.

Typically, when evaluating switch and router performance, manufacturers use the smallest legal frame size (Layer 2) to measure the frame per second (FPS) throughput rating to delineate equipment performance. The FPS rating is usually specified as an interface-to-interface rate, or as the total throughput capacity for the router or switch. It is measured by streaming frames through an interface pair and counting how many frames were forwarded through in one second. Based on this value, the device's throughput is then calculated.

In many instances, the term "packet" is used interchangeably with the term "frame," in which case, this measurement is expressed as packets per second (PPS). However, the two terms technically refer to different data transport structures. The term "frame" is commonly used to describe a structured collection of ordered bits used by a Layer 2 data transmission protocol to carry data. A frame has a base minimum size that includes header, error checking, and delineation information, with a variable data portion. The term "packet" refers to the link control information, Layer 3/Layer 4 header information, and the actual data. Layer 3 packets typically range from 4K to 64K in size. These packets are transported in Layer 2 frames. You should recall that when a packet exceeds the payload of a frame, the packet is fragmented into sections and transported in multiple frames to the packets destination. After all the frames have been delivered, the packet is reassembled and handed to the ULP.

When you are looking at PPS or FPS measurements provided in technical specifications disclosed in equipment documentation or product literature, be sure you know what packet/frame is being used to calculate the measurement. It is common today for vendors to use PPS to describe FPS (Layer 2) data transmission rates.

When calculating the FPS measurement, the entire length of the data-link transport frame must be used to get an accurate measurement. It is important to keep this in mind when measuring Ethernet performance because the frame preamble (the first 8 bytes) and the Inter-Frame Gap (IFG), which adds an additional 12 bytes to the frame, are not included as part of the minimum and maximum frame size. When calculating the PPS measurement, only the frame's data/information field is used. Needless to say, these two values will be different from one another.

PPS measurements describe the transmission and volume of the actual data being carried. FPS measurements describe the Layer 2 segment's transmission, throughput and, by extension, the segment's utilization performance. To calculate a throughput performance based on frame size (in bits), use the following formula:

$$\text{TFSB} = \text{TFSb} \times \text{FPS} = \text{T/S}$$

TFSB is the total frame size in bytes. TFSb is the total frame size in bits, which is multiplied by the FPS frame per rate. The result is the throughput per second rate (T/S). To derive the actual network utilization (based on the frame size), calculate the maximum FPS rate of the frame and compare the two results.

> **Note**
>
> nsec = 1×10^{-9} or .000000001
>
> usec = 1×10^{-6} or .000001
>
> msec = 1×10^{-3} or .0001
>
> sec = 1×10^{1} or 1

To calculate the FPS rate for a given frame size and trasmission rate, use the following calculations:

Step 1. Take the total frame size in bytes (TFSB) and multiply it by 8. That converts it to bits (TFSb). (For demonstration purposes, we are using a 64-byte frame on a 10Mbps Ethernet segment. Because Ethernet has an 8-byte preamble and a 12-byte IFG, we need to add an additional 20 bytes to the 64-byte frame.)

TFSB = 64 bytes + 8 bytes (PRE) + 12 bytes (IFG) = 84 bytes

TFSb = TFSB × 8 = 672 bits

Step 2. Multiply the TPSb by the appropriate bit time (bt) to get the frame transmission time (FTt). The bt value is the transmission time per bit, and PTt is the transmission time of a single packet. The following lists the bit times for the various Layer 2 LAN protocol transmission speeds:

1Mbps bt = 1 usec

4Mbps bt = 250 nsec

10Mbps bt = 100 nsec

16Mbps bt = 63.3 nsec

100Mbps bt = 10 nsec

1Gbps bt = 1 nsec

To calculate the frame transmission time, use the following calculation:

FTt = TFSb × bt = 672 bits × 100 ns = 67.200 nsec = 67.2 usec

Step 3. Now that we know how much time it takes to transmit a single frame (FTt), we can calculate how many frames can be transmitted in one second (FPS) by using this formula:

FPS = 1 sec ÷ FTt = 1 sec ÷ 67.2 usec = 14,880 FPS

The goal of any network transmission device is to be capable of transmitting at *wire rate*. Wire rate is the transmission of the smallest legal-sized data frame at the fast constant rate of transmission supported by the medium and transmission protocol, without adding any additional processing latency. Of course, the legal frame sizes vary depending on the Layer 2 protocol being used. For Ethernet, the minimum frame size is 72 bytes (64 bytes + 8-byte preamble), with a maximum of 1,526 (1,518 + 8-byte preamble) or 1,530 with 802.3ac due to the constraints of CSMA/CD. With Token Ring (which does not require collision detection), no minimum or maximum frame size is defined. Instead, the minimum frame size is calculated using the frame format data and LLC information, which is 43 bytes. The maximum frame size is dictated by the maximum amount of time a station can hold the token. So, on a 4Mbps ring, the maximum packet size is 4,472 bytes; on a 16Mbps ring, it is 17,800 bytes. FDDI, like Ethernet, supports a minimum and maximum frame size: 32-byte minimum (28 frame + 4 byte LLC-PDU, no data) and 4,500-byte maximum.

The reason the smallest legal frame size (64 bytes, in the case of Ethernet) expresses the FPS rating for port forwarding is because smaller frames are harder to process than larger ones. When looking at throughput utilization, the FPS rating is inversely proportional to the frame length. Mathematically speaking, you can move more small frames than large ones in a finite period.

Ideally, the goal of all switch and router designers is to design a product that will have no negative impact on the data transmission rate; this, however, is just not possible. All routers and switches require a certain amount of time to perform necessary functions. The period of time taken to perform these functions is known as latency.

To achieve "no latency" or "wire rate" performance, a device must be able to perform its frame processing (frame interrogation, such as CRC check, SAT lookup, and so on) during the time it takes to receive another packet. This processing on a 10Mbps store-and-forward Ethernet switch, for example, using a 64-byte frame, can be no greater than 67.2 usec, compared to a 1,518-byte frame that affords up to 1.2 milliseconds.

It should be noted that the smallest frame size is also the most inefficient. A smaller frame rate translates into a smaller data frame, resulting in a larger percentage of available bandwidth used to transmit framing information instead of actual data. Hence, larger frames, with their lower maximum FPS rates, use the available bandwidth more efficiently.

To give you some perspective on the number of frames a switch needs to process in order to achieve wire rate performance, and its relation to the switch's backplane capacity, let's consider the maximum frames per second (or packets per second) rate of a single Ethernet interface, which is 14,880 FPS (148,810 FPS for Fast Ethernet). In order for a switch with 24 Fast Ethernet ports to perform frame exchanges between an interface's ports at the maximum FPS, the switch would require a backplane capable of processing a minimum of 1,785,720 FPS. This would enable 12 simultaneous frame exchanges (one port sends and one port receives) to occur at a rate of 148,810 FPS. However, this backplane rate is quite restrictive, because it only permits all the ports to operate in half-duplex mode. If full-duplex operation between all the switch ports is required, the effective bandwidth requirements double, requiring a switch backplane capable of transferring 3,571,440 FPS.

Switch manufacturers usually state backplane capacity in terms of gigabits. So, in the previous example, the switch would need minimum backplane capacity of approximately 1.2 gigabits for half-duplex operation and 2.4 gigabits for full duplex and adequate port flooding support. If the switch does not have a backplane adequate enough to support the aggregate switch port throughput, it goes into what is known as a blocking state and effectively drops the frames. To calculate the required bandwidth capacity needed for your switch configuration, add the configured port speeds of the switch. If the number exceeds your switch's rated capacity, reconfigure the switch to avoid blocking.

Ideally, a switch should have enough backplane capacity to support full-duplex operation between all its ports. This is generally the case with most workgroup and enterprise-class switches on the market today. When LAN switches first came to market, they did not have adequate backplane (or processing capability) support to provide wire rate connections between all the switches ports simultaneously. This is also the case with many of today's inexpensive desktop-class switches. To accommodate data transmission in these cases, port blocking is employed when the switch's backplane becomes overloaded, to minimize frame loss. This problem also exists with many Gigabit Ethernet switch and uplink port implementations. A single Gigabit connection can transfer a maximum of 1,481,000 FPS, which is more than the minimum required backplane capacity for half-duplex operation of a 24-port Fast Ethernet switch. Implementations are getting faster, however, and with time wire rate Gigabit Ethernet will be available.

Caution should be taken when using vendor performance calculations to determine how to provision the ports on your switch. Rarely do switches perform at the rates their manufacturers claim. To be safe, use an adjusted backplane throughput rate. Assume that the switch will perform at 80 percent of its stated rate and then configure port speeds so utilization will not exceed this number.

Switch Forwarding Methodologies

Whereas the backplane capacity determines the switch's performance in terms of how much data will fit through the switch, the forwarding methodology dictates how long it takes to process each packet (the latency). It is the combination of these two elements that determines the switch's FPS rate. There are three types of frame forwarding methods employed by switch designers: *cut-through*, *store-and-forward*, and *hybrid* or *modified cut-through*. Most manufacturers of workgroup and enterprise LAN switches support all three forwarding methods either through manual configuration or automatically, based on traffic flow and traffic quality. However, there is a core of manufacturers that only build store-and-forward switches because this functionality is required for their proprietary enhancements.

Cut-through uses the same technology as a standard transparent bridge. The switch only looks at the frame up to the destination MAC address, which results in a very low latency. Total latency consists of the time it takes to examine the frame enough to get the destination address, perform the SAT lookup and whatever interframe spacing exists between the incoming frames. Cut-through switches are limited to switching between similar LAN media types (Ethernet to Ethernet, FDDI to FDDI, and so on) because the frame is merely looked at, then forwarded. Cut-through switches also filter frames like transparent bridges, so traffic that has destination addresses associated with the port on which the traffic was received is discarded. The downside to cut-through switching is that no error checking is performed, so misformed frames and collision fragments get forwarded on the same as "normal" frames.

The *store-and-forward* method is a more complicated and process-intensive forwarding scheme, and it consequently has a higher latency. A cut-through switch only looks at the frame up to the destination address, but a store-and-forward switch uses a four-step process to handle packets:

1. The switch reads the destination address of the frame.
2. The switch stores the frame in the port's frame buffer.
3. The address is compared against the SAT table so the destination port can be located.
4. Before the switch forwards the frame, it performs a Cyclic Redundancy Check (CRC) on the frame. If the frame is good, it is sent to the destination port; if the frame fails, it is dropped.

Although the same frame handling rules used by a cut-through are followed, the storing of the frame provides the facilities for greater control of frame flow. Because the frames are stored, the data within them can now be examined, and the switch can be directed to handle certain types of data one way and other types another. Misformed frames, runts (frames smaller than the minimum frame size), jabbers (frames that exceed the maximum frame size), and collision fragments are all discarded by a store-and-forward type switch. After the packet is filtered and deemed valid, it can then be forwarded on to its destination. Where the packet processing latency on a cut-through switch is defined largely on the time it takes to perform its SAT lookup, the frame processing latency on a store-and-forward switch varies depending on the size of the packet. Commonly, a store-and-forward switch's latency rating is measured using a 64Kbps packet size. Store-and-forward switching is the basis for many of the proprietary virtual LAN (VLAN) technologies. The store-and-forward method is also required to provide any switching between different Layer 2 network protocols (such as FDDI or Ethernet) on the same switch fabric. This enables the switch to effectively behave like a translating bridge.

Hybrid switches are, as the name implies, switches that use a combination of cut-through or store-and-forward methods. Hybrid switches are also known as modified cut-through or adaptive switches. Hybrid switches, by default, start in a modified cut-through mode. During the SAT lookup phase, the first 64 bytes of the frame are examined by the switch before forwarding the frame. This allows the switch to identify any runts or collision fragments and discard them. In the event of numerous collisions (Ethernet) or high frame error rate (Token Ring and FDDI), the switch will change its frame processing method from cut-through to store-and-forward, which will perform its own CRC checks of the frame in the buffer before sending it on. When the frame error rates are reduced, the switch will go back to cut-through mode and begin the monitoring process again. The frame processing latency of hybrid switches depends on the operating mode of the switch. In modified cut-through mode, the latency is at a fixed rate. In store-and-forward mode, the latency is variable, again depending on the size of the packet. The advantage of the hybrid switch is that it provides the data integrity checking of store-and-forward switching (if needed) and the reduced latency (more efficient packet processing) advantage of cut-through switching.

Source routing is also available on Token Ring switches as a forwarding method. In terms of operation and configuration, the same rule that applies to source route bridges applies when source routing is used with switches. The switch must be configured with proper LAN and port IDs and must be able to understand the RIF field contents.

Layer 2 LAN Switching Services for Layer 3

Along with significant performance improvements, Layer 2 switching has also provided additional functions to enhance the deployment and performance of Layer 3 protocols. Three of these functions are Virtual LANs (VLANs), Layer 2 traffic prioritization, and Layer 3 switching. These functions, like port trunking and full-duplex port operation, are made possible by adding additional logic features to the switching logic's ASIC chip.

VLANs

A switch that supports VLAN provides the capability to partition the switch's ports into artificial Layer 2 broadcast domains. A VLAN behaves exactly like a physical Layer 2 broadcast domain. Unicast, broadcast, and multicast traffic sent into the VLAN are only visible to ports and the hosts connected to those ports that are designated for that VLAN. Also, as with traditional Layer 2 broadcast domains, it is possible to bridge between VLANs.

Initially, VLANs were introduced on a proprietary basis by several different switch vendors. The technology was seen as a way of maximizing the usefulness of the switch's port density by enabling it to support multiple networks on the same switch in a secure manner. That capability was welcomed with open arms by many network administrators because early switch products were expensive and VLANs made it possible to provide switching to different Layer 3 segments with a single Layer 2 device. In terms of host management, VLANs promised tremendous flexibility when a workstation needed to be moved from one network to another network (utilizing different Layer 3 addressing), instead of "repatching" in the wiring closet. Using VLANs, the workstation's port could just be reassigned to the VLAN corresponding to the new network. Along these same lines, VLANs made it possible for workstations in different geographical locations within the enterprise to be on the same local network segment. VLANs also provided mechanisms to control broadcast and multicast traffic, which can be a limiting factor in growing switched LANs. These mechanisms made it possible to utilize high-speed switch ports as a packet-switched backbone link that shared the port's available bandwidth between different VLANs.

Of course, like many new technologies, VLANs had a rough start. Various processing chores needed to maintain the VLANs in the initial implementations sacrificed most, if not all, of the performance gained from using switches. This problem, however, was overcome with the development of faster ASICs. Another rather cumbersome problem was managing VLAN construction and membership; although this problem has been somewhat addressed, it is not yet fixed.

Port-Based VLANs

The simplest and most common method of addressing the problem is with port-based VLANs. With port-based VLANs, the switch's ports are grouped into VLAN groups consisting of two or more ports. All the end-stations, multiport repeaters, and router interfaces connected to the port group are members of the same broadcast domain. Figure 6.12 is an example of a switch configured with two port-based VLAN devices: ports 1 through 6 are VLAN 1, and 7 through 12 are VLAN 2. End-stations connected to these ports only see Layer 3 broadcast traffic that is specific to their VLAN port group. In order for the two VLANs to exchange traffic, a router is needed.

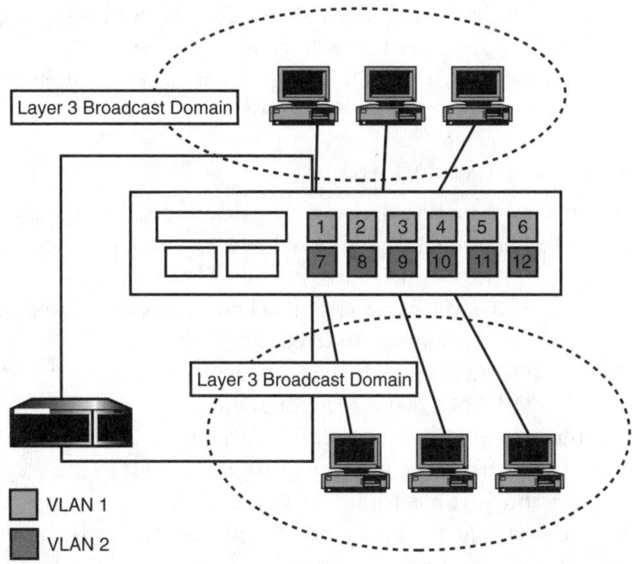

Figure 6.12 A switched backbone topology utilizing VLANs for Layer 3 segmentation.

The downside to port-based VLANs is mostly housekeeping chores. Port configurations must be manually tracked and labeled. There is also the added hassle of having to change the port allocations when networks expand and contract, or as end-stations are moved between VLANs. Special care must also be used when establishing links between switches. With most implementations, VLANs (by default) do not span across multiple switches automatically; they need to be configured on each switch accordingly.

MAC Address-Based VLANs

Another common VLAN construction method is MAC address-based VLANs. MAC-based VLANs determine the end-station's network access based on the station's individual MAC address. This method works well because (in most cases) every network device has its own universally unique MAC address. The MAC address VLAN approach is popular with enterprise-class network switches. A central configuration server is connected to the network, and all the enterprise switches communicate with it to learn VLAN configuration information. Then, when an end-station is connected to a port, the switch learns its address, checks with the server, and learns with which stations the newly connected station can communicate. The great thing about MAC-based VLANs is that they are very secure and you do not have to reconfigure the switch when end-stations are moved. The downside is that every node needs to have its MAC address entered into the VLAN configuration database. Otherwise, the station cannot communicate with anything on the network.

Upper Layer Protocol-Based VLANs

ULP-based VLANs use a common network layer delineation, such as protocol type, network or subnetwork address, or multicast address, to construct and manage membership. Policy-based VLANs, another form of Layer 3-based VLANs, decide membership using a set of Layer 2 and Layer 3 criteria. These approaches have some particular advantages if you want to partition off different network protocols to specific switch ports, or if you are using multicast address-aware applications, such as MBONE. They are implemented using VLAN matching or condition statements, which are similar in form and function to router ACLs, (covered in Chapter 10, "Configuring IP Routing Protocols on Cisco Routers"). The switch interrogates the incoming packets on its ports and looks for traffic that meets the VLAN qualification criteria. If traffic matching the VLAN criteria is seen, the port (or MAC address) is associated with the VLAN. After VLAN membership is established, the switch includes (that is, forwards) the newly assigned port in any unicast, broadcast, or multicast traffic distribution associated with the newly assigned VLAN.

Mind you, this is not Layer 3 routing; rather, the switch uses a distinct value in the Layer 2 frame, Layer 3 datagram, or Layer 4 message as a basis for adding a port (and the systems attached to it) to a specific VLAN.

ULP-based VLANs are a comparatively new and useful class of VLAN. Depending on VLAN implementation, VLAN membership can be based on IP, IPX, AppleTalk, and other popular LAN protocols. However, Layer 3-based VLANs operate at a slower processing rate. Examining each packet well into the datagram portion of the packet is time consuming in comparison to looking at a MAC address. Multicast VLANs consume even more switch resources than basic Layer 3 switch VLAN variations. This is because the switch not only needs to examine the Layer 3 portion of the packet, it also needs to maintain membership lists of the hosts involved in the multicast groups, along with maintaining distinct VLANs for each multicast group.

VLAN Configuration and Implementation

Regardless of the VLAN membership methodology, VLAN implementations represent an additional level of network complexity beyond the traditional plug-and-play "one network, one wire" approach. When VLANs first appeared, they were limited to a standalone switch, which meant that VLANs in large networks were really only useful for creating broadcast domain port groups, which could be connected to another switch or router that was part of the same Layer 3 network. Figure 6.13 provides a simple example.

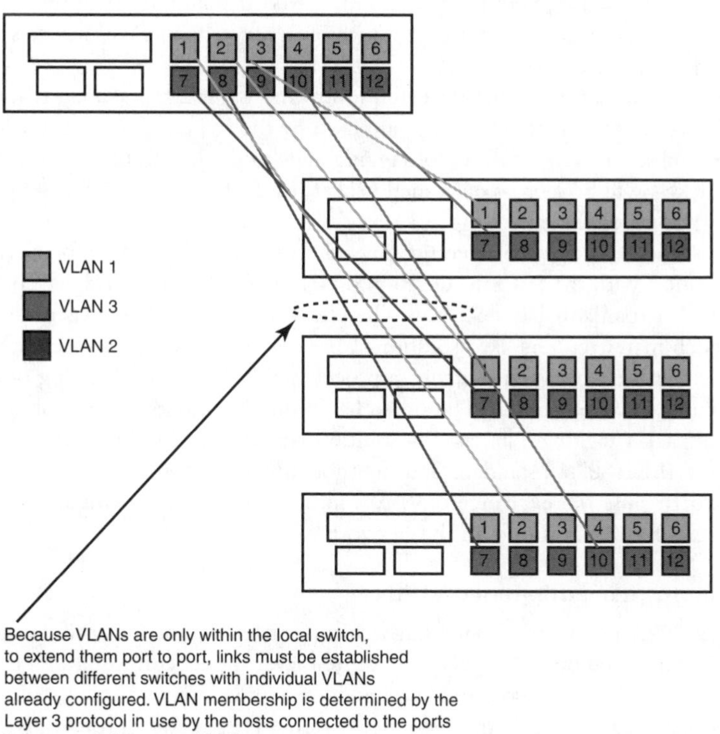

VLAN 1
VLAN 3
VLAN 2
Because VLANs are only within the local switch, to extend them port to port, links must be established between different switches with individual VLANs already configured. VLAN membership is determined by the Layer 3 protocol in use by the hosts connected to the ports that are part of the same VLAN.

Figure 6.13 Switches running independent VLANs connected with interconnection links.

When MAC and Layer 3-based VLANs became available, however, a means to distribute VLAN information between switches became necessary. Cabletron Systems' SecureFast and Cisco Systems' InterSwitch Link (ISL) were both very popular, but proprietary, and hence incompatible distributed VLAN solutions. Therefore, it was possible to deploy enterprise VLANs, provided you used switches that were all from the same switch vendor. The major source of incompatibility with SecureFast and ISL (and other proprietary VLAN distribution schemes) was the method they used to establish and describe VLAN membership, which was accomplished through the use of some form of packet (or frame) tagging.

There are two forms of packet tagging: implicit and explicit. Implicit packet tagging uses an existing packet attribute to associate a packet with a specific VLAN. A ULP-based VLAN is an example of using implicit tagging for VLAN identification. Although implicit frame tagging is easy to implement on a standalone switch to establish VLAN port membership, there is no easy way to ensure that an implicitly tagged frame will be forwarded correctly when it needs to leave the switch for delivery. To ensure proper delivery of implicitly tagged frame to the correct VLAN when it reaches a remote switch, some type of centralized configuration data must be used to determine a frame's VLAN assignment. For this reason, VLAN mechanisms utilize implicit tagging operate in conjunction with a VLAN configuration server. Switches installed on the network exchange configuration and VLAN mapping information. Although using a central server makes configuration easier, a server failure can have a severe impact on the network's operation.

With explicit packet tagging, an extension to the Layer 2 frame a tag is added to a packet as it enters the switch. The tags are used by the switches in the LAN to indicate VLAN membership. However, use of tagging adds to the length of the packet, so tagged packets will be seen as deformed packets by switches that do not support tagging and will be discarded.

With a marketplace full of incompatible solutions, a community effort was taken up in cooperation with the IEEE to develop a standard for "Virtual Bridged Local Area Networks," or standards-based VLANs. The effort began in 1996 between the major LAN switch manufacturers. Two solutions were considered. One was the exchange of VLAN information between switches, a process similar to the exchanging of dynamic routing tables between routers. The other, based on Cisco Systems' ISL solution, was the use of packet tagging. The tagging solution was selected by the IEEE to be the basis of the IEEE 802.1q standard. This method, incidentally, is also the more efficient of the two, because the exchange of VLAN information between switches would only increase the amount of traffic on the network.

IEEE 802.1q Standards-Based VLANs

The IEEE 802.1q standard provides an architectural model for the configuration, distribution, and operation of VLANs. It is a three-layer model consisting of a configuration layer, a distribution/resolution layer, and a mapping layer.

The *configuration layer* uses the Generic Attribute Registration Protocol (GARP) and Generic VLAN Registration Protocol (GVRP) to provide a signaling mechanism to exchange VLAN membership information between switches and other 802.1q-aware devices.

The distribution/resolution layer defines the 802.1q packet ingress and egress rules, which handle packet VLAN ID assignment and packet delivery. Under 802.1q, a packet's VLAN assignment is made based on the VLAN Identifier (VID) of the inbound port on which the packet is received. After the packet is associated with a VLAN, the switch (or network of switches) delivers the packet only to the ports associated with the packet's VLAN designation.

802.1q VLAN port membership is managed either through GVRP messages sent over interswitch links or through manual configuration. Regardless of the configuration method, it is essential that a single, unique VID is used to identify each VLAN.

802.1q VLANs handle port identification and packet designation using both implicit and explicit packet tagging. When a switch port is designated for VLAN membership, part of the configuration is its membership type assignment. The port can be assigned to be a tagged or untagged VLAN member. Typically, ports used for end-station terminations and interconnect links with non-802.1q-aware devices are defined as untagged members. Ports assigned to 802.1q-aware devices and 802.1q interswitch links are defined as tagged members. Through the use of explicit packet tagging, 802.1q maintains and distributes VLAN designation information between interswitch links that might have membership to multiple VLANs.

Under the 802.1 ingress rules, the packet is associated with the VLAN designated to the inbound port. After the egress port has been determined by the switch, the packet is either tagged or untagged, depending on the egress port's VLAN membership assignment. If the egress port is designated a tagged port, an explicit packet tag is added to the frame, the CRC is recalculated, and the frame is forwarded. If the egress port is an untagged member of the VLAN, no tag is added. If a tag exists, it is removed, and CRC is recalculated before it is forwarded out of the switch. Devices that are not 802.1q-aware do not understand the 802.1q tag, so they consider the frame corrupt and discard it. For this reason, it is important that, when you configure your 802.1q VLAN port assignments, you correctly designate the port as a tagged or untagged member.

The mapping layer provides the mechanisms for associating packets with a specific VLAN through the use of explicit and implicit packet tags. 802.1q implicit tags can be based on port, source, destination addresses, and so on. Whatever the implicit tag basis is, it must be associated with a unique VID. VID is used by the switch to determine the packet's VLAN membership. To implement explicit tagging, 802.1q defines a 4-byte field, which is appended to the packet after the source address field. The 802.1q tag is used by Ethernet, Token Ring, and FDDI. Because these three Layer 2 protocols differ in terms of format, the tag format varies between the three protocols, but contains the same core information, mainly the VID and the User Priority Field (UPF). The 802.1q tag consists of four elements that make up the VLAN tag field:

- Tag protocol identifier—Two bytes long, it is used by Token Ring, FDDI, and SNAP transmissions.
- User Priority Field (UPF)—Used in conjunction with the 802.1p traffic prioritization standard, this field provides ATM-like traffic classification services to Layer 2 LAN protocols.

- Canonical format indicator—A single-bit flag that is used by Token Ring. The flag specifies if the frame is actually a Token Ring frame or a Token Ring frame encapsulated in an Ethernet frame format.
- VLAN ID (VID)—Used by all the switches to identify VLAN membership.

Most switches shipping today that support VLANs support the IEEE 802.1q standard. Although 802.1q interpretability is generally assured between different vendors, a common enterprise configuration interface is still lacking. There are, however, efforts by most of the network management tool developers and some of the switch manufacturers to create common configuration tools, so keep an eye out.

VLAN Benefits

VLANs add an additional layer of complexity and provide a variety of benefits, particularly for enterprise networks. The most significant benefits are better overall Layer 2 bandwidth efficiency and easy Layer 3 network partitioning and expansion.

Because VLANs are artificially imposed Layer 2 broadcast domains, it is possible to partition networks into smaller Layer 3 network subnets without the additional complexity and cost of maintaining separate Layer 2 infrastructures to operate multiple discrete Layer 3 segments. By using VLANs, you can reduce Layer 2 collision domains and token contention problems, as well as control the extent of Layer 3 broadcast traffic. When it exceeds the normal operating percentage of 8 to 10 percent of Layer 3 segment's operating bandwidth, this kind of traffic impacts network performance and increases the chance of Layer 3 broadcast storms. VLANs can also reduce unneeded router processing overhead by eliminating the need to have geographically distant end-stations (with common resource access requirements) homed on different Layer 3 segments. Instead, dynamic VLANs can be created to have end-stations and servers in different locations function as if they were on the same local segment.

> **Note**
>
> Remember, when configuring 802.1q switches, each switch involved within a VLAN group must be configured using the same VID designation. In other words, if you want a port on switch B to belong to the same VLAN as another port on switch A, you need to make the VID for both of those ports identical between the two switches.

> **Note**
>
> Broadcast storms occur when the network is too large or when defective hardware or network configurations exist. Network broadcasts are essential for any network to operate, and when the network is of a reasonable size, broadcasts should not result in any significant impact on network performance. As Layer 3 networks increase in size, however, the broadcast domains (the possible number of hosts that can send or receive broadcast traffic) also increase. This results in higher network bandwidth utilization rates for broadcast traffic and less for "real" network traffic. Broadcast bandwidth utilization rates up to 25 percent are tolerable, but not optimal. When Layer 3 networks grow too large, broadcast storms can be triggered by minor network events and cause excessive broadcast traffic between hosts on top of an already high network broadcast rate. Broadcast storms will also occur regardless of the underlying Layer 3 broadcast when an end-station NIC, switch, or bridge port fails and randomly generates broadcast traffic, or if two or more bridge ports exist between two Layer 2 network segments and spanning tree is disabled.

In addition to better overall network performance, large-scale distributed VLANs effectively remove expansion limitations caused by physical cabling and equipment placement. VLANs not only provide the capability to run multiple Layer 3 network segments over the same switch, but also over the same high- and low-speed network links. For example, using a single or trunked group of Fast or Gigabit Ethernet links, it is possible to have all your Layer 3 segments operate over common network backbone links (see Figure 6.14).

When the need for additional segmentation or a new Layer 3 network arises, an additional VLAN can be added to the switched backbone. This capability also makes it possible for a LAN to operate different Layer 3 protocols in a segmented fashion. These switch ports can then be provisioned to only provide the Layer 3 network access the hosts need. This method not only allocates the Layer 2 bandwidth in a more efficient manner, but also enhances network security by allowing hosts to see only the Layer 3 traffic they need to see.

VLAN Usefulness

Although proprietary (and now standards-based) VLAN capability has been part of many switch manufacturers' feature sets for years, the implementation of VLANs has not become as prevalent as the implementing of "plain vanilla" Layer 2 switching. There are three main reasons for this.

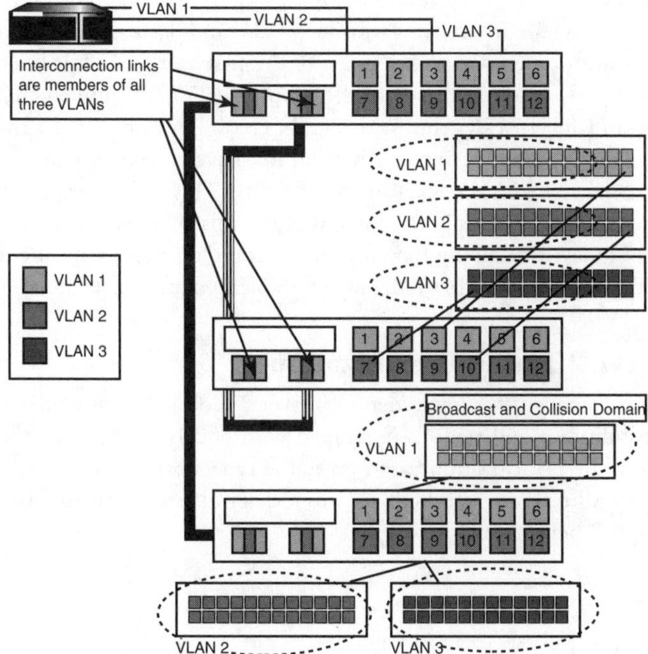

Figure 6.14 Multiple VLANs over common switched backbone.

The first reason is the complexity of using distributed VLANs. Deploying VLANs properly requires planning and, in many cases, major network redesign. When switching hit the marketplace, many manufacturers pushed Layer 2 switches as the panacea for large, congested, Layer 3 segmented networks. "Get rid of your routers, flatten your network," was their sales pitch. What most customers missed was that in order to correctly implement this idea, they needed to replace their shared networks with switched networks. Well, an entire switched network is quite expensive, but the idea sounded good. So, they bought some switches, attached multiport repeaters and MAUs to them, and then flattened their networks. For many, this resulted in networks with excessively large broadcast domains, with 40 percent of the network traffic being utilized for network broadcasts. So, when the same vendors came around a few years later and started talking about reducing broadcast domains and using multiport routers to manage multiple "virtual" Layer 3 networks, they got their hats handed to them.

The second reason VLANs are not as prevalent is that standards migration only affected those who really kept up on the evolving technology. Until the IEEE 802.1q standard was released, network administrators were forced to use only a single vendor's switching implementation, for compatibility reasons. Large-scale enterprise network administrators tend to keep away from non-standards-based solutions unless they have no other choice. Therefore, with no standards, no cross-vendor compatibility, and no real enhancement to the network that could not be achieved by other means, VLANs did not catch on with the individuals who could have really used them. Now, with standards-based VLANs available, utilization is starting to become more common, particularly in high bandwidth/speed data farms in WWW and database application services.

The third reason is perhaps an extension of the first and second. It has to do with the poor initial performance of most of the proprietary VLAN implementations. When the first store-and-forward switches with VLAN capability came on the market, many of them had the software functionality, but did not have switching engines powerful enough to process the VLAN packet flows efficiently. This lackluster performance early in the development of VLANs burned many of the initial buyers of the technology. The real irony today is watching the same issues come up again with the development of Gigabit Ethernet. The idea is solid, but the implementation falls short.

802.1p Layer 2 Traffic Prioritization

802.1p defines a standard for implementing Layer 2 CoS traffic prioritization. It provides the capability for end-stations (equipped with 802.1p-compatible NICs) and servers to set traffic prioritization based on traffic type, port, or VLAN. 802.1p provides eight different traffic classes (see Table 6.1) to specify different prioritization levels: 0 through 7, with 7 being the highest.

Table 6.1 **The IEEE 802.1p Traffic Class Recommendations**

User Priority Bits	Traffic Priority Class
111	Network Critical (7)
110	Interactive Voice (6)
101	Interactive Multimedia (5)
100	Streaming Multimedia (4)
011	Business Critical (3)
010	Standard (2)
001	Background (1)
000	Best Effort (0)

The 802.1p standard specifies a 3-bit header to indicate the packet's priority class. This header is part of the 802.1q 32-bit VLAN tag in the form of the user priority field. So, although 802.1p and 802.1q are defined as separate standards, this tandem relationship effectively makes 802.1p and 802.1q almost mutually exclusive to one another in terms of implementation. In order for the incoming packets to be placed in the appropriate delivery queue, both the end-station and the switch need to be able to discern different types of traffic when transmitting and receiving. Without the use of network-aware applications, traffic classification needs to occur by some other means at the network layer. This problem sends us back to 802.1q, which is well suited to perform the task of traffic identification and segmentation. 802.1q also has the added capability to distribute both priority definitions to other switches (through GARP). As a result, most switch vendors combine their 802.1p and 802.1q implementations, which results in them commonly being characterized together as 802.1q/p—leaving some networkers with the impression that CoS is just an add-on to 802.1q.

Architecturally, 802.1p is implemented as part of the switching logic (on the ASIC). When 802.1p is implemented, the switch partitions its port buffer memory into different forwarding queues. As prioritized traffic flows inbound, it is routed to the appropriate queue and forwarded based on priority. The idea of providing Layer 2 is not new. The 100VG-AnyLan (IEEE 802.12) with its demand-priority media-access protocol and Intel's High Priority Queue (HPQ) are other examples of Layer 2 CoS implementations. (Traffic prioritization does not go beyond the router, if you need to implement some type of Layer 3 Quality of Service [QoS] capability.) In both of these cases, new hardware and software were required to implement CoS. In some sense, 802.1p is no different, as it too requires that both the switches and the end-station's NICs be 802.1p-aware. This means that 802.1p might have limited value (you can implement CoS prioritizations based on VLAN traffic over ISLs) unless you have end-stations that can support 802.1p. It is not backward compatible. Of course, you might not require CoS, in which case, you should at least keep it in mind and look for it as a supported feature in your NIC and switch hardware selections.

With 802.1p traffic prioritization service now a standard, over time, it will become an important feature for NIC's workgroup and enterprise switches. Because it provides transport access for the latency-affected, convergence technologies (integrated voice, video, and data over the same wire) are starting to come out of Cisco Systems, Lucent Technologies, and Nortel (Northern Telecom).

Layer 3 Switching

Like 802.1p, Layer 3 switching promises, over time, to become a significant technology. Layer 3 switches, routing switches, and switch routers are different names for the same thing: hardware-based—specifically, ASIC-based—routing. The idea behind hardware-based routers is essentially to "route" Layer 3 datagrams at wire speed by developing a hardware chip to perform the same functionality that is achieved with software-based traditional routers. At the present time, several vendors are developing and marketing Layer 3 switches. Some of the more notable solutions are Cabletron Systems' SmartSwitch routers, Foundry Networks BIG Iron and ServerIron switches, and PacketEngines PowerRail routing switch. Cisco Systems, the undeniable king of software-based routers, is also marketing and developing ASIC-based switch router products in the form of their Routing Switch Modules (RSMs) on their Catalyst 4000, 5000, 6000, and 8000 series switches. However, Cisco's and all the other "switch router" vendors still have a ways to go before software routers become "a thing of the past."

This is largely due to the limitations intrinsic to hardware-based anything; there is only so much that can fit into a chip. So, although advances have been made in Layer 2 switching ASICs, the technology to develop a complete multiprotocol router chipset is still not available. Today, most Layer 3 switch routers offer IP and IPX, and some core filtering and packet processing functions. AppleTalk and the older legacy Layer 3 protocols have yet to make it into a production switch router (as of this writing). The switch routers in the marketplace today provide enough functionality to route IP/IPX network traffic, but few have developed a hardware-based routing suite that even comes close to providing the feature set the Cisco Internetwork Operating System offers. In reality, most vendors have had to move to software-based solutions for the majority of routing features, leaving only packet routing to be handled in hardware.

This is not to say that Layer 3 switch routing will not come into its own and deliver on its promise of wire rate routing. It just is not here today. Not to mention the fact that most Layer 3 routers have little to nothing in terms of any sort of WAN support (this is also changing somewhat).

ATM Switches

When ATM was introduced, its distinction from traditional LAN and WAN transmission protocols was immediately noticed. Its support of a diverse range of operating speeds, optimization towards multimedia applications through the use of small MTU size and Quality of Service (QoS) transport contracts, and dependency on switches for implementation made ATM stand out from the crowd. It was also believed by some of its strongest early advocates that ATM would evolve into the "all in one" LAN/WAN networking solution. ATM-specific applications would be developed, and mixed protocol LAN and WAN implementations would be replaced with ATM operating over different media and at different rates, depending upon the environment's needs.

In the end, however, traditional Layer 2 LAN transmission technologies, utilizing similar hardware switching technologies, evolved far beyond what ATM's creators had ever envisioned. Today, LAN protocols such as Ethernet have evolved to support more than adequate operating speeds and with further standards development (for example, IEEE 802.1p), can now provide Layer 2 traffic prioritization. There have also been efforts to develop QoS services for TCP/IP, which is used for all Internet data transport. QoS service for TCP/IP has become increasingly important as technologies to transmit voice over IP (VoIP) and real-time video over the Internet have matured. The Resource Reservation Protocol (RSVP) developed by the Internet Engineering Task Force (IETF) is one such method for providing QoS services for TCP/IP.

Given the current state of development of QoS services for traditional Layer 2 LAN transmission and Layer 3/4 transport technologies today, the usefulness of ATM in the LAN environment is questionable. Because ATM is a connection-oriented transmission protocol, it is basically incompatible in its native form with traditional connectionless protocol-based LANs. Because of the inherent interoperability issues between ATM and traditional or legacy Layer 2 and Layer 3-based LAN protocol functionality, several strategies have been developed to provide compatibility services between ATM and the legacy LANs. The first and most widely supported solution is the ATM Forum's LAN Emulation over ATM (LANE) protocol suite. The LANE suite provides all the services necessary to fully incorporate ATM connected end-stations into an existing traditional Layer 3 LAN network segment.

The utilization of ATM as a high-speed WAN protocol is growing, however, as the need for more and more bandwidth on the Internet backbone increases. As we learned earlier, LAN networks traditionally have been designed to operate using an 80/20 traffic utilization rule: 80 percent of the network traffic is within the local network, and 20 percent is outside the local network. Today, the 80/20 rule is more like 50/50 or 40/60. The need for access to resources outside the local network is growing annually. With this shift in network traffic utilization, and with IP being the dominant network transport protocol, the development of IP over ATM technology has been quite active for some time. The first result of this effort was the development of IETF's "Classical IP over ATM." Keeping in mind the now-legacy status of LANE and need for IP over ATM, we will look at LANE and then briefly discuss Classical IP and its derivatives later in this section.

Note

RSVP provides bandwidth reservation services for specific Internet application data flows. RSVP was initially developed by the University of Southern California Information Sciences Institute and the Xerox Palo Alto Research Center. The RSVP standards effort is being run by the IETF. RSVP provides three transport service levels: best-effort, rate-sensitive, and delay-sensitive. Transport is provided using RSVP tunnel interfaces, which are established between the gateways in the transport path between sending and receiving hosts, utilizing IP multicast and the Internet Group Membership Protocol (IGMP). For more information on RSVP, refer to the IETF Web site (`http://www.ietf.org`).

ATM Switch Architecture

ATM and LAN switch architecture are essentially the same. From an operational perspective, there are some big differences, because the ATM transport mechanism is so functionally different from traditional Layer 2 LAN and WAN transmission protocols. In Chapter 5's overview of the ATM protocol, the distinction between public and private ATM switches was introduced. The difference between them is that private ATM switches are used to provide ATM transport between privately managed and connected ATM switches, whereas public ATM switches are maintained and operated by a public network service provider, such as an Incumbent Local Exchange Carrier (ILEC) or Interexchange Carrier (IXC) which provides public ATM network access along the same lines as Transport-1 and ISDN carrier service.

This discussion focuses on private ATM switch architectures and features, because these switches are constructed specifically to provide ATM capabilities. Although functionally the same, public ATM service providers generally do not utilize application-specific switches. Instead, a general-purpose telecom carrier-class switch is used, which utilizes application-specific interface modules and software modules to perform specific tasks.

ATM private-class switches are comprised of the same architectural elements utilized by LAN/WAN switches:

- Physical layer interface
- Switch backplane
- Switch logic controller

Because ATM's founding premise was to provide transport over a large scaleable range of transmission speeds, a number of physical transmission media and interface types are available. Table 6.2 lists ATM's most commonly supported transmission media and their operating rates.

Table 6.2 **Common ATM Transport Implementations**

Transport Medium	Operating Rate	NNI or UNI Application
ATM over DS1 (Cat 3)	1.544	Public
ATM over UTP (Cat 3)	26.6	Private
ATM over DS3 (Cat 3)	44.736	Public
ATM over STP (Cat 3)	155.52	Private
ATM over SONET (SMF/MMF)	155.52	Public/Private
ATM over SONET UTP (Cat 3)	155.52	Private
ATM over SONET UTP (Cat 5)	155.52	Private
ATM over SONET (SMF)	622.08	Public/Private

SMF = Single mode fiber
MMF = Multimode fiber
STP = Shielded twisted pair
UTP = Unshielded twisted pair

What makes ATM switch ports significantly different from their LAN counterparts is that they only support one user network interface (UNI) or network-to-network interface (NNI) device per port. Commonly, ATM interface port densities are 8, 16, or 24 ports per switch or enterprise module, depending upon the supported port operating rate. It is also common for LAN switch manufacturers to provide a single 155Mbps ATM-UNI interface module that can be used to interconnect to another LAN switch or to an ATM switch backbone running LANE.

Historically, due to ATM's faster operating rates, ATM switch backplanes had significantly higher bandwidth ratings than traditional LAN switches. This is really not the case today. Like a LAN switch, the ATM switch backplane needs to have enough bandwidth to support the simultaneous operation of its ports at the highest operating rate. ATM interfaces do not support full-duplex operation, so the required switch backplane can be calculated quite easily by adding the number of ports, multiplying the result by their highest operating rate, and then dividing the sum by two. If the result is less than the switch's rated backplane, blocking will result when all the switch ports are active.

An ATM switch logic controller consists of the same basic elements found in a LAN switch logic controller: a CPU, a switch logic module, DRAM, and Flash. Due to ATM's fast operating rate and guaranteed QoS demands, cellprocessing is ideally performed in hardware. Therefore, the use of ASIC-based switch logic modules is quite common. However, where most ATM switches tend to fall short is in the area of buffer space. ATM switches commonly use a shared DRAM buffer to store cells they cannot process. If the switch comes under heavy load and the buffers overflow, the switch will drop cells. This, in turn, forces the transmitting applications to retransmit the lost packets, which contributes even further to the switch's overloaded condition. With ATM switches, buffer sizes often dictate how well the switch will perform. The general rule is "bigger is better." In the end, the speed of the ASIC and the backplane are irrelevant if the switch cannot actually process the cells.

ATM Cell Forwarding

ATM cells are forwarded using a technique derived from X.25 and Frame Relay known as *label swapping*. With label swapping, each cell has a logical connection identifier (LCID). In ATM's case, the Virtual Path Identifier (VPI) and the Virtual Channel Identifier (VCI) fields in the cell header are the cell's LCID. The VPI/VCI are used individually or collectively to determine which port the cell exits from. Each ATM switch in the connection path has its own mapping table that contains the local VPI/VCI mappings for each port and their relations to PVC and SVC connections handled by the switch. Cells are then forwarded between ATM switch ports based on their VPI/VCI. Once the cell reaches the ATM switch output port, a new VIP/VCI is inserted into the cell header and the cell is forwarded on to the next switch. This process is repeated until the cell reaches its final destination.

Because both the VPI and VCI are used in combination to make up the LCID, two types of forwarding can be performed. This circumstance has resulted in two different types of ATM switches: VP switches and VC switches. VP switches make forwarding decisions based only on the VPI value. VC switches alternatively forward cells based on both the VPI and VCI values. VP switches forward cells at a faster rate than VC switches, because only the VPI is considered when making forwarding decisions. Most public ATM carrier-class switches are VP-type switches, because the granularity afforded by VC-type switches is not required to transport cells across a carrier's ATM cloud. However, the additional examination of the VCI header is required at both ends of the public/private transit path and with most private ATM backbone implementations. Consequently, most private-class ATM switches are VC-type switches.

The label swapping approach works well for all packet- and cell-switching connection technologies because of their connection-oriented natures. As for the VPI and VCI, values are mostly arbitrary and only relevant to the local switch. Therefore, they can be either statically or dynamically allocated using signaling when ATM PVC or SVC connections are established.

LANE

With its fast operating rate and its optimization for carrying multimedia data (specifically, voice and streaming video), ATM is very attractive as a LAN backbone and high-speed server connection technology. It was for this kind of application that LANE was developed. Today, with Gigabit Ethernet and switched port trunking, the use of ATM as a LAN backbone solution is not nearly as attractive as it was a few years ago. Although ATM's popularity as a LAN backbone technology is fading, it is growing as a WAN backbone technology. The bandwidth limitations of Frame Relay are being realized with more network-intensive applications.

Before we look at how LANE works, a brief description of why it is needed is required. The major operational incompatibility between ATM and traditional LAN protocols is not specific to ATM only. This incompatibility exists with all Layer 2 WAN transmission protocols, and it is the inability to handle Layer 3 broadcast and multicast messages. This is not because of some operational limitation, but mainly because there are drastically different operational paradigms that govern the design of these protocols.

LAN transmission protocols operate over a shared transmission medium, so when a packet is sent, all the nodes need to examine it to see if it is destined for themselves. In a switched LAN environment, the switch logic performs this task by examining each of the packets sent to each of its ports, and then forwarding packets, when needed, to the port that is connected to the destination host. In the case of a broadcast packet, the switch (or bridge) floods the packet to all the switch's ports.

WAN protocols, on the other hand, are designed for synchronous or point-to-point operation, where the connection is always up, so to speak. Regardless of whether there is data to be transmitted, frames are always being sent across the link. With packet-based technologies such as Frame Relay that support point-to-multipoint connections, broadcasts need to be simulated by either the FR switch or application, so the broadcast is sent as a unicast to all the connected endpoints. Because WAN links operate either in routed Layer 3 segments or in a bridged capacity, the inability to handle broadcast and multicast traffic has little impact.

Because ATM is in essence a WAN transmission protocol, and works on the premise of point-to-point connections between endpoints, it has no capability to handle Layer 3 broadcasts and multicasts. However, what makes ATM different from traditional WAN protocols and what makes it even possible for it to be used in a shared-medium paradigm is the capability of a ATM host to send data to multiple end-stations over a single physical link, through the use of multiple virtual connections. This dynamic, simultaneous connection model is what comprises the "asynchronous" element of ATM. Although it is possible for an ATM host to establish asynchronously multiplex PVC and SVC connections between other ATM endpoints, these circuits are still only bidirectional point-to-point (or unidirectional point-to-multipoint) connections. LANE provides a set of services that simulate the broadcast and multicast functions needed for ATM-connected hosts to participate in a traditional LAN environment.

LANE's Function

LANE's function is to enable ATM-connected end-stations to interoperate within Layer 2/Layer 3 LAN segments with other end-stations connected by traditional LAN technologies. It performs this function by operating a collection of protocols that emulate IEEE 802.3 Ethernet and/or IEEE 802.5 Token Ring LAN in an ATM network environment. The goal is to have Layer 3 protocols function on ATM hosts just as if they were traditional LAN hosts.

LANE does not affect the actual operation of ATM, so no special signaling is required. Therefore, ATM switches providing cell relay do not need to be aware of its use. LANE operates as an overlay on top of ATM, so only the attached routers, LAN switches, and ATM-connected end-stations need to be aware of LANE's activities. This is different from traditional LAN switch VLANs, where the LAN switch is configured to provide network partitioning, and the clients are unaware of the fact that a switch and VLANs are even in use.

> **Note**
>
> Layer 3 broadcasts and multicasts are generally localized to the Layer 3 network segment and are not routed.

LANE operates as a client/server technology. Its basic function is to provide resolution services between MAC addresses and ATM addresses and to provide Layer 3 broadcast facilities. These are all required for the Layer 3 networking protocols to function as if they were operating in a connectionless Layer 2 protocol-based LAN environment. The emulation services are established and provided through the creation of virtual circuits between LANE clients and servers. These client/server interactions operate over the LAN Emulation User to Network Interface (LUNI). All the LANE operational functions are performed through the LUNI. These functions are initialization, registration, address resolution, and data transfer.

LANE's Components and Their Operation

Four components provide emulation services between ATM endpoints (end-stations or LAN switches or routers) and traditional LANs:

- LAN Emulation Client (LEC)
- LAN Emulation Server (LES)
- Broadcast and Unknown Server (BUS)
- LAN Emulation Configuration Server (LECS)

The LANE protocol also requires an emulated LAN (ELAN) to operate either as an Ethernet or Token Ring ELAN. It is possible to have more than one ELAN operate within a single ATM network. Connectivity between the different ELANS must be provided either by a translation bridge or by a router, in the same manner required by traditional LAN operation.

ATM endpoints using LANE exchange control messages and data by establishing bidirectional (send/receive) and unidirectional (receive only) point-to-point Virtual Circuit Connections (VCCs) between each other and the LANE servers. There are four basic kinds of VCCs: configuration, control, data direct, and multicast. Control and configuration VCCs are used to exchange messages between LANE clients and servers. Data direct VCCs are used by LANE clients to exchange Layer 3 datagrams. Multicast VCCs are established between LANE clients and the BUS to send and receive Layer 3 broadcast datagrams, and the initial unicast datagrams generated by the source client which have a Layer 3 address, but no ATM address.

Of the four operational LANE components, only three need to be replicated on a per-ELAN basis. The following is a brief description of each of the LANE components and their operation. It should be noted that all these components can be provided by any LANE-compatible ATM device (that is, a router, ATM switch, or ATM-equipped LAN switch). When implementing LANE from a network recovery standpoint, it is best to have the different LANE servers operating on different ATM devices, rather than on a single device. LANE operates the same with all compatible devices, but configuration will be specific to the device's manufacturing directions.

LAN Emulation Server

The LAN Emulation Server (LES) functions as an address resolution server for the entire ELAN. When an LEC joins the ELAN, or when it requires the ATM address associated with specific MAC address, the LES performs registration and lookup services. MAC-to-ATM address resolution is managed with the LAN Emulation Address Resolution Protocol (LEARP). The LES stores the MAC-to-ATM address mappings in a locally stored cache. If a request is made for an ATM address that the LES does not have, a LEARP query is made on the behalf of the LES by the BUS. If more than one ELAN exists, each ELAN within the ATM network must have its own LES.

Broadcast and Unknown Server

The Broadcast and Unknown Server (BUS) addresses ATM's inability to handle Layer 3 broadcast datagrams. It is responsible for sending broadcast and local multicast traffic to all LANE clients within a specific ELAN. It also performs Layer 2 broadcast functions such as LEARP queries. The BUS sends address queries, such as broadcast and multicast datagrams to LANE clients, in most cases through a unidirectional multicast VCC. Like the LES, each ELAN must have its own BUS.

LAN Emulation Configuration Server

The LAN Emulation Configuration Server (LECS) provides configuration for the LANE clients. The LECS is essentially a database server that contains information about all the ELANS in existence for a given ATM network. The LECS also has configuration information for all the LANE clients: which ELAN they belong to, the ELAN type and MAC address, the address of the LES, and so on. When the LANE client starts up, it establishes a configuration VCC with the LECS. If the LECS has configuration information for the requesting client, it provides the client with the needed information. If no configuration information exists, the LANE client configuration fails. A single LECS is all that is required to provide configuration information for all the ELANS within an ATM network.

The LAN Emulation Client

The LAN Emulation Client (LEC) exists on each ATM endpoint involved in the ELAN. A separate LEC must exist for each ELAN with which the ATM endpoint is involved. The LEC (with the assistance of the LAN emulation server) is responsible for performing ATM-to-MAC address translations. Each LEC client has both an ATM and emulated MAC (Ethernet or Token Ring) address for each connected ELAN. The LEC on ATM-equipped end-stations provides the interface used by the Layer 3 protocol to interact with other connected end-stations. LEC functionality is part of the ATM NIC, LAN switch, or router interface.

In order for the LEC to participate in the ELAN, three events must occur. First, the LEC needs to be configured for operation within a specific ELAN. This is accomplished by the LEC establishing a bidirectional configuration VCC with the LANE configuration server upon startup. After the LEC is configured, it must then establish a bidirectional control VCC with the LES and register its MAC address. This, in turn, initiates the creation of a unidirectional control VCC from the LES to the LEC, at which point the LEC can make and receive address resolution requests. The final event is the LEC establishing a bidirectional multicast VCC with the BUS and the BUS establishing a unidirectional multicast VCC with the LEC. This enables the LEC to send and receive Layer 3 broadcasts and local multicasts and respond to LEARP queries, which effectively completes the LEC membership into the LAN. After all the required VCCs are established, the LEC is part of the ELAN and can now exchange datagrams with other LECs and traditional LAN end-stations. Figure 6.15 illustrates the various VCCs needed for an LEC to operate in the ELAN.

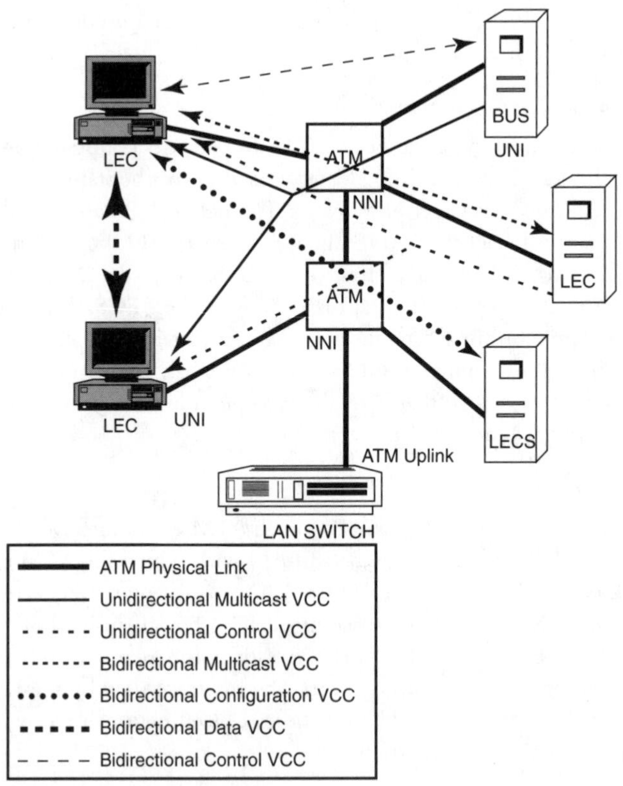

Figure 6.15 An LEC and its VCC relationships with the LANE servers.

Classical IP over ATM

The IETF's Classical IP over ATM was the first evolution of the IP over ATM effort. Classical IP, like LANE, treats ATM-connected devices as if they are all connected locally on the same Layer 2 segment. This leaves Layer 3 communication between segments to IP routers with ATM interfaces. This approach reflects IP's architectural Layer 2 indifference to the actual transmission of IP datagrams, which only requires that a datagram encapsulation process and an ATM-to-IP address resolution scheme be developed. RFC 1577 defines the operational mechanisms required for the implementation of Classical IP and ARP over ATM. Classical IP over ATM does not support broadcast or multicast traffic, so its usefulness is really limited to IP over ATM implementations that operate as point-to-point WAN links. Figure 6.16 illustrates the local perspective used for the connection of IP devices over ATM.

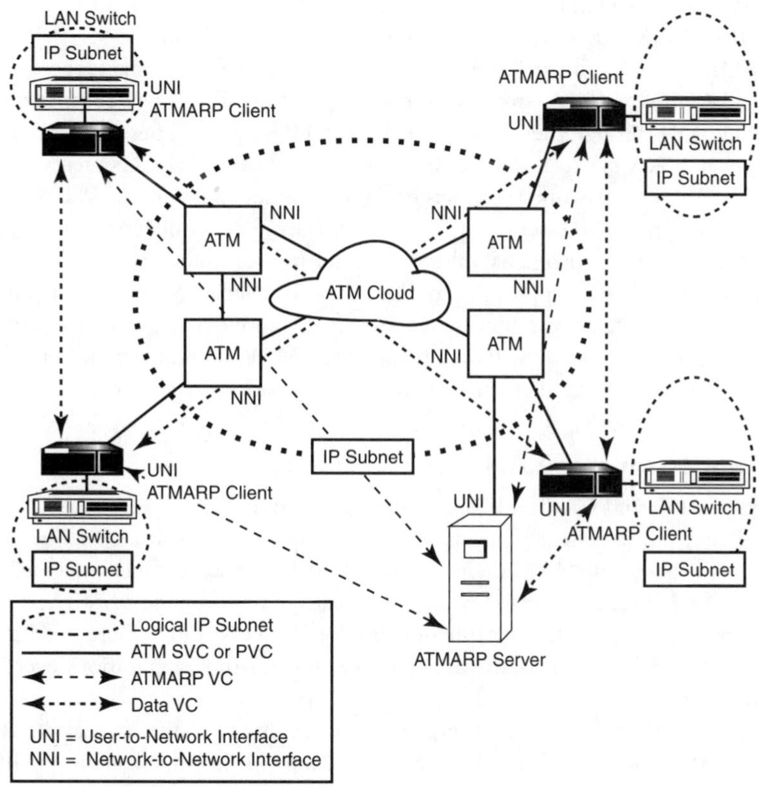

Figure 6.16 A Classical IP implementation over an ATM network.

The operation of Classical IP is straightforward. All the ATM end-stations must be directly attached to the same network segment and all must use a common IP network address space. IP datagram encapsulation is handled by ATM Adaptation Layer 5 (AAL5) by adding an LLC/SNAP header to the IP datagram. Data exchanges between ATM hosts are accomplished by establishing VCCs between each other over ATM PVC or SVC. Endpoint ATM addresses are resolved through an IP-to-ATM address resolution mechanism known as ATMARP. ATMARP is a modification of traditional ARP, so it will function in a non-broadcast environment. Inverse ATMARP functionality is also available for IP address resolution for circumstances when end-stations connected to the local segment know the ATM, but not the IP address of the station they want to exchange data with.

To accommodate ARP in a non-broadcast environment, an ATMARP server is established. Its purpose is to maintain a table with ATM-to-IP address mappings. This ATMARP server is used by all the ATM devices connected to the local IP subnet to resolve IP-to-ATM addresses (in a SVC environment) or VCs (in a PVC environment). The ATMARP server functionality is limited by the same operational scope used with traditional ARP—that is, the scope of the local IP subnet. The server constructs the ATMARP table through the exchange of ATMARP messages. These messages are exchanges of SVCs between the ATMARP server and the ATMARP clients. When the clients first join the network, they register their addresses and establish SVCs with the ATMARP. The ATMARP server then maintains these SVC connections for address verification and the exchange of address resolution request information.

Although the classical approach provides native IP over ATM services, it is limited by its lack of broadcast and multicast support. So, by employing point-to-multipoint VCCs and a Multicast Address Resolution Server (MARS), multicast and broadcast traffic can distribute local IP subnet multicast and broadcast datagrams.

The Limitations of LANE and Classical IP

Although LANE and Classical IP succeed in integrating ATM as just another Layer 2 protocol, they fail to provide a means of capitalizing on ATM's strengths. These strengths are, specifically, ATM's contract-based quality of service guarantees and its dynamic hierarchical routing and connection management functionality. There have been several attempts to develop functionality for IP networking where it is badly needed as the evolution of bandwidth-sensitive and real-time applications over IP continues.

Cisco Systems' Tag Switching Protocol and its evolution as the basis of the IETF's development of the Multi-Protocol Label Switching (MPLS) standard are the most noteworthy attempts to resolve many of the traffic flow management issues that ATM was developed to address, but are effectively lost when utilized with traditional connectionless Layer 3 protocols. Cisco's Tag Switching and MPLS efforts are not specific to ATM. They are instead a technology bridge that will be used between traditional connectionless IP networks and the public connection-oriented ATM. This adaptation will more than likely comprise most of the Internet and private networking backbones in the near future.

With MPLS packets, routing decisions are not made in the traditional manner. Instead, the first router (normally the edge router between the LAN and the Internet or private data network) decides the route the packet will take to reach its destination. This packet forwarding information is appended as a label, either within the Layer 2 packet header (if natively supported) or appended to the actual packet between the Layer 2 and Layer 3 headers. This label is then used by MPLS-supported routers to perform destination-based routing. Packets are forwarded between other MPLS routers based on comparisons between the packet's label and the MPLS router's Forwarding Information Base (FIB). FIB is generated by information learned from traditional dynamic routing protocols such as OSPF and BGP. After the packet leaves the MPLS cloud, the label is removed and the packet is handled by traditional means.

The advantage of MPLS is that it is a software-based mechanism that can work with most existing networking hardware and Layer 3 protocol technologies.

Summary

In this chapter, we covered a myriad of essential material dealing with the operation and implementation of LAN and ATM network switches. The material provided in this chapter offered an overview of the many topics discussed. You are encouraged to read additional material on these subjects. The following is a list of topics you should understand by the end of this chapter:

- The foundations of switch operations
- Transparent, translation, encapsulation, and source route bridging
- The spanning tree algorithm
- LAN and ATM switch architectures
- A comparison of bridge and router operation
- LANE and Classical IP over ATM, and MPLS
- VLANs
- Full-duplex switch port operation and switch port trunking
- The differences between store-and-forward and cut-through forwarding

This chapter concludes our discussion of Layer 2 networking. The next four chapters are dedicated to the implementation and configuration of multiprotocol enterprise routing using Cisco routers. Chapter 7 deals with basic operational and configuration aspects. Chapter 8 deals exclusively with the theory and operation of TCP/IP dynamic routing protocols. Chapter 9 provides additional advanced Cisco router configuration material, and Chapter 10 concludes the Cisco-centric material with the configuration of dynamic TCP/IP routing protocols.

Related RFCs

RFC 1483	Multiprotocol Encapsulation over ATM Adaptation Layer 5
RFC 1577	Classical IP over ATM
RFC 1626	Default IP MTU for Use over ATM AAL5
RFC 1755	ATM Signaling Support for IP over ATM
RFC 1954	Transmission of Flow Labeled IPv4 on ATM Data Links Ipsilon Version 1.0. P. (MPLS Information)
RFC 2098	Toshiba's Router Architecture Extensions for ATM (MPLS Information)
RFC 2225	Classical IP and ARP over ATM
RFC 2379	RSVP over ATM Implementation Guidelines
RFC 2382	A Framework for Integrated Services and RSVP over ATM

Additional Resources

Ali, Syed. *Digital Switching Systems: Systems Reliability and Analysis*. McGraw-Hill, 1997.

Cisco Systems, Inc. *Cisco IOS Switching Services*. Cisco Press, 1998.

Davie, Bruce, Paul Doolan and Yakov Rekhter. *Switching in IP Networks: IP Switching, Tag Switching, and Related Technologies*. Morgan Kaufman, 1998.

Lammle, Todd, Ward Spangenberg, and Robert Padjen. *CCNP Cisco LAN Switch Configuration Guide*. Sybex, 1999.

Lewis, Chris. *Cisco Switched Internetworks*. McGraw-Hill, 1999.

Metzler, Jim, and Lynn DeNoia. *Layer 3 Switching: A Guide for IT Professionals*. Prentice Hall, 1999.

Parkhurst, William. *Cisco Multicast Routing and Switching*. McGraw-Hill, 1999.

Perlman, Radia. *Interconnections: Routers and Bridges*. Addison-Wesley, 1992.

7

Introduction to Cisco Routers

THE GOAL OF THIS CHAPTER IS TO PROVIDE YOU WITH an overview of the Cisco router configuration options commonly used in enterprise networking environments, specifically:

- Configuration of IP, IPX, and AppleTalk protocols
- Basic protocol routing configuration
- Using the Cisco IOS command line and configuration modes
- Understanding memory usage
- Upgrading the IOS and disaster recovery
- Setting up router logging, accounting, and security

IP routing protocols (BGP, Open Shortest Path First (OSPF), and so on) and advanced configuration options (such as access control lists and Network Address Translation) will be covered in Chapter 9, "Advanced Cisco Router Configuration," Chapter 10, "Configuring IP Routing Protocols on Cisco Routers," and Chapter 11, "Network Troubleshooting, Performance Tuning, and Management Fundamentals." When you begin using Cisco routers, the hardest thing to master is the command structure. This chapter has plenty of examples and a configuration tutorial to help get you started.

Cisco Router Hardware

Cisco Systems began as a small startup in a San Francisco living room in 1984. Its founders were researchers at Stanford University who devised a "gateway server" to connect computers from different departments. At this time, Transmission Control Protocol (TCP)/IP was beginning to enter the networking community, and this husband and wife team (Len and Sandy Bosack) decided to take a chance and make a commercial version of their server. Cisco Systems' first generation of products were known as gateway servers, and the product line had four iterations:

- Advanced Gateway Server (AGS)
- Mid-Range Gateway Server (MGS)
- Integrated Gateway Server (IGS)
- Advanced Gateway Server Plus (AGS+)

These servers provided basic LAN-to-WAN and LAN-to-LAN routing utilizing Ethernet, Token Ring, and Serial Layer 2 technologies. Today Cisco's router product line is quite broad, and it provides routing solutions for every application.

Entry Level	Mid Range	High End
Cisco 16xx	Cisco 26x	Cisco 7xxx
Cisco 25xx	Cisco 36xx	Cisco 4xxx

Cisco 1600/1600-R small office stub LAN/WAN access router is Cisco's current IOS-supported entry-level modular router. Most versions come with a built-in 10BASE-T Ethernet interface and support for ISDN-BRI, synchronous Serial and Integrated DSU/CSU T carrier modular WAN interface cards (WICs), which are also compatible with the 2600 and 3600 series routers. The 1600 series uses the Motorola 68030 processor, running at 33MHz (6,000 packets per second, or PPS). The 1600 series routers operate using IOS version 11.1 and higher.

Cisco 2500 series is quite extensive; there are 32 different models. The 2500s, like the 1600s, are mostly used as small stub or end-node access routers for LAN-to-WAN access. They also have LAN-to-LAN (Token Ring, Ethernet) and Terminal Server (8 or 16 async ports) versions. The 2500 series is based on the Motorola 68030 processor, running at 20MHz (5,000PPS). This series supports IOS 10.x and higher.

The 2600 series was intended to be a modular replacement for the 2500 series. It uses a Motorola MPL 860 chip, running at 40MHz and can process up to 25,000PPS. It has a single 10-BaseT Ethernet interface and slots for two WICs and one expansion slot (for ATM, async and sync serial, and so on). It requires IOS 11.3 or higher. The 2600 also supports Voice over IP (VoIP).

The 3600 series routers (3620 and 3640) were the first of the third generation modular routers. The 3620 has two expansion slots, and the 3640 has four. They are based on the IDT R4700 processors running at 80MHz/16,000PPS and 100MHz/40,000PPS, respectively. They support interfaces for just about everything—ATM, Async and Sync Serial, HSSI, Ethernet, and Fast Ethernet—except fiber distributed data interface (FDDI) and Packet over SONET (POS). Both require Internetwork Operating System (IOS) 11.1 or higher.

The 4000s (4000, 4500-M, 4700, 4700-M) were the first generation of modular routers. The 4000 uses a 40MHz Motorola 68030, the 4500 uses an IDT Orion 100MHz, and the 4700 uses an IDT Orion 133MHz. The 4000 series routers come in a three-slot chassis, but they have also been built in line card form to operate in Cisco's Catalyst 5000 switches and Cabletron's 9000 series switches. They support up to 18 Ethernet interfaces, 6 Token Ring, 2 Fast Ethernet, 2 FDDI, 1 HSSI, 34 Async Serial, 16 Sync Serial, and ATM. They run IOS 9.14 and higher.

The 7x00 (7000, 7200, and 7500) series routers make up Cisco's high-end router line, and were Cisco's first generation of high-end routers. The 7000 and 7010 originally used separate route and switch (interface) processors. They now use a single processing unit called a *Route Switch Processor (RSP)*, and both models use Cisco's CxBus architecture. The 7200 series routers are the mid-line enterprise routers that use PCI bus architecture. The 7500 is the high end, sporting the CyBus architecture. The 7x00 series routers' processing speed is determined by the speed of the RSP (7000 and 7500) or Network Processing Engine (NPE) (for the 7200). The 7x00 series routers also support Cisco's *Versatile Interface Processor (VIP)*, which provides hot swap capability and Layer 2 switching capabilities. The 7x00 routers utilize IOS 11.x and higher, depending on the model.

Although it's not supported on every router platform, Cisco has a Layer 2 implementation for each of the following:

- Ethernet (10, 100, and 1,000Mbps)
- Token Ring (4 and 16Mbps)
- Asynchronous Serial (11,500Bps)
- Synchronous Serial (DS0, DS3)
- HSSI (High Speed Serial Interface)
- ISDN (BRI and PRI)
- ATM (DS3, 0C3)
- POS (0C3, 0C12)

Memory on Cisco Routers

Everything about Cisco routers revolves around memory. Memory is used to boot the router, store its operating system, perform the routing process, and store the router's configuration information. To perform these tasks, four types of memory are used:

- Read-only memory (ROM)
- Flash memory
- Non-volatile random access memory (NVRAM)
- Dynamic random access memory (DRAM)

ROM is used on the 1600, 2500, 2600, and 3x00 series routers to handle the router's bootstrap process, and it has just enough headroom to load the OS. In the case of the 2500, the ROM set also contains a limited version of the router's operating system, called the *Internetwork Operating System (IOS)*.

Flash memory is rewritable nonvolatile memory. In the early days of programmable hardware, floppy disks were used to store boot images and configuration files, as flash memory was excessively expensive. Flash is either mounted on the router's motherboard or installed in a router's PCMCIA slot(s). Flash memory is used on all Cisco routers to store the IOS. On 4x00 and 7x00 routers, along with the IOS flash, an additional flash chip known as the *bootflash* contains the bootstrap OS. By using the bootflash, the bootstrap application can be upgraded without a hardware change.

NVRAM is similar to flash in that it does not lose its contents when power is lost. It is used to store the router's configuration information.

DRAM is used for packet processing and IOS operation. On all router platforms (except the 7x00 series), DRAM is partitioned into primary and shared memory. There are two primary memory architectures used by Cisco routers:

- Read From Flash (RFF)—The Cisco 1600 and 2500 series (and older) routers use an RFF model. The IOS resides on flash, and IOS data structures are loaded into RAM as needed.
- Read From RAM (RFR)—The 1600-R, 2600, 36×0, 4×00, and 7x00 series routers are RFR routers. The IOS image is stored in compressed form on the router's flash. At boot, the IOS's whole image is decompressed from flash to DRAM where it is accessed by the router when needed.

Along with the IOS, primary memory is also used to store the router's configuration information and the data tables needed to process datagrams. For instance, if IP is being routed, the ARP table and IP route tables are stored in primary memory. If IPX is in use, the SAP and IPX route table would be stored. If AppleTalk is routed, the AARP, Zone Information Table, Name Binding Table, and AppleTalk Route Table would be stored.

The shared memory is used for datagram processing. By default, the DRAM is partitioned using a 75/25 split of the router's minimum DRAM configuration split. On 1600, 2500, and 4x00 series routers, the primary/shared partition cannot be adjusted. On the 2600 and 36x0 series routers, this partition can be readjusted to accommodate different interface configurations.

The 7x00 series does not use partitioned memory architecture. DRAM is used for IOS and routing table storage. Memory used for packet processing is on the interface controller or on the switch processor, depending on the router model.

Because the entire operation of the router is dependent on memory, it is essential that the router have enough to operate efficiently. On fixed DRAM partition routers, the DRAM and flash requirements are dictated by the type of IOS you want to support in terms of services. If you want multiprotocol firewall support, you need more DRAM and flash than you need if all you want to do is route IP. On adjustable partition and 7x00 series routers, flash requirements will also be dictated by IOS functionality. DRAM and/or switch processor memory requirements will depend upon the types of interfaces you plan to use. Be sure to check the memory requirements for the IOS and interfaces you want to use on your router before you order it.

Talking to Your Cisco Router (Through the Console)

Access to the router's console (in most cases) is required for access to the operating system for initial setup and configuration. After the router is online, Telnet can be used to access a virtual router terminal port.

After the router is online, *Simple Network Management Protocol (SNMP)* can be an alternative to the router's Command-Line Interface (CLI) to make changes and gather information about the router. Like Telnet, SNMP is dependent on TCP/IP for transport. Therefore, it requires TCP/IP to be enabled, in addition to its own protocol configuration. Once SNMP is configured and running on the router, an SNMP manager is used to send and receive commands. SNMP configuration will be covered in Chapter 10, "Configuring IP Routing Protocols on Cisco Routers," and a general overview is provided in Chapter 11, "Network Troubleshooting, Performance Tuning, and Management Fundamentals."

> **Note**
>
> Cisco also has a Windows-based tool called ConfigMaker which is often shipped with its low-end routers and is freely available on the Web at `http://www.cisco.com/warp/public/cc/cisco/mkt/enm/config/index.shtml`. This tool provides a GUI configuration interface to perform initial router configurations on Cisco 1600, 2500, 2600, 3600, and 4000 series routers (it also works with Cisco's Catalyst switches).
>
> The IOS also provides an AutoInstall feature, which uses BOOTP, RARP, or SLARP (SerialARP) to provide an IP address to an unconfigured router. The AutoInstall mechanism is evoked on all Cisco routers, as part of an unconfigured router's startup process. After the router has an IP address, a configured production router acts as a TFTP server from which the "newrouter" downloads its configuration.

All Cisco routers come with at least two Async Serial line ports, which are labeled "CONSOLE" and "AUX." The console port is special. It is the only means of direct access to the router's CLI, and it functions as the router's primary configuration access port when the router is unconfigured. It provides a serial display interface that is attached to send and receive router commands and status information. After you have access to the console port, you basically have control over the router. With this in mind, make sure that at the very least you use passwords and install the router in a secure place. The console port interface, depending on the Cisco router model, will either be a male RJ-45 or female RS232C interface, and will be on the side where the router's network interfaces are located (except for the 3600 series, which has console and AUX RJ-45 ports located in the front and its interfaces in the back). The console port is also the default output display for all the router's system messages.

Cisco provides a cable set with every router to attach a serial display terminal or a PC (running a terminal emulator) to the router. The cables are modular and support different configurations for different uses. The set includes RJ45-M to RJ45-M Cisco serial cable, which connects to the RJ45 port on the router. Because cables tend to get lost, the color matchings for the CAT 5 twisted pair and the Cisco cable are listed along with the cable/head pin match translations.

This is a straight-through cable, so both ends are the same. To construct a Cisco cable out of a standard CAT 5 Ethernet cable, you need to crimp the CAT 5 strands following the Cisco RJ-45 head pin assignments. Table 7.1 provides this information.

Table 7.1 **Cisco Console Cable Pinouts**

CAT 5 Standard Cable	CAT 5 RJ45 Head	Cisco Cable	Cisco RJ45 Head
Blue	Pin 1	Brown	Pin 8
White Blue	Pin 2	Blue	Pin 7
Orange	Pin 3	Yellow	Pin 6
White Orange	Pin 4	Green	Pin 5
Green	Pin 5	Red	Pin 4
White Green	Pin 6	Black	Pin 3
Brown	Pin 7	Orange	Pin 2
White Brown	Pin 8	Gray	Pin 1

Note

The AUX port (which is available on most Cisco router models) is a standard async port and can be configured to do a number of things, such as provide a second console port for out-of-band (dial-in) access to the router, provide remote network access over PPP or SLIP, or function as a backup link for a dedicated WAN connection.

RJ45-M (male) to RS232C-F (female) adapter (25-pin, labeled *Terminal*) connects your PC's COM port to the Cisco serial cable. See Table 7.2 for the RJ45 head to DB25/RS232 pin translation.

Table 7.2 **RJ45-M to DB25 Pin Translation**

RJ45-M	DB-25-F
Pin 1 Blue	to Pin 4
Pin 2 Orange	to Pin 20
Pin 3 Black	to Pin 2
Pin 4 Red	to Pin 7
Pin 5 Green	to Pin 7
Pin 6 Yellow	to Pin 3
Pin 7 Brown	to Pin 6
Pin 6 White	to Pin 5

RJ-45-M to DB-9-M/RS232 (9-pin, terminal cable) is used instead of the RS-232-F if you have a DB-9 serial port. See Table 7.3 for the RJ-45 head to DB-9 pin translation.

Table 7.3 **RJ-45-M to DB-9-F Pin Translation**

RJ45	DB9-F
Pin 1 Blue	to Pin 7
Pin 2 Orange	to Pin 4
Pin 3 Black	to Pin 3
Pin 4 Red	to Pin 5
Pin 5 Green	to Pin 5
Pin 6 Yellow	to Pin 2
Pin 7 Brown	to Pin 6
Pin 8 White	to Pin 9

RJ45-M to DB25/RS232 (25-pin, modem cable) is used for connecting a modem to the console port for out-of-band access and with the terminal cable to connect to the console port on Cisco 4000 and 7000 series routers. See Table 7.4 for the RJ-45 head to DB-25 pin translation.

Table 7.4 **RJ-45-M to RS-232C-M Pin Translation**

RJ45-M	DB-25-M
Pin 1 Blue	to Pin 6
Pin 2 Orange	to Pin 9
Pin 3 Black	to Pin 3
Pin 4 Red	to Pin 7

continues

Table 7.4 **Continued**

RJ45-M	DB-25-M
Pin 5 Green	to Pin 7
Pin 6 Yellow	to Pin 2
Pin 7 Brown	to Pin 20
Pin 8 White	to Pin 4

Use the cable set to build the cable appropriate to your situation, which in most cases will mean attaching a PC serial COM port with terminal emulation software to the router. The AUX port is a standard Cisco Async Serial port. It is not used for initial configuration and system messaging. It can be configured to act as a secondary console port, but it is not one by default.

After you have the correct console cable for your router, you will need to use a terminal emulation package to access the CLI. Any terminal emulation package that allows access to the serial port directly (without a modem) will do. If you are using a Windows system, you can use HyperTerminal. On a Macintosh, a package called *Z-Term* is popular.

Now, your cables are connected between the router and your PCs, and the terminal package of your choice is up and running. With the router's power on (and with no existing configuration), you can enter a carriage return, and you should see the following:

```
% Please answer 'yes' or 'no'.
Would you like to enter the initial configuration dialog? [yes/no]:
```

If the initial dialog or some kind of output appears, you are ready to move on to configuring your router. However, if there is no output on the terminal, reset the router's power and see if the boot banner comes up. It will look something like this:

```
System Bootstrap, Version 11.0(10c), SOFTWARE
Copyright© 1986-1996 by Cisco Systems
2500 processor with 14336 Kbytes of main memory


Notice: NVRAM invalid, possibly due to write erase.
```

If nothing appears, check your terminal package and cable connections. In most cases, the problem will end up being the cable combination you are using. The following are the different cable configurations used with most popular computer platforms:

- A PC will use the Cisco "terminal" DB-9-F or RS-232C-F and the Cisco console cable.
- A Macintosh will use a Macintosh modem cable, Cisco terminal adapter, and the Cisco serial cable.
- A UNIX system will use the same cable set as a PC, but may require a male-to-male gender changer on certain hardware platforms (for example, Sun's).

Note

I recommend Columbia University's terminal emulation package, C-Kermit. C-Kermit is available as both a free and a commercial package and is available for almost every computing platform ever created. Information is available at `http://www.columbia.edu/kermit/`, and the software can be downloaded from `ftp://watson.cc.columbia.edu/`. C-Kermit comes in precompiled form for Macintosh (MAC-Kermit), DOS (MS-Kermit), and the most popular UNIX flavors. Though going out and getting a new communications program may seem like a pain, I highly recommend C-Kermit because it is quite flexible, and there is excellent documentation available on the Web, as well as in book form. C-Kermit is a must if you use a terminal program a great deal, and if you configure routers differently.

After C-Kermit is installed and running, you need to configure it to use the appropriate communications settings. Cisco console port (and all default Cisco Async Serial ports) settings are as follows:

```
VT100 9600/8/1/none
```

These settings are the default on almost all terminal programs, but it always pays to make sure the settings are correct. Here are the steps needed to configure C-Kermit to function as a connected serial VT100 terminal:

Step 1. Starting at the DOS prompt, type `D:\ckermit> kermit` to start the C-Kermit application. This is the default application prompt where configuration commands are entered:

```
[D:\CKERMIT] MS-Kermit>
```

Step 2. This command specifies which COM port to use:

```
[D:\CKERMIT] MS-Kermit> set port com1
```

Step 3. This command sets the port's speed:

```
[D:\CKERMIT] MS-Kermit> set speed 9600
```

Step 4. This command sets the terminal emulation:

```
[D:\CKERMIT] MS-Kermit> set term type vt100
```

Step 5. This command sets the stop bit:

```
[D:\CKERMIT] MS-Kermit> set stop 1
```

Step 6. This command sets parity:

```
[D:\CKERMIT] MS-Kermit> set parity none
```

Step 7. This command tells the program not to look for a modem carrier detect signal:

```
[D:\CKERMIT] MS-Kermit> set carrier off
```

Step 8. This command starts the terminal session:

```
[D:\CKERMIT] MS-Kermit> connect
```

To end a terminal session on DOS, use Alt+X; on a UNIX system, use Ctrl+C; on a Macintosh, use the menu: Special, Reset, Terminal.

If your cable type and connections are correct, check to make sure the serial port you're using is the correct one and that you have the access privileges to use it (Windows NT and UNIX users).

Cisco IOS

Cisco IOS is the standard operating system for all Cisco routers. It is a deeply feature-rich, yet efficient, routing-centric OS. Aside from creating the router industry, what has made Cisco so popular among network managers and administrators is its solid implementation of standard and proprietary routing protocols and its reliability and security enhancements. Cisco IOS is the standard by which other routing implementations are measured in terms of protocol implementation stability, IETF (RFCs), IEEE, and ANSI hardware and software standards implementations.

The IOS interface was based on the TOPS-20 command shell interface, which was a user-centric, help-oriented shell. The IOS command interface is now widely emulated on a variety of hardware vendors' configuration interfaces. In fact, it is now quite common for vendors to claim (as a selling point) that their network hardware sports a "Cisco-like" configuration interface or that they are in the process of creating one for their product line. The driving reason for this copycat interface approach is the demand of customers for all network equipment to have a similar interface so retraining is not needed. Sadly, most of these emulation attempts fall short, not because the interface is bad (though in some cases it is), but rather because they just do not support all the options and protocols that Cisco IOS does. The current IOS supports the following:

Network Protocols	**Hardware Protocols**
IPv4 and v6	Ethernet
AppleTalk	FDDI
IPX	Token Ring
VINES	HDLC
DECnet	PPP
XNS	ATM
Apollo DOMAIN	POS
CLNS	Frame Relay
Layer 2 bridging	SRB, RSRB, and so on

IOS Versions and Releases

The IOS is platform hardware-centric, and each hardware platform has its own specific version. For example, even though the Cisco 3640 and the Cisco 3620 are in the same product family, each uses a different IOS. The commands and features are the same; you just need the IOS binary image for the right processor, which will differ depending on the router model. In addition, not all versions of the IOS support the same feature set (networking protocols, security enhancements, and so on). Although there are numerous subtle variations on the IOS depending upon the hardware platform, the IOS essentially comes in three flavors:

- IP only—Comes in several variations (IP, IP Plus, IP 40, IP 40 Plus, to name a few). These variations provide additional services such as data encryption, firewalls, and Network Address Translation (NAT).
- IP/IPX/AT/DEC—Provides multiprotocol support.
- ENTERPRISE—Provides support for all the IOS enhanced features.

The IOS feature set has a direct bearing on the amount of memory your router requires. When purchasing a router or an IOS upgrade, make sure you have enough DRAM and flash memory to support the IOS you want to run.

When downloading images, be sure you get the correct platform, feature set, and version. You should also be aware that along with hardware and feature set support, IOS versions are released under different classifications. The differences in the various releases' classifications are as follows:

- Major release (MR), according to Cisco, "delivers a significant set of platform and feature support to market." There have been five major IOS releases: 8.0, 9.0, 10.0, 11.0, and 12.0. Only 10.0 through 12.0 are still supported. An IOS MR evolves from an early deployment release to a general deployment status.
- Early deployment release (ED) is an enhancement to a MR, providing support for a new hardware or software feature.
- Limited deployment (LD) is the halfway mark of an IOS version between MR and general deployment.
- General deployment (GD) indicates that the IOS has reached a mature stable phase and is recommended for general use.
- Deferred releases (DR) are old EDs that have been replaced by a newer ED.

Note

You can download IOS versions off the Cisco Web site if you have a support contract and a Cisco Connection Online (CCO) account. You should also check the release notes associated with an IOS release so you can be aware of any problems or workarounds needed for your configuration.

As a rule, unless you need some feature or support only provided by an initial MR or EDR, wait until the IOS has reached limited deployment status. Do not be the first kid on the block to deploy a MR or ED into a production network; it will only cause you pain. In most cases, there are no major issues with these releases, but there are usually plenty of little ones. Although MR and ED are good for testing how new releases will work in your network, it is best to wait until an IOS has reached a LD or better state. 11.x is Cisco's currently shipping IOS. 12.X is currently in LD. Examples in this book will use IOS versions 11.2x and 12.0.4. Full documentation for all the Cisco IOS versions is available at Cisco Web site: `http://www.cisco.com/univercd/cc/td/doc/product/software/index.htm`.

The IOS Command Line and Configuration File Creation

The first experience most people have with configuring Cisco routers is using the `<setup>` command, or what is commonly known as the *initial configuration dialog*.

```
--- System Configuration Dialog ---

At any point you may enter a question mark '?' for help.
Use ctrl-c to abort configuration dialog at any prompt.
Default settings are in square brackets '[]'.
Would you like to enter the initial configuration dialog? [yes]:
```

When the router comes out of the box and you turn it on for the first time, you will see this dialog. The purpose of the IOS `<setup>` application is toprovide an easy way to get the router configured with a basic configuration. Once established, you can continue to refine and edit the configuration. The reason the `<setup>` command is invoked at startup is because there is no startup configuration (`startup-config`) in the router NVRAM.

All Cisco routers have at least two configuration files on Cisco routers: startup and running. The *startup configuration* is the configuration file the router stores in NVRAM and is read into RAM when the router boots up. The *running configuration* is a copy of the router's stored configuration that is loaded into the router's RAM when the router boots up. The running configuration and the startup configuration are identical until a configuration change is made.

The actual configuration is not a "text file" *per se*, as one would perceive it to be from using the IOS. Actually, it is a collection of binary data structures. The IOS acts as a translator, converting these structures into text representations and vice versa when the router's configuration is being created.

There are two methods for installing and creating a router configuration file:

- Using the setup script
- Using the Cisco IOS configuration mode(s)

Both of the methods are dependent on interaction with IOS CLI shell. The IOS CLI, more commonly referred to as the *EXEC shell*, is accessed by establishing a terminal session through a direct or virtual terminal line. A direct line is often the router's serial console port. You can also get to the command line via the asynchronous serial ports, like the AUX port or an async port on an access-server. Once IP is configured on the router, the router is also accessible over what is known as a *Virtual Terminal Line (VTY)*. VTYs are Telnet or rlogin sessions that are established over VTY interfaces provided by the IOS. Connections to VTY interfaces are established using the IP addresses configured on the router's interfaces. There are three EXEC modes available through the interface shell, discussed in the following sections. Each mode supports password protection, but by default there are no passwords.

User EXEC

The user EXEC mode is basically a user shell. Access to the user EXEC is protected by the password set for the type of access or by the router's authentication mechanism (see the section "IOS Authentication and Accounting" later in this chapter for more details). The IOS by default uses a two-tier security mode. The security mode defines which commands are available for use from the EXEC shell. The user EXEC shell is security level 0, which basically allows the user to establish connections from the EXEC shell to other TCP/IP hosts (such as Telnet or `rlogin`) or start up a network transport session (PPP, SLIP, and so on). The following code lists the available user EXEC commands:

```
access-enable          Create a temporary Access-List entry
access-profile         Apply user-profile to interface
clear                  Reset functions
connect                Open a terminal connection
disable                Turn off privileged commands
disconnect             Disconnect an existing network connection
enable                 Turn on privileged commands
exit                   Exit from the EXEC
help                   Description of the interactive help system
lat                    Open a lat connection
lock                   Lock the terminal
login                  Log in as a particular user
logout                 Exit from the EXEC
name-connection        Name an existing network connection
pad                    Open an X.29 PAD connection
ping                   Send echo messages
ppp                    Start IETF Point-to-Point Protocol (PPP)
resume                 Resume an active network connection
rlogin                 Open a rlogin connection
slip                   Start Serial Line IP (SLIP)
systat                 Display information about terminal lines
telnet                 Open a Telnet connection
terminal               Set terminal line parameters
```

```
tn3270                        Open a tn3270 connection
traceroute                    Trace route to destination
tunnel                        Open a tunnel connection
where                         List active connections
```

The user EXEC shell provides a session establishment point for access server (terminal server) users and a way to enable a privileged EXEC mode. The <ping> and <traceroute> applications are also available in user EXEC. User EXEC mode uses a > prompt to distinguish itself. To enter the privileged EXEC mode, use the <enable> user EXEC command. When in privileged EXEC the prompt will change from > to #:

```
Router> enable
Password:*******
Router#
```

Privileged EXEC

The privileged EXEC mode enables complete control over the router. This mode, commonly called *enable mode* because it is invoked with the user EXEC command <enable>, has the highest IOS security level (15) by default. Privileged EXEC mode can (and should) be password protected, but it is not by default. The enable passwords are set using the <enable password> and <enable secret> commands. Enable secret is the preferred mode of password protection because it is stored in the configuration information in an encrypted form. The IOS can also be configured to use more than one privileged EXEC security level. When multiple privilege levels are defined, EXEC commands can be assigned to specific levels that can require further password authentication. This is useful when you want users to only have access to specific commands (see the section "IOS Authentication and Accounting" later in this chapter for details).

> **Note**
>
> In this book, IOS commands will be formatted using < > to border the command text. If a command has options, they will be described in [ ]. As an example, look at the IOS command <debug [IP] [ospf] events>. debug is the command, IP is an IOS/router interface or service, ospf is an option or process pertaining to the interface or service, and events specifies what is being examined by the debug command.

> **Note**
>
> There is only one IOS shell. When you move from one EXEC mode to another, you are not starting a new EXEC shell (which is the common perception, derived from people with experience using UNIX). Mode changes (user to privileged, for example) operate along the lines of a command execution which, when successful, provides additional command access. This can be an important point for people who are users to the shell model. While shell users normally type <exit> to get to the previous level, in Cisco this ends the EXEC session completely. You should use <disable>, which closes the privileged EXEC mode.

The default privileged EXEC mode(15) is quite powerful; think of it as root or the administrator account. It should be password protected and restricted from general user access. The reason it is so powerful is because once you are in privileged EXEC, you effectively have complete control over the router. All router commands are available, and any router element can be viewed and configuration mode invoked.

The privileged EXEC command `<show>` is used to see the status and configuration of any aspect of the router. Some common uses of `<show>` are:

- `<show interfaces [type] [number]>`—Shows information about router interfaces
- `<show running-config>`—Shows the router's current configuration
- `<show startup-config>`—Shows the router's NVRAM stored configuration

To enter the configuration EXEC mode, use the `<configure [option/source]>` privileged EXEC command:

```
Router#configure terminal
```

To exit out of configuration EXEC mode, the key combination Ctrl+Z is used. Additionally, to exit out of the privileged EXEC mode and return to the user EXEC mode, use the `<disable>` command, which is illustrated in the example below:

```
router#disable
router>
```

Configuration EXEC

Configuration EXEC mode is only for creating and modifying the router's configuration files. It's entered from a privileged EXEC mode with a security level of 1 or higher. The `<configure>` command's default security setting is level 15, part of the IOS's default privileged EXEC command set. The configuration EXEC's command set is used only for router configuration. There are four configuration options.

- `<configure terminal>` provides the only direct access used to the configuration EXEC CLI. For most beginners, this is the most widely used configuration method, after `<setup>`. For simple changes and getting started with configuration, this is the way to go. No router status information can be retrieved while in `<configure terminal>` mode. Therefore, if you want to check something about the router's configuration, you need to exit out. All configuration commands must be entered one command at a time. If you want to disable an interface or service, use `<no>` before the command. Here is an example of using `<configure terminal>` and the `<no>` option. To exit out of configuration mode, the key command Ctrl+Z is used.

  ```
  Router#config terminal
  Enter configuration commands, one per line.  End with CNTL/Z.
  Router(config)#hostname test-router
  test-router(config)#no hostname test-router
  Router(config)#^Z
  Router#
  ```

> **Note**
>
> On Cisco routers, all configuration changes are active changes, so when you make them they are in effect. When changes occur, they are made to the running configuration, but they leave the startup untouched. If you make a major configuration mistake, just reboot the router.

Notice how the router's hostname changed as soon as the command was executed.

- `<configure memory>`—Reloads the startup-config into DRAM, overwriting any changes made to the running-config that are contained in the original startup-config (just in case you made a mistake). Additional changes made to the running-config, which are not contained in the startup-config, will not be altered.
- `<overwrite-network>`—Erases the running-config before reloading the startup-config. This returns configuration to the router to its boot or last saved state.

Note

In early versions of the IOS, the `<configure>` command was used to manipulate the router's configuration, and the `<write>` command was used to save and display the configuration. In IOS version 12.x, the `<copy>` and `<show>` commands have replaced or duplicated certain operations. Here are the command equivalents:

- To print the running configuration to the terminal window, use either:

  `<write terminal>`

  or

  `<show running-config>`
- To save the running configuration to NVRAM (so it can be the startup configuration if the router is rebooted), use either:

  `<write memory>`

  or

  `<copy running-config startup-config>`
- To configure the router using a Trivial File Transfer Protocol (TFTP) server, use either:

  `<configure network>`

  or

  `<copy tftp running-config>`
- To delete the startup configuration, use either:

  `<write erase>`

  or

  `<erase startup-config>`
- On 7x00 only, use:

  `<erase nvram:>`

<configure network> loads a file from a defined source, then loads it into DRAM and appends it to the running configuration. When managing large configuration elements, such as ACLs or static user accounting, this is the configuration method to use. In IOS 11.x, the command uses TFTP as the network protocol in interactive mode to specify the file source and destination:

```
IOS11.2-GW#config network
Host or network configuration file [host]?
Address of remote host [255.255.255.255]? 192.168.0.2
Name of configuration file [IOS11.2-GW-confg]? new.gw.conf
Configure using new.gw.conf from 192.168.0.2? [confirm]
```

In IOS version 12.x, the <configure network> command is aliased to the <copy> command, which uses a Universal Resource Locator (URL) to define the source file. In addition to TFTP (the default), other file sources can be specified, such as flash:\\, ftp:\\, or rcp:\\.

```
IOS12.x-GW#configure network tftp://192.168.64.46/acl2
Host or network configuration file [host]?
This command has been replaced by the command:
         'copy <URL> system:/running-config'
Address or name of remote host [192.168.64.46]?
Source filename [acl2]?
Configure using tftp://192.168.64.46/acl2? [confirm]
```

We will take a closer look at using this option in the section "Configuring Your Router with <copy> and TFTP" later in this chapter.

The configuration EXEC commands use a tiered command structure, which is apparent when you use the <configure terminal> CLI mode. When you first enter the configuration EXEC CLI, you are in *global configuration mode*. Global mode commands are used to set most of the router's global operational parameters and basic network protocol services. Some examples of these settings are the following:

- Time and date settings
- Security privileges and enable passwords
- User accounts and authentication
- Configuration register settings
- Boot parameters

From global mode, you can enter an interface or process-specific subcommand mode. Subcommand modes are used for configuring specific interfaces and router processes. Each mode has a command set specific to its subprocess and/or a process-specific subset of the global configuration command set. The most commonly used subcommand modes are the following:

<interface>

Async	Async interface
BVI	Bridge-group virtual interface
Dialer	Dialer interface
Ethernet	IEEE 802.3
FDDI	ANSI X3T9.5
Fast Ethernet	Fast Ethernet IEEE 802.3
Port-channel	Ethernet channel of interfaces
Group-Async	Async group interface
Serial	Serial
Loopback	Loopback interface
Null	Null interface
Tunnel	Tunnel interface

<router>

bgp	Border Gateway Protocol (BGP)
egp	Exterior Gateway Protocol (EGP)
eigrp	Enhanced Interior Gateway Routing Protocol (EIGRP)
igrp	Interior Gateway Routing Protocol (IGRP)
ospf	Open Shortest Path First (OSPF)
rip	Routing Information Protocol (RIP)

<line>

0-6	Async Serial line
aux	Auxiliary line
console	Primary terminal line
vty	Virtual terminal line

Of course, after you have configured the router, you will want to save the changes. It is a common mistake to believe that configuration changes are saved, because they become active as soon as they are applied. This is not the case. To save changes made to the router's configuration, the `<write memory>` or `<copy running-config startup-config>` privileged EXEC commands must be used. Both of these commands copy the running configuration to the router's NVRAM. It also does not hurt to have a copy of the configuration on something other than the router. Again, TFTP is good for doing this.

Navigating the IOS Command Line

As far as command-line interfaces go, the IOS's is decent.

True to its TOPS-20-style interface, the IOS has extensive command help, which is context-sensitive and assists with command syntax. It does not provide information on the proper use of a command or the command's dependency on other commands. So if you know what you want to do, in-line help can be a great help. If you don't, well, it's not.

Command References for IOS 9.x and Up

Cisco provides a wealth of information on its Web site. Unfortunately, the same problem plagues the Web site as the IOS's help system: if you don't know what you are doing, it's hard to find the right information. Here are the locations of the IOS command references for IOS 9.x and up. Like all Cisco documentation, they are well written and usually quite clear.

Command Reference for IOS 9.X—`http://www.cisco.com/univercd/cc/td/doc/product/software/ssr921/rpcr/index.htm`

Command Reference for IOS 10.X—`http://www.cisco.com/univercd/cc/td/doc/product/software/ios100/rpcs/index.htm`

Command Reference for IOS 11.1—`http://www.cisco.com/univercd/cc/td/doc/product/software/ios111/supdocs/sbook/index.htm`

Command Reference for IOS 11.2—`http://www.cisco.com/univercd/cc/td/doc/product/software/ios112/sbook/index.htm`

Command Reference for IOS 12.0—`http://www.cisco.com/univercd/cc/td/doc/product/software/ios120/12cgcr/index.htm`

You can invoke command help at any time in every EXEC mode (even when you are in the middle of the command) by typing a question mark or part of a command with a question mark. When used by itself, <?> will list all of the available commands:

```
Router#?
Exec commands:
  access-enable    Create a temporary Access-List entry
  access-profile   Apply user-profile to interface
  access-template  Create a temporary Access-List entry
  bfe              For manual emergency modes setting
  clear            Reset functions
  clock            Manage the system clock
  configure        Enter configuration mode
  connect          Open a terminal connection
  copy             Copy configuration or image data
```

When used in mid-command, <?> will list the possible IOS commands that contain the letters typed before the <?>:

```
Router(config)#ip rou?
route  routing
Router(config)#ip rou<tab>
Router(config)#ip route
```

The <?> command can also be used after a complete word in a multiword command to provide the next command option:

```
Router(config)# ip ?
Global IP configuration subcommands:
  access-list            Named access-list
  accounting-list        -Select hosts for which IP accounting information is
kept
  accounting-threshold   Sets the maximum number of accounting entries
  accounting-transits    Sets the maximum number of transit entries
  address-pool           Specify default IP address pooling mechanism
  alias                  Alias an IP address to a TCP port
  as-path                BGP autonomous system path filter
  bgp-community          Format for BGP community
  bootp                  Config BOOTP services
  classless              Follow classless routing forwarding rules
  community-list         Add a community list entry
  default-gateway        Specify default gateway (if not routing IP)
  default-network        Flags networks as candidates for default routes
  dhcp-server            Specify address of DHCP server
```

The command line also supports UNIX-like word completion by using the Tab key and command abbreviation. These options are helpful for people who hate to type (like me) or who are bad spellers. For example, the `<copy running-con fig startup-config>` command can be typed as:

```
<cp run start>
```

Here is another example, to display the running configuration:

```
<show run>
```

Another win for bad typists is the IOS's command-line history and command-line editing feature. Both options are set using the `<terminal>` command. To enable history and editing in privileged EXEC mode use:

```
Router#terminal editing
Router#terminal history size 100
```

The `size` option allows you to specify how many commands to save in the memory buffer. To see all the terminal settings, use `<show terminal>`. The up arrow and Ctrl+P, and the down arrow and Ctrl+N keys are used to scroll through the command history. The up arrow and Ctrl+P move back toward the session's first command. The down arrow and Ctrl+N move ahead toward the session's last command (and the EXEC prompt). Command history is available in all EXEC modes. Included in Table 7.5 is a list of line editing commands, the majority of which you will forget. The important ones are left arrow and Ctrl+B to move the cursor left, right arrow and Ctrl+F to move the cursor right, and Backspace and Ctrl+H to delete the line recursively.

Table 7.5 **Line Editing Commands**

Command	**Description**
Ctrl+P, up arrow	Previous command in history buffer
Ctrl+N, down arrow	Next command in history buffer
Ctrl+B, left arrow	Move CLI cursor to the left
Ctrl+H, right arrow	Move CLI cursor to the right
Ctrl+A	Move to beginning of command line
Ctrl+E	Move to the end of the command line
Ctrl+K	Delete line back from cursor position
Ctrl+U	Cut line and store in terminal buffer
Ctrl+Y	Paste line in stored terminal buffer
Ctrl+D	Delete character under cursor
Esc+F	Move forward one word
Esc+B	Move back one word

The IOS command line also supports a variation of the UNIX pager display, where <more> is automatically invoked:

```
Router#?
Exec commands:
  access-enable    Create a temporary Access-List entry
  access-profile   Apply user-profile to interface
  access-template  Create a temporary Access-List entry
  bfe              For manual emergency modes setting
  clear            Reset functions
  clock            Manage the system clock
  configure        Enter configuration mode
  connect          Open a terminal connection
  copy             Copy configuration or image data
  debug            Debugging functions (see also 'undebug')
  disable          Turn off privileged commands
  disconnect       Disconnect an existing network connection
  enable           Turn on privileged commands
  erase            Erase flash or configuration memory
  exit             Exit from the EXEC
  help             Description of the interactive help system
  lat              Open a lat connection
  lock             Lock the terminal
  login            Log in as a particular user
  logout           Exit from the EXEC
  mrinfo           Request neighbor and version information from a multicast
 --More--
```

<more> uses the Spacebar to scroll to the next page and <cr> to see the next line. You can type any character to return to the EXEC prompt.

Configuring Your Router with *<copy>* and TFTP

Configuration EXEC mode is fine for simple adds and changes. For large changes (such as static routing table or ACL maintenance), the command line is just not practical. In addition to the <write network> command, the IOS also provides <copy>. The command uses a "from-to" syntax style similar to the DOS copy command. The <copy> command is used to copy files on and off the router's NVRAM, DRAM, and flash "file systems" (running or startup configuration, or the IOS). To see the contents of the router's memory file systems, the <show> and <dir> (12.x only) commands are needed.

> **Note**
>
> Cisco, in later versions (12.x) of the IOS, uses the term file system to describe memory (originally with only the 7x00 series routers but now with all).

Along with the router's memory file system, a variety of source and destination targets are supported:

```
Router#copy
  /erase          Erase destination file system (not available in all IOS
                  versions)
  flash:          Copy from flash: file system
  mop             Copy from a MOP server (not available in all IOS versions)
  ftp:            Copy from ftp: file system
  nvram:          Copy from nvram: file system
  rcp:            Copy from rcp: file system
  null:           Copy to null: file system (not available in all IOS
                  versions)
  tftp:           Copy from tftp: file system
  xmodem          Copy from xmodem server (not available in all IOS versions)
  ymodem          Copy from ymodem server (not available in all IOS versions)
  running-config  Copy from current system configuration
  startup-config  Copy from startup configuration
```

The following sections take a close look at some of the possible source and destination options and usage combinations.

Using <*copy*> to Erase File Systems

The `<copy /erase>` command is introduced in IOS version 12.x, and when used in conjunction with the `[null]` file system will erase a memory partition. For example, to erase NVRAM, the command `<write erase>` would be used. The `<copy>` command approach would use the following:

```
Router#copy /erase null: nvram:startup-config
Destination filename [startup-config]?
Erasing the nvram filesystem will remove all files! Continue? [confirm]
[OK]
Erase of nvram: complete
Router#
```

The `<erase>` privileged EXEC command will perform the same task. The command differs slightly between IOS 11.x and 12.x, reflecting the context change to memory as a "file system":

IOS 11.3:

```
Router#erase ?
  flash           Erase System Flash
  startup-config  Erase contents of configuration memory
```

IOS 12.x:

```
Router#erase ?
  flash:          Filesystem to be erased
  nvram:           Filesystem to be erased
  startup-config  Erase contents of configuration memory
```

Using <*copy*> with Flash Memory

The [flash:] target is used to represent all flash memory installed on a router's motherboard and installed in PCMCIA slots (if no partitioning is used). If you want to see the contents of a specific PCMCIA flash card, it must be indicated using the slot [slot0:]. You can copy to or from flash cards using [slot:] as the target. Here are some examples:

```
7200router#copy run slot0:
Destination filename [running-config]?
5810 bytes copied in 0.816 secs
```

Now let's see what is on the PCMCIA card in slot0 using the <show [filesystem]:[partition]:> privileged EXEC command and the <dir [filesystem]:[partition]:> privileged EXEC command. The show command output is different in IOS version 12.x than in previous versions, so both command outputs are shown.

Using IOS version 12.x <show>:

```
7200router#sh flash
1    .D unknown  3CC2EF20  4DC648    21  4965832 Jun 16 1999
      11:27:14 c7200js-mz.113-6.bin
2     .. unknown  6665A22D  B7DC14    23  6952268 Jun 16 1999
      11:47:03c7200-js40-m
      z.120-4.bin
3     .. unknown  60D41E19  B7E8EC    14     3159 Jun 20 1999
      09:34:20 running-config
8656660 bytes available (11921644 bytes used)
7200router#
```

Using IOS version 11.x <show>:

```
2500Router#sh flash
System flash directory:
File  Length   Name/status
  1   9212736  c2500-enter40-l.113-9.bin
[9212800 bytes used, 7564416 available, 16777216 total]
16384K bytes of processor board System flash (Read/Write)
2500Router#
```

Using `<dir>` (available for all routers in version 12.x; available on 7x00 series only in earlier IOS versions):

```
7200router#dir /all
Directory of slot0:/
  1  -rw-     4965832   Jun 16 1999 11:27:14  [c7200-js-mz.113-6.bin]
  2  -rw-     6952268   Jun 16 1999 11:47:03  c7200-js40-mz.120-4.bin
  3  -rw-        3159   Jun 20 1999 09:34:20  running-config
20578304 bytes total (8656660 bytes free)
7200router#
7200router#dir slot0:
Directory of slot0:/
  2  -rw-     6952268   Jun 16 1999 11:47:03  c7200-js40-mz.120-4.bin
  3  -rw-        3159   Jun 20 1999 09:34:20  running-config
20578304 bytes total (8656660 bytes free)
7200router#
```

One of the advantages of using PCMCIA flash cards is using them to move configuration and IOS files from router to router. This example installs a new static route table from a PCMCIA card in `slot0:` to the running configuration:

```
3620router#sh slot0:
PCMCIA Slot0 flash directory:
File   Length        Name/status
  1    1132          static-192.168.x
  2    4498744       c3620.113.2.bin
  3    3659          3620-config
  4    4080564       c3620.112.bin
  5    1132          3620a-config
[13085648 bytes used, 3691568 available, 16777216 total]
16384K bytes of processor board PCMCIA Slot0 flash (Read/Write)
cisco-3620#copy slot0:static-192.168.x running-config
Destination filename [running-config]?
3159 bytes copied in 0.336 secs
cisco-3620#
```

When copying any file, unless no previous file exists, copied information is appended to the existing file.

Using <*copy*> with TFTP

Even though later IOS releases have added copy support for TCP/IP, FTP, and RCP, the standard used for offloading and uploading IOS images and configuration files has been TFTP. TFTP is a UNIX routine used in conjunction with BOOTP, which provides a file transfer mechanism for loading OS and configuration information for diskless workstations. TFTP has no security, and files can be copied back and forth with no authentication. It uses IP's UDP port (69) for transport. On UNIX systems, TFTP is run from **/etc/inetd.conf** or as a boot service started from **/etc/rc** or equivalent.

With both methods, the TFTP root directory must be indicated and must be world readable and writable. A file must exist in the TFTP root directory with the same name as the file you want to write.

After the router has an IP address, TFTP can be used to load the rest of the router's configuration, and if needed you can boot your router off an IOS image stored on the server. Using <copy> with a TFTP server works the same as copying from a memory file system. Here are some command examples. To copy the running configuration to a TFTP server, type the following:

```
Router#copy running-config tftp
Remote host []? 192.168.0.2
Name of configuration file to write [router-config]?
Write file router-confg on host 192.168.0.2? [confirm]
Building configuration...
Writing router-config !! [OK]
Router#
```

Keep in mind that if you want to save different versions of the same router's configuration (a good idea) or save a different router's configuration on the same TFTP server, you should change the name of configuration file to write accordingly. Here is an example of using TFTP with the <configure network> privileged EXEC command:

```
Router#configure network
Host or network configuration file [host]?
Address of remote host [255.255.255.255]? 192.168.0.2
Name of configuration file [router-confg]? 2600c.config
Configure using mis-ir26 from 192.168.0.2? [confirm]
Loading mis-ir26 from 192.168.0.2 (via Ethernet0): !
[OK - 784/32723 bytes]
2600Router#
```

TFTP will be reviewed again in the section "Upgrading Your Router's IOS."

Note

TFTP is also available for Windows and Macintosh as an add-on application. Cisco provides a free server on its Web site: `http://www.cisco.com/pcgi-bin/tablebuild.pl/tftp`. Another good server is available from SolarWinds at `http://www.solarwinds.net`. A great Macintosh TFTP server is available as shareware from `http://www.2gol.com/users/jonathan/software`.

Basic Cisco Router Configuration

There are hundreds of different IOS elements that can be configured on a given router, all of which are fully documented by Cisco. To help you get started, IOS has the <setup> helper script. The problem with <setup> is that quite often, the setup configuration ends up being the production configuration, leaving the router only half configured and a potential security risk.

The goal of this section is to show you how to configure some of the basic Layer 2 and Layer 3 protocols. The examples are structured so you can follow along on a router to get a feel for working with the command line. To follow along, you need (at minimum) a router with a serial and Ethernet interface and multiprotocol version IOS 11.2.x or greater.

Here is the scenario (see Figure 7.1): You have a small company (WidgetCo) with two offices (Ridgefield and Concord) that have Ethernet LANs with a dedicated T1 connecting them. The Ridgefield office also has a T1 to the Internet. The Concord office contains Finance, which uses a UNIX accounting system accessed with dumb VT100 terminals. Sales and Marketing are in Ridgefield, and they have PCs running NT with a Novell LAN and Macintoshes. The Ridgefield office also contains Research, which uses UNIX, and the Management team, which uses PCs and Macintoshes.

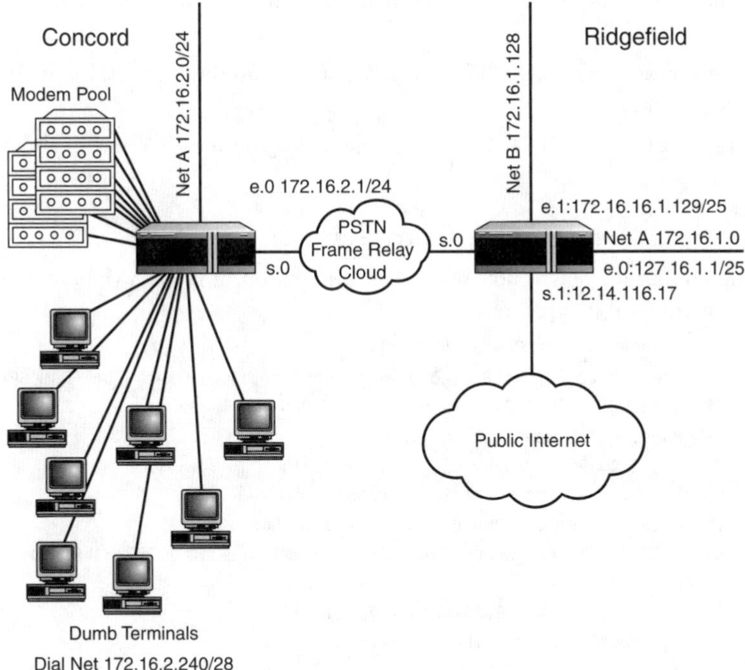

Figure 7.1 The WidgetCo network map.

The WidgetCo network has a Cisco 2600 in the Concord office with one Ethernet, 16 Async Serial ports, and one integrated T1 serial interface. The Ridgefield office has a Cisco 2514 with two Ethernet and two serial interfaces.

Helpful IOS Commands

Understanding how the router is configured and how its interfaces are behaving is very important. Earlier, you were introduced to some of the variations of the `show` command. Here are some additional usages that are quite handy:

- `<show interfaces>`—Displays the status of the router's interfaces. This command can also show a specific interface by adding the target to the end of the command: `<show interface [target]>`.
- `<show interfaces accounting>`—Provides throughput statistics for a router's interfaces.
- `<show memory failures alloc>`—Shows router processes that have failed due to memory constraints.
- `<show memory>`—Lists all of the router's memory usage statistics.
- `<show version>` and `<show hardware>`—Provide summary information on the router's installed interfaces, hardware, memory, and IOS version.
- `<show protocols>`—Displays information about all of the router's active Layer 3 protocols.
- `<show processes cpu>`—Provides information on the router CPU usage and load averages.
- `<terminal monitor>`—Displays console messages on a VTY terminal display.

Configuring Your Router from Scratch

Assuming you have a new router with no configuration in NVRAM, here is an example of what you will be greeted with:

```
Cisco Internetwork Operating System Software
IOS (tm) C2600 Software (C2600-I-M), Version 11.3(2)XA1, PLATFORM SPECIFIC
RELEASE SOFTWARE (fc2)
TAC:Home:SW:IOS:Specials for info
Copyright © 1986-1998 by Cisco Systems, Inc.
Compiled Fri 03-Apr-98 07:00 by rnapier
Image text-base: 0x80008084, data-base: 0x8051BDD8
cisco 2610 (MPC860) processor (revision 0x100) with 12288K/4096K bytes of
memory.
Processor board ID JAB021703TR (3788815090)
M860 processor: part number 0, mask 32
Bridging software.
```

```
X.25 software, Version 3.0.0.
1 Ethernet/IEEE 802.3 interface(s)
1 Serial network interface(s)
16 terminal line(s)
32K bytes of non-volatile configuration memory.
4096K bytes of processor board System flash (Read/Write)
Notice: NVRAM invalid, possibly due to write erase.
         --- System Configuration Dialog ---
Would you like to enter the initial configuration dialog? [yes/no]:
```

This has little value to you, so answer <no> or use Ctrl+C to terminate. This routine, after a brief cleanup, drops you in user EXEC mode. From EXEC, you move on to <enable> privileged EXEC mode and finally the CLI configuration EXEC mode <configure terminal>. Before you start configuring interfaces, you need their type, slot, and port information. Port information can be retrieved by using the <show running-config> privileged EXEC command. Here is the output from the Concord router:

```
Current configuration:
!
version 11.3
no service password-encryption
!
!
!
!
!
interface Ethernet0/0
 no ip address
 shutdown
!
interface Serial0/0
 no ip address
 shutdown
!
ip classless
!
line con 0
line 33 48
line aux 0
line vty 0 4
```

This router has one Ethernet interface and one serial interface.

Configuring Interfaces

There are two class C IP networks used by WidgetCo—one for each location. Using variable-length subnet mask (VLSM), you can break up the space to provide address space for all the LAN and Remote Access dial networks. Each network will be able to support up to 126 hosts:

Ridgefield:

Net A `172.16.1.0 /25 or 255.255.255.128`

Net B `172.16.1.128 /25`

Concord:

Net A `172.16.2.0 /24`

To configure the WAN link between Concord and Ridgefield, you have some choices. The first one is to subnet one of the class C address spaces even further and use some of the address space. Because Ridgefield is already subnetted, the Concord address space 172.16.2.0 /24 could be subnetted like this:

```
172.16.2.0 /25
172.16.2.128 /26 or 255.255.255.192
172.16.2.192 /27 or 255.255.255.224
172.16.2.224 /28 or 255.255.255.240
172.16.2.240 /28 or 255.255.255.240
172.16.2.248 /30 or 255.255.255.240
172.16.2.252 /30 or 255.255.255.240
```

The `/25` segment could be used for the LAN, and either of the `/30` spaces could be used for the WAN link addressing. WidgetCo also plans to provide dial-in access at the Concord office, so one of the `/28` spaces could be used for dial-in address allocation, leaving Concord with plenty of addresses to add additional networks in the future.

The second option is to use `<IP unnumbered>` on the WAN interfaces. The IOS `<IP unnumbered>` option allows an interface to share the IP address of another interface. This effectively creates a virtual bridge, whereby the LAN interfaces appear to be directly connected to one another. Because the serial interfaces are borrowing the IP addresses of their reference interfaces to exchange IP datagrams, this method is only practical for point-to-point serial links.

Both of these approaches have limitations. The subnetting option segments the address space into spaces that may not have practical use, especially if the LAN segments have enough hosts to utilize the larger address spaces. The unnumbered option can present problems with certain IP routing protocols. The unnumbered interface is not accessible with Telnet using a VTY line or directly monitorable with SNMP as the interface has no IP address.

As it turns out, the Concord office is quite large so `<ip unnumbered>` is a better option. Now let's examine the LAN and WAN interface configurations for the Concord and Ridgefield offices.

The Concord side:

```
Router(config)#hostname Concord-GW
Router(config)# enable secret *****
Concord-GW(config)#int e0
Concord-GW(config)#description Concord LAN-GW
Concord-GW(config-if)#ip address 172.16.2.1 255.255.255.128
Concord-GW(config-if)#no shut
Concord-GW(config-if)exit
Concord-GW(config)#int s0/0
Concord-GW(config)#description point to point to Ridgefield
Concord-GW(config-if)#ip unnumbered ethernet 0
Concord-GW(config-if)#encapsulation frame relay
Concord-GW(config-if)#frame-relay interface-dlci 16
Concord-GW(config-if)#frame-relay lmi-type ansi
Concord-GW(config-if)#no shutdown
Concord-GW(config-if)exit
Concord-GW(config)# ip route 0.0.0.0 0.0.0.0 172.16.1.1
```

The Ridgefield side:

```
Router(config)#hostname Ridge-GW
Ridge-GW(config)# enable secret #####
Ridge-GW(config)#int e0
Ridge-GW(config)# description Ridgefield LAN-GW-Netb
Ridge-GW(config-if)#ip address 172.16.1.1 255.255.255.128
Ridge-GW(config)#int e1
Ridge-GW(config)# description Ridgefield LAN-GW-Neta
Ridge-GW(config-if)#ip address 172.16.1.129 255.255.255.128
Ridge-GW(config-if)#no shut
Ridge-GW(config-if)exit
Ridge-GW(config)#int s0
Ridge-GW(config)# description point to point to Concord
Ridge-GW(config-if)#ip unnumbered ethernet 0
Ridge-GW(config-if)#encapsulation frame-relay
Ridge-GW(config-if)#frame-relay interface-dlci 16
Ridge-GW(config-if)#frame-relay lmi-type ansi
Ridge-GW(config-if)#no shut
Ridge-GW(config-if)exit
Ridge-GW(config)#int s1
Ridge-GW(config)#description Internet-GW
Ridge-GW(config-if)#ip address 12.14.116.17 255.255.255.252
Ridge-GW(config-if)#encapsulation ppp
Ridge-GW(config-if)exit
Ridge-GW(config)# ip route 0.0.0.0 0.0.0.0 12.14.116.18
```

Another default setting is DNS resolution. However, for resolution to work properly, the router needs to know where the DNS server is or have a local host table. To set the DNS server, use `<ip name-server [ip address]>`. To enter host addresses, use `<ip host [hostname] [ip address]>`. To disable DNS resolution so the IOS will not try to resolve your typing mistakes, use the global configuration EXEC command `<no ip domain-lookup>`.

WidgetCo has one DNS server on the Ridgefield network, but the Finance terminal users hate to type and fear that if the point-to-point link goes down, they will be unable to work. So, you need to add the hosts they connect to:

```
Concord-GW(config)# ip name-server 172.16.2.100
Concord-GW(config)# ip host cashbox 172.16.2.20
Concord-GW(config)# ip host lostit 172.16.2.22
Concord-GW(config)# ip host postman 172.16.2.30
```

Note

By default, the IOS uses HDLC for the synchronous serial interface encapsulation. In addition, ATM and x.25 are also available for use on serial interfaces.

The `<ip route [network] [mask] [gateway] [distance]>` command is used to set the router's default routes. For the Concord network, the default gateway is the Ethernet interface on the Ridgefield router as it is the next hop. In Ridgefield's case the remote end of its Internet link is its default gateway. Using this approach, Concord sends all its Internet-bound traffic to the Ridgefield gateway, where it is forwarded on to the ISP's router. IP route configuration will be covered in depth in Chapter 10.

Configuring AppleTalk and IPX

IP, however, is not the only protocol used by WidgetCo. AppleTalk and IPX are also required. Unlike IP, IPX and AppleTalk are not enabled by default. To enable AppleTalk, the `<appletalk routing>` command is used. IPX follows suit with the `<ipx routing>` command. Both commands must be entered into the routers' configuration. AppleTalk and IPX protocol routing must be enabled before the interfaces can be configured.

AppleTalk and IPX are simple to configure, requiring only a network address (and zone names, for AppleTalk) to start the process.

The Concord side:

```
Concord-GW(config)#
Concord-GW(config)#description Concord LAN-GW
Concord-GW(config-if)#ip address 172.16.2.1 255.255.255.0
Concord-GW(config-if)#ipx network 4000
Concord-GW(config-if)#ipx type-20-propagation
Concord-GW(config-if)#ipx encapsulation sap
Concord-GW(config-if)#appletalk cable-range 31-31
Concord-GW(config-if)#appletalk zone S&M
Concord-GW(config-if)#appletalk glean-packets
Concord-GW(config-if)#appletalk protocol rtmp
Concord-GW(config-if)#exit
Concord-GW(config)#int s0
Concord-GW(config)#description point to point to Ridgefield
Concord-GW(config-if)ipx network 45
Concord-GW(config-if)ipx type-20-propagation
Concord-GW(config-if)appletalk cable-range 2-2
Concord-GW(config-if)appletalk zone crossover
Concord-GW(config-if)exit
Concord-GW(config)#
```

Note

In addition to their native protocols, AppleTalk and IPX can be routed with Cisco's proprietary routing protocol EIGRP. Configuration of multiprotocol routing with EIGRP will be covered in Chapter 10.

The Ridgefield side:

```
Ridge-GW(config)#
Ridge-GW(config-if)# description Ridgefield LAN-GW
Ridge-GW(config-if)#ipx network 8990
Ridge-GW(config-if)#ipx type-20-propagation
Ridge-GW(config-if)#ipx encapsulation sap
Ridge-GW(config-if)#appletalk cable-range 33-34
Ridge-GW(config-if)#appletalk zone research
Ridge-GW(config-if)#appletalk zone FrontOffice
Ridge-GW(config-if)#appletalk glean-packets
Ridge-GW(config-if)#appletalk protocol rtmp
Ridge-GW(config-if)#exit
Ridge-GW(config)#int s0
Ridge-GW(config-if)# description point to point to Concord
Ridge-GW(config-if)ipx network 45
Ridge-GW(config-if)ipx type-20-propagation
Ridge-GW(config-if)#appletalk cable-range 2-2
Ridge-GW(config-if)#appletalk zone crossover
Ridge-GW(config-if)exit
Ridge-GW(config)#
```

All that is required to enable IPX or AppleTalk is a network address. After a network address is set, the nodes can do the rest. From a practical perspective, however, some additional configuration options may be required depending on the interface type or for performance reasons. In the WidgetCo configuration, some of the PC users are using IPX (because it is faster) to access the NT servers. In order for NetBIOS broadcasts to be forwarded across all the network links, `<ipx type-20-propagation>` needs to be enabled. You might also find with IPX that SAP and RIP updates can consume too much bandwidth and need to be reduced. Reduction can be accomplished by using the interface configuration EXEC command `<ipx update interval>`:

```
Ridge-GW(config-if)#ipx update interval rip changes-only
Ridge-GW(config-if)#ipx update interval sap changes-only
```

In the example above, the command instructs the router to only send RIP and SAP updates when changes occur. This approach reduces traffic on WAN links and places the RIP/SAP information distribution responsibility on the local servers.

Note

To see information about the interfaces that are running AppleTalk and IPX, two commands are available. To view interface-specific information, use `<show appletalk interface [interface slot/port]>` for AppleTalk and `<show ipx interface [interface slot/port]>` for IPX. If an interface summary is all you need, you can use `<show appletalk interface brief>` for AppleTalk and `<show ipx interface brief>` for IPX.

Viewing Routing Tables

After your interfaces are configured, you may want to see the routing tables of the various protocols you have running on your router. Like everything else on a Cisco router, if you want to see something, a variation of the `<show>` command is used. To see the IP routing table, use the privileged EXEC command `<show ip route>`. Here is the IP routing from the Concord router:

```
Concord-GW#sh ip route
Codes: C - connected, S - static, I - IGRP, R - RIP, M - mobile, B - BGP
       D - EIGRP, EX - EIGRP external, O - OSPF, IA - OSPF inter area
       N1 - OSPF NSSA external type 1, N2 - OSPF NSSA external type 2
       E1 - OSPF external type 1, E2 - OSPF external type 2, E - EGP
 i - IS-IS, L1 - IS-IS level-1, L2 - IS-IS level-2, * - candidate
default U - per-user static route, o - ODR
Gateway of last resort is 172.16.1.1 to network 0.0.0.0
     172.16.0.0/16 is subnetted, 2 subnets
C       172.16.1.0/25 is directly connected, Serial0/0
C    172.16.2.0/24 is directly connected, Ethernet0/0
S*   0.0.0.0/0 [1/0] via 172.16.1.1
Concord-GW#
```

To see AppleTalk routing information, two privileged EXEC commands are used, `<show appletalk route>` and `<show appletalk zone>`. Here is Ridgefield's AppleTalk routing information:

```
Ridge-GW#show appletalk route
Codes: R - RTMP derived, E - EIGRP derived, C - connected, A - AURP
       S - static  P - proxy
3 routes in internet
The first zone listed for each entry is its default (primary) zone.
C Net 33-34 directly connected, Ethernet0, zone research
          Additional zones: 'FrontOffice'
C Net 2-2 directly connected, serial0, zone crossover
R Net 31-31
```

Here is the WidgetCo AppleTalk zone information:

```
Ridge-GW#sh apple zone
Name                                Network(s)
FrontOffice                        33-33
research                            33-33
crossover                   2-2
S&M                        31-31
Total of 4 zones
Ridge-GW#
```

The IPX routing table is displayed using the <show ipx route> command at the privileged EXEC prompt:

```
Ridge-GW#sh ipx route
Codes: C - Connected primary network,    c - Connected secondary network
       S - Static, F - Floating static, L - Local (internal), W - IPXWAN
       R - RIP, E - EIGRP, N - NLSP, X - External, A - Aggregate
       s - seconds, u - uses, U - Per-user static
2 Total IPX routes. Up to 1 parallel paths and 16 hops allowed.
No default route known.
R     4000 (SAP,        Se0
C         45 (PPP),           Se0
C       8990 (SAP),           Et0
Ridge-GW#
```

The <show [protocol] route> command is the first command you should use when network access problems arise—because if there is no route, there is no way to get there.

Configuring Async Lines

The last configuration task on the WidgetCo network is to set up the async serial lines in the Concord office, of which there are 16. Async lines 33 to 39 are used for dumb terminals, lines 40 and 41 for serial-line printers, and lines 42 to 48 are used for an IP (SLIP/PPP) dial pool.

Configuring Dumb Terminal Service

To set up the terminal lines, you need to define how you want the lines to behave when users are connected to them. In WidgetCo's case, the lines will be used for VT100 terminal access via Telnet to UNIX hosts. In case you were wondering where the line numbers came from, they were displayed along with the rest of the interface information when you ran the <show running-config> at the beginning before you started to configure the routers:

```
ip classless
!
line con 0
line 33 48
line aux 0
line vty 0 4
```

To start, as with any Cisco interface, you need to move into the interface's subconfiguration mode (while in the configuration EXEC):

```
Concord-GW(config)#line 33 39
```

After you're there, you want to define the transport protocols that can be used with the line(s). Because you are only using IP and Telnet, you should use the `<transport [direction] [protocol/all]>` command and specify Telnet only:

```
Concord-GW(config-line)#transport input telnet
Concord-GW(config-line)#transport output telnet
```

The `<privilege level>` command sets the EXEC mode to which the line defaults:

```
Concord-GW(config-line)#privilege level 0
```

The `<login [service]>` sets the type of password checking used by the line for access authentication. The `<login local>` option requires a local username and password account to exist on the router (see the section "IOS Authentication and Accounting" for more details):

```
Concord-GW(config-line)#login local
```

If the lines were going to be directly attached to equipment console ports instead of dumb terminals, you might have to change the lines' operating speed or parity. To configure the lines' operational behavior, use the following commands:

```
Concord-GW(config-line)#speed 9600
Concord-GW(config-line)#databits 8
Concord-GW(config-line)#stopbits 1
Concord-GW(config-line)#parity none
Concord-GW(config-line)#terminal-type vt100
Concord-GW(config-line)#length 24
Concord-GW(config-line)#width 80
```

The following command sets the line's escape character, which is Ctrl+Shift+6 by default. The escape character can be set by entering the character or its ACSI number.

```
Concord-GW(config-line)#escape-character q
```

The `<session-limit>` command controls the number of concurrent Telnet sessions that can be run on a line (the command sequence Ctrl+Shift+6+X suspends a session). The `<notify>` command has the line notify the user when his/her attention is needed at an open session. `<session-timeout>` sets the line's idle time out:

```
Concord-GW(config-line)#session-limit 5
Concord-GW(config-line)#notify
Concord-GW(config-line)#session-timeout 15
```

Note

There are applications that provide ASCII tables for all the major computing platforms. A good Web resource is `http://www.efn.org/~gjb/asciidec.html`.

Note

In addition to providing async serial access to the network, async ports can also provide access to serial devices from the network with Telnet. This provides a great way to access console ports of equipment over a dial-out modem. To access a device using Telnet, open a Telnet session to port 2000 plus the port's line number. For example, a modem connected on line 44 would have 2044 as its port.

Now your terminal lines are set up and ready to be attached to dumb terminals. As these are the only network access points for some of the WidgetCo finance department users, you are not restricting their access to the user EXEC. If the need arises, you could set up the terminals to just provide Telnet access by `<autocommand connect>` to the line configuration.

Configuring LPR Service

IOS supports a variation of the Berkeley line printer daemon (LPD) over async serial lines, which is used for TCP/IP remote printing on UNIX systems. To setup LPD, first the async port(s) must be configured to accept data flows without the chance of an EXEC session starting. This is done using the `<no exec>` command. The port should transport all input and output types. The `<telnet transparent>` commands sets the line to act as a permanent data pipe so the printer data will be received by the printer without mistakenly being interpreted as line control data. Serial line printers usually like hardware flow control, whereas laser printers prefer software flow control. Check with your printer's documentation for the preferred setting.

To set up LPR service on lines 40 and 41:

```
Concord-GW(config)#line 40 41
Concord-GW(config-line)#no exec
Concord-GW(config-line)#no vacant-message
Concord-GW(config-line)#transport input all
Concord-GW(config-line)#telnet transparent
Concord-GW(config-line)#stopbits 1
Concord-GW(config-line)#flowcontrol hardware
Concord-GW(config-line)#exit
```

After the line(s) are configured, the `<printer [name] [line #] [option]>` command is used to define the LPD printer name. The `<newline-convert>` option fixes problems with single character line terminators, which are not handled correctly by all UNIX systems:

```
Concord-GW(config)#printer cpop line 40 newline-convert
Concord-GW(config)#printer ccrunch line 41 newline-convert
```

After the router is configured, the UNIX or NT system that wants to use the printer will have to be configured. Here are some sample UNIX **/etc/Printcap** settings from `cashbox.widgetco.com`:

```
cornpops|Printer on Cisco termserver port 40:\
              :lp=:rm=192.168.17.1:rp=cpop:\
              :sd=/var/spool/lpd:\
              :lf=/var/log/lpd-errs:

cptncrunch|Printer on Cisco termserver port 40:\
              :lp=:rm=192.168.17.1:rp=ccrunch:\
              :sd=/var/spool/lpd:\
              :lf=/var/log/lpd-errs:
```

To print from NT you will, of course, need the appropriate drivers for the printer.

Configuring IP Dial-In (PPP/SLIP) Service

IP dial-in access is commonplace today (in some form) on most networks. Dial-in access provides an inexpensive way for users to access network services (mail, Internet, and so on) remotely and is seen as an indispensable tool for sales and marketing types. The major drawbacks to dial-in are security and, most recently, speed.

From a network security standpoint, a *dial-in pool* is a hole in your network which must be protected. The IOS's basic security model (old-mode) revolves around local user accounts and cleartext passwords (which are what WidgetCo uses). The IOS's enhanced security model (new-mode) is covered in the "IOS Authentication and Accounting" section later in the chapter.

In terms of speed, async dial-in is slow, and it's not going to get any faster. A Cisco 2511 with USR Sporter 33.6K modems will operate at around 2 to 4Kbps. Using USR (56K) I-modems moves the data rate to 4 to 6Kbps. These rates in comparison to dedicated access servers (Cisco AS5200 and USR TotalControl) are off by 2 to 4Kbps. So if fast access is required, you probably want to use an access server. Nevertheless, if inexpensive and reasonable access is acceptable (and what you can afford), this will be just fine.

IP (dial-in/dial-out) modem pool access has two elements:

- Creating the dial pool
- Configuring the line interfaces

Setting up the dial access has two parts:

- Creating the `<group-async>` interface
- Configuring the lines to support the interface

> **Note**
>
> More information on configuring LPD on UNIX and NT systems can be found on the Web. The following is a short list of sites, but there are more.
>
> Basic **printcap** information:
>
> `http://www.lantronix.com/htmfiles/ts/techtip/faq/rtel004.htm`
>
> `http://support.wrq.com/techdocs/7045.htm`
>
> `http://outoften.doc.ic.ac.uk/local/system/lpd.html`
>
> LPD access for Macintoshes:
>
> `http://www.umich.edu/~rsug/netatalk/`
>
> LPD service on Windows NT:
>
> `http://bmewww.eng.uab.edu/HOWTO/NT-basedLPservice/`

The <group-async> interface needs to be created first. The interface functions like any other router interface; it requires Layer 3 addressing (<ip unnumbered>) and a Layer 2 framing protocol <encapsulation ppp>. The <group-range> defines the async lines that make up the group:

```
Concord-GW(config)#interface group-Async 1
Concord-GW(config-if)#ip unnumbered Ethernet0
Concord-GW(config-if)#ip tcp header-compression passive
Concord-GW(config-if)#encapsulation ppp
Concord-GW(config-if)#async mode interactive
Concord-GW(config-if)#async default routing
Concord-GW(config-if)#flowcontrol hardware
Concord-GW(config-if)#group-range 42 48
```

Alternatively, if you do not have an access-server router, you can also use the AUX line on your router for dial-in access (this is great for a backdoor access to your network). The configuration elements are the same, only the interface type changes. Instead of creating a <group-async> interface, a single <async> interface is built (this dial pool also supports AppleTalk):

```
Ridge-GW#(config)#interface async 1
Ridge-GW#(config-if)#ip unnumbered Ethernet0
Ridge-GW#(config-if)#async default routing
Ridge-GW#(config-if)#async mode interactive
Ridge-GW#(config-if)#appletalk cable-range 40-40 40.89
Ridge-GW#(config-if)#appletalk zone outofband
```

The <group-async>/<async> interface is also responsible for providing the IP address source for the dial-in connections. Addressing can be provided from a local address pool using <peer default ip address pool [pool-name]>:

```
Concord-GW(config-if)#peer default ip address pool 17-pool
```

The actual pool is created with the <ip local pool [poolname] [starting address ending address]> command:

```
Concord-GW(config)#ip local pool 17-pool 192.168.17.248 192.168.17.252
```

Or from a adjacent DCHP server using <peer default ip address pool dhcp>:

```
Concord-GW(config-if)#peer default ip address pool dhcp
```

If you plan to use DHCP, you also need to tell the router which server to use:

```
Concord-GW(config)#ip dhcp-server 192.168.17.25
```

You need to choose one of the two addressing mechanisms; you cannot use both. One last thing on addressing: regardless of the assignment method, if possible have the addressing be part of the same address pool used by the router's gateway interface (such as the interface you used with <ip unnumbered>).

When configuring the line(s), the important settings are the lines' speed and <modem [access type]>. You can get dial-in working with just those two settings. Granted, there is no security, but you can get a PPP session running. In the WidgetCo example, you are using local security username accounting that is specified with the <login local> setting. The modems are also accessible (with Telnet) from the network for dial-out use with <modem InOut>:

```
Concord-GW(config)#line 42 48
Concord-GW(config-line)#login local
Concord-GW(config-line)#transport input all
Concord-GW(config-line)#modem InOut
Concord-GW(config-line)#speed 115200
Concord-GW(config-line)#flowcontrol hardware
```

The AUX port configuration for Ridgefield's IP/AppleTalk dial-in is almost identical to the line configurations for Concord, except only dial-in is permitted and the line speed (which is limited on the AUX port) is set to 38400:

```
Ridge-GW#(config-line)#login local
Ridge-GW#(config-line)#modem dialin
Ridge-GW#(config-line)#transport input all
Ridge-GW#( (config-line)#speed 38400
Ridge-GW#(config-line)#flowcontrol hardware
```

Creating Dial-In User Accounts

After the modem pool is configured, all that is left is user account creation. Because you are using local user authentication, the users are added to the router's configuration through global configuration EXEC mode using <username [name] password [password text]>. This method works fine for adding a few users, but a more practical method is to load the user table using the privileged EXEC <copy> command from a file on a TFTP server. The file should contain one user per line, using the same command syntax format used when entering users through the configuration CLI:

```
!
username fred password fred,4r
username ralph password ou812
username george password dollar
!
end
```

Incidentally, the same syntax rules apply to all network loadable configuration files: one command per line, using the identical CLI syntax.

> **Note**
>
> If you use AAA instead of local authentication, there is no need to specify an authentication method. By default, the port will use the globally defined authentication method for login.

Disaster Recovery

The routers are up, everything is great, and BANG!—disaster strikes. The enable password is lost, or a power surge zapped the router and the IOS image is corrupted. It doesn't matter how it happened. You just need to get the router back online, and to do that you need it up, and you need access.

Setting the Bootstrap Behavior

A Cisco router's bootstrap behavior is set using what is known as the *configuration register (conf-reg)*. On the early Cisco routers (such as AGS), the `conf-reg` was set with jumpers on the router's system board. On the later Cisco models (Cisco 1000 and up), the conf-reg can be set in global configuration mode using the `config-register` command:

```
Router# configure terminal
Router(config)#config-register 0x2102
Router(config)#^Z
Router#
```

Alternatively, if access to global configuration mode is unavailable, the ROM Monitor (`rommon`), which is similar in function to a PC's BIOS mode, can be used. To access `rommon`, you need to be attached to the console port and you need to issue a terminal break sequence through your terminal emulator software. Check your package's documentation for the correct key sequence; C-Kermit's break key is Alt+B. In most cases, break is disabled once the router is up and running. So the easiest way to get into `rommon` is to power-cycle the router and enter the break sequence when the router begins the boot process after the "bootstrap" banner has been displayed:

```
System Bootstrap, Version 11.0(10c), SOFTWARE
Copyright © 1986-1996 by Cisco Systems
2500 processor with 14336 Kbytes of main memory
Abort at 0x1098FEC (PC)
>
```

Note

Keep in mind that whenever you make a `conf-reg` change, it will need to be set back to "running" `conf-reg` setting after you have corrected the problem. This can be done using the <config-register> command in global configuration mode. You can, by the way, avoid <rommon> altogether, and change the `conf-reg` setting in global configuration mode if you have access.

Disaster recovery on Cisco routers using `rommon` and `conf-reg` settings is straightforward because all Cisco routers use the same basic boot sequence:

1. The router is powered up.
2. It checks its `conf-reg` settings.
3. It checks the NVRAM for boot loader information.
4. It loads the IOS.
5. It loads the NVRAM config.

Because the `conf-reg` is the first boot element checked, it is possible to configure the router to do a number of different things just by changing its `conf-reg` settings. If possible, before making any `conf-reg` changes, be sure that you know what your router's current settings are. `conf-reg` information is listed as the final system variable in the output of the `show version` or `show hardware` privileged EXEC command.

```
Router#sh hardware
Cisco Internetwork Operating System Software
IOS (tm) 4500 Software (C4500-IS-M), Version 11.3(5), RELEASE SOFTWARE (fc1)
Copyright © 1986-1998 by Cisco Systems, Inc.
Compiled Tue 11-Aug-98 04:16 by phanguye
Image text-base: 0x600088F8, data-base: 0x60710000
ROM: System Bootstrap, Version 5.2(7b) [mkamson 7b], RELEASE SOFTWARE (fc1)
BOOTFLASH: 4500 Bootstrap Software (C4500-BOOT-M), Version 10.3(7), RELEASE
SOFT
WARE (fc1)
Router uptime is 3 minutes
System restarted by power-on
System image file is "flash:c4500mz.113-5.bin", booted via flash
Cisco 4500 (R4K) processor (revision D) with 32768K/4096K bytes of memory.
Processor board ID 05475810
R4700 processor, Implementation 33, Revision 1.0
G.703/E1 software, Version 1.0.
Bridging software.
X.25 software, Version 3.0.0.
4 Ethernet/IEEE 802.3 interface(s)
128K bytes of non-volatile configuration memory.
16384K bytes of processor board System flash (Read/Write)
4096K bytes of processor board Boot flash (Read/Write)
Configuration register is 0x2002
Router#
```

Setting the *conf-reg* Value in ROM Monitor

There are two different versions of the ROM monitor used on Cisco routers. On the 2500 series, routers use a very terse ROM monitor compared to the 1600, 3x00, 4x00, and 7x00 series routers. When in `rommon` on a 2500 series, the rommon prompt looks like this:

```
>
```

To get help, you can type `?` or `H`, which will list the available ROM monitor commands. To set the `conf-reg` value on a Cisco 2500, type the `O/R` command combination and the desired `conf-reg` setting (with no line spaces):

```
>O/R0x2002
```

Then, initialize the router:

```
>I
System Bootstrap, Version 11.0(10c), SOFTWARE
Copyright © 1986-1996 by Cisco Systems
```

This will boot the 2500 series router using the new `conf-reg` setting.

On the 1600, 2600, 3x00, 4x00, and 7x00 series, ROM monitor is slightly more friendly. The prompt looks like this (the number in the prompt is the command history value):

```
rommon 3 >
```

Help is available by typing `?` or `help`. The available commands vary slightly depending upon the router series. To set the `conf-reg` type `<confreg>`, which starts up the configuration summary utility, type:

```
rommon 11 > confreg
    Configuration Summary
enabled are:
load rom after netboot fails
ignore system config info
console baud: 9600
boot: image specified by the boot system commands
      or default to: cisco2-C4500
do you wish to change the configuration? y/n [n]:
```

Alternatively, you can just enter the `conf-reg` setting directly and `<reset>` the router:

```
rommon 12 > confreg 0x2002
You must reset or power cycle for new config to take effect
rommon 13 >reset
System Bootstrap, Version 5.2(7b) [mkamson 7b], RELEASE SOFTWARE (fc1)
Copyright © 1995 by Cisco Systems, Inc.
```

Configuration Register Settings

Here are the most common (needed) `conf-reg` settings and their effects on the router. These settings should work on any Cisco router above the 1000 series (unless indicated).

- Register setting 0x2010—Configures the router to check NVRAM and boot directly into ROM monitor.
- Register setting 0x8000—Configures the router (2500,4000,7x00) and loads the router into diagnostic mode. Setting 0xA042 is used on 4500 and 4700 series routers.
- Register setting 0x2101—Loads the IOS from ROM, and the NVRAM configuration is loaded into RAM as the running configuration. This is useful for performing IOS upgrades or if the flash IOS image is corrupted. Not all routers have a ROM-based IOS to boot from, so this option only works on those that do (2500, 4000, 7x00). If this setting is used on routers that do not have an IOS in ROM, RFF IOS-based routers will boot with flash in read/write mode. RFR IOS-based routers will just read the IOS in flash, which is always in read/write mode. See the section "Upgrading Your Router's IOS" for more information.
- Register setting 0x2102—This is the Cisco default `conf-reg` setting. The router loads the IOS from flash or the boot system source specified in NVRAM. After the IOS is loaded, the startup configuration is loaded from NVRAM to RAM as the running configuration. By default, console break option is disabled once the IOS is loaded, so if you want to have it enabled, use `conf-reg` setting 0x2002 instead.
- Register setting 0x2141—Tells the router to load the IOS from ROM (Cisco 2500, 4x00, and 7x00) and ignore the configuration in NVRAM. Because the startup configuration is not copied to DRAM, the router starts in setup mode. This is great if the IOS image is damaged, or if you need to perform an IOS upgrade on an RFF and have no local TFTP server.
- Register setting 0x2142—The router loads the IOS from flash and ignores the NVRAM (just like setting 0x2141). This is the Cisco recommended setting to use for password recovery. A choice exists between two recovery processes: brute force or subtle. This brute force method erases the router configuration completely. If you have a backup of the router configuration or do not need any of the configuration information, this setting works fine.

To change the `conf-reg` setting after the router is up and running use the following steps:

1. In `rommon`, use the `<o/r>` or `<confreg>` command to set the `conf-reg` to 0x2142, then reset the router.
2. The router will load the startup configuration process. Answer `no` or type Ctrl+C.

3. Enter privileged mode, and erase NVRAM:

```
Router#erase startup-config
```

4. Enter global configuration mode and reset the `conf-reg` for normal operation:

```
Router#config t
Router(config)#config-register 0x2102
Router#^z
Router#reload
```

When the router reloads, you can use startup mode (the default when no configuration is in NVRAM) or the command line to reconfigure the router.

The subtle approach is used if you want to only change a bad or damaged password. The process is a little more complicated, but still quite easy. Using the same 0x2142 `conf-reg` setting, you boot the router, kill the startup process, and enter privileged EXEC mode.

1. Copy the startup configuration stored in NVRAM into DRAM:

```
Router#copy startup-config running-config
00:17:45: %SYS-5-CONFIGI: Configured from memory by console
Router#
```

2. Enter global configuration mode and change the enable line or username password that needs resetting:

```
Router#configure terminal
Enter configuration commands, one per line.  End with CNTL/Z.
Router(config)#enab
Router(config)#enable secret anypass
Router(config)#^Z
Router#
```

After the password is reset, you also need to re-enable (with the `<no shutdown>` command) all the router's interfaces that were in use in your configuration.

3. After the interfaces have been re-enabled, reset the conf-reg to the appropriate operational mode, <copy> the running configuration to NVRAM, and <reload> the router.

Upgrading Your Router's IOS

TFTP is the easiest way to upgrade your router's IOS. It is recommended that you have the TFTP server on the same local segment, if possible. On RFF routers (1600, 2500 series), the router boots with the flash in read-only mode, and because it reads data structures from the IOS stored in flash, it cannot be altered while the router is using them. To perform an IOS upgrade, you need to configure the router to load its IOS off the TFTP server instead of using the flash stored image.

On RFR routers (1600-R, 2600, 36x0, 4x00, and 7x00 series), the router's flash is in read/write mode all the time because the IOS is loaded into DRAM at boot. IOS can be upgraded while the router is in operation.

There is, however, an element of risk to this approach. If the IOS image is corrupt, or if there are incompatibilities between your configuration and the new IOS, you could have some problems. To play it safe on both RFR and RFF router models, use TFTP to boot the IOS image you plan to upgrade on your router. Then, upgrade it.

Setting Up TFTP Boot

By default, the IOS uses the first IOS image it finds on the flash filesystem. The <boot system [source] [filename]> global configuration command is used to specify the IOS load source. This configuration variable is read by the boot loader before the IOS is loaded. To configure the router to boot from TFTP, use the <boot system> variation:

```
<boot system tftp [filename] [IP address]>
```

Here is an example:

```
Router#config terminal
Enter configuration commands, one per line.  End with CNTL/Z.
Router(config)#boot system tftp c2500-enter40-1.113-9.bin 192.168.0.2
Router(config)#^Z
Router#
```

When you reboot the router, it will load the specified IOS version (c2500-enter40-1.113-9.bin) from the defined TFTP server (192.168.0.2). When the router is loading the IOS, you will see the following:

```
Loading c2500-enter40-1.113-9.bin from 192.168.0.2 (via Ethernet0):
!!!!!!!!!!!!!!!!!!!!!!!!!!!!!!!!!!!!!!!!!!!!!!!!!!!!!!!!!!!!!!!!!!!!!!!!!!!!!
!!!!!!!!!!!!!!!!!!!!!!!!!!!!!!!!!!!!!!!!!!!!!!!!!!!!!!!!!!!!!!!!!!!!!
!!!!!!!!!!!!!!!!!!!!!!!!!!!!!!!!!!!!!!!!!!!!!O!!!!!!!!!!!!!!!!!!!!!!!!!!!
!!!!!!!!!!!!!!!!!!!!!!!!!!!!!!!!!!!!!!!!!!!!!!!!!!!!!!!!!!!!!!!!!!!!!!!!!
!!!!!!!!!!!!!!!!!!!!!!!!!!!!!!!!!!!!!!!!!!O!!!!!!!!!!!!!!!!!!!!!!!!!!!!!!
!!!!!!!!!!!!!!!!!!!!!!!!!!!!!!!!!!!!!!!!!!!!!!!!!!!!
[OK - 9212736/13868534 bytes]
```

An ! indicates that a 64-bit UDP packet has been successfully transferred, and O indicates a missed packet.

Installing the New IOS

After the router is up and running, the IOS can be loaded using the <copy> privileged EXEC command:

```
Router#copy tftp flash
System flash directory:
No files in System flash
[0 bytes used, 16777216 available, 16777216 total]
Address or name of remote host [255.255.255.255]? 192.168.0.2
Source file name? c2500-enter40-l.113-9.bin
Destination file name [c2500-enter40-l.113-9.bin]?
Accessing file 'c2500-enter40-l.113-9.bin' on 192.168.0.2...
Loading c2500-enter40-l.113-9.bin .from 192.168.0.2 (via Ethernet0): ! [OK]
Device needs erasure before copying new file
Erase flash device before writing? [confirm]
Copy 'c2500-enter40-l.113-9.bin' from server
  as 'c2500-enter40-l.113-9.bin' into Flash WITH erase? [yes/no]yes
Erasing device... eeeeeeeeeeeeeeeeeeeeeeeeeeeeeeeeeeee.......
Loading c2500-enter40-l.113-9.bin from 192.168.0.2 (via Ethernet0):
!!!!!!!!!!!!!!!!!!!!!!!!!!!!!!!!!!!!!!!!!!!!!!!!!!!!!!!!!!!!!!!!!!!!!!!!!!!!!!!
!!!!!!!!!!!!!!!!!!!!!!!!!!!!!!!!!!!!!!!!!!!!!!!!!!!!!!!!!!!!!!!!!!!!!!!!!!!!!!!
!!!!!!!!!.....
[OK - 9212736/16777216 bytes]
Verifying checksum... OK (0x99D9)
Flash device copy took 00:04:27 [hh:mm:ss]
```

If you have enough flash space, it is possible to store different IOS versions and use the <boot> command to define which IOS version will be loaded at boot time. If you do not have enough flash, you can always use TFTP.

Some Additional IOS Options

If you add additional flash to a router that already has an IOS image in flash, the new flash will be available as a second partition. This is fine if you want two different IOS images on separate partitions. But if you need a larger flash space to load a large IOS image, this will not work. To change this partition, boot the router using TFTP and erase the flash partition with the IOS image. After the partition is erased, both flash chips will appear as a single partition.

If, however, you want to partition your flash and load two IOS versions, you need to boot with TFTP. First, erase the flash partition:

```
Router#erase flash
System flash directory:
File  Length   Name/status
  1   9212736  c2500-enter40-l.113-9.bin
[9212800 bytes used, 7564416 available, 16777216 total]
Erase flash device? [confirm]
```

```
Are you sure? [yes/no]: yes
Erasing device...
eeeeeeeeeeeeeeeeeeeeeeeeeeeeeeeeeeeeeeeeeeeeeeeeeeeeeeeeeeeeeeeeeee .
..erased
Router#
```

Then, in global configuration mode, partition the flash:

```
Router#config t
Enter configuration commands, one per line.  End with CNTL/Z.
Router(config)#partition flash 2
Router(config)#^Z
Router# show flash
System flash directory, partition 1:
File  Length   Name/status
No files in System flash
[0 bytes used, 8388608 available, 8388608 total]
8192K bytes of processor board System flash (Read/Write)
System flash directory, partition 2:
No files in System flash
[0 bytes used, 8388608 available, 8388608 total]
8192K bytes of processor board System flash (Read/Write)
```

With motherboard-mounted flash, the number of partitions depends on the number of flash slots. After the partitions are in place, you can TFTP the IOS images. When you copy the images, you will be asked to choose which partition will be the destination:

```
Router#copy tftp flash
Partition   Size    Used      Free      Bank-Size  State          Copy Mode
  1         8192K     0K      8192K     8192K      Read/Write     Direct
  2         8192K     0K      8192K     8192K      Read/Write     Direct
[Type ?<no> for partition directory; ? for full directory; q to abort]
Which partition? [default = 1]
System flash directory, partition 1:
No files in System flash
[0 bytes used, 8388608 available, 8388608 total]
Address or name of remote host [192.168.0.2]?
Source file name? c2500-d-l.120-3.bin
Destination file name [c2500-d-l.120-3.bin]?
Accessing file 'c2500-d-l.120-3.bin' on 192.168.0.2...
Loading c2500-d-l.120-3.bin from 192.168.0.2 (via Ethernet0): ! [OK]
```

After your images are loaded, you need to indicate which image to use. If you do not specify an image, the first flash partition will be used:

```
Router#config t
Enter configuration commands, one per line.  End with CNTL/Z.
Router(config)# boot System flash c2500-d-l.120-3.bin
Router(config)#^Z
Router#
```

When partitioning is used, it will appear in the configuration file, listing the number of partitions and the size of each partition:

```
!
version 11.3
service timestamps debug uptime
!
hostname Router
!
boot System flash c2500-d-l.120-3.bin
!
partition flash 2 8 8
```

If flash chips of different sizes are used, each partition will reflect the size of the chip.

On 1600, 36x0, and 7x00 routers, PCMCIA flash cards are used along with motherboard-mounted flash chips. PCMCIA flash cards can be partitioned into more than one partition. Also, when working with PCMCIA flash cards, the commands differ between the modular routers (1600 and 3600) and 7x00 series routers.

On the modular routers, the <show> command is used to see what is on the flash card. The files on the PCMCIA card in slot0 are the following:

```
Router3600#sh slot0:
-#- ED --type----crc--- -seek--nlen -length- -----date/time------name
1   .. image    5621FA2E  69803C   21  6782908 Jun 06 1999 02:41:21 c3600-
js-mz.120-4.bin
1212356 bytes available (6783036 bytes used)
```

Remember, on 7x00 series routers (and on all routers running IOS version 12.x), the <dir> command is also available. To see what is on the PCMCIA card in slot0:

```
Router7200#dir slot0:
Directory of slot0:/
  1  -rw-     6782908   Jun 06 1999 02:41:21  c7200-js-mz.120-4.bin
7995392 bytes total (1212356 bytes free)
Router7200#
```

Another difference is how flash is erased. On a modular router, <erase> is used:

```
Router3600#erase slot0:
Erase flash device, partition 0? [confirm]
Are you sure? [yes/no]: yes
```

```
Erasing device...
eeeeeeeeeeeeeeeeeeeeeeeeeeeeeeeeeeeeeeeeeeeeeeeeeeeeeeeeeeeeeeee .
..erased
Router3600#
```

On a 7x00 series router, <format> is used instead of <erase> when working with flash:

```
Router7200#format slot0:
Format operation may take a while. Continue? [confirm]
Router7200#
```

Aside from these differences, working with PCMCIA flash cards is the same as working with motherboard-mounted flash, including the ability to have all of the router's flash accessed as a single partition.

When Something Goes Wrong with IOS

If something happens to your router's IOS image, there are recovery options besides sending your router back to Cisco. On the 2500, 4x00, and 7x00, there are two IOS versions. On the 2500, there is a stripped-down version in ROM. If the conf-reg is set for normal operation (0x2102 or 0x2002), the ROM version is used when the flash or other indicated boot method fails. When in ROM IOS, you can reload your flash IOS copy or modify your <boot system> options.

On the 4x00 and 7x00, there is an IOS in flash and boot-IOS stored in bootflash. To see what version of the IOS you have in bootflash, use the privileged <show bootflash> command:

```
router#sh bootflash
-#- ED --type----crc--- -seek--nlen -length- -----date/time------name
1   .. unknown  E8E9F469  1BE584   25  1565956 Jun 20 1997 13:02:09 c7200-
boot-mz.111-13a.CA1
1841788 bytes available (1566084 bytes used)
router#
```

As with the 2500 series, if there is a failure in the boot process, the bootstrap IOS version is used.

With Cisco's later modular routers (1600, 2600, and 3600 series), there is no ROM IOS or bootflash. Instead, they have added the ability to xmodem or TFTP (2600 only) an IOS image into the router's DRAM. To use this option, you need to be in rommon and connected to the console port, which is where you will be if there is an IOS load failure. The xmodem method loads the IOS image from the system you're using to display the console over the console serial connection. This means you need to use a communication package that can perform xmodem (or ymodem) file transfers. Sorry, C-Kermit is not well suited for this. On PCs, I recommend ProCOMM. On Macintosh, use Z-term (it's shareware and easy to get on AOL or the Web).

After you are on the console and have the IOS image you want to use, type at the rommon prompt:

```
rommon 3 > xmodem -yr c3620-i-mz.113-9.bin
```

This executes an xmodem download using the ymodem (1K) transfer protocol (ymodem is much faster than xmodem, if you can use it) to boot the router, using the downloaded image after the transfer is finished. You also need to set the name of the file you are downloading, which in this case is `<c3620-i-mz.113-9.bin>`. After you have entered the command, you will be greeted with the following:

```
Invoke this application only for disaster recovery.
Do you wish to continue? y/n  [n]:  y
Ready to receive file  ...
```

After you reply `yes`, rommon tells you it's ready to receive the file. At this point you should start downloading the IOS image with your communications package. An IP-only IOS image (about 2.5MB) will take an hour to download. When completed, you will get an acknowledgement message, then the IOS image will decompress and load itself into DRAM and the router will come online:

```
### Send (Y) c3620-i-mz.113-9.bin: 2771280 bytes, 52:41 elapsed, 876 cps,
91%
Download Complete!
program load complete, entry point: 0x80008000, size: 0x2a4834
Self decompressing the image :################################
#############################################################################
#############################################################################
############################ [OK]
```

After the router is up, you can reload an IOS image to flash or adjust your boot parameters.

Configuring the Router's Clock

Cisco routers, like all computers, keep track of time. While the time of day is unimportant to the router, it is of some importance to us, especially if system logging, accounting, and SNMP are enabled. When you configure your router, you need to set its clock, either manually each time the router reboots, or by using the *Network Time Protocol (NTP)*.

Note

Cisco 7x00 series routers use a hardware clock. On all other models when the system boots up the clock is set to March 1, 1993.

IOS uses *Universal Time Coordinated (UTC)*, more commonly known as *Greenwich Mean Time (GMT)*. This means that if you want your logging and accounting information to reflect local time, you need to set your time zone and its offset from UTC/GMT. A good Web site to verify your time zone and check your offset is `http://aa.usno.navy.mil/aa/faq/docs/Worldtzones.html`. If you live in the U.S., the offsets are as follows:

Eastern Standard Time (EST)	-5
Central Standard Time (CST)	-6
Mountain Standard Time (MST)	-7
Pacific Standard Time (PST)	-8

To set your time zone, use the global configuration command `<clock timezone>`:

```
Router#config terminal
Enter configuration commands, one per line.  End with CNTL/Z.
Router(config)#clock timezone EST -5
Router(config)#^Z
```

If you live in a place where daylight savings time is used, you also need to set your daylight savings time zone using the global configuration command `<clock summer-time>`. The convention is `<time region (for example, Eastern) Daylight Time>`:

```
Router#config t
Enter configuration commands, one per line.  End with CNTL/Z.
Router(config)#clock summer-time EDT recurring
Router(config)#^Z
```

To set the router's system clock, use the privileged EXEC command `<clock set>`. The convention is `hour:min:second day month year`:

```
Router#clock set 12:00:15 14 June 1999
```

The router's clock must be set every time the router boots (or reboots). Because this can be inconvenient, IOS supports NTP to perform this function. To use NTP, the router must be able to access a stratum 2 or better NTP server. In many cases, a site's ISP will provide access to a timeserver for its customers. If this service is unavailable, there are several public timeservers available over the Internet.

> **Note**
>
> For more information about NTP, check out `http://www.eecis.udel.edu/~ntp/`. This Web site provides some great information on NTP, plus NTP software and a list of public timeservers.

IOS provides the facility for the router to act as an NTP client, an NTP server, or an NTP peer. As an NTP client, the router contacts the defined NTP server, then verifies and synchronizes its system clock. As an NTP server, the router can be used by other systems on the network as their NTP server. In a peer scenario, the router and the NTP source exchange time synchronization information.

To configure the router as an NTP client in global configuration mode, use the `<ntp server>` command and the interface command `<ntp broadcast client>`:

```
3600router#config terminal
Enter configuration commands, one per line.  End with CNTL/Z.
3600router(config)#ntp server 192.168.0.5
3600router(config)#ntp server 128.4.1.1
3600router(config)#interface FastEthernet1/0
3600router(config-if)#ntp broadcast client
3600router(config-if)#^Z
3600router#
```

This configures the router to use a local timeserver (192.168.0.5) and an authoritative Internet timeserver (128.4.1.1) for time synchronization. The router uses Fast Ethernet interface 1/0 to listen for NTP broadcasts from the local NTP server. To see if NTP is working correctly, use the privileged EXEC command `<show ntp status>`:

```
3600router#sh ntp status
Clock is synchronized, stratum 3, reference is 130.140.55.216
nominal freq is 250.0000 Hz, actual freq is 250.0000 Hz, precision is 2**24
reference time is BB0FC7AA.95E93186 (14:25:46.585 EDT Mon Jun 14 1999)
clock offset is 7.3012 msec, root delay is 21.21 msec
root dispersion is 65.20 msec, peer dispersion is 37.78 msec
3600router#
```

To see the status of the NTP server relationships between the router and its defined NTP servers, use the privileged EXEC command `<show ntp associations>`:

```
local-AS#sh ntp associations
      address       ref clock      st  when  poll reach  delay  offset   disp
*~130.140.55.216  128.8.10.6     2   196   256  377     3.2    7.30    2.1
 ~130.140.67.12   130.140.55.4   4   252  1024  377    13.2  -46.59   11.4
+~130.140.55.4    128.118.25.3   3   201   256  377     3.5  -40.32    4.8
 * master (* master (synced), # master (unsynced), + selected, - candidate,
~ configured)
```

Note

To use NTP to update the 7x00 hardware clock, add the global configuration command `<ntp update calendar>`. You can also use the privileged EXEC command `<show calendar>` or the `<show clock>` command to display the router's time and date settings.

One last thing about time: even though the router has a sense of time, the IOS messages do not. To enable timestamping on router messages, you need to use the global configuration command <`service timestamps log`> for logging messages and <`service timestamps debug`> for debug messages. Timestamping logging messages is important and should be enabled, because it provides the time and date the event occurred. Timestamping is helpful for debug messages as well, but in most cases is not required, and when enabled just adds to the size of the debug message.

IOS Message Logging

Keeping up with what is going on with your router is important. *Logging* provides this simple, but essential facility. In most cases, log information is simply status data, such as changes in the router's interface status, modifications to running configuration, and debugging output. When things are operating smoothly, this data is nice to have. When a problem comes up, however, this data can be quite valuable. The IOS uses UNIX's syslog logging system to generate IOS logging messages. The IOS provides four methods for viewing logging information:

- Console—The router's console port
- Monitor—The router's system monitor, a VTY "console" message display
- Trap—Syslog output to a remote syslog server running on UNIX or NT
- Buffer—A place to store a list of logging events in the router's DRAM

By default, all console and monitor methods are enabled, buffer is disabled, and trap (while set up) needs to be configured to know where to send its messages.

The trap and DRAM buffer logging methods provide a way to store logging information. Of the two, the trap method is more useful for logging because the messages are stored on a remote server. The buffer approach stores the messages in DRAM, which is lost when the router is shut down or rebooted. Storing messages in DRAM can also affect the router's performance if your router is tight on shared DRAM (which is why it is disabled by default). If your router has enough memory to spare, saving logging information on the router will not affect performance. It is helpful to have the messages accessible on the router, especially when you are trying to debug a problem.

Before enabling local logging, check your router's memory usage (see the general info section for command help) after the router has been in production for a little while. This will give you an accurate picture of the router's actual memory use, so you can make sure that you have enough DRAM to allocate for logging buffer storage.

Note

To view messages sent to the monitor, use the privileged EXEC command <`terminal monitor`>.

Setting Up Buffered Logging

To set up buffered logging, you first need to enable it. Then the local DRAM allocation and logging event history needs to be set. All logging parameters are set in global configuration mode:

```
local-AS(config)#logging on
local-AS(config)#logging buffered 64000
local-AS(config)#logging history size 250
```

In the example above, all logging has been enabled, a local buffer of 64K has been allocated, and up to 250 logging messages will be stored in the buffer (the maximum is 500). The buffer logfile is a rotating one, so once the message count has reached its limit, it starts to overwrite itself, deleting the last message in the file as new messages are added. To view logging information, use the privileged EXEC command `<show logging>`:

```
local-AS#sh logging
Syslog logging: enabled (0 messages dropped, 0 flushes, 0 overruns)
    Console logging: level informational, 17 messages logged
    Monitor logging: level debugging, 234 messages logged
    Trap logging: level informational, 744 message lines logged
        Logging to 130.140.55.216, 7 message lines logged
    Buffer logging: level debugging, 810 messages logged
Log Buffer (64000 bytes):
00:00:08: %LINK-3-UPDOWN: Interface Ethernet0/0, changed state to up
00:00:08: %LINK-3-UPDOWN: Interface Ethernet0/1, changed state to up
```

Cisco uses the syslog level classification to define the severity of logging messages:

Message Severity Level	Meaning	Explanation
0	Emergencies	System is unusable
1	Alerts	Immediate action needed
2	Critical	Critical conditions
3	Errors	Error conditions
4	Warnings	Warning conditions
5	Notifications	Normal but significant conditions
6	Informational	Informational messages
7	Debugging	Debugging messages

Each level sends messages recursively. If the warning's level is set, all message levels below it are sent as well, so the message level defines what classes of messages will be sent. The default logging configuration has all router event messages logged. This is achieved with the logging level set to `debugging`:

```
Router#sh logging
Syslog logging: enabled (0 messages dropped, 0 flushes, 0 overruns)
    Console logging: level debugging, 15 messages logged
    Monitor logging: level debugging, 0 messages logged
    Trap logging: level informational, 19 message lines logged
    Buffer logging: disabled
```

It is possible to define which logging level will be sent to which logging method (console, monitor, trap, or buffer). The global configuration command <logging [method] level> is used to set the logging method's message level. This can be quite helpful, especially if you use debug mode often. When using the default logging settings, all the debug messages are sent to the console. This makes using the console to enter commands difficult, particularly if there is a lot of debug message feedback. To avoid this, set the console's logging level to informational; this way, debug messages are set to the system monitor and the console is clear:

```
local-AS#config t
Enter configuration commands, one per line.  End with CNTL/Z
local-AS(config)#logging console informational
local-AS(config)#^Z
local-AS#
```

Another case where this is helpful is with the buffer logging method. Sending debug messages to the buffer will overwrite it rather quickly. If you plan to use buffered logging at all effectively, set its level at most to informational. Some care should be taken when using logging on routers with heavy processing load. Logging, like all router functions, uses memory and processor resources that can be used for processing datagrams. If these resources are tight, logging and other nonessential services will affect router performance.

Setting Up Trap Logging

Sending router messages to a remote host running syslog is more useful than local logging methods, but it requires a little more work. In addition to configuring the router, you must set up a host to process the syslog messages. Syslog comes with most standard UNIX implementations and is available as an add-on service for Windows NT. Some resources for NT syslog are the following:

http://www.kiwi-enterprises.com/

http://www.adiscon.com/NTSLog/main.asp

http://www.cls.de/syslog/eindex.htm

Here is an IOS trap logging configuration example:

```
Local-AS#config t
Enter configuration commands, one per line.  End with CNTL/Z.
local-AS(config)#logging 130.140.55.216
local-AS(config)#logging trap informational
local-AS(config)#logging source-interface Ethernet 0/0
local-AS(config)#^Z
local-AS#
```

The router configuration is easy; it is the logging server that requires a little work. To start you need to run syslog. On UNIX this is accomplished by starting the syslogd daemon with specific flags:

OS	Daemon	Logfile Root Location
BSDI	/sbin/syslogd -n	/usr/var/log
SunOS	/usr/etc/syslogd -n	/var/log
Linux	/usr/sbin/syslogd -n	/var/log
Solaris	/usr/sbin/syslogd -n	/var/log

The syslog daemon gets its configuration information from /etc/syslog.conf. This file will need to be edited to reflect how you want syslog to handle your router's log messages.

Syslog decides how to handle a message by looking at the incoming message's facility and severity level pair. Each message has a facility level pair associated with it. The message's value pair is defined by the program or service issuing the message, using syslog's defined facility and severity levels. The IOS uses syslog's standard severity values to classify messages and uses syslog facility local7 by default. There are 10 different syslog facility definitions:

`auth`	Authorization system
`cron`	Cron/at facility
`daemon`	System daemons
`kern`	Kernel
`local0-7`	Local use
`lpr`	Line printer system
`mail`	Mail system
`news`	Usenet news
`syslog`	Syslog itself
`user`	User process
`uucp`	UNIX-to-UNIX copy system

IOS allows you to configure the router to use any defined syslog facility by using the global configuration command `<logging facility>`:

```
local-AS#config t
Enter configuration commands, one per line.  End with CNTL/Z.
local-AS(config)#logging facility local6
local-AS(config)#^Z
local-AS#
```

After the message's facility level pair has been determined, it checks the syslog.conf file to see which action it should take. There are five syslog-defined actions:

- Log to a file (/usr/log/<filename>)
- Forward the message to another syslog process on another host (@hostname or @ip address)
- Write the message to a specified user's operator window. (user,username)
- Write the message to all users' operator windows (*)

Here is a sample syslog file that uses facility local7, local6, and local5 to collect logging messages for Internet gateway routers, internal routers, and access servers:

```
#/etc/syslog.conf
#
#Logging for all Network Gateways
local7.emerg                                @logmaster.x2net.net
local7.alert;local7.crit;local7.err         *
local7.warning                              root,help
local7.notice                               root,netadmin,help
local7.info                                 /var/log/cisco-gw.info
local7.debug                                /var/log/cisco-gw.debug
#
#Logging for all internal routers
local6.emerg                                @logmaster.x2net.net
local6.alert;local7.crit;local7.err         *
local6.warning                              root,help
local6.notice                               root,netadmin,help
local6.info                                 /var/log/cisco-int.info
local6.debug                                /var/log/cisco-int.debug
#
#Logging for all access-servers
local5.emerg                                @logmaster.x2net.net
local5.alert;local7.crit;local7.err         root
local5.warning                              root,help
local5.notice                               root,netadmin,help
local5.info                                 /var/log/cisco-as.info
local5.debug                                /var/log/cisco-ast.debug
```

Log messages with a high severity level are sent to the master syslog server and to users' operator windows, whereas general router information and debug messages are sent to files. The files can be stored or sent on for review. It is a good idea to use different syslog facilities for different routers, to make log management easier. It also helps to use syslog facilities that are not predefined for a specific use (kern, user, mail, and so on) make the logfiles that are easier to read. Here is some sample logfile output from syslog facility local6:

```
Jun  1 15:31:09 130.140.56.9 20: 00:01:43: %AT-6-ONLYROUTER: Ethernet1:
AppleTalk interface enabled; no neighbors found
Jun  1 15:49:24 130.140.56.9 21: 00:19:58: %SYS-5-CONFIGI: Configured from
console by admin on console
Jun  1 15:53:12 130.140.56.9 22: 00:23:45: %SYS-5-CONFIGI: Configured from
console by admin on console
```

Log Management

Logs are useless unless someone looks at them. One easy way to deal with logfiles is to mail them to yourself. This can be done with a simple UNIX shell script:

```
#!/bin/sh
# This script mails logfiles to group of users or mail archive.
# This script should be run once a day or once a week from root
# cron@11:59pm
# depending on the logfile. Here are some sample crontab entrys
# Once a day crontab entry
# 55 23 * * * /usr/local/scripts/maillog.sh > /dev/null
# Once a week crontab entry run on Friday
# 55 23 * * 5 /usr/local/scripts/maillog.sh > /dev/null
# All that needs to be defined is the users and the logfiles name
#
# Edit these as needed
users="user1@mailhost user2@mailhost"
logtitle="gw routers .info log"
logfile=/usr/var/log/cisco.info
#
#
mail=/usr/ucb/mail

echo Moving logfile.
if [ -f $logfile ]; then
                mv $logfile $logfile.old
fi
echo Mailing logfile.
$mail -v -s "$logtitle" $users < $logfile.old
echo Cleaning up.
```

```
touch $logfile
rm $logfile.old
echo "Done, have a nice day!"
```

The report, which is mailed to you, will look like this:

```
Subject: gw routers log
Date: Wed, 16 Jun 1999 23:55:00 -0400 (EDT)
From: System Administrator <root@any.com>
To: mjm@any.com
Jun 16 07:48:08 130.140.56.10 747: 21:15:54: %SYS-5-CONFIGI: Configured from
console by martin on console
Jun 16 07:51:58 130.140.56.10 748: 21:19:43: %SYS-5-CONFIGI: Configured from
console by martin on console
Jun 16 07:54:04 130.140.56.10 749: 21:21:49: %SYS-5-CONFIGI: Configured from
console by martin on console
Jun 16 07:59:45 130.140.56.10 750: 21:27:30: %SYS-5-CONFIGI: Configured from
console by martin on console
Jun 16 08:00:41 130.140.56.10 751: 21:28:26: %SYS-5-CONFIGI: Configured from
console by martin on console
Jun 16 11:43:20 130.140.56.10 752: 1d01h: %SYS-5-CONFIGI: Configured from
console by martin on console
Jun 16 11:44:06 130.140.56.10 753: 1d01h: %SYS-5-CONFIGI: Configured from
console by martin on console
Jun 16 11:44:20 130.140.56.10 754: 1d01h: %AAAA-3-TIMERNOPER:
AAA/ACCT/TIMER: No periodic update but timer set.
Jun 16 11:44:20 130.140.56.10 755: -Traceback= 6016B84C 6016C97C 6016DBCC
601677D8 60170A20 601A53A8 601DAE8C 601DAE78
Jun 16 14:26:34 130.140.56.10 756: 1d03h: %SYS-5-CONFIGI: Configured from
console by martin on console
Jun 16 14:26:34 130.140.56.10 757: 1d03h: %AAAA-3-TIMERNOPER:
AAA/ACCT/TIMER: No periodic update but timer set.
```

This script can also be used to mail accounting and authentication logfiles. All you need to do is edit the users and logfile values, then add the script to root's crontab file. If your pager supports email, you can have the reports sent to your pager as well.

IOS Authentication and Accounting

IOS supports two modes of system authentication: old-mode and new-mode. Old-mode uses a static, locally stored user table or *Terminal Access Controller Access Control System (TACACS)* for user authentication. TACACS (introduced in IOS version 10.0) is a security software suite that provides a configurable modular system for providing authentication, accounting, and authorization. The service is provided by running a service or process daemon on a Windows NT or UNIX system. The server stores the user authentication and authorization data, and logs user and system accounting events. There are three versions of TACACS supported by IOS:

- TACACS (CiscoSecure 1.x)
- Extended TACACS (available in old-mode and new-mode)
- TACACS+ (CiscoSecure 2.x) (available only in new-mode and preferred by Cisco)

New-mode is also referred to as *AAA (authentication, accounting and authorization)*. In addition to user authentication, AAA provides facilities for system and IOS command accounting and authorization. These additional control facilities allow you to keep accounting logs on user access to the router (a must for dial-in), and, if needed, to configure logged, restricted EXEC command access. AAA support is essential in access-server environments and can be quite useful in situations where many people are responsible for maintaining the network. AAA also provides the capability to use Remote Authentication Dial-In User Service (RADIUS) and Kerberos V5 as an alternative to TACACS. However, if you plan to use AAA new-mode authorization and accounting, you will want to use TACACS+. While Cisco provides reasonable support for RADIUS authentication, there is limited support for the authorization and accounting services.

RADIUS (introduced in IOS version 11.2) is an open protocol security authentication and accounting system. Originally developed by Livingston for its portmaster access servers, RADIUS is now maintained by Merit and other commercial vendors and is considered the de facto remote access server (RAS) authentication method. What makes RADIUS so popular is its support for different authentication mechanisms, such as secure ID and smartcards, user profiling, and strong accounting support. RADIUS, sadly, is an IP-only tool and requires users to have more than one account if they require different access services.

Kerberos (introduced in IOS version 11.2) is a trusted third-party authentication system developed at the Massachusetts Institute of Technology (MIT). Kerberos is named after Cerberus, the three-headed dog who guarded the gates of Hades, and it uses the Data Encryption Standard (DES) encryption algorithm for transmitting authentication data. With Kerberos, both users and systems must have an identity, known as a *principle*, stored on a centralized authentication and database server, known as the *Key Distribution Center (KDC)*. All access verification questions are deferred to the KDC (which is why Kerberos is called a "trusted third-party" system). The collection of users and systems that trust the Kerberos KDC to authenticate is known as a *Kerberos realm*.

On the system side, authentication information is exchanged using a key called a *servtab* to encrypt messages between the system and the KDC. On the client side (which, from the Kerberos perspective, is the unsecured element), however, authentication is a little more complicated. A simple view of the Kerberos user authentication process looks like this:

1. A user logs in, and a request is sent to the KDC, where the user's access rights are verified.
2. If the user is valid, a ticket-granting ticket (TGT) and a session key (SK) (which is used as a one-time password [OTP] to decrypt the TGT) are sent back in encrypted form, using the user's password as the source of the encryption key. The user is now authenticated locally. To access other systems that are part of the realm, another ticket is needed.
3. To get a ticket (which has a limited life span), the user generates a ticket request for the remote host it wants to access. The TGT and an authenticator are sent to the Ticket Granting Server (TGS). The request is verified, and a ticket and session key (an OTP to decrypt the ticket) for the requested system are returned.
4. After the ticket is acquired, the user logs in to the system. The system then verifies that the ticket and authenticator match, and if so, access is permitted.

In addition to authentication, Kerberos also supports application data encryption for Telnet, rlogin, rsh (remote shell), and rcp (remote copy). There are two versions of Kerberos in use today: version 4 and version 5. Version 4 is the most common because it is distributed in some form with most UNIX distributions. It is also popular because it can be used outside the United States. KerberosV4 is not without its shortcomings; the most obvious is its lack of support for fully qualified hostnames (which can be corrected with major changes to the code). KerberosV5 added full hostname support, the ability to choose the encryption method by adding encryption algorithm identifier tags to messages, and authentication forwarding (which allows a server to act on behalf of a client for certain application functions). The Cisco IOS only has support for version 5.

IOS's Kerberos support provides kerberized application access (Telnet, rlogin, and so on) from the USER exec and access-server authentication. Accounting must still be provided with RADIUS or TACACS.

The following are online resources for RADIUS, TACACS and Kerberos:

RADIUS:

`http://www.merit.edu/aaa/basicsvr.html`

`ftp://ftp.merit.edu/radius/releases/radius.3.6B.basic.tar`

`http://www.merit.edu/radius/releases/radius.3.6B.basic.tar`

TACACS:

http://fxnet.missouri.org/individuals/wes/tacplus.html

ftp://ftp-eng.cisco.com/pub/xtacacs/

http://www.easynet.de/tacacs-faq/

CiscoSecure 2.0 (TACACS+)

http://www.cisco.com/warp/public/480/tacplus.shtml

http://www.cisco.com/warp/public/480/cssample2x.html

CiscoSecure 1.0 (TACACS+)

http://www.cisco.com/warp/public/480/cssample.html

Kerberos:

http://gost.isi.edu/info/kerberos/

http://web.mit.edu/kerberos/www/

http://www.pdc.kth.se/kth-krb/

http://www.cybersafe.com/

Using Old-Mode Authentication

By default, the IOS uses AAA old-mode authentication. Under AAA old-mode, each dedicated and virtual line interface that requires authentication must be configured using the `login` command. Here is an example:

```
local-AS#config t
Enter configuration commands, one per line.  End with CNTL/Z.
local-AS(config)#line vty 0 4
local-AS(config-line)#login ?
  local   Local password checking
  tacacs  Use tacacs server for password checking
  <cr>
local-AS(config-line)#login local
local-AS(config-line)#exit
local-AS(config)^Z
local-AS#
```

In the example above, you set up old-mode authentication on the VT ports on the router. You now need to create user accounts for each user who requires VT access to the router. This is the same process you used to create user accounts for the WidgetCo dial-in solution:

```
local-AS#config t
Enter configuration commands, one per line.  End with CNTL/Z.
local-AS(config)#username admin password apassword
local-AS(config)#^Z
local-AS#
```

After the username is created, you can log in to the router using old-mode authentication:

```
$telnet 192.160.56.5
Trying 192.160.56.5...
Connected to 192.160.56.5.
Escape character is '^]'.
User Access Verification
Username: admin
Password:
local-AS>
```

If you want to use a TACACS server, set up the interface and specify the address of the TACACS server:

```
local-AS#config t
Enter configuration commands, one per line.  End with CNTL/Z.
local-AS(config)#line AUX 0
local-AS(config-line)#login tacacs
local-AS(config-line)#exit
local-AS(config)#tacacs-server host 192.160.56.1
local-AS(config)#^Z
local-AS#
```

Note

All user accounts and line passwords are in cleartext form unless password encryption is enabled (by using the global configuration command `<service password-encryption>`). Also, there is no facility in the IOS for users to change their passwords, which substantially limits the security value of local user accounting.

Using AAA New-Mode

New-mode is enabled (along with authentication, accounting and authorization; hence, the name AAA) with the global configuration command <aaa net-mode>:

```
local-AS#config t
Enter configuration commands, one per line.  End with CNTL/Z.
local-AS(config)#aaa new-mode
local-AS(config)#^Z
local-AS#
```

Unlike old-mode, when new-mode authentication is enabled, it is up to the administrator to define the authentication options the router will use.

New-mode authentication supports both a local static user table and the security authentication protocols mentioned earlier. If you use a remote security protocol, it is a good idea to create a local backup account in case your authentication server fails and you need to access the router.

It's possible to have privileged EXEC and configuration EXEC modes authenticated remotely as well. This latter option does not work well with RADIUS, so unless you do not have Cisco's TACACS, it should be avoided. It is just as easy to build an enable password file and update your router monthly using TFTP.

One other thing about new-mode is that, unlike old-mode, it can apply authentication to all line interfaces (including the console) when using the "default" list.

Setting Up Login Authentication

Always create the local administrative account first. Then, enable <aaa new-mode>. After AAA has been enabled, the authentication service type and list are defined:

```
Router#config t
Enter configuration commands, one per line.  End with CNTL/Z
Router(config)#username root! password anypass
Router(config)#aaa new-mode
Router(config)#aaa authentication login default radius local
```

In this example, the authentication list uses <radius> as the primary authentication means, and the <local> static username list is the secondary. To set up authentication for PPP service, add the configuration line:

```
Router(config)#aaa authentication ppp ppp-authen radius local
```

If the router is accessible to the public, you may want to change the default Username: and Password: prompts. This can be done with the global configuration EXEC command:

```
Router(config)#aaa authentication password-prompt >
Router(config)#aaa authentication username-prompt #
```

Now the router's login looks like this:

```
User Access Verification
#root!
>
Router>
```

Remember, to remove any IOS configuration option, use <no> preceding the command. To reset the username and password prompts back to the default settings, the <no> option is used:

```
Router(config)# no aaa authentication password-prompt >
Router(config)# no aaa authentication username-prompt #
```

Now the router's login is back to its default:

```
User Access Verification
Username: root!
Password:
Router>
```

At this point in the configuration, the router will only use local authentication, because the RADIUS server has not been defined. Three configuration commands are needed to set up RADIUS:

```
Router(config)#radius-server host 192.168.0.10 auth-port 1645 acct-port 1646
Router(config)#radius-server key agoodkey
Router(config)#ip radius source-interface eth0/0
```

For TACACS+ the commands are as follows:

```
local-AS(config)#tacacs-server host 192.168.0.11
local-AS(config)#tacacs-server key anotherkey
ip tacacs source-interface eth0/0
```

The first two commands are required for authentication to work. The <[service]-server host> command defines the IP address of the server and, if RADIUS is used, the UDP port numbers employed for the accounting and authorization server. The <[service] -server key [word]> command specifies the shared text key that the RADIUS server and the router use for authentication. The < IP [service] source-interface> command is optional; it is used to define which interface will be identified with all of the requests. This last option makes it easier to configure, because the server only needs to know about one interface.

Setting Up RADIUS

RADIUS is a client/server communications protocol used in authentication and accounting. Clients (routers, access servers, and so on) forward user connection information to the RADIUS server, where the user's access is verified. User information or *profiles* are stored on the RADIUS server as a collection of attribute-value pairs that describe the user's access privileges. These attribute-values are in part defined as part of the RADIUS protocol. There are about 50 standard attributes, plus an innumerable amount of vendor-specific attributes.

If you are planning to use a free RADIUS version, you will be running it on UNIX. The best source for the code is at Merit. Many UNIX implementations come with prebuilt RADIUS implementations that are just fine. However, you cannot adjust how the daemon behaves if you do not have the source code, and RADIUS does support some different compiling options. Depending on the version of RADIUS you have, the binary will be located in **/etc/raddb/radiusd** or **/usr/private/etc/radiusd**. If the binary code is not in either of these places, you can use the UNIX command `<locate>` to find it. The directory where RADIUS looks for its configuration files is built into the daemon. The default (again it depends on the version) is **/etc/raddb** or **/usr/private/etc/raddb**. The default can be changed using a flag when you start the RADIUS daemon, which is usually part of the system's startup scripts.

To start RADIUS from **/etc/rc.local**, use

```
if [ -f /etc/raddb/users -a -f /usr private/etc/users ]; then
        echo -n ' radiusd';     radiusd
fi
```

RADIUS stores its logfile in its home directory, **etc/raddb**. Accounting files are stored in **/var/account/radius**. This, like its home directory, can be changed in the source code or with a flag.

After the daemon is built and added to the system startup scripts (and if need be, added to **/etc/services**), only two files need to be created: **/etc/raddb/clients** and **/etc/raddb/users**. The client file tells the radius daemon which hosts to provide authentication and accounting services to. It uses a tab-delimited format:

```
[IP address or hostname]<tab>[Key]
#client key type
192.168.18.20     ***hello12     type=nas
192.168.18.21     paslam69;        type=nas
foo-bar.widgetco.com         way----out             type=nas
```

The key is used for authentication and encrypting the passwords exchanged between the RADIUS server and the client. The key is limited to 15 characters in length. The `type` field is not supported on all RADIUS versions; it is used to specify the type of client. There are several attributes for different vendors. To see if your RADIUS version supports the `type` option, check the README file.

The RADIUS users file requires a little more work. The users file stores the user profile information. The profile tells RADIUS how to authenticate the user and, once authenticated, what services they have access to. Profile entries are created using a collection of *(attribute = value)* pairs to specify the different user profile elements. Fortunately, RADIUS provides the capability to create a default user, which can be set up to reflect a common service access profile. The default can save you the headache of having to create a separate profile for each user.

The first entry is the realm name. The realm name is added after the username, preceded by the `@` symbol at the username or login prompt: `anyuser@WIDGET.COM`. The authentication protocol defines what type of authentication protocol (PAP, CHAP, and so on) should be expected (the `<-DFLT>` option will support any type). The authentication type specifies the kind of authentication to use. The `NULL` realm is used to define what authentication type should be used if no realm is specified. This is not required, but comes in handy with users who forget things.

Note

One of RADIUS's strong points is its support of different authentication mechanisms (Kerberos, secure ID, and so on). To provide access to these functions, RADIUS uses the concept of realms. You do not need to use this if you provide username and password information in the /etc/raddb/users user profile. If you want to use an alternative user accounting method, you first need to define the realm and method in the /etc/raddb/authfile. To use realm authentication, the user profile needs the attribute "Authentication-Type = Realm" which tells RADIUS to check the /etc/raddb/authfile for the authentication method. Here is an authfile example:

Realm	Authentication Protocol	Authentication Type
WIDGET.COM	-DFLT	UNIX-PW
DEVELOP.WIDGETCO.COM	-DFLT	MIT-Krb
NULL	-PW	UNIX-PW

Note

There are two basic types of users: session and dial-in. Session users require some form of terminal access. Dial-in users are looking for authentication, data-link, and network addressing information. RADIUS supports a standard set of attributes and (depending on the version) a bunch of vendor-specific ones. These attributes are defined in the dictionary file stored in the RADIUS home directory. Here are some sample profiles you can use to get started (they can also be used for the default user profile depending on your needs).

Standard username/password login:

```
Anyuser Password = "test",
       Service-Type = Administrative-User,
       Login-Service = Telnet,
       Class = Telnet-User,
```

Dynamically assigned address PPP user:

```
PPPuser Authentication-Type = Realm
        Service-Type = Framed,
        Framed-Protocol = PPP,
        Framed-IP-Address = 255.255.255.254,
        Framed-IP-Netmask = 255.255.255.255,
        Framed-Routing = None,
        Framed-MTU = 1500,
        Framed-Compression = Van-Jacobson-TCP-IP,
       Class = PPP-dial,
```

Statically addressed SLIP user:

```
SLIPuser Authentication-Type = Realm
        Service-Type = Framed,
        Framed-Protocol = SLIP,
        Framed-IP-Address = 192.168.18.24,
        Framed-IP-Netmask = 255.255.255.0,
        Framed-Routing = None,
        Framed-MTU = 1045,
        Framed-Compression = Van-Jacobson-TCP-IP
   Class = SLIP-dial,
```

Setting Up Authorization

What is the difference between authentication and authorization? *Authentication* is used to permit or deny access to the router. *Authorization* provides the ability to restrict access to services once the user is on the router. Authorization requires `<aaa new-mode>` to be activated before any command configuration can take place. This option is oriented more toward use with access-server routers than with traditional network gateways. Authorization, mind you, is not required when using AAA with access servers; it's just there to provide you with more control over users' actions if you need it.

Caution should be taken when using authorization, because mistakes can lead to wasted time messing with `rommon`. Make sure before you implement authorization services that authentication is fully configured and tested.

There are three authorization commands used for restricting access: two for controlling EXEC command access, and one for controlling network events:

- Starting an EXEC shell—`<aaa authorization exec [primary authen] [secondary authen]>`.
- Specific EXEC commands—`<aaa authorization command [privilege level 0-15] [primary authen] [secondary authen]>`. This option must be used in conjunction with the IOS's `<privilege>` command, which is used to assign specific privilege levels for commands.
- Starting a network event (SLIP, PPP, and so on)—`<aaa authorization network primary authen] [secondary authen]>`.

For example, to have RADIUS verify whether the user profile has rights to use PPP, use:

```
<aaa authorization network radius>
```

An alternative to this would be to permit a PPP session if the user has been authenticated:

```
<aaa authorization network if-authenticated>
```

If you only want certain TACACS+ users to have access to privileged mode, you would need to use:

```
<aaa authentication login default tacacs+ local>
```

and

```
<aaa authorization commands 15 tacacs+>
```

The authorization process works by having the router check the user's authentication method and access privileges. If they meet the configured authorization requirements, access to the command is permitted. This means you have to configure the users' RADIUS/TACACS user profiles to reflect the specific types of services they need. This can be quite a task, so you must decide if the work is worth the return.

Configuring Accounting

Accounting is one of those things you just need to do, especially with services like dial-in, where security risks are high. Accounting gives you the ability to track service abuse and login failures, generate usage statistics, and so on. If you are not using some kind of system accounting on at least your high-risk system, you really should do so. There are two types of IOS accounting: user and operations. IOS provides access information on network sessions (PPP, SLIP, ARA) and outbound connections (such as Telnet, rlogin). Operational accounting, on the other hand, tracks the information pertaining to router-centric activities. The IOS splits operational accounting between two accounting systems:

- `<system>` accounting—Provides system event information (similar to logging information)
- `<command>` accounting—Keeps track of EXEC shell commands usage

IOS accounting requires a TACACS+ or RADIUS server to process the accounting records. The records are collections of attribute-value pairs (similar to those used to create user profiles). Here is an example of a RADIUS exec accounting record:

```
Wed Jun  9 12:27:09 1999
        NAS-IP-Address = 192.168.17.1
        NAS-Port = 2
        NAS-Port-Type = Virtual
        User-Name = RAD-USER
        Called-Station-Id =
        Calling-Station-Id = 192.168.17.1
        Acct-Status-Type = Stop
        Acct-Authentic = RADIUS
        Service-Type = Shell-User
        Acct-Session-Id = 00000005
        Acct-Terminate-Cause = Idle-Timeout
        Acct-Session-Time = 601
        Acct-Delay-Time = 0
```

In order to get accounting started, you need to be using `<aaa new-mode>` security and have TACACS+ or RADIUS configured on the client (router) and server ends. Accounting is configured using variations of the `<aaa accounting [service] [notice] [RADIUS/TACACS+]>` command.

If you are using RADIUS, you can only collect accounting information for network, outbound connections, and EXEC sessions, providing essentially only user accounting information. If you need to collect "system accounting" information about the router, you will have to use TACACS+. This provides support for collecting accounting information about the router using the <system> and <command> options.

```
local-AS(config)#aaa accounting network start-stop radius
local-AS(config)#aaa accounting command start-stop radius
local-AS(config)#aaa accounting exec start-stop radius
local-AS(config)#aaa accounting system start-stop tacacs+
local-AS(config)#aaa accounting commands start-stop tacacs+
```

There are three accounting notice options:

- <start-stop> sends a notice to the accounting server when the session starts and ends.
- <wait-start> performs the same function as <start-stop>, except the actual session does not start until the accounting notice is acknowledged by the accounting server.
- <stop-only> provides the most basic accounting information and only sends a notice when the session is completed.

To see all the active sessions on the router, you can use the <show accounting> privileged EXEC command:

```
local-AS#show accounting
Active Accounted actions on tty0, User flasmaster Priv 1
 Task ID 26, EXEC Accounting record, 00:00:19 Elapsed
 taskid=26 timezone=UTC service=shell
local-AS#
```

The output is displayed using the attribute-value pairs of the accounting system you are using, so the output will vary depending whether you are using TACACS+ or RADIUS.

Summary

This chapter covered plenty of material to provide you with the background you'll need to implement Cisco routers in a multiprotocol environment:

- Cisco console access
- IOS command basics
- Basic interface configuration
- Basic IP, AppleTalk, and IPX protocol configuration
- Using rommon and disaster recovery
- Configuring the router's clock
- Upgrading the IOS
- Configuring IOS logging
- Configuring IOS authentication and accounting

In Chapter 9, "Advanced Cisco Router Configuration," we will cover additional Cisco IOS configuration options, such as ACLs and WAN interface configuration. Chapter 10 will cover IP dynamic routing configuration.

Related RFCs

RFC 868	Time Protocol
RFC 1179	Line Printer Daemon Protocol
RFC 1350	TFTP Protocol Revision 2
RFC 1510	The Kerberos Network Authentication Service (V5)
RFC 2138	Remote Authentication Dial-In User Service
RFC 2139	RADIUS Accounting

Additional Resource

Da Cruz, Frank, and Christine M. Gianone. *Using C-Kermit.* Digital Press, 1993.

8

TCP/IP Dynamic Routing Protocols

IN CHAPTER 2, "THE NETWORKER'S GUIDE TO TCP/IP," we reviewed the different protocols that make up the TCP/IP protocol suite. In the section covering Internet Protocol (IP), we reviewed the following:

- Classful IP addressing
- Classless IP addressing
- The IP datagram routing process
- Static and dynamic routing

Then, in Chapter 3, "The Networker's Guide to AppleTalk, IPX, and NetBIOS," we looked at AppleTalk and Internetwork Packet Exchange (IPX) protocol suites and the protocols that make these respective suites work. All these protocol suites provide the same fundamental set of communication services and can operate over the same transport media. TCP/IP's distinction is that it can scale to an almost limitless size, whereas IPX and AppleTalk have size and operational limitations that make them undesirable for use in large-scale enterprise and global networks.

The flexibility of the TCP/IP protocol suite is a result of IP's connectionless datagram delivery process and the diversity of IP dynamic routing protocols. Although TCP/IP was designed to operate on a global scale, the actual methodology of how TCP/IP is implemented has changed as the global Internet has grown. IP's original dynamic routing protocols have evolved, and new protocols have been created to meet IP's evolving needs. This diversity in the implementation of IP and the variety of its associated dynamic routing protocols reflect the specific requirements for IP datagram delivery. The goal of this chapter is to provide you with a more in-depth look at TCP/IP's delivery fundamentals and the dynamic routing protocols that support this process. By looking at each of the routing protocols, you learn their strengths and weaknesses and, most importantly, how to use them correctly. This chapter builds on the concepts covered in Chapter 2. In this chapter, we review and examine the following:

- Dynamic routing concepts and terms
- TCP/IP interior gateway protocols
- TCP/IP exterior gateway protocols

In the following section, we review some of the general concepts and terms associated with IP routing. This section also relies on material covered in Chapter 2, specifically:

- IP datagram delivery process
- IP address space
- VLSM (variable-length subnet masks)
- CIDR (classless interdomain routing) and classful addressing

An Introduction to General Routing Concepts and Terms

Networking has its own language, consisting of terms with very specific meanings. The goal of this section is to lay the groundwork and give you the fundamentals of the TCP/IP routing protocols, then address the particulars of each protocol in its own section.

Dynamic Routing Protocol Basics

The basic job of a *dynamic routing protocol* is to find the best, most efficient route to forward IP network traffic. The dynamic routing protocol can also find a secondary route in the event that the best route is lost. Enterprise networks commonly use multiple network paths to interconnect segments. Routing protocols provide two main advantages:

- The capability to balance the traffic load between these paths and keep traffic flowing in the event that a path is lost
- The capability to easily add routes (such as new network segments) to the network

IP network routing can be categorized into two distinct types of routing:

- Intranetwork routing
- Internetwork routing

Intranetwork routing, or *interior routing*, exchanges route information between routers within defined routing processes. An intranetwork can have single or multiple routing processes. Protocols that perform intranetwork routing are known as *interior gateway protocols (IGPs)*. The Routing Information Protocol (RIP) and Open Shortest Path First (OSPF) are popular IGPs.

Internetwork routing has two dimensions: First, in the context of the global Internet, internetwork routing is the exchange of routing information between large, centrally administered networks known as *autonomous systems (ASs)*. A particular type of routing protocol known as *Exterior Gateway Protocol (EGP)* performs this task. EGP was the first protocol created to perform this function. Today, Border Gateway Protocol (BGP) version 4 is the most commonly employed protocol for this task. Refer to the section "TCP/IP Exterior Gateway Protocols" later in this chapter for more information on EGP and BGPv4.

The other, more useful dimension of internetwork routing is the process of running multiple distinct routing protocol instances or a collection of different IGP routing protocols to distribute routing information within a network hierarchy. A common form of this model is when a single root network acts as a common backbone, and child networks branch off that backbone. The distinction between this type of network from, say, an enterprise intranetwork is utilization of different routing protocols within each child, so that each child network administers its own routing policy (the function of deciding what networks are available to the network and/or the scope of IP network spaces that are managed under a single administrative authority). The root network can act on behalf of all child networks in the context of the global Internet.

Note

Under the OSI model, intranetwork routing is known as intradomain routing.

Note

Under the OSI model, internetwork routing is known as interdomain routing.

Deciding to Use a Routing Protocol

If you're wondering whether you should use a routing protocol, the answer is a qualified yes. Routing protocols have come a long way since Routing Information Protocol (RIP) version 1 was released as part of the BSD UNIX TCP/IP suite. In the early days of computer networking, computer networks were more like host-to-host interconnections than actual networks. As LAN (Ethernet, Token Ring, and so on) and UNIX workstation technologies evolved, small clusters of hosts connected over a shared physical network segment (a hub or switch) replaced the mainframe in some cases. Under the host-to-host internetwork model (see Figure 8.1), redundancy was achieved by each site (intranetwork) having its own connections to every other site. The idea was that if a site lost its direct connection it would still be able to reach its destination by sending its data to another site that still had a connection to the destination site. Because each of the sites had point-to-point links of similar speeds, the only thing the routing protocol had to do was provide a way to get there.

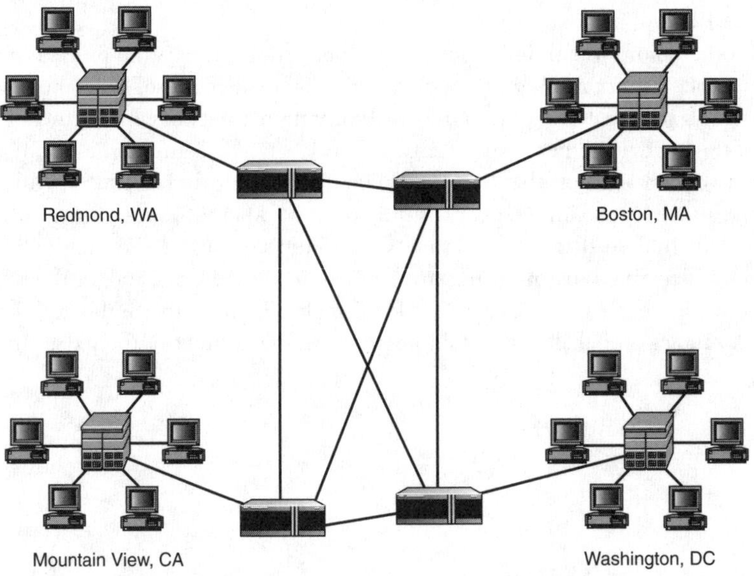

Figure 8.1 The host-to-host paradigm.

Today's intranetworks consist of hundreds of IP subnets and support thousands of hosts, as shown in Figure 8.2. Communication between hosts can happen at gigabit speed. WAN intranetworks today can operate at Fast Ethernet speeds and greater. Gone are the days of direct point-to-point links; now, intranetworks connect to single or multiple Internet backbone networks that provide for extensive network diversity and redundancy.

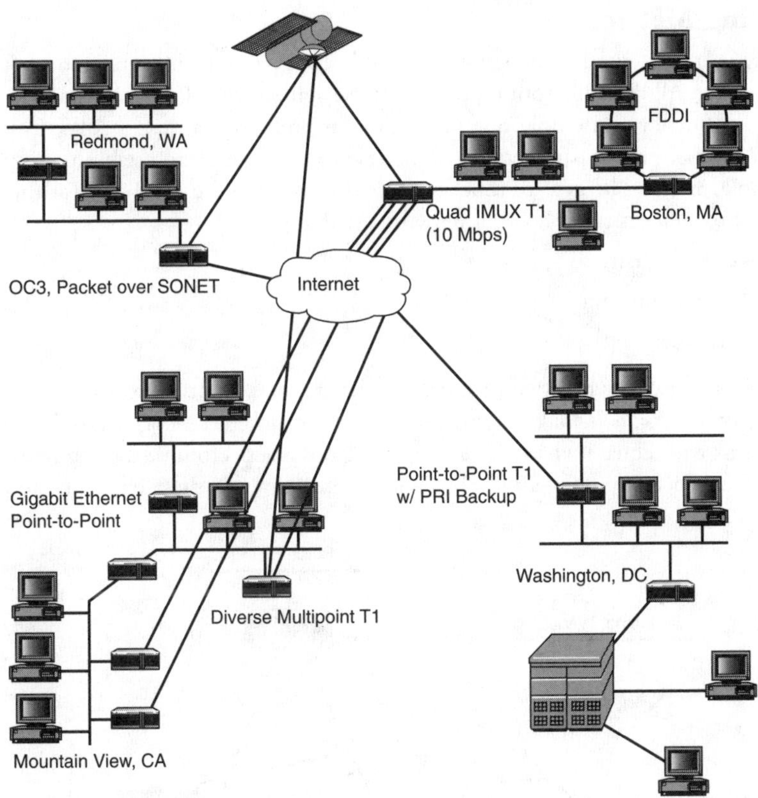

Figure 8.2 The network-to-network paradigm.

Dynamic routing protocols can support the demands of modern networking. These protocols are more than a process for determining how to reach other networks. Along with route discovery, the dynamic routing protocol needs to exchange routing information with other designated routers in its process or domain, and determine the optimum path and/or manage loadsharing across paths, based on information received from other members of the routing process or domain. Finally, the dynamic routing protocol must be able to react to and change router behavior in response to link failures, and adjust for load and topology changes within the scope of its process.

Now, here is the qualifier: Not every network needs to use a dynamic routing protocol. There are pros and cons, and as a network administrator, you need to make this decision. The rest of this chapter provides you with the information you need in order to make it.

Routing Metrics

Routing metrics are used by dynamic routing protocols to establish preferences for a particular route. All dynamic routing protocols use some form of route metrics. In certain cases, a protocol can use only one metric. Other protocols can use a collection of metrics, or a set of metric parameters may be calculated together to establish a single, but adjustable, metric. The goal of route metrics is to provide the capability for the routing protocol to support the following network attributes:

- Route diversity
- Route redundancy
- Load balancing

Route diversity exists when two or more unrelated access points or paths exist for the same network. *Route redundancy* exists when two or more access points to the same network exist with equal metrics. *Load balancing* is the practice of distributing network traffic evenly across multiple links. In Figure 8.3, all three of these elements are possible.

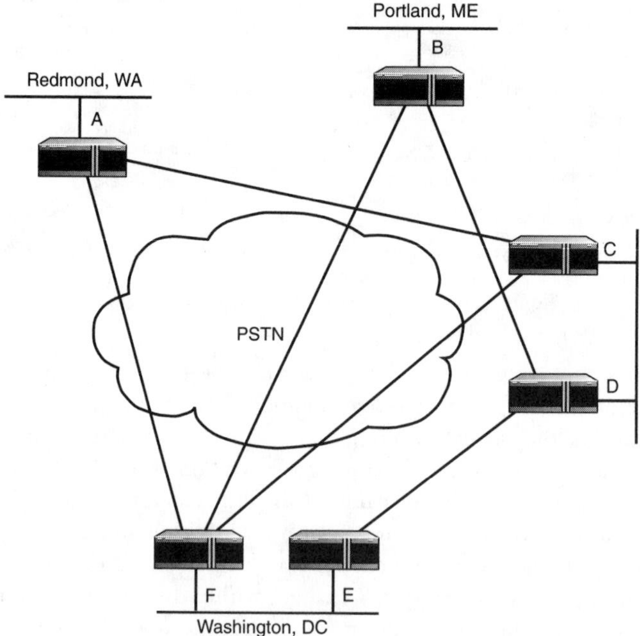

Figure 8.3 A route-diverse network.

All the access routers support route diversity because there are at least two access points to each network. Route redundancy and load balancing is available between the Washington and Princeton networks. The single largest advantage in using a dynamic routing protocol is the gain of route redundancy and load balancing capability. Without the use of a dynamic routing protocol, neither of these facilities would be available.

To get a better understanding of how route metrics work, let's take another look at Figure 8.3. For starters, if the link between F and C is a 56K line and the link between E and D is a T1, and we want to move a 4Mb file, we want our data to move across the E and D link. But what if the link between F and A is a 10Mbps link and the link between A and C is an OC3 (155Mbps)? Then, we would want our data to move across the F to A, A to C link.

If RIP was the routing protocol for this network, our data would have been sent across the F to C or E to D link. RIP determines the best route based on hop count. If, on the other hand, we had been running IGRP, our file transfer would have been sent across the F to A, A to C link, because IGRP uses lowest bandwidth and fastest path to decide the best route. Metrics tell the dynamic routing protocol (and the router) which route is the best route.

The following list describes the most common routing metric variables:

- Hop count—As the most basic metric, hop count is the number of intermediate systems (routers) between the router and the destination router. The hop count between Washington and Portland from Router E is 2, from Router F is 1.
- Bandwidth—This metric reflects the interface's ideal throughput. For example, a serial interface on a Cisco router has a default bandwidth of 1.544Mbps, and Ethernet has a default bandwidth of 10Mbps.
- Load—Although the bandwidth metric sets the ideal, the load metric varies based on the actual usage (traffic).
- Delay—Delay is the total time needed to move a packet across the route. The shortest time is the best route.
- Reliability—Reliability estimates the chance of a link failure and can be set by an administrator or established by the given protocol.
- Cost—This metric sets the preference for a given route. The lower the cost, the more preferable the route. Most routers have a default value for each type of interface. The interface's default cost is directly related to its speed.

Note

Keep in mind that packet routing is a hop-by-hop event. These metrics are set on the interfaces on the router. The routing protocol then sends those values to the other routers who use them to determine the best route from their viewpoint.

Network Convergence

Convergence is the process of bringing all the routing tables of all the routers in the network (or that belong to the process) to a state of consistency.

Routing information is distributed between physically connected routers as broadcast or multicast messages. Network information is designated in a router-to-router (hop-to-hop) fashion the same way that IP datagrams are delivered. Using Figure 8.4 as an example network, let's walk through the convergence process from Router A's perspective.

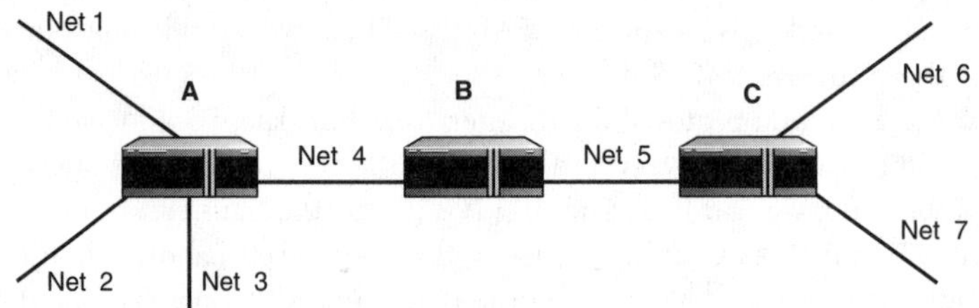

Figure 8.4 A simple convergence.

Router A has four networks attached to it. Router B is directly attached to Router A. Router A sends information about all of its directly connected networks to Router B. Router B then tells Router C about all of the networksit knows about: its own directly connected networks, and all of Router A's directly connected networks. Router C tells Router B about its directly connected networks and Router B tells Router A about Router C's networks.

Convergence time is how long it takes for routers to learn the network topology and/or changes in the network topology (for example, a failed link or the addition of a new network segment). When a change takes place, the network is in a state of flux, and it is possible that some routers will forward traffic to paths that are not available or no longer exist. In large networks, it is preferable to use a routing protocol that has a fast convergence time.

Distance Vector Protocols

All routing protocols use an algorithm or set of algorithms to calculate anddistribute routing information. A *distance vector protocol (DVP)* is a dynamic routing protocol that uses an algorithm based on the work of Bellman, Ford, and Fulkerson known as *vector distance* or the *Bellman-Ford algorithm*. RIPv1, RIPv2, and Cisco's Interior Gateway Routing Protocol (IGRP) are considered DVPs.

The idea behind the DVP algorithm is that each router on the network (routing domain or process) compiles a list of the networks it can reach and sends the list to their directly connected neighbors. The routers then create routing tables based on what they can reach directly and indirectly, using their neighbor routers as gateways. If multiple paths exist, the router only keeps the best one. This route is chosen based on the protocol's particular metric for determining the best route. Routing protocols each use different route metrics to choose the best route.

Link State Protocols

Link state protocols (LSPs) are based on a different algorithm than DVPs. LSPs are based on a graph theory algorithm called *Dijkstra's algorithm*, named after its creator E.W. Dijkstra. The algorithm works like this: The premise is that a collection of points is seen as a tree; one point is the base and the shortest path to all other points is calculated using the other points as the pathway. The cost of the path is the sum of the path costs between the source point and the terminus point. Whereas DVPs are interested in the shortest route, LSPs are interested in the operational status of all of the links in the network. Each host compiles a "map" of the routing domain from its own point of view.

From a practical perspective, LSPs and DVPs have some similarities. In both protocols, the router comes online and finds other routers as "neighbors" that are directly accessible from its connected network interfaces and part of the same process. The router then sends out a message known as a link state announcement (LSA) from each of its network interfaces. The LSA contains information about the router's interfaces: their cost metric, operational state, and which routers are reachable from them. The directly accessible routers then forward the LSA to other neighbors. LSAs are saved together as a link state database. The router uses the database to compute its own routing table.

The important differences between DVPs and LSPs are the following:

- LSPs only exchange link information, not routing tables the way that DVPs do.
- LSPs only send LSA changes (such as routing updates) when an actual change in the network occurs. DVPs send updates at a fixed time whether changes in the network have happened or not.

OSPF (Open Shortest Path First) is an IP dynamic routing protocol that uses link state (Dijkstra's algorithm) as its routing algorithm.

Variable Length Subnet Mask (VLSM) Routing

One of the more complex concepts associated with IP datagram routing is classless interdomain routing (CIDR) and the proper use of IP subnetting and VLSM. In Chapter 2, CIDR and VLSM were introduced. The goal here is to revisit these concepts in the context of IP routing protocols. You should recall from Chapter 2 that the IPv4 address space can be provisioned using either the classful or classless method:

- *Classful* uses natural bit mask network boundaries to define IP networks of different sizes. There are three usable classful network ranges: Class A, Class B, and Class C. The class of an IP address can be determined by looking at the first 8 bits of the binary address.

First 8 bits:	Address Class	Natural Network Mask
0XXXXXXX	Class A	255.0.0.0
01XXXXXX	Class B	255.255.0.0
011XXXXX	Class C	255.255.255.0

  These classful spaces can be subnetted into smaller IP subnets to allow for more efficient use of address space. The catch is that the major address space must be subnetted using only one subnet mask. For example, a Class B network can be subnetted using a Class C netmask, thereby creating 255 IP subnets, each of which can support 254 hosts.

- *Classless*, as in classless interdomain routing (CIDR), is network addressing without address classes (Class A, Class B, and so on). Instead, IP address groups are created. These groups can be of any size (within the limits of 32-bit address space) based on a power of 2. Classless addressing is the most common model used today with routers, and classful addressing is used with host computers. Classless network masks are defined by a bit representation that indicates how many bits are used for the network address.

The important thing to remember about classless and classful addressing is that the netmask representations change, but the netmask itself stays the same. A 255.255.255.0 classful netmask is the same amount of binary 1s on the netmask field as the classless representation of /24. Both mean that 24 1s will be used to mask the address space. The change is in how the IP address space is allocated and provisioned as variable address groups rather than as ridged classed subnets.

These changes affect the way IP addresses are interpreted and the way route entry for the network is created. With classful addressing, each classful network must have a routing table entry. With classless addressing, a network address group can be expressed as a single entry.

Classless addressing enables the use of *variable-length subnet masks (VLSM)*. VLSM is a tremendous win for routers and the way routing tables are created. With VLSM, you can variably subnet a major network address space or further subnet an already subnetted address space. For example, an ISP assigns you four Class C address spaces, 192.119.17.0 to 192.119.20.0, and each network can support 254 hosts each. Instead of adding the four Class C networks to its routing tables, the ISP will just add one route entry: 192.119.17.0 /22. Now, let's say you have nine networks spread out over three locations for which you have to provide addresses, and you are also using shared Ethernet repeaters, so the maximum amount of users you want to have per network is about 50 hosts. You subnet the 192.119.17.0 /22 into network spaces you can use efficiently (as shown in Table 8.1).

Table 8.1 **VLSM Subnetting for the Network Example in Figure 8.5**

Network Address	Netmask
192.119.17.0	/26
192.119.17.64	/26
192.119.17.128	/26
192.119.17.192	/26
192.119.18.0	/26
192.119.18.64	/26
192.119.18.128	/26
192.119.18.192	/26
192.119.19.0	/26
192.119.19.64	/30
192.119.19.68	/30
192.119.19.72	/30

Now, let's use these network partitions to construct a network. In Figure 8.5, the assumption is that all networks have a network root of 192.119 unless otherwise marked.

You could have followed classful subnetting rules to create this network (because you have only one subnet mask in the major address space), but you would run into a small issue regarding the point-to-point links. These would have been made using the remaining three subnets, which would have depleted any reserved address space you could have used for dial-in or further network expansion (not to mention that it would have been a waste of 180 host addresses).

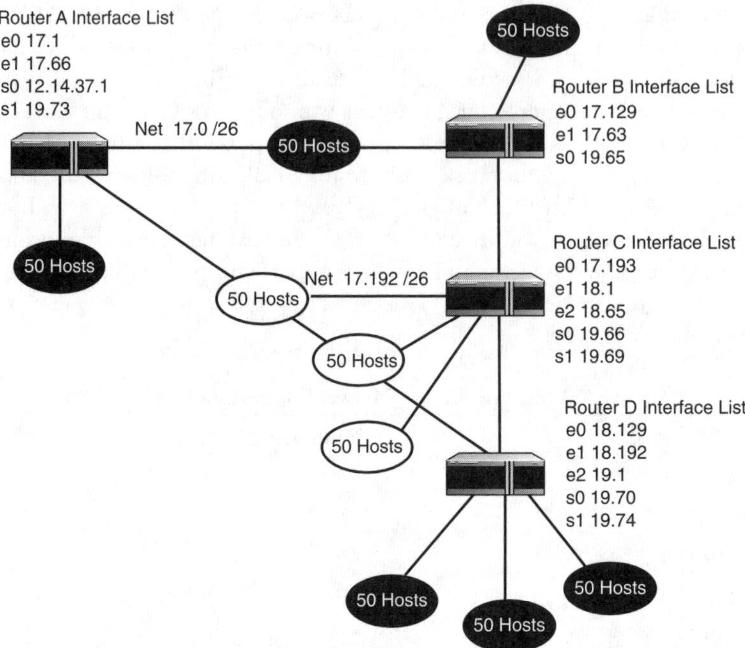

Figure 8.5 Network map for 192.119.17.0 /21 (a VLSM network).

By using VLSM to allocate the address space, you use the address space more efficiently and gain the ability to summarize the routes in your routing table. Because this network is small and you are using shared Ethernet, the additional network load that would be created by a dynamic routing protocol might not be desirable. The ability to use summarized routes means a shorter routing table, resulting in faster route lookups. Tables 8.2 and 8.3 show the route tables from Routers A and B.

Table 8.2 **Routing Table from Router A**

Network	Netmask	Gateway
17.0	/26	Direct
17.64	/26	Direct
19.72	/30	Direct
19.0	/26	17.129
	19.74	
19.64	/29	17.129
	19.74	
18.0	/24	17.129
	19.74	
17.129	/25	17.12
	19.74	

Table 8.3 **Routing Table from Router B**

Network	Netmask	Gateway
17.0	/26	Direct
17.128	/26	Direct
19.64	/26	Direct
17.192	/26 19.66	17.1
17.64	/26 19.66	17.1
18.0	/24 19.66	17.1
19.68	/29 19.66	17.1

Rather than enter routes for each subnet, summarized entries are made for the network address spaces that are contiguous. The network address space needs to be contiguous so the netmask will allow the correct host bits to be recognized. To get a better understanding of how this works, you need to look at the addresses the way the router does (as binary numbers). In this example, Router A uses 192.119.17.64 /29 to summarize the 192.119.17.64 /30 and 192.119.17.68 /30 networks:

```
192.119.17.64 /30
Address:    11000000.01110101.00010011.01000000
Mask:       11111111.11111111.11111111.11111100
192.119.17.68 /30
Address:    11000000.01110101.00010011.01000100
Mask:       11111111.11111111.11111111.11111100
192.119.17.64 /29
Address:    11000000.01110101.00010011.01000000
Mask:       11111111.11111111.11111111.11111000
```

With a bit mask of /30, there are only two usable addresses for hosts. In the case of 192.119.17.64 /30, the addresses 192.119.19.65 and 192.119.19.66 are usable addresses. The 192.119.19.64 and 192.119.19.67 addresses are reserved for the network loopback and network broadcast address. Route summarization works because routers are only concerned with the next hop, rather than the whole path. The router just forwards the traffic to the next hop. When the next hop gets the packet, it uses its routing table to decide where to forward the packet next. Because the spaces are contiguous, moving the netmask back still allows the proper distinction between the network and host addresses. The /29 mask allows for 3 bits to be used for host address space. So, effectively, any traffic that has a destination address that falls within the allowed host space will be forwarded.

Note

When a router forwards an IP datagram, only the network address portion of the destination address matters. The host address portion is only important when the IP datagram reaches the destination network segment. A common problem with using VLSM to segment IP address spaces is that the subnet masks on end-stations are set incorrectly. If this condition exists it is possible that IP datagrams can be discarded or lost because the host portion of the destination address might be interpreted as part of the network portion. Keep this in mind when summarizing IP address space statically on systems that do not directly support VLSM.

Routing protocols that support CIDR and VLSM use summarization when possible to reduce routing table sizes. Not all routing protocols support VLSM, however, so if VLSM support is something you need, make sure the protocol you choose supports it. Although VLSM can be a wonderful way to stretch limited address space and reduce routing table sizes, it also creates conditions that can result in network performance issues. For example, large Layer 3 broadcast domains (greater than 254 hosts) can be exploited with ICMP tools such as ping to create broadcast storms. Another common problem isthat some network administrators think that because they have large IP subnet spaces, they can forget Ethernet collision domain rules (no more than 50 to 150 hosts per Ethernet segment) and create large networks of shared hosts.

Discontinuous IP Subnets

Although route summarization is a good thing overall, there are circumstances when route summarization is undesirable. This is most often the case when a network has discontinuous subnets (see Figure 8.6). *Discontinuous subnets* exist when a routing domain with two networks using the same classful network address space are connected across a network that uses a different classful address space. This creates a problem because some routing protocols perform route summarization using the network's highest boundary or natural class mask. Not all routing protocols send netmasks in their updates. Rather, they use the natural address classful boundary, or if the network is directly connected, they use the subnet mask that is set on the interface (this allows classful address subnetting to work).

Because the router uses the netmask to determine the network and host portion of the address, it forwards the traffic correctly. However, if the subnet masks are not correct on all of the hosts/routers, the data will not be forwarded correctly because they will use their configured subnet masks to discern the network and host portions of the address. The problem created with discontinuous subnets is that because some routing protocol summarizes at the natural mask, a routing loop can result. A *routing loop* is a condition where the router forwards datagrams back to itself.

This summarization condition is illustrated in Figure 8.6. Router A thinks that all traffic destined for network 144.128.0.0 is local to it, and Router B thinks that all traffic for 144.128.0.0 is local to it. But, 144.128.0.0 has been subnetted further than its natural mask. In this case, the routing protocol cannot support discontinuous subnets. To correct this problem, route summarization needs to be disabled. Unfortunately, disabling route summarization comes with problems of its own because the router will be required to maintain routes for each IP subnet in the network. On a small 5 or 10 subnet network, this is not a real problem, but if you have a VLSM-segmented Class B address space, your routing tables can become quite large and cause poor router performance if the router is short on memory or processing speed. When possible, you should try to avoid discontinuous subnets. If this is not an option, using a different routing process or routing protocol and route redistribution to merge the different IP address spaces into a working policy is also an alternative.

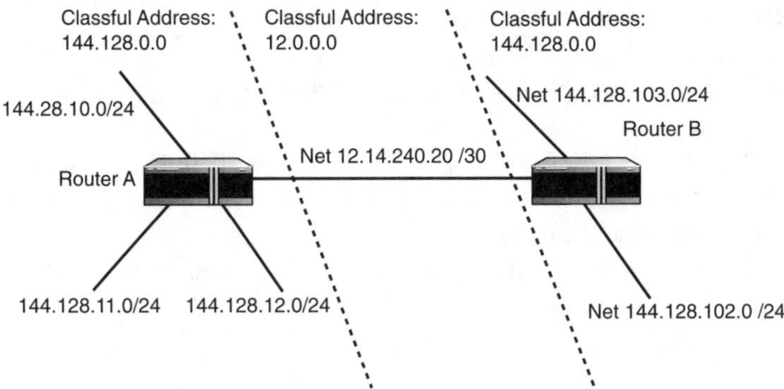

Figure 8.6 A discontinuous network.

Route Redistribution

It is not uncommon to have more than one routing protocol operating in a large internetwork/intranetwork. Not all IP routing protocols are supported by network hardware vendors. Some vendors even have their own proprietary protocols (for example, Cisco's IGRP and EIGRP). Having multiple routing domains enables you to overcome discontinuous subnet issues. Whatever the reason, *route redistribution* is the process of having a routing protocol distribute to other members of its routing process domain routes that were learned from a source outside of the domain. Setting up route redistribution will be covered in Chapter 10, "Configuring IP Routing Protocols on Cisco Routers."

TCP/IP Static Routing

The battles over "Which is better, static or dynamically built routing tables?" will continue as long as there is a choice. The decision is up to the administrator, because he or she has to live with whatever process is used to manage route collection. The idea of static routing is a very "IP-centric" concept. Most popular networking protocols provide routing information dynamically and offer little or no manual manipulation of network routes. Static routing has strong benefits and, unfortunately, some strong weaknesses. The intent here is to illustrate them both so you have something to compare when you look at the dynamic routing protocols.

Strengths of Static Routing

Using static routes to manage a routing domain is a very popular thing to do. Dynamic routing protocols represent advanced TCP/IP knowledge to most people, so administrators who are new to networking tend to steer clear of them. Most people who use TCP/IP have at least some experience with static routing in setting a default gateway. Lack of experience has perpetuated the use of static routing even in situations where it is not the best solution.

There are five commonly acknowledged strengths of static routing:

- Ease of use
- Reliability
- Control
- Security
- Efficiency

The reason static routing is so popular is because of its ease of use, reliability, and control. UNIX/NT system administrators (who usually get stuck with managing the IP network) are the most common advocates of the use of static routes. System administrators love its simplicity almost as much as they love having control over what goes on with the systems they administer. Static routing can be set up on most modern operating systems' TCP/IP implementations.

An administrator can build a master route table for the network and distribute it to end-stations so the routes are added to the local table when the system boots up. If a change is needed, the file can be copied over the network to the end-stations, the system is rebooted, and the new route is added. Although ease of use is always an advantage, the bigger win with using static routing table entries is the control and reliability they provide. Because static routing tables are entered manually, unless a mistake is made or a hardware failure occurs, the tables are absolutely correct. Dynamic routing protocols can fail or be misconfigured. They also suffer from network convergence latency and potential routing loops.

It is commonly held that static routing provides an element of security. Fundamentally, this is true. An end-station cannot reach a network to which it has no route. This approach is not exactly security in the formal sense but rather, "security through obscurity." It accomplishes the same end goal that a traditional security approach would—limiting network availability through the use of access control lists (ACLs). But, it does so by an obscure means (omitting routes from certain workstations), instead of a secure one (creating an access list that only allows certain hosts to access the network segment and reporting when an attempt is made by an end-station that is not permitted).

Although static routing might fall short on the security side, it makes up forit in efficiency. If static routing is anything, it is network and router friendly. Dynamic routing protocols add to network traffic and use memory and processor cycles on the router. With static routing, none of the router resources are used for route acquisition and no additional network traffic is generated by sending and receiving dynamic routing updates.

Weaknesses of Static Routing

Static routing, sadly, has two weaknesses that are difficult to overcome, the first of which becomes evident to most network administrators when their networks start to grow. Static routing does not scale. If you have 20 hosts connected to an Ethernet hub and a 1.544Mbps Internet link, static routing is just fine. If you have 30 high-speed routers and 128 IP subnets, static routing is not going to work. Building 30 different routing tables to route traffic to 128 different subnets will not work in a clean and efficient manner.

The second weakness is rather ironic: What makes static routing so reliable in small static networks is what makes them impractical to use in large networks. Static routes are *static*! Static routing is not adaptable. When a link fails, there is no means to redirect the traffic being forwarded to the link. The absolute nature of static routes is why most routers have a default gateway or route of last resort. In the event that the dynamic routing protocol fails, it forwards the traffic to the default gateway.

When static routes are used improperly, they can, at the very least, affect network performance and, at worst, they can cause routing loops or data loss due to traffic hemorrhages. With that in mind, use of static routes in complex networks should be looked at in terms of growth of the network and what effects will result if they are lost or removed from the routing table.

Note

Do not underestimate the amount of bandwidth used by routing protocols! In older shared segment networks, bandwidth is precious. Protocols, like RIP, constantly send routing updates, even after the network has achieved convergence.

Note

If you plan to use static routes in your network design, be sure to document on your network map which routers contain which routes and to what devices. This will be a large asset when you or someone else (who did not design the network) is troubleshooting network problems.

As a rule, static routes have a role in every routing environment. They should, however, be used as part of a sane routing strategy. Chapter 10 looks at different ways to use static routing in enterprise network environments.

TCP/IP Interior Gateway Protocols

At this point, you should have a solid understanding of how IP datagrams are routed, what the basic elements of dynamic routing protocols are, and why they are necessary. This section provides you with an overview of how RIP, IGRP, EIGRP, and OSPF work. Implementation and configuration of these protocols are covered in Chapter 10.

Routing Information Protocol, v1 and v2

RIP has the largest support base of any of the dynamic routing protocols. Almost every Layer 3 network device on the market supports at least one version or both versions of RIP. RIP is also available on most of the popular operating systems (UNIX and Windows NT), giving end-stations the capability to support multiple network gateways. RIP's universal support and easy setup make it very common in small, static point-to-point and redundant path LAN networks.

RIP comes in two flavors: v1 and v2. RIPv1 is a basic single metric distance vector protocol. RIPv1 provides support only for classful addressing and is limited to a network radius of 15 hops (Layer 3 address segments) from the first router to the last.

RIPv2 fixes the security holes that existed in RIPv1, and provides support for routing authentication, route summarization, CIDR, and VLSM support. Both RIPv1 and RIPv2 use the same message format. However, RIPv1 cannot read RIPv2 messages, so if both versions of the protocols are running, compatibility options need to be enabled. RIPv1 uses UDP (User Datagram Protocol) broadcasts (source and destination port 520) to transport messages. RIPv2 uses the multicast address 224.0.0.9 (also to port 520) to transport messages.

Note

Today, RIP is considered by some to be antiquated and ineffective, particularly RIPv1. Despite this belief, RIP is still common, especially in old Novell NetWare shops, because IPX uses RIP for route management. Another popular use for RIP is to redistribute routing information from systems that only support RIP into another routing protocol. RIP's real value to us is that it illustrates all the basic elements of a dynamic routing protocol.

Note

Extended RIP (ERIP) is also available on some platforms. ERIP extends the network diameter to 128 hops.

RIP uses only one metric to determine the best route: hop count. The term *hop* refers to the unit of distance between forwarding points, such as end-stations or routers. Corresponding to RIP's limited network topology, RIP supports hop counts with a range from 1 (directly connected) to 16 (unreachable network). This distance limitation constrains RIP's usage in large enterprise networks. But, RIP was not designed and should not be used to manage routing in an enterprise. RIP was designed to be used in networks of limited size using similar Layer 2 transport, which means that it does work well in small LAN networks.

RIP is a distance vector protocol, based on the vector distance algorithm (described in the "Distance Vector Protocols" section earlier in the chapter). Distance vector messages are sent as value pairs. The first value is the vector or destination. The second value is the distance or hop count. After the router comes online, it builds its local routing table and then joins the RIP process by sending a RIP message with the request command flag set. The router listens for responses from the other directly connected routers, which contain the other routers' routing tables.

Using the example shown in Figure 8.7, Router E's initial routing table would look like what is shown in Table 8.4.

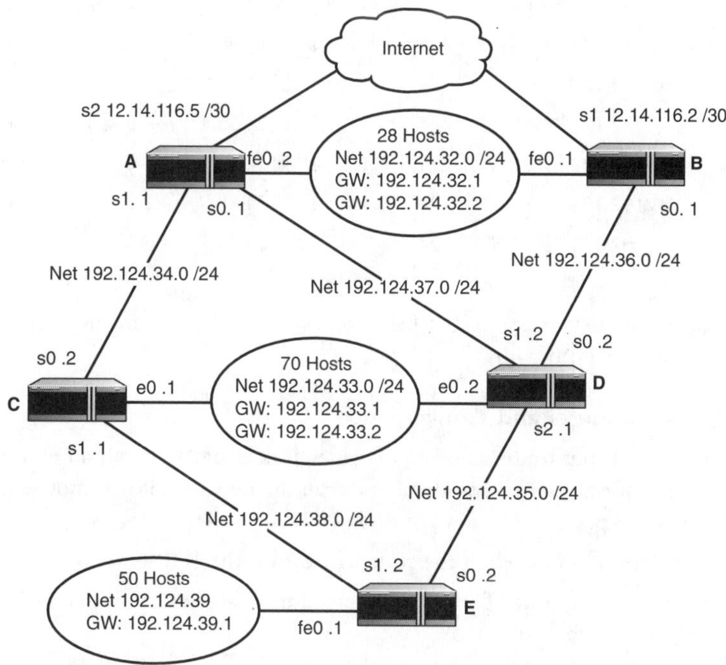

Figure 8.7 An example network using RIP.

Table 8.4 **Router E's Initial Routing Table**

Source	Network	Netmask	Gateway	Metric	Interface
Local	192.124.35.0	/24	192.124.35.2	0	s0
Local	192.124.38.0	/24	192.124.38.2	0	s1
Local	192.124.39.0	/24	192.124.39.1	0	fe

Router E then takes the routing tables from its directly connected neighbors and builds a routing table for the whole network. After Router E has received its updates from Routers C and D, its routing table looks like Table 8.5.

Table 8.5 **Router E's Enhanced Routing Table**

Source	Network	Netmask	Gateway	Metric	Interface
RIP	192.124.32.0	/24	192.124.35.1	3	s0
RIP	192.124.38.1			3	s1
RIP	192.124.33.0	/24	192.124.35.1	2	s0
RIP	192.124.38.1			2	s1
RIP	192.124.34.0	/24	192.124.38.1	2	s0
Local	192.124.35.0	/24	192.124.35.2	1	s0
RIP	192.124.36.0	/24	192.124.35.1	2	s0
RIP	192.124.37.0	/24	192.124.35.1	2	s0
Local	192.124.38.0	/24	192.124.38.2	1	s0
Local	192.124.39.0	/24	192.124.39.1	1	s0

RIP only keeps the best route in its table. Routes of equal value are also kept and traffic is forwarded over both paths.

RIP Routing Messages and Convergence

When a router sends out routing update messages, it does not send out an entire routing table, but rather a list of vectors and distances to all the networks it has knowledge of.

There are two types of RIP message updates:

- Requests are sent out when the router first joins the RIP process.
- Responses are sent out of all the connected interfaces every 30 seconds (unless the default timing is adjusted).

Note

The IP subnet is only sent in RIPv2. "Source" indicates where and how the route was learned. "Metric" is the hop count to the network from the router. "Interface" indicates which interface is adjacent to the next hop or network.

RIP will also send an update in the event of a topology change; this is known as a *triggered update*. Convergence of the routing tables is a router-to-router event, so the convergence time of this network is 30 seconds per router. Total convergence time for Figure 8.8 is two minutes.

RIPv1 and RIPv2 use the same message format. As you can see, RIPv2 uses the unused fields in the RIPv1 message to hold the additional data. Up to 25 routes can be sent in a RIP message (total message size is 512 bytes). The message fields for Figure 8.8 are defined as follows:

- Command—The command flag defines the message type: request or response.
- Version—Specifies the version of RIP that created the message: version 1 or version 2.
- Address Family Identifier—States which protocol is using RIP for routing calculations. The ID for IP is 2.
- Route Tag—This is a version 2-specific field. It is reserved for use in redistribution. RIPv1 does not use this field.
- IP Address (the vector)—This can be an IP address of a host, IP classful network address, IP classful subnet address, or a default route.
- Subnet Mask: RIPv2 only—The netmask of the IP address.
- Next Hop: RIPv2 only—This is used as a hint; if an alternative same or lower hop gateway exists on the same subnet it will be sent along in the update in this field.
- Metric—The distance (hop count) between the router that is sending the update and the destination network.

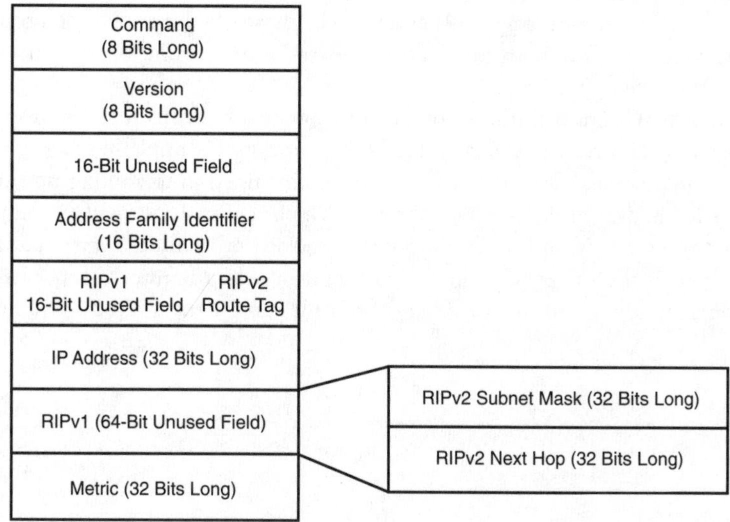

Figure 8.8 The RIPv1 and RIPV2 message formats.

All routers on the network keep track of their update times and the ages ofthe routes in their tables. If a router has not received an update message from a particular network after 180 seconds, the route is marked as invalid. After the route has been marked as invalid, it is in a hold-down state. While the route is in this state, the router continues to forward traffic using that route. During this period, the router is also trying to find a route with a higher metric to use as an alternative. If a higher metric alternative is not found, and there still has not been an update from the router, the route is flushed from the table and users are notified by Internet Control Message Protocol (ICMP) that their datagrams cannot be delivered.

Route Poisoning

To prevent routing the routing loops that are endemic to DVPs, RIP employs route poisoning. *Route poisoning* is a method of inflating route metrics contained in the update messages that are sent to neighbor routers. This is done so the routes that are exchanged between locally connected routers are not interpreted as alternative routes. If this were allowed to happen, routing loops would occur in the event of a router failure. RIP uses two types of route poisoning:

- Split horizon
- Poison reverse

In Figure 8.9, the Fast Ethernet hub that Router E is attached to has just died. Router E has the route to 192.124.39.0 in hold-down mode and is looking for an alternative route. Router C has just sent its update and says that it can reach 192.124.39.0 with a hop count of 2. Router E is happy; it changes its routing table and starts sending from the 192.124.38.0 network that is destined for the 192.124.39.0 to Router C, and Router C starts sending it back to Router E. This starts the beginning of a routing loop that continues until both routers have raised the hop count for the route to 16, known as the *Count to Infinity*.

To avoid the Count to Infinity, routers implement split horizon and poison reverse. Both methods achieve the same goal by different means. By implementing split horizon, the router deletes the network information contained in its routing update, and that it has learned from the interface through which it is transmitting the routing table update. In other words, an interface's route message to other routers after split horizon will not contain the routes it learned from them. So, when Router C sends its update to Router E, it deletes the vector distance information for the 192.124.39.0 network.

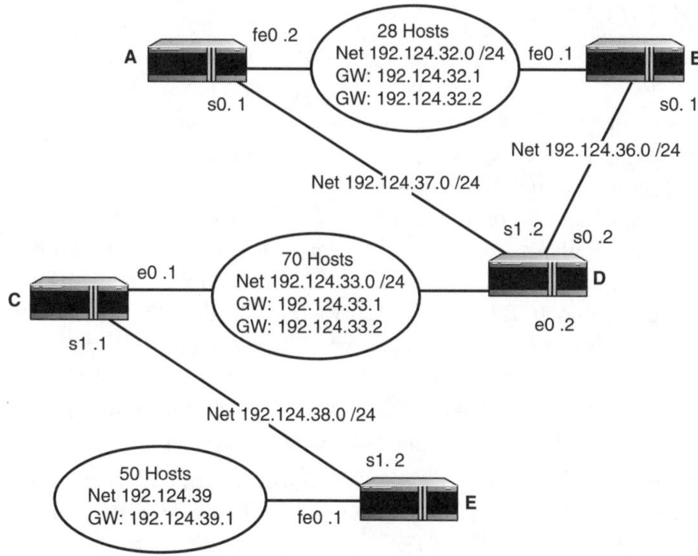

Figure 8.9 Route poisoning example.

Poison reverse does just what its name indicates. When the router sends out an update, it sets the hop count to 16 (because RIP only supports a maximum network diameter of 15) for all the routes its has learned from the interface through which it is transmitting the routing table update. Both methods achieve the same goal by different means. Both of these route poisoning strategies work well with RIP because its vector distance algorithm is interested in only one thing: finding the lowest hop count path.

RIP's Value Today

RIP is old, it's slow, and it is very limited in terms of tunable metrics. On the other hand, it works with just about anyone's network product and is easy to set up. If RIP is used in environments for which it was designed, it works quite well. The problem is that small routing diameter networks with similar bandwidth links are not how networks are built today. RIP is fine for small networks, small LAN segments connecting to the Internet, or small point-to-point office networks. In conjunction with a more advanced protocol (via route redistribution), RIP can enable multipoint gateway services for end-station systems and allow legacy network equipment to interoperate with equipment that supports newer protocols. Because RIP was enhanced to support VLSM and authentication, it has received a new lease on life and will probably be around for some time to come.

IGRP and EIGRP

Interior Gateway Routing Protocol (IGRP) and *Enhanced Interior Gateway Routing Protocol (EIGRP)* are Cisco Systems' proprietary protocols, which means they are only supported on Cisco routers. This can be a severely limiting factor because interoperability with other non-Cisco routers would require running an additional routing protocol and redistributing it into the IGRP or EIGRP process.

IGRP was created as a more robust alternative to RIPv1, though operationally, IGRP behaves in a similar manor. It is a distance vector protocol. Each router sends its routing table out of each of its connected interfaces and implements split horizon to perform route poisoning. IGRP, also like RIPv1 (but not RIPv2), provides no support for VLSM. When IGRP was first released, it was a big win over RIPv1 (even with RIPv2's added support of VLSM, it still offers some major advantages if VLSM is not a requirement). RIPv1's operation was similar and IGRP eliminated many of RIPv1 operational limitations. IGRP's highlights include the following:

- Support for multiple route metrics—Where RIPv1 uses the hop count metric to determine a route's desirability, IGRP supports four metrics—bandwidth, delay, load, and reliability—all of which are tunable by the administrator.
- Support for ASs and multiple routing process domains—IGRP has support for external routes, routing domain segmentation within the AS, and classful route summarization. RIP can only operate as a single routing domain.
- Support for larger network diameters—IGRP can support a network of up to 255 hops in size versus RIP's 15 and ERIP's 128.

The downside is that IGRP has longer distances between routing updates, which are sent out at 90-second intervals. Although this system uses less network bandwidth for routing information, network convergence is slower. If fast convergence times are a core requirement, EIGRP or OSPF is a better choice.

EIGRP is a hybrid protocol that uses the *diffusing update algorithm (DUAL)*. Where distance-vector protocols require route poisoning to correct possible routing loops, DUAL uses diffusing computations to perform route calculations. Route computations are "shared" across routers. Routing tables are still exchanged between routers (such as RIP and IGRP), but updates are only sent when changes in the network topology occur.

EIGRP also uses the same routing metrics as IGRP. EIGRP provides support for network summarization, VLSM, route redistribution, and multiprotocol routing (IPX and AppleTalk). The "E" for enhanced has mostly to do with how EIGRP computes, distributes, and processes its routing table. EIGRP is commonly referred to as a hybrid or *distance state protocol* because it has none of the shortcomings of a distance vector protocol and provides all of the advantages (fast convergence, low bandwidth consumption) of a link state protocol. If EIGRP has a shortcoming it is the fact that is only supported by Cisco products. IGRP and EIGRP follow an almost identical configuration process. IGRP and EIGRP metrics and other configuration options will be covered in Chapter 10.

OSPF

Open Shortest Path First (OSPF) is a link state protocol. Routers using distance vector protocols exchange routing information, in the form of routing tables listing all the networks in which they have knowledge. Routers running link state protocols exchange link state information about the network in which they are directly connected. Each then uses this information to construct a map of the entire network topology from its perspective.

OSPF offers several advantages over distance vector protocols:

- Better reliability—OSPF routers construct their own routing table describing network reachability from their perspective form within the network. All network routes are stored in the topological or link state database, where they can be retrieved quickly in the event that a routing change must be made. RIP, as you might recall, discards all routes except the best one, making it dependent on a routing update in order to construct a new route in the event that the old route is no longer viable.
- Fast convergence—After the network has converged and all the routers have constructed their own routing tables and network maps, updates are sent out only when changes in the network topology occur. When changes occur, they are flooded across all network links so each router can adjust for the change in the network topology immediately.
- Unlimited network size—No maximum hop distance is enforced with OSPF. Network size is constrained only by available router processing power and Layer 2 medium limitations.
- VLSM support—OSPF provides full support for VLSM.
- Type of Service routing—OSPF supports Layer 4 Quality of Service routing. ToS/QoS was part of the original TCP/IP suite, but was never compatible with RIP. This lack of support resulted in lack of implementation. ToS/QoS is experiencing something of a revival now that network-based applications are becoming more and more performance sensitive.
- Low bandwidth usage—OSPF uses multicast instead of local network broadcasts to send out OSPF messages. This limits the impact of OSPF messages to only OSPF-speaking hosts.
- Dynamic load balancing and route selection—OSPF uses a cost metric to determine route desirability. Cost is based on the link speed. Links of equal cost are load balanced, and in cases where there are links of unequal cost between similar destinations, the lowest cost link is preferred. If the lowest cost link is down, the next lowest cost link will be used.

The price that must be paid for these advantages is the requirement for more processing power and memory than is needed for distance vector protocols. OSPF is an Open Systems protocol and an open standard (that is, in the public domain). It was introduced by the Internet Engineering Task Force (IETF) in 1989 in RFC 1131 (OSPF v1), and made a standard in RFC 2328 (OSPF v2). OSPF was intended to be an updated Open Systems alternative to RIP and to provide an Open Systems alternative to Cisco's proprietary IGRP and EIGRP protocols.

OSPF is designed to operate in both large and small-scale networks. The network diameter (the number of subnets and routers) has a direct relation to how OSPF is implemented. OSPF has different configuration options and utilities to address operational issues that arise in both large and small network environments.

Implementing OSPF requires some forethought and planning; a poorly implemented OSPF solution can affect the efficiency of the network. The following are major OSPF elements you need to understand when designing an OSPF implementation:

- The OSPF area hierarchy
- Router designations
- OSPF network types
- Inter/intra routing

The OSPF Area Hierarchy

OSPF's only limiting aspect is the memory required by the router to compile the shortest path distances needed for the routing table. On a small network consisting of 20 or so routers, these computations are manageable, but on a network of 100 or more, these calculations can cripple a small router with limited CPU and memory.

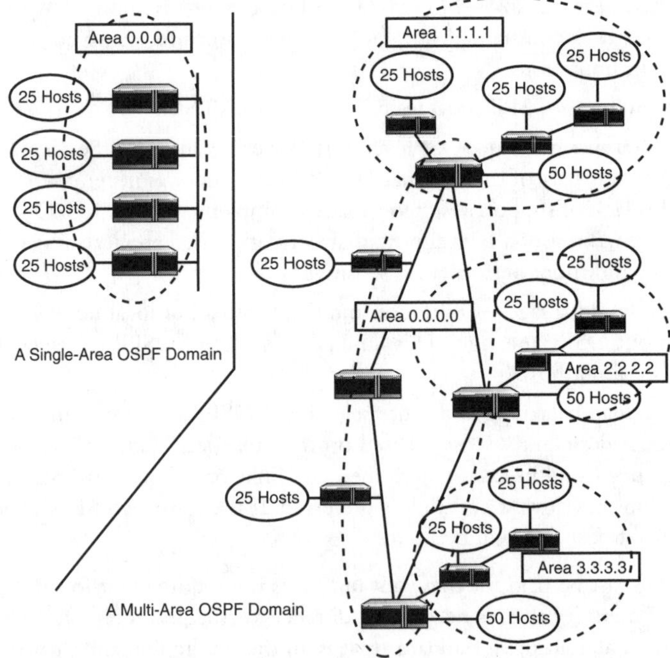

Figure 8.10 Single- and multi-area OSPF network topologies.

To address computational issues, OSPF uses a hierarchical network segmentation structure. These structures are called *areas*, OSPF's convention for logically grouping a collection of networks. Every OSPF network has at least one area, known as a *root* or *backbone area*. The backbone area, commonly designated as area 0 or area 0.0.0.0, is a special OSPF area used to connect all other areas of the OSPF network domain. In a single-area OSPF network (see Figure 8.10), only the backbone area is used. In a multi-area network (again, see Figure 8.10), all areas must have one interface connected to the backbone area. Because this is not always possible, OSPF provides the capability to create virtual links between disconnected areas and the backbone network.

The backbone network is used to exchange inter-area routing information. The idea behind segmenting the network into areas is that different routers can compute shortest path information for sections of the network, instead of requiring each router to compute the shortest path information for the entire network.

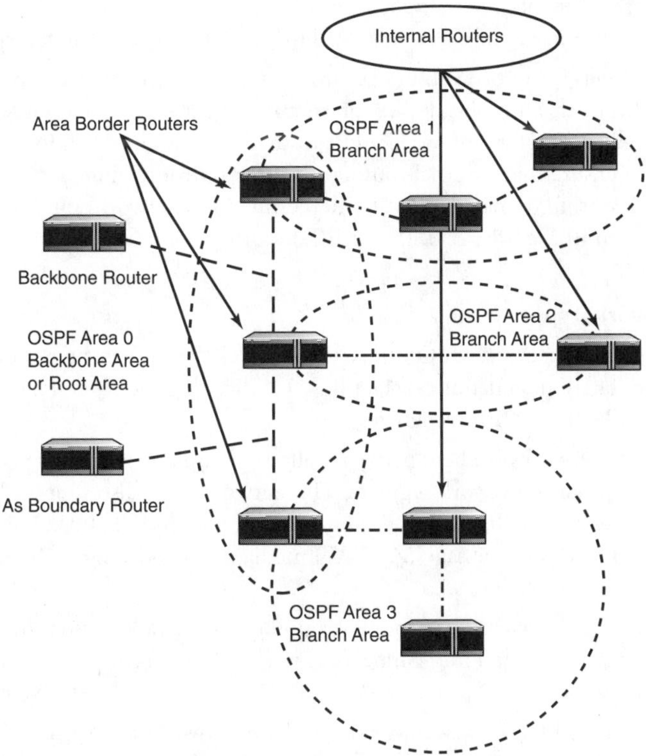

Figure 8.11 The OSPF router class tree.

Router Designations

There are four different designations for routers in an OSPF network (as shown in Figure 8.11):

- Backbone routers—These routers maintain complete routing information for all the networks (all the areas or domains) that are connected to the backbone.
- Area border routers—These are routers that connect two or more areas (inter-area routing). These routers usually sit on the network backbone and connect the branch areas to the root area. They are different from backbone routers because they traditionally do not redistribute other AS routing information and belong to only one AS. These routers only maintain information about the backbone and the areas to which they are attached.
- Internal routers—These routers are involved only in routing their internal area (intra-area routing) and only maintain information about the area in which they operate. All out-of-area traffic is forwarded to the border router, because only network summaries of other areas within the network are maintained locally.
- Autonomous system (AS) boundary routers—These routers have interfaces on the network that are outside the internetwork routing domain. These routers commonly redistribute internal and external routing domain information. Their job is to ensure that external routing information is distributed through the internal routing domain so that traffic destined for networks outside the domain is delivered in accordance with the EGP routing policy.

OSPF Network Types

Along with router classification, OSPF also designates the different types of networks it operates on. These designations dictate how OSPF sends messages. There are three types of networks that OSPF recognizes:

- Broadcast networks—These networks allow multiple routers to be connected to the same physical network segment. The network must also support broadcast and multicast capabilities. OSPF uses multicast addresses 224.0.0.5 and 224.0.0.6 to send and receive messages. All LAN media types and some WAN media types support these capabilities.
- Point-to-point networks—A point-to-point Layer 2 link is used toconnect a pair of routers. A serial line connection is a common example of a point-to-point network.
- Nonbroadcast multi-access networks—These networks allow multiple routers to be connected to the same network segment, but do not support the broadcast and multicast capabilities needed to send OSPF messages. In these cases, messages are sent using unicast, and additional configuration is needed so routers know who to exchange messages with.

A typical multi-area OSPF implementation will consist of mostly broadcast networks and nonbroadcast multi-access networks. In large WAN internetworks, it is common for the backbone area to be a collection of meshed Frame Relay PVCs (permanent virtual circuits) with the WAN/LAN gateway routers acting as area border routers/backbone routers.

In Figure 8.12, the ABRs need to know who their neighbor routers are so they can exchange routing updates. So, for example, Router D needs to know that routers A, B, and C are its neighbors and it should send unicast LSAs to their respective IP addresses. Another approach uses all broadcast networks with MAN/LAN Layer 2 technologies.

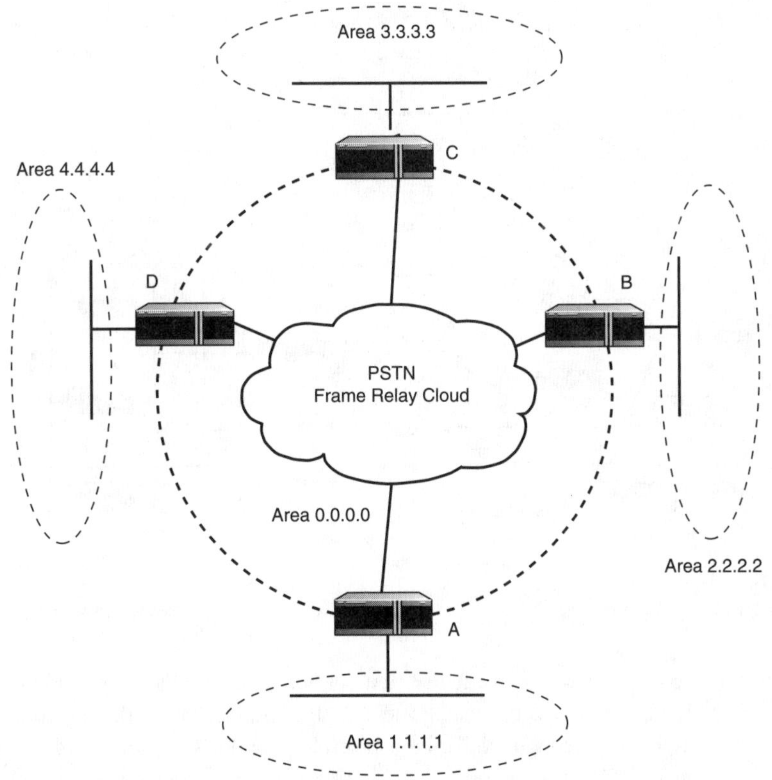

Figure 8.12 A multi-area OSPF implementation using broadcast and nonbroadcast networks.

> **Note**
> MAN refers to a metropolitan area network.

The backbone network is comprised of LAN and MAN/WAN links again, with area border routers/backbone routers connecting the stub areas to the backbone. In Figure 8.13, all the Layer 2 transport facilities support broadcast, so all the networks exchange information without any designation of a "neighbor" router. The "dark fiber" in this diagram means that the telephone company (LEC, or local exchange carrier) is leasing the fiber connection to the customer, and it's the customer's responsibility to manage the circuit.

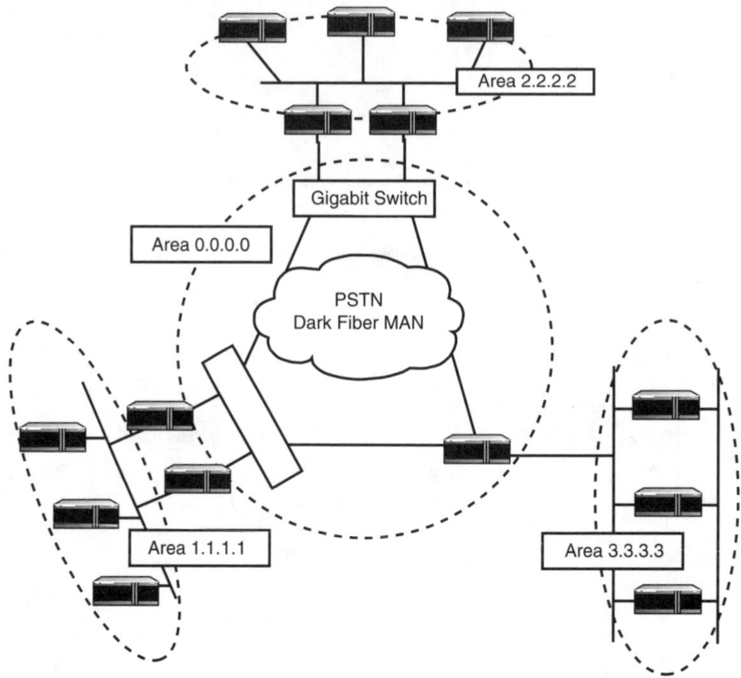

Figure 8.13 A multi-area OSPF implementation using MAN/LAN broadcast networks.

Regardless of how the areas are constructed, it is essential that all areas be either physically or virtually connected to the backbone area. It is only through the backbone network that inter-area message exchange is possible. The ABR routers facilitate this exchange by sending OPSF messages (LSAs) between their connected networks. Because all areas have ABR connected to the backbone, all LSAs are sent to all the networks in the routing domain.

Another way OSPF provides for exchanging information is by forming adjacencies. *Adjacencies* are relationships formed between routers that provide a way for routers to exchange routing information. The advantages of adjacencies are that they reduce the amount of link calculation and route table construction needed for a router to build a complete routing table. Routers become adjacent under different circumstances:

- If a link type is point-to-point, the routers on each end will become adjacent in order for routing information to be distributed between the connected networks.
- If a router is a designated router or a backup designated router on a broadcast or nonbroadcast multi-access network, adjacency is formed with all of the routers within the common network.
- If a link is a virtual link, an adjacency will be formed with the backbone network, so backbone information can be exchanged.

The most common instance of adjacencies is in networks where a designated router can be used.

On networks where more than one router is used to provide connectivity, a designated router will be elected to represent the router group. The *designated router* (see Figure 8.14) will compute the routing table and exchange routing messages with other routers within the area. There is also a *backup designated router*, which performs the same functions if the primary fails.

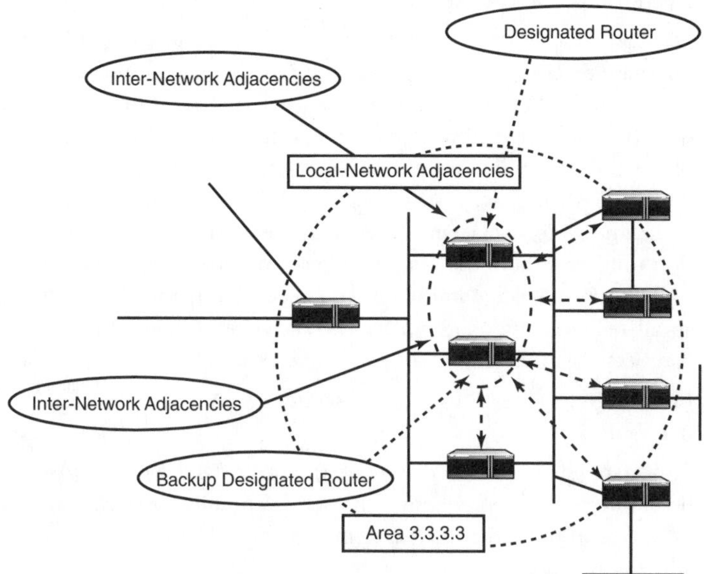

Figure 8.14 OSPF router designations and adjacencies.

The designated router then sends OSPF messages that announce the link states for the networks adjacent to it. The designated router in turn distributes link states for networks it learns about from other routers on the domain. The advantage of using the designated router is that route and message processing are handled by two routers instead of by every router on the network.

OSPF Messages

In order for OSPF to operate, a great deal of information about the status of the network must be sent, processed, and maintained. The OSPF message header and different types of messages that are used to exchange information about the OSPF domain include:

- Version—Indicates the OSPF version being used. Version 2 is the current version.
- Packet type—This indicates the type of message being sent. There are four message types:
  - Hello—Messages are used to create adjacent neighbor relationships. Depending upon the type of network, different hello protocol rules are followed. Hello messages are exchanged between routers that have become adjacent. Once adjacency is established, routers will exchange routing information with one another. In situations where there is a designated (and backup) router, hello messages will be exchanged between the designated router and the other routers on the network.
  - Database description—This message is sent in response to a hello request. Database description messages are used to exchange topological database information. The topological or link state database is the result of the link state (Dijkstra) algorithm calculation performed by the router. The result of the calculation lists every possible route to every network in the routing domain. The database is used to identify possible routing loops, redundant paths, and so on. Adjacent routers exchange topological database information until both have the same information. Once they are identical, the routers will exchange link state requests and link state updates to learn about networks they did not have knowledge of before their adjacency had been established.
  - Link state request—This message requests LSAs from other OSPF neighbor routers.
  - Link state update—This message responds to (and sends) LSAs to neighbor router link state requests. Link state announcements are what OSPF uses to send the actual link state information. There are five link state advertisements:
    - Router links advertisements—These are sent by all OSPF routers. They are used to describe the state and cost of the connected interface within the given area.

- Network links advertisements—Designated routers on multi-access networks generate this message. The messages list all routers connected to the network.
- Summary links advertisements—ABRs send these messages; they provide summary information for networks inside and outside the OSPF network.
- Autonomous system external links advertisements—These LSAs are sent by AS boundary routers. They describe network destinations that are external to the OSPF network.
- Link state acknowledgment—These are sent to the LSA sender to acknowledge that the LSA has been received.

- Packet length—The total size of the message in bytes.
- Router ID—The ID number of the router sending the message. Usually, the IP address of the interface that is sending the message.
- Area ID—The ID number of the destination area.
- Checksum—Cyclical redundancy check (CRC) of the message. Excludes the authentication fields.
- Authentication type and value—OSPF supports process authentication. The type field specifies what type is being used, and the value field is the authentication key.

OSPF messages are sent either by using a technique known as *flooding* or directly to a node address. Flooding is accomplished by having each router forward an OSPF message out of each of its connected interfaces. This is the most common means of propagating network changes throughout the OSPF routing domain. If a message is intended for all the routers in the OSPF domain, the message is sent to the multicast address 224.0.0.5. In cases where router adjacencies exist, OSPF messages intended for the designated router or backup designated router are sent using the multicast address 224.0.0.6.

After an OSPF message is received, it must be verified before it can be processed. There are eight steps to OSPF message verification:

1. The IP packet CRC checksum is verified.
2. The destination IP address must be for the router interface that received the message or one of the designated multicast addresses.
3. The IP protocol ID must identify that the message is an OSPF message (89 is OSPF's IP protocol ID).
4. The source IP address must be checked to make sure the message did not originate from the receiving station.
5. The OSPF version number is verified.
6. The OSPF header CRC checksum is checked.
7. The area ID must be either that of the interface or the backbone area.
8. The OSPF authentication type and value must correspond and be correct.

If any of the steps do not meet the expected criteria, the message is discarded.

Routing Table Creation, Path Costs, and Path Types

All the OSPF messages are stored and compiled to create a topological or link state database on the router. The router then uses this information to construct a routing table. The LSAs contained in the OSPF messages are added to the link state database as they are received. LSAs are not discarded as new updates arrive. This allows the router to keep an updated link map of the entire routing domain.

The first step in building a routing table is the creation of the *shortest path tree (SPT)*, as shown in Figure 8.15.

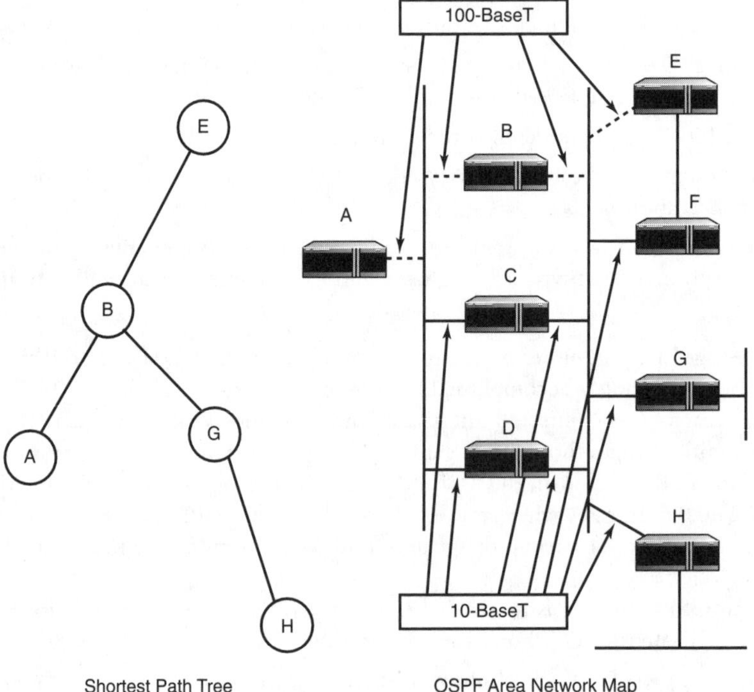

Figure 8.15 Shortest path tree.

The tree is created by examining the link state/topology database and building a tree reflecting the shortest path to each network from the router's point of view. The path consists of the series of hops that must be traversed in order to reach the destination network. The router constructs the SPT by calculating the shortest path (from itself) and the additive cost for the path for each of the networks in the area. The cost of the path is the sum of the path costs between the source point and the terminus point. (OSPF costs are determined by dividing the link bandwidth by 100 million.)

A Network Path Cost Model	Network Bandwidth Path Cost(in Bits)
100Mbps	1
16Mbps	6
10Mbps	10
4Mbps	25
1.544Mbps	65
64K	1562
56K	1786

The cost metric is what OSPF uses to determine the best route. OSPF will naturally determine the cost based on the Layer 2 type of the link. Link costs can be adjusted to reflect an administrator's preference for a route. Although it is not generally a good idea to start adjusting cost metrics, there might be a need because OSPF's cost metric is only valid up to 1Gbps. As Gigabit Ethernet becomes more and more common, an administrator may want to adjust the costs for all the links in the OSPF domain, so preference between high-speed paths is possible. If adjustments are not made longer, transit paths may be preferred over shorter ones because the additive cost of the longer path may be lower.

The SPT is calculated (and routing entries are created) from a local-to-remote perspective. SPT calculations and (route entries) are first done for the router's connected networks, then transit networks that are connected to its attached networks, and finally the stub networks that branch off the transit networks.

In a single-area OSPF domain, the results of the SPT are used to create route entries. In a multi-area OSPF domain, the intra-area calculations are completed first, then the inter-area paths are examined, calculated, and added. The last route entries entered are for networks external to the OSPF domain and for routes redistributed from other routing processes.

The manner in which a route entry is created is indicated by the path types. There are four OSPF path types:

- Intra-area paths—Paths to networks that are directly connected to the router.
- Inter-area paths—Paths to networks that are in a different area, but are within the boundaries of the OSPF domain.
- Type 1 external paths—Routes to destinations outside the OSPF domain that have an OSPF cost associated with them.
- Type 2 external paths—Routes to destinations outside the OSPF domain that have no OSPF cost. Type 2 is the default type used for all external routes.

Note

When designing your network, keep in mind how path costs are assigned, how they are used, and what kind of manipulation may be needed to determine the best path.

OSPF's Value

Although the workings of OSPF are quite complex, the actual configuration and operation are easy from an administrative point of view. OSPF does require some pre-implementation planning, but its speed and scalability make it an ideal choice for large enterprise internetworks. The other element that makes OSPF attractive is that it is an open standard and is widely implemented on most networking hardware. This is important because no vendor has a perfect solution for every problem.

TCP/IP Exterior Gateway Protocols

Exterior routing protocols (ERPs) are used primarily in Internet backbone networks that connect internetworks. ERPs send information the same way RIP does: as a vector (the AS number) and as a distance (the destination network information). A simple way of looking at ERPs is as a means for distributing static routes between internetworks.

Autonomous Systems and Exterior Routing

Routing between autonomous systems or exterior routing is strictly an Internet convention. What was once a practice undertaken by a chosen few is now (with more and more networks connecting to the Internet) becoming a commonplace activity. ERPs were created to provide a more efficient way to build and manage Internet backbone routing tables. The idea was to segment the Internet into separate "ubernetworks" called autonomous systems (ASs). An AS is, according to RFC 1267, *A Border Gateway Protocol 3 (BGP-3)*, Lougheed and Rekhter, Network Working Group 1991:

> "A set of routers under a single technical administration, using an interior gateway protocol and common metrics to route packets within the AS and using an exterior gateway protocol to route packets to other autonomous systems."

Coincidentally, this idea has translated well into the way the modern global Internet has evolved.

> **Note**
>
> It is common to confuse an internetwork and an AS because the terms are often used interchangeably. The difference is that an internetwork uses one or more IGP protocols to exchange routing information within a closed network. An AS uses both IGP(s) and ERPs to exchange routing information within the context of an open network (like the Internet). In addition, an AS network is registered with an Internet registration authority (ICANN, AIRN, or RIPE).

There are three types of ASs:

- Stub—Stub ASs reach other networks through one gateway. Routing for stub AS is commonly achieved with static routes. Stub ASs are uncommon today, because most IP addresses are provided by ISPs and NSPs, and those addresses fall under their own AS.
- Multihomed nontransit—Large private enterprise networks, with multiple Internet access points, commonly operate as ASs. Because these networks operate their own network backbone infrastructures, they do not want to have traffic other than their own traversing their network backbone. This is accomplished by only advertising routes to networks within their own AS.
- Multihomed transit—A multihomed transit network allows traffic belonging to other networks to traverse across its network in order to reach its destination. In addition, it also announces other AS routes and will allow ASs to exchange routing messages across its network.

IGP and ERP Differences

Conceptually, the idea behind AS-to-AS routing or exterior gateway routing and IGP routing operate under the same operational premise: the hop-to-hop connectionless routing paradigm, where IGP route path decisions are made at each hop. ERPs direct traffic between ASs in the same way. Because each protocol uses a different operational model, however, there are some distinctions between their approaches. An easy way to remember the distinctions between EGP and IGP protocols are the three Cs:

- Context—IGPs and ERPs both operate in different routing contexts. IGP protocols are interested in building and exchanging routing information. ERPs are interested in network reachability information.
- Connectivity—IGP networks are connected physically to one another. Routing tables are built on direct connectivity and local adjacency, using the hop-to-hop routing paradigm common to IP. ERPs exchange network reachability information. Information is exchanged by designated peers. Routes are based on AS paths that need to be traversed in order to reach the destination network.
- Choice—IGP protocols support route metrics that enable the router to choose the best route path. ERPs do not support routing metrics. ERPs use policy routing to manage traffic behavior. These distinctions reflect the different operational paradigms. They are not alternative solutions for the same problems. IGPs are designed to provide a dynamic means of constructing routing tables that enable datagram forwarding to occur in the most efficient and reliable manner. ERPs provide the capability to route IP datagrams between distinct open internetworks in a dynamic and manageable fashion. Figure 8.16 illustrates the roles both protocols take in IP datagram delivery.

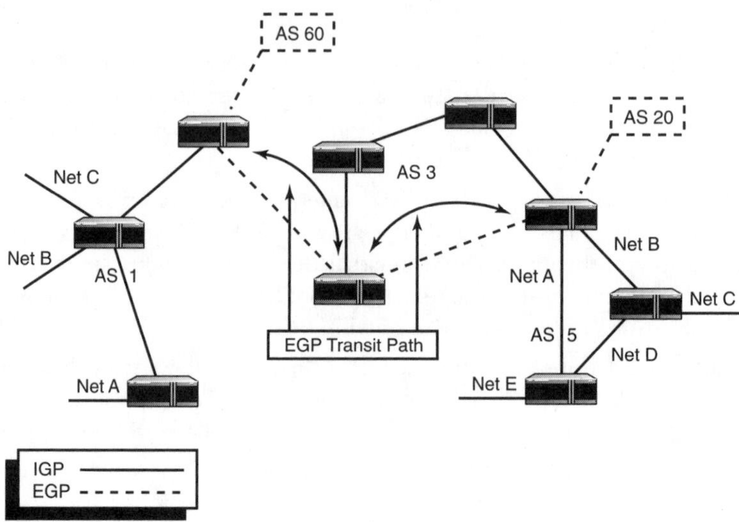

Figure 8.16 The IGP to ERP handoff.

A host on Network A in AS 1 wants to deliver a packet to a host on Network 5 in AS 5. The destination address is outside the local network, so the datagram is forwarded to the gateway. The gateway has a route for the destination network, which is part of AS 5, and is reachable via AS 3, which is a multihomed transit AS. The datagram is then forwarded across AS 3's transit path. Once it is forwarded to AS 5's gateway, AS 5's IGP process determines the rest of the path. If AS 5 were also available via AS 60, a policy would have been required on AS 1's and AS 5's routers to set the path preference.

BGP Version 4

The *Border Gateway Protocol (BGP-4)* is the protocol that is most commonly used to exchange routing information between ASs. According to RFC 1267:

> "The Border Gateway Protocol is an inter-autonomous system routing protocol… The primary function of a BGP speaking system is to exchange network reachability information with other BGP systems. This network reachability information includes information on the full path of ASs that traffic must transit to reach these networks."

BGP is based on the vector distance algorithm. Routing information is sent using the AS number as the vector and the gateway address as distance. BGP systems exchange network reachability information with other BGP systems. These systems do not need to be directly accessible to each other; the only requirement is that the two systems are reachable across the network. Instead of using broadcast or multicast network facilities that are dependent on direct connectivity, BGP uses a TCP unicast (port 179), peer-to-peer-based message exchange process to distribute routing information. These exchanges are established through a multistage communication process known as the *BGP finite state machine* (see Figure 8.17).

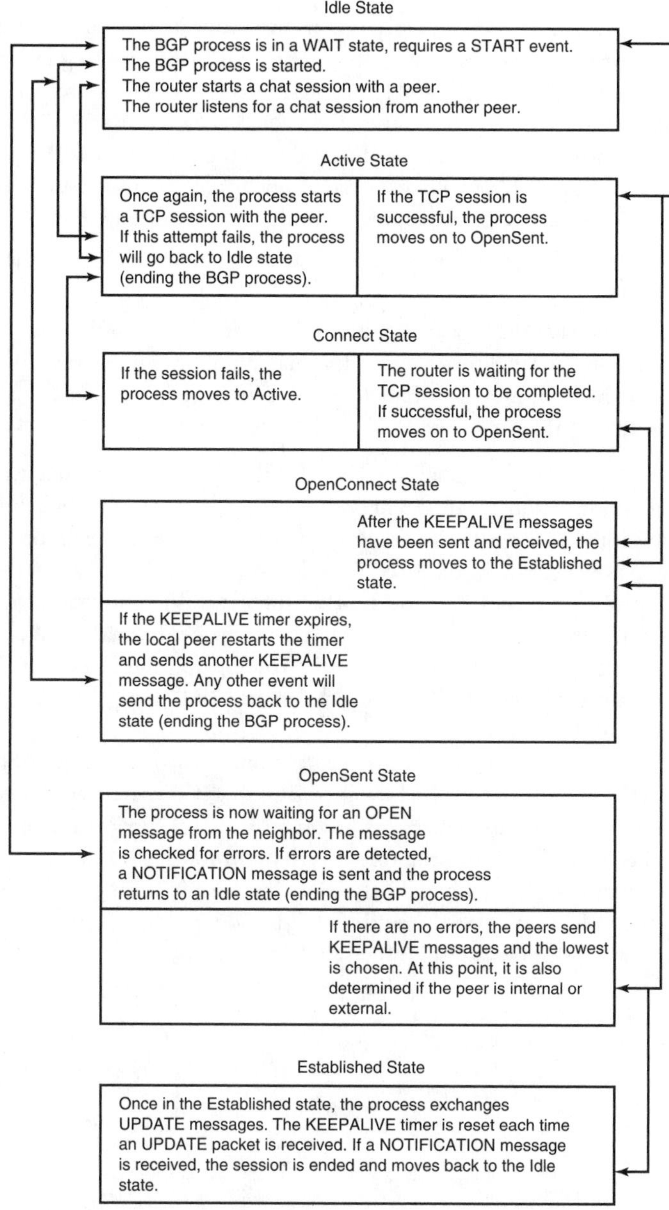

Figure 8.17 The BGP finite state machine.

All BGP messages use a 19-byte header consisting of a 16-byte marker, a 2-byte message length field, and a 1-byte type field. There are four types of BGP messages:

- `OPEN`—The `OPEN` message contains generic information needed to establish peer-to-peer BGP communication. The fields are version, AS number, hold time, BGP identifier, and options.
- `UPDATE`—The `UPDATE` messages are used to send routing update messages. When two peers first establish their BGP session, they exchange routing tables. Network changes are sent as single update messages as they are realized. Along with adding and withdrawing routes, the `UPDATE` message also sends along path attribute information, which is used to indicate preferred routes. Routing information, also known as *network layer reachability information (NLRI)*, is sent as tuples: netmask (using CIDR representation) and network address.
- `NOTIFICATION`—These messages are used to communicate errors in the peer-to-peer exchange.
- `KEEPALIVE`—`KEEPALIVE`s are sent to keep the peer-to-peer session open. The hold down value in the `OPEN` message sets the maximum amount of time that can lapse between BGP messages. The `KEEPALIVE` message is sent out at a rate appropriate to that of the hold time to keep the session running.

After the peer session has been established, reachability information is exchanged. The peers exchange information about the networks they can reach. However, unlike IGPs, which use the source as the gateway, BGP reachability information includes the full path of all the ASs that must be traversed in order to reach the destination network. This distinction is what makes IGP and EGP so different. BGP "routes" provide a list of points the traffic must pass through in order to reach the destination. Whereas IGP directs traffic to a gateway, unaware of the entire route path, BGP uses routing policies to determine route preference, facilitate load balancing, and manage route recovery in the event that a BGP peer is lost.

BGP and EGP protocols in general have a very limited scope of use. They are used almost exclusively by ISPs and NSPs, but since BGP-4, EGPs have been increasing in popularity among network administrators because of their flexibility. You as a network administrator need to be aware of the role they have in routing traffic across the Internet. Configuring BGP will be addressed in Chapter 10, "Using IP Routing Protocols on Cisco Routers." Policy routing will be addressed in Chapter 9, "Advanced Routing Options and Configuration."

Summary

At this point, you should have a good understanding of routing concepts, including the strengths and weaknesses of the protocols, how they're used and why. We focused on the open protocol standards EIGRP and IGRP, OSPF, IGP and ERP, RIP and BGP. You should be familiar with the factors that affect dynamic routing, such as metrics and convergence, and how each protocol adjusts to changes in the network's configuration and state.

Related RFCs

RFC 827	Exterior Gateway Protocol (EGP)
RFC 1058	Routing Information Protocol
RFC 1131	OSPF Specification
RFC 1265	BGP Protocol Analysis
RFC 1266	Experience with the BGP Protocol
RFC 2082	RIP-2 MD5 Authentication
RFC 2178	OSPF Version 2
RFC 1700	Assigned TCP/IP Port Numbers
RFC 1721	RIP Version 2 Protocol Analysis
RFC 1771	A Border Gateway Protocol 4 (BGP-4)

Additional Resources

Halabi, Bassam. *Internet Routing Architectures*. Cisco Press, 1997.

Moy, John T. *OSPF: Anatomy of an Internet Routing Protocol*. Addison-Wesley, 1998.

Stewart, John W. *BGP4, Inter-Domain Routing in the Internet*. Addison-Wesley, 1998.

Tanenbaum, Andrew S. *Computer Networks*. Third Edition. Prentice Hall, 1996.

9

Advanced Cisco Router Configuration

IN CHAPTER 7, "INTRODUCTION TO CISCO ROUTERS," we focused on providing you with the essential skills you need to configure and maintain Cisco routers. This chapter covers enterprise-related configuration tasks such as

- Controlling network traffic flow
- Router link redundancy and backup
- Network address translation
- Tunnel (VPN) configuration
- WAN interface configuration

Keep in mind that these services are not necessarily exclusive to enterprise networking, but they do represent, to many, the next step on the internetworking evolutionary ladder.

Access Control Lists

Originally, access control lists (ACLs) were used for their packet-filtering capabilities, most commonly for IP security filtering. As the IOS software has evolved, the usefulness of ACLs has also grown (depending on the protocols and features supported by IOS) to allow filtering capabilities for all the IOS-supported Layer 3 protocols (and MAC addresses), and traffic TOS and QOS identity services for traffic control tools.

The ABCs of Access Control Lists

An ACL (or access-list) is a series of action statements used for examining network protocol traffic. There are two possible action statements:

- `permit`
- `deny`

ACLs perform different tasks, depending on how they are applied. After the list is applied, the router's network traffic is compared to the list of action statements recursively (from the top down). When a match is found, the evaluation stops and the associated action is taken. If no explicit match is found, a default action is taken. The implicit default action for all access control lists is `deny any`.

When constructing access control lists, the order of the action statements is crucial. Traffic matches are made recursively, so ideally you want statements at the top of the list to be specific and become increasingly general as the list progresses. In this same vein, place entries in descending order (most likely to least likely). The goal is to have packets processed as quickly as possible out of the ACL. Poorly placed statements or an excessively long ACL will produce undesirable results, ranging from restrictive or ineffective routing to degradation of overall performance.

> **Note**
>
> A slight caveat to this rule exists when constructing IP ACLs. The IOS reorders IP ACL entries based on their network address, using a low-to-high ordering scheme. For example, if you have an ACL that filters addresses using the IP network addresses:
>
> 10.30.100.0 /24
>
> 172.116.45.0 /24
>
> 192.168.34.0 /24
>
> The ACL will be ordered using entries from the 10.30.100.0 /24 network first and the 192.168.34.0 /24 last, regardless of the order in which they are entered during the configuration. The same rule holds true for ACL entries created using only one IP address space. ACL entries for hosts with addresses from the beginning of the address range will be closer to the top of the ACL. So, if an ACL permits the hosts
>
> 192.168.30.22
>
> 192.168.30.200
>
> 192.168.30.3
>
> 192.168.30.14
>
> 192.168.30.198
>
> the ACL will be ordered starting with 192.168.30.3 and ending with 192.168.30.200. Regardless of the order in which the actual entries are made, the router will reorder the addresses during the construction of the ACL.

Originally, different types of access control lists were distinguished by the number associated with each list. Access control lists are created with the global configuration EXEC command `<access-list [id number]>`. Each protocol has a specified ACL number range. Table 9.1 lists the relevant number ranges associated with each of the different Layer 2 and Layer 3 protocols. In later IOS versions, named access control lists were added for IP, IPX, and some of the other protocols. The configuration of named ACLs is slightly different from standard ACLs. A named ACL is configured in two parts:

1. The named list is created in global configuration using the command `<[ip|ipx] access-list [standard|extended] [acl name]>`.
2. Entries are added using a subprocess command `<[permit¦deny] [source address]>`.

Table 9.1 **ACL ID Filter Ranges**

Protocol	**Number Range**
IP standard	1-99 and 1,300-1,999 (expanded range, IOS 12.x and later only)
IP extended	100-199 and 2,000-2,699 (expanded range, IOS 12.x and later only)
IPX standard	800-899
IPX extended	900-999
IPX SAP	1,000-1,099
AppleTalk	600-699
48-bit MAC address	700-799
Extended 48-bit MAC address	1,100-1,199

In Table 9.1, you will notice that for most protocols, ACLs come in two basic flavors: standard and extended. Standard lists provide basic matching. For example, IP standard access control lists perform test matches against the source network information contained in the packet. IPX standard lists, however, can match against both the source and network address information. Extended lists provide a great degree of matching flexibility for not only source and destination network addressing, but also for protocol type, transport source, and destination ports.

Applying Access Control Lists

As mentioned in the previous section, access control lists are not just for packet filtering. ACLs are used to provide Boolean logic matching for a variety of tools. Due to their diverse functions, ACLs are never applied as just an "access-list." Rather, an ACL (using the ACL ID number) is associated with the filtering or logic command that is specific to an application.

When ACLs are used for protocol-specific packet filtering, they are applied to interfaces as access-groups. Each interface can support a single, standard or extended, inbound and outbound access-group. This variability applies to all enabled protocols. Each access-group is applied as an interface configuration subcommand `<[protocol] access-group [access-list number] [in|out]>`. If no filter direction is specified, the access-group will be applied as an inbound filter.

When creating the ACL, take into account where the ACL will be placed in the interface's path. Traditionally, access-groups used for security are inbound filters, on the assumption that you are defending against unauthorized packets from outside your LAN. When inbound filters are applied, all traffic flows to the interface. During the matching process, the inbound list is first compared to the source address of the incoming packet. If the packet matches an address on the list, it is processed further. If no match is made, it is processed by the default action (usually `deny`). Outbound filters are also used for security filtering in scenarios where the router is acting as a firewall and outbound access is restricted or requires authentication. More often, though, they are used as traffic filters, to discard unneeded traffic and permit everything else. With outbound filters, the traffic flows through the interface, a route lookup is performed, and then the packet is processed through the filter. If a match is made, the packet is forwarded on, otherwise it is discarded. The first match of an outbound filter is made against the source address of the outgoing packet. The issue to be aware of is that the type of list and match tests used to process packets changes, depending on which side of the interface it's applied to.

> **Note**
>
> IOS also supports advanced IP access control lists of two types: dynamic and reflexive. Dynamic ACLs provide lock and key IP traffic filtering. Lock and key filtering allows you to configure a gateway router to act as a firewall. Before a user can send traffic out through the router, he must authenticate first. After being successfully authenticated (and assuming the destination host is also permitted) temporary access through the gateway is granted.
>
> Reflexive lists are created with named ACLs. Where they differ from regular ACLs is that, rather than having a static list of matching statements that leave specific ports open, reflexive lists add ACL entries for specific hosts on a dynamic and temporary basis. These entries are based on upper-layer session activity that is defined as part of the reflexive ACL. This kind of configuration is particularly advantageous from a security perspective, because it defends against most network-based security assaults (when used in conjunction with extended ACL matching statements). By default, all traffic is denied. Only traffic that matches the dynamic entries will pass through the router, and these entries are transaction-specific and always changing.

Let's look at an example. You need to restrict access to local hosts from the external hosts connected to the Internet. Therefore, you need to create an ACL to apply to the outbound interface of the Internet gateway router. This task can be accomplished using either an outbound or an inbound list. With an inbound filter, you can do one of the following:

- Use a standard access-list that specifies which external hosts are permitted to access the local network
- Use an extended list that permits all external hosts to access only specified local hosts

You can also accomplish the same thing by using an outbound filter. A standard access-list could be created that would specify which local hosts could access external hosts. Alternatively, an extended list could be used that permits all local hosts to have access to specified external hosts only.

ACLs used for filtering IP route announcements and dynamic routing protocol redistributions are applied as distribute-lists. As the name implies, distribute-lists are used for controlling the distribution of routing information announcements transmitted out of the routers' interfaces. Distribute-lists are applied using the dynamic routing protocol configuration subcommand `<distribute-list [ip acl number] [in|out] [interface¦protocol]>`. Distribute-lists will be covered in the "Filtering with Distribution Lists" section in Chapter 10, "Configuring IP Routing Protocols on Cisco Routers."

A standard IP access-list is used when you want to restrict a host's Telnet access to the router's virtual terminals. This list is known as an access-class. An access-class is applied as a line configuration subcommand `<access-class [acl number] in>`. When applying ACLs to VTYs, be sure to apply ACL to all the VTYs. It is common to make the mistake of just applying the ACL to the first VTY interface, `vty 0`. To apply the list to all the VTYs, use the following:

```
Router#config t
Enter configuration commands, one per line.  End with CNTL/Z.
Router(config)#line vty 0 4
Router(config-line)#access
Router(config-line)#access-class 9 in
Router(config-line)#^Z
Router#
```

> **Note**
>
> In situations where you are filtering "internal" host access as an alternative to using an ACL, it is possible to route undesirable packets to the null 0 interface. The null 0 interface is a pseudo-interface available on all Cisco routers. To route traffic to null, use the `<ip route>` command and replace the gateway address with `null 0`. Here is a example that routes to null 0 any packet with a destination address using any of the unregistered IP address spaces:
>
> ```
> hobo(config)#ip route 10.0.0.0 255.0.0.0 null0
> hobo(config)#ip route 172.16.0.0 255.266.0.0 null0
> hobo(config)#ip route 192.168.0.0 255.255.0.0 null0
> ```

The in directional statement is not necessary, because all access control lists are implicitly applied as inbound filters. Adding the in statement does not hurt, however, and explicitly applying the filter direction is a good habit to get into when installing access-groups on interfaces.

The activation of a dial-on-demand routing (DDR) interface is determined by certain kinds of traffic the router sees as interesting. Different types of interesting traffic (for each protocol, if needed) are defined with ACLs and applied to the DDR interface with the interface configuration subcommand `<dialer-group [group number]>`. The dialer group number corresponds to the dialer-list. The dialer-list is created in global configuration EXEC mode using the `<dialer-list [dialer list number] protocol [ip|appletalk|ipx] list [acl]>` command.

IP access-lists are also used for traffic redirection through policy routing, and for queuing lists through the route-map configuration subcommand `<match ip address [acl number]>`.

Creating Access-Lists

Now let's get our hands dirty and create some standard access-lists. We will start creating a standard IP access-list. Standard ACLs are built line by line, using the global configuration command `<access-list [1-99|1300-1999] [permit|deny] [source address] [wildcard mask]>`:

```
asbr-a2#config t
Enter configuration commands, one per line.  End with CNTL/Z.
asbr-a2(config)#access-list 1 permit 192.168.5.3
```

In the example above, we created an action statement that says "Permit any traffic that comes from the IP host 192.168.5.3. All other traffic will be denied." It is also possible to make this entry this way:

```
asbr-a2(config)#access-list 1 permit 192.168.5.3 0.0.0.0
```

This method uses a wildcard mask. Think of a wildcard mask as an inverted subnet mask. Like subnet masks, wildcard masks are expressed as four octets. With subnet masks, however, the 1 bits are used to mask the network portion address. Only the bits covered by the mask are looked at; the others are ignored. The wildcard mask uses the same approach, but instead of masking the network address, the host space is masked. In other words, a 0 bit indicates that an exact match is required and any 1 bits are ignored. The advantage to using wildcard masks is greater filtering flexibility. To calculate the wildcard mask of a network address, subtract the network's subnet mask from 255.255.255.255. The yield is the wildcard mask for the address space. In the example above, the 0.0.0.0 wildcard mask requires that each bit in the source address match the address specified in the action statement. If you wanted to permit all traffic from the 192.168.5.0 /24 network, you would use the following:

```
asbr-a2(config)#access-list 1 permit 192.168.5.0 0.0.0.255
asbr-a2(config)#int serial 0
asbr-a2(config-if)# ip access-group 1 in
```

Alternatively, let's say you want to restrict some network addresses and permit all others. To accomplish this, IOS provides the any keyword, a wildcard for creating global permit or deny actions. As with the host keyword, the same effect can be created by using specific IP addresses and IP netmask addresses. Here is an example that creates an ACL using the any keyword and using only IP addresses. Both achieve the same result:

```
asbr-a2(config)#access-list 1 deny 192.168.5.0 0.0.0.255
asbr-a2(config)#access-list 1 permit any
asbr-a2(config)#int serial 0
asbr-a2(config-if)# ip access-group 1 in
asbr-a2(config)#access-list 1 deny 192.168.5.0 0.0.0.255
asbr-a2(config)#access-list 1 0.0.0.0 255.255.255.255
asbr-a2(config)#int serial 0
asbr-a2(config-if)# ip access-group 1 in
```

This access list denies traffic originating from 192.168.5.0 /24, but permits traffic originating from any other network. Remember, all access control lists have deny any as the implicit action statement for any unqualified traffic. It is recommended that you use this fact to your advantage when creating ACLs and avoid using explicit deny statements unless required.

Note

When adding a single host entry in an ACL, it is not necessary to specify the wildcard mask 0.0.0.0, because any address not accompanied by a wildcard mask is assumed to be 0.0.0.0. It is also possible to add a single host entry using the host ACL keyword. Here is a example:

```
asbr-a2(config)#access-list 8 permit host 192.168.30.98
```

Note

Before we move on to extended ACLs, some specific words are needed on the care and feeding of ACLs. First, when you make an addition to an ACL, it cannot be undone. It is forever there in stone. If you try to delete the line using the <no> command, the entire ACL will be deleted. (This is not true of named access-lists; it is possible to delete line entries without deleting the entire list.) Second, every addition made to an ACL from the configuration command line is appended to the end of the list. Considering the sensitivity that ACLs have to the order of matching statements, appending a matching statement to the end of the list is not always desirable.

For these reasons, many authorities (including Cisco) recommend that you use a text editor and the `<copy tftp running-config>` command to create and install your ACLs. You will find this method much saner and a whole lot more efficient than trying to build ACLs with the command line. By the way, just in case you missed it, when you add anything to an ACL, it is appended to the bottom. This means that if you do not delete the ACL (with the `<no access-list [acl number]>` global configuration command) before you copy your new ACL into the running-config, all your changes will be appended to the bottom of the old ACL.

Extended IP Access Control Lists

Extended IP access-lists require a little more care and feeding than standard IP lists, because they provide additional matching capability. Standard ACLs are straightforward, and this characteristic lends them to being utilized with distribution, dialer, and route-map matching lists. Extended lists are used to create traffic filters, and the feature that distinguishes this function is the operation/port-value pair match. Extended ACLs are created line by line using the global configuration command <access-list [100-199|2000-2699] [permit|deny] [protocol] {[any|host] [source|wildcard mask]} {[any|host] [destination|wildcard mask]} [op] [port-value] [options...] >.

First, let's create an extended ACL without operator/value matches:

```
asbr-a2(config)#access-list 100 permit ip any 192.168.5.0 0.0.0.255
asbr-a2(config)#interface s0
asbr-a2(config-if)# ip access-group 100 in
```

This extended ACL permits all IP traffic from any source to any host on network 192.168.5.0 /24. All other traffic is discarded. Using ip as the protocol effectively permits all IP protocol traffic. Alternatively, you can create filters that are protocol-specific. The protocol choices are illustrated in Table 9.2. Extended lists, like standard lists, provide the option to use any in place of a network address. If you want to permit a host address instead of a network address range, the keyword host is used along with the host's IP address.

Table 9.2 **IP ACL Protocol Filtering Keywords**

Protocol	ACL Keyword
Enhanced Interior Gateway Routing Protocol	eigrp
Interior Gateway Routing Protocol	igrp
Open Shortest Path First	ospf
Generic Route Encapsulation Tunneling Protocol	gre
IP in IP Tunneling Protocol	ipinip
Internet Control Message Protocol	icmp
Transmission Control Protocol	tcp
User Datagram Protocol	udp
Internet Group Management Protocol	igmp
Protocol Independent Multicast	pim

Operator/port-value pair matches are available when ACL entries use a specific protocol type. When you create packet filter ACLs, you will generally use specific TCP and UDP operator/port-value pair matches. Table 9.3 lists the various operators available for each of the IP protocol options.

> **Note**
>
> Table 9.3 does not list the operator matches for ICMP, which are quite extensive. For more information, you can use the IOS inline help to get a list of the command options: <access-list 100 permit icmp any any ?>. For additional information, see the references listed at the end of the chapter.

Table 9.3 **Extended IP ACL Operators**

ACL	OperatorFunction	Protocol Availability
`established`	Match established connection-oriented VC connections	TCP
`eq`	Match on a transport port number	TCP, UDP
`gt`	Match packets with a greater port number	TCP, UDP
`lt`	Match only packets with a lower port number	TCP, UDP
`neq`	Match only packets not on a given port number	TCP, UDP
`range`	Match only packets in the range of port numbers	TCP, UDP
`dvmrp`	Distance Vector Multicast Routing Protocol, used with `mrouted` for MBONE routing	IGMP
`host-query`	Multicast host query	IGMP
`host-report`	Multicast host report	IGMP
`pim`	Protocol Independent Multicast	IGMP
`trace`	Multicast trace	IGMP
`log`	Log matches against this entry	All IP protocols
`log-input`	Log matches against this entry, including input interface	All IP protocols
`precedence`	Match packets with given precedence value	All IP protocols
`tos`	Match packets with given TOS value	All IP protocols

Match availability depends on the kind of protocol being used to create an ACL entry. In the case of TCP and UDP, there are varieties of matches that can be used, whereas routing and tunneling protocols are quite limited. TCP and UDP use transport layer service port numbers for port-value matching. For example, let's create a traffic filter ACL that permits inbound SMTP mail delivery and DNS service:

```
asbr-a2(config)#access-list 102 permit tcp any any eq 25
asbr-a2(config)#access-list 102 permit tcp any any eq 53
asbr-a2(config)#access-list 102 permit udp any any gt 1024
asbr-a2(config)#access-list 102 permit tcp any any gt 1024
asbr-a2(config)#interface s0
asbr-a2(config-if)# ip access-group 102 in
```

By using the operator/port-value pair, only inbound mail and DNS zone transfers are permitted. The `gt 1024` statements permit the local users to access external Internet hosts. Table 9.4 provides a list of commonly filtered TCP and UDP service ports.

Table 9.4 **Common Internet Known Service Ports**

Service ID	Service Name	Transport Layer Protocol and Port Number
`Bgp`	Border Gateway Protocol	TCP 179
`Bootp/dhcp`	Dynamic Host Configuration Protocol	UDP 67
`Cmd`	UNIX R commands	TCP 514
`Domain`	Domain Name Service	TCP and UDP 53
`Rsh`	BSD Remote Shell	TCP 512
`Ftp`	File Transfer Protocol	TCP 21
`Ftp-data`	FTP data connections	TCP 20
`Gopher`	Gopher protocol	TCP 70
`Http`	Hypertext Transport Protocol	TCP 80
`Irc`	Internet Relay Chat	TCP 194
`Imap v2`	Interactive Mail Access Protocol	TCP 143
`Imap v3`	Interactive Mail Access Protocol v3	TCP 220
`Klogin`	Kerberos login	TCP 543
`Kshell`	Kerberos shell	TCP 544
`Kdc`	Kerberos server	TCP and UDP 750
`Kreg`	Kerberos registration	TCP 760
`Kpasswd`	Kerberos password service	TCP 751
`Kpop`	Kerberos POP	TCP 1109
`Eklogin`	Kerberos encrypted `rlogin`	TCP 2105
`Login`	BSD remote login	TCP 513
`Lpd`	Berkeley printer daemon	TCP 515
`Nntp`	Network News Transport Protocol	TCP 119
`Nfs`	Network File System	TCP and UDP 2049
`Netbios-dgm`	NetBIOS datagram service	UDP 138
`Netbios-ns`	NetBIOS name service	UDP 137
`Netbios-ss`	NetBIOS session service	UDP 139
`Ntp`	Network Time Protocol	UDP 123
`Pop2`	Post Office Protocol v2	TCP 109
`Pop3`	Post Office Protocol v3	TCP 110

Service ID	Service Name	Transport Layer Protocol and Port Number
Radius	RADIUS authentication server	UDP 1645
Rip	Routing Information Protocol	UDP 520
Radacct	RADIUS accounting server	UDP 1646
Snmp	Simple Network Management Protocol	UDP 161
Snmptrap	SNMP traps	UDP 162
Sunrpc	Sun Remote Procedure Call	UDP 111
Smtp	Simple Mail Transport Protocol	TCP 25
Socks	SOCKS	TCP 1080
Shttp	Secure Hypertext Transport Protocol	TCP 443
Tacacs	TACACS Access Control System	TCP AND UDP 49
Tftp	Trivial File Transfer Protocol	UDP 69
Telnet	Telnet	TCP 23
Time	Time	TCP 37
Uucp	UNIX-to-UNIX Copy program	TCP 540
X11	X Window system display	TCP 6000-6004

Traditionally, it was believed that filtering below TCP/UDP port 1024 would provide adequate protection from external intruders, so many sites have an inbound packet filter that looks something like this:

```
access-list 101 permit tcp any host 192.168.0.5 eq www
access-list 101 permit tcp any 192.168.0.0 0.0.0.255 eq telnet
access-list 101 permit tcp any 192.168.0.0 0.0.0.255 eq domain
access-list 101 permit udp any 192.168.0.0 0.0.0.255 eq domain
access-list 101 permit tcp any host 192.186.0.23 eq smtp
access-list 101 permit tcp any 192.168.0.0 0.0.0.255 eq ftp
access-list 101 permit tcp any 192.168.0.0 0.0.0.255 eq ftp-data
access-list 101 permit tcp any 192.168.0.0 0.0.0.255 gt 1024 established
access-list 101 permit udp any 192.168.0.0 0.0.0.255 gt 1024
```

Note

Named access-lists permit operator/port-value pair matches to be specified for both the source and destination address designation. Only extended named ACLs permit these matches to be performed in conjunction with the destination address.

With today's Internet, however, this approach does not quite offer adequate protection. This list does greatly restrict inbound access, but it does not provide any defense against denial of service and SMURF attacks, and leaves the majority of service access partially unrestricted, open to exploitation. Unfortunately, these ports need to remain unrestricted because they are used randomly to initiate sessions. Even if you only accept inbound traffic from known hosts, there is still the possibility of a hacker using address masquerading to access your site. For these reasons, firewalls are often utilized to provide a single defense point, which allows you to tailor a security filter ACL around a specific host and address range instead of the entire network. Let's take a closer look at these defensive procedures.

Defense Against IP Denial of Service Attacks

A denial of service attack exploits the three-way handshake used by TCP to establish transport sessions. To establish a TCP session, host-a initiates the session by sending a TCP packet with the SYN bit enabled and an Initial Sequence Number (ISN), which is used to track the order of all subsequent exchanges. The destination host-b receives this packet and returns a TCP packet with its own SYN and ISN, along with an acknowledgement (ACK) of the sending host's SYN. This is commonly referred to as the SYN-ACK step. When the initializing host (host-a) receives the SYN-ACK, it sends an ACK of the destination host's (host-b) ISN. An attack is launched by a third host (host-c), which impersonates host-a. Host-c sends datagrams as host-a to host-b, and host-b in turn replies to the real host-a. Host-a, meanwhile, will not respond to any of the host-b datagrams, but they will continue to fill host-a's buffers until they overflow and the system crashes. The trick is that host-c needs to predict host-b's ISN, but because this is a systematically increasing number, all the attacker needs to do is sniff the initial SYN-ACK from host-b. This trick is not too hard to do with the right tools on today's Internet.

To defend against these attacks, the IOS (11.2 and higher) provides TCP intercept on 4000 and 7x00 series routers. When enabled, the router acts as a bridge between the source and destination hosts. If the connection is valid, the data exchange proceeds between the servers until the session ends. If the session is an attack, the router times out the connection, and the destination server is never involved with the transaction. The software can also operate passively, permitting transactions to directly occur between hosts. These transactions are monitored, however, and in the event that there is a connection failure or a partially open session for an excessive period, the router terminates the connection. TCP intercept is enabled as a global configuration command using `<ip tcp intercept list [ACL]>` and `<ip tcp intercept mode [intercept|watch]>`. The ACL defines the network and/or hosts that TCP intercept should interact with. Here is a configuration example that configures the router to passively monitor all TCP transactions:

```
!
ip tcp intercept 100
ip tcp intercept mode watch
!
!
```

```
access-list 100 permit tcp any 172.16.30.0 0.0.0.255 eq 514
access-list 100 permit tcp any 172.16.30.0 0.0.0.255 eq 513
access-list 100 permit tcp any 172.16.30.0 0.0.0.255 eq 512
access-list 100 permit tcp any 172.16.30.0 0.0.0.255 eq 443
access-list 100 permit tcp any 172.16.30.0 0.0.0.255 eq 25
access-list 100 permit tcp any 172.16.30.0 0.0.0.255 eq 80
access-list 100 permit tcp any 172.16.30.0 0.0.0.255 eq 22
access-list 100 permit tcp any 172.16.30.0 0.0.0.255 eq 21
access-list 100 permit tcp any 172.16.30.0 0.0.0.255 eq 23
access-list 100 permit tcp any 172.16.30.0 0.0.0.255 eq 514
access-list 100 permit tcp any 172.16.30.0 0.0.0.255 eq 514
```

Host sessions over slow WAN connections and TCP transactions that have long windows between acknowledgements can be interpreted as attacks. So some consideration should be given to the types of data transactions being handled by the router before implementing `<tcp intercept>`.

Address masquerading is quite simple to accomplish. A system without access assumes the IP address of a system with access. System services that function without authentication (that is, NFS, `rlogin`, TFTP) are particularly vulnerable to masquerading, because the IP address is used as the authentication identifier. To defend against address masquerading, you need to deny external network packets that contain your local network address in the source address field. In addition, you should filter out inbound source packets that might originate from the unregistered IPv4 address space. Here is an example of that kind of denial. The local network address range is 12.14.52.0 /22. The 10.0.0.0, 172.16.0.0, and 192.168.0.0 ranges are the "unassigned" IP address ranges, which should always be filtered out just as a standard precaution:

```
!
interface serial 0
ip address 12.14.116.1 255.255.255.0
ip access-group 99 in
!
access-list 99 deny   12.14.52.0 0.0.0.255
access-list 99 deny   12.14.53.0 0.0.0.255
access-list 99 deny   12.14.54.0 0.0.0.255
access-list 99 deny   12.14.55.0 0.0.0.255
access-list 99 deny   10.0.0.0 0.255.255.255
access-list 99 deny   172.16.0.0 0.0.255.255
access-list 99 deny   192.168.0.0 0.0.255.255
access-list 99 permit any any
```

Note

Here are some of the better security/hack information sites on the Web:

`http://www.cert.org/`

`http://www.rootshell.com/`

`http://www.phrack.com/`

SMURF attacks use IP-directed broadcasts to overload the network segment. Usually, an ICMP message (like a ping) or IP datagram with the network's broadcast address as the destination address (and a forged or bogus source address) is sent to every host on the segment, which forces all the recipient hosts to respond. A continuous flood of these packets will effectively "melt" the segment, allowing no real traffic to be passed. These attacks can be prevented by denying ICMP echo replies as part of your inbound security filter:

```
access-list 106 deny icmp any any echo-reply
access-list 106 permit icmp any any
```

In addition, by disabling the router interface's capability to forward IP-directed broadcasts, you can prevent the propagation of such traffic. This is accomplished with the interface configuration subcommand <no ip directed broadcasts>.

AppleTalk Access-Lists

ACL filtering for AppleTalk is primarily oriented toward the restriction of user access to network resources. AppleTalk ACLs make it possible to restrict access to nodes, printers, file servers, networks (all or part of a cable-range), and zones. This flexibility provides administrators with a great deal of control over what resources are accessible to users, zones, and networks. You might want to restrict this kind of traffic on your network because it's chatty; it takes up a lot of bandwidth, relatively speaking. AppleTalk network traffic filtering is more complex than IP traffic filtering due to AppleTalk's use of dynamic addressing and its logical naming scheme.

AppleTalk ACLs are built using a variety of matching criteria, which are listed in Table 9.5.

Table 9.5 **AppleTalk ACL Filtering Matches**

ACL Match Operator Data Value	Description
`1-65279`	Match on a AppleTalk node address
`network`	Match on a Phase 1 AppleTalk network address
`includes`	Match on a Phase 2 AppleTalk network address
`nbp`	Match on a name binding protocol (object, type, zone)
`within`	Match on a Phase 2 AppleTalk network address
`zone`	Match on a AppleTalk zone name

AppleTalk ACLs are created with the same line-by-line entry format used to create IP ACLs, and you should follow the same approach for ACL creation, editing, and installation—in other words, using a text editor. The global configuration command for creating AppleTalk ACL action statements is <access-list [600-699] [permit|deny] [match operator] [match data value]>. Aside from care and feeding, however, AppleTalk ACLs are very different from their IP and IPX counterparts, particularly when it comes to filtering AppleTalk's named space.

In Chapter 3, "The Networker's Guide to AppleTalk, IPX, and NetBIOS," we reviewed AppleTalk's naming entities. The NBP provides the mechanisms needed to map and distribute `object:name@zone` information throughout the network. ZIP is used for the creation and distribution of zone information between routers and nodes, both of which compile their own Zone Information Tables (ZITs). To filter these services, two different ACLs are required. NBP and network filtering (which conceptually resembles IP source/destination filtering) are applied to an interface as an inbound or outbound `<appletalk access-class>` command. Both NBP `object:type@zone` filtering and AppleTalk network address filtering are applied with the `<appletalk access-class [acl]>` command. It is also possible for both types of filtering to be performed on the same list.

Because zone information distribution filtering is accomplished through two different operations, it also uses two different filter applications. The `<getzonelist-filter>` is used to suppress zone information from being sent to user workstations when ZIP requests are made. When the `<getzonelist-filter>` is in place, only the zones permitted in the ACL will be sent in the ZIP response. To suppress the zone name propagation between routers, the `<zip-reply-filter>` is used. Only the zone names permitted in the ACL are sent in response to ZIP network-to-zone-name requests. These filters, like `<appletalk access-groups>`, are applied as interface configuration subcommands. Now that you have flipped back to Chapter 3 and reviewed the AppleTalk section, let's create an AppleTalk traffic filter.

> **Note**
>
> Because ACLs implicitly deny all traffic that is unspecified, generic ACL matches for AppleTalk are provided for dealing with unspecified network information (that is, no explicit match is applicable to them in the ACL). These ACLs perform roughly the same purpose as the any IP access-list keyword. This being AppleTalk, however, one operator would not do; three are required. The `other-access` match operator is used with a `permit` or deny statement to control the flow of packets containing network or cable-range information not explicitly matched by an ACL action statement. The `additional-zones` match operator is used for handling unspecified zones, and `other-nbps` is used for handling unspecified NBP packets.

Creating an AppleTalk ACL Filter

Figure 9.1 shows a three-segment AppleTalk network tied together with two half-routers. The central office only wants its main office zone and file servers and printers available to the users in the outland zone.

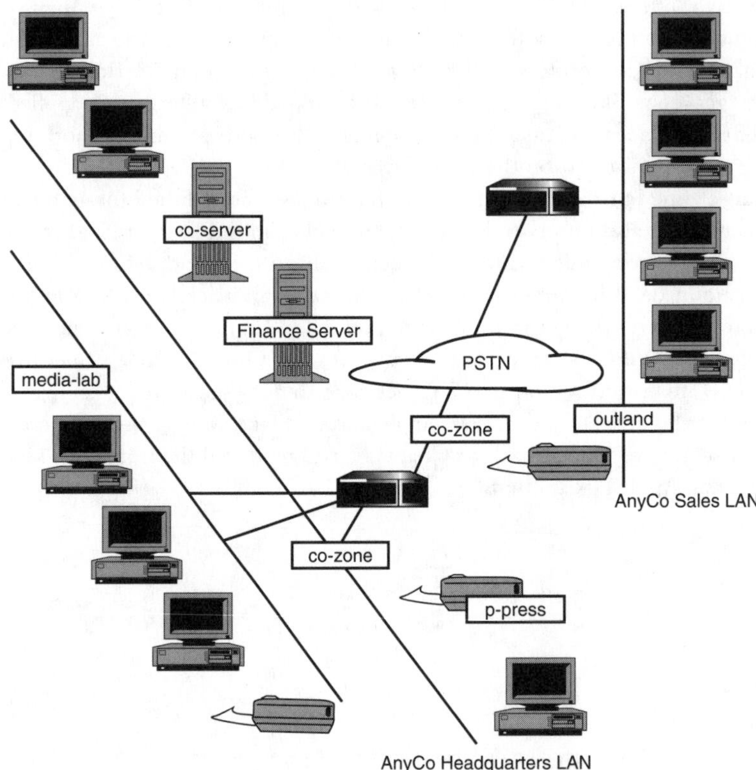

Figure 9.1 The AnyCo corporate AppleTalk network.

Here is the AppleTalk ACL being applied as an outbound access-group on the central office half-router:

```
!
interface ethernet 0
appletalk cable range 3-8 3.174
appletalk zone co-zone
appletalk zone media-lab
appletalk getzonelist-filter 608
!
!
interface serial 0
appletalk cable range 10-10 10.139
appletalk zone co-zone
```

```
appletalk access-group 601 out
appletalk zip-reply-filter 608
!
access-list 601 permit nbp 1 object CO server
access-list 601 permit nbp 1 type AFPServer
access-list 601 permit nbp 1 zone co-zone
access-list 601 permit nbp 2 object Finance Server
access-list 601 permit nbp 2 type AFPServer
access-list 601 permit nbp 2 zone co-zone
access-list 601 permit nbp 3 object p-press
access-list 601 permit nbp 3 type LaserWriter
access-list 601 permit nbp 3 zone co-zone
access-list 601 permit other-access
access-list 601 deny other-nbps
!
access-list 608 permit zone co-zone
access-list 608 permit zone outland
access-list 608 deny additional-zones
```

The ACL applied as the <appletalk access-group 601 out> permits the servers "co-server" and "Finance Server" and the printer to be accessible from the "outland" network. To suppress the zone announcements to the router and other workstations, ACL 608 was applied as the <appletalk getzonelist-filter [acl]> and the <appletalk zip-reply-filter [acl]>. The numbers used with the <permit nbp> action statement are sequence numbers that tie the object:name@zone elements together in the list.

IPX Access-Lists

The virtues of IPX filtering cannot be underestimated. With the IPX protocols, the filtering of the protocol's chattiness on some networks becomes a necessity, not an option. IPX, like IP, has two types of lists: standard and extended (IOS also supports named IPX lists). Standard lists are created using the global configuration command <access-list [800-899] [permit|deny] [source] [destination]>. The source and destination address match values can be either the IPX 32-bit network address or 96-bit network and node address. Here is an example standard IPX ACL that allows any packet to reach network 45:

```
asbr-a2(config)#access-list 810 permit -1 45
```

Note

One word of caution: If you plan to filter zone list announcements between routers, make sure that all the directly adjacent routers use the same filter. Otherwise, the ZITs will vary from router to router, resulting in inconsistent zone announcements and a great deal of user distress.

Extended lists match on source and destination address, and they provide the capability to filter on the basis of IPX protocol type and socket numbers. The extended IPX ACL command format is as follows:

```
<access-list [900-999] [permit|deny] [protocol] [source] [source socket]
[destination] [destination-socket]>
```

Table 9.6 lists the available IPX protocol matches, and Table 9.7 lists the available socket matches.

Table 9.6 **IPX Extended ACL Protocol Match Operators**

ACL Protocol Match Operator	Data Value Description
`Any`	Match on any IPX protocol packet
`Ncp`	Match on any NetWare Core Protocol packet
`NetBIOS`	Match on any IPX NetBIOS packet
`Rip`	Match on any IPX Routing Information Protocol packet
`Spx`	Match on any Sequenced Packet Exchange packet
`Sap`	Match on any Service Advertising Protocol packet

Table 9.7 **IPX Extended ACL Socket Match Operators**

ACL Socket Match Operator	Data Value Description
`All`	Match on any socket
`Cping`	Match on Cisco IPX ping
`Diagnostic`	Match on diagnostic
`Eigrp`	Match on IPX EIGRP
`Ncp`	Match on NetWare Core Protocol
`NetBIOS`	Match on IPX NetBIOS
`Nlsp`	Match on NetWare Link State Protocol
`Nping`	Match on standard IPX ping
`Rip`	Match on IPX RIP
`Sap`	Match on SAP
`Trace`	Match on traceroute

Now, let's create an extended ACL that filters inbound SAP and RIP announcements:

```
asbr-a2(config)#access-list 902 deny sap any sap any sap
asbr-a2(config)#access-list 902 deny rip any rip any rip
asbr-a2(config)#access-list 902 permit any any all any all
asbr-a2(config)#interface s0
asbr-a2(config-if)#ipx access-group 902
```

Remember, when you apply access-filters, inbound filters match packets coming into the interface and discard packets that fail. Outbound filters process the packet, and then a match is performed. In addition to standard and extended IPX access-lists, the IOS supports several additional IPX traffic filtering options. Refer to the IOS documentation for further details.

Displaying ACL Information

To display ACLs and ACL logging information, use the user EXEC command <show [protocol] access-list [acl number]>. If no ACL protocol (such as AppleTalk, IPX, IP, and so on) type or number is specified, all the lists will be displayed. If you plan to use ACLs for security purposes, it is also a good idea to enable accounting on the interfaces that are using ACL security filters. Accounting is available for IP and IPX.

IP accounting is enabled as an interface configuration subcommand <ip accounting>. To enable ACL violations, use the <ip accounting access-violations> variation of the command. IP accounting information is displayed with the user EXEC commands <show ip accounting> and <show ip accounting access-violations>.

IPX accounting is enabled using the interface configuration subcommand <ipx accounting>. To display IP accounting information, use <show ipx accounting>.

You can also log hits on ports by adding <log> at the end of the stanza. The access-list and logging information count is viewed by using <sh access-list [acl number]>.

Policy Routing

Policy routes provide an intelligent alternative to IP static routes. A static route's entry is limited to specifying a single gateway based on the destination address of a datagram. Policy routes can, among other things, direct traffic to different gateways based on various criteria, such as Layer 3 source or destination addresses, or Layer 4 destination ports. Policy routing is often used to provide gateway redirection services based on quality of service requirements, and for Layer 4 redirection services for use with HTTP and other application layer caching servers.

Policy routes are constructed out of two IOS elements: a route-map and one or more standard or extended IP ACLs. Route-maps behave in a manner similar to an outbound traffic filtering ACLs. Packets are processed through the router's interface. The IP datagram is processed through the route-map and then compared to a series of action statements. If no match is found, the packet is processed normally. If a match is found, the route-map will make the appropriate changes to the datagram as specified in the route-map. The route-map's capability to change the actual datagram distinguishes it from a filtering ACL.

A route-map action statement is also different from an ACL. An ACL action statement results in one of two actions: `permit` or `deny`. Route-map action statements are constructed as action pairs. The first action is the `<match>` statement, which uses an ACL to provide packet matching criteria. The second action is the `<set>` statement, which performs the desired processing action. Route-maps, in addition to their use in constructing policy routes, are also used for route redistribution processing.

Creating a Policy Route

The first step in creating a policy route is to construct the route-map. Route-maps are configured in two steps:

1. First, the criteria ACLs need to be created. As mentioned earlier, these are used by the route-map's `<match>` action to provide the packet evaluation criteria.
2. The second step is to create the route-map itself. Route-maps are created with the global configuration command `<route-map [map name] [permit|deny] [sequence number]>`. The map name serves as the route-map identifier, which associates the route-map with the policy route statement.

Policy routes, like ACLs, use deny as their implicit action. If no action statement is declared, the map sequence will be interpreted as a deny action.

The sequence number serves two roles:

- Sequence identification
- Sequence order

Route-map statements can be entered in any order, but they will be processed and matched in sequence, starting with the lowest-numbered statement. When building map statements, it is not a bad idea to leave numbers between statements (like old BASIC programming) so if you want to add a statement later you can just slide it in.

Let's create a simple map to redirect Web traffic to a Web cache server:

```
jung(config)#route-map http permit 10
jung(config-route-map)#match ip address 113
jung(config-route-map)#set ip next-hop 172.16.84.2
```

> **Note**
>
> The deny action has a different effect depending on the application of the route-map. If the map is being applied as a policy route and a datagram matches a `deny` statement, the datagram is processed normally (outside of the policy route). If the map is used for filtering a route redistribution, a deny match will not redistributed.

In the example above, the <match ip address [acl number]> route-map configuration subcommand is used to reference ACL 113 for matching criteria. TCP packets that meet this criterion are forwarded to the Web cache server with the IP address of 172.16.84.2. All other traffic is implicitly denied—in other words, is routed normally. The <match ip address> operator is used to define policy route matching criteria. Table 9.8 lists the set operators available for use with policy routing.

Table 9.8 **Route-Map Operators for Use with Policy Routing**

Route-Map Set Operator	Action Description
`set interface`	Directs the datagram out of the specified interface if no destination route exists in the router's routing table. Used for traffic redirection policy routing.
`set ip next hop`	Directs the datagram to the specified gateway address. Used for traffic redirection policy routing.
`set ip precedence`	Sets the precedence 0 to 7 (7 being the highest) value in the type of service field in the IP header. Used for quality of service policy routing.
`set ip tos`	Sets the type of service bits (1 to 15) in the IP header's type of service field. Used for QoS policy routing.

After the route-map is created, it needs to be applied to the gateway interface that serves as the outbound gateway for the network traffic you want to redirect. Because policy routes filter as the traffic is moving outbound, the map's placement needs to be in front of the gateways you are redirecting to. Keep this in mind when you are creating and installing your policy route-map. The map is installed with the <ip policy route-map [map name]> interface configuration subcommand:

```
jung(config)#int e0
jung(config-if)#ip policy route-map http
```

If you need to monitor your policy route's behavior, a debug command is available: <debug ip policy>.

Gateway Redundancy

Redundancy is a large part of enterprise network design. When designing enterprise networks, single points of network failure should be avoided whenever possible, particularly when network hardware is involved. You should recall from Chapter 6 that Layer 2 redundancy can be accomplished quite easily by connecting duplicate backbone links. The spanning-tree algorithm used in most bridges and switches today by default will disable one of the links, leaving it in a suspended state. In case of a failure, the suspended link is enabled and service is restored. At Layer 3, it is possible to have gateways and nodes share routing information. AppleTalk and IPX behave this way by default. With IP, it is a little more difficult, because some end-node configuration is needed. However, most major operating systems provide support for RIP to operate in a "snooping" mode to build a local dynamic routing table (some also support OSPF in this mode). Unfortunately, the feasibility of having every IP host run a routing protocol is quite limited, because of the technical support required, additional processing overhead, and peripheral support availability needed to implement and manage such processes.

There is really no efficient alternative for providing IP gateway assignments, so statically assigned gateways continue to be the most common method of providing a default gateway for external network access. They are often preferred over the possibility of running a dynamic protocol on an end-station. The static default is universally supported, easy to maintain, and has no possibility of failing—unless, of course, the router goes down. Regardless of how much redundancy you build into your network infrastructure, when the router fails the end-stations are stranded and cut off from the rest of the network until the router is recovered.

Gateway failures are perhaps the largest single point of failure for any enterprise network. Cisco and other vendors have developed router redundancy protocols to address this problem. The idea behind router redundancy protocols is simple. Two or more gateway routers collectively share a virtual IP and MAC address, which is set as the default gateway for the network segment. The routers sharing the IP address use an election protocol to determine who is the gateway and who, in effect, stands by to take over when a failure occurs. There are three such protocols used to perform this service:

- Cisco Systems' Hot Standby Router Protocol (HSRP)
- Digital Equipment Corporation's IP Standby Protocol (IPSTB)
- A new open standard called Virtual Router Redundancy Protocol (VIRP)

IPSTB and VIRP are not supported currently in the Cisco IOS, so we will only focus on HSRP. IPSTB and VIRP are supported by other router vendors, so you should at least be aware of their existence. IPSTB and VIRP are not compatible with Cisco's HSRP, and Cisco has no plans to support VIRP (the newer and open standard) in the future. More information on IPSTB and VIRP is available in RFC 2338.

HSRP

HSRP works by sharing a virtual IP address between two routers, where one is the primary and the other the standby. When HSRP is active, all traffic and operational activity (ping responses, VTY sessions, and so on) is handled by the primary router. HSRP is activated using the `<standby [group identifier] ip [virtual ip]>` interface configuration subcommand. In situations where only one HSRP instance is in use on a router, the group identifier is unnecessary. If more than one HSRP group exists, a group identifier (which can be from 1 to 255) is required for each instance. The identifier is used by both of the routers to identify them as part of the HSRP pair. In addition, the group ID is used to identify HSRP status messages sent between the routers and is part of the virtual MAC address (0000.0c07.ac**), the last two integers being the HSRP identifier.

Figure 9.2 depicts a standard HSRP application: two LAN gateway routers configured as an HSRP pair. Network A contains users; Network B is acting as a DMZ network which serves as an interconnection point for the LAN's various public and private outbound gateways. HSRP is configured on both router interfaces.

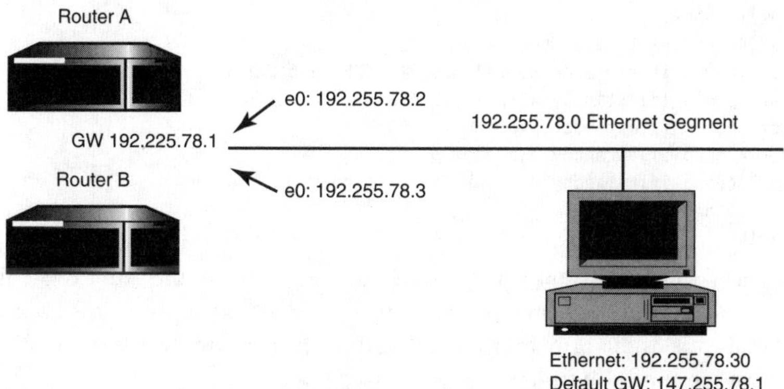

Figure 9.2 A standard HSRP implementation.

The best approach for setting up HSRP is to configure one router at a time. The first router configured should be the primary router. One of the major drawbacks of HSRP is that the standby router stays in a passive state unless the primary fails. Because routers are expensive, it is common to back up a high-end router with a router that has more moderate performance capabilities. For example, if a 4700 is the primary router, a 2500 or 2600 series would be a good standby router choice (this is the case in the example).

> **Note**
>
> In HSRP's original form, only one HSRP group could be configured on a given interface. Certain IOS versions and router platform combinations support multiple HSRP groups (each with a different virtual IP address) on the same interface. This enables you to set up some end-stations to use one virtual gateway and others to use another, providing some level of load sharing. This approach, by the way, is the operational model utilized by the VIRP standard, which Cisco is claiming to implement in upcoming versions of its IOS.

To ensure that the proper router acts as the primary, the <standby priority [0-255]> command provides the capability to set the router's selection preference. The default is 100, and the router with the highest priority value becomes the primary router. Let's look at the configurations on husserl and hegel. Because husserl will be the primary router, we will configure it first:

```
husserl(config)#int Fast Ethernet 0
husserl(config-if)#ip address 147.225.78.2 255.255.255.0
husserl(config-if)#standby ip 1 147.225.78.1
husserl(config-if)#standby 1 priority 200
husserl(config-if)#standby 1 prempt
husserl(config-if)#standby 1 track Fast Ethernet 1
husserl(config-if)#exit
husserl(config)#int Fast Ethernet 1
husserl(config-if)#ip address 147.225.91.2 255.255.255.0
husserl(config-if)#standby 4 ip 147.225.91.1
husserl(config-if)#standby 4 priority 200
husserl(config-if)#standby 4 prempt
husserl(config-if)#standby 4 track Fast Ethernet 0
husserl(config-if)#^Z
husserl#
```

The <standby [group id] prempt> command ensures that in the event of a failure (and the standby router becomes primary), when the intended primary is back online, it resumes its role as the primary router. When the husserl router's configuration is finished, we move on to hegel:

```
hegel(config)#int ethernet 0
hegel(config-if)#ip address 147.225.78.3 255.255.255.0
hegel(config-if)#standby ip 1 147.225.78.1
hegel(config-if)#standby 1 priority 195
hegel(config-if)#standby 1 track ethernet 1
hegel(config-if)#exit
hegel(config)#int ethernet 1
hegel(config-if)#ip address 147.225.91.3 255.255.255.0
hegel(config-if)#standby 4 ip 147.225.91.1
hegel(config-if)#standby 4 priority 195
hegel(config-if)#standby 4 track ethernet 0
hegel(config-if)#^Z
hegel#
```

With this configuration, because HSRP is running on both interfaces, we want to make sure that one router acts as the primary for both groups. This is achieved by adding the <standby [group identifier] track [interface type:slot:port]> command. The standby track tells the router to monitor its sister interface. If this interface goes down, the HSRP member interface's priority drops by 10. When the standby router's priority drops, it takes over. This is particularly important when running HSRP on both interfaces. If your standby tracking priority and prempt settings are not properly configured, you can have a situation where one router's interface is acting as the primary for one group and the other is acting as the primary of the other. This can be desirable if both routers have equal capabilities. If you have a high-end router for your primary and a low-end router for backup, however, you will want to ensure that the primary router is the correct one.

It is also possible to set up authentication for HSRP groups. This option is helpful when a router is participating in several HSRP groups. The password is a clear text password, which needs to be shared between the two group members. Authentication is enabled, like other HSRP options, as a variation of the standby command `<standby [group identifier] authentication [password]>`. To display status information about HSRP groups running on the router, the user EXEC command `<show standby>` is provided. Here is the output from the router `hegel`:

```
hegel# show standby

Ethernet0 - Group 1
  Local state is Active, priority 195
  Hellotime 3 holdtime 10
  Next hello sent in 00:00:01.360
  Hot standby IP address is 147.225.78.1 configured
  Active router is local
  Standby router is 192.168.78.2
  Standby virtual mac address is 0000.0c07.ac01

Ethernet0 - Group 4
  Local state is Active, priority 195
  Hellotime 3 holdtime 10
  Next hello sent in 00:00:01.360
  Hot standby IP address is 192.168.91.1 configured
  Active router is local
  Standby router is 192.168.91.2
  Standby virtual mac address is 0000.0c07.ac04
```

Network Address Translation

Internet access today (from both the home and office) is almost as commonplace as the telephone. With the demand for access to the Internet today almost doubling in size on a yearly basis, IPv4 address space is getting tight. Additionally, before Internet access became widespread. The TCP/IP protocol was widely used on many private enterprise networks. Since these networks were private (no Internet access), many administrators used "illegal" IP address ranges. These "illegal" address ranges, were IP addresses "assigned" to companies, by IANA for use on the public Internet. Other network administrators use the IANA provided "unregistered" IP address spaces to address their networks. Of course, in either instance, when these networks then wanted to connect to the Internet they were forced into the very unappealing and painful task of re-addressing the networks. Network Address Translation (NAT) was primarily developed to address these troublesome issues. NAT provides the capability to have a group of end-stations utilize an "illegal" or "unregistered" IP address space. Sharing a group (or single address, if necessary) of "registered" IP addresses to access hosts (that is, web, ftp, database servers and so on). NAT, in its basic form, functions as described next.

An internal (privately addressed) host establishes a connection with an external (publicly addressed) host. When the session is initially opened, the NAT gateway router assigns an externally routable IP address to the internal host establishing the external session (a one-to-one IP address translation). This address is then used to readdress the source address of all the outgoing IP datagrams sent from the internal host. As far as the external host is concerned, this external address is the real IP address of the internal host. So, when the NAT gateway router receives traffic from the external host with a destination address being used by an internal host, the incoming IP packets are readdressed again. This time the IP datagram's destination address is changed to reflect the actual internal address of the host. Along with translation, it is the job of the router to keep track of which host is assigned which address, how long they have had the address, and how long since the last time they have used the address. After a defined period of inactivity, external addresses are returned to the "address pool" and are reassigned to hosts who need them.

A variation of NAT known as Port Address Translation (PAT) is also often used to provide addressing translation for the SOHO PPP dial connections. PAT allows a single IP address to be used to manage multiple Layer 4 sessions (a one-to-many IP address translation). PAT is often represented as NAT on equipment that typically operates over PPP asynchronous links, where the IP address is assigned randomly to the router each time the link is connected.

NAT is commonly used in a fashion similar to the PPP/PAT scenario by small regional ISPs to provide addressing to dedicated attachment customers. The ISP designates a small block of routable addresses (instead of a single IP address), which are shared by the customers' end-stations to establish Internet connections. This is an attractive approach for ISPs that typically provide mail, Web hosting, DNS, and so on for small businesses that do not have their own Information Technology (IT) staffs. Instead of having the ISP dedicate a large address range block from their (limited) registered IP address space, a small address range (typically a /28 or /29 address range) is provided for host addressing through NAT. The customer then uses an unregistered Class C or Class B address range to provide local host addressing. These solutions are particularly well suited to NAT's functional design, which is geared toward scenarios where a limited amount of hosts are communicating outside the local network at any given time.

> **Note**
>
> PAT uses a single public IP address to represent all the nonpublic hosts. Each nonpublic transaction originates on a different UDP/TCP port number. Therefore, it appears that the single address is the originator of all the requests.

NAT configuration and operation are quite simple. The router is configured with an outside and inside interface. The outside interface is connected to the external Internet or intranetwork using a publicly routable IP address. The inside interface is connected to the local network that is using an "illegal" or unregistered address range. NAT translations replace the source address of the packet sent by an inside host with an address taken from a pool of externally valid addresses. The router keeps track of these translations as they occur. When a response is sent back, the router replaces the incoming packet's destination address with the inside address of the host that originated the request.

The fact that the router rewrites the Layer 3 addressing information contained in the IP datagrams makes a NAT router distinct from a traditional router. Most routers only perform Layer 2 readdressing, which is a necessary part of the packet forwarding process.

Configuring NAT

Enabling NAT on a router is a three-step process:

1. Creation of the address translation pool or static entries
2. Creation of the ACL that specifies which inside addresses to translate
3. Enabling NAT on the inside and outside interfaces

The IOS NAT implementation can provide translation services for unregistered inside addresses to registered outside addresses (the most common application of NAT). Additionally, it can provide overlapping NAT translation. Overlapping translation occurs when the inside address range is officially registered to another publicly accessible network. NAT is also capable of providing round-robin TCP load sharing across multiple hosts. With load sharing, a virtual host is created that is announced as the legitimate address of the service. The NAT router serves as the gateway to the virtual host, and as service requests come in, they are directed to the "real" hosts that respond to the service request.

Note

In scenarios where PAT is used, in addition to inside/outside address translation information, the transport layer (TCP/UDP) source and destination port information are retained in the NAT translation table as well.

Inside Address Translation

Inside address translation is provided statically or dynamically. Static NAT mapping translates an inside unregistered host address to an outside registered address. Static mappings are used when an inside host needs to be accessed by external hosts, as with an SMTP mail server. Static entries are created with the global configuration command <ip nat inside source static [inside ip] [outside ip]>. Dynamic entries are inside-to-outside address translations made on a temporary basis. A pool of outside addresses is allocated, and addresses are drawn from the pool as translations are required. Dynamic inside-to-outside translations remain until they expire (after 24 hours) or are cleared by an administrator. Here is an example of a static translation entry:

```
sartre(config)#ip nat inside source static 192.168.4.36 12.14.116.5
```

The global configuration command <ip nat pool [name] [starting outside address range] [ending outside address range] [prefix-length]> is used to create the outside NAT translation pool. The pool name is an identifier needed for the global configuration command <ip nat inside source list [acl] pool [pool name]>, which establishes dynamic NAT translation. The outside address range should ideally be allocated within a classful or classless network range.

The [prefix-length] or [netmask] sets the network mask that should be associated with the addresses in the pool. In this example, the local ISP has assigned the address range 12.14.116.16 /28 for the creation of the outside address pool:

```
sartre(config)# ip nat sartre-NAT 12.14.116.17 12.14.116.31 prefix-length 28
```

When dynamic translation is in use, addressees are translated on a first come, first served basis. If all the addresses in the pool are allocated, inside hosts will be sent an ICMP host unreachable message. If your outside address pool is small, it can be advantageous to modify the translation expiration from the default 24 hours to a more reasonable (smaller) time range. This is accomplished with the global configuration command <ip nat translation timeout [seconds]>. Another available option is to enable PAT. As stated earlier, PAT enables multiple inside TCP and UDP requests to be associated with a single outside address. PAT is enabled by adding the [overload] flag to the <ip nat inside source list [acl] pool [pool name] [overload]> configuration command.

Note

Two approaches can be used to allocate NAT address ranges. The first approach is to allocate the NAT outside pool range from the same network space used by the gateway interface. This makes announcing routing for the NAT pool easy because the gateway and NAT pool addresses can be routed with the same network announcement. The other approach is to assign an address range different from that used by the gateway interface. With this approach, announcements for the NAT pool are also directed to the gateway interface, but require an additional network announcement. If a dynamic routing protocol is being used to announce the publicly accessible routes, a loopback interface can be installed, using an address from the pool range.

Dynamic Source List Creation

The dynamic source list is a standard IP access-list that defines which inside addresses are eligible for NAT translation. Only addresses defined in the ACL will be translated. ACL entries are entered on a line-by-line basis using the global configuration command <access-list [1-99] permit [source network address/host address] [wildcard mask]>.

This approach allows the administrator to define NAT translation service on a host-by-host or network-wide basis. If all local networks are eligible, the "any any" ACL <access-list [1-99] permit any any> can be used. When the list is created, dynamic translation is enabled using the <ip nat inside source list [alc] pool [pool name]> command, referring to the ACL number and outside address pool name. This example utilizes the translation pool we created in the previous section:

```
sartre(config)#ip access-list 4 permit any any
sartre(config)#ip nat inside source list 4 pool sartre-NAT overload
```

Due to the small outside address pool, overload is enabled for more efficient use of the translation address pool.

NAT Interface Specification

When the source translation eligibility and outside address pool list definitions are completed, the inside and outside NAT translation interfaces must be defined. Interfaces are defined with the interface configuration subcommand <ip nat inside> and <ip nat outside>:

```
sartre(config)# interface s0
sartre(config-int)#ip address 12.14.116.5 255.255.255.252
sartre(config-int)# ip nat outside
sartre(config-int)#exit
sartre(config)# interface e0
sartre(config-int)#ip address 192.168.1.1 255.255.255.0
sartre(config-int)# ip nat inside
sartre(config-int)#exit
```

When both interfaces are configured, NAT is up and running. To see NAT translation table and operational statistics, the user EXEC commands <show ip nat translations> and <show ip nat statistics> are available. To clear NAT translations, you can use the privileged EXEC command <clear IP net translations *>.

NAT's Shortcomings

Although NAT is useful, it is not without shortcomings, the largest of which is the fact that NAT is CPU intensive. If you are planning to deploy NAT on any sizable scale (a pool of 128 addresses or more), you will want to use a 36x0 series or higher router to minimize any user-perceivable performance decrease. In terms of actual deployment, if possible do not run NAT on your gateway router. In most cases, the gateway router is performing security filtering and other CPU-intensive tasks. Adding the additional NAT processing will only decrease the gateway's performance. Use a separate router to provide NAT services, and it will dedicate that router's CPU to a single task and make NAT/performance-related troubleshooting easier and much less intrusive. It also provides you with the capability to do some additional filtering (which is needed if you use NAT as part of a router-based firewall solution).

Not all IP-based applications behave well with NAT, particularly those that are dependent on hostname/IP address verification and third-party authentication, such as Kerberos and AFS. NAT also has trouble with certain ISO load sharing and queuing schemes, depending on the IOS version the router is using (11.0 releases are prone to this problem). Most IP applications behave just fine, however, though it is always wise to test all your application requirements before you deploy.

Cisco Tunneling

The tunnel interface is a virtual point-to-point link tied to a physical source interface. That is to say, there is a virtual interface that corresponds to a logical interface, and between two interfaces of this type, traffic is wrapped and delivered across a public/private internetwork as if it were a point-to-point link. The tunnel interface has three components:

- The passenger protocol—The network protocol you are sending over the tunnel, such as AppleTalk, IP, and so on.
- The carrier protocol—The protocol that encapsulates the data. Generic Route Encapsulation (GRE) is the most commonly used protocol and is needed if multiple protocols are being handled. Cisco also supports Cayman (for AppleTalk over IP), EON (for CLNP over IP), and NOS.
- The transport protocol—This is IP (Internet Protocol), which handles the delivery of the data between the physical links tied to the tunnels.

The disadvantage of tunneling is performance. The encapsulation and de-encapsulation of the LAN protocols is time consuming, and processor and memory intensive. With Cisco IOS 11.1, GRE tunneling is supported on all Cisco 1600 and IP routers. Tunneling's greatest advantage is that it allows multiprotocol virtual private networks to be connected with a single access protocol, which (if required) can be sent across public backbone networks in a secure manner. VPNs offer tremendous cost savings to companies that need private connectivity between offices. See Figure 9.3 for a point-to-point tunneling topology example.

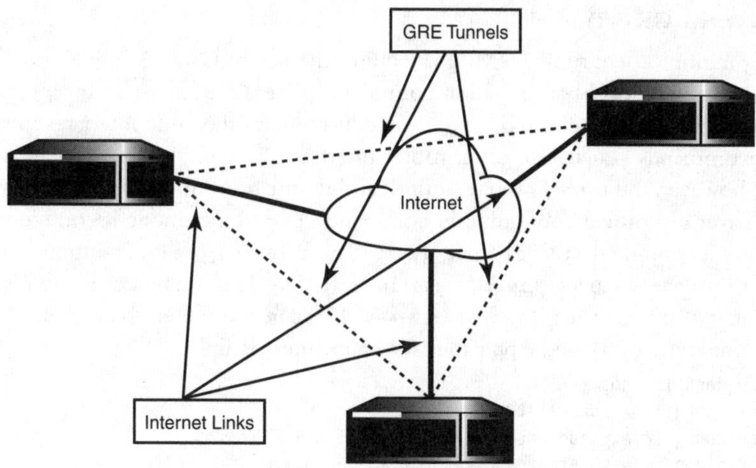

Figure 9.3 A point-to-point tunneling example.

There are two types of VPNs: encrypted and unencrypted. Unencrypted VPNs encapsulate data inside of a tunneling protocol packet. With an encrypted VPN, the protocol datagram reaches the VPN gateway, where it is encrypted and placed inside another protocol datagram for delivery. When the datagram reaches the destination router, the packet is decrypted and forwarded on to its destination host. Several open and private VPN encryption standards exist. Secure Shell (SSH), Point-to-Point Tunneling Protocol (PPTP), SOCKS, and IPSec are all popular secure tunneling methods. All these tunneling approaches use some combination of the 40-bit data encryption standard and RSA public key encryption. The Cisco IOS supports VPNs through both encrypted (DES-40, DES-56, DES-128-bit) and unencrypted tunnels.

Note

DES is a private key `crypto` algorithm based on a 56-bit encryption key, in which both the sender and receiver share a common key. RSA is a public key encryption system in which two keys are used: a public key used to encrypt data, and a private key used to decrypt data. There are no specific minimum and maximum RSA key lengths. However, most applications use key sizes ranging from 56 to 1,024 bits in length. Data encryption is a complex subject far beyond the scope of this book. See the references at the end of the chapter for resources providing information on data encryption.

Note

Administrators beware: Tunneling places additional load on the router's memory and CPU. Encrypted tunnels add even more processing load. Although tunneling (both encrypted and unencrypted) is supported on almost the entire Cisco router platform, this does not mean that you should implement tunneling (especially, encrypted tunnels) on Cisco router models. For most real-world tunneling applications, you will want to use at a minimum a 36x0 series router and the right software version and feature set.

Unencrypted GRE Tunnels

Creating multiprotocol tunnels (MPTs) requires the use of GRE. A tunnel interface is created using the global configuration command `<interface tunnel [interface number]>`. After the tunnel interface is created, it employs the same interface configuration subcommands available to other router interfaces. To complete the tunnel configuration, however, the tunnel's mode, source, and destination must be set, plus any other desired Layer 3 protocol configuration commands. The tunnel mode is configured using the subcommand `<tunnel mode [aurp|cayman|dvmrp|gre]>`. The tunnel source is set with `<tunnel source [ip address|interface]>`. The tunnel destination is set with `<tunnel destination [remote gateway ip address]>`. Here is an example of a GRE tunnel moving IP and AppleTalk across an Internet link:

```
router(config)#config t
router(config)#int tunnel 100
router(config-if)#ip address 147.225.96.1 255.255.255.252
router(config-if)#appletalk cable-range 45-45 45.149
router(config-if)#appletalk zone test-tunnel
router(config-if)#tunnel source ethernet 0
router(config-if)#tunnel destination 147.225.69.1
router(config-if)#tunnel mode gre ip
```

Encrypted GRE Tunnels

To create encrypted GRE tunnels, you need an enterprise plus or plus 40 version of the IOS software. The tunnel interface configuration is the same as it is in unencrypted tunnels. However, there are added commands for the exchange of DES encryption keys, as well as for the creation and application of the encryption maps, which are used to define what traffic will be encrypted.

After the tunnel interface is created, the first encryption step is to create the DES key and define the DES encryption algorithms you will use. Here is an example:

```
bumb(config)#crypto gen-signature-keys vpn
Generating DSS keys ....
 [OK]

bumb(config)#crypto algorithm 40-bit-des
```

Note

Encrypted GRE tunnel endpoints need to be configured at the same time because the router administrators need to actively verify the DES key exchange.

Note

Enterprise IOS comes with one encryption engine: the IOS `crypto` engine. However, the software encryption engine (as mentioned earlier) places additional load on the router when used. In addition, two IOS `crypto` engines, the Cisco RSP7000 and 7500, have the VIP 2 `crypto` and Encryption Service Adapter (ESA) `crypto` engines available. The ESA is also available on the 7200. When using encryption with these routers, using the VIP2 or ESA `crypto` engines will result in better performance than using the IOS `crypto` engine.

After the DES keys have been created, they need to be exchanged between the routers that will make up the encrypted tunnel. This exchange is done in global configuration mode, with both router administrators on the telephone so the DES key information can be verified. One side is configured as active and the other as passive. The passive router will wait until the active router connects. After it connects, the active key information will be displayed on both routers' consoles. After you both verify that the same key information is being displayed, passive will agree to accept active's key. The process then repeats itself for active to accept passive's key. To configure the router to be the passive DES key exchange partner, use the global configuration command `<crypto key-exchange passive [tcp port number]>`. To enable an active exchange, use the command `<crypto key-exchange [passives interface ip address] [tcp port number]>`. Both the active and passive routers need to use the same TCP port number. Here are examples of the passive and active key exchange commands used on routers `hobo` (passive) and `bum` (active):

```
hobo(config)#crypto key-exchange passive 8080
bum(config)#crypto key-exchange 172.16.44.5 8080
```

Now let's create and apply our `crypto` maps. The `crypto` map uses an extended ACL to match traffic. Because we are encrypting GRE traffic only, our ACL looks like this (we are using the 172.16.0.0 /16 for all private network traffic):

```
bum(config)#access-list 100 permit gre 172.16.0.0 0.0.255.255 72.16.0.0 0.0.255.255
```

A `crypto` map is configured the same way as a route-map. The `crypto` map is created globally `<crypto map [map-name]>`. The map name is used to associate the map with an interface, using the interface subcommand `<crypto map [map-name]>`. The following example creates the `crypto` map on the router `bum`, using the ACL created above, and applies it to the tunnel and the tunnel source interface:

```
bum(config)#crypto map vpn 10
bum(config-crypto-map)#set peer hobo
bum(config-crypto-map)#match address 100
bum(config-crypto-map)#set algorithm 40-bit-des
bum(config-crypto-map)#exit
bum(config)#interface tunnel 100
bum(config-if)#crypto map vpn
bum(config-if)#exit
bum(config) interface serial0
bum(config-if)#crypto map vpn
```

Displaying Tunnel Information

To monitor and display encrypted and unencrypted tunnel configurations, the following IOS commands are provided:

- `<show interface tunnel [number]>`—Tunnels are just like any other router interface. This command will display the source and destination addresses, tunnel protocol, and standard input and output statistics.

- `<show crypto algorithms>`—Displays the router's supported algorithms.
- `<show crypto engine connections active>`—Displays the active encrypted connections.
- `<show crypto engine brief>`—Shows a summary of the router's `crypto` engine, `crypto` engine info, key serial number, name, type, and so on.
- `<show crypto map>`—Displays the router's available `crypto` maps.
- `<show crypto mypubkey>`—Displays the router's public DES key.
- `<show crypto pubkey [name|serial number]>`—Displays a peer's public key or all the public keys stored on the router.

Cisco Router Wide Area Interface Configuration

In Chapter 5, "WAN Internetworking Technologies," we examined the various signaling and framing protocols associated with the following:

- Bell Labs/AT&T T-Carrier standard
- Integrated Services Digital Network (ISDN)
- Packet-switched network technologies: Frame Relay, ATM, and SONET

Although some of these technologies have been implemented in certain examples in this book, no direct attention has been paid to IOS interface configuration steps needed to implement these transport protocols. This is the goal of this section. Although it is not possible to address all possible configuration options, the intent here is to provide you with some configuration examples to get you started.

Transport-1 (DS1)

The AT&T/Bell Labs Transport-1 standard is the easiest and most often used WAN transport. Connectivity is provided through a V.35 synchronous serial (external DSU/CSU) or integrated DSU/CSU serial interface. As with all serial interfaces, encapsulation, framing, and line signaling protocols must be also be specified. Cisco, as mentioned earlier, uses HDLC for the default framing protocol. Alternatively, PPP (with or without authentication) can also be used. Serial interfaces are configured using the global configuration command `<interface serial [slot/port]>`. Framing and line signaling are only configured if the serial interface is an integrated DSU/CSU. Otherwise, framing and signaling are configured on the external DSU/CSU.

> **Note**
> On Cisco 4xxx and 7xxx routers, integrated DSU/CSU interfaces are called `<controllers>`. On 16xx, 25xx, and 36xx series routers, they are referred to as `<service-modules>`.

Transport-1 supports two framing and signaling combinations: D4-ESF/B8ZS, which is the default, and D4-SF/AMI. You must specify the type of framing and signaling you want to use when you order your circuit from your RBOC or CLEC. Signaling and framing are configured on integrated controllers using the interface configuration subcommands `<service-module t1 [framing] [esf|sf]>` and `<service-module t1 [linecode] [AMI|B8ZS]>`. When using an integrated controller, the number of timeslots or T1 channels must be specified. Channels are a carryover from the early days of data communications. At one time, "channelized" T1 service was easily available. With channelized or fractional T1 service, it's possible to order a T1 and only pay for using part of the circuit. The service was sold in 128K blocks. Today, most carriers only offer fractional data rate service using Frame Relay. Point-to-point dedicated service is sold only as dedicated 56K (DS0), T1, and T3. Channel specification is performed using the interface configuration subcommand `<service-module t1 timeslots [1-24|all]>`.

If the circuit is a dedicated 56Kbps, only one channel is specified; if a T1 is in use, either `<1-24>` or `<all>` can be used. You might also be required to provide timing for your circuit; if this is the case, only one side is configured as the clock source. To configure the integrated controller to provide the network clock, use the interface configuration subcommand `<service-module t1 clock source>`. By default, the controller expects the circuit to provide the clock. Here is an integrated DSU/CSU control configuration example:

```
persephone(config)# interface serial0/0
persephone(config-if)#ip address 172.16.30.1 255.255.255.252
persephone(config-if)#no shutdown
persephone(config-if)#encapsulation ppp
persephone(config-if)#service-module t1 framing esf
persephone(config-if)#service-module t1 linecode b8zs
persephone(config-if)#service-module t1 timeslots 1-24
persephone(config-if)#service-module t1 clock source
persephone(config-if)#ppp authentication chap
```

Configuring PPP Encapsulation and Authentication

PPP is IOS's open systems alternative Layer 2 encapsulation protocol. PPP provides data-link encapsulation for synchronous and asynchronous serial and ISDN BRI interfaces. Like HDLC, PPP can transport any IOS-supported Layer 3 network protocol. To enable PPP encapsulation, the interface configuration subcommand `<encapsulation ppp>` is used.

One of the many virtues of PPP is its built-in protocol level authentication. This makes it the ideal protocol for use with DDR and remote access over dial networking scenarios. PPP supports two authentication mechanisms: Password Authentication Protocol (PAP) and Challenge-response Authentication Protocol (CHAP). These protocols do not participate in the authentication; they only transport the authentication credentials. The host/router is responsible for determining if actual access is permitted.

The PAP authentication protocol provides a mechanism for PPP peers to send authentication information to one another. When a PPP connection is established, the initiating router sends a "clear text" username and password to the host router. If the username and password are correct, the session establishment is finished and the link is established. To enable PAP authentication on an ISDN BRI, async, or dialer interface, use the following steps:

1. Enable PPP encapsulation.
2. Enable PPP PAP authentication using <ppp authentication pap>.
3. Configure the PAP username and password the interface will send to the host router. This is done with the <ppp pap sent-username [username] password 7 [password]> command.

If you have <AAA new-mode> authentication and authorization enabled for PPP, the username and password can be verified by whatever mechanism you have indicated—RADIUS, TACACS, or local. If you have old-mode (the IOS default) authentication, the username and password must be configured on the host router. Here is an example using an DDR async interface:

```
asbr-a2(config)#int async 1
asbr-a2(config-if)#ip address 172.16.44.5 255.255.255.252
asbr-a2(config-if)#dialer in-band
asbr-a2(config-if)#dialer map ip 172.16.44.6 modem-script usr
asbr-a2(config-if)#ppp authentication pap
asbr-a2(config-if)#ppp pap sent-username test password test
```

CHAP's function is the same as PAP's: to provide an authentication credential exchange mechanism. Where CHAP is different is that instead of sending the authentication information to the host router just once, both the local and remote router randomly authenticate each other throughout the life of the session.

CHAP authentication requests are referred to as challenges. With CHAP authentication enabled, routers use their hostnames or a text string set on the interface as an authentication key. Instead of a password, a CHAP secret is configured on both routers. To enable CHAP authentication on an ISDN BRI, async, or dialer interface, use the following steps:

1. Enable PPP encapsulation.
2. Enable PPP CHAP authentication using <ppp authentication chap>.
3. Configure the CHAP secret and hostname that the router will use to issue and verify CHAP challenges.

A CHAP authentication challenge can be initiated by a router on either side of the connection. The CHAP challenge begins by having one of the routers transmit an authentication key (the hostname or other text key) to the remote router. This key is then encoded by the challenged host, using the shared secret, and transmitted back to the challenge initiator. The key is decoded using the key, and if there is a match, the CHAP authentication challenge is successful. It is also possible to have RADIUS or TACACS provide the authentication key and secret information. Here is an ISDN dialer example using CHAP with the credential information configured locally on the router:

```
asbr-a2(config)#username asbr-a1 password anydaynow
asbr-a2(config)#int bri 0
asbr-a2(config-if)#ip address 172.16.101.6 255.255.255.252
asbr-a2(config-if)#encapsulation ppp
asbr-a2(config-if)#ppp chap password 7 anydaynow
asbr-a2(config-if)#ppp chap hostname asbr-a2
```

Displaying PPP-Related Configuration Information

To view the array of PPP link state, negotiation, and authentication messages, IOS provides the following <show> and <debug> commands:

- `<show ppp multilink>`—Displays PPP multilink interface information
- `<debug ppp authentication>`—Displays the PPP authentication dialog
- `<debug ppp negotiation>`—Displays the PPP Layer 2 negotiation dialog
- `<debug ppp packet>`—Displays PPP packet information on the router console
- `<debug ppp error>`—Displays PPP authentication, link quality, and assorted PPP error messages on the router console

ISDN

Cisco supports ISDN BRI and ISDN PRI interfaces. BRI interfaces are supported on the 16xx, 25xx, 26xx, 36xx, 4xxx, and 7xxx series routers. They are commonly used for establishing dial-on-demand routing (DDR) links. ISDN PRIs are supported on the 4xxx, 36xx, and 7xxx series routers using the multichannel interface processor. Because the ISDN BRI is the more common implementation of ISDN, the focus is on its configuration.

With ISDN, the circuit provisioning is the hardest part. In most cases, when an ISDN BRI is ordered, it is attached to a modem. Most local exchange carriers or competitive local exchange carriers understand this implementation, so when you say, "I have a Cisco router," there is a good chance that confusion will follow.

One way of making this process much easier is to use the NIUF (North American ISDN Users Forum) compatibility packages and provisioning guidelines. These are available on the NIUF Web site at `http://www.niuf.nist.gov/niuf/docs/428-94.html`. The guidelines provide descriptions of different ISDN BRI provisioning packages, which you specify to your carrier to ensure that your provisioning is correct. NIUF computability package M, S, or T (variations on alternate voice/data on two B channels and one D channel) work quite well for most ISDN modems and Cisco router applications.

General ISDN Configuration

Before the ISDN BRI interface can be used, the router needs to set the ISDN switch type, the BRI's service profile identifiers (SPIDs), and local directory numbers (LDNs). The ISDN switch type can be configured globally or as an interface configuration subcommand. When a multi-BRI interface card is used, one ISDN switch type is supported for all the BRI interfaces on the card. This can cause a slight problem if you are provisioning ISDN BRI circuits from different LEC or CLEC central offices with different ISDN switch types. The problem can be avoided, however, if your circuit provisioning is done in accordance with an ISDN "standard" such as National ISDN (in the U.S.) or NET3 (in the U.K. and Europe) instead of specific switch types. The ISDN switch type is set using the global configuration command `<isdn switch-type [switch type]>`. Table 9.9 lists the switch type options available.

Table 9.9 **Cisco Router Supported ISDN TelCo Carrier Switch Types**

IOS Setting	Switch Type
`basic-1tr6`	1TR6 switch type for Germany
`basic-5ess`	AT&T 5ESS switch type for the U.S.
`basic-dms100`	Northern DMS-100 switch type
`basic-net3`	NET3 switch type for the U.K. and Europe
`basic-ni`	National ISDN switch type
`basic-ts013`	TS013 switch type for Australia
`ntt`	NTT switch type for Japan
`vn3`	VN3 and VN4 switch types for France

After the switch type is specified, the ISDN BRI SPIDS and LDNs are configured using the BRI interface configuration subcommand `<isdn spid[1|2] [spid] [LDN]>`. Not all ISDN BRI provisioning uses two SPIDS; in these cases, only SPID1 is configured. Here is an ISDN example (setting the switch globally), using a Northern Telecom DMS-100 switch and two SPIDS:

```
descartes(config)#isdn switch-type basic-dms100
descartes(config)#interface bri 0
descartes(config-if)#isdn spid1 80055534340101 5553434
descartes(config-if)#isdn spid2 80055512120101 5551212
```

ISDN BRI DDR Configuration

Although it is not always the most cost effective networking solution, ISDN DDR is the most flexible. In this scenario, the BRI interface has two data channels, both of which operate at 56 or 64Kbps.

> **Note**
>
> Although ISDN bearer channels are 64Kbps each, it is sometimes necessary to configure them at 56Kbps for transport compatibility across T-carrier links. If the ISDN call path is not entirely all "ISDN digital," transport problems can arise. Using the 56Kbps channel rate works around this problem, especially when using ISDN over long distance lines.

For most applications, only a single channel is needed. If more bandwidth is required, however, both BRI channels can be utilized by using multilink PPP. Multilink PPP provides load balancing and packet fragmentation capabilities over multiple asynchronous serial and ISDN interfaces. Multilink ISDN configurations are implemented either as a single ISDN BRI interface or as a dialer interface, which permits two or more BRI interfaces to be used collectively to transport data as a single logical interface. The dialer interface is a parent logical interface that manages the physical interface(s). Datagrams are delivered to the dialer interface, which in turn directs the data flow across the physical interfaces under its control. The dialer interface also controls the multilink session, bringing up interfaces and tearing down interfaces as the data flow demands. In single ISDN BRI configurations, the BRI interface operates as a dialer interface. In a multi-ISDN BRI configuration, a separate dialer interface is created, and the BRI interfaces are added to it by way of a dialer rotary group.

Let's examine the configuration process for configuring a single BRI multilink PPP interface:

```
leibniz#config t
Enter configuration commands, one per line.  End with CNTL/Z.
leibniz(config)#interface bri0
leibniz(config-if)#isdn switch-type basic-ni
leibniz(config-if)#ip address 172.16.96.5 255.255.255.252
leibniz(config-if)#isdn spid1 0835866201 8008358662
leibniz(config-if)#isdn spid2 0835866401 8002358664
leibniz(config-if)#no shut
```

After the <no shutdown> command is entered, the BRI should come up and negotiate the terminal endpoint identifier (TEI) with the LEC/CLEC TelcCo carrier switch. The TEIs are the ISDN B channel's Layer 2 addresses. The TEI negotiation messages will be displayed on the router's console:

```
22:41:34: %LINK-3-UPDOWN: Interface BRI0:1, changed state to down
22:41:34: %LINK-3-UPDOWN: Interface BRI0:2, changed state to down
22:41:34: %LINK-3-UPDOWN: Interface BRI0, changed state to up
22:41:38: %ISDN-6-LAYER2UP: Layer 2 for Interface BR0, TEI 87 changed to up
22:41:38: %ISDN-6-LAYER2UP: Layer 2 for Interface BR0, TEI 88 changed
```

The default is to negotiate TEI when the interface comes online, but it is possible to have this occur when the first ISDN call is placed. This behavior is set using the interface subcommand <isdn tei [first-call|powerup]>. To check the state of the ISDN BRI interface, the <show isdn status> user EXEC command is used:

```
leibniz#sh isdn status
Global ISDN Switchtype = basic-ni
ISDN BRI0 interface
        dsl 0, interface ISDN Switchtype = basic-ni
    Layer 1 Status:
        ACTIVE
    Layer 2 Status:
        TEI = 87, Ces = 1, SAPI = 0, State = MULTIPLE_FRAME_ESTABLISHED
        TEI = 88, Ces = 2, SAPI = 0, State = MULTIPLE_FRAME_ESTABLISHED
```

```
Spid Status:
    TEI 87, ces = 1, state = 5(init)
        spid1 configured, spid1 sent, spid1 valid
        Endpoint ID Info: epsf = 0, usid = 2, tid = 1
    TEI 88, ces = 2, state = 5(init)
        spid2 configured, spid2 sent, spid2 valid
        Endpoint ID Info: epsf = 0, usid = 4, tid = 1
Layer 3 Status:
    0 Active Layer 3 Call(s)
Activated dsl 0 CCBs = 0
The Free Channel Mask:  0x80000003
Total Allocated ISDN CCBs = 0
```

After the interface is configured, the next step is to configure multilink PPP and create the dialer map:

```
leibniz(config-if)#encapsulation ppp
leibniz(config-if)#ppp multilink
leibniz(config-if)#ppp authentication chap
leibniz(config-if)#ppp chap password agoodpass
leibniz(config-if)#dialer idle-timeout 300
leibniz(config-if)#dialer load-threshold 128 inbound
leibniz(config-if)#dialer map ip 172.16.96.6 name descartes broadcast 8008358861
leibniz(config-if)#dialer map ip 172.16.96.6 name descartes broadcast 8008358863
leibniz(config-if)#dialer-group 1
leibniz(config)#dialer-list 1 protocol ip list 100
leibniz(config)#access-list 100 permit tcp any any eq 80
leibniz(config)#access-list 100 permit tcp any any eq 110
leibniz(config)#access-list 100 permit tcp any any eq 443
leibniz(config)#access-list 100 permit udp any any eq 54
leibniz(config)#access-list 100 permit icmp any any echo
leibniz(config)#ip route 0.0.0.0 0.0.0.0 172.16.96.6
```

The first two dialer interface commands `<encapsulation ppp>` and `<ppp multilink>` set the interface's operational properties. The `<dialer load-threshold [1-255] [inbound|outbound]>` command sets the interface traffic level that must be met before another link is enabled. The threshold `<128 inbound>` specifies that if the first channel's inbound load exceeds 50 percent of the circuit's bandwidth, to bring up the other link. The IOS uses a scale of 0 to 255 to set many of its option variables—128 is the midpoint, which equals 50 percent. The `<dialer idle-timeout [seconds]>` specifies how long the link can remain idle before the call is terminated. The dialer interface is built in three steps:

1. The dialer map is created with `<dialer map [L2/L3 protocol] [destination address] [options] [LND number]>`. A dialer map entry is needed for each LND.
2. Specify which `<dialer group>` number is used to identify the `<dialer rotary-group>` and `<dialer-list>` associated with the interface. The dialer-list indicates which ACL should be used to match interface traffic, to determine if it is "interesting" enough to start a call. The dialer rotary-group is used to assign BRI interfaces to a `<dialer>` interface in a multi-BRI configuration.

3. Create the ACL. When creating the ACL, think about the kinds of traffic the link will handle. Many configurations use the blanket statement `<access-list 100 ip any any>`. This ACL, however, can give you a very large telephone bill, because anything will bring the link up. The ACL example above will only activate the line when a DNS, Web, secure Web, or POP mail request is made. Now let's look at implementing BRI backup for dedicated synchronous serial links.

The following is a configuration example for multi-BRI dialer implementation. The configuration process is identical to the single BRI, except for the dialer interface and the dialer rotary-group:

```
!
interface Dialer1
ip address 172.16.96.5 255.255.255.252
dialer map ip 172.16.96.6 name descartes broadcast 8005551212
encapsulation ppp
dialer in-band
dialer idle-timeout 300
dialer load-threshold 20 either
dialer-group 1
ppp multilink
ppp authentication chap
ppp chap password 7 03055C04090B314D5D1A
!
interface BRI2/0
no ip address
encapsulation ppp
no keepalive
isdn spid1 91468160100101 6816010
isdn spid2 91468160110101 6816011
dialer load-threshold 20 either
dialer rotary-group 1
!
interface BRI2/1
no ip address
encapsulation ppp
no keepalive
isdn spid1 91468160100101 6816010
isdn spid2 91468160110101 6816011
dialer load-threshold 20 either
dialer rotary-group 1
!
ip route 0.0.0.0 0.0.0.0 172.16.96.6
!
access-list 100 permit tcp any any eq www
access-list 100 permit tcp any any eq pop3
access-list 100 permit tcp any any eq 443
access-list 100 permit icmp any any echo
access-list 100 permit tcp any any gt 1000
dialer-list 1 protocol ip list 100
```

ISDN BRI Interface Backup Configuration

IOS uses HSRP to provide LAN gateway redundancy, but in most cases, it is the WAN link that presents the greatest possibility of failure. In scenarios where multiple external paths exist, this vulnerability to failure is minimized somewhat. But, because most WAN-like failures are due to PSTN switch and network failures, unless these circuits have been engineered for complete path redundancy, when one link fails, the other very likely will go down as well. Complete path redundancy is quite costly and difficult, unless your site is serviceable from multiple LEC/CLEC COs. To defend against such failures, the IOS provides the capability to assign a `<backup>` asynchronous serial or ISDN interface to be activated in the event of a failure or if additional bandwidth is needed.

To configure the router to use link backup, the primary interface is configured first. You specify which interface will be used as the backup and under what conditions the backup link will be enabled and disabled. Three interface configuration subcommands are used:

- `<backup interface [type/slot/port>` specifies the backup interface.
- `<backup load [enable load|never] [disable load|never]>` defines the primary load requirements that need to be met to bring up and bring down the link. This command can also be used to configure the backup interface to come online during high bandwidth demand periods. For example, the configuration entry `<backup load 90 40>` configures the interface to come online when the primary interface has a utilization load of 90 percent and disable when the load returns to 40 percent.
- `<backup delay [enable delay|never] [disable delay|never]>` specifies the time between the primary link failure and the backup link activation, and how long the backup link will remain active once the primary link has been restored.

Now let's examine a "backup only" configuration:

```
leibniz(config-if)#ip address 172.16.44.1 255.255.255.248
leibniz(config-if)#backup interface bri0
leibniz(config-if)#backup delay 15 60
leibniz(config-if)#backup load never never
leibniz(config-if)#ex
leibniz(config)#int bri 0
leibniz(config-if)#ip address 172.16.44.2 255.255.255.248
leibniz(config-if)#isdn switch-type basic-ni
leibniz(config-if)#isdn spid1 0835866201 8008358662
leibniz(config-if)#isdn spid2 0835866401 8002358664
leibniz(config-if)#ppp multilink
leibniz(config-if)#dialer load-threshold 20 either
leibniz(config-if)#dialer string 8005551212
leibniz(config-if)#dialer-group 1
leibniz(config-if)#ex
leibniz(config)#dialer-list 1 protocol ip list 100
```

```
leibniz(config)#access-list 100 permit ip any any
leibniz(config)#ip route 0.0.0.0 0.0.0.0 bri 0 172.168.44.3
leibniz(config)#ip route 0.0.0.0 0.0.0.0 serial 0 172.168.44.4
```

In this configuration, the ISDN BRI interface is configured to direct dial, rather than use a dialer map. When dial backup is enabled, the backup interface is effectively controlled by the primary interface; it cannot be used for any other purpose. This condition is easily verifiable by looking at the interface configuration with the `<show interface [type/port/slot]>` command. The partial `<show interface>` command output for the primary and backup interfaces illustrates this interdependence:

```
Serial0 is up, line protocol is up
  Hardware is HD64570
  Internet address is 172.16.44.1/24
  Backup interface BRI0, failure delay 15 sec, secondary disable delay 60 sec,
  kickin load not set, kickout load not set
  MTU 1500 bytes, BW 1544 Kbit, DLY 20000 usec, rely 255/255, load 1/255
  Encapsulation HDLC, loopback not set, keepalive set (10 sec)
  Last input 00:00:01, output 00:00:01, output hang never
  Last clearing of "show interface" counters never
  Queuing strategy: fifo

BRI0 is standby mode, line protocol is down
  Hardware is BRI
  Internet address is 172.16.44.2/24
  MTU 1500 bytes, BW 64 Kbit, DLY 20000 usec, rely 255/255, load 1/255
  Encapsulation PPP, loopback not set
  Last input 00:01:38, output 00:01:38, output hang never
  Last clearing of "show interface" counters never
  Input queue: 0/75/0 (size/max/drops); Total output drops: 0
  Queuing strategy: weighted fair
```

Note that the two interfaces are both addresses from the same IP address subnet (this is not permitted, unless dial backup is in use). Although these interfaces are joined at the hip, so to speak, both interfaces must be configured appropriately, as though they were each the primary interface. This is accomplished by setting the default route for each interface, instead of just setting a single next hop address. If this approach is not followed, when the primary interface fails, the default gateway will be lost and the traffic will be dropped by the router because it does not know where to send it.

> **Note**
>
> If a dynamic routing protocol is used, this will not occur because protocol will readjust its forwarding behavior for the link failure.

Let's examine how the router behaves using this configuration. The following code output illustrates the IP routing table of our example router, leibniz, under normal operation, where the primary interface only is listed and is set as the gateway of last resort. The router leibniz's routing table under normal conditions looks like this:

```
Codes: C - connected, S - static, I - IGRP, R - RIP, M - mobile, B - BGP
       D - EIGRP, EX - EIGRP external, O - OSPF, IA - OSPF inter area
       N1 - OSPF NSSA external type 1, N2 - OSPF NSSA external type 2
       E1 - OSPF external type 1, E2 - OSPF external type 2, E - EGP
       i - IS-IS, L1 - IS-IS level-1, L2 - IS-IS level-2, * - candidate default
       U - per-user static route, o - ODR
       T - traffic engineered route

Gateway of last resort is 172.168.44.4 to network 0.0.0.0

S     172.16.0.0/16 [1/0] via 172.16.44.0
C     192.168.191.0/24 is directly connected, Ethernet0
S*    0.0.0.0/0 [1/0] via 172.168.44.4, Serial0
```

When the primary link fails, the BRI interface takes its place and also takes the default gateway role, because a distinct static entry is part of the configuration. This is illustrated in the following code sample. It is important to note that if the BRI default route entry was not specified, no gateway of last resort would exist and none of the traffic would be forwarded by the router, because it would not know where to send it. The router leibniz's routing table with the backup line active and a new default gateway looks like this:

```
Codes: C - connected, S - static, I - IGRP, R - RIP, M - mobile, B - BGP
       D - EIGRP, EX - EIGRP external, O - OSPF, IA - OSPF inter area
       N1 - OSPF NSSA external type 1, N2 - OSPF NSSA external type 2
       E1 - OSPF external type 1, E2 - OSPF external type 2, E - EGP
       i - IS-IS, L1 - IS-IS level-1, L2 - IS-IS level-2, * - candidate default
       U - per-user static route, o - ODR
       T - traffic engineered route

Gateway of last resort is 172.168.44.3 to network 0.0.0.0

         172.16.0.0/16 is variably subnetted, 2 subnets, 2 masks
C       172.16.44.0/24 is directly connected, BRI0
S        172.16.0.0/16 [1/0] via 172.16.44.0
C      192.168.191.0/24 is directly connected, Ethernet0
S*     0.0.0.0/0 [1/0] via 172.168.44.3, BRI0
```

When the primary link is restored, it resumes its place as the gateway of last resort (see the following code example). The backup interface remains in the routing table until the call is terminated and it returns to its standby state. The router leibniz's routing table during the transition from the backup link to the primary link looks like this:

```
Codes: C - connected, S - static, I - IGRP, R - RIP, M - mobile, B - BGP
       D - EIGRP, EX - EIGRP external, O - OSPF, IA - OSPF inter area
       N1 - OSPF NSSA external type 1, N2 - OSPF NSSA external type 2
       E1 - OSPF external type 1, E2 - OSPF external type 2, E - EGP
       i - IS-IS, L1 - IS-IS level-1, L2 - IS-IS level-2, * - candidate default
       U - per-user static route, o - ODR
       T - traffic engineered route
```

```
Gateway of last resort is 172.168.44.3 to network 0.0.0.0

         172.16.0.0/16 is variably subnetted, 2 subnets, 2 masks
C       172.16.44.0/24 is directly connected, BRI0 is directly connected, Serial0
S       172.16.0.0/16 [1/0] via 172.16.44.0
C       192.168.191.0/24 is directly connected, Ethernet0
S*      0.0.0.0/0 [1/0] via 172.168.44.3, BRI0
                [1/0] via 172.168.44.4, Serial0
```

When the transition period is over, the BRI is placed back into a standby state. When the backup link is down, the BRI and secondary default gateway are purged from the routing table and the router returns to its pre-failure state.

Displaying ISDN Configuration Information

Knowledge of the operational information is essential to using ISDN, perhaps even more than with any other protocol or interface type. When testing ISDN configurations, use <show controllers bri [slot/port]> and do not use the following debug commands all at once. If you do decide to use all the ISDN debug commands at the same time, be prepared for a screen full of noise. If you make a mistake, you can disable the debug commands using <no debug all>.

- <debug isdn events>—Shows the Layer 3 operational events
- <debug q921>—Displays the Layer 2 operational events
- <debug dialer>—Displays dialer operational events

For general behavioral and session information, use the following commands:

- <show isdn active>—Shows which BRI links are active and for how long
- <show isdn history>—Displays BRI call history—who has called, how long the connection was up, and so on
- <show isdn status>—Displays the BRI Layer 1, 2, 3 status, SPID(s), and TEI

Frame Relay Configuration

Frame Relay (FR) connections are configured on synchronous V.35 serial or integrated DSC/CSU serial controller interfaces. Frame relay is commonly provisioned over a T1 digital transport circuit, operating at the primary rate of 1.544Mbps. The actual transport link is either a permanent or switched virtual circuit (PVC/SVC), with the former being the most common provisioning implementation.

Because FR PVC is still the most common implementation, we will focus our examples on FR PVC configuration. Configuring a FR PVC has three basic steps:

1. The interface must be enabled to use FR encapsulation, using the interface configuration <encapsulation [type]> subcommand:

   ```
   jupiter(config)# interface serial0
   jupiter(config-if)#encapsulation frame-relay
   ```

2. After encapsulation is set, the Data-Link Connection Identifier (DLCI) and Local Management Interface (LMI) type need to be configured. The DCLI is set using the FR interface configuration subcommand `<frame-relay interface-dcli [dcli]>`. The LMI type can be configured dynamically (on IOS versions 11.2 and later) by the router when the interface is activated or the LMI can be set explicitly in the routers FR interface configuration. The default LMI is Cisco. IOS supports three LMI types: ansi (most often used), cisco, and q933a. Let's look at Jupiter's DCLI and LMI configuration:

```
jupiter(config-if)#frame-relay interface-dcli 35
jupiter(config-if)#lim-type ansi
```

3. The last step is the configuration of the Layer 3 protocol address assignment. FR supports four different network topologies: point-to-point, meshed and partially meshed point-to-point, and multipoint. We will look at meshed and multipoint topologies in a moment. If you are only running IP over a FR point-to-point connection, it is possible to configure the interface to use either unnumbered IP or static IP address assignments. With unnumbered IP, one router appears as a DTE and the other as a DCE. You should recall that the IP address of one of the router's IP configured Ethernet interfaces is used to exchange IP datagrams across the link. Alternatively, if you are using the point-to-point FR connection for multiprotocol transport, it is best to assign Layer 3 addresses for all the protocols. Here is an example of a point-to-point interface using AppleTalk and IP and static Layer 3 to DCLI mappings:

```
jupiter(config-if)#ip address 172.16.33.1 255.255.255.252
jupiter(config-if)#appletalk cable-range 6678-6678 6678.1
jupiter(config-if)#appletalk zone WAN
jupiter(config-if)#frame-relay map ip 172.14.33.2 892 broadcast
jupiter(config-if)#frame-relay map appletalk 6678.2 892 broadcast
```

The `<broadcast>` option indicates that network broadcasts should be forwarded over the interface.

Note

Although FR at the primary rate, DSO, is by far the most common FR implementation, recently carriers have started to offer FR PVCs and SVCs rates beyond the primary rate, up to DS3, showing that PVCs and SVCs have finally come into their own.

Note

Remember that the DLCI can be either locally significant or assigned by the FR provider. The DLCI address is used for PVC identification, which is mapped to a Layer 3 protocol address. The function is normally performed dynamically by default. It is possible to create static Layer 3 to DCLI translation maps on the interface using `<frame-relay map [protocol type] [remote PVC interface protocol address] [dcli] [options]>`.

Frame Relay Topologies

The underlying idea behind FR is that transport paths and their required bandwidth are not dedicated—which is typically the case with data links—but instead are virtual transport paths (using a shared bandwidth pool) established between network termination points on the FR switch network.

When compared to dedicated bandwidth (i.e., point to point T1 carrier circuit) circuits, Frame Relay has its shortcomings, such as higher transport latency, a greater potential for transport-related packet loss and retransmissions, and the potential side effects from over-subscription of the cloud's available transmission bandwidth. If the FR provider's network is implemented (that is, not oversubscribed) and managed correctly, however, these shortcomings are easily overshadowed by the cost-effective and bandwidth-efficient topology options FR's "virtual circuit" implementation model provides.

Because packet and cell-switched circuit terminations are "virtual links" between two termination points, established across a shared transport, the possibility exists to establish multiple virtual circuit (VC) terminations on a given point (that is, a single router interface). This multipoint capability makes FR (and other packet and cell-switched technologies) very attractive to enterprise network designers. It allows construction of point-to-point and point-to-multipoint partially and fully meshed WANs with a high degree of flexibility. Additionally, these topologies can be implemented at carrier and hardware costs that are substantially less than what would be incurred if a similar topology was implemented using dedicated circuits.

Figure 9.4 illustrates the possible multi-VC topologies that are supportable using FR transport: point-to-multipoint (non-mesh), point-to-point (full mesh), and point-to-point (partial mesh).

To support multi-PVC meshed and partially meshed FR implementations, IOS uses serial sub-interfaces to provide the Layer 3 and DCLI access points needed to establish the "links" between the different remote end-point routers.

> **Note**
>
> Due to the "virtual" nature of packet-based transport networks (FR, X.25, and ATM), the network is often referred to as a cloud.

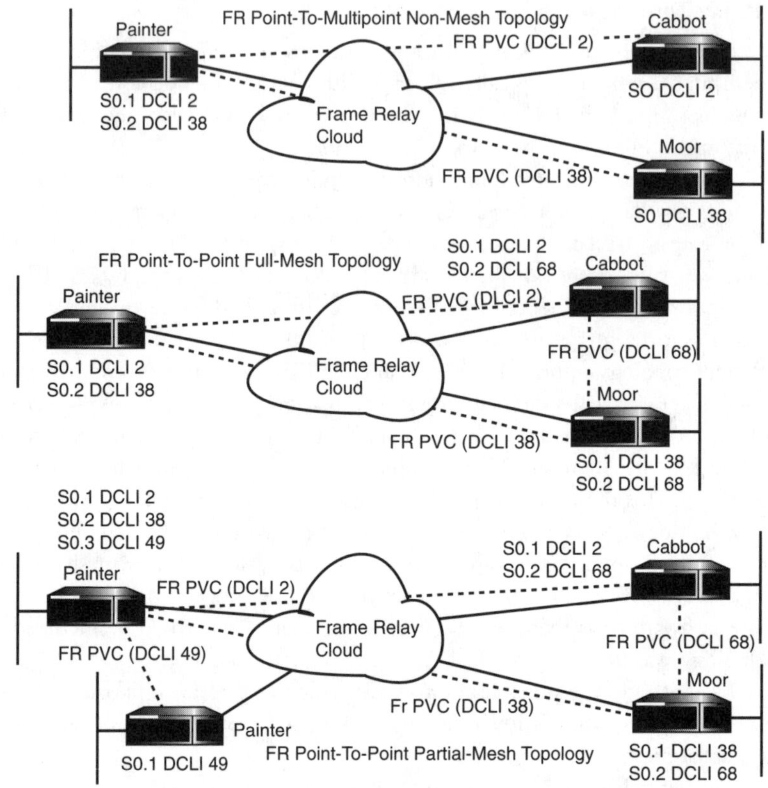

Figure 9.4 Point-to-point and point-to-multipoint FR topologies.

Before you begin to establish serial sub-interfaces, you need to configure the primary serial interface's encapsulation and LMI information. After the primary serial interface is configured, serial sub-interfaces can be established using the interface configuration subcommand `<interface serial [physical slot/port sub-interface number [multipoint|point-to-point]>`. After the sub-interface is established, the Layer 3 address information needs to be configured and the sub-interfaces DCLI needs to be configured using the interface configuration sub-command `<frame-relay interface dcli [dcli]>`. Here is a configuration example from a fully meshed FR topology supporting IP and AppleTalk:

```
vulcan(config)#interface s0
vulcan(config-if)#encapsulation frame-relay
vulcan(config-if)#lim-type ansi
vulcan(config-if)#exit
vulcan(config)#interface s0.34
vulcan(config-subif)#ip address 172.16.240.1 255.255.255.252
```

```
vulcan(config-subif)#frame-relay interface-dcli 34
vulcan(config-subif)#appletalk cable-range 89-89
vulcan(config-subif)#appletalk zone cloud
vulcan(config-subif)#frame-relay map appletalk 34 broadcast
vulcan(config-subif)#exit
vulcan(config)#interface s0.28
vulcan(config-subif)#ip address 172.16.240.4 255.255.255.252
vulcan(config-subif)#frame-relay interface-dcli 28
vulcan(config-subif)#appletalk cable-range 899-899
vulcan(config-subif)#appletalk zone cloud
vulcan(config-subif)#frame-relay map appletalk 28 broadcast
vulcan(config-subif)#exit
```

Note

Using FR also places network bandwidth availability, performance, and network reliability (and a good portion of infrastructure costs) in the hands of the carrier or FR provider. This can either be a blessing or a curse depending on your service contract, FR provider, and actual availability and performance requirements. Caveat emptor.

Note

Sub-interfaces are available on LAN interfaces as well. They are commonly used in routing enterprise network backbone traffic where tagged VLANs (802.1Q or Cisco's proprietary Inter-Switch Link [ISL]) are used to carry virtual network segments across a common transport media segment. Instead of using a single router interface to feed each VLAN, a single high-speed interface can be used with multiple sub-interfaces. To configure the sub-interface on LAN interfaces, use the global configuration command `<interface slot/port [sub-interface number]>`. Here is the Fast Ethernet configuration example:

```
leibniz(config)#interface fa0/0.1
leibniz(config-subif)#ip address 172.16.48.1 255.255.255.0
leibniz(config-subif)#exit
leibniz(config)#interface fa0/0.2
leibniz(config-subif)#ip address 172.16.49.1 255.255.255.0
leibniz(config-subif)#exit
leibniz(config)#interface fa0/0.3
leibniz(config-subif)#ip address 172.16.100.1 255.255.255.0
leibniz(config-subif)#exit
leibniz(config)#interface fa0/0.4
leibniz(config-subif)#ip address 172.16.32.1 255.255.255.0
leibniz(config-subif)#exit
```

With certain FR topology implementations utilizing sub-interfaces with FR, along with the Layer 3 addressing and DCLI information, some additional interface configuration options might be required. The first of which might be the configuration of an FR map. FR maps are needed if inverse ARP (Layer 3 address to DCLI mapping, which is enabled by default) is disabled or not supported by the Layer 3 protocol being implemented. In these cases, the mapping needs to be done in order for the network traffic to be transmited over the correct VC. In the previous configuration example, FR maps are used to associate different AppleTalk networks with specific sub-interfaces.

The second option allows you to specify whether or not the interface will function as a FR point-to-point or FR multipoint interface (there is no default). The point-to-point designation is used when a sub-interface has a single DCLI. Multipoint designation is used when multiple DCLIs are associated with the sub-interface, or if the FR provider is providing multipoint grouping.

By using the "multipoint" designation, you are instructing the router that all the associated DCLIs are allowed to exchange frames (using inverse ARP or frame-relay maps to handle the DLCI to Layer 3 address mapping). This distinction is what makes it possible for one router to function as a multipoint hub in point-to-multipoint topology implementations where the multipoint router has point-to-point PVCs to two or more endpoints. These single endpoints then exchange packets with each other through the multipoint router. However, although it is possible to exchange packets between the DCLIs, by default only unicast packets are sent. FR does not natively support broadcast and multicast transmissions. Therefore, in order for the messages to be properly relayed, the multipoint router must send a unicast copy of these packets to each endpoint. To enable broadcast simulation, add the `[broadcast]` keyword to the end of the `<frame-relay interface dcli>` or `<frame-relay map>` interface configuration command.

> **Note**
>
> With multipoint grouping, a collection of FR endpoints function as if they are connected over a local shared transport. The FR provider provisions the multipoint group (up to 32 nodes) by associating all the endpoint members' local DCLIs to a multipoint group. Each endpoint is assigned an address reflecting a common network address (the same IP subnet or equivalent) with the DCLI functioning in effect as a MAC address. From a conceptual perspective, the endpoints function as DTE peers connected to a common virtual DCE.

> **Note**
>
> Not all Layer 3 protocols can operate over a point-to-multipoint topology. The reason for this is that not all protocols will permit datagrams to be relayed out of the same interface from which they are received. In addition, the multipoint sub-interface designation also enables "split-horizon" (disabled by IOS when frame-relay encapsulation is selected) on any IP routing protocol.

Point-to-Point VC Topology

Typically, multi-VC FR implementations (partial and full mesh) are configured using partially or fully meshed point-to-point sub-interfaces. Although quite effective, they are costly in comparison to multipoint implementations because each link requires its own PVC. In addition to cost, the actual configuration is fairly complex, requiring that some real thought be paid to addressing the various point-to-point PVC sub-interfaces. Good documentation is a big help when debugging problems (so be sure to write some). Figure 9.5 illustrates a fully meshed FR topology using point-to-point PVC. Following the illustration are interface configuration examples from two of the routers (batman and robin).

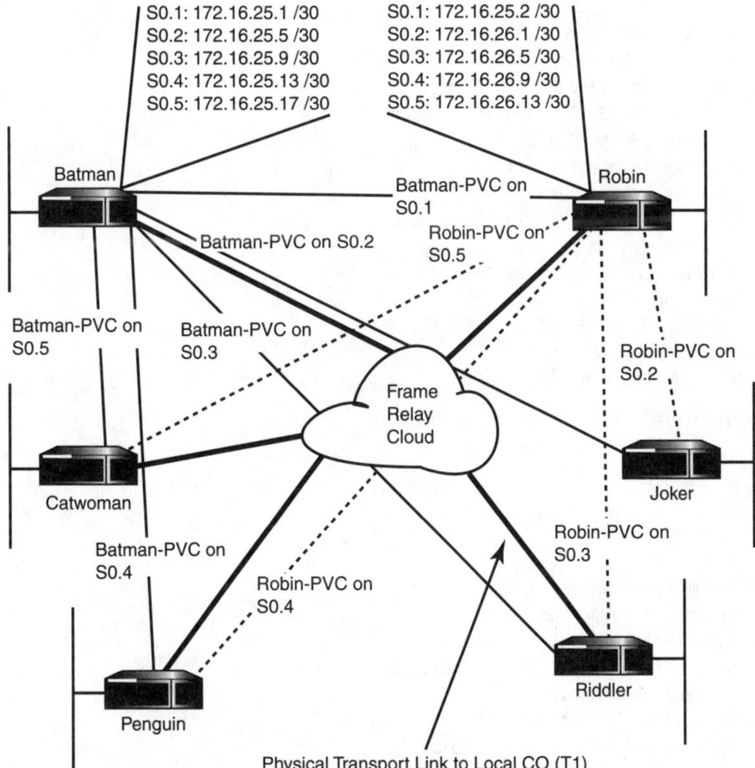

Figure 9.5 A full-mesh FR point-to-point PVC topology.

Here is the interface configuration from the router batman:

```
!
hostname batman
!
interface serial0/0
 no ip address
 encapsulation frame-relay
frame-relay lmi-type ansi
interface serial0.1 point-to-point
 ip address 172.16.25.1 255.255.255.252
 description PVC to robin
 frame-relay interface dcli 45
interface serial0.2 point-to-point
 ip address 172.16.25.5 255.255.255.252
 description PVC to Joker
 frame-relay interface dcli 89
interface serial0.3 point-to-point
 ip address 172.16.25.9 255.255.255.252
 description PVC to Riddler
 frame-relay interface dcli 90
interface serial0.4 point-to-point
 ip address 172.16.25.13 255.255.255.252
 description PVC to penguin
 frame-relay interface dcli 47
interface serial0.5 point-to-point
 ip address 172.16.25.17 255.255.255.252
 description PVC to Catwoman
 frame-relay interface dcli 89
```

Here is the router interface configuration from robin:

```
!
hostname robin
!
interface serial0/0
 no ip address
 encapsulation frame-relay
frame-relay lmi-type ansi
interface serial0.1 point-to-point
 ip address 172.16.25.2 255.255.255.252
 description PVC to batman
 frame-relay interface dcli 45
interface serial0.2 point-to-point
 ip address 172.16.26.1 255.255.255.252
 description PVC to Joker
 frame-relay interface dcli 67
interface serial0.3 point-to-point
 ip address 172.16.26.5 255.255.255.252
 description PVC to Riddler
 frame-relay interface dcli 89
interface serial0.4 point-to-point
 ip address 172.16.26.9 255.255.255.252
```

```
  description PVC to penguin
  frame-relay interface dcli 53
 interface serial0.5 point-to-point
  ip address 172.16.26.13 255.255.255.252
  description PVC to Catwoman
  frame-relay interface dcli 78
```

Multipoint VC Topology

Multipoint VC implementations (both grouped and point-to-multipoint), although more cost effective, are just not as common as point-to-point. This is partly because not all FR providers provide multipoint service, because it requires a degree of configuration and management involvement beyond the standard PVC service model. It is also because point-to-point requires far more involvement with the FR provider than most network engineers are comfortable with (this is even more so when the endpoints involved are international). Figure 9.6 illustrates a FR multipoint group implementation based on the topology used in the previous point-to-point example. The illustration is followed by the interface configurations from the routers batman and robin.

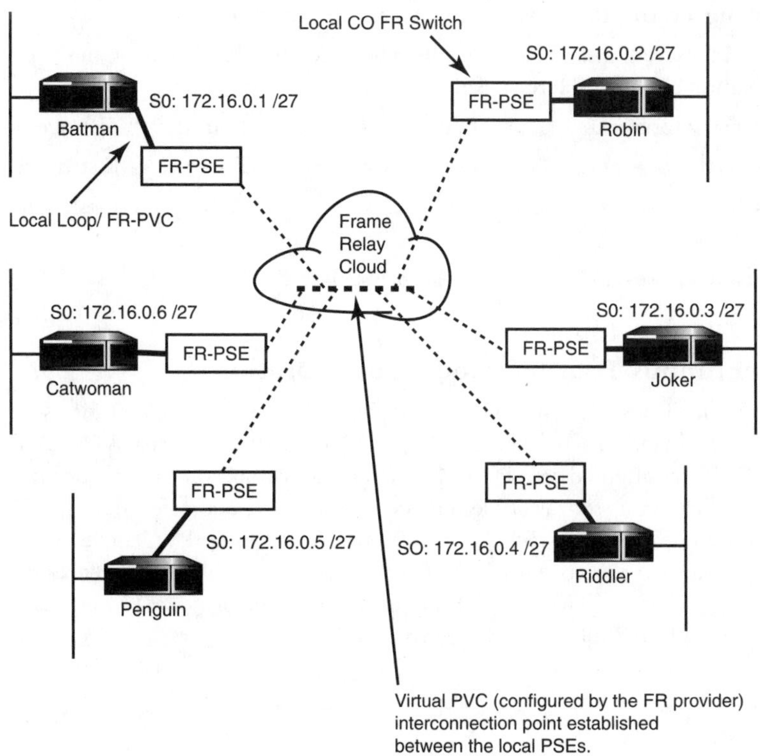

Figure 9.6 A multipoint FR example.

Here is the multipoint interface configuration from the router batman:

```
hostname batman
!
interface serial0
encapsulation frame-relay
frame-relay lmi-type ansi
interface serial 0.1 multipoint
ip address 172.16.44.1 255.255.255.248
frame-relay interface-dlci 99 broadcast
```

Here is the multipoint interface configuration from the router robin:

```
!
hostname robin
!
interface serial0
encapsulation frame-relay
frame-relay lmi-type ansi
interface serial 0.1 multipoint
ip address 172.16.44.2 255.255.255.248
frame-relay interface-dlci 34 broadcast
```

Displaying Frame Relay Configuration Information

IOS provides several user EXEC commands to monitor FR behavior and configuration elements. Here are some of the more essential ones.

- `<show frame-relay pvc>`—Displays information about the PVC usage statistics
- `<show frame-relay lmi>`—Displays information about the LMI statistics
- `<show frame-relay map>`—Displays Layer 3 to DCLI address translation map entries
- `<show frame-relay traffic>`—Displays FR traffic statistics

Asynchronous Transfer Mode Configuration

Asynchronous Transfer Mode (ATM) is implemented in two different forms on Cisco routers. One form is a synchronous serial interface with an external ATM data service unit (ADSU), configured to use ATM-DXI encapsulation. Another is a dedicated ATM interface processor (available on 4xxx, 36xx, and 7xxx series routers). ATM, like other cell/frame switching technologies, is provisioned using PVC and SVC. SVCs are only supported on routers with ATM interface processors. Slight differences exist between configuring ATM on synchronous serial and dedicated ATM processors, and for this reason both will be covered separately.

ATM-over-Serial Configuration

ATM configuration on synchronous serial is performed in the following manner:

1. The protocol addresses should be assigned, and the interface should be enabled.
2. ATM encapsulation must be enabled on the serial interface, using the interface configuration subcommand <encapsulation atb-dxi>:

```
persephone(config)# interface serial0/0
persephone(config-if)#ip address 172.16.30.1 255.255.255.252
persephone(config-if)#ipx network 89990
persephone(config-if)#no shutdown
persephone(config-if)#encapsulation atm-dxi
```

3. When enabled, the PVC's virtual channel identifier (VCI), virtual path identifier, and ATM encapsulation type are configured using <dxi pvc [vpi] [vci] [snap|nlpid|mux]>. Cisco's ATM implementation will support Layer 3 network protocol transport using a distinct PVC for each network protocol with <multi-plex> ATM encapsulation. Alternatively, all the Layer 3 protocols can be multiplexed across a single PVC using <snap> (the default for IOS 10.3 and later) encapsulation. Remember, the VCI and VPIs are only used locally, but do need to be distinct for each PVC.
4. Create Layer 3 protocol to ATM VCI/VPI maps. The maps associate a VCI/VPI with the remote Layer 3 address at the far end of the link. Protocol maps are created with the interface configuration subcommand <dxi map [appletalk|ip|ipx] [layer 3 remote address] [vpi] [vci] [broadcast]>. Layer 3 to VCI/VPI maps are required with all ATM PVC configurations.

Now let's finish up the rest of our ATM-over-serial configuration:

```
persephone(config-if)#dxi pvc 0 2 mux
persephone(config-if)#dxi pvc 1 3 mux
persephone(config-if)#dxi map ip 172.16.30.2 0 2 broadcast
persephone(config-if)#dxi map ipx 89990.cc45.6783.345f 1 3 broadcast
```

ATM-over-Interface Processors

When configuring ATM over traditional serial interfaces, the PVC is configured on the ADSU, so only the VPI and VCI need to be created. On Cisco routers with ATM processors, PVCs need to be established between the ATM switch and the router.

To create an ATM PVC, use the interface processor subcommand <atm pvc [vcd] [vpi] [vci] [aal-encapsulation] [peak average burst]>. The [vcd], or virtual circuit descriptor, is the PVC ID used by the ATM switch and router to distinguish different PVCs. The [aal-encapsulation] command designates the ATM adaptation layer and encapsulation used with the PVC. The encapsulation choices are the same ones available with serial configuration. You should recall there are three ATM adaptation layers: AAL1, AAL3/4, and AAL5, the last of which is the best choice for most LAN/WAN environments.

Let's look at a basic PVC configuration:

```
orion(config)#interface atm0/0
orion(config-if)#172.16.56.2 255.255.255.252
orion(config-if)#atm pvc 2 2 31 all5mux
```

When the interface is configured, it is also necessary to create Layer 3 to VCI/VPI mappings. However, with ATM interface processors, instead of creating and applying individual map entries as we would if this were an ATM serial interface, Layer 3 to VCI/VPI map entries are made into a map-list.

Map-lists are created in the same fashion as ACLs: one entry per line. In the following example, we will finish configuring the ATM interface on orion, by creating and applying the ATM map-group. A map-list is created in global configuration mode using `<map-list [map list name]>`. After the list is created, entries are added using the map-list subcommand `<[protocol] [protocol address] [atm-vc|atm-nasp] [VCD|NASP address] [broadcast]>`:

```
orion(config)#map-list jupter
orion(config)(config-map-list)# ip 172.16.56.1 atm vc 2 broadcast
orion(config)(config-map-list)#exit
orion(config)#interface atm0/0
orion(config-if)#map-group jupter
orion(config-if)#exit
orion(config)#
```

After the map-list is created, it is applied to the interface—oddly enough, as a map-group—with the interface processor subcommand `<map-group [map-list name]>`.

Cisco supports ATM SVCs in fully meshed (point-to-point) and multipoint topologies. The idea behind SVCs is to provide a "bandwidth on demand" service model. SVCs are commonly used in large-scale classical IP-over-ATM (CIP) scenarios. SVCs, despite their signaling configuration requirements, are easier to manage in large-scale deployments. Figure 9.7 illustrates this complexity.

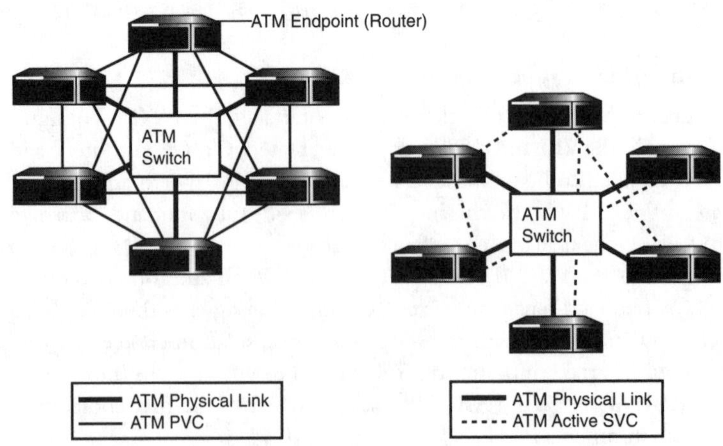

Figure 9.7 ATM PVC topology versus ATM SVC topology.

CIP with PVCs requires individual virtual circuits to be established between every router involved. SVCs are only established when needed, as opposed to the PVC scenario, where the virtual circuit is always active until removed from the interface processor. In an SVC topology (and PVC), the routers involved relate to one another as if all were connected to the same local LAN segment. The routers establish SVC connections using the Q.2931 signaling protocol between themselves and their local ATM switch. It is the local ATM switch, however, that is actually responsible for all SVC call setups and teardowns. To facilitate this, SVC calls are established over signaling PVC that is established between ATM end-point device and its adjacent ATM switch. SVC's between ATM connected end-points are established using network service access point (NASP) addresses. Since, SVC calls need to be established when the end-point has layer 3 traffic to transmit to another end-point, some type of NASP to Layer 3 address mapping needs to be provided, this mapping is accomplished using map-lists. Each SVC router uses a map-list containing Layer 3 to NASP addressing mappings for its brother routers. These addresses are sent to the local ATM switch, which in turn establishes the SVC (with a remote ATM switch) and manages the call level routing, although the router forwards packets through the established SVC.

In order to use SVC, follow these steps:

1. Establish the signaling PVC between the router and the local ATM switch using the interface processor configuration subcommand `<atm pvc [vcd] [vpi] [vci] qsall>`.
2. Configure the local NSAP address. The NSAP address is a 40-digit hexadecimal string. It is entered using the following interface processor subcommand `<atm nsap-address xx.xxxx.xx.xxxxxx.xxxx.xxxx.xxxx.xxxx.xxxx.xxxx.xx>`.
3. Configure the map-list containing the remote Layer 3 to NSAP addressing mappings.

Here is an example of an SVC router configuration:

```
interface atm 2/0
ip address 172.16.40.1 255.255.255.248
map-group mars
atm nsap-address ff.ffx0.00.ffx679.ffff.ffff.6709.6795.fff6.5567.34
atm pvc 5 5 5 qsall
!
!
map-list mars
ip 172.16.40.2 atm-nsap ff.ffx0.00.ffx679.ffff.ffff.6709.6795.8891.35
ip 172.16.40.3 atm-nsap ff.ffx0.00.ffx679.ffff.ffff.6709.6795.8591.36
ip 172.16.40.4 atm-nsap ff.ffx0.00.ffx679.ffff.ffff.6709.6795.3821.37
ip 172.16.40.5 atm-nsap ff.ffx0.00.ffx679.ffff.ffff.6709.6795.7781.38
```

In the previous example, the map-list "mars" is used to provide a remote network address to NSAP mappings for all its related SVC neighbors.

Classical IP over ATM

ATM Address Resolution Protocol (AARP) is provided as part of the IETF's CIP scheme to provide "classic" ARP services for ATM NASP-to-IP address translation. Therefore, in SVC and PVC ATM IP network environments, it is possible to use ATM ARP as an alternative to map-lists. When deploying ATM ARP, one router is configured as an AARP server and the others are configured as AARP clients. The server builds its NASP-to-IP table by sending inverse ARP requests to the connected clients to learn their IP and NASP addresses.

To configure an ATM processor to function as an AARP server, the following interface configuration subcommands are used:

```
persephone(config)#interface atm 3/0
persephone(config-if)#ip address 172.16.100.2 255.255.255.0
persephone(config-if)#atm nsap-address
ff.ffx0.00.ffx679.ffff.8901.6709.6795.ff66.5467.67
persephone(config-if)# atm arp-server nasp self
persephone(config-if)#atm arp-server time-out 15
```

The <atm arp-server nasp [nasp address]> command is used to configure both ARP clients and servers. When the router is the ARP server, the <self> option is used in place of the NASP address. The <arp-server time-out [min]> command specifies how long an entry can remain in the AARP table before the entry is deleted. On the client side only, the <atm arp-server nasp [nasp address]> command is required to enable AARP:

```
tisiphone(config)#interface atm 3/0
tisiphone(config-if)#ip address 172.16.100.15 255.255.255.0
tisiphone(config-if)#atm nsap-address
ff.9901.00.ffx679.ffff.8901.6709.6725.5568.5467.84
tisiphone(config-if)# atm arp-server nasp
ff.ffx0.00.ffx679.ffff.8901.6709.6795.ff66.5467.67
tisiphone(config-if)#
```

ATM LANE is also supported on ATM interface processors. Routers can be configured as both LANE servers and clients. Due to LANE's inherent complexity, LANE server and client configuration is beyond the scope of this book. Please refer to the Cisco documentation for LANE configuration information.

Displaying ATM Configuration Information

Here is a brief list of ATM status commands. Like all <show>-based commands, they can be retrieved using <show atm|dxi ?> from the user EXEC shell.

- <show dxi map>—Displays information about the DXI-maps configured on the serial ATM interface
- <show dxi pvc>—Displays information about the serial DXI PVC
- <show atm map>—Displays the list of ATM static maps to remote Layer 3 addresses

- <show atm traffic>—Displays global inbound and outbound ATM traffic information
- <show atm vc>—Displays information about all the PVCs and SVCs on the router
- <show atm interface atm [slot/port]>—Displays ATM interface processor information

Packet over SONET

A Packet over SONET (POS) interface is available for Cisco 7x00 series routers. POS interface is a single (SONET stub) OC-3155Mbps single-mode fiber SONET-compatible interface. The POS interface configuration process is similar to the one followed when configuring other integrated WAN interface controllers.

In addition to the Layer 3 address configuration, the interface's framing type and clocking source must be set. An MTU size adjustment option is also available. The POS interface is configured by using the global configuration command <interface posi [slot|port]>. After the interface processor is selected, the remaining configuration commands are POS interface specific subcommands (aside from Layer 3 addressing). The POS interface's default settings are SDH framing, remote clock, and the default OC-3 interface MTU, 4,470 bytes. Therefore, you may only need to perform Layer 3 address configuration:

```
thor(config)#interface pos 1/0
thor(config-if)#ip address 172.16.55.1 255.255.255.252
```

To configure the POS interface controller to transmit clock information, use the command <config-if>posi internal-clock.

To use STM-1 framing, use the command <config-if>posi framing-sdh.

To adjust the MTU size to reflect the LAN packet media (which is 1,500 in most cases), use <config-if>mtu 1500.

Summary

The goal of this chapter was to increase your Cisco IOS configuration knowledge on subjects directly applicable (but not exclusive) to enterprise networking concerns. You should now have a familiarity with the following subjects:

- The creation and application of access control lists
- Policy route creation with route-maps
- Router redundancy protocols, and the configuration of HSRP
- Implementing network address translation
- Cisco tunneling and GRE VPNs
- Layer 2 WAN protocol and WAN interface configuration

In Chapter 10, we explore implementing static and dynamic TCP/IP routing protocols on Cisco routers. Implementing an IP routing scheme can be a complex task. Chapter 10's focus is to review the various IP routing options supported by the IOS, and to provide a Cisco IOS command overview and give some common implementation examples.

Related RFCs

RFC 1170	Public Key Standards and Licenses
RFC 1332	The PPP Internet Protocol Control Protocol (IPCP)
RFC 1483	Multiprotocol Encapsulation over ATM Adaptation Layer 5
RFC 1490	Multiprotocol Interconnect over Frame Relay
RFC 1577	Classical IP and ARP over ATM
RFC 1618	PPP over ISDN
RFC 1631	The IP Network Address Translator (NAT)
RFC 1700	Assigned TCP/IP Port Numbers
RFC 1701	Generic Routing Encapsulation (GRE)
RFC 1702	Generic Routing Encapsulation over IPv4 networks
RFC 1827	IP Encapsulating Security Payload (ESP)
RFC 1829	The ESP DES-CBC Transform
RFC 1853	IP in IP Tunneling
RFC 1919	Classical versus Transparent IP Proxies
RFC 1928	SOCKS Protocol Version 5
RFC 1932	IP over ATM: A Framework Document
RFC 1948	Defending Against Sequence Number Attacks
RFC 1968	The PPP Encryption Control Protocol
RFC 1994	PPP Challenge Handshake Authentication Protocol (CHAP)
RFC 2171	MAPOS—Multiple Access Protocol over SONET/SDH Version 1
RFC 2172	MAPOS Version 1 Assigned Numbers
RFC 2176	IPv4 over MAPOS Version 1
RFC 2226	IP Broadcast over ATM Networks
RFC 2338	Virtual Router Redundancy Protocol

Additional Resources

Chappel, Laura, Editor. *Advanced Cisco Router Configuration*. Cisco Press, 1999.

Cisco Systems, Inc. *Cisco IOS Dial Solutions*. Cisco Press, 1998.

———. *Cisco IOS Network Security*. Cisco Press, 1998.

———. *Cisco IOS WAN Solutions*. Cisco Press, 1998.

Garfinkel, Simson, and Gene Spafford. *Practical UNIX and Internet Security*. O'Reilly & Associates, 1996.

Schneier, Bruce. *Applied Cryptography,* Second Edition. John Wiley & Sons, 1996.

Scott, Charlie, and Paul Wolfe and Mike Erwin. *Virtual Private Networks*. O'Reilly & Associates, 1998.

10

Configuring IP Routing Protocols on Cisco Routers

THIS CHAPTER DISCUSSES THE IMPLEMENTATION OF STATIC AND dynamic IP routing on Cisco routers in an enterprise network environment. To start, we evaluate route distribution methods and explain the general configuration elements that apply to configuring dynamic routing protocols on Cisco routers. Then, we review basic steps for configuring and monitoring IP static routing, RIP, IGRP, EIGRP, AURP, and OSPF. For easy reference, each section starts with a command summary list and utilizes (whenever possible) the same core network topology (testnet, our example network) illustrated in Figure 10.1. This chapter concludes with the implementation of BGPv4 and an overview of IP route redistribution.

The best way to learn something is to do it. Therefore, as you did in Chapter 7, "Introduction to Cisco Routers," you should recreate the configuration examples using a single router or pair of routers. The section examples revolve around the `asbr-a1/2` and `asbr-b1/2`. Loopback interfaces can be used to simulate additional network interfaces. To create a loopback interface, from global configuration mode, type the following:

```
Router#config t
Router(config)#interface loopback 1
Router(config-if)#
```

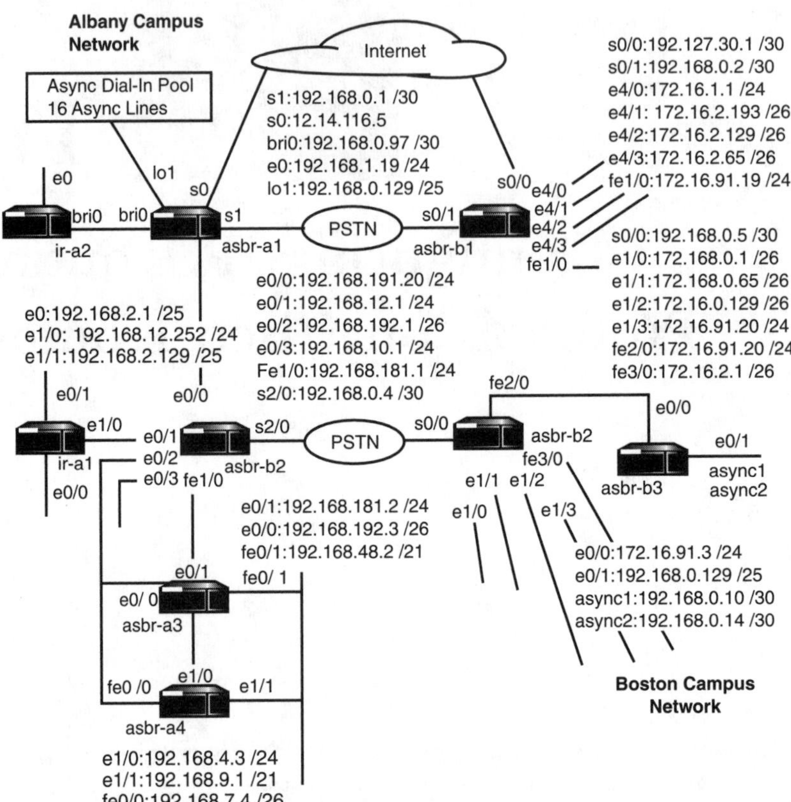

Figure 10.1 The testnet network map.

The Boston network will use networks with the classful root of 172.17.0.0, and Albany will use 192.168.0.0 for its IP address space. Classless/VLSM addressing will be used with the protocols that can support it.

Choosing the Right Protocol

The task of enabling IP route advertisement varies in complexity, depending on the operational scope and size of the network. Static routing, which is byfar the easiest and most problem-free method, is tedious to manage in large networks and provides no recovery facility when link failures occur. Alternatively, dynamic routing protocols address the shortcomings of static routing, but do, however, come with their own operational concerns.

Unfortunately, when selecting an IP announcement methodology, there is no single correct answer. There is, however, a wrong answer, which is to select a method that will not be able to meet the networks operational requirements or scale appropriately to enable future growth. The way to avoid this problem is to examine the network's topology, availability, and performance requirements through a series of questions:

- What kind of network is this (ISP backbone/multipolicy, single GW LAN/WAN, multipoint LAN/WAN, enterprise backbone/single-policy, and so on)?
- What is the network diameter (more or fewer than 15 hops, how many routers, and so on)?
- Is CIDR/VLSM addressing support required?
- Does the network use redundant or multiple paths between network segments?
- What type of equipment will be used to route traffic on the network, and is a standards-based routing protocol required?
- What are the performance requirements of the routers in the network? For example, is convergence time a factor?

After you have a list of requirements, you can review your available options and determine the solution appropriate for your networking environment. Table 10.1 provides a brief feature comparison of the popular IP routing options supported on the Cisco IOS.

Table 10.1 **Comparison of Routing Protocol Features**

Protocol Feature	**RIP v1/v2**	**IGRP**	**EIGRP**	**OSPF**	**Static**	**BGP**
Supports classful addressing	yes	yes	yes	yes	yes	yes
Interior Gateway Protocol	yes	yes	yes	yes	no	no
Exterior Gateway Protocol	no	no	no	no	no	yes
Supports classless addressing	yes (V2)	no	yes	yes	yes	yes
Supports load sharing	no	yes	yes	yes	no	yes
Supports authentication	yes (v2)	no	yes	yes	no	yes
Easy implementation	yes	yes	yes	no	no	no
Routing algorithm	DV	DV	DUAL	LS	none	DV
Supports weighted metrics	no	yes	yes	yes	no	no
Fast convergence	no	yes	yes	yes	yes	yes
Uses broadcasts for route updates	yes (v1)	yes	no	no	no	no
Uses multicast for routing updates	yes (v2)	no	yes	yes	no	no
Supports large network diameters	no	no	yes	yes	yes	yes

DV=Distance Vector; LS=Link State; DUAL=Diffusing Update Algorithm

Route Selection

The router will use all the available sources of reachability information to construct the most accurate and efficient routing table possible, based on the available information. Each information source is assigned an administrative distance, which is used to determine a route's integrity. Table 10.2 lists the IOS's default administrative distance values for its supported IP routing protocols.

Table 10.2 **IOS Administrative Distances**

Protocol	**Distance**
Connected interface	0
Static route	1
EIGRP summary route	5
BGP (external)	20
EIGRP (internal)	90
IGRP	100
OSPF	110
RIP	120
EIGRP (external)	170
BGP (internal)	200
Unknown	255

The lower the administrative distance value, the more trusted the route. So, you can see in situations where multiple routing protocols are being used for route advertisement, the router will prefer information provided from certain protocols over others.

One of the big advantages of employing this advertisement hierarchy is that it gives you another way to manage traffic flow in dynamic routing. By using static routes (which have a lower administrative distance than any dynamic protocol), you can overrule dynamic announcements in multipath networks to specify the route path to specific hosts. By the same token, it is possible to set a static route to use a higher administrative distance so a dynamic route is preferred and the static route is only used in the event that dynamic route announcement is lost.

Displaying General Routing Information

There are several IOS commands used for controlling and displaying information about IP routing.

Display commands:

```
<show ip route>
<show ip route connected>
<show ip route [address/hostname]>
<show arp>
```

```
<show ip protocol>
<show ip masks>
<show ip masks [network address]>
<traceroute>
```

Control commands:

```
<clear ip route *>
<clear ip route [network] [mask]>
<clear arp>
```

Displaying IP Network Information

In Chapter 7, "Introduction to Cisco Routers," the <show ip route> user EXEC command was introduced. Now, let's take a closer look at the IOS's route table display:

```
BBR-172#sh ip route
Codes: C - connected, S - static, I - IGRP, R - RIP, M - mobile, B - BGP
       D - EIGRP, EX - EIGRP external, O - OSPF, IA - OSPF inter area
       N1 - OSPF NSSA external type 1, N2 - OSPF NSSA external type 2
       E1 - OSPF external type 1, E2 - OSPF external type 2, E - EGP
       i - IS-IS, L1 - IS-IS level-1, L2 - IS-IS level-2, * - candidate
default
       U - per-user static route, o - ODR
       T - traffic engineered route

Gateway of last resort is 172.16.191.220 to network 0.0.0.0

     172.16.0.0/16 is variably subnetted, 17 subnets, 6 masks
S        172.16.80.0/24 [1/0] via 172.16.191.220
B        192.168.91.0/24 [20/0] via 192.168.0.2, 6d14h
O        172.16.48.0/21 [1/0] via 172.16.191.140
O        172.16.161.128/25 [1/0] via 172.16.191.253
C        192.168.0.1/29 is directly connected, Serial0/2
C        172.16.12.0/24 is directly connected, Ethernet3/1
O IA     172.16.160.96/30
           [110/1563] via 172.16.191.220, 4d21h, FastEthernet2/0
C        172.16.192.0/21 is directly connected, Fddi6/0
B        192.168.176.0/21 [20/0] via 192.168.85.75, 6d14h
C        172.16.181.0/24 is directly connected, Ethernet3/0
0        172.16.186.0/24 [1/0] via 147.225.12.250
0        172.16.185.0/24 [1/0] via 147.225.12.250
C        172.16.191.0/24 is directly connected, FastEthernet2/0
S*    0.0.0.0/0 [1/0] via 147.225.191.220
```

The <show ip route> command output displays all the known network routes and the following information (from left to right):

- The source of the route. The legend for all available protocols is displayed at the top of the route table.
- The network and netmask. The netmask can be displayed using bitcount (shown), decimal (default), or hexadecimal. To set the netmask display format, use the <ip netmask-format [format]> command. The command can be used for a temporary <terminal> session <terminal ip netmask-format [format]> or used to permanently set the mask format display in configuration EXEC line mode command:

```
router# config t
Enter configuration commands, one per line.  End with CNTL/Z.
router(config)# int vty 0 4
router(config-line)#ip netmask-format bit-count
Router(config-line)#^Z
Router#
```

- The route's administrative distance and routing metric.
- The network's next hop gateway. If the route is learned dynamically, the route's age or last update time and the interface that the route was learned on may also be listed.
- The gateway of last resort, if one has been set.

Keep in mind that the routing table displays the best routes, not necessarily all available routes. Only the route with the best administrative distance/metric will appear in the table. In cases where routes of equal cost to the same network are available, both will be listed. Routes to directly connected networks will not appear for disabled interfaces, and neither will static routes that are dependent on the disabled interface. All dynamic routing protocols also require that at least one active interface exists and that its address corresponds to a network being announced by the protocol. Although it would seem odd to announce a route for a network that is not active, it is quite common for dial-on-demand and dial access-servers to use network address space on interfaces that are not available all the time. Static routes are usually used to provide reachability information in these situations. An alternative is to use a loopback interface, which acts as a placeholder for the network, and allows it to be announced dynamically.

In addition to the basic <show ip route> form of the command, the <show ip route connected> form of the command can be quite useful. <show ip route connected> displays all the network routing information about all the active router interfaces:

```
Router>sh ip route connected
C    192.168.0.0/24 is directly connected, Serial0
     192.168.1.0/26 is subnetted, 3 subnets
C       192.168.1.64 is directly connected, Ethernet0
C       192.168.1.192 is directly connected, Loopback2
C       192.168.1.128 is directly connected, Loopback1
Router>
```

The <show ip route connected> command will display all the routing entry information for the specified network or host:

```
Router>sh ip route 172.16.191.1
Routing entry for 172.16.191.0/24
  Known via "static", distance 20, metric 0
  Routing Descriptor Blocks:
  * 192.168.0.2
      Route metric is 0, traffic share count is 1

Router>
```

This command variant is quite useful when debugging route announcement issues that arise when you use dynamic routing protocols and route redistribution.

The <show ip masks [address]> command is also invaluable when you're employing classful or classless subnetting. Errors in subnetting large address spaces are common. This command lists all the mask variations used with a given IP address space:

```
router#show ip masks 172.16.0.0
Mask            Reference count
255.255.255.255 4
255.255.255.252 1
255.255.255.248 2
255.255.255.128 1
255.255.255.0   6
255.255.248.0   3
```

A problem commonly associated with dynamically addressed networks is incorrect ARP table entries. For example, a node's IP address can change or a new IP address is assigned to a node that already had an address on the same segment (an expired DHCP lease, perhaps). These are prime targets for bad ARP entries. Since the router plays such an important role in data delivery for all the connected nodes, it is essential that the router's ARP table be correct. A good place to start looking when nodes are suddenly unreachable and no hardware failure exists is the router's ARP tables. Like the IP routing table, the ARP table is viewable in user EXEC mode. To display the ARP table, the <show arp> command is used:

```
Router>show arp
Internet  192.168.191.220        -  00e0.1ef2.15a1  ARPA   Ethernet2/0
Internet 192.168.191.253         0  00c0.4906.9488  ARPA   Ethernet2/0
Internet 192.168.191.240         0  00e0.1e34.a758  ARPA   Ethernet2/0
Internet 192.168.191.141        37  00e0.b08d.ccf0  ARPA   Ethernet2/0
Internet 192.168.191.142        41  00e0.1e5b.6e80  ARPA   Ethernet2/0
Internet 192.168.190.136         -  00e0.1ef2.15b1  ARPA   Ethernet3/0
Internet 192.168.190.137         -  00e0.1ef2.15b1  ARPA   Ethernet3/0
Internet 192.168.190.138         -  00e0.1ef2.15b1  ARPA   Ethernet3/0
Internet 192.168.190.139         -  00e0.1ef2.15b1  ARPA   Ethernet3/0
```

A great tool for displaying information about the dynamic IP routing protocols running on the router is <show ip protocols>. This command can display a verbose or terse report. Verbose is the default, and it displays all the running processes, their update status, the networks being announced, and any neighbor routers:

```
ASBR-34#sh ip protocols
Routing Protocol is "bgp 66"
  Sending updates every 60 seconds, next due in 0 seconds
  Outgoing update filter list for all interfaces is
  Incoming update filter list for all interfaces is
  IGP synchronization is enabled
  Automatic route summarization is disabled
  Neighbor(s):
    Address          FiltIn FiltOut DistIn DistOut Weight RouteMap
    172.16.85.75
  Routing for Networks:
    172.16.12.0/24
    172.16.80.0/24
    172.16.161.128/25
    172.16.191.0/24
    172.16.192.0/21
  Routing Information Sources:
    Gateway         Distance      Last Update
    172.16.85.75        20       6d16h
  Distance: external 20 internal 200 local 200
```

The terse report is retrieved by using the <show ip protocols summary> command. It only lists the protocol's process and, if applicable, the process ID:

```
Router#sh ip protocols summary
Index Process Name
0     connected
1     static
2     ospf 45
3     rip
Router#
```

Managing IP Routing Information

Problems do arise, and when they do, it is best to start slowly by verifying that what you believe to be happening is in fact happening. The <show> commands work well in this regard. In most situations where IP routing is the suspect, the problem is lack of a route, or a bad route or ARP entry.

To flush the entire routing table, the privileged EXEC command `<clear ip route *>` is used. To remove an individual IP route entry, use

```
<clear ip route [address] [mask]>
```

When the route table is flushed, the router will recreate the table starting with its directly attached interfaces. If a dynamic protocol is in use, the flush will trigger an update immediately. Static route entries that are loaded from the configuration file at boot will also be restored shortly after the table has been flushed. Be aware that while the router rebuilds its route table, connectivity will be lost. The connectivity outage will vary in length from a few seconds to several minutes, depending on the size of network and the routing table.

The `<clear arp>` privileged EXEC command clears the router's ARP table. If a situation requires the entire IP route table to be cleared, it is not a bad idea to clear the ARP table as well, preferably *before* you flush the IP route table. This way, you will ensure that the router has accurate ARP entries for its adjacent neighbors before the route table is cleared. The ARP table will reconstruct itself quickly, since it is only dependent on connectivity with nodes that are directly accessible to its connected interfaces.

Because dynamic protocols refresh route entries automatically, only the most drastic situations will require the entire route table to be flushed. In most cases, only a single failed route is the problem, such as a dynamic route in a hold-down state (waiting to expire) or a static route that points to a nonexistent gateway. To remove an invalid dynamic route from the routing table, use `<clear ip route [network] [mask]>`. It is also possible to clear a single ARP table entry with `<clear arp [mac address]>`. Removal of an invalid static route must be done from global configuration EXEC mode. After you're in configuration mode, place a `<no>` in front of the same command string you would use to create the route:

```
Router#config t
Enter configuration commands, one per line.  End with CNTL/Z.
Router(config)#no ip route 192.160.0.4 255.255.255.252 172.16.0.6 20
Router(config)#^Z
Router#
```

After the static route is removed from the configuration, the IOS will purge it from the IP routing table.

Note

When changing any IP routing behavior, it is best to perform changes from the router's console. VTY line connectivity can be interrupted while some changes are taking effect. Consequently, if the change fails, all VTY connectivity might be lost, leaving the router inaccessible.

Managing Static Routing

Display commands:

```
<show ip route>
<show interface [type/slot/port]>
```

Global configuration commands:

```
<ip route [network] [mask] [gateway] [administrative distance]>
<ip classless>
<ip subnet-zero>
<ip forward-protocol udp [port number]>
<no ip source-route>
```

Interface configuration subcommands:

```
<ip helper address [ip address]>
<bandwidth>
<mtu>
<no ip redirects>
<no ip unreachables>
```

Control commands:

```
<copy tftp [route-table-name]>
```

Although static routing is not an ideal total solution for IP route announcement in large networks, it is a very effective solution for small (single gateway) networks and provides needed gateway and route redirection services. Static routes are also essential to announcing networks where the access links are unstable or temporary (as with dial-up connections).

Static routes are set in the Cisco IOS using the `<ip route>` configuration EXEC command. As noted in the previous section, static routes are set in the router's configuration file. The IOS's handling of static routes is different than, say, a UNIX or NT workstation that has static entries in its startup configuration file. With the IOS, static route entries are managed as an IP routing process, so they are reloaded after routing table flush. Static routes entered on an end-station would typically need to be re-entered or reloaded (by rebooting the system) after the routing table has been flushed.

Configuring Default Routing

If the IP network is closed (network reachability is limited to explicitly announced networks defined in the routing table), a default route is not needed. A typical example is a closed or private network where there is no Internet access or where access is provided through a proxy server or firewall. In closed network architectures, typically, traffic destined for unannounced networks is discarded and the user is notified with an ICMP message. To disable ICMP notifications and ICMP redirection attempts, the `<no ip unreachables>` and `<no ip redirects>` commands can be set on the router's connected interfaces. These options also provide an additional level of network security by limiting the capability of IP traffic to be redirected across a path that may insecure.

In most situations, however, a default route is needed. The most common one is where a single point Internet connection exists or where it is undesirable to exchange routing information but reachability information is required.

In Chapter 7, we first used the <ip route> command to set the default route on the Concord and Ridgefield routers. In this case, we are setting the default route for `asbr-a2` to forward all traffic that has no explicit route out `asbr-a1`'s dedicated Internet link through Fast Ethernet 0/0 (refer to Figure 10.1). If the link fails, it should forward all of the traffic out interface s2/0 to `asbr-b2`. Notice on the route entry for 192.168.0.6, we've added a number for administrative distance of 20:

```
asbr-a1#config t
Enter configuration commands, one per line.  End with CNTL/Z.
asbr-a1(config)#ip route 0.0.0.0 0.0.0.0 192.168.0.6 20
asbr-a1(config)#ip route 0.0.0.0 0.0.0.0 192.168.191.20
asbr-a1(config-if)#^Z
asbr-a1#
```

When interface Fast Ethernet 0/0 is up, the router designates 192.168.191.19 as its gateway of last resort:

```
asbr-a2#sh ip route
Codes: C - connected, S - static, I - IGRP, R - RIP, M - mobile, B - BGP
       D - EIGRP, EX - EIGRP external, O - OSPF, IA - OSPF inter area
       N1 - OSPF NSSA external type 1, N2 - OSPF NSSA external type 2
       E1 - OSPF external type 1, E2 - OSPF external type 2, E - EGP
       i - IS-IS, L1 - IS-IS level-1, L2 - IS-IS level-2, * - candidate
default
       U - per-user static route, o - ODR

Gateway of last resort is 192.168.191.20 to network 0.0.0.0

     192.168.181.0/25 is subnetted, 1 subnets
C    192.168.12.0/24 is directly connected, Ethernet1/0
C    192.168.181.128 is directly connected, Ethernet1/3
C    192.168.10.0/24 is directly connected, Ethernet1/2
C    192.168.191.0/24 is directly connected, FastEthernet0/0
     192.168.0.0/24 is variably subnetted, 2 subnets, 2 masks
C    192.168.0.4/30 is directly connected, Serial2/0
C    192.168.192.0/21 is directly connected, Ethernet1/1
S*   0.0.0.0/0 [1/0] via 192.168.191.20
```

If interface Fast Ethernet 2/0 fails, 192.168.0.6 (accessible through interface serial2/0) is designated as the gateway of last resort:

```
asbr-a2#sh ip route
Codes: C - connected, S - static, I - IGRP, R - RIP, M - mobile, B - BGP
       D - EIGRP, EX - EIGRP external, O - OSPF, IA - OSPF inter area
       N1 - OSPF NSSA external type 1, N2 - OSPF NSSA external type 2
       E1 - OSPF external type 1, E2 - OSPF external type 2, E - EGP
       i - IS-IS, L1 - IS-IS level-1, L2 - IS-IS level-2, * - candidate
default
       U - per-user static route, o - ODR

Gateway of last resort is 192.168.0.6 to network 0.0.0.0
     192.168.181.0/25 is subnetted, 1 subnets
C    192.168.12.0/24 is directly connected, Ethernet1/0
C    192.168.181.128 is directly connected, Ethernet1/3
C    192.168.10.0/24 is directly connected, Ethernet1/2
     192.168.0.0/24 is variably subnetted, 2 subnets, 2 masks
C    192.168.0.4/30 is directly connected, Serial2/0
C    192.168.192.0/21 is directly connected, Ethernet1/1
S*   0.0.0.0/0 [20/0] via 192.168.0.6
asbr-a2#
```

By setting the administrative distance on the static route pointing to `asbr-b2`, the route only appears in the routing table when a route with a lesser administrative distance does not exist (in other words, when the router's interface is down). In this case, the default static route (which uses the default administrative distance of 0) directs all remote traffic out through the local Internet connection via `asbr-a1`.

The problem with this solution is the intrinsic nature of static routes. The route fail-over only occurs if the interface on the router fails. If the next hop interface is unavailable but the local router interface is still on line, the router will continue to forward traffic to the unavailable interface. This solution does offer some redundancy over having nothing at all, because statistically, the Fast Ethernet hub will probably fail (disabling the interface) before the next hop router's hardware interface will.

Using Static Routing in the Enterprise

Static routing, although not the best method, is often used for route advertisement in enterprise networks because it is stable and removes the possibility of IP traffic being misdirected by some runaway, misconfigured, or broken dynamic routing process. Static routing is nevertheless difficult to scale when you manage networks with large routing tables.

One way to scale a static routing configuration is to manage the static tables like access-lists, using <copy> and TFTP. Because the most tedious task in maintaining large static routing tables is adding and removing static routes, copying those changes to the <running-config> is more efficient than making single line changes in configuration mode. This approach also has the major advantage of having a copy of the routing table separate from the router configuration; it's easier to load the table to a different router in case of a hardware failure.

For asbr-a2 to reach all the testnet networks, 15 static routes are required. Here is the static route config file for asbr-a2. Like the default gateway scenario, a secondary route is provided with a higher administrative distance, to be used in case of a link failure:

```
!   -To reach networks connected to asbr-b1 via asbr-a1 (primary) and
asbr-b2
!  (backup)
ip route 172.16.3.0 255.255.255.0 192.168.191.20
ip route 172.16.3.0 255.255.255.0 192.168.0.6 40
ip route 172.16.2.192 255.255.255.192 192.168.191.20
ip route 172.16.2.192 255.255.255.192 192.168.0.6 40
ip route 172.16.2.64 255.255.255.192 192.168.191.20
ip route 172.16.2.64 255.255.255.192 192.168.0.6 40
ip route 172.16.2.192 255.255.255.192 192.168.191.20
ip route 172.16.2.192 255.255.255.192 192.168.0.6 40
ip route 172.16.2.128 255.255.255.192 192.168.191.20
ip route 172.16.2.128 255.255.255.192 192.168.0.6 40
! Equal hop networks off asbr-b1 and asbr-b2
!
ip route 172.16.91.0 255.255.255.0 192.168.191.20
ip route 172.16.91.0 255.255.255.0 192.168.0.6
ip route 192.168.0.128 255.255.255.128 192.168.191.20
ip route 192.168.0.128 255.255.255.128 192.168.0.6
ip route 192.168.0.9 255.255.255.252 192.168.191.20
ip route 192.168.0.9 255.255.255.252 192.168.0.6
ip route 192.168.0.13 255.255.255.252 192.168.191.20
ip route 192.168.0.13 255.255.255.252 192.168.0.6
!  -To reach networks connected to asbr-b2 via asbr-b2 (primary) and asbr-a1
!  (backup)
ip route 172.16.1.0 255.255.255.192 192.168.0.6
ip route 172.16.1.0 255.255.255.192 192.168.191.20 40
ip route 172.16.1.64 255.255.255.192 192.168.0.6
ip route 172.16.1.64 255.255.255.192 192.168.191.20 40
ip route 172.16.1.128 255.255.255.192 192.168.0.6
ip route 172.16.1.128 255.255.255.192 192.168.191.20 40
ip route 172.16.1.192 255.255.255.192 192.168.0.6
ip route 172.16.1.192 255.255.255.192 192.168.191.20 40
ip route 172.16.2.0 255.255.255.0 192.168.0.6
```

```
ip route 172.16.2.0 255.255.255.0 192.168.191.20 40
! To reach networks of asbr-a1 via asbr-a1 (primary) and asbr-b2 (backup)
ip route 192.168.160.96 255.255.255.252 192.168.191.20
ip route 192.168.160.96 255.255.255.252 192.168.0.6 40
! Default routes
ip route 0.0.0.0 0.0.0.0 192.168.0.6 40
ip route 0.0.0.0 0.0.0.0 192.168.191.20
end
```

To load the static table, use the `<copy tftp running-config>` privileged EXEC command:

```
asbr-a2#copy tftp run
Host or network configuration file [host]?
Address of remote host [255.255.255.255]? 192.168.191.202
Name of configuration file [asbr-a1-confg]? asbr-a2.static
Configure using asbr-a2.static from 192.168.191.202? [confirm]
Loading asbr-a2.static from 192.168.191.202 (via Ethernet0): !
[OK - 1984/32723 bytes]
asbr-a2#
```

After the routing table is loaded, the IP route table looks like this:

```
asbr-a1#sh ip route
Codes: C - connected, S - static, I - IGRP, R - RIP, M - mobile, B - BGP
       D - EIGRP, EX - EIGRP external, O - OSPF, IA - OSPF inter area
       N1 - OSPF NSSA external type 1, N2 - OSPF NSSA external type 2
       E1 - OSPF external type 1, E2 - OSPF external type 2, E - EGP
       i - IS-IS, L1 - IS-IS level-1, L2 - IS-IS level-2, * - candidate
default
       U - per-user static route, o - ODR

Gateway of last resort is 192.168.191.20 to network 0.0.0.0

     192.168.181.0/25 is subnetted, 1 subnets
C       192.168.181.128 is directly connected, Ethernet1/3
C    192.168.10.0/24 is directly connected, Ethernet1/2
     192.168.160.0/30 is subnetted, 1 subnets
S       192.168.160.96 [1/0] via 192.168.191.20
     172.16.0.0/16 is variably subnetted, 10 subnets, 2 masks
S       172.16.1.128/26 [1/0] via 192.168.0.6
S       172.16.2.128/26 [1/0] via 192.168.191.20
S       172.16.1.192/26 [1/0] via 192.168.0.6
S       172.16.2.192/26 [1/0] via 192.168.191.20
S       172.16.1.0/26 [1/0] via 192.168.0.6
S       172.16.2.0/24 [1/0] via 192.168.0.6
S       172.16.3.0/24 [1/0] via 192.168.191.20
S       172.16.91.0/24 [1/0] via 192.168.191.20
```

```
                      [1/0] via 192.168.0.6
S       172.16.1.64/26 [1/0] via 192.168.0.6
S       172.16.2.64/26 [1/0] via 192.168.191.20
C    192.168.191.0/24 is directly connected, Ethernet0
     192.168.0.0/24 is variably subnetted, 3 subnets, 3 masks
S       192.168.0.0/24 [1/0] via 192.168.0.4
C       192.168.0.4/30 is directly connected, Serial2/0
S       192.168.0.128/25 [1/0] via 192.168.191.20
                         [1/0] via 192.168.0.6
S*   0.0.0.0/0 [1/0] via 192.168.191.20
C    192.168.192.0/21 is directly connected, Ethernet1/1
```

In the event that the Fast Ethernet 0/0 on `asbr-2a` fails, the routing would adjust itself to only use 192.168.0.6 as the next hop gateway:

```
asbr-a1#sh ip route
Codes: C - connected, S - static, I - IGRP, R - RIP, M - mobile, B - BGP
       D - EIGRP, EX - EIGRP external, O - OSPF, IA - OSPF inter area
       N1 - OSPF NSSA external type 1, N2 - OSPF NSSA external type 2
       E1 - OSPF external type 1, E2 - OSPF external type 2, E - EGP
       i - IS-IS, L1 - IS-IS level-1, L2 - IS-IS level-2, * - candidate
default
       U - per-user static route, o - ODR

Gateway of last resort is 192.168.0.6 to network 0.0.0.0

     192.168.181.0/25 is subnetted, 1 subnets
C       192.168.181.128 is directly connected, Ethernet1/3
C    192.168.10.0/24 is directly connected, Ethernet1/2
     192.168.160.0/30 is subnetted, 1 subnets
S       192.168.160.96 [40/0] via 192.168.0.6
     172.16.0.0/16 is variably subnetted, 10 subnets, 2 masks
S       172.16.1.128/26 [1/0] via 192.168.0.6
S       172.16.2.128/26 [40/0] via 192.168.0.6
S       172.16.1.192/26 [1/0] via 192.168.0.6
S       172.16.2.192/26 [40/0] via 192.168.0.6
S       172.16.1.0/26 [1/0] via 192.168.0.6
S       172.16.2.0/24 [1/0] via 192.168.0.6
S       172.16.3.0/24 [40/0] via 192.168.0.6
S       172.16.91.0/24 [1/0] via 192.168.0.6
S       172.16.1.64/26 [1/0] via 192.168.0.6
S       172.16.2.64/26 [40/0] via 192.168.0.6
     192.168.0.0/24 is variably subnetted, 3 subnets, 3 masks
S       192.168.0.0/24 [1/0] via 192.168.0.4
C       192.168.0.4/30 is directly connected, Serial2/0
S       192.168.0.128/25 [1/0] via 192.168.0.6
S*   0.0.0.0/0 [40/0] via 192.168.0.6
C    192.168.192.0/21 is directly connected, Ethernet1/1
asbr-a1#
```

If the asbr-2a interface came back up, 192.168.191.20 would become the default route again.

As you can see, it is possible to manage IP route announcements using only static routing. However, this example also shows the work required to manage static tables sanely.

Configuring Classless Routing

Cisco IOS provides support for both classful and classless addressing. Depending on the IOS, version support for both is enabled by default. To verify if classless support is enabled, use the privileged EXEC command `<show running-config>` and check the configuration file for the configuration command `<ip classless>`. If it appears, it is enabled. If not, and classless support is required, then in global configuration EXEC mode, enter the command `<ip classless>`. When enabled, variable subnetting is permitted of classless address spaces.

One easy way to use VLSM is to break up the address space in blocks of four, because the theory behind classless addressing is any address space that is a multiple of two (for example, 2 × 2 = 4 = /30 mask, 2 × 4 = 8 = /29 mask, andso on) minus the network and broadcast address. Although this is correct in theory, in order to maintain computability with classful subnetting, by default, IOS does not permit the usage of network addresses that end on natural zero boundaries. This is done because these addressees can be mistaken for network broadcast addresses because the netmask is not sent along in the packet's address information. To enable the use of "zero boundary" network addresses, which is particularly useful when working with CIDR's created out of Class C addresses, IOS provides the global configuration EXEC command `<ip subnet zero>`. Here is an example of what happens when a zero boundary address is used without enabling `<ip subnet zero>`:

```
Router#config t
Enter configuration commands, one per line.  End with CNTL/Z.
Router(config)#int e0
Router(config-if)#ip address 192.168.0.1 255.255.255.252
Bad mask /30 for address 192.168.0.1
Router(config-if)#
With the <ip subnet zero> option enabled, the address is permitted:
Router#config t
Enter configuration commands, one per line.  End with CNTL/Z
Router(config)#ip subnet-zero
Router(config)#int e0
Router(config-if)#ip address 192.168.0.1 255.255.255.252
Router(config-if)#^Z
Router#
```

Configuring IP Control Services

IOS supports a number of options to adjust IP's behavior. Some of these commands enhance the security of IP, and others enhance IP's performance:

- `<ip forward-protocol udp [port number]>`—This command allows IP UDP broadcasts to be forwarded between IP network segments that are connected to each other across the router. IP (Layer 3) broadcasts are not forwarded between networks by default. This option allows UDP broadcast service requests like DHCP and BOOTP to be relayed to servers on remote networks. This command must be used with `<ip helper address [address]>`, which is set on the router interface that is performing the forwarding. The address specifies where the request should beforwarded. To configure DHCP/BOOTP forwarding:

```
Router(config)#ip forward-protocol udp 67
Router(config)#int e0
Router(config-if)#ip helper-address 192.168.1.24
```

- `<no ip source-route>`—IP source routing is an option built into the IP protocol suite for testing (source routing is indicated in the IP datagram's header). It provides the capability to dictate the route path that a datagram will travel. Source routing, like ICMP redirects, can be exploited to create potential security problems. Unless required, source routing should be disabled.
- `<ip mtu [bytes]>`—Each router interface has a default MTU size, which is used for determining if IP fragmentation is needed. The IOS supports MTU path discovery, which allows the devices along the route path to adjust for differences in maximum MTU sizes. Ideally, an MTU of 1500 is preferred, because it provides for the maximum PDU size allowable for Ethernet, which will eliminate the need for resizing. Certain Layer 2 protocols support MTU sizes larger than 1500, which means that when packets are exchanged between the two different protocols, fragmentation occurs. Whenever possible it is best to reduce interfaces with larger MTU sizes down to 1500. The MTU value is part of the IGRP and EIGRP route metric set, but has no direct role in determining a route's vector (metric cost).
- `<bandwidth [kilobytes]>`—The `<bandwidth>` is a configurable parameter on each of the router's interfaces used for setting the interface's data transfer rate in kilobytes per second. This value is also used by IGRP, EIGRP, and OSPF to calculate route metrics. An interface's MTU and bandwidth values are displayed as part of the `<show interface [type/slot/port]>` command output:

```
Router#show interfaces e0
Ethernet0 is administratively down, line protocol is down
  Hardware is Lance, address is 0010.7b37.b27c (bia 0010.7b37.b27c)
  Internet address is 192.168.0.1/30
  MTU 1500 bytes, BW 10000 Kbit, DLY 1000 usec,
     reliability 255/255, txload 1/255, rxload 1
```

Configuring Dynamic IGP and EGP IP Routing Protocols

All dynamic IP routing protocols are configured as IOS subprocesses, much the same way a router interface is configured. The `<router [protocol] [process id]>` global configuration command enables the process:

```
asbr-a1#config t
asbr-a1(configure)#router ospf 89
asbr-a1(config-router)#network 172.16.0.0 0.0.255.255 area 0.0.0.0
asbr-a1(config-router)#
```

After a process is enabled, it does not become active until a `<network [IP network address]>` statement that corresponds to a configured router interface is added to the subprocess.

After the process is configured, you need to add each directly connected network that you wish the protocol to announce. If the network is not addedto the process, it will not be announced by the protocol. Occasionally, you might want to receive, but not send, network announcements on an interface (and you still want the network announced by the routing process). This is common when a routing protocol is used to manage a "private" network that is attached to an Internet gateway, which is also using a routing protocol. In situations like this, the interface can be run in passive mode using the routing configuration subprocess command `<passive-interface [interface]>`:

```
asbr-a1#config t
asbr-a1(configure)#router ospf 89
asbr-a1(config-router)# passive-interface s1
```

After the interface is in passive mode, it will only receive routing announcements. When a router interface has been configured to operate in passive mode, it will only accept routing announcements. This option is useful when you only want the router to obtain reachability information, but not announce. Initially, this might seem odd, this is a common practice in situations where data services are being provided by an outside network (for example, Internet Gateway or News Service feed). In these situations, the provider will use a static route to reach your gateway, but otherwise, has no need for your routing information. However, by having the gateway listening to the provider's route announcements, the router will then incorporate the reachability information into its routing update that it transmits to the other routers. This way, when the provider makes changes to its network, your network is dynamically updated instead of you having to add or remove static routes.

Note

The IGP process ID is often the same number as the network's autonomous system (AS) number. This is a practice and not a requirement; if your network does not have an assigned AS number, any number will do. When using RIP, no process ID is required because only one RIP process is supported. This is in contrast to IGRP, EGRP, and OSPF, where multiple processes can be supported.

Additionally, at least one interface must be configured with an IP address in order for any routing process to be created.

Here is an example of using the <passive-interface> command to manage an internal and external route announcement. The Albany campus network uses asbr-a1 to announce of the local network to its local ISP (refer to Figure 10.1). It then uses OSPF to manage announcements internally with the Boston network.

Here are the RIP and OSPF configurations used on asbr-a1:

```
router rip
 version 2
 network 12.0.0.0
 network 192.168.160.0
 network 192.168.191.0
 network 192.168.161.0
 passive-interface e1
 passive-interface s1
 neighbor 12.14.116.66
 no auto-summary

router ospf 89
 network 12.14.116.64 0.0.0.3 area 0.0.0.0
 network 192.168.0.0 0.0.0.3 area 0.0.0.0
 network 192.168.191.0 0.0.0.255 area 0.0.0.0
 network 192.168.160.96 0.0.0.3 area 0.0.0.0
 network 192.168.161.128 0.0.0.128 area 0.0.0.0
 passive-interface s1
```

By enabling the interfaces as passive, only the active interfaces announce all the networks that are part of the process. The RIP process announces the local network to the ISP using interface s1. The e1 interface is also active, so it can send/receive RIP updates with asbr-a2. None of the internal WAN networks are included in the RIP process. The OSPF process suppresses announcements on interface s1 and includes asbr-a1's WAN link, which exchanges network announcements with the Boston office.

Using RIPv1 and RIPv2

Global configuration commands:

```
<router rip>
<key chain [key chain name]>
<key [id number]>
<key-string>
<default key-string>
<access-list [1-99]>
```

Global router subprocess commands:

```
<neighbor>
<timers basic [update] [invalid] [holddown] [flush]>
<distance>
```

```
<auto-summary>
<version [1] [2]>
<offset-list [acl] [in/out] [metric] [interface type/slot/port]>
```

Global interface subprocess commands:

```
<ip rip send version [1/2/1 2]>
<ip rip receive version [1/2/1 2]>
<ip rip authentication mode [md5] [text]>
```

Control commands:

```
<debug rip>
<debug rip events>
```

RIP is an IGP, distance vector-based routing protocol defined in RFC 1058. RIP uses a single routing metric—hop count—to determine the best route path. RIP, you may recall, has two versions: 1 and 2. The main differences are that version 1 uses UDP broadcasts and has no support for authentication and VLSM. RIP version 2 was added to Cisco IOS version 11.1, and it uses multicast for sending routing updates, as well as supports MD5 and text key authentication and VLSM. The IOS, by default, sends version 1 updates but will receive both version 1 and 2 updates. RIP's strength as a protocol is that it is supported in some form by virtually every network hardware vendor and operating system platform.

In the previous example, RIP made the network announcements between the local Albany network and its ISP. The Albany network uses both classful and classless addressing, and since netmasks need to be sent in the update messages, RIP version 2 was required. You can force this in the configuration by using the `<version [1] [2]>` subcommand. Using the version command disables RIPv1 support altogether. Alternatively, if you need to support some legacy hardware that only speaks RIPv1, you can specify on each interface what message types to support:

```
asbr-a1(config)# interface e0
asbr-a1(config-if)#ip rip send version 1 2
asbr-a1(config-if)#ip rip receive version 2
```

This configuration would only accept RIPv2 messages but send both v1 and v2 updates. Let's examine the RIP configurations used on `asbr-a1` and `asbr-a2`:

```
asbr-a1#config t
Enter configuration commands, one per line.  End with CNTL/Z.
asbr-a1(config)#router rip
asbr-a1(config-router)#version 2
asbr-a1(config-router)#network 12.0.0.0
asbr-a1(config-router)#network 192.168.160.0
asbr-a1(config-router)#network 192.168.161.0
asbr-a1(config-router)#network 192.168.191.0
asbr-a1(config-router)#
```

```
asbr-a2config t
Enter configuration commands, one per line.  End with CNTL/Z.
asbr-a2(config)#router rip
asbr-a2(config-router)#version 2
asbr-a2(config-router)#network 192.168.181.0
asbr-a2(config-router)#network 192.168.191.0
asbr-a2(config-router)#network 192.168.192.0
asbr-a2(config-router)#network 192.168.10.0
asbr-a2(config-router)#network 192.168.12.0
```

After all the network advertisement statements have been added, RIP is up and running. Here is the IP route table from `asbr-a1`:

```
asbr-a1#sh ip route
Codes: C - connected, S - static, I - IGRP, R - RIP, M - mobile, B - BGP
       D - EIGRP, EX - EIGRP external, O - OSPF, IA - OSPF inter area
       N1 - OSPF NSSA external type 1, N2 - OSPF NSSA external type 2
       E1 - OSPF external type 1, E2 - OSPF external type 2, E - EGP
       i - IS-IS, L1 - IS-IS level-1, L2 - IS-IS level-2, * - candidate
default
       U - per-user static route, o - ODR

Gateway of last resort is not set

R    192.168.12.0/24 [120/1] via 192.168.191.19, 00:00:26, Ethernet0
     192.168.192.0/26 is subnetted, 1 subnets
R       192.168.192.64 [120/1] via 192.168.191.19, 00:00:26, Ethernet0
R    192.168.181.0/24 [120/1] via 192.168.191.19, 00:00:26, Ethernet0
R    192.168.10.0/24 [120/1] via 192.168.191.19, 00:00:26, Ethernet0
     192.168.161.0/25 is subnetted, 1 subnets
C       192.168.161.128 is directly connected, Loopback1
C    192.168.191.0/24 is directly connected, Ethernet0
     12.0.0.0/30 is subnetted, 1 subnets
C       12.14.116.4 is directly connected, Serial1
asbr-a1#
```

IOS allows you to adjust how the RIP process behaves and interprets route advertisements.

Note

RIP version 1 does not send subnet masks in its update messages. So, networks are announced using the address space's classful root. Routers then apply their locally configured subnet mask to interpret the address space. RIP version 2 sends both the network address and mask in its updates. However, RIPv2 summarizes at network boundaries, announcing only the classful root. RIPv2 uses the subnet masks on the corresponding interfaces to construct the subnet mask elements of its update messages. If you use classless addressing with a large classful space, this can present problems if discontinuous subnets are used. The RIP subprocess command <no auto-summary> disables network summarization.

Adjusting Distance and Metrics

To adjust the administrative distance applied to all route advertisements received by the RIP process, use the router subcommand `<distance [10-255]>`.

If you need to change a route's metric, a standard access list is used in combination with the RIP subcommand `<offset-list [acl] [in/out] [metric] [interface type/slot/port]>`. The Albany network has three access points to the 192.168.192.64 /26 network: `asbr-a2`, `asbr-a3` and `asbr-a4`. In order to make all three routes have the same cost, the advertisement for the 192.168.192.64 network must be adjusted on `asbr-a2`. First, a standard IP access-list is created:

```
asbr-a2(config)#access-list 1 permit 192.168.192.64 0.0.0.0
```

Then, in the RIP subprocess, the metric adjustment is applied. This adjustment forces the metric for IP address 192.168.192.64 from 1 to 3 only on announcements sent to `asbr-a1` from `asbr-a2`:

```
asbr-a2(config-router)#offset-list 1 out 2 ethernet 0
```

To verify that the adjusted route metric is being announced, enable RIP debugging with the `<debug ip rip>` privileged EXEC command and examine the RIP update message.

```
RIP: sending v2 update to 224.0.0.9 via Ethernet0 (192.168.191.19)
     192.168.10.0/24 -> 0.0.0.0, metric 1, tag 0
     192.168.12.0/24 -> 0.0.0.0, metric 1, tag 0
     192.168.20.0/24 -> 0.0.0.0, metric 16, tag 0
     192.168.192.64/26 -> 0.0.0.0, metric 3, tag 0
     192.168.181.0/24 -> 0.0.0.0, metric 1, tag 0
```

Adjusting RIP Process Timers

RIP's default is to send a message update every 30 seconds (the router's entire routing table) whether there is a change or not. The IOS RIP updates can have up to a 4.5 second variation in frequency; this is to prevent all of the routing tables from sending updates at the same time.

In addition to the `[update]` timer, there is the `[invalid]` timer, which sets how long a route can remain in the routing table without being updated. If the `[invalid]` timer expires, the route is set to a metric of 16 and the `[flush]` is started. Upon expiration, the route is purged from the table.

IOS also supports a `[hold down]` timer, which is triggered when an update has a different metric than one previously recorded. These timers can be adjusted with the RIP routing subprocess command `<timers basic [update] [invalid] [holddown] [flush]>` to reduce message update times. In our present example, because the RIP process is only used for advertising routes to the ISP and changes in the topology are rare, setting the timers for updates that are more infrequent is preferable over the default timers. All values are set in seconds:

```
asbr-a1(config-router)#timers basic 120 240 180 300
```

Note

This option is also available with IGRP.

In situations where you have networks that consist of nonbroadcast media, or where it is desirable only to exchange routing updates between two hosts connected over broadcast-supported media, IOS provides the router subprocess command `<neighbor [ip address]>`. The ISP provider for the Albany network uses Frame Relay over HDLC, so a neighbor must be specified:

```
asbr-a1(config-router)#neighbor 12.14.116.66
```

Configuring Authentication

When you employ any routing protocol in a publicly accessible environment, it is wise to use authentication to verify with whom you are exchanging route advertisements. RIP version 2 supports clear text and MD5 authentication. Both routers must exchange the same password regardless of the authentication method. To use authentication, each router needs to create a `<key chain>` and key and then enable authentication on the appropriate interfaces. To set up a key chain, use the following:

```
asbr-a1(config)#key chain test
asbr-a1(config-keychain)#key 1
asbr-a1(config-keychain)#key-string atestkey
asbr-a1(config-keychain)#accept-lifetime 12:00:00 31 dec 1998 infinite
asbr-a1(config-keychain)#send-lifetime 12:00:00 31 dec 1998 infinite
```

The example above creates a key-chain called "test" and a single key with the password "atestkey". A key chain containing a key with "atestkey" as a password must also be created on all routers that `asbr-a1` wishes to exchange route announcements. After the keys are in place, the interface is configured:

```
asbr-a1(config)# interface s1
asbr-a1(config-if)#ip rip authentication key-chain test
asbr-a1(config-if)#ip rip authentication mode md5
```

Note

The `<neighbor>` command is a general routing protocol configuration command that is usable with all IGP routing protocols.

Note

This key generation process is also used to generate MD5 keys for use with OSPF MD5 authentication.

Examining the RIP Process

To verify RIP operation and troubleshoot possible configuration problems, the privileged EXEC commands <show ip protocols>, <debug ip rip>, and <debug ip rip events> can be used. <show ip protocols> provides status information on the process state. <debug ip rip events> displays process event messages about RIP message updates. <debug ip rip> displays the contents of RIP update message being sent. To disable the debug commands, use the <no> command in front of the debug command or <no debug all>.

Using IGRP/EIGRP

Global configuration commands:

```
<router igrp [process id]>
<router eigrp [process id]>
<appletalk routing eigrp [process id]>
<ipx router eigrp [process id]>
```

Global router subprocess commands:

```
<neighbor>
<timers basic [update] [invalid] [holddown] [flush]>
<distance>
<no auto-summary>
<network [IP address¦ip network number]>
<offset-list [acl] [in/out] [metric] [interface type/slot/port]>
<distribute-sap-list [IPX access-list number] [in¦out]>
<variance [multiplier]>
<traffic-share [balanced¦min]>
```

Global interface subprocess commands:

```
<ip eigrp-bandwidth-percent [percent]>
<appletalk eigrp-bandwidth-percent [percent]>
<ipx bandwidth-percent eigrp [process id] [percent]>
<ip summary-address eigrp [ip address] [netmask]>
```

Display commands:

```
<show ip protocols>
<show ip eigrp topology>
<show ip eigrp traffic>
<show ip eigrp neighbors [options]>
<show ip eigrp traffic>
```

Interior Gateway Routing Protocol (IGRP) and Enhanced Interior Gateway Routing Protocol (EIGRP) are Cisco Systems' proprietary dynamic routing protocols. IGRP, like RIP, is a distance vector protocol that broadcasts its router table out of all its interfaces at regular (adjustable) intervals. Unlike RIP, IGRP supports unequal path traffic sharing, uses its own transport protocol (similar to UDP) to send its update messages, and supports a network diameter of 224 hops compared to RIP's 15.

Two elements make IGRP a significant improvement over RIP. The first is its support of five routing metrics compared to RIP's single hop-count metric:

- Bandwidth (K1)
- Delay (K2)
- Reliability (K3)
- Load (K4)
- MTU (K5)

The second is its significantly faster convergence time, which uses "flash" updates to immediately propagate changes to the network topology whenever they occur.

EIGRP, as I'm sure you have guessed, is an enhancement to Cisco's IGRP protocol and represents a change in the routing algorithm. IGRP, as a true distance vector protocol, sends updates as tuples or distance pairs containing the network (vector) and path cost (distance). EIGRP uses the DUAL algorithm (Diffusing Update Algorithm), developed at SRI International. DUAL is considered a hybrid protocol because it employs the distance vector and route metric basis established for IGRP, but also uses link state information collection and announcement features. EIGRP (like OSPF) establishes relationships with other EIGRP neighbors, through the use of a "hello" protocol. After a neighbor relationship has been established, routing information is exchanged via unicast or multicast transport—using EIGRP RTP (Reliable Transport Protocol)—depending on the type of messages being exchanged. DUAL, unlike distance vector and link state protocols, uses a system of diffusing calculations shared across multiple routers to compute routing calculations. This is a different approach than what is used with distance vector and link state protocols, where each router performs its own route calculations.

Note

IGRP and EIGRP, straight out of the box, determine a route's metric using the additive sum of all the segment delays plus the bandwidth value of the slowest interface in the path. In a network comprised of same-bandwidth links, hop-count is used to determine desirability. In mixed media networks (with links of differing bandwidth), the lowest metric path is chosen. IGRP and EIGRP calculate an interface's bandwidth using the interface's `<bandwidth>` setting in KBPS divided by 10 to the 7th power or 10,000,000. All five metrics can be enabled using `<metric weight [tos] [k1] [k2] [k3] [k4] [k5]>`. This is not, however, advisable. The defaults are `tos=0 k1=1 k2=0 k3=1 k4=0 k5=0`. If k5 is set to 1, reliability and load will be used in addition to bandwidth and delay for calculating the route metric.

EIGRP's other enhancements include the following:

- Multiprotocol routing support (IP, IPX, and AppleTalk)
- CIDR support for classless summarization
- Fast convergence and partial updates, so only changes in topology are announced and only routers affected by a network changes are forced to re-compute their routing tables
- VLSM support

From a configuration perspective, IGRP and EIGRP are quite similar, so our examples will use EIGRP. An indication will be made if an option is only available with EIGRP. One difference should be noted outright. After convergence is achieved, EIGRP will only send out routing updates when topology changes occur. IGRP, on the other hand, uses timed, regular updates (regardless of changes in topology) and flash updates (when changes occur between scheduled updates). To adjust IGRP timers, use the router configuration subcommand <timers basic [update] [invalid] [holddown] [flush]>. IGRP has a 90-second default update interval, 270-second invalid timer, and 630-second flush timer. Like RIP, if you adjust the times on one router, you have to adjust them on all the routers.

Basic EIGRP (IGRP) Configuration

To get started, we will configure EIGRP on asbr-a2. The EIGRP process is started with the global configuration EXEC command <router eigrp [process id]>, and IGRP is started with <router igrp [process id]>. The process ID must be a number between 1–65,535. Multiple EIGRP and IGRP processes can operate on the same router. This permits you to segregate routing policies inside the internetwork and then use redistribution to announce between processes (a concept often used in large corporate networks where network management responsibility is decentralized). The process ID is used to identify the separate routing processes, so the process ID must be the same for each router that participates in the same EIGRP/IGRP process. Because EIGRP/IGRP is dependent on the process ID (PID) being uniform across all the routers, the PID is often referred to as an AS number. This does not mean you need an AS number to EIGRP/IGRP; any number will do, it just needs to be consistent across all the routers that are meant to exchange routing information:

```
asbr-a2(config)#router eigrp 99
```

After the process is enabled, IP network announcements are entered using the `<network [ip address|ip network number]>` router configuration subcommand. Announcements are entered using the network number or IP address of the interface:

```
asbr-a2(config-router)#network 192.168.191.20
asbr-a2(config-router)#network 192.168.12.0
asbr-a2(config-router)#network 192.168.192.0
asbr-a2(config-router)#network 192.168.10.0
asbr-a2(config-router)#network 192.168.181.0
asbr-a2(config-router)#network 192.168.0.4
asbr-a2(config-router)#no auto-summary
```

No matter how the announcement entry is made, it will appear in the router's configuration as classful network address summaries:

```
router eigrp 99
 network 192.168.191.0
 network 192.168.12.0
 network 192.168.192.0
 network 192.168.10.0
 network 192.168.181.0
 network 192.168.0.0
 no auto-summary
```

How the announcements are displayed in the configuration has no bearing on how they will be announced; that is determined by which protocol is used. If IGRP is in use, the networks will be announced classfully, depending on the subnet mask of the router for mask interpretation. Remember that when you subnet class A and B address spaces, you must use a consistent subnet maskfor the entire address space. If EIGRP is in use, the network will be announced along with their subnet masks. EIGRP by default uses classful auto-summarization; this should be disabled ifVLSM and a single classful spaceis being used for addressing. Disabling classful auto-summarization can be accomplished with the `<no auto-summary>` router configuration subcommand. The configuration outlined thus far represents the minimum required configuration to enable IP EIGRP/IGRP routing.

Note

IGRP/EIGRP will only send routing information updates on interfaces with addresses that correspond to those announced in its configuration. In situations where a large classful address space is used, this may not be desirable, especially if multiple IGRP/EIGRP processes are announcing different segments of the network that have the same classful root address (for example, 172.16.0.0 is a class B root address). In these situations, the `<passive-interface>` command can be used to suppress routing message updates from being sent out of an interface as part of the particular process.

Before we look at the EIGRP specific configuration options, a quick word about default routing: IGRP and EIGRP both use the `<ip default network [ip network number]>` global configuration EXEC variable to set and distribute default network information. This setting can be set on one gateway router and redistributed to the rest of the members of the IGRP/EIGRP routing process, or it can be set on each router in the process. The idea behind the default-network is to send external traffic not to a specific gateway, but rather to a default network where the gateway can be determined (in the case of multiple network exit points) by the locally attached router. It should be noted that a default gateway setting using the `<ip route 0.0.0.0 0.0.0.0 [ip remote gateway]>` may be required on the actual outbound gateway that is attached to the default network.

EIGRP Specific Configuration Options

EIGRP's support of IP summary-address announcements has an impact on CIDR supernets. Although it is recommended that you disable classful summarizations or avoid classful summarization altogether, if you use CIDR supernetting with a class C address space (not uncommon with sites that use small regional ISPs for Internet access), you will need to use summary announcements in order to have EIGRP announce the superset correctly. IP summarizations are configured using the interface configuration subcommand `<ip summary-address eigrp>`. Here is an example: The ISP for the Albany network (refer to Figure 10.1) has provided four class C addresses to support a network expansion that will be attached to `asbr-2a`. The addresses were assigned as a single /22:

```
192.168.4.0 /22 = 192.168.4.0, 192.168.5.0, 192.168.6.0, 192.168.7.0
```

Moreover, we want to maintain this address provisioning. Under normal operation, EIGRP would announce each class C network as a separate route entry, even though the address block is being used as a contiguous CIDR superblock. By using an aggregate summary address, the 192.168.4.0 /22 network is announced correctly as a supernet:

```
asbr-a2(config)#int e2/0
asbr-a2(config-if)#ip address 192.168.4.1 255.255.252.0
asbr-a2(config-if)#ip summary-address eigrp 99 192.168.4.1 255.255.252.0
```

Using the `<show ip protocol>` command we can see the summarization listed (notice that auto-summarization has been disabled):

```
Routing Protocol is "eigrp 99"
  Outgoing update filter list for all interfaces is not set
  Incoming update filter list for all interfaces is not set
  Default networks flagged in outgoing updates
  Default networks accepted from incoming updates
  EIGRP metric weight K1=1, K2=0, K3=1, K4=0, K5=0
  EIGRP maximum hopcount 100
  EIGRP maximum metric variance 1
  Redistributing: eigrp 99
```

```
Automatic network summarization is not in effect
Address Summarization:
  192.168.4.0/22 for Ethernet0/0
Routing for Networks:
  192.168.191.0
  192.168.12.0
  192.168.192.0
  192.168.10.0
  192.168.181.0
  192.168.0.0
  192.168.4.0
Routing Information Sources:
  Gateway         Distance      Last Update
Distance: internal 90 external 170
```

One of EIGRP's biggest enhancements is the support of IPX and AppleTalk routing. Enabling EIGRP AppleTalk routing is done using the `<appletalk routing eigrp [process id]>` command. Let's look at a configuration example:

```
Router#config t
Enter configuration commands, one per line.  End with CNTL/Z.
Router(config)# appletalk routing eigrp 99
```

To enable EIGRP IPX routing, the global configuration command `<ipx router eigrp [process id]>` is used. When enabled, IPX networks are added using an IPX version of the EIGRP `<network [ipx network address]>` or `<network all>` command:

```
Router(config)#ipx router eigrp 99
Router(config-ipx-router)# networks all
```

IPX EIGRP functionality is similar to IP EIGRP; it supports redistribution of IPX, RIP and NLSP, route filtering, and IPX-specific adjustments such as variable SAP announcements (interface subcommand `<ipx sap-incremental-eigrp>`) and SAP filtering with the IPX EIGRP subcommand `<distribute-sap-list [IPX access-list number] [in|out]>`. Each of the protocol-specific EIGRP processes run distinctly, so different PID should be used when configuring the separate instances.

Enabling Load Sharing and Balancing with EIGRP/IGRP

EIGRP is particularly sensitive to bandwidth consumption when it sends routing update information. The amount of bandwidth that can be used to send updates is adjustable (for all EIGRP protocol implementations). The bandwidth allocation for EIGRP is set as an interface configuration subcommand:

```
<ip eigrp-bandwidth-percent [percent]>
<appletalk eigrp-bandwidth-percent [percent]>
<ipx bandwidth-percent eigrp [process id] [percent]>
```

IGRP and EIGRP provide the capability to distribute network traffic load over links of unequal costs for load balancing. IGRP supports asymmetric load balancing over four unequal-cost paths. EIGRP supports up to six. The router configuration subcommand `<variance [multiplier]>` is used to enable this feature, and its default setting is 1. The variance value is used to recompute the metrics of routes of lesser value than that of the primary route. For example, if the variance was set to 3, any route with a metric 3 times greater (the lower the number, the better the metric) will be used to forward traffic.

There is also a way to balance the traffic by using the router configuration subcommand `<traffic-share [balanced|min]>`. This configuration helps the router determine if the traffic should be balanced (default) or should favor the lowest path cost first.

Adjusting Administrative Distances

EIGRP and IGRP handle metrics the same way, but compute administrative distances differently. IGRP uses a single administrative distance of 100. This value is adjustable using the `<distance [1-255]>` router configuration subcommand. EIGRP uses three administrative distances:

- Internal distance is used for routes announced from the EIGRP routing process, and its default setting is 90. The `<distance [1-255]>` subcommand will adjust this value to set which route will be preferred if multiple routing protocols are providing the same announcement.
- Summary distance is used for routes generated by summary addresses statements, and it has a default of 5.
- External distance is used when you redistribute routes from OSPF, and its default is 170. The values are adjustable using the `<distance eigrp [internal distance] [summary distance] [external distance]>` EIGRP configuration subcommand.

Monitoring IGRP and EIGRP

The IOS provides various informational commands to help you monitor your EIGRP processes, but not as many for IGRP. To display basic summary information, use the `<show ip protocols>` command:

```
Routing Protocol is "eigrp 99"
  Outgoing update filter list for all interfaces is not set
  Incoming update filter list for all interfaces is not set
  Default networks flagged in outgoing updates
  Default networks accepted from incoming updates
  EIGRP metric weight K1=1, K2=0, K3=1, K4=0, K5=0
  EIGRP maximum hopcount 100
  EIGRP maximum metric variance 1
```

```
Redistributing: eigrp 99, igrp 99
Automatic network summarization is not in effect
Address Summarization:
  192.168.4.0/22 for Ethernet0/0
Routing for Networks:
  192.168.191.0
  192.168.12.0
  192.168.192.0
  192.168.10.0
```

There are four <show ip eigrp> subcommands:

- <show ip eigrp neighbors [options]> provides operational information on each of the neighbors the router knows about:

```
asbr-a2#show ip eigrp neighbors ?
  <1-65535>  AS Number
  Ethernet   IEEE 802.3
  Loopback   Loopback interface
  Null       Null interface
  Serial     Serial
  detail     Show detailed peer information
  <cr>
```

- The neighbor's IP address and accessible interface are listed, along with the period since the router has heard from the neighbor (uptime).
- <show ip eigrp interfaces> and <show ip eigrp traffic> provide EIGRP traffic statistics:

```
asbr-a2#show ip eigrp interfaces ?
  <1-65535>  AS Number
  Ethernet   IEEE 802.3
  Loopback   Loopback interface
  Null       Null interface
  Serial     Serial
  detail     Show detailed peer information
  <cr>

asbr-a2#show ip eigrp traffic ?
  <1-65535>  AS Number
  <cr>
```

- <show ip eigrp topology> provides summary information on the EIGRP topology information source database:

```
asbr-a2#show ip eigrp topology ?
  <1-65535>         AS Number
  A.B.C.D           Network to display information about
  active            Show only active entries
  all-links         Show all links in topology table
  pending           Show only entries pending transmission
  summary           Show a summary of the topology table
  zero-successors   Show only zero successor entries
  <cr>
```

IOS also supports various debugging options for IGRP and EIGRP:

```
asbr-a2#debug ip igrp
  events         IGRP protocol events
  transactions   IGRP protocol transactions

asbr-a2#debug ip eigrp
  <1-65535>       AS number
  neighbor        IP-EIGRP neighbor debugging
  notifications   IP-EIGRP event notifications
  summary         IP-EIGRP summary route processing
```

Using OSPF

Display commands:

```
<show ip protocols>
<show ip ospf [local process id]>
<show ip ospf neighbor>
<show ip ospf database>
<show ip ospf virtual-links>
```

Interface configuration commands:

```
<ip ospf cost>
<ip ospf demand-circuit>
<ip ospf network [broadcast or non-broadcast]>
<ip ospf network point-to-multipoint>
<ip ospf authentication-key [password]>
<ip ospf priority [0-255]>
```

Note

As with all debugging commands, use these commands with caution. To disable a debugging command, use <no> preceding the command or use <no debug all>.

OSPF configuration subprocess commands:

```
<router ospf [local process id]>
<network <IP address/network number> <reverse network mask> <area id>
<area [area id] range [ip network address] [standard network mask]>
<distance [10-255]>
<distance ospf [external/inter-area/inter-area] [10-255]>
<area [area id] stub>
<area [area id] nssa>
<area [transit area id] virtual-link [router id]>
<redistribute [process] subnets>
<no ospf auto-cost>
<no ospf auto-cost-determination>
<passive-interface>
<neighbor [ip address] [priority (0-255)]>
```

Additional commands:

```
<ip classless>
<ip subnet-zero>
```

Implementing OSPF requires some up-front planning. Unlike other IGPs, the network topology plays a role in how OSPF functions. OSPF divides a large internetwork or AS into a collection of centrally connected hierarchical segments called *areas*. Each area should consist of no more than 50 routers, and must have one router that provides access to the backbone area. The *backbone area* (0.0.0.0) is used for exchanging Link State Announcements (LSAs) and acts as a transit network for all inter-area datagram traffic. All IP traffic exchanged between areas traverses the backbone network.

OSPF Router Type Definitions

You should recall that there are four classifications of OSPF routers:

- Backbone Routers—These routers maintain complete routing information for all the networks (all the areas or domains) that are connected to the backbone.
- Area Border Routers (ABRs)—These routers connect one or more areas to the backbone area, and only maintain information about the backbone and the areas they are attached to. Any router that has interfaces attached to the backbone and at least one area is an ABR.
- Internal Routers (IRs)—These routers are only involved in intra-area routing. IRs only contain information about the area they operate in. All extra-area traffic is forwarded to the area border router because only summaries of other areas within the network are maintained locally.
- Autonomous System Boundary Routers (ASBRs)—These routers are used to exchange routing information between OSPF and EGP. Any OSPF router that redistributes routing information is considered an ASBR.

Network topology plays a role in OSPF performance because each router maintains a database on the state of the network. OSPF was designed to be implemented in large-scale internetworks with 30 or more routers.

This does not mean OSPF will not function properly in small network environments; on the contrary, it will work well. Actually, in situations where a standards-based routing protocol is required, OSPF is a significantly better choice than RIP, especially when it comes to speed and network utilization.

Ideally, the backbone routers should bear the brunt of the processing by constructing a complete topology map from the LSAs sent by the ABRs that have established adjacencies with the IR. Therefore, the more network topology and addressing lends itself to subdivision and summarization, the more efficient and scalable OSPF becomes. OSPF's tightly structured model, where each router class performs a function, gives OSPF the capability to efficiently manage the route processing load efficiently.

Configuring OSPF

Configuring OSPF is similar to setting up IGRP and EIGRP. To get the OSPF process started, it requires a local process ID in the range of 1 to 65,535 (just like IGRP and EIGRP). In addition, like IGRP and EIGRP, it is possible to run multiple OSPF instances. This should only be done in special circumstances and generally avoided. Each OSPF process builds its own topological database and has its own routing algorithm process, so running multiple OSPF instances adds unneeded load to the router.

Where OSPF differs from a configuration standpoint is in the way the network announcements are entered. OSPF is a true classless protocol. When configuring RIPv2, EIGRP, and so on, networks are entered (and listed in the configuration) using their natural classful boundaries. OSPF provides the capability to enter the network announcements in their classless form, along with a *network mask* to interpret the network address. One thing about the netmask: It is entered in reverse form, just like access lists. Also, like access lists, the same calculation formula is used to calculate the reverse mask:

```
 255.255.255.255
-255.255.252.0   (Subnet mask)
_________________
   0.  0.  3.255 (Reverse mask)
```

The capability to enter the network and mask gives you a lot of flexibility in terms of how networks are announced. With other IGP protocols, network summarization can present problems, especially in large internetworks where classless addressing is used extensively. To address this problem, summarization is disabled with the router configuration subcommand `<no auto-summary>`. OSPF does not summarize unless you configure it to.

In large LANs where subnetting is used and the possibility of discontinuous subnets exists, auto-summarization can be a pain. In a large internetwork, however, summarization is an efficiency gain, since summarization reduces the number of routes required in the routing table. On large internetworks, whenever possible, you want to deploy your network address space using classless boundaries that can be treated as CIDR supernets. This makes it easy to use CIDR addressing to summarize network announcements. Establishing an effective network addressing hierarchy is essential for OSPF to function efficiently.

> **Note**
>
> A situation where multiple OSPF processes might be desirable is when the router is acting as gateway between two separate internetworks. A common example would be if OSPF was being used as an EGP by a network service provider and as an IGP by a client. The client gateway router would run both an internal and external OSPF process. The router would construct its routing table using announcements from both processes.

With this in mind, the test network has been readdressed to take advantage of address summarization. Figure 10.2, in addition to illustrating the network's readdressing, also shows the OSPF area partitioning. OSPF area IDs can be any number between 1 and 4,294,967,295 or a variation of the IP network address in dotted quad form. The Boston LAN networks use area ID 172.16.0.0 and the Albany LAN networks use area ID 192.168.0.0.

To get our feet wet, let's configure OSPF on `asbr-b1` and `asbr-a1`. The network's topology, as well as the IP addressing hierarchy, has an effect on how well OSPF functions. Because all inter-area traffic must flow across the backbone network, it makes sense to have all the WAN and backbone LAN networks reside in area 0.0.0.0 and to treat each of the respective LANs as their own area. When you are designing the network topology and/or partitioning the network into areas, it is a good idea to list which contains each of the routers and lists which router interface is in which area. This helps you visualize the traffic flow.

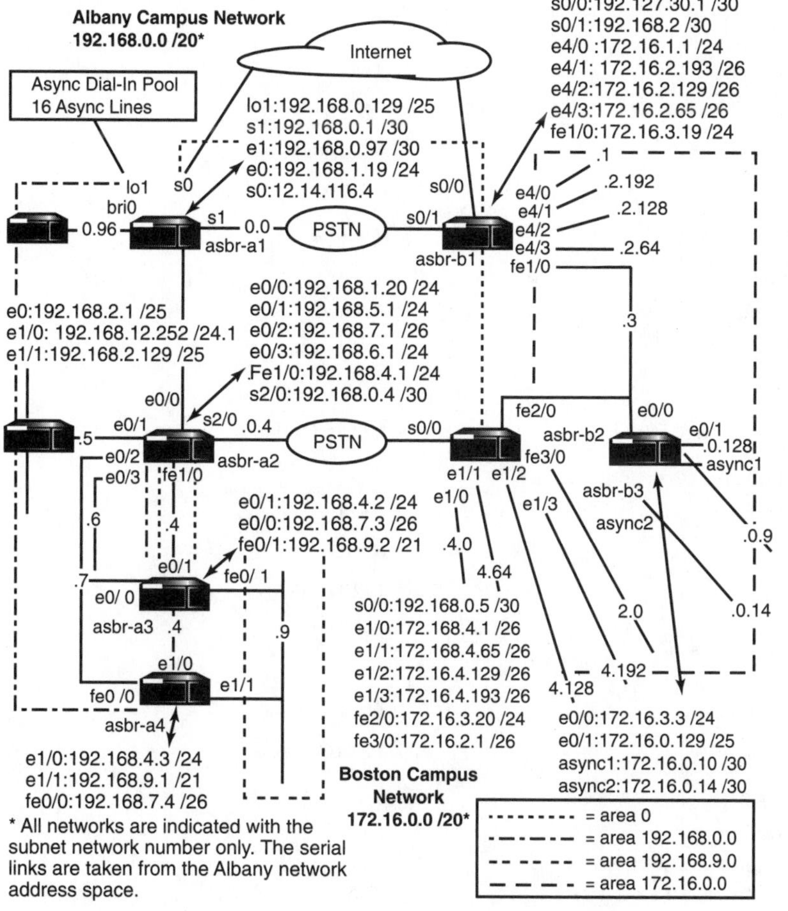

Figure 10.2 The testnet using CIDR summarization.

Table 10.3 **Router Interface Summary for the OSPF Networks in Figure 10.2**

	ALBANY
	ASBR-A1
Interface	**Area**
s1	0.0.0.0
e0	0.0.0.0
bri0	192.168.0.96
lo1	192.168.0.0
ASBR-A2	
Interface	**Area**
e0/0	0.0.0.0
e0/1	192.168.0.0
ASBR-A2	
Interface	**Area**
e0/2	192.168.0.0
e0/3	192.168.0.0
fe1/0	0.0.0.0
s2/0	0.0.0.0
ASBR-A3	
Interface	**Area**
e0/0	192.168.0.0
e0/1	0.0.0.0
fe0/1	192.168.9.0
ASBR-A4	
Interface	**Area**
e1/0	0.0.0.0
e1/1	192.168.9.0
fe0/0	192.168.0.0
IR-A1	
Interface	**Area**
All Int	192.168.0.0
IR-A2	
Interface	**Area**
All Int	192.168.0.0

Boston

ASBR-B1

Interface	Area
s0/1	0.0.0.0
fe1/0	0.0.0.0
e4/0	172.16.0.0
e4/1	172.16.0.0
e4/2	172.16.0.0
e4/3	172.16.0.0

ASBR-B2

Interface	Area
s0/0	0.0.0.0
fe2/0	0.0.0.0
e1/0	172.16.0.0
e1/1	172.16.0.0
e1/2	172.16.0.0
e1/3	172.16.0.0

ASBR-B3

Interface	Area
e0/0	0.0.0.0
e0/1	172.16.0.0
async1	172.16.0.0
async2	172.16.0.0

IR-B1

Interface	Area
All Int	172.16.0.0

IR-B2

Interface	Area
All Int	172.16.0.0

For OSPF, at least one interface must be configured on the router in order to enable the process. The OSPF process is enabled using the configuration EXEC command `<router ospf [local process id]>`. The PID is local only to the router, and OSPF uses the largest IP address configured on the router as its PID. If this interface is removed or shut down, the OSPF process will select a new ID and resend all its routing information. There is no way to specify which interface to use as the PID. The IOS will, however, choose a loopback interface over any other regardless of the IP size.

One method you can use to control the OSPF process IDs is to assign addresses from the unregistered space 192.168.254.0 /24 to loopback interfaces on each of the routers in the process. Each router would get a different incremental address, and this way you could directly associate an OSPF ID with each router in your network:

```
asbr-a1#config t
asbr-b2(config)# int loopback 1
asbr-a2(config-if)#ip address ip address 192.168.254.1 255.255.255.0
```

Entering OSPF Route Announcements

OSPF network entries are made with the OSPF configuration subcommand `<network <IP address/network number> <network mask> <area id>`. As mentioned before, with OSPF there is more than one way to configure network announcement entries. You will find that the best method is to use the network address and its corresponding mask. This way only the network space you want announced gets announced. It also forces you look at how you are using CIDR and VLSM beforehand, to ensure that you have your addressing straight. Here is the configuration for `asbr-b2` using this approach:

```
asbr-b2#config t
Enter configuration commands, one per line. End with CNTL/Z.
asbr-b2(config)#router ospf 202
asbr-b2(config-router)#network 192.168.0.0 0.0.0.3 area 0.0.0.0
asbr-b2(config-router)#network 172.16.1.0 0.0.0.255 area 172.16.0.0
asbr-b2(config-router)#network 172.16.2.192 0.0.0.63 area 172.16.0.0
asbr-b2(config-router)#network 172.16.2.128 0.0.0.63 area 172.16.0.0
asbr-b2(config-router)#network 172.16.2.64 0.0.0.63 area 172.16.0.0
asbr-b2(config-router)#network 172.16.3.0 0.0.0.255 area 0.0.0.0
```

An alternative to this is to use the network addresses of the connected interfaces. When using this approach, the netmask/wild card bits are set to 0.0.0.0. Router `asbr-a1` is configured using this approach:

```
asbr-a1(config)#router ospf 57
asbr-a1(config-router)#network 192.168.3.19 0.0.0.0 area 0.0.0.0
asbr-a1(config-router)#network 192.168.0.1 0.0.0.0 area 0.0.0.0
asbr-a1(config-router)#network 192.168.0.97 0.0.0.0 area 192.168.0.0
```

Now that both routers are configured, let's check the routing table on `asbr-a1` to make sure we are in fact getting announcements from `asbr-b2`:

```
asbr-a1#sh ip route
Codes: C - connected, S - static, I - IGRP, R - RIP, M - mobile, B - BGP
       D - EIGRP, EX - EIGRP external, O - OSPF, IA - OSPF inter area
       N1 - OSPF NSSA external type 1, N2 - OSPF NSSA external type 2
       E1 - OSPF external type 1, E2 - OSPF external type 2, E - EGP
       i - IS-IS, L1 - IS-IS level-1, L2 - IS-IS level-2, * - candidate
default
```

```
       U - per-user static route, o - ODR

Gateway of last resort is 12.14.116.6 to network 0.0.0.0

     192.168.160.0/30 is subnetted, 1 subnets
C       192.168.160.96 is directly connected, Ethernet0/0
     172.16.0.0/32 is subnetted, 4 subnets
O IA    172.16.2.129 [110/65] via 192.168.0.2, 00:00:51, Serial0/0
O IA    172.16.2.193 [110/65] via 192.168.0.2, 00:00:51, Serial0/0
O IA    172.16.1.1 [110/65] via 192.168.0.2, 00:00:56, Serial0/0
O IA    172.16.2.65 [110/65] via 192.168.0.2, 00:00:41, Serial0/0
C    192.168.191.0/24 is directly connected, Ethernet0/1
     192.168.0.0/30 is subnetted, 1 subnets
C       192.168.0.0 is directly connected, Serial0/0
     12.0.0.0/30 is subnetted, 1 subnets
C       12.14.116.4 is directly connected, Loopback1
S*   0.0.0.0/0 [1/0] via 12.14.116.6
```

According to the routing table, everything is fine; the 172.16.2.x /26 and 172.16.1.0 /24 network spaces are being announced as inter-area routes just as they are supposed to be.

As with RIP, EIGRP, and so on, it is possible to use just a natural address mask. This approach saves typing, but should be used with caution. If the address space is being used in different OSPF areas, routes will not be announced correctly. Let's reconfigure the `asbr-b1` router to see what can happen and how to avoid these problems.

Because the Boston network uses only address space from 117.12.0.0 /16 and 192.168.0.0 /24, we can save ourselves a lot of typing by just entering networks classfully:

```
asbr-b2(config)#router ospf 202
asbr-b2(config-router)#network 172.16.0.0 0.0.255.255 area 0.0.0.0
asbr-b2(config-router)#network 192.168.0.0 0.0.0.255 area 0.0.0.0
asbr-b2(config-router)#network 172.16.0.0 0.0.255.255 area 172.16.0.0
```

Again, let's look at the `asbr-a1` routing table:

```
asbr-a1>sh ip route
Codes: C - connected, S - static, I - IGRP, R - RIP, M - mobile, B - BGP
       D - EIGRP, EX - EIGRP external, O - OSPF, IA - OSPF inter area
       N1 - OSPF NSSA external type 1, N2 - OSPF NSSA external type 2
       E1 - OSPF external type 1, E2 - OSPF external type 2, E - EGP
       i - IS-IS, L1 - IS-IS level-1, L2 - IS-IS level-2, * - candidate
default
       U - per-user static route, o - ODR

Gateway of last resort is 12.14.116.6 to network 0.0.0.0

     192.168.160.0/30 is subnetted, 1 subnets
C       192.168.160.96 is directly connected, Ethernet0/0
```

```
         172.16.0.0/32 is subnetted, 4 subnets
O           172.16.2.129 [110/65] via 192.168.0.2, 00:05:05, Serial0/0
O           172.16.2.193 [110/65] via 192.168.0.2, 00:05:05, Serial0/0
O           172.16.1.1 [110/65] via 192.168.0.2, 00:05:05, Serial0/0
O           172.16.2.65 [110/65] via 192.168.0.2, 00:05:05, Serial0/0
C        192.168.191.0/24 is directly connected, Ethernet0/1
         192.168.0.0/30 is subnetted, 1 subnets
C           192.168.0.0 is directly connected, Serial0/0
         12.0.0.0/30 is subnetted, 1 subnets
C           12.14.116.4 is directly connected, Serial1/0
S*       0.0.0.0/0 [1/0] via 12.14.116.6
```

Do you see anything wrong? (Refer back to the previous route table.) Now, all the networks are being announced as area 0.0.0.0. Let's look at the running-config on asbr-b1:

```
!
router ospf 202
 network 172.16.0.0 0.0.255.255 area 0.0.0.0
 network 192.168.0.0 0.0.0.255 area 0.0.0.0
!
ip classless
!
line con 0
line 33 48
```

The second address and area entry was never accepted because it was considered part of the first announcement entry. If you change the entry to

```
asbr-b2(config)#router ospf 202
asbr-b2(config-router)#network 192.0.0.0 0.255.255.255 area 0.0.0.0
asbr-b2(config-router)#network 172.16.0.0 0.0.255.255 area 172.16.0.0
```

Then, all the 172.16.0.0 /16 networks appear as inter-area routes. This is not correct either because we want 172.16.3.0 /24 to be part of the backbone so we can have asbr-b3 function as an ABR. So, let's add 172.16.3.0 /24 to the OSPF configuration:

```
asbr-b2(config-router)#network 172.16.3.0 0.0.0.255 area 0.0.0.0
% OSPF: "network 172.16.3.0 0.0.0.255 area 0.0.0.0" is ignored. It is a
subset
of a previous entry.
```

Again, OSPF sees the subnet as part of the broad 172.16.0.0 /16 network. So, route announcement entry is ignored. To avoid this problem, always start with the specific networks first and then use the broader network space entries:

```
asbr-b2(config)#router ospf 202
asbr-b2(config-router)#network 172.16.3.0 0.0.0.255 area 0.0.0.0
asbr-b2(config-router)#network 192.168.0.0 0.0.0.255 area 0.0.0.0
asbr-b2(config-router)#network 172.16.0.0 0.0.255.255 area 172.16.0.0
```

To make sure that the correct interfaces are in the OSPF areas, you can use the user EXEC command <show ip ospf>:

```
sbr-b2#sh ip ospf
 Routing Process "ospf 202" with ID 172.16.2.193
 Supports only single TOS(TOS0) routes
 It is an area border router
 Summary Link update interval is 00:30:00 and the update due in 00:28:24
 SPF schedule delay 5 secs, Hold time between two SPFs 10 secs
 Number of DCbitless external LSA 0
 Number of DoNotAge external LSA 0
 Number of areas in this router is 2. 2 normal 0 stub 0 nssa
    Area BACKBONE(0.0.0.0)
        Number of interfaces in this area is 2
        Area has no authentication
        SPF algorithm executed 3 times
        Area ranges are
        Link State Update Interval is 00:30:00 and due in 00:28:17
        Link State Age Interval is 00:20:00 and due in 00:18:10
        Number of DCbitless LSA 0
        Number of indication LSA 0
        Number of DoNotAge LSA 0
    Area 172.16.0.0
        Number of interfaces in this area is 4
        Area has no authentication
        SPF algorithm executed 2 times
        Area ranges are
        Link State Update Interval is 00:30:00 and due in 00:28:22
        Link State Age Interval is 00:20:00 and due in 00:18:22
        Number of DCbitless LSA 0
        Number of indication LSA 0
        Number of DoNotAge LSA 0

asbr-b2#
```

This output tells us there are two interfaces in the backbone area and four interfaces in area 172.16.0.0, which is just how we configured it to be. As you can see, there is more than one way to configure announcements, and each has its own merits. The first two approaches are quite effective for most configurations. The third approach should be avoided unless you feel comfortable with address subnetting and IP masking.

OSPF Address Summarization

In the examples above, all the networks are announced individually. With OSPF, network summarization is an option permitted on ABRs. To configure route summarization on an ABR, the `<area>` OSPF subcommand is used. The `<area>` command is used for configuring several different options; this command is shown below. OSPF route summarization is limited to networks within the same OSPF area. To illustrate this, we are going to reconfigure `asbr-1a` to summarize 117.16.2.x /26 network spaces as 117.16.2.0 /23. Currently, these routes appear on `asbr-a1` as follows:

```
     172.16.0.0/32 is subnetted, 4 subnets
O IA    172.16.2.129 [110/65] via 192.168.0.2, 00:24:23, Serial0/0
O IA    172.16.1.1/32 [110/65] via 192.168.0.2, 00:37:25, Serial0/0
O IA    172.16.2.193 [110/65] via 192.168.0.2, 00:24:23, Serial0/0
O IA    172.16.2.65 [110/65] via 192.168.0.2, 00:24:23, Serial0/0
```

After all the involved networks are configured to be announced, the `<area>` command is used to establish the summarization network range:

```
asbr-b2(config-router)#area 172.16.0.0 range 172.16.2.0 255.255.255.0
```

Now, all the 172.16.2.x /26 networks are summarized as one /24 network:

```
     172.16.0.0/16 is variably subnetted, 2 subnets, 2 masks
O IA    172.16.1.1/32 [110/65] via 192.168.0.2, 00:37:25, Serial0/0
O IA    172.16.2.0/24 [110/65] via 192.168.0.2, 00:03:09, Serial0/0
```

Summarization can be done with any classless or classful address range, provided it falls on the correct boundaries. Now, this may not seem like much of a savings on a small network, but on a large internetwork where you are subnetting several Class A or B sized address spaces, every route counts. As you can see from the examples above, OSPF does require some forethought in terms of addressing segmentation and area planning. Failure to do so will often result in problems down the road, which can range from the annoying to the disasterous.

Adjusting OSPF Distance and Metrics

OSPF has a default administrative distance of 110 and uses a single route metric: route cost. This information, along with the route's age, route source, and accessible interface, is displayed as part of an OSPF-generated route entry in the route table. All regular OSPF routes are indicated with an "O"; an additional legend marker is to used indicate the type of OSPF route:

```
asbr-a1>sh ip route
Codes: C - connected, S - static, I - IGRP, R - RIP, M - mobile, B - BGP
       D - EIGRP, EX - EIGRP external, O - OSPF, IA - OSPF inter area
       N1 - OSPF NSSA external type 1, N2 - OSPF NSSA external type 2
       E1 - OSPF external type 1, E2 - OSPF external type 2, E - EGP
       i - IS-IS, L1 - IS-IS level-1, L2 - IS-IS level-2, * - candidate
default
```

```
       U - per-user static route, o - ODR

Gateway of last resort is 12.14.116.6 to network 0.0.0.0

     192.168.160.0/30 is subnetted, 1 subnets
C       192.168.160.96 is directly connected, Ethernet0/0
     172.16.0.0/32 is subnetted, 4 subnets
O       172.16.2.129 [110/65] via 192.168.0.2, 00:05:05, Serial0/0
O       172.16.2.193 [110/65] via 192.168.0.2, 00:05:05, Serial0/0
O       172.16.1.1 [110/65] via 192.168.0.2, 00:05:05, Serial0/0
O       172.16.2.65 [110/65] via 192.168.0.2, 00:05:05, Serial0/0
C    192.168.191.0/24 is directly connected, Ethernet0/1
     192.168.0.0/30 is subnetted, 1 subnets
```

The default administrative distance for all OSPF routes (which is 110) can be adjusted using the router subcommand <distance [10-255]>. It is also possible to adjust the distances of external, inter-area, and intra-area routes using the OSPF configuration subcommand <distance ospf [external/inter-area/inter-area] [10-255]>.

OSPF Route Types

An "O" type route indicates that the route is directly connected to the backbone area. "IA" routers indicate that the route is an inter-area. Backbone and inter-area route costs reflect the cost of the entire route path. "E1" indicates that route is an external, type 1 route. Type 1 metrics show the total cost of route path between the source and destination. All directly connected external networks use type 1 metrics since the actual cost to reach the network is the cost of the internal path to the ASB router. The "E2" marker is used for type 2 routes. Redistributed routes are announced as type 2 metrics. The route cost associated with a type 2 metric only reflects the cost to the source of the external route.

You should recall from Chapter 8, "TCP/IP Dynamic Routing Protocols," that OSPF uses a structured scale for establishing interface costs based on bandwidth, (see Table 10.4).

Table 10.4 **OSPF Interface Bit Value and Cost**

Network Bandwidth (in bits)	Bandwidth (in KBit/s)	Path Cost
100,000,000	100,000 (Fast Ethernet)	1
16,000,000	16,000 (16Mb Token Ring)	6
10,000,000	10,000 (Ethernet)	10
4,000,000	4,000 (4Mb Token Ring)	25
1,544,000	1,544 (serial T1)	64
64,000	640 (single B-ISDN)	1562
56,000	560 (single T-channel)	1786

continues

Table 10.4 **Continued**

Network Bandwidth (in bits)	Bandwidth (in KBit/s)	Path Cost
33,600	33.6 (codex modem)	3354
28,000	2.8 (28K baud modem)	2604
14,400	1.4.4 (14.4 baud modem)	5208
9,600	9.6 (9,600 baud modem)	10417
1,200	1.2 (1,200 baud modem)	83333

The importance of setting the interface's bandwidth correctly can not be stressed enough. Setting this parameter correctly ensures that the route costs are properly calculated. The route cost is the sum of all interface costs in the route path, divided by 100,000,000. The sum total is then rounded to the highest number, yielding the route cost. If your network uses interfaces that support bandwidth speeds above Fast Ethernet (in other words, Gigabit Ethernet, POS, Smart Trunk Interfaces), you might want to set OSPF interface costs manually:

- Disable OSPF auto-cost using `<no ospf auto-cost>` for IOS version 12.x and later, or `<no ospf auto-cost-determination>` with IOS 11.x.
- Set each interface's cost with the interface configuration EXEC command `<ip ospf cost>`.

Stub Networks

You will also notice on the `<show ip route>` route type code legend OSPF NSSA Type 1 (N1) and Type 2 (N2) routes. *NSSA* stands for *not so stubby area*. An *OSPF stub area* is an area with one or more outbound access paths that only receive internal routing information and cannot redistribute external routing information. Traffic destined for networks external to the local internetwork or AS use a default route which is statically set or generated by the ASBR and distributed as part of the OSPF announcement (see the section "Route Control and Redistribution," for more details). Commonly, a stub area is a small office/home office (SOHO) or remote office router connected to the internetwork over a DDR connection.

To minimize the amount of routing information sent across the link, the area is configured as a stub. To configure an area as a stub on the ABR, the command `<area [area-id] stub no summary>` is used. The DDR BRI interface on `asbr-a1` is configured as an OSPF stub area:

```
router ospf 67
 network 192.168.0.97 0.0.0.0 area 192.168.0.96
 area 192.168.0.96 stub no-summary
```

An NSSA network is very similar. An NSSA area only has routing information about the internal internetwork provided by the OSPF process. The difference is that an NSSA area can redistribute external routes into the general OSPF routing policy, so they can be forwarded to the rest of the internetwork. If the remote router attached to `asbr-a1` was a remote office that was connected to the larger internetwork, we could reconfigure area 192.168.0.96 on `asbr-a1` as an NSSA area and have the routing information redistributed into the OSPF process. The OSPF routing subcommand `<area [area id] nssa>` is used to designate an area as an NSSA:

```
asbr-a1(config-router)#area 192.168.0.96 nssa
```

The configuration on `ir-a2` looks like this:

```
rip 89
network 192.168.64.0
network 192.168.65.0
network 192.168.66.0
network 192.168.67.0

router ospf 34
network 192.168.0.98 0.0.0.0 area 192.168.0.96
area 192.168.0.96 nssa
redistribute rip subnets
```

All the routes redistributed from the NSSA network will appear as NSSA type 2 routes (N2). When configuring an area as an NSSA, all the routers in the area must be configured to recognize the area as an NSSA.

Adjusting OSPF's Behavior

It is not always possible to construct a network topology that allows for proper router/area segmentation. It is possible to connect an ABR across a non-area-zero transit area. A virtual link is established between a true ABR and the "virtual ABR" via the OSPF subprocess configuration command `<area [transit area id] virtual-link [router id]>`. Each side of the link uses the router id of the remote end. Let's examine a virtual link configuration between `ir-a1` and `asbr-a2` (the `<show ip ospf>` command will display the router's OSPF ID):

```
hostname ir-a1
!
router ospf 78
network 192,168.12.252 0.0.0.0 area 192.168.0.0
network 192.168.2.0 0.0.0.255 area 0.0.0.0
area 192.168.0.0 virtual-link 192.168.254.5

hostname asbr-a2
!
```

```
router ospf 78
 network 192.168.0.5 0.0.0.0 area 0.0.0.0
 network 192.168.1.19 0.0.0.255 area 0.0.0.0
 network 192.168.4.1 0.0.0.0 area 192.168.0.0
 network 192.168.6.1 0.0.0.0 area 192.168.0.0
 network 192.168.5.0 0.0.0.266 area 192.168.0.0
 network 192.168.7.1 0.0.0.63 area 192.168.0.0
 area 192.168.0.0 virtual-link 192.168.254.4
```

OSPF, like RIP, EIGRP, and IGRP, supports passive mode on any interfaces. To configure an interface to only receive OSPF messages, use the general router configuration subprocess command <passive-interface>.

OSPF Authentication

OSPF supports both text password and MD5 key authentication to verify route message exchanges between routers. Authentication needs to be enabled as part of the OSPF process using the OSPF configuration subcommand <area authentication> for clear text or <area authentication [message-digest]> to use MD5 key authentication (see the section, "Configuring Authentication" for MD5 key generation instructions).

All the testnet area 0.0.0.0 routers use clear text authentication; the password is "quickkey". Let's take a look at asbr-a1 and asbr-a2's configurations:

```
hostname asbr-a1
!
interface Ethernet0
 ip address 192.168.1.19 255.255.255.0
 no ip directed-broadcast
 ip ospf authentication-key quickkey
!
!
interface Serial1
 ip address 192.168.0.1 255.255.255.0
 no ip directed-broadcast
 ip ospf authentication-key quickkey
!
router ospf 67
 network 192.168.0.97 0.0.0.0 area 192.168.0.96
 network 192.168.1.0 0.0.0.255 area 0.0.0.0
 network 192.168.0.0 0.0.0.3 area 0.0.0.0
 network 192.168.0.128 0.0.0.128 area 192.168.0.0
 area 0.0.0.0 authentication
 area 192.168.0.96 nssa

hostname asbr-a2
!
```

```
interface Ethernet0
 ip address 192.168.1.20 255.255.255.0
 ip ospf authentication-key quickkey
!
interface Serial0
 ip address 192.168.0.5 255.255.255.252
 ip ospf authentication-key quickkey
!
router ospf 78
 network 192.168.0.5 0.0.0.0 area 0.0.0.0
 network 192.168.4.1 0.0.0.0 area 192.168.0.0
 network 192.168.7.1 0.0.0.0 area 192.168.0.0
 network 192.168.6.1 0.0.0.0 area 192.168.0.0
 network 192.168.5.1 0.0.0.0 area 192.168.0.0
 network 192.168.1.20 0.0.0.0 area 0.0.0.0
 area 192.168.0.0 virtual-link 92.168.254.4 authentication-key quickkey
 area 0.0.0.0 authentication
```

After authentication has been enabled for an OSPF area, all message exchanges halt until all of the authentication keys have been set on the involved interfaces. Do not forget that all the keys have to be the same in an area.

Designated and Backup Designated Router Election

It is possible to configure which routers will be the designated and backup designated routers within an area by using the `<ip ospf priority [0-255]>` interface command, the highest priority value winning the election. By default, the IOS sets all OSPF interfaces with a priority of 1. When configuring your router, set higher priority values on the router interface you want to act as DR and BDR. It is also possible to exempt an interface from the DR/BDR election by setting its priority to zero. This, like most of the configuration options available with OSPF, requires planning. Simply changing a router's interface will not make it the areas DR or BDR if they already exist. Only a re-election will make the change. In situations where a DR and BDR already exist and you want to force an election, simply shut down the DR/BDR interface momentarily; this will force an election which will be won by your re-prioritized interface.

> **Note**
>
> The use of virtual interfaces is highly discouraged and generally reflects poor planning on the part of the network designer. It was included only to provide transitional capability, not features.

Using OSPF with Nonbroadcast and DDR Networks

In Chapter 8, "TCP/IP Dynamic Routing Protocols," we reviewed how OSPF classifies connections into three categories:

- Broadcast
- Point-to-point
- Nonbroadcast multiaccess networks

When you use a transport medium that does not support broadcasts (Frame Relay being the most common type), the routers involved need to have a means to exchange OSPF messages. To get OSPF messages to exchange properly, you need to configure the interface to conform to the type of network connection (point-to-point, multipoint, or point-to-multipoint) you have in place. If you have a point-to-point or multipoint Frame Relay connection, you can configure the interface as a broadcast or nonbroadcast network using the interface configuration subcommand `<ip ospf network [broadcast or non-broadcast]>`. If you configure a Frame Relay interface (or any other media type, such as Ethernet, Token Ring, and so on) as a nonbroadcast network type, you need to specify a neighbor with which to exchange OSPF messages. The neighbor is specified with the OSPF configuration subcommand `<neighbor [ip address] [priority (0-255)]>`:

```
interface Serial0
ip address 192.168.0.1 255.255.255.252
encapsulation frame-relay
frame-relay lmi-type ansi
ip ospf network non-broadcast
!
ospf 87
network 192.168.0.1 0.0.0.3 area 0.0.0.0
neighbor neighbor 192.168.0.2 priority 2
```

It is possible to fool point-to-point and multipoint frame-relay interfaces into acting like broadcast media by configuring them as `<ip ospf network broadcast>` interface types. Provided that each link is directly connected, this approach will work fine. With this scenario, only routers with non-zero priority values need to have `<neighbor>` routers specified.

In situations where Frame Relay is being used in a point-to-multipoint scenario, the `<ip ospf network [point-to-multipoint]>` network type is specified. Under these circumstances, neighbor routers are not used. A point-to-multipoint scenario uses a single IP address space, creating effectively a virtual subnet, where the multipoint Frame Relay router acts as a traffic bridge between disconnected point-to-point routers (see Figure 10.3).

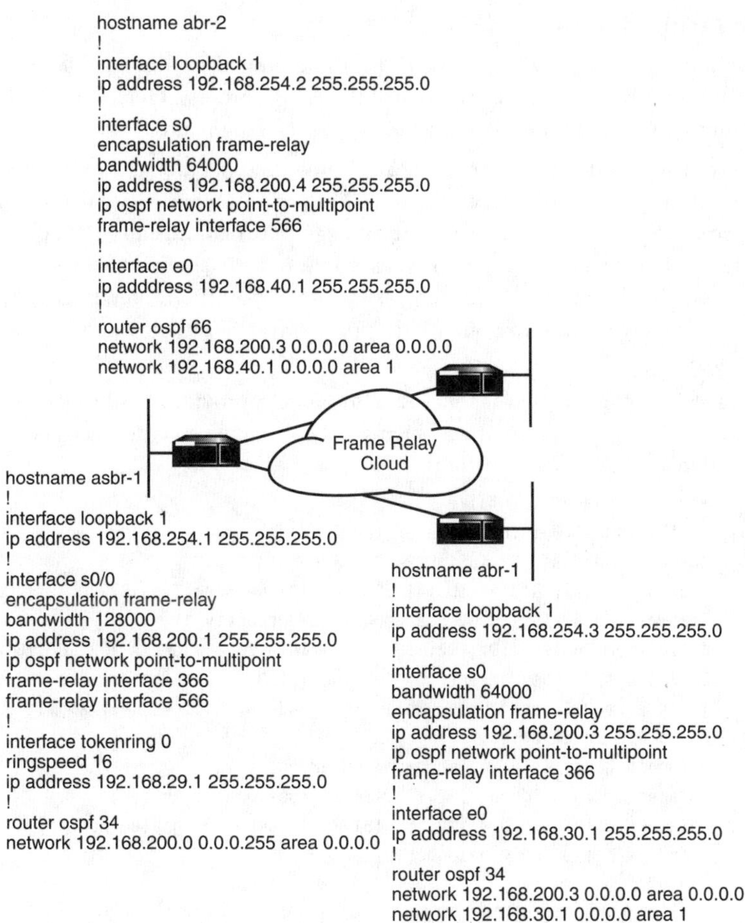

hostname abr-2
!
interface loopback 1
ip address 192.168.254.2 255.255.255.0
!
interface s0
encapsulation frame-relay
bandwidth 64000
ip address 192.168.200.4 255.255.255.0
ip ospf network point-to-multipoint
frame-relay interface 566
!
interface e0
ip adddress 192.168.40.1 255.255.255.0
!
router ospf 66
network 192.168.200.3 0.0.0.0 area 0.0.0.0
network 192.168.40.1 0.0.0.0 area 1
Frame Relay Cloud
hostname asbr-1
!
interface loopback 1
ip address 192.168.254.1 255.255.255.0
!
interface s0/0
encapsulation frame-relay
bandwidth 128000
ip address 192.168.200.1 255.255.255.0
ip ospf network point-to-multipoint
frame-relay interface 366
frame-relay interface 566
!
interface tokenring 0
ringspeed 16
ip address 192.168.29.1 255.255.255.0
!
router ospf 34
network 192.168.200.0 0.0.0.255 area 0.0.0.0
hostname abr-1
!
interface loopback 1
ip address 192.168.254.3 255.255.255.0
!
interface s0
bandwidth 64000
encapsulation frame-relay
ip address 192.168.200.3 255.255.255.0
ip ospf network point-to-multipoint
frame-relay interface 366
!
interface e0
ip adddress 192.168.30.1 255.255.255.0
!
router ospf 34
network 192.168.200.3 0.0.0.0 area 0.0.0.0
network 192.168.30.1 0.0.0.0 area 1

Figure 10.3 A point-to-multipoint Frame Relay mesh.

When using DDR links with OSPF, it is quite common to filter out OSPF messages from seeming like interesting outbound traffic to bring up the link. It is, however, desirable to have the routing tables across the routing domain as accurate as possible. With the interface command `<ip ospf demand-circuit>`, it is possible to have the local end router suppress periodic OSPF messages that contain no topology changes. If an OSPF message does contain topology change information and the DDR link is brought up, the messages are passed on to the remote router. When the link is up for normal data use, OSPF messages are exchanged normally. This feature is enabled on the network access router, not the remote router.

OSPF Monitoring Commands

Throughout the section, we have used various commands to display different types of information relating to the functioning of the OSPF process. Here is a command and result summary of the more useful `<show ip ospf>` commands:

- `<show ip ospf [local process id]>` provides operational information on the OSPF process—in other words, what kind of OSPF process is being deployed, area information, link state update and age information, and SPF algorithm statistics. This command can also be executed just as `<show ip ospf>`. In this form, it will list all the local OSPF processes running on the router. Because you should only be running one OSPF process, you can use just this command form.
- `<show ip ospf interface [type/slot/port]>` provides OSPF-related information about a specific interface, such as its cost, network type, adjacencies, and designated router information:

```
asbr-a1>ip ospf int fa1/0
FastEthernet1/0 is up, line protocol is up
Internet Address 192.168.19.20/24, Area 0.0.0.0
Process ID 101, Router ID 147.225.181.19, Network Type BROADCAST, Cost: 1
Transmit Delay is 1 sec, State DROTHER, Priority 1
Designated Router (ID) 192.168.181.2, Interface address 192.168.181.2
Backup Designated router (ID) 192.168.181.3, Interface address 192.168.181.3
Timer intervals configured, Hello 10, Dead 40, Wait 40, Retransmit 5
Hello due in 00:00:03
Neighbor Count is 6, Adjacent neighbor count is 2
Adjacent with neighbor 192.168.181.2  (Designated Router)
Adjacent with neighbor 192.168.181.3  (Backup Designated Router)
Suppress hello for 0 neighbor(s)
```

- `<show ip ospf neighbor>` provides information about the router's relationship to the other routers. Who are the designated and backup designated routers and what are those the routers' adjacency states and interface priorities?

```
asbr-b2>sh ip ospf neighbor

Neighbor ID     Pri   State        Dead Time   Address         Interface
12.14.116.5       1   FULL/  -     00:00:34    192.168.0.1     Serial0/0

<show ip ospf database> displays the routers entire link state (topological)
database.

             OSPF Router with ID (192.168.254.1) (Process ID 57)


                       Router Link States (Area 0.0.0.0)

Link ID         ADV Router      Age      Seq#        Checksum Link count
12.14.116.5     12.14.116.5     1109     0x80000011 0x2CAB    3
```

```
172.16.2.193     172.16.2.193    144      0x8000000D 0xC737    3

                Summary Net Link States (Area 0.0.0.0)

Link ID         ADV Router      Age      Seq#       Checksum
172.16.1.1      172.16.2.193    756      0x80000007 0xC135
172.16.2.65     172.16.2.193    756      0x80000007 0x3481
172.16.2.129    172.16.2.193    757      0x80000007 0xB1C3
172.16.2.193    172.16.2.193    757      0x80000007 0x2F06
192.168.160.96  12.14.116.5     1043     0x80000006 0x4BE7

                Router Link States (Area 192.168.0.0)

Link ID         ADV Router      Age      Seq#       Checksum Link count
12.14.116.5     12.14.116.5     1042     0x80000006 0xBDD6    1

                Summary Net Link States (Area 192.168.0.0)

Link ID         ADV Router      Age      Seq#       Checksum
172.16.1.1      12.14.116.5     1043      0x80000006 0x475D
172.16.2.65     12.14.116.5     1290      0x80000006 0xB9A9
172.16.2.129    12.14.116.5     1290      0x80000006 0x37EB
172.16.2.193    12.14.116.5     1290      0x80000006 0xB42E
172.16.91.0     12.14.116.5     1290      0x80000006 0xC978
192.168.0.0     12.14.116.5     1290      0x80000006 0x14E9
192.168.191.0   12.14.116.5     1291      0x80000006 0xCAA6
```

- <show ip ospf virtual-links> will display information about the state of virtual links running on the router.

Note

IOS also provides an extensive <debug> toolkit for viewing the various database and functional elements that make up the OSPF process:

```
asbr-a2#debug ip ospf ?
  adj             OSPF adjacency events
  events          OSPF events
  flood           OSPF flooding
  lsa-generation  OSPF lsa generation
  packet          OSPF packets
  retransmission  OSPF retransmission events
  spf             OSPF spf
  tree            OSPF database tree
```

Note

As with all debug commands, these should be used with caution and disabled when no longer required because they can impact router performance. If you are experiencing difficulties with OSPF, start out with the <debug ip ospf events> command.

Using AURP

Display commands:

```
<show appletalk aurp events>
<show appletalk aurp topology>
<show appletalk routes>
<show appletalk zone>
Configuration commands:
<interface tunnel [number]>
<appletalk protocol AURP>
<tunnel source [ip address]>
<tunnel destination [ip address]>
<tunnel mode [aurp]>
```

In Chapter 3, "The Networker's Guide to AppleTalk, IPX, and NetBIOS," we discussed the function of AppleTalk's AURP (AppleTalk Update Based Routing Protocol), which is an enhancement to the RTMP (Routing Table Maintenance Protocol) that extends AppleTalk networks across large IP internetworks. AURP works by essentially encapsulating AppleTalk routing and data packets into TCP/IP UDP packets.

IOS implements AURP using `<tunnel>` interfaces. To establish the tunnel, a router on each of the LANs creates a tunnel interface. When the tunnel is created, it is bound to a "real" router interface that serves as the tunnel's source address. This "source" address is used by the remote end to send data to the tunnel interface. Along with a tunnel "source", a tunnel "destination" address is configured. The "destination" address is the "real" interface where all the tunnel's outbound traffic is addressed. When configuring an AURP tunnel, the source interface is the "IP only" interface on the router. The destination interface is the "IP only" interface of the remote router. Figure 10.4 illustrates the different "interfaces" involved with configuring the AURP tunnel.

AppleTalk must be enabled on the router(s) the tunnel interfaces are configured on because the AppleTalk network needs to be directly accessible through one of the tunnel router's interfaces.

To monitor AURP, the <show appletalk aurp events> and <show appletalk aurp topology> commands are available in the user EXEC. The <show appletalk route> and <show appletalk zone> user EXEC commands are also useful when setting up AURP.

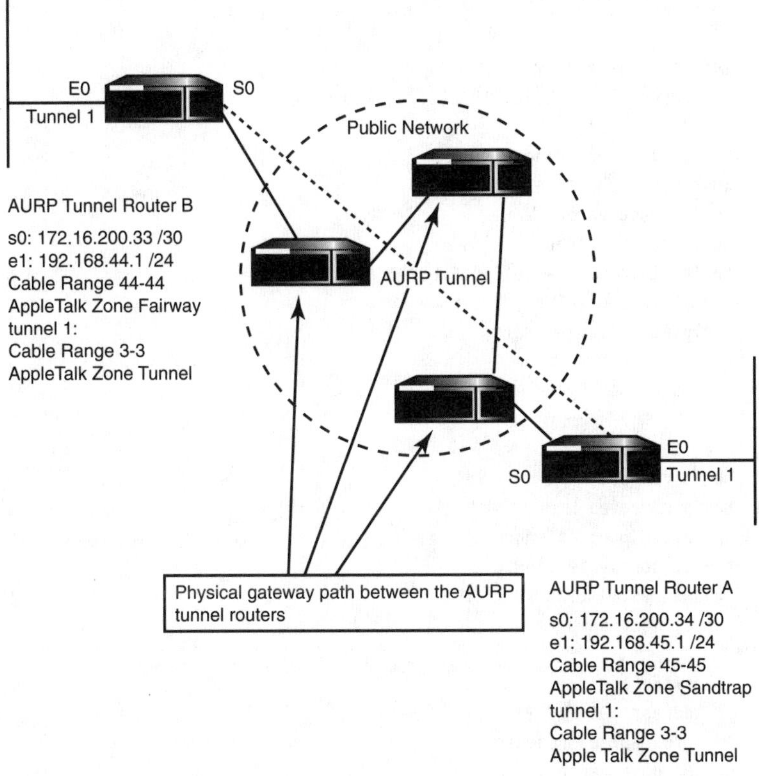

Figure 10.4 AURP tunnel interfaces.

Using our sample network, illustrated in Figure 10.5, let's configure a AURP tunnel between routers asbr-a1 and asrb-b1:

```
hostname asbr-a1
!
interface Tunnel1
 no ip address
 no ip directed-broadcast
 appletalk protocol aurp
 tunnel source 192.168.0.1
 tunnel destination 192.168.0.2
 tunnel mode aurp

 interface serial1
 ip address 192.168.0.1
 no ip directed-broadcast
 no ip route-cache
```

```
 no ip mroute-cache

 interface Ethernet0
 ip address 192.168.191.19 255.255.255.0
 no ip redirects
 no ip directed-broadcast
 ip route-cache same-interface
 no ip route-cache
 no ip mroute-cache
 appletalk cable-range 65015-65015 65015.117
 appletalk zone Albany Backbone
 appletalk glean-packets

 hostname asbr-b1
!
interface Tunnel1
 no ip address
 no ip directed-broadcast
 appletalk protocol aurp
 tunnel source 192.168.0.2
 tunnel destination 192.168.0.1
 tunnel mode aurp

 interface serial1
 ip address 192.168.0.2
 no ip directed-broadcast
 no ip route-cache
 no ip mroute-cache

 interface Fast Ethernet1/0
 ip address 172.16.3.19 255.255.255.0
 no ip redirects
 no ip directed-broadcast
 ip route-cache same-interface
 no ip route-cache
 no ip mroute-cache
 appletalk cable-range 60001-60001 60001.125
 appletalk zone Boston Backbone
 appletalk glean-packets
```

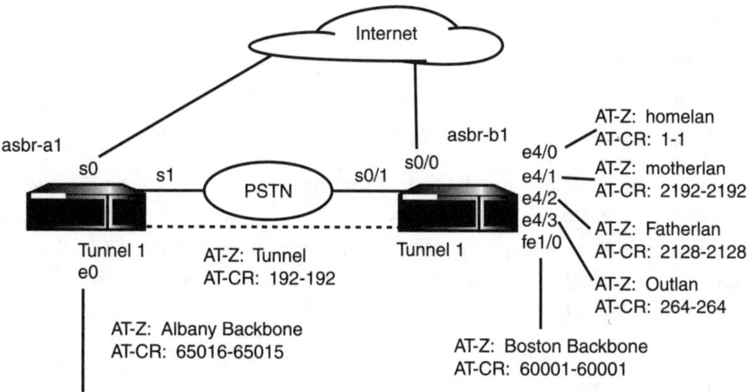

Figure 10.5 AppleTalk network segments on `asbr-a1` and `asbr-b1`.

Using BGP

Display commands:

```
<show ip bgp>
<show ip bgp summary>
<show ip bgp neighbor>
<show ip bgp peer-group>
```

Configuration commands:

```
<router bgp [as number]>
<network [ip network address] mask [subnet mask]>
<neighbor [ip address] remote-as [remote-as number]>
<neighbor [peer-group name] peer-group>
<neighbor [peer-group]remote-as [local-as number]>
<neighbor [ip address|peer-group] remote-as [as-number]>
<neighbor [ip address|peer-group] soft reconfiguration>
<neighbor [ip address|peer-group] weight [weight]>
<neighbor [ip address|peer-group] version [1-4]>
<neighbor [ip address|peer-group] [password]>
<neighbor [ip address] [peer-group name]>
<neighbor [ip address|peer-group] ebgp-multihop>
<timers bgp [keepalive] [holdtime]>
<distance [external] [internal] [local]>
<no auto-summary>
<neighbor [ip address|peer-group] [distribute-list] [acl][in|out]>
<redistribute [protocol] [Process ID] [metric] [weight]>
<neighbor [ip address] filter-list [acl number] weight[0 65535]>
```

Control commands:

```
<clear ip bgp [*|ip address]>
```

Additional commands:

```
<ip as-path access-list [acl number] [permit|deny] [string]>
```

For many users, the usefulness of BGP is questionable. BGP, althoughconsidered a routing protocol, could be more aptly described as a network reachability protocol. Its focus is not on announcing how to reach a destination, but rather on how to get somewhere that can reach the destination. With this goal in mind, let's review the three types of routing BGP is used for:

- Inter-autonomous system routing—A collection (two or more) routers from different autonomous systems exchange BGP information with the goal of maintaining a complete reachability table for the entire intranetwork.
- Intra-autonomous system routing—Used by peer-routers connected to the same autonomous system to exchange external AS information.
- Pass-through autonomous system routing—Used to exchange BGP reachability information between a collection of routers belonging to the same or different ASs across an uninvolved third-party AS. In this case, the ASs exchanging reachability information are dependent on the third-party AS's IGP protocol to transport the BGP traffic.

As stated earlier, almost all applications of BGP occur within the context of the public Internet. Most private enterprise networks do not have these types of requirements, and those that do can safely accomplish their goals with OSPF or EIGRP. However, with businesses' growing dependency on the Internet and the hyper-growth of many corporate intranets, the usefulness of BGP is finally reaching the private network. It is particularly applicable to environments where access to multiple private/public networks links is required. Because BGP was designed to manage the Internet, it is powerful, and as you can imagine, can be quite complex to configure. With this in mind, we will look at some basic inter-autonomous and intra-autonomous BGP configurations.

Inter-Autonomous BGP

For most BGP applications, the goal is to provide network-reachability information for your local network between two or more ISP gateways. In this situation, your goal is not to configure a large BGP routing policy. All you want is to have Internet access redundancy. An easy way to achieve this goal is to use *ships in the night (SIN) routing*. SIN routing exists where a router is running multiple dynamic protocols, without redistributing them into one another. The effect is that only the SIN router processes a complete internal and external network routing table. The other internal network routers use the SIN router as the default gateway, so they blindly forward all of their external traffic to the SIN router, which then uses its full routing table to decide the best next hop. When configuring BGP for the SIN routing context, you want to set up the BGP process on your internal network's gateway router instead of configuring BGP on each of your Internet access gateways.

Figure 10.6 illustrates a SIN routing router topology configuration example.

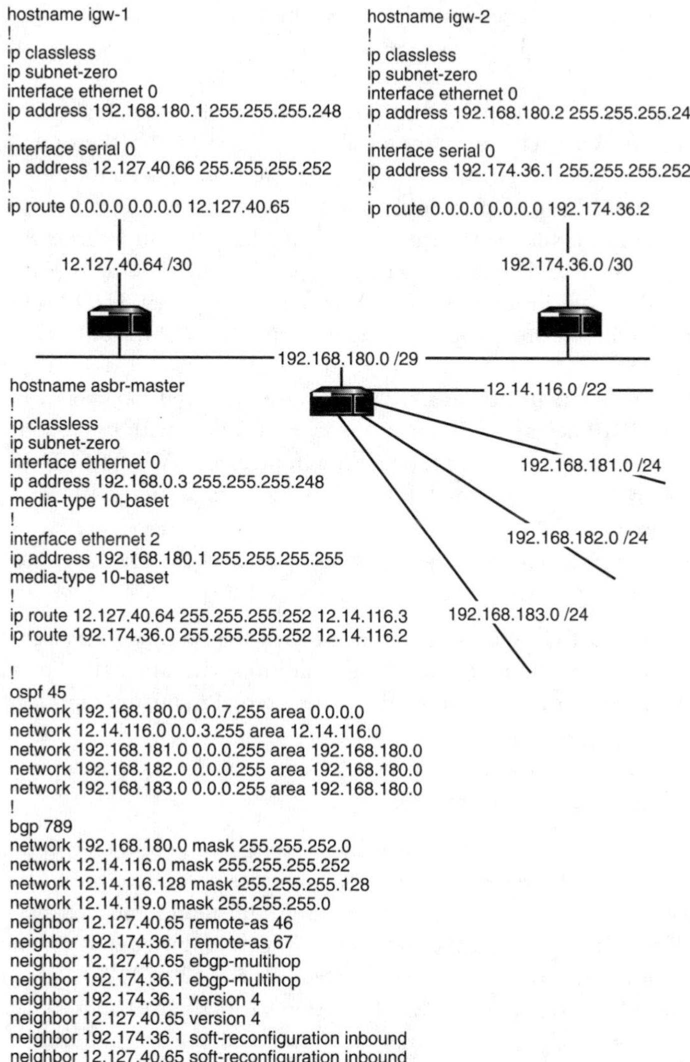
hostname igw-1
!
ip classless
ip subnet-zero
interface ethernet 0
ip address 192.168.180.1 255.255.255.248
!
interface serial 0
ip address 12.127.40.66 255.255.255.252
!
ip route 0.0.0.0 0.0.0.0 12.127.40.65
hostname igw-2
!
ip classless
ip subnet-zero
interface ethernet 0
ip address 192.168.180.2 255.255.255.248
!
interface serial 0
ip address 192.174.36.1 255.255.255.252
!
ip route 0.0.0.0 0.0.0.0 192.174.36.2
12.127.40.64 /30
192.174.36.0 /30
192.168.180.0 /29
12.14.116.0 /22
192.168.181.0 /24
192.168.182.0 /24
192.168.183.0 /24
hostname asbr-master
!
ip classless
ip subnet-zero
interface ethernet 0
ip address 192.168.0.3 255.255.255.248
media-type 10-baset
!
interface ethernet 2
ip address 192.168.180.1 255.255.255.255
media-type 10-baset
!
ip route 12.127.40.64 255.255.255.252 12.14.116.3
ip route 192.174.36.0 255.255.255.252 12.14.116.2
!
ospf 45
network 192.168.180.0 0.0.7.255 area 0.0.0.0
network 12.14.116.0 0.0.3.255 area 12.14.116.0
network 192.168.181.0 0.0.0.255 area 192.168.180.0
network 192.168.182.0 0.0.0.255 area 192.168.180.0
network 192.168.183.0 0.0.0.255 area 192.168.180.0
!
bgp 789
network 192.168.180.0 mask 255.255.252.0
network 12.14.116.0 mask 255.255.255.252
network 12.14.116.128 mask 255.255.255.128
network 12.14.119.0 mask 255.255.255.0
neighbor 12.127.40.65 remote-as 46
neighbor 192.174.36.1 remote-as 67
neighbor 12.127.40.65 ebgp-multihop
neighbor 192.174.36.1 ebgp-multihop
neighbor 192.174.36.1 version 4
neighbor 12.127.40.65 version 4
neighbor 192.174.36.1 soft-reconfiguration inbound
neighbor 12.127.40.65 soft-reconfiguration inbound

Figure 10.6 A multipoint access Internet gateway model.

The example shows both ISP routers being directly accessible from the internal gateway router. The SIN approach can also be followed using a single internal/Internet access router where the internal network and the multiple Internet links are all homed into the same router. This approach does have some drawbacks. The first is the fact that there is no hardware redundancy; if the internal gateway fails, you are completely down. To overcome this and still maintain a single gateway for end-stations and servers, use HSRP (or VIRP) with two gateway routers. The second drawback has to do with security. Internet gateways are managed by you and in part by your ISP. It is usually not a good idea to have your ISP accessing your internal network gateway router.

After the connectivity has been established between the internal gateway router and the Internet gateway routers, you want to establish a local BGP process on your gateway router. This is started using the global configuration EXEC command `<router bgp [as number]>`. If you are planning to use BGP to provide reachability information over the Internet, you will need a registered AS number from an Internet Addressing Authority (IAA), such as the American Registry for Internet Numbers (ARIN). For unannounced BGP (for internal use only), any number between 32768 and 64511 will do. For this example, we are going to announce networks 12.14.116.0 /22 and 192.168.180 /22.

Before we start configuring our BGP process, we first need to establish routes to our remote-AS BGP neighbors. In most cases, you will peer with the router on the other end of your Internet link. In this case, the routers are (ATT) 12.127.40.65 and (PSI) 192.144.36.1. We can accomplish peering with static routes:

```
ASBR-Master#config t
ASBR-Master(config)#ip route 12.127.40.64 255.255.255.252 12.14.116.3
ASBR-Master(config)#ip route 192.174.36.0 255.255.255.252 12.14.116.2
```

Now, let's enable our BGP process. The process ID is the AS number, which, in this case, is 789. The process ID is used for communication with other BGP processes. If you are planning to run BGP on more than one internal router, be sure that the same AS is used:

```
ASBR-Master(config)#router bgp 789
ASBR-Master(config-router)#
```

After the process starts, we add the networks we want announced. Like OSPF, BGP uses netmasks as part of the network number announcements. This means that you can announce parts of networks as VLSM subnets, as CIDR address ranges, or as classful networks. All that matters is that the addressing has correct bit boundaries and that you want them announced. For example, we do not want the 12.14.117.0 and 118.0 spaces announced and we are using 12.14.116.0 in various VLSM address segments. So, we will only announce the networks we know should be accessed from the Internet:

```
ASBR-Master(config-router)#network 12.14.116.0 mask 255.255.255.252
ASBR-Master(config-router)#network 12.14.116.128 mask 255.255.255.128
ASBR-Master(config-router)#network 12.14.119.0 mask 255.255.255.0
ASBR-Master(config-router)#network 192.168.180.0 mask 255.255.252.0
```

Now, let's configure our remote neighbors, using the BGP configuration subprocess command <neighbor [ip-address] remote-as [as number]>:

```
ASBR-Master(config-router)#neighbor 12.127.40.65 remote-as 46
ASBR-Master(config-router)#neighbor 192.174.36.1 remote-as 67
ASBR-Master(config-router)#neighbor 192.174.36.1 version 4
ASBR-Master(config-router)#neighbor 12.127.40.65 version 4
ASBR-Master(config-router)#neighbor 192.174.36.1 soft-reconfiguration inbound
ASBR-Master(config-router)#neighbor 12.127.40.65 soft-reconfiguration inbound
```

The <neighbor [ip address] version 4> command forces which version of BGP will be used, instead of using the default negotiation process which starts at version 4 and moves downward until a version match is found. The <neighbor [ip address] soft-reconfiguration inbound> command indicates that changes to the inbound BGP configuration should be enabled automatically. Generally, whenever a change is made to the BGP configuration, the connection between neighbors needs to be reset. This is accomplished with <clear ip bgp *> (to reset all sessions) or <clear ip bgp [neighbor ip address]> to reset a specific session. With our configuration, our outbound connections need to be reset whenever we make a change.

After BGP routing is established with our Internet access providers, we can also establish BGP adjacencies with other ASs, if we desire. To establish a peer session with BGP neighbors that are not accessible over a direct connection, the <neighbor [ip address] ebgp-multihop> BGP configuration subprocess command must added to the BGP configuration in addition to the <remote-as> declaration. In this example, our BGP router is the default gateway for the internetwork. Once the externally destined traffic arrives, it will decide which ISP gateway to forward the traffic to based on its local routing table. Alternatively, we can redistribute the BGP derived routing information into the local IGP process.

When using BGP to provide reachability information about networks within the AS (in perhaps a large global private internetwork), two approaches can be taken:

- The first is to create BGP peer groups, which will exchange internal BGP routing information and provide gateway services for the local internetworks.
- The second is to redistribute the BGP process into the IGP process and have the IGP routers determine the destination path.

The IBGP approach will be discussed in the next section. BGP redistribution will be covered in the section "Route Control and Redistribution." In most cases, you want to avoid BGP-to-IGP and IGP-to-BGP redistribution.

Intra-Autonomous BGP Configuration

In the inter-autonomous BGP example, we used a single "distribution" router to act as a gateway between the BGP and IGP processes. In large private enterprise network environments, it is not uncommon for each location to have an Internet gateway and one or more internal access gateways to some or all of the other internal network sites. Our testnet example uses just such a network model. When configuring Internal BGP (IBGP), all the IBGP routers need to have peering sessions with all the other IBGP routers.

Let's examine the IBGP/EBGP configurations on `asbr-a1` and `asbr-a2`. For this example, the test network uses AS number 30001. Both Albany and Boston use an IGP protocol to manage local route announcement. BGP is used to advertise routes over the private WAN links and with the ISPs. `asbr-a1/a2` and `asbr-b1/b2` use identical BGP configurations. Here are the BGP configurations for `asbr-a1` and `asbr-b1`:

```
hostname asbr-a1
!
autonomous-system 30001
!
router bgp 30001
network 192.168.0.0 mask 255.255.255.252
network 192.168.0.4 mask 255.255.255.252
network 192.168.0.128 mask 255.255.255.128
network 192.168.1.0 mask 255.255.255.0
network 192.168.2.0 mask 255.255.255.0
network 192.168.3.0 mask 255.255.255.0
network 192.168.4.0 mask 255.255.255.0
network 192.168.5.0 mask 255.255.255.0
network 192.168.6.0 mask 255.255.255.0
network 192.168.7.0 mask 255.255.255.0
network 192.168.8.0 mask 255.255.255.0
network 192.168.9.0 mask 255.255.255.0
network 12.14.116.4 mask 255.255.255.252
neighbor 192.168.0.6 remote-as 30001
neighbor 192.168.0.6 filter-list 30001 4 weight 45000
neighbor 12.14.116.5 remote-as 8446
neighbor 12.14.116.5 weight 40000
neighbor 192.127.30.1 remote-as 899
neighbor 192.127.30.1 ebgp-multihop
neighbor 192.127.30.1 weight 30000
neighbor 192.168.1.20 remote-as 30001
neighbor 192.168.1.20 filter-list 30001 4 weight 42000
no auto-summary
```

```
!
ip as-path access-list 4 permit 30001
```

```
hostname asbr-b1
!
autonomous-system 30001
!
network 192.127.30.0  mask 255.255.255.252
network 172.16.0.0 mask 255.255.240.0
neighbor 192.127.30.1 weight 40000
neighbor 192.127.30.1 remote-as 899
neighbor 12.14.116.5 remote-as 8446
neighbor 12.14.116.5 weight 30000
neighbor 12.14.116.5 ebgp-multihop
neighbor 192.168.0.5 remote-as 30001
neighbor 172.16.2.30 remote-as 30001
neighbor 192.168.0.5 filter-list 4 weight 45000
neighbor 172.16.2.30 filter-list 4 weight 42000
no auto-summary
!
ip as-path access-list 4 permit 30001
```

Routers `asbr-a1/a2` and `asbr-b1/b2` are exchanging external BGP and internal BGP routing information about their local networks. As you can see, both IBGP and EBGP peer sessions (also referred to as *speakers*) are configured using the `<neighbor [ip address] remote-as [as number]>` command. What demonstrates the existence of IBGP peer session is that the remote AS number is identical to that of the local AS number. Because there are four IBGP routers in AS-30001 that have EBGP peer sessions and provide similar network reachability information, administrative weights are used to set preferences for which routes the speakers' broadcasts will be sent to.

Note

To prevent routing loops, external network information learned from IBGP peers is not readvertised to other IBGP peers.

Weights are used to ensure that the external routes provided through the locally adjacent peer are preferred over the remote peer. This same approach is used with the IBGP routers using a special kind of access list for filtering `as-path` information. `as-path acl` is created in general configuration EXEC mode using `<ip as-path access-list [acl number] [permit|deny] [string]>`. The filter is then applied to a neighbor using the `<neighbor [ip address] filter-list [acl number] weight [0 - 65535]>`.

Because all the AS-30001 routers are using similar configurations, a BGP peer-group could be created to reduce some of the configuration overhead. BGP peer-groups provide a way of configuring a core set of BGP parameters that are used by all of the peer-group members. Peer-groups are created in the BGP configuration sub-process:

```
asbr-b2(config-router)#neighbor as30001 peer-group
asbr-b2(config-router)#neighbor as30001 remote-as 30001
```

The command shown above sets up the peer-group and establishes it as an IBGP peer-group. The following command establishes all the members of the IBGP peer-group:

```
asbr-b2(config-router)#neighbor 192.168.0.5 as30001
asbr-b2(config-router)#neighbor 192.168.0.6 as30001
asbr-b2(config-router)#neighbor 192.168.0.1 as30001
asbr-b2(config-router)#neighbor 192.168.0.2 as30001
asbr-b2(config-router)#neighbor 192.168.1.19 as30001
```

The following command enables MD5 authentication (and the password) to be used when establishing BGP peer sessions:

```
asbr-b2(config-router)#neighbor as30001 password as-30001
```

It is also possible to enable filtering with `<neighbor [peer-group name] filter-list [acl] [in|out] weight>`. Additional options include router distribution, route-maps, version, advertisement interval—any configuration option that can be applied group-wide. It is also possible to override a peer-group option by configuring an option locally.

After they are configured, neighbor routers just create the same peer-group within their local BGP configuration using `<neighbor [peer-group name] peer-group>`. Any member of the peer-group can make contributions to the group-wide configuration.

> **Note**
>
> An alternative to using weights is to use the administrative distances to determine route preference. With BGP, three distances are used: external, internal, and local. To change the distances from their default settings, use the BGP configuration subcommand `<distance [external] [internal] [local]>`. The range, in case you forgot, is 1 to 255.

BGP Reflectors

An alternative to having a group of routers configured for external peer relationships is to have one router form an external peer relationship and then act as a BGP reflector. IBGP routers do not re-advertise external routes learned from IBGP peers to other IBGP peers. This is done so routing loops are not created. This was the reason each of the IBGP peers needed its own EBGP session to construct a full routing table. When a reflector is used, it acts as a routing information hub which all the IBGP peers connect to, removing the need to have each of the IBGP peers have peer sessions with all the IBGP routers within the AS. In situations where more than one reflector is needed (or desired), the involved IBGP routers will require a cluster id for routeidentification. In other scenarios where only one reflector is used, the cluster is identified by the reflector router's ID. Figure 10.7 illustrates an EBGP/IBGP reflector configuration using our test network.

Here is the IBGP reflector configuration from `asbr-a1`:

```
hostname asbr-a1
!
autonomous-system 30001
!
router bgp 30001
bgp cluster-id 30001
network 192.168.0.0 mask 255.255.255.252
network 192.168.0.4 mask 255.255.255.252
network 192.168.0.128 mask 255.255.255.128
network 192.168.1.0 mask 255.255.255.0
network 192.168.2.0 mask 255.255.255.0
network 192.168.3.0 mask 255.255.255.0
network 192.168.4.0 mask 255.255.255.0
network 192.168.5.0 mask 255.255.255.0
network 192.168.6.0 mask 255.255.255.0
network 192.168.7.0 mask 255.255.255.0
network 192.168.8.0 mask 255.255.255.0
network 192.168.9.0 mask 255.255.255.0
network 12.14.116.4 mask 255.255.255.252
neighbor 12.14.116.5 remote-as 8446
neighbor 192.127.30.1 remote-as 899
neighbor 192.127.30.1 ebgp-multihop
neighbor 192.168.0.2 remote-as 30001
neighbor 192.168.0.5 remote-as 30001
neighbor 192.168.0.5 remote-as 30001
neighbor 192.168.0.5 route-reflector-client
neighbor 192.168.1.20 route-reflector-client
```

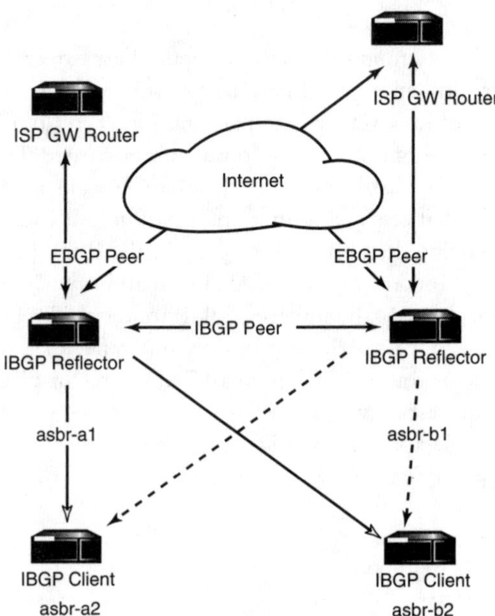

Figure 10.7 The EBGP and IBGP peer relationships in our test network.

Router `asbr-b1` uses an identical configuration. Notice that `asbr-a1` and `asbr-a2` have a full peer relationship and are not configured as reflector clients. As for the client configuration, IBGP sessions are only established with the reflector(s).

Monitoring the BGP Process

As with the IGP protocols, BGP has a set of IOS commands that reveal various operational details about the process. Here is a command list and brief command result summary:

- `<show ip bgp>`—This command displays BGP summary route information, the route's weight, path source, router's ID, cluster attributes, and so on.
- `<show ip bgp summary>`—This is the most often used of all the BGP informational commands. This command yields summary statistics about the running BGP process, and, most importantly, provides the connection status of all the neighbor peer sessions: the activity state, bytes transferred, up/down status, version, and so on.

- <show ip bgp peer-group>—This command provides summary information on the peer-group configuration, particularly what options have been inherited, and/or added or overwritten.
- <show ip bgp neighbors>—Another very useful tool, this command provides detailed information on the status and configuration of the router's configured peers.

Debugging commands are also available. <debug ip bgp> enables all of them, although this is not recommended. <debug ip bgp events> echoes BGP operational events—peer session establishment and communication—and other housekeeping. <debug ip bgp updates> displays the route update information and exchanges between peers. All the <show> and <debug> commands are displayed using the root of the command and a question mark:

```
asbr-a2#debug ip bgp ?

  events      BGP events
  keepalives  BGP keepalives
  updates     BGP updates
  <cr>

asbr-a2#show ip bgp ?
  A.B.C.D           Network in the BGP routing table to display
  cidr-only         Display only routes with non-natural netmasks
  community         Display routes matching the communities
  community-list    Display routes matching the community-list
  dampened-paths    Display paths suppressed due to dampening
  filter-list       Display routes conforming to the filter-list
  flap-statistics   Display flap statistics of routes
  inconsistent-as   Display only routes with inconsistent origin ASs
  neighbors         -Detailed information on TCP and BGP neighbor connections
  paths             Path information
  peer-group        Display information on peer-groups
  regexp            Display routes matching the AS path regular expression
  summary           Summary of BGP neighbor status
  <cr>
```

Route Control and Redistribution

IOS, for better or worse, supports the redistribution of both static and dynamically derived routing information through dynamic routing protocols. You should take notice of the introductory qualification of "for better or worse." The "better" is the capability to use route redistribution to have different dynamic routing protocols announce each other's routing information. Redistribution is helpful whenmultiple dynamic routing protocols are used to manage different parts of a large internetwork.

This condition is often found in large corporate networks where different sites (and sometimes departments) are managed individually, but need to interact with the larger corporate whole. Here, two-way route redistribution is a tremendous win.

Internet or private global network connections are another context where distinct dynamic routing protocols are in use to enforce the separate routing polices of the two networks. With single-point connections, static routes can be used to send packets destined to external networks to the Internet gateway. In situations where multiple Internet and private network links exist (as an alternative to using BGP), it is possible to use filtered redistribution to limit the inbound and outbound network announcements between the different routing policies using distribution lists and route-maps.

Static route redistribution is often used in dial-access and DDR networks where dynamic protocols are not always effective. Instead of adding static routes on every router on the network, a single router can redistribute a collection of static routes. Static route redistribution can also be used to limit network announcements inside internal networks. By using multiple dynamic routing processes to manage different sections of the network, and then statically redistributing on a selective basis, you will only announce the networks that need to be announced between the separate routing processes.

The "worse" is the potential for routing loops and the difficulty surrounding the translation of different routing protocol metrics and distances. Here are some guidelines to help you avoid the "worse" side of using redistribution:

- Ask yourself, "What is the advantage to using redistribution (in a given situation) over using a static route?" In most cases, a static route or default route can be used to provide a network announcement. If you have only a few networks to redistribute, use static routes. Ideally, you only want to be running a single routing protocol. In situations where legacy equipment might reduce your choices in terms of which protocols you can use, it might be more advantageous to use a single protocol or static routing (to address legacy hardware issues) than to add the additional processing and network traffic associated with multiple protocol redistribution.
- If possible, avoid redistributing between classless and classful protocols if you are using VLSM. Classful protocols do not include network address masks in their messages. Instead, they apply the mask of the interface that receives the route announcements. Consequently, only VLSM networks that have compatible masks will be advertised, and the others will not be redistributed.
- When using redistribution with multigateway networks (or HSRP) it is essential that the gateway routers use metrics and distances that favor one router over the other. Earlier in this chapter, we discussed how a route's administrative distance determines its desirability in comparison to the same route from another source. The lower the administrative distance, the more desirable the route. When redistributing routes from one protocol to another, the redistributed routes are

assigned a default administrative distance. If the same network information is redistributed by more than one router in the routing process, a routing loop will result. To avoid this while still maintaining redundancy, adjust the administrative distance for the redistribution on the routers so one is preferable to the other. This will result in having the "preferable" routes used, and in the event that the preferred router fails, the "less-preferable" routes will be available.

Basic Redistribution

Basic route redistribution is enabled with the routing protocol configuration subcommand `<redistribute [source] [process id] [metrics]>`. The redistribution metrics can be set specifically for the redistributed routes in the `<redistribution>` command stanza or by using a `<default-metric [metrics]>` router subprocess declaration. In either case, the metric settings reflect those used specifically by the protocol, such as hop-count for RIP, cost for OSPF, and so on. Figure 10.8 illustrates a simple redistribution example. ASBR-20 is redistributing RIP, OSPF 20, and EIGRP 22 so `ir-1a`, `ir-1b`, and `ir-3c` can reach each other's networks.

As you can see in Figure 10.8, the redistribution configuration is similar for all the protocols. The major difference is that each protocol applies its own metrics to the redistributed routes. OSPF uses cost, which you can estimate adding the sum costs of all the interfaces. In our example, the interfaces directly connecting the RIP and EIGRP networks are Ethernet, which have an OSPF cost of 10. In fact, all the involved interfaces are Ethernet, so the cost was determined by adding the interface cost for each gateway a datagram would have to pass through to reach its destination. A similar approach was followed in determining the RIP metric: the average hop-count was used as the metric. The EIGRP (same for IGRP) metric you should recall is as follows:

- Bandwidth in KB/s
- Delay (in tenths of milliseconds)
- Reliability 1-255
- Load 1-255
- MTU

When configuring EIGRP metrics for redistribution, use the redistribution source interface's default metrics. These can be retrieved (you should know this, by the way) using the `<show interface [type/slot/port]>` command. Unless metrics are specified, the redistributing protocol will use a default value, which in most cases identifies the routes as directly connected. This makes each redistributed route appear to have the same path metric, which basically defeats the advantage of using a dynamic routing protocol to determine the best route path. With simple redistributions, using the default will generally work fine. In situations where multiple gateways and redistribution are combined, however, the use of metrics is strenuously recommended.

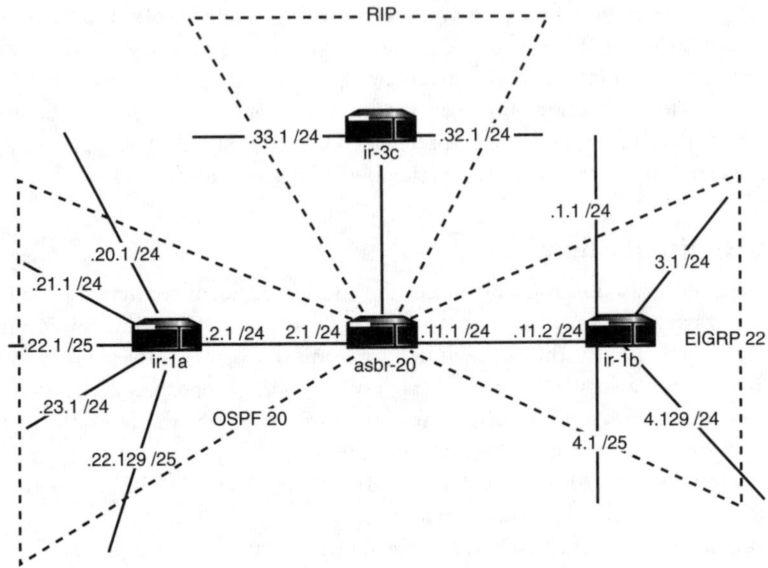

Figure 10.8 An example network running three routing protocols that require redistribution.

OSPF requires a little more specificity than the other protocols when redistributing routes. The `<metric-type [1 or 2]>` is an option that enables the redistributed routes to be identified as OSPF external type 1 or type 2 routes. The `<subnets>` command is specific only to OSPF; it specifies that subnet information will be redistributed along with the network information. If `<subnets>` is not used, the netmask information is not sent, only the classful network address is sent. If VLSM is used with large classful A or B space, and if `<subnets>` is not specified, redistribution will not work. As a rule, the`<subnets>` command should be part of any redistribution (both static and dynamic) declaration. When an OSPF router redistributes any routing information, it becomes an ASBR. As an ASBR, the router can also redistribute default route information. To enable OSPF to generate and announce a default route statement, use the OSPF configuration subcommand `<default-information originate always>`. This command should only be configured on the router that functions as the network's default gateway.

As an alternative to SIN routing, here is an OSPF/BGP redistribution using default route distribution:

```
router ospf 45
 redistribute static subnets
 redistribute bgp 23 metric 1600 metric-type 2 subnets
 network 192.168.1.0 0.0.0.255 area 0.0.0.0
 network 192.168.2.96 0.0.0.3 area 192.168.0.0
 network 192.168.5.0 0.0.0.255 area 192.168.0.0
 default-information originate always
!
router bgp 23
 network 172.16.0.0
 network 192.168.0.0 mask 255.255.240.0
 neighbor 12.14.116.2 remote-as 45550
 neighbor 12.14.116.2 version 4
 neighbor 12.14.116.2 soft-reconfiguration inbound
```

The `<ip default-network [ip address]>` used by IGRP and EIGRP generates a default network statement.

Controlling Redistribution

In some cases, it is not always desirable to redistribute all the routing announcements from one protocol into another. To restrict the flow of dynamic route announcements between routers belonging to the same routing policy or between protocol redistributions, IOS provides route filtering. Route filtering is particularly advantageous for administrators who need to keep strict control over the route announcements exchanged between routers.

Filtering with Distribution Lists

The route filter's responsibility is to suppress unwanted routing information from being redistributed, entered in or advertised out of the routing table. Although the result is the same as redistribution, this process behaves differently if a link state or distance-vector protocol is being used. When you use a route filter with a distance vector protocol, the filter affects not only the router the filter refers to, but also all the routers that receive advertisements from the filtered router. This is the result of the distance vector using their routing tables as the basis for the network announcements. When filtering is employed with a link state protocol, however, the filter will only affect the local routing table. Link state protocols use link state announcements to construct their routing tables, and these are unaffected by route filters.

Routing filters are configured in two parts, starting with the filter list, which is a standard access list. Access list entries are entered using only the network address:

```
asbr-a1(config)#access-list 2 permit 129.0.0.0
asbr-a1(config)#access-list 2 permit 192.168.0.0
asbr-a1(config)#access-list 2 permit 192.168.10.0
```

After the list is created, it is applied to the routing protocol configuration with the routing configuration subcommand `<distribute-list [acl number] [in|out] [protocol|interface]>`. Inbound filter lists are applied to interfaces. They permit or deny the flow of route announcements through a router interface. Inbound filters are used for suppressing route propagation with distance vector protocols or the addition of route announcements on a local router when used with a link state protocol. They are often used to correct a side effect of route redistribution known as *route feedback*. Route feedback occurs when routing information is advertised in the wrong direction across a redistributing router. Inbound filters are also useful from a security perspective, if you are only routing to or from a specific set of networks. Applying an inbound route filter will suppress any unauthorized networks from being injected into the routing table.

Here are inbound filter examples using OSPF and RIP (see Figure 10.9). `ABR-a57` is filtering announcements from `ir-23` and `ir-24`. Only the 192.168.x.0 networks need to be accessible throughout the network. The 10.0.x.0 networks are used for local testing and only need to be available to the networks directly connected to `ir-33` and `ir-34`. The first router configuration example shows OSPF using an inbound distribution list to suppress the addition of the 10.0.x.0 routes on the local routing table. The next example shows the same thing with RIP. The RIP usage of the inbound `<distribute-list>` command also prevents the 10.0.x.0 routes from being announced to `ABR-a58` and `ABR-a56`:

```
hostname ABR-a57
!
access-list 1 permit 192.168.21.0
access-list 1 permit 192.168.22.0
access-list 1 permit 192.168.23.0
access-list 1 permit 192.168.24.0
access-list 1 permit 192.168.25.0
access-list 1 permit 192.168.26.0
access-list 1 permit 192.168.27.0
!
ospf 57
network 172.16.192.0 0.0.0.255 area 57
network 172.16.60.28 0.0.0.3 area 0.0.0.0
network 172.16.60.32 0.0.0.3 area 0.0.0.0
distribute-list 1 in FastEthernet0/0
```

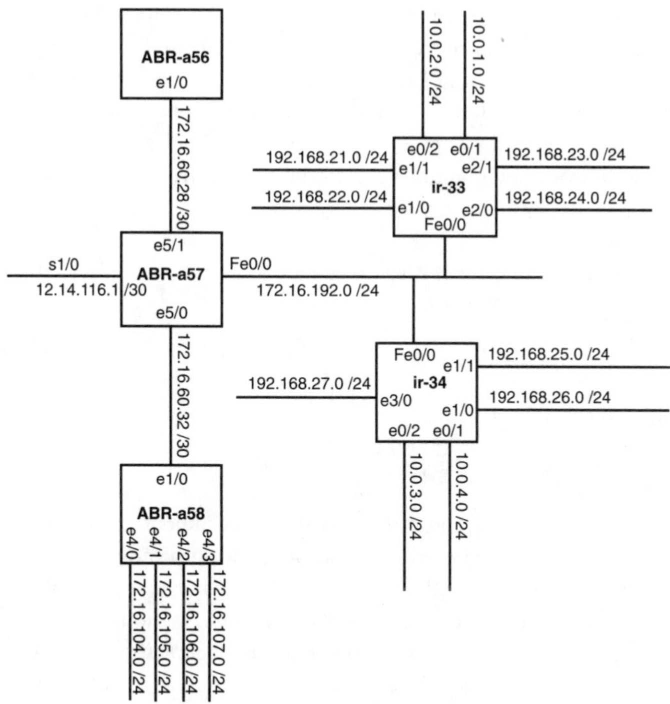

Figure 10.9 A campus backbone network example for using the `<distribute-list>` route distribution/suppression command.

```
hostname ABR-a57
!
access-list 1 permit 192.168.21.0
access-list 1 permit 192.168.22.0
access-list 1 permit 192.168.23.0
access-list 1 permit 192.168.24.0
access-list 1 permit 192.168.25.0
access-list 1 permit 192.168.26.0
access-list 1 permit 192.168.27.0
!
rip
version 2
network 172.16.0.0
distribute-list 1 in FastEthernet0/0
```

Outbound filters on distance-vector protocols can be used for protocol redistributions and for filtering the network announcements sent out of a router's interface. Take for instance our example illustrated in Figure 10.7. In the RIP example above, the goal was to limit the network accessibility to the 10.0.x.0 networks. Because `ABR-a57` is also an Internet gateway, we might want the 10.0.x.0 users to have Internet access (assuming our usage of the 10.0.0.0 network address was permissible), but still want to limit their Internal access. So, instead of using an inbound filter on the `ABR-a57` Fe0/0 interface, which prevents the 10.0.x.0 routes from being added to the `ABR-a57` routing table, outbound filters can be applied to interface e5/0 and e5/1 to suppress the 10.0.x.0 network from being announced to the rest of the private network:

```
hostname ABR-a57
!
access-list 1 deny 10.0.0.0
access-list 1 permit any
!
rip
version 2
network 172.16.0.0
distribute-list 1 out Ethernet5/0
distribute-list 1 out Ethernet5/1
```

Link state protocols, however, can only use outbound filters for limiting the network announcements being redistributed into another routing protocol. Again, let's use Figure 10.7. In this scenario, OSPF is being used to manage all the internal routing (including the test networks) for the network. This information is then redistributed into BGP, which is being used to announce reachability for the network segments. However, because the test networks are using "unregistered" IP address space that cannot be announced publicly, they need to be filtered from the redistribution. To prevent their announcement, they are being suppressed using an outbound distribution list:

```
hostname abr-a57.

!
router ospf 45
network 172.16.192.0 0.0.0.255 area 57
network 172.16.60.32 0.0.0.3 area 0.0.0.0
network 172.16.60.28 0.0.0.3 area 0.0.0.0

 default-information originate always
 distribute-list 2 out bgp 23
!
bgp 890
redistribute ospf 45
neighbor 12.14.116.2 remote-as 2889
neighbor 12.14.116.2 password gimmyroutes
neighbor 12.14.116.2 soft-reconfiguration inbound
neighbor 12.14.116.2 version 4
neighbor 12.14.116.2 distribute-list 4 out
neighbor 12.14.116.2 distribute-list 3 in
!
access-list 2 deny 10.0.1.0
access-list 2 deny 10.0.2.0
```

```
access-list 2 deny 10.0.3.0
access-list 2 deny 10.0.4.0
access-list permit any
!
access-list 4 permit any
!
access-list 3 permit any
```

The OSPF distribution list redistributes all the networks except the 10.0.x.0 test networks. The distribution lists on the BGP process do not restrict announcements in either direction. When using redistribution to announce networks, however, particularly to the Internet, it is a good idea to implement at the very minimum a basic filter so when the need arises to restrict something from the redistribution it is simply an access-list change.

Filtering with Route-Maps

In Chapter 9, we used route-maps to configure policy routing. Route-maps can also be used to control the flow of route redistributions and apply metrics to redistributed routes for IGP and EGP protocols. Here is a simple IGP example that does both.

`asbr-2b` is redistributing EIGRP 45, the local routing process for the Boston network into OSPF 87, which is the routing process for Albany network. The problem is that `asbr-b1` is redistributing OSPF 87 into EIGRP 45, so `asbr-b2` needs to filter any local Albany routes that are being announced from the redistribution running on `asbr-b1`. The filtering is being done by a route-map named "grinder."

Let's take a look at how the route-map "grinder" is created. First, a standard IP ACL containing all of the Albany local routes needs to be created:

```
asbr-b2(config)#Access-list 1 permit 192.168.4.0
asbr-b2(config)#Access-list 1 permit 192.168.5.0
asbr-b2(config)#Access-list 1 permit 192.168.6.0
asbr-b2(config)#Access-list 1 permit 192.168.7.0
asbr-b2(config)#Access-list 1 permit 192.168.9.0
asbr-b2(config)#Access-list 1 permit 192.168.0.0
asbr-b2(config)#Access-list 1 permit 192.168.1.0
asbr-b2(config)#Access-list 1 permit 192.168.2.0
asbr-b2(config)#Access-list 1 permit 192.168.12.0
asbr-b2(config)#Access-list 1 permit 12.14.116.0
asbr-b2(config)#Access-list 2 permit 172.16.0.0 0.0.255.255
```

Note

Route-maps are used extensively for controlling BGP routing announcements. This application, however, is way beyond the scope of this book. For more information on using route-maps and BGP, check the additional resources section at the end of the chapter.

Now, we configure the route-map. The first line permits any route announcement that matches any address in access-list 2 to be redistributed with an OSPF type 2 metric:

```
asbr-b2(config)#route-map grinder permit 10
asbr-b2(config-route-map)#match ip address 2
asbr-b2(config-route-map)#metric-type 2
```

The second line configures the route-map "grinder" to deny any route announcement that matches any address in the access list 1:

```
asbr-b2(config)#route-map grinder 10
asbr-b2(config-route-map)#match ip address 1
```

After the map is completed, it just needs to be applied to the OSPF process:

```
asbr-a2(config-router)#$redistribute eigrp 99 route-map grinder
```

Here are the basic match and set statements you can use to control route redistributions:

```
<match ip address [acl]>
<match metric [metric]>
<match ip route-source [acl]>
<set metric [metric values used by the redistributing protocol]>
<set metric type [type-1|type-2]>
```

Keep in mind that route-maps, like access-lists, use deny as their default action, so you need to declare permit statements when configuring the map statements.

Summary

The focus of this chapter has been the configuration of static and dynamic IP routing on Cisco routers using the Cisco IOS. At this point, you should have a familiarity with the following concepts:

- Managing static routing
- Implementing RIP, OSPF, and EIGRP
- Adjusting routing preferences with administrative distances
- Implementing AURP
- Implementing BGP
- Route redistribution

In Chapter 11, "Network Troubleshooting, Performance Tuning, and Management Fundamentals," we look at network management, troubleshooting, and performance fundamentals. The chapter's goal will be to introduce basic network management and diagnostics concepts. Covered topics include SNMP, RMON, hardware and software network management tools, network management models, network performance baselining, and common network problem identification.

Related RFCs

RFC 1245	OSPF Protocol Analysis
RFC 1246	Experience with the OSPF Protocol
RFC 1397	Default Route Advertisement in BGP2 and BGP3 Versions of the Border Gateway Protocol
RFC 1467	Status of CIDR Deployment in the Internet
RFC 1518	An Architecture for IP Address Allocation with CIDR
RFC 1519	Classless Inter-Domain Routing (CIDR): An Address Assignment and Aggregation Strategy
RFC 1582	Extensions to RIP to Support Demand Circuits
RFC 1586	Guidelines for Running OSPF over Frame Relay Networks
RFC 1587	The OSPF NSSA Option
RFC 1721	RIP Version 2 Protocol Analysis
RFC 1774	BGP-4 Protocol Analysis
RFC 1817	CIDR and Classful Routing

Additional Resources

Ballew, Scott M. *Managing IP Networks with Cisco Routers*. O'Reilly & Associates, 1997.

Fischer, Paul. *Configuring Cisco Routers for ISDN*. McGraw-Hill, 1999.

Lewis, Chris. *Cisco TCP/IP Routing Professional Reference*. McGraw-Hill, 1998.

Wright, Robert. *IP Routing Primer*. Cisco Press, 1998.

11

Network Troubleshooting, Performance Tuning, and Management Fundamentals

THIS CHAPTER DEALS WITH PERHAPS THE MOST DAUNTING TASKS any network administrator will face, network performance and network management. To those new to networking, there is a common misconception that designing and building the network are its most difficult aspects. What they soon learn after the network is completed is that building was the fun part. Now that it is in place, the work begins. The work consists of keeping the network up and running and providing a consistent level of reliability and performance. Computer networks today are essentially the nervous system of the business. They connect everything together and without them, everything just stops working (in most cases).

The problem is that computer networks are not stationary systems like file and database systems. Computer systems have finite limitations that can be monitored and in many cases corrected in a reasonably straightforward manner. If a system becomes sluggish, its processing and memory resources can be monitored. If its storage systems become too full, data can be moved off to tape or disks can be added. This is not to say that system management is an easy task. Rather, there are cause and effect relationships that can be identified. Computer networks, on the other hand, are much more difficult. They are a collection of complex, interdependent systems. Physical cabling, routers, switches, repeaters, and end-stations are all equal partners in the operation of the network. Keeping track of all these elements, where they are, how they are configured, and how they are functioning, and most importantly, being able to correct problems when they arise, are all within the broad scope of network management.

Network Analysis and Performance Tuning

Using a management and monitoring system on your network is the best way to identify network faults and performance issues. These systems, however, are not really designed to resolve problems after they are identified. In addition, their capability to detect performance-related issues is really limited by your own ability to identify performance issues. This is to say, they are data collection systems, not fault and performance problem-resolution systems. The network administrator and/or network operations staff performs this function. You can have all the network management data in the world, but if you cannot understand the information that is being collected in relation to how your network functions, it is useless. This is where network analysis and performance tuning come in.

Actually, the purpose of network analysis is twofold. First, in order to implement any effective monitoring and management system, you need to have an understanding of how your network functions. Your understanding should encompass the following:

- WAN link and LAN segment bandwidth utilization
- WAN and LAN protocol usage
- Average and peak utilization network response times
- Network traffic flow patterns between segments, gateways, and servers
- Network application services
- Layer 2 and Layer 3 packet statistics

Through the use of network analysis techniques, you can establish a set of network performance baselines. Network baseline data is used for a variety of network management-related tasks. For example, when setting up network monitoring, baseline data is used for setting management and monitoring probe alarms. Although default values ("norms") for most monitoring variables exist, these might not always apply to your network environment. By using network baseline data to set monitoring alarm thresholds, your network management system will be more effective and provide more useful information.

The second purpose of network analysis is to resolve network fault conditions and to improve overall network performance through performance tuning. Performance tuning is making configuration and design adjustments to the network to improve its performance based on the results of management and analysis data. Again, here is where network baseline data is helpful. If the network users complain of poor network performance, without any performance baseline data it is difficult to pinpoint a cause. Therefore, instead of identifying the actual problem, additional resources are implemented which might or might not actually fix the problem. There are some contexts where additional network resources are needed, but implementing them without identifying the actual problem you are trying to correct can sometimes make the problem worse. Throwing money at performance problems is an expensive approach to resolving them. Your approach to network analysis depends largely on whether you are determining the network's performance baseline or resolving a network performance or fault condition.

Tools

Although network management systems and vendor-specific GUI-based configuration tools are the most commonly associated networking tools for network data collection and diagnostics, they are not the only tools needed to effectively manage your network. Packet analyzers or packet sniffers, time domain reflectors, bit error rate testers, and software tools like ping and traceroute are each valuable tools that you will at some point need in order to resolve a problem on your network. However, the most valuable tool of all, on any network, is proper documentation. Without adequate documentation, your network will become an unmanageable beast before your eyes. Documentation helps you and others visualize the role of each component in the network and provides a common reference point for configuration, technical contact, and service and warranty information.

Documentation

This point cannot be emphasized enough: *Good documentation is very important.* Without it, you are lost. One of the most common misconceptions about computer networks is that they are too complicated to manage. This is not the case. A network that is well designed and properly documented is easily managed, and when problems do arise, they are often easy to pinpoint and resolve.

Now, the word "documentation" is often misunderstood and is, in some circles, a "dirty" word. So, let's start with what documentation is not. It is not notes scribbled on napkins and pizza boxes, or a notebook filled with sketches. Proper network documentation is a set of documents that describes the following:

- The topology of your network, each of its hardware components, and the protocols in use to transmit, transport, and manage data delivery.
- The site's (or sites') cabling infrastructure and labeling scheme, kind of cabling used (copper and fiber), types of connectors, patch panels, average cable length between the end-stations and the interconnect panel, and any certification documentation.
- The servers on the network, their operating system, the function they perform, network address information, administrator information, and so on.
- A management document that describes the general operational policies regarding the network. This document would also describe the management system, documentation policy, format, and contact information. This document is usually associated with a company's computer and physical security policy and a disaster recovery plan.

The actual format and method you follow to document your network is up to you. Your company might already have a standard and format they use for this kind of information. The key is to make sure it is accurate, up to date, readable, and easily accessible.

To get you started, here are some basic ideas about the kinds of information your documentation should contain.

Topology Maps

An updated network topology map is essential. This map should depict (at least) the cabling plant and the bridges, switches, hubs, routers, and so forth on the network. The map should indicate the physical locations of the equipment, network segments, and device and interface addressing information. In most cases, it is easier to develop maps that depict specific elements. For instance, one map might document physical device interconnections, another might describe the network logically from a protocol addressing perspective. One map might depict the equipment racks by device and location in the computer or cable distribution room. The idea is to map out the network in such a way that anyone can find anything by looking at the correct map. Mapping is a common feature of many management systems, and some are better at it than others. These systems also support automatic device discovery and can relay status information. Though they are helpful to display and manage the network, they are not suitable for constructing all the maps necessary for a solid documentation set.

Device Information

Each network component should have a device information sheet. This can be an Access or Oracle database, a set of index cards, even a set of Web pages. The idea is to construct a uniform fact sheet about each device on the network from end-station to routers. You should develop a universal identity scheme for all the networked devices. This can be a hostname, internal audit, or lease serial number; you can even use the device's MAC address. The ID needs to be unique to the device and that one that will ideally never be duplicated or become obsolete (such as an employee or department name). This tag is what you can use to sort through the information. Then, for each device, you want to record some subset of the following information:

- Universal ID
- Hostname/system name
- Installation location
- Network address(es) and supported protocols
- MAC address(es)
- Telco circuit ID(s) (if appropriate)
- Manufacture and model
- Vendor information and contact numbers

- Service contract number
- Local technical contact
- Data port assignments (what end-devices are connected to its ports)
- Software version
- Dependent or adjacent devices
- Description of primary function

This information can be used when resolving network-related problems. The device tag should be associated with any trouble ticket or change event that the device is involved in.

Change Log

All changes should be documented. Network patches, device moves, software and hardware upgrades, and device module additions and removals should be in the log. The log entry should describe the type of change, devices involved and/or affected, and any performance evaluation or testing data. It is also wise to reference a trouble ticket or incident record that is associated with the actual change or was the motivation for the change. The goal of the change log is to provide enough information so those who come after you will understand what has been done, why it was done, and, most importantly, what it might have directly or indirectly affected.

Documentation takes work, but it is essential to maintain a network of any size. The work pays off when you try to fix a problem and all the information you need is available and not in some technician's head (who is at home asleep).

Protocol Analyzers

The function of protocol analyzers is to capture and display network protocol data. They are extremely powerful tools, and are indispensable for troubleshooting and analyzing networks. Protocol analyzers perform real-time analysis or offline analysis of some or all of the network segment traffic. Through the use of packet filters and triggers, the analyzer can capture data based on application traffic type, Layer 2 or Layer 3 address, transaction type, or specific error. In either real-time or offline mode (by saving the data collection to a file), collected traffic samples can be analyzed and displayed in statistical or graphical form, using either generic or custom analysis templates. The generic analysis function displays summaries of errors (runts, jabbers, beacons, ring resets, collisions, and broadcasts). The analysis will also display data on packet sizes, protocol type, and application type distribution. In many ways, protocol analyzers and RMON probes are similar in terms of functionality, packet collection capabilities, and analysis. In fact, many protocol analyzers on the market can operate as RMON probes. It should be noted, however, that protocol analyzers have been around much longer than RMON probes and offer a greater degree of flexibility and functionality than RMON probes when it comes to troubleshooting. The major difference between them is that protocol analyzers are geared toward real-time operation.

Protocol analyzers exist for FDDI, Ethernet, Token Ring, and WAN network interfaces. The analyzer is essentially a computer with special software and hardware that provides the analysis and capturing functionality. Analyzers function much like regular end-stations, but the difference is that a regular end-station listens to the network segment and only copies packets destined for itself. A protocol analyzer copies all the packets on the segment that pass across its network interface. In the earlier discussion of protocol analyzers, it was mentioned that this was known as "promiscuous mode." LAN protocol analyzers generally use standard network interfaces for data collection and operate in a non-intrusive mode. WAN protocol analyzers, however, need to be in some part of the transmission path. To use a WAN analyzer, there will be some minor service disruption. To listen to the transmission path, WAN analyzers come with passthrough interface cards that sit between the computing device (for example, a router) and the DSU/CSU. This allows the analyzer to see all of the traffic on the transmission path and keep the link in operation. The service intrusion comes when the analyzer is installed and removed.

As mentioned earlier, the most popular protocol analyzer is the Network Associates (formally Network General) Expert Sniffer and Distributed Sniffer products. The Expert Sniffer comes as a software/NIC bundle either for laptops and desktop PCs or as a reconfigured system using a Compaq, Toshiba, or Dulch computer. The Distributed Sniffer System (DSS) product functions as an RMON collection system or as a real-time analyzer. The DSS uses a collection of analyzers installed on different network segments. The analyzers are then available through a central console. The Expert and Distributed Sniffer System protocol analyzers provide a large collection of reporting and capturing capabilities. There are also a number of third-party developers that have created extensions and add-on scripts for enhanced collection and reporting capabilities.

Along with Network Associates, a number of software and hardware protocol analyzers are manufactured by the following vendors: Telecommunications Techniques (LANHAWK), IBM (LAN Manager), FTP software (LANwatch Network Analyzer), Microsoft (included with Windows NT), and Hewlett-Packard (Network Advisor).

Time Domain Reflectors

Time Domain Reflectors (TDR) devices are used for diagnosing cable failure and associated problems. TDRs work by sending a test signal down the cable segment and listening for an "echo" or reflection. The TDR signal is sent at a specific amplitude and rate; if no echo is detected, the cable is free of errors. If an echo is detected, the TDR scope can determine the type of problem based on the type of reflection or Cable Signal Fault Signature (CSFS) detected. Depending on the type of cable fault, different CSFSs are generated. Along with the type of cable fault, TDR scopes tell you how far down in the segment the fault is located.

There are TDRs available for both copper and fiber cable types. A typical TDR scope has two components: a TDR scope and a cable terminator. The scope is on one end and the terminator is connected to the other end of the segment. Do not use a TDR device with equipment attached at the other end of the cable. This can cause damage to the TDR and the attached equipment. Many TDR scopes that are on the market can also perform limited segment monitoring and packet analysis. This can be handy for resolving frame error type problems. These devices are not, however, substitutes for a protocol analyzer.

Bit Error Rate Testers

Bit Error Rate Testers (BERTs) are used to verify and test telecom transmission circuits. These tools are commonly used by Local Exchange Carriers/Competitive Local Exchange Carriers (LEC/CLEC) for testing T1/DS1/T3 and SONET/SDH circuits. The most common BERT in use today is the Telecommunications Techniques T-BERT 244 analyzer. BERTs operate by sending different test patterns across the transmission circuit. There are two types of patterns: fixed and pseudorandom. Fixed patterns send short continuous patterned bit sequences. The most common are the mark (all ones) and space (all zeros) patterns. Various patterns of ones and zeros are used to test different circuit types. Pseudorandom patterns are long random patterned bit sequences. When a group of bits is randomly examined, the stream appears to be random. It is only when the entire stream is examined and is repeated that the stream appears as an actual pattern. Pseudorandom patterns are used for circuit stress testing. Fixed testing is used for testing fault conditions. When performing fault testing, a string of different fixed patterns is sent in a continuous multiple pattern stream.

BERT testing can be intrusive or non-intrusive, depending on how the BERT is placed in the circuit path. For non-intrusive testing, the BERT taps into the transmit and receive pairs and monitors the traffic stream. With intrusive testing, one or two BERTs can be used to transmit and receive test patterns. When two BERTs are used, they replace the DSU/CSU at the circuit's ends and exchange bit streams. When only one BERT is used, the remote DSU/CSU is placed in loopback mode and the BERT sends and receives the test pattern signals.

Unless you are managing a large number of your own dedicated telecommunications circuits, the actual usefulness of a BERT is limited. If you have many inbound circuits, however, a BERT can be useful for doing your own stress and diagnostic testing.

Packet Internet Groper (Ping)

Ping began as a simple IP-based UNIX tool for testing host reachability. Over time, it has evolved, like many other "simple" UNIX tools, to other operating systems and even to support reachability testing for other Layer 3 protocols (IPX, AppleTalk, and so on). As an IP application, ping functions by sending ICMP `echo_request` messages at a target destination host. When the target host receives these packets, it responds to the source host with ICMP `echo_reply` messages.

The initial UNIX implementation of ping simply indicated if the host was reachable. Today, most UNIX-based implementations of ping transmit an infinite stream of ICMP messages at the target host. This action is terminated using the Ctrl+C keystroke combination. Alternatively, ping implementations that come with Cisco IOS and Windows by default only transmit a short burst (The IOS version sends five ICMP requests and Windows version sends four.). The ping command output example below illustrates the basic information you should expect to be returned from most ping implementations:

```
thor# ping 192.168.55.1
PING 130.140.55.1 (130.140.55.1): 56 data bytes
64 bytes from 192.168.55.1: icmp_seq=0 ttl=30 time=1.164 ms
64 bytes from 192.168.55.1: icmp_seq=1 ttl=30 time=0.918 ms
64 bytes from 192.168.55.1: icmp_seq=2 ttl=30 time=0.983 ms
64 bytes from 192.168.55.1: icmp_seq=3 ttl=30 time=0.939 ms
64 bytes from 192.168.55.1: icmp_seq=4 ttl=30 time=0.945 ms
64 bytes from 192.168.55.1: icmp_seq=5 ttl=30 time=0.946 ms
64 bytes from 192.168.55.1: icmp_seq=6 ttl=30 time=0.932 ms
^C
—- 192.168.55.1 ping statistics —-
7 packets transmitted, 7 packets received, 0% packet loss
round-trip min/avg/max = 0.918/0.975/1.164 ms
thor#
```

> **Note**
>
> In most cases, ping comes as part of an operating system's standard TCP/IP implementation. As a rule, UNIX ping implementations are more feature-rich than their Windows and Macintosh (which has no native implementation) counterparts.
>
> There are, however, a number of great add-on tools available on the Internet. Here is a short list of some ping tools for Mac and PC.
>
> For Windows Systems, check out the following:
>
> | | |
> |---|---|
> | VisualRoute | `http://www.visualroute.com/` |
> | Enhanced Ping | `http://SolarWinds.Net/Ping/` |
> | !TraceRoute | `http://SolarWinds.Net/TraceRoute/` |
> | Tray Ping | `http://home.t-online.de/home/je.weiland/trayping.html` |
>
> For Macintosh Systems, check out the following:
>
> | | |
> |---|---|
> | MacTCP Watcher | `http://www.stairways.com/info/products.html` |
> | AG Net Tools | `http://www.aggroup.com/` |

Although the display might vary depending on the implementation, ping displays some kind of visual queue indicating a transmitted packet (some are more verbose than others). The standard output generally indicates the ICMP message size, its sequence number, its Time To Live (TTL) value, and its round-trip delivery time, which is the time from when the messages was sent and a reply was received. When the command execution is complete, ping provides summary information detailing the success rate (packets sent, received, and lost) and the round-trip time averages (min, avg, max) in milliseconds. All ping implementations use default ICMP message size. UNIX versions tend to utilize a 56-byte packet, with an 8-bit ICMP header, the IOS implementation uses a 100-byte message and Windows ping uses a 32-byte ICMP message. The message size is user-definable with flags in the command line syntax. Because ping tracks are both successful and failed ICMP requests, they can also be used to measure packet loss between the test host and destination target (that is, another host or gateway) and round-trip delivery time. In addition to host reachability information, ping also provides valuable data for measuring packet loss and transit link latency. The following code is an output example from the Cisco IOS ping implementation, where the ping attempt has failed:

```
emmet> ping 192.168.30.1
Sending 5, 100-Byte ICMP Echoes to 192.168.30.1, timeout is 2 seconds:
.....
Success rate is 0 percent (0/50
emmet>
```

Note

Ping implementations vary from platform to platform in terms of their command line interface (CLI), supported features, and protocols. When using ping on a system for the first time, check the documentation to see what options are available. On a UNIX system, this can be done using the `<man ping>` command. On Cisco IOS, use `<ping ?>`, and on Windows systems, type `C:\ping <cr>` and the CLI usage and options are displayed.

Using Ping

The ping command can executed from the command line in both the Windows NT and UNIX environments, using the following command syntax <ping [hostname/IP address>. Depending on the ping implementation, either a group or stream of ICMP messages will be sent to the specified target host. When debugging user network problems, this should be the first thing you ask them to do. If the ping test is successful, move on to the problem specifics. Quite often, a user will complain that he cannot get email, when it turns out that he cannot reach anything on the network because he is not physically connected or the IP address information is not properly configured. When using the ping command the host's IP implementation needs to be functioning correctly. This can easily be accomplished by pinging the loopback interface that uses the IP address 127.0.0.1:

```
%ping 127.0.0.1
PING 127.0.0.1 (127.0.0.1): 56 data bytes
64 bytes from 127.0.0.1: icmp_seq=0 ttl=255 time=0.478 ms
64 bytes from 127.0.0.1: icmp_seq=1 ttl=255 time=0.161 ms
^C
—- 127.0.0.1 ping statistics —-
2 packets transmitted, 2 packets received, 0% packet loss
round-trip min/avg/max = 0.161/0.319/0.478 ms
%
```

If this ping test fails, the host's IP implementation is not functioning correctly or not configured (an all too common problem with hosts configured using DHCP). Check the host's IP configuration information. On a UNIX host, this is done using the <ifconfig -a> command, which will list all of the configured interfaces and if they are up or down. On a Cisco router, you can use the <show interface [slot/port]> command. On a Windows NT system, use the CLI command shell and type <ipconfig /all>. On a Windows 95/98 system, from the CLI DOS shell or from the RUN dialog in the Start menu, execute the command <winipcfg>. On a Macintosh, there is no native TCP/IP configuration verification method (aside from the TCP/IP Control Panel), so to verify TCP/IP operation you need to use a third-party tool like MacTCP Watcher.

Note

It is possible to use the ping command to generate a list of the reachable hosts on an IP subnet. The list is generated by pinging the broadcast address of the IP subnet. This forces all the hosts on the subnet to reply to the ICMP request. This option should be used with caution, especially on CIDR "supernet" address spaces because it generates tons of network traffic. In fact, using the ping command in this fashion can be deadly if used in "fast packet" mode, which sends a burst of messages as fast as possible at one time, or if a large ICMP message size is configured. For this reason, filtering out ICMP requests is a wise idea when configuring your router.

In addition to performing basic host reachability tests, ping can be used in UNIX and Windows shell scripts to run basic performance checks and perform long-term host availability monitoring. Here is an example of a performance monitoring tool that performs a set of pings to a host using different ICMP message sizes. Scripts like this are great for testing for packet loss and measuring transit path performance. This script runs as a UNIX Cron job and mails the test results:

```
#!/bin/sh
# This script runs as a CRON job sending a group ICMP packets
# of various sizes. The summary results of these transactions are then
# sent via SNMP mail to an administrator.
#
# This tool is well suited for measuring WAN and LAN transit link
# performance between file and other high access/high availability
# devices.
#
# To enable ping fast packet "flooding" add the -f flag to the execution # command.
#
# Some editing is needed for the script to operate correctly.
# 1. Define the target host.
# 2. Define the number of ICMP requests.
# 3. Set the UNIX binary paths.
# 4. Set the temp directory path.
# 4. Set the mailtarget (who is getting the report).
#
# 6. Define the message size. The following is a list of suggested
# sizes:
# 56-bits
# 256-bits
# 512-bits
# 768-bits
# 1000-bits
# 1500-bits
# Keep in mind the number of ICMP messages and their size affect
# the runtime of the script and the expected round-trip results. Larger
# packets will take longer to process. Larger packet sizes (above 512)
# also may overrun the application buffer. So when configuring the
# script for sending large packet sizes, keep the number of requests
# low.
#
# This is set in the Check Script command section of the script.
#
# To enable the fast packet option if desired, by adding the -f
# flag to the command line.
 #
# For best results run the script at different points during normal
# usage hours. Keep track of the data to establish a performance
# baseline
.
# Large shifts in performance should be investigated.
```

```
#
# Ping Variables
# Set the Target
target=
# Set the ICMP message count, 20 is the default.
pc=20
# Set the UNIX binary paths
ping=/bin/ping
mail=/usr/bin/Mail
rm=/bin/rm
tail=/usr/bin/tail
cat=/bin/cat
# Set the temp path
temp=/tmp
# Set the mail target
mailtarget=anyuser@any.com
#Do not change these variables
logfile=/tmp/load.data

# Beginning of the script.
# Clear all log and temporary files.
rm /tmp/logfil*
#
#
# The script runs tests at 56, 256, and 512. More test can be added;
# just copy the test command set, add it to the bottom of the Check
# Script run, and set the ICMP message size.
# Start of the Check Script Command Section
#
# 56k check
# Set the ICMP message size, if you want to use fast packet add the -f
# before the -c flag. ping -f -c $pc -s $size
size=56
# Here is the ping command
ping -c $pc -s $size $target > $logfile.t
tail -n 2 $logfile.t > $logfile.rept
echo $size-Bit Loss Statistics > $temp/head.temp
# Generate Report
cat $temp/head.temp $logfile.rept > $logfile.$size
rm $logfile.t $logfile.rept $temp/head.temp
#
# 256k check
# Set the ICMP message size, if you want to use fast packet add the -f
# before the -c flag. ping -f -c $pc -s $size
size=256
# Here is the ping command
ping -c $pc -s $size $target > $logfile.t
tail -n 2 $logfile.t > $logfile.rept
echo $size-Bit Loss Statistics > $temp/head.temp
# Generate Report
cat $temp/head.temp $logfile.rept > $logfile.$size
```

```
rm $logfile.t $logfile.rept $temp/head.temp
#
# 512k check
# Set the ICMP message size, if you want to use fast packet add the -f
# before the -c flag. ping -f -c $pc -s $size
size=512
# Here is the ping command
ping -c $pc -s $size $target > $logfile.t
tail -n 2 $logfile.t > $logfile.rept
echo $size-Bit Loss Statistics > $temp/head.temp
# Generate Report
cat $temp/head.temp $logfile.rept > $logfile.$size
rm $logfile.t $logfile.rept $temp/head.temp
#
# End of the Check Script Command Section
# report mailer section
cat $temp/load.data.* > $logfile
cat $logfile | mail -s " Packet Loss Report from $target" $mailtarget
# End Of Script
```

For performing host monitoring, ping can be used in a script to periodically send messages to a group of hosts. This example also runs as a UNIX Cron job. It sends a set of ICMP messages to a set of hosts. If the hosts fail to send an ICMP echo-request, the script sends an email notice. This script works well for off-site notification if you have the capability to receive email on your pager.

```
#!/bin/sh
# This is a simple ping check script that is run from a UNIX
# cron job. Ideally, you want to run this script on two different
# hosts that can send mail alerts in the event of a host failure.
#
# To have the script run every five minutes, use the following Cron
# entry.
#
#0,5,10,15,20,25,30,35,40,45,50,55 * * * * /usr/local/etc/ping.sh
#
#Enter the hosts you want to ping here. The host entry can be an
# IP address or hostname. Entries should be separated by a space.

hosts="testhost testhost2 192.168.33.1"

# Enter the paths for ping and mail for your UNIX environment here.

ping=/bin/ping
mail=/usr/bin/Mail

# If any of the hosts fail to respond the script will mail
# a notification message to you.

for i in $hosts;
do
```

```
        $ping -c 3 $i >/dev/null 2>&1
        if [ $? != 0 ]; then
                echo $i is down|$mail your@mailaddress-here
        fi
done
```

Although ping is not a glamorous tool in comparison to other network tools, it is flexible (as you can see from the previous examples). When using ping as a standalone tool or in a script, keep in mind that a reachability failure or loss of packets does not automatically indicate that the host is down or there are physical layer problems with one of your gateways or transport links. A failed ping attempt or one with a high level of dropped packets can also indicate a routing problem or an excessive level of network traffic.

Traceroute

To check the route to the destination host or to verify that all of the gateways in the transit path are up, an additional application, commonly called traceroute, is used. Traceroute, like ping, uses ICMP echo-request messages to trace the gateway path one hop at a time. It accomplishes this by sending three ICMP messages with variable TTL counters. The first ICMP message, which is sent to the host's default gateway, has a TTL of 1. The message is received by the gateway, and it decreases the TTL counter to zero and returns an ICMP host unreachable notification message. Then traceroute sends an ICMP message with a TTL of 2, to reach the next gateway. This process continues, incrementing the TTL by one, each time a successful ICMP host unreachable message is received. This continues until the final destination host is reached. This occurs when the traceroute program receives an ICMP `echo_response` from the destination host.

As traceroute maps the transit path, it displays the gateway (hostname or address) and the round-trip delivery time averages for the ICMP messages sent to each gateway. The output from a UNIX-based traceroute implementation follows:

```
%traceroute www.yahoo.com
traceroute to www.yahoo.com (204.71.200.74), 30 hops max, 40 byte packets
 1  enss (192.207.123.1)  4.871 ms  1.988 ms  2.334 ms
 2  (137.39.3.89)  6.034 ms  13.311 ms  6.728 ms
 3  (146.188.177.134)  5.914 ms  5.232 ms  5.042 ms
 4  (146.188.177.149)  6.619 ms  12.762 ms  7.101 ms
 5  (206.132.150.129)  7.227 ms  6.767 ms  6.327 ms
 6  (206.132.253.97)  6.392 ms  117.242 ms  70.468 ms
 7  (206.132.151.22)  133.405 ms  75.601 ms  89.595 ms
 8  (206.132.254.41)  239.353 ms  239.590 ms  241.420 ms
 9  (208.178.103.58)  291.413 ms *
10  www9.yahoo.com (204.71.200.74)  73.694 ms  68.444 ms  67.123 ms
```

Traceroute is available for UNIX (it is native on some UNIX implementations; for others, it is available as source on the Internet), Windows NT (native), Windows 95/98, and Macintosh (as a shareware add-on), and as part of Cisco IOS.

Network Baselining and Problem Identification

Network performance baselining is the measurement of key performance aspects of the network's operation. Data collection is done in sampled increments over a fixed period of time so that an accurate picture of the network's traffic flow patterns can be established. The idea is that regular baselining is continuously performed, with comparisons made to previous performance baselines.

Ideally, baselining should begin when the network is placed into production. However, baselining can be helpful no matter when you start. The following are just some of the reasons for performing network baselining:

- Ensure that your design is suitable for your user requirements. Most data networks grow out of necessity, not planning. A business is started, it moves into office space, it purchases some computers, it needs a LAN, and so on. This is not what is considered to be a designed network. Rather, it's just a collection of computers connected together. A sort of "orange juice can and string" network—you can hear sound (move data) on it, and you are happy. This is a very common approach in the start of new businesses, since networked communication between computers is usually the only design goal. So if it works, that's usually all that matters.
- Later on, the business grows and network-dependent applications start to become important to business function that the network's initial reactive design approach begins to show its limitations. Unfortunately, this is also usually the time that a business can least afford any major business disruption, such as a new network implementation or a painful network upgrade. So, when the time comes to actually design the network and deploy it from the ground up, it is essential that you have some accurate understanding of how the old network functions.
- Until the users are on the new network, however, using the new servers, and so on, no real method for verifying the suitability of the design exists. It is possible, though, to run test simulations and performance baselines on gateways, WAN links, and servers before they are moved into production, to examine their behavioral characteristics. Then, after the network is in production, you can test data for comparison to your initial production baseline.
- Verify that your network design has been implemented properly. Unless you have installed the entire network yourself, (which is just not possible in large network implementations), you cannot be sure that all the configurations will be error free. In many cases, an initial performance baseline will usually reveal any configuration or implementation problems.

- Measure the effects of a topology or hardware addition upgrade. As mentioned above, any change to the network can have a positive or negative effect on the overall performance. As network changes are implemented, their effects need to be evaluated. Quite often, a change can have both positive and negative effects; with thorough testing, these effects can be isolated and corrected.
- Maximize the utilization of your WAN and Internet links. The telecommunication circuits used for WAN and Internet links are quite expensive. Evaluating their utilization can help you determine if your link is too small or not large enough.

Ideally, any baseline study will last usually two or more days, depending on the number of Layer 2 segments you need to sample. Before the use of network switching became popular, baselining was straightforward. Now, to get a truly accurate picture of network performance sampling, you need to analyze each segment. Switches discard runts and jabbers and influence the behavior of Layer 3 broadcasts.

In selecting a baselining period, you should take into account the time of the month that you are conducting the study. For example, if your business is calendar-driven, and at the end of every month the network is swamped with everyone using email and database applications to meet deadlines, this is when you want to run your baseline. This same approach should be used to pick your sampling times. If the office opens at 8:00, start your first sample at 8:30, when everyone is checking email and getting started for the day. After lunch is also another prime sampling period. Another approach is to take random samples every hour. Moreover, if you run your backups at night over the network, be sure to collect samples during this period as well. Deciding how and when to take your samples is largely dependent on how the network is used. If your business is a "9 to 5" shop, some well-timed samples will tell you what you need to know. If it's a 24-7 environment, the once-an-hour method is for you. Several data measurements are used to determine the network performance baseline. These measurements can be categorized into four basic data categories, discussed in the sections that follow.

WAN Link, LAN Segment, and Node Layer 2 Bandwidth Utilization

Utilization is a measurement of the percentage of time the transmission medium is in an active state (transmitting frames, for example). Two types of utilization measurements exist: Layer 2 utilization, expressed in Frames Per Second (FPS) and Layer 3 utilization, expressed in Packets Per Second (PPS). Utilization is measured by calculating the number of frames/packets transmitted over a one-second interval.

> **Note**
> The terms "packet" and "frame" are often used interchangeably. Frame refers to Layer 2 transmission messages, and packet refers to a Layer 3 datagram; packets are transported in frames.

When looking at network performance in terms of utilization, three components need to be examined: composite bandwidth utilization, broadcast bandwidth utilization, and composite bandwidth utilization distribution. These three components determine if the network is performing up to its expectation.

The average utilization is a measurement of the network throughput based on average packet size and frame rate. The network segments average utilization measurement is calculated by recording the network segments traffic throughput (consisting of both, good and bad frames) over a specified period of time (usually a 30- to 60-minute interval). The average utilization measurement is a single value determined by protocol analyzer or network monitoring tool used to collect the traffic sample. The analyzer/monitor determines the average utilization measurement, by calculating the mean average throughput of sample data collected.

The peak and low utilization measurements are determined using the same network sample data used to determine the network average utilization. However, the peak and low measurements are concerned with utilization in terms of network throughput over time, instead of single utilization value. Therefore, the peak and low utilization measurements are typically expressed through a "utilization versus time" graph that plots the network segments utilization over the duration of the traffic sample period.

Because all three of these measurements are based on traffic samples versus long-term traffic monitoring that, in order to get a good determination of the network segments actual average, peak, and low utilization measurements, you need to perform several network segment traffic collections. You should perform them during different times of the day, over an extended calendar period, and then developing your own histograph comparing the results.

As you begin to plot your network's utilization, you will begin to see consistent utilization trends over time.

Note

The first value is the composite bandwidth. The composite bandwidth is the amount of available segment bandwidth capacity that is being used for data transmission. The composite value is determined by examining the frame/second rate of the traffic on the network segment, and calculating utilization based on the mean average packet size and transmission rate. Actually, it consists of three values:

- The mean average utilization rate
- The peak utilization usage rate
- The low utilization usage rate

Each of these values (mean/peak/low) is derived by taking repeated FPS/PPS measurements over a period of time.

Note

Any unexplained utilization shifts typically indicate a hardware or software component problem on the segment. This should be investigated to determine the cause of the problem.

After you start to develop a utilization profile, you then want to compare your actual utilization measurements to what the expected operational "norms" are in order to determine if you need to make any network backbone or user segment topology changes to improve performance. When doing so, keep in mind that average, peak, and low utilization measurements vary from network to network, depending on traffic flow patterns, applications, and number of users.

In regards, to the "normal" average network utilization measurement values, industry-standard utilization averages can be used as baselines for comparison. With the understanding that your network's utilization should fall in close relation to the baseline average to ensure optimum performance. For example, on a shared 802.3 Ethernet segment, the average utilization should be around 30 percent of the available bandwidth, with 55 percent being the maximum peak bandwidth utilization rate and a 1 percent collision rate. On an 802.5 Token Ring, the average utilization is 40 percent, with a peak of 65 percent. Table 11.1 provides some baseline average and peak utilization guidelines for Ethernet, Token Ring, and FDDI.

Table 11.1 **The LAN Protocol Utilization Baseline Measurement Guidelines**

Transmission Medium	Transmission Speed	Average Utilization	Peak Utilization
Shared Ethernet	10Mbps and 100Mbps	30 percent	80 percent
Switched Ethernet	Half-Duplex	80 percent	90 percent
Switched Ethernet (including Gigabit Ethernet)	Full-Duplex	97 percent†	97 percent†
Token Ring	4/16Mbps	40 percent	65 percent
Switched Token Ring	Half-Duplex	80 percent	90 percent
Switched Token Ring	Full-Duplex	97 percent†	97 percent†
FDDI	100Mbps	60 percent	90 percent

† *In the case of full-duplex transmission links, the bandwidth utilization rate is the same for both the transmit and receive pairs.*

Now, if your network's average utilization rates are on the high side (5 to 15 percent over the average). This does not necessarily mean that you have a problem. Rather, additional sampling and analysis should be done to determine if these rates represent a real performance problem or just the natural operational rate of your users and their applications. You see the baselines, while accurate are skewed towards the worst case operational scenario, rather then "normal" operational conditions. Particularly in the case of Ethernet, which of all the LAN transmission protocol is the most prone to performance problems as the network segments diameter increases in size. The baseline utilization guidelines for shared Ethernet calls for a 30 percent average utilization. This baseline value is calculated under the assumption that the majority of the segment traffic utilizes Ethernet's most inefficient frame size (64 byte frames) with a segment diameter of 1,024 nodes on a single segment. Although the scenario is possible, it is rather unrealistic.

A more realistic scenario is a shared Ethernet segment utilizing an average packet size between 500 and 1,000 bytes in length. With a segment diameter node count between 10 and 150 stations per segment, a utilization rate of 40 to 60 percent with a peak of 80 percent, and a 1 percent collision rate is acceptable. The reason is that the size of the Ethernet frame and the number of nodes on the segment affect the overall efficiency of the segment. The larger the Ethernet frame size is, the more efficiently the available bandwidth is used. The higher the Ethernet node count is on the segment, the greater the decrease in throughput. A segment with 10 nodes can operate with a utilization rate of between 90 and 97 percent. At 100 nodes, the utilization is between 70 and 90 percent. At 250 nodes, the utilization falls between 60 and 80 percent.

The point here is that you need to monitor your network's utilization, average frame size, and transmission rate to determine what should be your network's average utilization should be. An Ethernet network that typically utilizes larger frame sizes will operate more efficiently, at a higher utilization rate then an Ethernet network using smaller packet sizes, operating at a lower utilization rate. In the end, you need to monitor the segments utilization and traffic size distribution consistently, to really know if your network is performing efficiently.

Token Ring and FDDI's utilization rates are based on maximum ring size and station counts, using a Token Holding Timer reflective of these ring sizes (Token Ring = 260 stations and FDDI = 500 stations). Again, similar to Ethernet, Token Ring and FDDI's averages are anything but average. Most installations use much smaller station diameters (between 30 and 100 stations) to make station management bearable. When establishing an average utilization for token-based transmission protocols, the Token Rotation Timer (TRT) is a good indication of how well the ring performs. An average TRT between 5 and 120 usec (for Token Ring), or less than 1.5x the target TRT (for FDDI), indicates an optimum-sized and well-performing ring.

This brings us to the second measurement, the percentage of the composite bandwidth used for network broadcasts. Network broadcasts are a percentage of all network traffic. This percentage, however, should be as low as possible. As a rule, broadcasts should make up no more than 8 to 10 percent of the total traffic transmitted on the network. Broadcast rates higher than this (15 percent or more) can indicate either a hardware problem of some kind (for example, broadcast storms), or that the Layer 3 network(s) address range is too broad and the Layer 3 network needs to be partitioned into smaller address segments using gateways. What is important to keep in mind is that any percentage of network traffic above the norm is bandwidth that could be utilized for transmitting real data. If you are running multiple Layer 3 protocols, your broadcast rate might be slightly higher than the 8 to 10 percent acceptable rate.

Note

The different percentages reflect the effect on the segment's efficiency using Ethernet's smallest and largest frame size.

The third measurement—the composite bandwidth utilization distribution—is the breakdown of the composite bandwidth used by each node on the network segment. Nodes utilizing more than 60 percent of the composite segment bandwidth should be identified and their function on the network segment should be assessed. High utilization is common for servers and gateways. On standard end-nodes, however, this usually indicates a software or hardware problem. When examining bandwidth utilization on a node-by-node basis, it is also helpful to look at the percentage of broadcasts generated by each node.

Each of these measurements can be performed with hardware- or software-based protocol analyzers or a packet-capturing RMON probe.

Network Throughput

One of the major performance goals of any network is to provide the highest possible throughput in the shortest amount of time. Network throughput is measured by performing transaction timing analysis. This analysis measures the actual time it takes to transmit and receive information on the network, using three time measurements: transfer time, transaction time, and event time.

Transfer time is the bit transfer rate over a one-second period. On LANs, the minimum throughput for a shared segment LAN is about a 200Kbps host-to-host transmission rate. Over a WAN link, a host-to-host throughput rate minimum of 50Kbps (on a T1/DS1 circuit) is considered average. When examining the transmission rate, the operative word is consistency. Networks and/or network segments that consistently provide transmission rates below these averages are oversubscribed. This often indicates the need for a larger circuit and/or higher bandwidth transmission path.

Transaction time is the amount of time it takes to process a network event. The actual time measurement is the period between the first packet being sent and the last packet being received, ending the transaction.

Note

When looking at network broadcast measurements, it is important to have an understanding of the types of applications in use—specifically, how they utilize the network to transmit user data. It should also be determined if the high level of broadcast traffic is having any adverse effect on the actual performance of the network. There are a number of applications that use broadcasts to transmit user data. For example, many stock price ticker applications (used in brokerage houses) use UDP broadcasts to send and receive price updates. Another example is Windows-based PCs that use NETBUI with NetBIOS applications. In these situations, a high level of broadcast traffic is expected and should not impact network performance. Although these cases are the exception, they do exist.

The transaction itself can be a file transfer of a file server, a login/authentication process, or even a database query. Any network-related transaction that is performance sensitive can be used. The goal is to get a real-time measurement for these transactions for long-term benchmarking evaluation. The transaction time measurement is going to be relative to the transaction being timed. So, the actual expected performance will need to be determined to see if the application is performing properly. This, however, is a system management issue, so you will need to get this value from your system or database manager. Again, consistency is the idea here; you want the transaction time to be constant each time the measurement is taken. Large deviations should be investigated, in relation to network transfer times and event times. If an increase in transaction time is noted, and the transfer and event times have remained constant, the increase will generally be related to a systems issue, such as increased demand or limited resources, which should be addressed by the systems administrator.

Event time is the amount of time between the transmission of a request and the transmission of the response to the event. Like transaction time, this measurement is a benchmark for long-term analysis, and consistency in measurement is the goal. Event time is dependent on the performance of the system that is being asked to respond. This value looks at the time it takes for the event to be acknowledged and initiated, not the actual time it takes to complete the transaction. Long event times, particularly over WAN links, often indicate a saturated link in the transmission path or an oversubscribed service.

When measuring network throughput, you want to perform a variety of tests using different hosts and test file sizes. To perform throughput tests, you need to use a protocol analyzer or packet snooping application that can support filtering and time stamping. Because you only want to examine the traffic between your test station and target station, you need to use the packet timestamps to calculate the transaction and event times.

Network Errors

Knowing the amount and type of errors generated on the network segment is important for determining the segment's performance, and for properly troubleshooting performance problems and network outages, when they occur. During the network baselining process, a certain percentage of network errors are collected. Although network errors represent some type of abnormal transmission event, not all errors are an indication of a hard failure. In most cases, errors are indications of operational disturbances in the normal data transmission flow (which might even be necessary for proper operation). In some cases, network errors can be addressed by the transmission protocol without an administrator's intervention. Errors, in other cases, are an actual by-product of the data transmission process (Ethernet Collisions, for example) that are needed for the transmission process to function normally.

Under normal operation, network error rates should be a low percentage of the overall bandwidth utilization (1 to 3 percent). A high rate of errors, (being a consistent error rate of using 5 percent or more to the segments bandwidth) is an indication of problem that requires further investigation to identify the error source. Typically, a high-error percentage rate is an indication of failing hardware or software driver, a mis-configuration in circuit provisioning (WAN links), or a violation of transmission cable length specifications. Each of these impacts the performance of the device with which it is associated. However, excessive error rates also degrade network performance on the whole, by consuming bandwidth that would otherwise be used for sending data. Additionally, depending on the type of network error, the segment might evoke "recovery processes" or "transmission back-off algorithms" in an attempt to stabilize the segment. Both of these events effectively halt data transmission during the initiation process, further degrading performance.

With the exception of WAN transmission equipment, most error conditions are only visible through the use of a network probe and a network management/analysis tool. Intermittent or unexplained network performance drops overall degradation in network performance (for example, excessive local network transaction times, unexpected service disconnects, and so on) and are operational indications of the existence of an error condition and, unless you have deployed error recording and event alarming, might be your only indication of a problem. Listen to your users and take performance complaints seriously. Although the majority of complaints might be nothing, the ones that are serious, if identified early, can be corrected before they gravely impact network service.

Now, lets take a look at some of the different types of transmission and delivery-related errors.

Note

Another common network error not related to data transmission is duplicate Layer 3 (network layer) end-station addresses, particularly with IP. Under normal operating conditions, IPX and AppleTalk are relatively immune to duplicate addressing problems. However, because IP addresses can be statically assigned and, in many cases, changed by users with ease, duplicate addressing with IP is common. A packet sniffer can be used to track down duplicate network address users using the system's MAC addresses. If you do not know the location of the devices or the MAC addresses associated with the duplicate address, however, tracking down the location of the mis-addressed station is difficult. This is why good network documentation is so important.

Ethernet Errors

When looking at Ethernet performance, you will come across four types of errors: collisions, errored or corrupted frames, illegal frame, and jabbers. Collisions are a natural byproduct of the CSMA/CD medium-access method employed by Ethernet to manage access to the transmission medium. Collisions result when two stations are on the same segment attempt to transmit data on the medium simultaneously. Under normal operation, a moderately populated segment (20 to 200 end-stations) should have a collision rate of around 1 to 2 percent of the overall segment traffic. Higher density segments (200 to 500 end-stations) should expect collision rates of around 3 to 5 percent.

If the collision rate for the segment exceeds normal operational values, the end-stations involved should be identified and examined for possible faults. High collision rates can be the result of a bad hub or transceiver, a faulty NIC or connector, or a cable length violation.

When errored or corrupted frames are transmitted over the wire, they are recorded as CRC errors by protocol analyzers and management probes. To ensure data integrity, Ethernet employs Cyclical Redundancy Checking. Before an Ethernet frame is transmitted, the NIC performs a CRC on the "header and data" portion of the frame and appends the result, to the end of the Ethernet frame. When the station receives a frame, it performs its own CRC check on the header and data portion and compares it to the CRC transmitted along with the Ethernet frame. If they match, the frame is processed further; if they do not, a CRC error results and is sent to the network—the frame is discarded. Often, a CRC is due to an alignment error in the bit placement of the frame. An occasional CRC error can be attributed to environmental or corrupted ULP data. CRC error rates above 2 percent of network traffic indicate a problem, often a bad NIC, transceiver or cable connection.

Note

On switched Ethernet segments that are operating in full-duplex mode (a single end-station attached to switch port or an inter-switch link), there should be no collisions. This is because full-duplex operation disables the collision detection aspect of CSMA/CD used by Ethernet in order for the end-station to send and receive data simultaneously. If you do see collisions on a link you believe is configured for full-duplex operation, chances are that, either the end-station or the switch port is misconfigured.

As part of Ethernet's CSMA/CD implementation, Ethernet has defined minimum and maximum packet sizes. The minimum packet sizes are 72 bytes, 64-bytes data, including the frame header, the data PDU and the frames CRC, plus an 8-byte preamble. The maximum packet sizes are 1,526 bytes (802.3) or 1,530 bytes (802.3ac) (1,518-byte, including the CRC and header, plus an 8-byte preamble).

Ethernet frames that are less than the minimum size are known as *runts*, and excessively long packets are known as *jabs*. Illegal packet sizes are caused by a bad NIC or corrupted NIC driver. Another indication of a bad NIC is a condition called *jabbering*, which occurs when a NIC sends continuous stream bits that are longer than 1,526/1,530 bytes in length. Jabbering can bring the entire segment to halt because the jabbering station typically cannot hear collisions on the segment, and it just continues transmitting streams of random bits, consuming all the segment's available bandwidth.

Token Ring Errors

Token Ring has two types of error conditions: soft and hard. Soft errors are error conditions of an intermittent nature that only impact data transmission on a temporary basis. When a soft error is encountered on a Token Ring system, an error recovery process is initiated. This process begins with the detecting station collecting error type information and then generating a report indicating the type of error (and the station at fault, if possible) and sends it to the Ring Error Monitor (REM). REM is collects ring error information to troubleshoot error conditions on the ring. Each station keeps track of the errors it detects on the ring and, when an error is detected, it increments it's internal soft error counter. The remainder of the recovery process depends on the category and type of soft error that occurs.

Token Ring soft errors are divided into two categories: isolating and non-isolating. Isolating soft errors (ISEs) are soft errors that can be attributed to a specific Token Ring fault domain. A fault domain is a logical area of the ring that consists of the station that has detected the error, it's nearest active upstream neighbor (NAUN), and the physical ring segment that separates the two stations. Isolating error's can be often be attributed to a specific ring station within the fault domain, because the nature of the soft error is such that it is detected by the next station in the ring's logical order, which detects and sends the error report. Alternatively, Non-isolating soft errors (NSE) can be detected by any station on the ring, making fault isolation difficult. However, form a error reporting standpoint NSE are handled in a manor similar to ISE, where the detecting station collects error data, and generates and transmits an error report to REM.

> **Note**
>
> Almost all references to Ethernet frame sizes do not include the frame preamble when calculating the minimum and maximum Ethernet frame sizes. Hence, the standard minimum and maximum packet sizes are cited as 64 bytes and 1,518 bytes. Technically, however, although the frame preamble information is discarded when the frame reaches its destination, it is part of the frame when it travels across the wire. Therefore, when measuring the performance of the segment in terms of transmission utilization and time, the entire frame must be accounted in the calculation.

Not all ISE's and NSE's soft errors impact the rings performance equaly. The soft errors severity dictates level of level of the ring recovery effort needed to stabilize the ring. There are three soft error ring recovery procedure's, Type 1, Type 2, and Type 3:

- Type 1—Do not require any ring recovery effort, aside from reporting. A Type 1 error might result in an ULP retransmission of the data contained in packets that might be lost as a result of the error. A high rate of Type 1 errors associated with a particular station are an indication of a faulty NIC, a bad lobe connector, a bad horizontal, or patch cabling.
- Type 2—Type 2 soft errors require a ring-purge procedure to be invoked by the Active Monitor as a recovery effort. The ring-purge procedure attempts to reset the ring back to normal repeat mode. It is initiated by the active monitor on the detection of a lost token, frame, or a disruption in the ring's timing. The active monitor sends the ring-purge frame and awaits its return. If it returns, it checks the ring-purge frame for errors and, if there are none, generates a free token. If an error is detected, it transmits another ring-purge frame and checks it again for errors. If the active monitor fails to receive the ring-purge frame or continues to receive "errored" ring-purge frames until the ring-purge timer expires, the active monitor then evokes the token-claiming process which results in the re-election of the active monitor. Type 2 soft errors result in some type of error relating to the validity of the frame or token, and are commonly attributed to ring reconfiguration or a possibly a faulty NIC.
- Type 3—Type 3 soft errors require the election of a new active monitor and a ring-purge procedure. They are either a result of the loss of the ring's active monitor, or result in loss of the active monitor.

The level of ring recovery associated with each soft error must also be distinguished. In addition to being categorized as either an ISE or a NSE Token Ring soft error, each Token Ring soft error is further classified as a Type 1, Type 2, or Type 3 error. The following is a list describing the most common Token Ring soft errors (including their classification and recovery type):

- Line error—An isolating, Type 1 error. It indicates that the frame has a CRC error.
- Frame copy error—A non-isolating, Type 1 error. It indicates that a station copied a frame with its source address.
- Multiple monitors—An isolating, Type 1 error. It indicates that the active monitor received a ring-purge frame that it had not sent.
- Receiver congestion error—A non-isolating, Type 1 error. It indicates that a station was unable to receive a frame that was addressed to its address.

- Burst error—An isolating, Type 2 error. It indicates that a signaling error occurred. Burst errors are a common result of a station being added or removed to the ring. Occurrences of burst errors when no reconfigurations are occurring are often an indication of a cabling problem.
- Abort delimiter transmitted error—An isolating, Type 2 error. It indicates that the ring station detected a corrupted token or some type of unrecoverable internal error, which requires it to abort its transmission.
- Lost token error—An isolating, Type 2 error. It indicates that the active monitor's "Any_Token" time expired. If the active monitor fails to detect a token or frame once every 10usec, it purges the ring.
- Lost frame error—A non-isolating, Type 3 error. A lost frame error indicates that a station has sent a frame and has not received it back. The error is generated if the ring station (functioning as a standby monitor) fails to receive a token once every 2.6 seconds. This is tracked locally by the station with the Good_Token timer. An error is also genterated if the station is not polled once every 15 seconds by the ring's active monitor (also tracked locally with the Receive_Notfication timer).
- Frequency error—A non-isolating, Type 3 error. It indicates that a station received a frame that is out of clock frequency with the rest of the ring. As mentioned in Chapter 5, one of the functions of the active monitor is to provide the ring's master clock for transmission timing. Token Ring uses a line clock rate of 8MHz (for 4Mbps Token Ring) and 32MHz (for 16Mbps Token Ring). Although the ring signal (maintained by the frame/token) is retimed as the token is passed from station to station, due to signal attenuation from the transmission medium and varying component tolerances the frames, timing shifts slightly out of phase with the master transmission clock (maintained by the active monitor) as it traverses the ring. This phase shift is known as *jitter.* Token Ring's ring station density, cable type, and length design specifications are provided, taking into account a jitter rate that can be compensated for through the use buffer memory on the Token Ring NIC. This buffer is known the *elastic buffer.* Every Token Ring chipset contains an elastic buffer that corresponds to the NIC's operating rate. However, the elastic buffer is only used on the station that is designated as the ring's active monitor. The active monitor uses the elastic buffer to reset the ring's phase as the frame/token passes around the ring. When the jitter build-up on the ring exceeds the level that can be compensated for, a frequency error results.

Token Ring hard errors represent a class of errors that directly impacts the operation of the ring, typically, some type of hardware failure. Upon the detection of a hard error, the station evokes the Token Ring beaconing process in order to recover the ring. What makes hard errors different from soft errors is the element of an actual failure in the physical transmission infrastructure (for example, the complete failure of a NIC, lobe cable, or horizontal cable).

The beacon process is, in essence, a failure alarm that is sent by the station detecting the error to the rest of the stations on the ring. The beacon frame consists of three information elements: generating a host address, the host's NAUN address, and the hard error beacon type. The hard error beacon type provides some indication of the type hard error that has evoked the beacon process and the status of the beacon process. The following lists the hard error beacon types:

- Signal loss error—Indicates that a cable fault is the source of the error condition
- Streaming signal—Indicates that one of the stations in the fault domain is in a failure state
- Ring recovery—Indicates that the error has been isolated and the ring is in recovery

After the beaconing process is initiated, the other stations on the ring enter beacon repeat mode. In most cases, the station generating the beacon is not the station at fault. Rather, it is the beaconing station's NAUN. These two stations make up the hard error fault domain. The recovery process that is the end-result of the beacon process is two-fold, fault identification and ring reconfiguration.

Fault identification is achieved by both the error-generating station and it's NAUN by performing phases 0 (lobe Test) and 2 (Duplicate Address Verification Test) from the ring station insertion process. If either of these tests fail on either station, the failed station removes itself from the ring. After the failed station is out of the ring, reconfiguration can take place through automatic, or, if necessary, manual means. This process begins with the transmission of a beacon frame indicating ring recovery and the invoking of the beacon removal process. The following is a list of the common hard errors and their possible cause:

- Signal loss error—Indicates a cabling fault somewhere between the station and the NAUN. This could be a faulty MAU port, cable short, bad cable termination, or a bad connector.
- Streaming signal (bit streaming)—Indicates that the NIC at fault is overwriting the token and the frames it receives.
- Streaming signal (frame streaming)—Indicates that the NIC at fault is sending a continuous stream of tokens, frames, or abort delimiters.
- Internal error—Indicates that the NIC detected an unrecoverable error, which caused the detecting station to remove itself from the ring and force a ring re-initialization.
- Frequency error—Indicates that a station detected a discrepancy between its local adapter clock and the ring clock provided by the active monitor. This initiates an active monitor election. In the event that the active monitor election cannot be resolved, the ring re-initializes.

In most cases, when a hard error condition exists, you might need to isolate the stations in the fault domain from the rest of the ring to determine the actual cause of the failure. Isolation of the faulty station does not correct the problem; it only prevents it from impacting the operation of the rest of the ring. Hard error verification tests also do not always identify and remove the station responsible for the hard error condition. In situations where a hard error occurs, both stations are re-inserted back into the ring because they both passed the self tests. The best course of action is to remove the stations and test their NICs and cabling components. Additionally, when examining Token Ring failures, keep in mind that in addition to error conditions described above, station insertion and operation problems can also be the result of software driver problems or perhaps even something as simple as a NIC configured with the wrong operating speed.

T-Carrier Errors

WAN digital transmission circuits, like their LAN counterparts, are subject to intermittent transmission disruptions and failures. Unlike LANs, however, WAN circuits operate under a QoS agreement, on the side of the carrier/service provider. The average QoS objectives for digital transmission circuits range between 99.96 and 99.98 percent of error-free transmission for yearly service availability. To meet this QoS range, the transport must operate within a range of no more then 6 to 9 seconds worth of transmission errors on a given day, with a maximum of no more then 4 to 5 severely errored seconds per day. (A severely errored second is a one-second period where errors (no data) are consecutively transmitted.) In practical terms, this means that at least 96 to 98 percent of the frames transmitted over a one-second period have at least one error.

T-Carrier error types differ, depending on the framing format and transmission medium. Table 11.2 lists the error types associated with the D4 and ESF framing formats transmitted different transmission mediums.

Table 11.2 **WAN Frame Type Transmission Errors**

Transmission Medium	D4 Framing Format	ESF Framing Format
Copper-based mediums	Bipolar violations	Bipolar violations
	Frame errors	Frame errors
	Timing errors	CRC errors
	Signal errors	Timing errors
		Signal errors
Microwave/satellite	Frame errors	Frame errors
	Bipolar violations*	Bipolar violations*
	Timing errors	CRC errors
	Signal errors	Timing errors
		Signal errors

Transmission Medium	D4 Framing Format	ESF Framing Format
Fiber-optic cable	Frame errors Bipolar violations* Timing errors Signal errors	Frame errors Bipolar violations* CRC errors Timing errors Signal errors

* *Bipolar violations are present on copper interconnect sections of the transport segment.*

Digital transport errors are measured by the occurrences of bit-placement errors in the frames that comprise the entire bit-stream. These measurements are made using a BERT or a Protocol Analyzer outfitted with WAN analysis module. Circuit testing and monitoring can be performed through obtrusive (service effecting) or unobtrusive (non-service effecting) means. Obtrusive testing is done by using Bit Error Rate Tests (BERTs) that send different bit-stream patterns over the transmission span and measure the results. BERTs are performed by the carrier before the circuit is placed in service to validate its proper operation. They are also used to locate failures in the transmission path when hard failures occur. Unobtrusive testing collects error statistics on the bit-stream flow. If you do not have a BERT or an equivalent tool (which is often the case, due to their expense), you can use the DSU/CSU to collect transmission error statistics. The following list provides a description of each of the possible WAN error types:

- Bipolar violations (BPV)—Occur when two one-bit pulses of the same polarity occur in a row. For data transmission over copper transport, bipolar signaling is used. Binary zeros are represented by an absence of voltage. Binary ones are represented by alternating + and – DC voltage signals on the line. BPVs do not occur on fiber optic, microwave, or satellite transmissions because they use unipolar signaling. Fiber optic and radio wave transmission systems correct BPVs, if detected. However, BPVs that are "missed" are represented in the form of bit errors (for example, an incorrectly placed bit representation).
- Frame errors—Frame errors are discrepancies in what is expected to be the frame pattern and what is actually received. Frame error measurement is based on the counting of the framing bits contained in the frame, rather than examining the entire frame.
- CRC errors—To provide additional data validity checking capabilities, Extended Superframe (ESF) provides CRC-6 error checking. The CRC-6 tag is calculated by the transmitting equipment by adding the sum of all the ones and zeros and dividing it by a prime number. Then, this sum is added to the end of the data frame. The receiving equipment at the end of the transmission span (within the circuit path) performs the same calculation and compares the result. If the two sums do not match, bit errors have occurred during the transmission over the span. The CRC calculations of the data frame are recalculated each time the frame passes through a digital cross-connect, which interconnects the transit segments that make up the entire data transmission circuit.

- Timing frequency errors (TFEs)—TFEs are "slips" in the circuit timing. These slips are manifested as the addition or deletion of bits in the data stream. TFEs result from a poor circuit timing source.
- Signal errors—Result when the signal frequency level of the circuit drops above or below the tolerances required for proper transmission. Signal levels vary from segment to segment, depending on the length of the transmission medium. For proper operation, however, the signal level must be maintained on each segment consistently. Faulty circuit transmitters, poor or open cross-connects, an improper LBO setting on the CSU, or a cable length violation are all possible reasons for signal errors.

Retransmission Errors

Retransmissions are the most common ULP layer error detected. They occur when an application does not receive a proper response (or no response at all) within a specific time window. The occurrence of an occasional network retransmission does not signify any problem per se. Hosts that generate excessive amounts of network retransmissions, however, usually have some kind of physical layer problem (for example, cabling issues, a bad NIC, or corrupted or incompatible network drivers). Retransmission problems should be investigated aggressively because high retransmission rates consume bandwidth and contribute to overall performance loss. Additionally, if the problem is cabling-related, excessive collisions and other packet type errors might be generated as well.

Now that you have an understanding of the various types of errors that can and do occur, it is time to move on to troubleshooting those errors.

Developing Troubleshooting Skills

Some individuals are born troubleshooters. For the rest of us, however, troubleshooting (like everything else) is a skill that is learned. The development of troubleshooting skills is essential to the success of any network administrator. A large part of the development of these skills happens naturally over time as your understanding and experience with the different technological aspects of computer networking grow. Although technical knowledge is important, however, your ability to ask the right questions when problems arise, and to be able to listen and understand the answers to those questions, is the most important (and often the hardest) troubleshooting skill any technical professional must develop.

Problem resolution commonly takes place in either a proactive or reactive mode. The ultimate goal of proactive problem resolution is to identify problems and resolve them before they impact service performance. Reactive problem resolution (or fire-fighting mode) deals with problems as they arise. Most IT and MIS shops operate to some degree in both proactive and reactive mode (some, sadly, are more reactive than proactive). The basic nature of information systems support requires this dual mode approach.

Network troubleshooting also lends itself to this dual mode resolution process. Proactive monitoring of network performance, essential equipment, and transmission links with SNMP and other tools is highly recommend, if the network plays any significant role in the operation of the business enterprise. Where reactive problem resolution comes into play is in the identification and resolution of user problems.

As a network administrator, when dealing with user issues you quickly realize that *everything* is on the network. Because everything is dependent on the network, it is often the first suspect. In terms of support, it is your job, in many cases, to identify the real problem and assist those actually responsible for correcting the problem (which, in some cases, might be you). The following are some guidelines you might find helpful when troubleshooting a problem. Their focus is to suggest some ways for looking at and identifying problems:

1. Look at the big picture and then figure out what perspective you have. Ask yourself, "Do I refer to all carbonated beverages as soda, pop, or Coke?" Say things like, "Is there a problem with the network?" These generalizations are commonly used in human communication. They serve as a type of shorthand used under the assumption that the speaker and the party being addressed both understand the meaning. This kind of communication is common when users report problems. Users tend to generalize and use generic claims like, "The network is down," and assume that you immediately know the problem. To realize that this is happening, you must look at the big picture. Figuring out what you are looking at involves asking the user effective questions that actually describe the problem. Although this might be frustrating at times, without accurate information, it is not possible to effectively troubleshoot a problem.
2. Ask the right questions. When you talk to the user about the problem, ask the right questions and listen to the answers. It can save you time. Because the problem occurs with the user, in most cases, he or she is your best source of information. Some examples of the kinds of questions you should ask are as follows:
    - Can you describe the actual problem being experienced? (Be sure to get a list of applications and services used when the problem occurred.)
    - Has the problem occurred previously?
    - Is it a chronic problem? If so, how long have you been experiencing it?
    - If it is chronic, does it occur in along with another event that might or might not be related?
    - Is it an isolated problem, or is it being experienced by other users in your local area?
    - Can you replicate the problem?
    - Have any changes been made to your workstation recently?

3. After you figure out what you are looking at, be sure to map it out. When getting the details of the problem from the user, take notes, draw a picture, do whatever you need to get the facts straight. Returning to a user later to ask the same questions again is unprofessional. After talking with the user, identify the various components that are related to the problem. For example, if the user is having problems sending SMTP mail, identify the components involved in sending mail from the user's workstation. Then, check your notes and see what information you have that might pertain to components involved in the operation. For instance, when you asked the user if he had problems sending all mail messages, just messages to users within the company, or just users outside the company, he indicated that the problem was only with sending messages to users outside the company. This condition could relate to the SMTP relay host and perhaps the firewall (if one were in use).
4. Don't get lost in the big picture. When examining the components related to a process, define clear start and stop points. Every process has a beginning and an end. Sometimes, these points are not easy to see, but they are there. Although this seems obvious, quite often, the most obvious problems are the hardest to spot.
5. Document, document, document. When working on a problem, especially a performance-related one, be sure to document the results of your tests, changes you make, and so on. Proper documentation makes it easy to retrace steps after you have a solution. Problem resolution documentation serves as reference material for the future, so, if a similar problem arises, you have something to work from. Documentation also makes it easy for you identify and undo a change you might have made trying to fix the problem that ended up creating another problem. Many individuals also find that the documentation process itself aids in their understanding the problem and solution even better.
6. Mind the details. Always remember that computer networking is about details. Each component, from a patch cord to a router, plays a role in overall functioning of the network. A bad component, however benign, can affect the performance of the network. Remember, when looking at a process, be sure to identify all the elements involved. Then, when you are troubleshooting, assume that each of the elements could be at fault and check each element. A bad network card can bring down a whole network segment. Often, it's the last thing anyone looks for.

Network Management Fundamentals

Before computer networks emerged, there were mainframe time-sharing systems. These systems were standalone supercomputers that shared memory, processing, and storage resources between a certain number of users. Users would access the mainframe through dumb terminals that were connected to the server via serial controllers or modems. This computing model is still used today with some UNIX systems and client/server computing. But, large mainframe time-sharing systems required more attention and management than most UNIX systems today. Mainframe management was performed by systems administrators. These administrators would monitor the mainframe's processor and memory utilization, I/O controllers, line terminals and modems, multiplexers, channel banks, dedicated circuits, disk, tape, and card reading systems. All these systems required constant monitoring to continue functioning properly and effectively. What system resources were over-utilized or under-utilized? What was the system's load average and peak load? Were the local and remote access lines adequate to meet the access demand? The administrators would work with system operators and analysts to resolve system resource failures and reprovisions or add system resources.

Managing these complex systems was a monumental task. In most cases, separate management and configuration systems were required to manage all the different system resources. These systems were comprised of custom tools to extract data from all of the subsystems, and this complexity was expensive in terms of development and personnel costs. It was also very difficult to integrate new resources or products that were made by different manufacturers.

When LAN and WAN networks began to emerge in the 1980s with the introduction of the personal computer, similar configuration, management, and monitoring began to appear for computer networks. These new technologies were proprietary. Standards bodies, manufacturers, network managers, and network providers realized that there was a need for "universal" standards-based automated network management.

> **Note**
>
> In the 1980s, there were no Internet service providers (ISPs). Instead, the "Internet" was a collection of regional networks managed by government and research entities. Internet standards were developed by the Internet Activities Board/Internet Architecture Board (IAB) and the Internet Engineering Task Force (IETF).

Network Management Functions and Architectures

Network management is a complex undertaking. Even the name "network management" is confusing, because unlike systems management, the circumstances constantly change. Network management consists of separate and distinct tasks that contribute to the management of a complex collection of interdependent systems. The primary network management functions include the following:

- Physical infrastructure, interconnection cabling and patch panel design, installation and management, end-station patching and network hardware installation. Cable testing and length validation also fall into this category.
- Device configuration, bridge, router, switch, and repeater configuration. Backup, archiving, and documenting device configurations. Creating and updating network topology maps, showing device relationships, installation location, and other basic configuration information (IP address, device type, manufacture, and so on). Proper network documentation is essential when diagnosing network problems and performing performance testing.
- Link and services monitoring, network performance baselining, and periodic performance revaluation. Proactive and reactive hardware, link and network service failure detection. Network security monitoring.

In small office networks (generally 20 to 50 nodes), these duties are often performed by a solitary network administrator. In this kind of scenario, spreadsheet record-keeping or station and topology mapping with graphics packages like SysDraw or VISIO is possible, because one person will always be able to find and control the documentation. Router and other device configurations can be saved to a TFTP server. Link and service monitoring, if they are done at all, are accomplished with custom checking scripts, or with a free or commercial monitoring package and an email-enabled pager.

In large enterprises, a network environment that consists of hundreds or thousands of nodes, some kind of Network Management System (NMS) is generally required. These systems are either maintained by the IT or MIS department, or outsourced to network management firms like Electronic Data Systems (EDS). In either case, large-scale NMSs are expensive and complicated. They require expensive hardware and software components and, most importantly, trained staff to interpret the management data and do problem resolution. No matter how great your management system is, it cannot interpret the data it collects or fix your network when it goes down. That must be done by the network administration staff.

When determining your network's management system requirements, it is essential that you define what the NMS's operational scope will be. Without defining what your network management needs are, you can easily spend too much time and money developing a system that can perform many functions that you will not need. For example, the use of custom monitoring scripts using ping is a perfectly acceptable method for verifying if an IP-based device is up on the network. Ping might not tell you if it is operating correctly, but it will verify that it is online. This solution, for most scenarios, is adequate. If the script fails, it can send you email notification of the failure. Alternatively, you can spend some significant money on a management system that will poll the device and verify that all of the services are functioning correctly, and attempt to restart it if it's not. When restart fails, the system will notify you with an email. The point is that you need to determine what your network's operational and service requirements are and then use a network management approach that will accommodate those requirements in a sane, cost-effective manner.

The broad scope of activities that fall under network management often become blurred together. Network management system developers have a tendency to promise much more than they can deliver—easy configuration, automatic data collection, problem resolution, and so on. Effective network management is not easy; in fact, it can be downright tedious and frustrating at times. NMSs are only a tool, a very helpful tool when implemented correctly. They are not the only tool necessary for maintaining the efficient operation of the network and, in many cases, not the most important.

NMS Architecture

An NMS consists of the network hardware components that need to be managed running a software or firmware-based management interface or agent, a network management protocol, and a network management console.

An NMS can manage essentially any type of hardware device, such as a router, bridge, DSU/CSU, network server, or user end-station. These devices are classified into two categories: managed and unmanaged nodes. Managed nodes have the ability to perform basic testing and diagnostics on themselves and report their operational status to a management entity. The tests are performed using a proprietary or standards-based agent. An agent is software that runs on the device and provides management functionality by collecting and relaying management data about the device to the management system. The agent can also be used as an interface to the system's proprietary configuration and testing interface, permitting the NMS to perform remote device configuration and testing through the network management entity. Unmanaged nodes are devices that cannot directly support a management agent. These devices utilize a management proxy interface. A management proxy acts on the behalf of unmanageable devices to provide management information to NMS entities. A proxy interface can be an end-station or hardware device that runs an NMS agent and has interfaces (usually an RS-232) that connect to non-manageable devices. The proxy interface runs software that can communicate with the unmanageable node's configuration and management interface. The software translates NMS management information requests into commands that the device's local management interface can understand, which, in turn, relays its responses to the NMS.

The Network Management Protocol (NMP) defines the structure used by the management agents and the management entities to format, transmit, and exchange management information. There are several experimental, proprietary, and standards-based protocols that can be used to send messages between reporting and data-collection devices. The most well known and supported is the IP's Simple Network Management Protocol (SNMP).

Depending on the size and scope of the NMS, there can be one or several Network Management Consoles (NMC). The NMC is a computer that operates one or more management entities. These management entities perform basic management functions. The functions are essentially query interfaces that collect information using the NMP. This data is then stored in specialized databases that are used for performance analysis, problem tracking and resolution, trouble ticketing, and inventory control. The databases are relational; they can generate management reports on network utilization, availability, and inventory. Most NMCs also perform automatic device discovery and topology mapping. Figure 11.1 illustrates the basic NMS relationship structure.

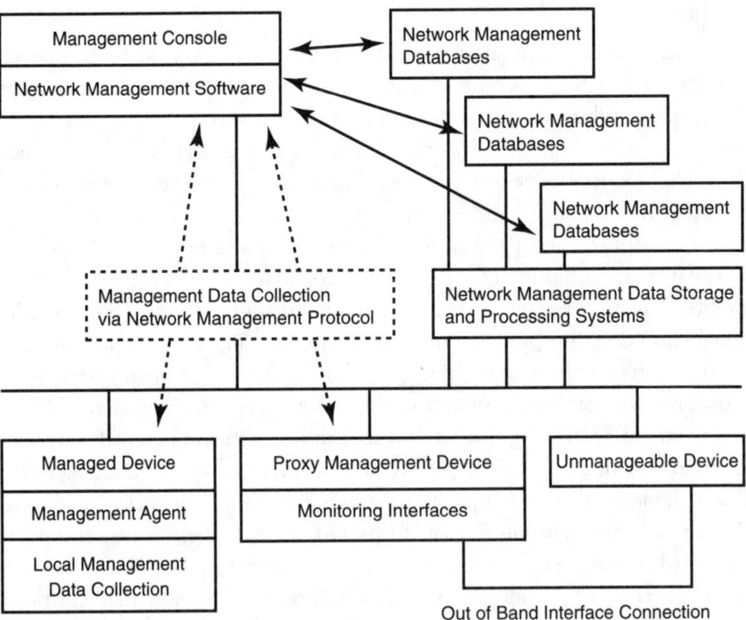

Figure 11.1 The basic NMS architecture model.

OSI Network Management Model

The International Standards Organization (ISO) developed an Open System Interconnection (OSI) network management model to provide a common reference for understanding and developing network management protocols and systems. The model defines five functional management areas:

- Fault management—This is the function of detecting, isolating, logging, and resolving network conditions that prevent the network from operating in a normal condition. Fault management elements would involve the detection and notification of abnormal conditions through the use of device queries using availability and performance thresholds to determine ranges of normal and abnormal operation. After a problem is identified, it is entered into a trouble ticketing system. The trouble ticketing system provides problem documentation, status reporting, and resolution documentation.
- Configuration management—In large-scale network environments consisting of perhaps thousands of manageable nodes and devices, configuration and version control is essential. Incompatible versions of software can lead to many network and general interoperability-related problems. Configuration management systems poll network devices and create a database of the devices' hardware configuration and software versions. Additionally, these systems provide facilities for performing software upgrades and installing new software packages. After an accurate configuration inventory is available for all the nodes on the network, the database can be consulted when diagnosing system performance issues. These systems have become popular for large scale Windows NT installations, which do not support remote administration easily.

  Configuration management also entails managing network resources. Adding or removing nodes, manual traffic redirection, and service prioritization during periods of high network utilization are examples of when this kind of management would occur.
- Accounting management—The computer network is a common enterprise-wide resource. It is common to bill various business cost centers or customers based on network utilization. Accounting management in these situations is also used for equipment depreciation and amortization, staff-related costs, and interdepartmental billing. This information can also be used for apportioning network resources between users. The historical data can be used to substantiate the need and/or establish a basis for justifying capital expenditures for new resources.

- Security management—This is the function of implementing and enforcing network and computer security. The network represents a substantial security risk for any enterprise. Networking equipment, management systems, and access points must be adequately secured. This entails the physical security for network equipment, the partitioning and securing of public and private data cabling, and the use of data encryption to transmit data between nodes and over public access links. The administration of a trusted third party or one-time password authentication system is part of this function, as well as the creation of an adequate security log and system event database.
- Performance management—This process evaluates the performance of the network. Ideally, the network should be operating in a manner that permits the connected nodes to process network transactions at their relative capacity. In order to ensure that network performance is maintained at an acceptable performance level, it is common to have service level agreements (SLAs) that govern network performance and availability. Performance and availability are measured using variables such as the following:
  - Number of CRC errors
  - Bit error rates
  - Bandwidth utilization rates
  - Ratio between broadcast and "real" network traffic
  - Protocol utilization and percentage rates
  - Percentage of packet loss and packet retransmission
  - Average transmission rate

Network performance validation begins when the network is first deployed. During the initial deployment network, utilization data should be collected. This data is used first to assess if the original design assumptions for network utilization were correct, or they were over or under expectations. After the first round of performance validation, and depending on the outcome, adjustments might need to be made. It is important to realize that no matter how much testing and modeling is done before a deployment, real user traffic is the only real test of network performance. If the network capacity is inadequate, additional resources or a topology change might be needed. If the network has excess capacity, cost savings might be gained by scaling back under-utilized WAN and remote access capability. After these adjustments are made, a second round of validation testing should begin. If no further adjustments are needed, this data could determine the network's average utilization and peak utilization rate. These values can then be used as the network's performance baseline. Future performance sampling can be compared against this baseline to determine if configuration changes are needed or if additional resource re-allocation is required.

The initial baseline can also be the basis for performance SLAs and for evaluating performance affecting network problems. Long-term monitoring of essential network access points such as network backbone interfaces gateway interfaces, remote access servers, and file and database servers is also quite desirable. Often, shifts in network usage can be easily identified by examining histograms of important network resources.

The area of performance management has always been a concern for network administrators of large shared transmission medium and WAN networks. With the arrival and mass deployment of network switches and high bandwidth WAN links, the importance of performance management has been reduced somewhat. It's not that monitoring is no longer needed, but that network switches make the collection of network segment performance data samples very difficult. The growing importance of network-intensive applications will drive new approaches to network sampling, however, because these new applications make performance management even more vital.

SNMP

SNMP is the de facto protocol for network management. It is part of the TCP/IP Internet protocol suite. Its predecessors are the Simple Monitoring Gateway Protocol (SMGP) and the OSI Common Management Information Protocol (CMIP). In 1988, the Internet Activities Board began an investigation to determine what would be the Internet standard management model. SMGP, which was already in use, had been passed over for the OSI's CMIP model. The problem was that, in the CMIP IP version, CMIP over TCP (CMOT) was not ready for production deployment. Therefore, for an interim period SMGP became the management standard. Then, a transnational framework called SNMP was developed to provide a migration from SMGP to CMOT. In 1990, SNMP was adopted as the standard management protocol and management model for use on the Internet.

The SMMP management model uses managed devices, agents, and network-management systems. Managed devices are nodes running an SNMP agent. These devices collect management data, then format and transmit using the SNMP protocol. SNMP agents are processes that run on the managed device. The agent maintains the local management's knowledge base of information for the device they are operating on. The agent creates this local management knowledge base by using the device's Management Information Base (MIB) as a collection template. MIB is essentially a data map for collecting management information. The agent, running on the managed device, collects management data using the MIB as a data collection template. The network-management system sends and receives SNMP messages to and from the managed devices. The agents then interpret the network management station's data requests, which request certain MIB-defined data objects, and return the requested data in SNMP form to the network management station.

There are three versions of SNMP. The standard is SNMP v1, as defined in RFCs 1212, 1215, and 1213. SNMP v1 has the broadest support of all three versions. It supports a community-based security model, which operates by having the devices and the manager employ a common logical name for community identification, authentication, and access, which can be either read-only or read and write. By default, the community name "public" is used for read-only, and "private" is used for read and write access. The community name is set by the network administrator. It is also possible for a managed device to support more than one community name. The community name is the only form of authentication for SNMP v1 to send and receive SNMP messages. The community name must be common. If no name is set, some clients will exchange SNMP messages with any network management station. This open security model has some significant security implications. Additionally, many administrators deploy devices without changing the default community names, making it possible for anyone to send and receive SNMP information from the device. As a rule, you should always change the default community names of SNMP supportable network devices if you enable SNMP.

These security problems led to the development of SNMP v2 and SNMP v3. Both versions are draft standards. SNMP v2 began as an enhanced version of SNMP v1 with additional security enhancements (authentication and access-control services) and improved management capabilities (support for distributed management, variable thresholds for alarms, and event messaging when defined thresholds are exceeded). At this point, there are more than 25 RFCs pertaining to the development of SNMP v2. Disagreements over the best method for implementing security and administrative enhancements led to the development of three experimental variations of the SNMP v2: SNMP v2u, SNMP v2*, and SNMP v2c. The current draft standard is based on SNMP v2c, which is essentially the SNMP v1 security model with some added management and monitoring functionality. This draft has resulted in an updated MIB, revised data structure rules, additional commands, and new SNMP protocol data unit (PDU) definitions. The SNMP v2 draft standard is defined in RFCs 1902 through 1910. SNMP v2 extensions have been largely adopted by most vendors today, but SNMP v1 is still the functional standard.

SNMP v3 became a draft standard in 1998, and it is based on the security models developed for SNMP v2u and SNMP v2*. It provides DES encrypted message exchange, user authentication, a tiered view access model and an updated SNMP PDU that supports the new authentication scheme. SNMP v3 is defined in RFCs 2271 through 2275.

SNMP v1 was originally developed as a TCP/IP protocol that used UDP connectionless service transport. With SNMP v2, IPX and AppleTalk were also defined for SNMP message transport, and SNMP v3 provides additional functionality and expands the usefulness of SNMP. However, vendor support for the new versions of SNMP has been inconsistent: Network managers based on newer versions of SNMP are generally backward compatible, but managers based on earlier versions do not work with the newer version. Because support for SNMP v3 is just becoming available, and the current proposed standards do not change the commands and data format substantially, the rest of our SNMP discussion focuses on SNMP v1 and v2, which currently have the broadest support.

SNMP Commands

SNMP gets the "simple" part of its name from the simplicity of its command structure. SNMP uses the request/response paradigm. The manager sends query commands to the managed devices and they in turn send responses. SNMP v1 defines five operational commands:

- `GetRequest`—The `Get` operation is used by the management station to retrieve single or multiple management data variables from a managed device.
- `GetNextRequest`—The `GetNext` operation is used by the management station to retrieve the next management data variable relative to the MIB location of the last `Get` request. This command is used to read management data in sequential order. It provides a means of retrieving tabular data nested in the MIB that cannot be directly accessed through the use of the `Get` operation.
- `SetRequest`—The `Set` command is used by the management station to issue change commands to the managed device. `Set` commands can only be issued to change settings defined in the managed device's MIB. Additionally, they must be issued by a management station that has a community name defined for read and write access.
- `GetResponse`—`GetResponse` commands are used by the agent running on the managed device to send responses to the management station's `Get` and `GetNext` commands.
- `Trap`—`Trap` messages are unsolicited messages sent from the managed device to the management station. Part of the agent configuration is the definition of the host(s) to send `Trap` messages. These messages are used to relay the occurrence of specifically defined events such as a link failure or reboot.

With the addition of SNMP v2 authentication, access control, and manager-to-manager data exchange capabilities, SNMP v2 also added two management station request operations:

- `GetBulkRequest`—The `GetBulkRequest` operation is used by the management station in a manner similar to the `GetNextRequest`, except that `GetBulkRequest` is much more efficient. `GetBulkRequest` allows the management station to request management data recursively from the relative position of the last `GetRequest`, without any data size specification. As much data will be sent as can fit in the PDU. With the `GetNextRequest` command, the data size needs to be specified along with the request, so if actual data size exceeds the specified size, there can be errors. This command was added to accommodate data extraction from large table sizes, which were implemented in SNMP v2.

- `InformRequest`—This command was developed to allow communication between management stations. With the `InformRequest` command, management stations can send `Trap` messages to each other about locally monitored events. This facility enables distributed local management stations to monitor local networks, which in turn relay management data and events to a master network management station.

With SNMP, the message is the management. The agents and management stations are merely command and control interpreters. They perform collection and display functions, but the actual management is all performed using SNMP. To this end, all data request and responses must be made using specific network layer addressing. SNMP agents do not blindly start sending SNMP messages after the device is functioning. Management data must be requested by a verifiable host with the proper access credentials. Management stations that support SNMP v1 and SNMP v2 do data verification to determine which SNMP version is supported by the agent. After the supported SNMP version is determined, the management station uses that version to structure management data requests.

SNMP Components

Now, let's look at the different components of SNMP. As a network management protocol, it defines the way network management information should be structured and interpreted. The basis of this functionality is SNMP's MIB. A MIB is the rule base for defining management data structures. The MIB is used by the management station and agents to format management data requests. SNMP defines the formatting of the management data structures in the Structure of Management Information (SMI) definition. For SNMP v1, the SMI is described in RFC 1155. The SMI for SNMP v2 is described in RFC 1902. Since SNMP is a hardware-independent network management protocol, it needs to accommodate the different data representation generated by different types of devices. To facilitate this kind of communication, SNMP uses Abstract Syntax Notation One (ASN.1). All network management data structure requests and responses are formatted using ASN.1. In order to transport these management messages, they must be encoded into a format that will permit their transmission between managed devices and management stations. SNMP defines this format as a set of data encoding rules known as the Basic Encoding Rules (BER).

SMI

The SMIs for SNMP v1 and SNMP v2 define the hierarchical rules for describing management data structures. The top of this structure is the MIB, which is essentially a container for specific management data structures, known as managed objects. Managed objects or MIB objects are the data mappings used by agents to collect management data. A managed object exists for each management data variable that is collected by the agent. Each object can contain a single data type and value or a list of data types and values. The SNMP v1 and SMTPv2 SMIs define two groups of data types: simple and application-wide. ASN.1 is used for the actual representation and implementation of the SMI-defined data structures.

BER

The SNMP BER defines the process used to convert the ASN.1 format into a binary format for transmission in the SNMP PDU. This serialized data is then converted back into ASN.1 format when it reaches its destination. The actual SNMP-PDU data is in BER hexadecimal form. The BER conversion is a function of the SNMP process operating on the managed device. Only a subset of BER is used for SNMP; its usage is a legacy of the CMIP standard, which uses the CCITT X.209 specification. From a functional perspective, BER employs a least to most significant bit ordering scheme (little-endian), using 8-bit octets. ASN.1 data types are encoded using three variable length fields. The first field, defined as the tag field, indicates the ASN.1 data type. The second field, the length field, indicates how many octets make up the actual ASN.1 data type. The third field is the value field, and it consists of *x* amount of octets that make up the actual ASN.1 data.

ASN.1

ASN.1 is a platform and implementation-independent data description language developed by the ISO for unambiguous data representation. ASN.1 presents data in an abstract form using a highly structured syntax that only describes the ASN.1 data types and the assignment of values to those data types. ASN.1 is not interested in how the data will be processed, only with its correct presentation. ASN.1 is a very complex language that offers a broad set of data type definitions. This makes it an ideal tool for representing data on any system type. The SNMP SMI defines only a small group of ASN.1 data types for use with SNMP. ASN.1 uses the following basic structure to define data types and values:

```
data type value name | data type identifier ::= data value or {data type identifier
➥(data value)}
```

Data type name labels can be specifically defined through the use of a unique data identifier that appears in initial lowercase, uppercase form (for example, `sysContact`). The data type identifier is either a named data type expressed in title form (such as `MacAddress`) or as an ASN.1-defined data type, expressed as an ASN.1 keyword in uppercase form (for example, `INTEGER`). Data values are expressed as alphanumeric characters in all lowercase form (for example, `linkstatus 0`).

> **Note**
>
> ASN.1 is a programming language, and a rather complicated one at that. Its complete syntax, grammar, and implementation are far beyond the scope of this book. Unless you are planning to develop your own MIBs, an intimate knowledge of ASN.1 is not required. Table 11.3 lists some of the more commonly used ASN.1 keywords used with SNMP MIB's. The goal in this section is to provide you with a basic understanding of how SNMP messages are constructed. Additional references on SNMP are listed at the end of the chapter.

Table 11.3 **ASN.1 Keywords Used with SNMP**

BEGIN	OCTET	STRING	OBJECT
IDENTIFIER	SEQUENCE	OF	NULL
END	INTEGER	DEFINITIONS	DESCRIPTION

To implement the SMI MIB, an ASN.1 module is used. The ASN.1 module is a collection of ASN.1 data representations that are related to a common global data set. The ASN.1 module provides the ability to reference specific data objects in an ordered manner. The ASN.1 module is constructed with an ordered list of SMI MIB objects. The objects contained in the module can be imported and exported between different modules.

These "MIB objects" are created with an ASN.1 macro. All SMI managed objects are described using the OBJECT-TYPE macro (ASN.1 macro names appear in all uppercase). A macro is a special data type that supports clauses. Clauses are a defined subset of data definitions that are used to describe the macro-defined data type. The clauses enforce the use of specific set information to define the proprieties of the macro-defined data type. Every time the macro is used, some or all of these properties must be defined. The `OBJECT-TYPE` macro uses following format:

```
data type value name OBJECT-TYPE
    SYNTAX data type identifier
    UNITS
    ACCESS
    STATUS
    DESCRIPTION
    REFRENCE
    INDEX
    DEFVAL
    ::= {data value}
```

- `SYNTAX`—Defines the ASN.1 data type (simple, application wide [tagged], or structured [list or table]) which is used to define the object.
- `ACCESS`—Defines the level of access for the object (read-only, read-write, and so on).
- `STATUS`—Defines the definition status of the object (current, obsolete, and so on).
- `DESCRIPTION`—Details the role or function of the managed object and how to use it.
- `REFERENCE`—This is an optional definition used when the object is derived from a previously defined object. It points to the object that was used as a basis for the current object.
- `INDEX`—When an object contains a list or tabular data values, `INDEX` clauses are used to distinguish data values in different tables and rows.

- `DEFVAL`—Another optional definition used to provide information regarding the default value to be used with this object.

Three main ASN.1 data types are used for formatting SMI data type definitions within MIB objects:

- Universal—ASN.1 universal "simple" data types are unique values associated with an ASN.1 data type. There are four SMI data types defined for SNMP v1. The SNMP v2 SMI added one additional data type. As indicated above, simple data types are expressed using ASN.1 keywords. All five types are described in Table 11.4.

Table 11.4 **ASN.1, (SMI v1 and SMI v2) Simple Data Types**

Data Type	Description
`INTEGER`	A signed number between –2,147,483,648 and 2,147,483,648. ASN.1 specifies no `INTEGER` size limit. The SMI, however, limits it to 32 bits in length.
`OCTET STRINGS`	A sequence of octets, which are 8 bits in length. The octets can represent ASCI or binary data.
`OBJECT IDENTIFER`	A data type used for identifying the location of a data type. (We will look at object identifiers further when we discuss SNMP MIBs).
`NULL`	Both a data type and a data value. It is used as a placeholder or to indicate that a data type does not contain any valid data
`BIT STRING`	The `BIT STRING` data type was added in the SMI for SNMP v2. It is a sequence of zero or more bits that represent a data value. The value can be binary or hexadecimal.

- Tagged—ASN.1 tagged data types provide the capability to create new data types, which are actually just specific definitions of universal (`INTEGER` and `OCTET STRING`) ASN.1 data types. Tagged data types must be either globally unique within the ASN.1 implementation or within a given ASN.1 module. There are four types or classes of tags defined by ASN.1 for this purpose: universal tags, which are defined to identify globally unique data types; application-wide tags, which are used within an ASN.1 module; context-specific tags, which are used to describe specific structured data types; and private tags, which are used for special purpose data types. In addition to the creation of classed data types, tagged data types also support the use of specific semantics to defined specific data value ranges. This is known as subtyping. To create the SMI-defined, SNMP-specific, application-wide data types, ASN.1 application-class tags are used with subtyping. These are defined in Table 11.5.

Table 11.5 **SMI-Defined Application-Wide Data Types**

Data Type	Description
`IpAddress ::= [APPLICATION 0] IMPLICIT OCTET STRING (SIZE(4))`	Stores an IP address in four 8-bit octets.
`Counter ::= [APPLICATION 1] IMPLICIT INTEGER (0..4294967295)`	An undefined variable counter that can store increments up to 429,967,295. After it reaches the maximum defined value, it returns to zero and counts up again.
`Gauge ::= [APPLICATION 2] IMPLICIT INTEGER (0..4294967295)`	An undefined variable that can be any range between 0 and 429,967,295. The value can increase or decrease between the set ranges. However, after the minimum or maximum of the range is reached, the value will remain at that value until the value shifts toward the opposite range. Like a speedometer with a range between 0 and 80, the car can be traveling at 90, but the speedometer reads 80.
`TimeTicks ::= [APPLICATION 0] IMPLICIT INTEGER (0..4294967295)`	A counter that increments in hundredths of a second from a specified event. The counter maximum is 497 days, at which point the counter returns to 0.
`Opaque::= [APPLICATION 0] IMPLICIT INTEGER (0..4294967295)`	This data type can be used to store data types that are outside of the SMI definition. The data can be any random information type that can be represented in the integer range between –2,147,483,648 to 2,147,483,648.

- Structured—ASN.1 structured (or constructed) data types provide the capability to represent a collection of one or more simple data types. There are two ASN.1 structured data types used for this function: `SEQUENCE` and `SEQUENCE OF`. The `SEQUENCE` type is used for creating an ordered list or list of lists (essentially a table). ASN.1 `SEQUENCE` definitions use the following format:

```
ASN.1 data type name ::= SEQUENCE

    {
    data identifier data value,

    }
```

The `SEQUENCE OF` type is used for the creation of an ordered list of data values that are associated with the same data type. This is a function similar to creating an array in C. Here is an example of the basic format and a simple array:

```
ASN.1 data type name ::= SEQUENCE OF
ASN.1 data type name  ASN.1 data type identifier ::= {}

UselessArray ::= SEQUENCE OF INTEGER
UselessArray uselessArray ::= {23,78,898,122,45}
```

Structured data types make it possible for MIB objects to contain lists and tables. The `SEQUENCE` data type is used by the SMI for the creation of one-dimensional columnar data structures. The `SEQUENCE OF` data type is used for the creation of two-dimensional tabular data structures. Here is an example of a MIB object using a `SEQUENCE` data type:

```
— Define the object that will use the SEQUENCE

someSequence OBJECT-TYPE
    SYNTAX TestSequence
    ACCESS read-only
    STATUS obsolete
    DESCRIPTION
        "A sequence example"
    ::= (exampleMib 1)

— Then define the SEQUENCE

TestSequence ::= SEQUENCE
     {
      exampleOne  OCTET STRING (SIZE(6))
      exampleTwo  OCTET STRING (SIZE(6))
      exampleThree  OCTET STRING (SIZE(6))
      }

—Then each of the objects that are contained in the SEQUENCE need to be defined —
(here is just one).

exampleOne OBJECT-TYPE
    SYNTAX OCTET STRING (SIZE(6))
    ACCESS read-write
    STATUS obsolete
    DESCRIPTION
        "an array value"
    ::= {someSequence 1}
```

MIB Structure, Object Identifiers, and MIB Types

As stated previously, the MIB module is a "data map" or collection of management data objects, which is used by the management station to structure management requests. These requests return data variables about the status of the specific device characteristic that was queried. The SNMP protocol facilitates those requests. There are three types of MIB modules used to collect management data: standard, experimental, and enterprise. Regardless of the MIB type, all are constructed using the same hierarchical structure. Additionally, each uses the same basic hierarchy and global identification mechanism.

The MIB module uses an inverted tree hierarchy to organize the MIB and its objects. Figure 11.2 provides a simple view of the MIB hierarchy.

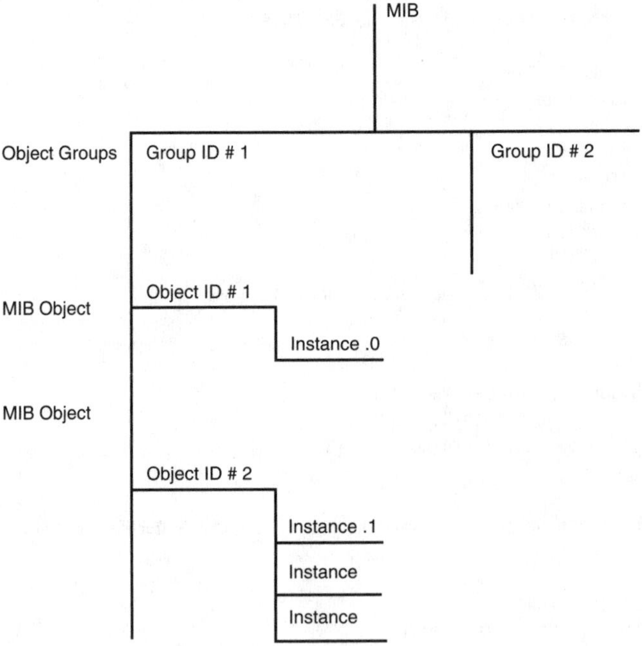

Figure 11.2 The MIB hierarchy.

The "top" of the MIB is the root of the tree. The SMI specifies that MIBs contain information about their creator, version, and any dependency on other MIBs for their functionality, etc. The MIB's interrelated managed objects are grouped together into object-groups. Each object-group has its own MIB-unique identifier. Additionally, each of the managed objects in the object-group each has its own object-identifier

(OID). The objects themselves can be either scalar or tabular. Scalar objects are typically simple ASN.1 data types that contain a single value or instance. Data variables contained in managed objects are referred to as instances. Scalar object instances use an instance identifier of ".0" to identify themselves. Tabular objects can contain one or more data instances. When requesting an object that contains multiple instances, the instance identifier ".1" is used. With tabular objects, the `INDEX` clause defines what the instance identifier is for each of the tabular data values.

The MIB module's type indicates its use and its identification in the global MIB naming mechanism. Standard type MIB modules are defined by standards bodies (mainly IETF/IAB) for the management of standards-based services and functions. The MIB-2 and RMON-MIB modules are perhaps the most recognizable and widely implemented. Almost every Internet standards-based technology has a standard MIB module. MIB modules exist for 802.3 Ethernet, ATM, BGP, DNS, SONET, OSPF, and Frame Relay, to name just a few. Devices that support SNMP and utilize these technologies will generally implement support for some or all of the standard's defined MIB module groups.

Experimental MIB modules are essentially "MIBs in progress." During the standards-based MIB development process, MIBs are assigned temporary MIB identifiers for testing. Once the MIB becomes a standard, it is assigned a unique, standard MIB identifier.

Enterprise MIB modules are developed for specific devices by manufacturers. These MIB modules permit manufacturers to develop management capabilities for their products' specific features. A device-specific MIB will often implement objects and object-groups from standard MIB modules. Although all SNMP supported devices will support standard MIB modules to some extent, it is the vendor-specific MIB module that provides the additional control and management capabilities that are often needed to manage the device effectively. When using a management system, you typically need to install your equipment vendor's MIBs into the management system's MIB repository. In most cases, these MIBs are downloadable from the vendor's Web site.

The MIB type also contributes to its global ID. In accordance with the SMI, all of the MIB types are uniquely identified through the use of the ISO/ITU-T global naming tree, which is illustrated in Figure 11.3. To access any MIB object data value, its global OID is used to retrieve the information. This can be accomplished by using one of two methods. The OID can be called using a text string naming the whole tree path or using the dotted integer form. To access the sysContact object in the MIB-2 module using the named form, the request would look like this:

```
iso.identfied-orgnaization.dod.internet.mgmt.mib-2.system.sysContact.0
```

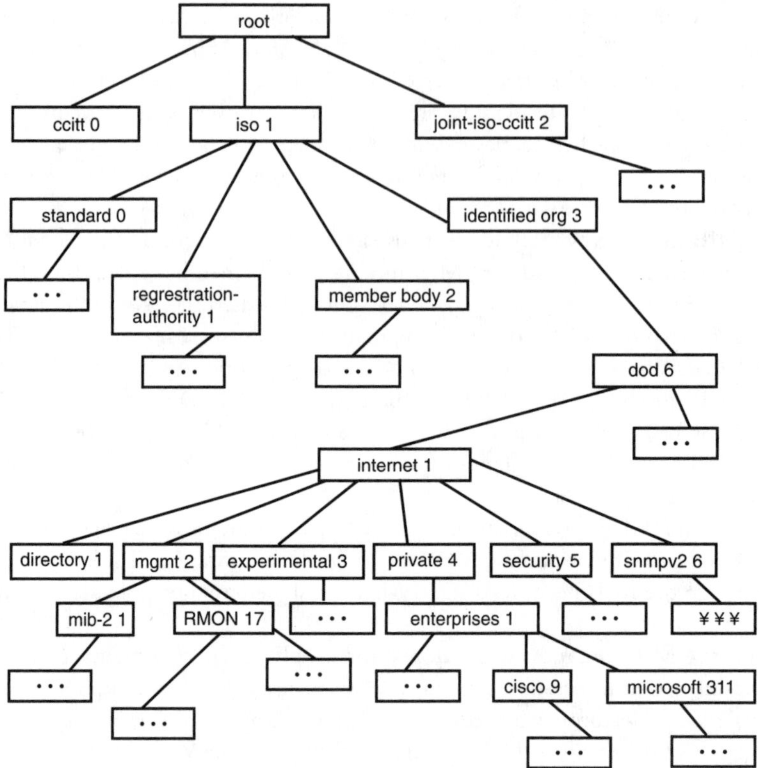

Figure 11.3 The SNMP view of the ISO/ITU-T global naming tree (including MIB-2).

In dotted integer form, it would look like this:

```
1.3.6.1.2.1.1.4.0
```

It is also possible to mix the two forms:

```
iso.identfied-orgnaization.dod.internet.mgmt.1.1.4.0
```

In the ISO/ITU-T scheme, the management subtree is where all the IETF/IAB standard MIB modules reside. The experimental subtree contains all of the in-process MIB modules. The private.enterprises subtree contains all of the custom MIB modules created by device manufacturers, management station vendors, and software developers. Each MIB developer is assigned a subtree ID. This ID serves as the root location vendor's MIB.

> **Note**
>
> Most SNMP managers and some agent implementations provide a MIB "walking" tool that allows you to browse MIB modules. The network management tools section lists some of the available tools.

The SNMP Message Formats

The SNMP message is used to transport the BER-encoded, SMI-defined ASN.1 objects and values. The message has two parts: the message header (see Figure 11.4) and the SNMP PDU. SNMP v1 and SNMP v2 use the same message header formats (see Figure 11.5), but different PDUs. The SNMP v1 SMI defines a PDU for data operations and for Trap operations (see Figure 11.6). The SNMP v2 SMI defines a single PDU for both data and Trap operations and a separate PDU (see Figure 11.7) for the `GetBulkRequest` operation. First, let's examine the SNMP v1 and v2 message header. Then, we will examine the various SNMP message and Trap PDU formats.

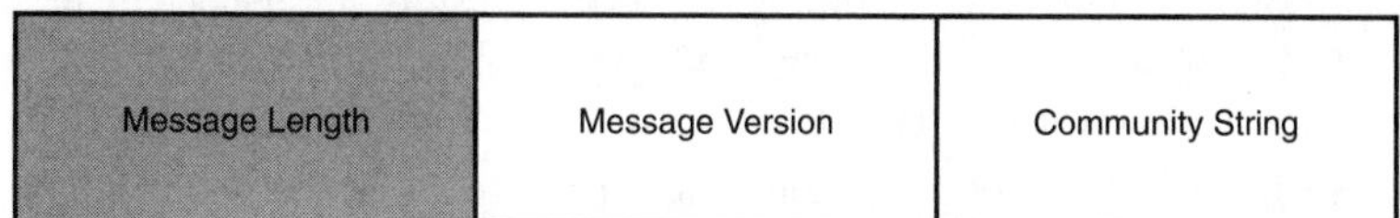

Figure 11.4 SNMP v1 and SNMP v2 message headers.

- Version Number—This field specifies which version of SNMP the message is formatted in. This provides computability determination between the manager and agent. If either receives a message formatted in a version they do not support, the message will be discarded.
- Community Name—This defines the community that the message is intended for. Both the manager and the agent must share a common community name. If the community name does not match the agents, the message will be discarded.

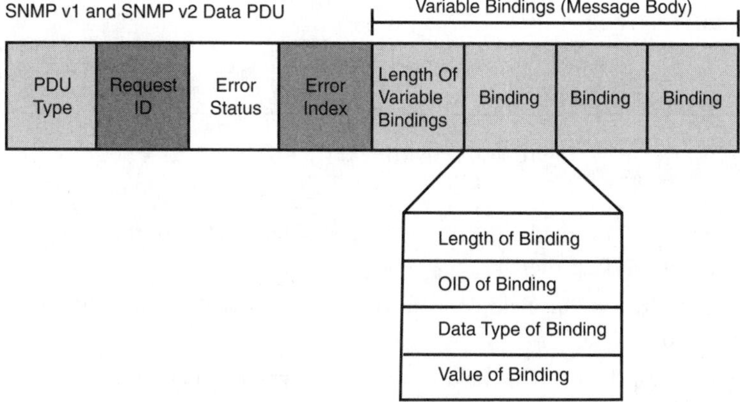

Figure 11.5 SNMP v1 and SNMP v2 data PDU.

- PDU Type—This field specifies the type of SNMP operation data contained in the message body. With SNMP v1, the accepted values are `getRequest`, `getNextRequest`, `getResponse`, and `setRequest`. With SNMP v2, the accepted values are `getRequest`, `getNextRequest`, `getResponse`, `setRequest`, and `trap`.
- PDU Length—Describes the size of the PDU (in octets) after the PDU length field.
- Request ID—This field is used by the management station for matching send requests with received requests. It is not used by the agent. However, the agent needs to include the Request ID in its response to the management station.
- Error Status—This field is used by the agent to notify the management station that the request was a failure. Different values are supported to indicate the type of error. If no error occurred, the field is set to zero.
- Error Index—This field (if possible) associates the error with a specific variable binding that is contained in the PDU message body.
- Message Body—When management data is requested, it is done so using the global OID of the management object. The agent in response sends the value it has for the specified OID. This can be either an ASN.1 data type or a NULL. It is also possible to specify the value of data type being requested. These messages are known as variable bindings. Each variable binding consists of four fields: length, OID, type, and value. When the variable binding is sent as a request form to the management station, only the length and OID fields will contain data. The amount of variable bindings contained in a message is undefined. Instead, the size of the SNMP message is defined by the agent and is typically 64 bytes in size. The message size is, however, an adjustable variable.

SNMP v1 Trap PDU

Enterprise	Agent Address	Generic Trap Type	Specific Trap Code	Time Stamp	Variable Bindings (Message Body)

Figure 11.6 SNMP v1 Trap PDU header.

- Enterprise—This field contains the variable-length OID of the object that has generated the Trap message.
- Agent Address—This field contains the network address of the station that has generated the Trap.
- Generic Trap Type—This field identifies the type of Trap message being sent (`linkUp`, `linkDown`, `coldStart`, `warmStart`, `enterpriseSpecific`, and so on).

- Specific Trap Type—This field is used to indicate the type of Trap being generated if it's not a generic Trap. If the generic Trap type is set to enterpriseSpecific, this value is checked and compared to Traps listed in the enterprise-specific MIB, which is indicated in the Enterprise Trap header field.
- Time Stamp—This field indicates the time (in hundredths of a second) of the Trap event relative to the system's bootstrap. It uses the MIB-2 `sysUpTime` variable at the time of the Trap generation to derive this value.
- Message Body—This contains the variable bindings associated with the Trap event.

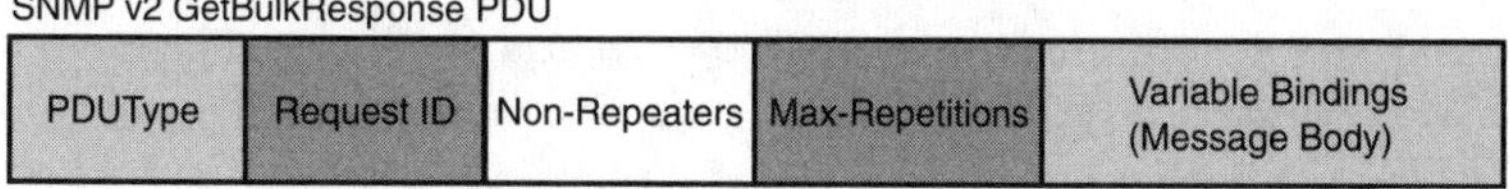

Figure 11.7 SNMP v2 GetBulkResponse PDU header.

- PDU Type—Specifically identifies this SNMP message as a `GetBulkResponse` operation.
- Request ID—This field is used by the management station for matching send requests with received requests. It is not used by the agent. However, the agent needs to include the Request ID in its response to the management station.
- Non-repeaters—This field indicates how many data variables should be retrieved in the request.
- Max-repeaters—This field indicates how many variables beyond the specified amount should be retrieved.

The Non-repeaters and Max-repeaters fields are used for shaping the scope of the bulk message request. This is needed since the request is blind and specifies no set data size limitation.

SNMP Agents and Management Tools

A variety of SNMP-based network management tools are in the marketplace today. The major players in this market are the following:

- Hewlett-Packard's OpenView—What began as an SNMP management tool has grown into a network management environment. OpenView provides SNMP monitoring, software distribution, asset tracking, and remote system administration. As the oldest of the distributed network management systems, it is also the most flexible and broadly supported. OpenView will operate with all the major relational database products (Oracle, Informix, and Sybase) and the Remedy helpdesk and trouble ticketing system. OpenView comes in various forms ranging from enterprise total management systems to a Windows-based workgroup management console. Information on OpenView is available at http://www.openview.hp.com.

- Tivoli Systems—This package provides enterprise and midsize (1,000 nodes or fewer) management solutions. Tivoli was acquired by IBM, who then handed them its NetView/AIX management solution. Tivoli's enterprise management solution is modular, providing different modules for asset management, help desk, change, remote administration, directory and accounting services, and security management. Tivoli uses both SNMP and proprietary device agents. Its midsized product, IT Director, provides software distribution, network monitoring, and remote administration. Tivoli's enterprise solution, like OpenView's, utilizes an external RDBMS for data collection and analysis. Information on Tivoli is available at `http://www.tivoli.com`.
- Cabletron's Spectrum—Where Tivoli and OpenView provide management services for just about everything, Spectrum's focus is primarily to perform enterprise network monitoring. Its strength is fast performance utilizing SNMP standard technologies. It uses its own proprietary database, a centralized management server, and remote management display consoles. Spectrum is available in both UNIX and Windows NT versions. Cabletron also markets a "management console" package called Spectrum Element Manager, which provides a decent "laptop" SNMP and RMON management console. Information on Spectrum is available at `http://www.spectrummgmt.com`.
- Castle Rock's SNMPc—This package is a Windows-based SNMP management suite. The management server and console both operate on Windows NT and provide support for RMON I, and integrated email and paging notification. It comes in both enterprise and workgroup versions. SNMPc is also part of the CiscoWorks for Windows management system for Cisco routers and switch products. CiscoWorks provides configuration, real-time monitoring, and topology mapping services. CiscoWorks for Windows comes with SNMPc. The package, however, can also interoperate with OpenView for Windows. Information on SNMPc is available at `http://www.castlerock.com`.

This is by no means an extensive list of the commercial SNMP and network management products on the market. There are a large number of network management firms marketing SNMP development tools, network probes, and MIBs. A number of public license and shareware tools are also available on the Web. Here are a few of the most useful and reliable:

- MRTG (Multi-Router Traffic Grapher)—This is a must-have tool for doing traffic flow monitoring on routers, switches, bridges, and servers; anything that supports SNMP MIB-2 can be monitored with this tool. It was created by Tobias Oetiker and Dave Rand and is available for UNIX and Windows NT.

The tool is basically a Perl script that polls SNMP-aware devices and constructs a graphical representation of the traffic flow on the devices' interfaces based on the data it collects. These graphs are then formatted into HTML pages that can be made available with any HTTP server. MRTG is available at `http://ee-staff.ethz.ch/~oetiker/webtools/mrtg/mrtg.html`.

- Scotty—Scotty is a TCL-based SNMP network development environment built on top of the Tool Command Language (TCL). Scotty has two components: the TNC Tcl extension, which is used to access management objects and develop SNMP management scripts, and Tkined, which is a graphical network editor that can develop network topology maps and provide visual displays for SNMP alarm events. Scotty is a tool that is definitely oriented toward people with programming experience. It is dependent on TCL and TK (the TCL graphics element) and runs and compiles on most popular UNIX systems. Scotty is available at `http://wwwhome.cs.utwente.nl/~schoenw/scotty`.
- SNMP Watcher—This is a Macintosh-based SNMP management console. It is available at `http://www.dartmouth.edu/netsoftware/snmpwatcher`.
- SMAP—SMAP is a Macintosh SNMP graphing tool. It supports dynamic node discovery and provides graphical traffic and device status visualization. It supports MIB-II, RMON, AppleTalk, IP, and IPX. SMAP is available at `http://www.kagi.com/drm/SMAP.html`.
- GxSNMP—This is the GNOME SNMP manager for Windows NT/95. It provides basic collection and mapping service and it's free. GxSNMP is available at `http://gxsnmp.scram.de`.
- nTOP—Top is a process visualization tool for UNIX. nTOP performs a similar function for networks by listing hosts and their traffic statistics (utilization, traffic matrixes, IP protocols, and so on). The package is freely available for UNIX and is shareware for Windows NT. nTOP is available at `http://www-serra.unipi.it/~ntop`.

Configuring an SNMP Agent on a Windows NT System

With Windows NT 4.0 systems, SNMP is installed as a service. To add the SNMP service, you need to login with an administrator account:

1. Select Start, Settings, then Control Panel.
2. Double-click the Network control panel and click the Services tab.
3. After you select the Services tab, click the Add button. You are presented with the Select Network Service dialog box. Using the up and down arrows, scroll down to SNMP Service and click OK. Now, NT loads the service. You are asked for the Windows NT 4.0 CD to install the service, unless you have a copy of the i386 directory already installed on your hard disk (this is, by the way, a good idea).

After the service is installed, you are presented with the SNMP Agent properties window (see Figure 11.8). You are asked to configure the MIB-2 system-group configurable values: `sysContact`, `sysLocation`, and `sysServices`.

- `sysContact`—Responsible for the system
- `sysLocation`—Defines where the system is physically located
- `sysServices`—Indicates what system services are supportable on the system

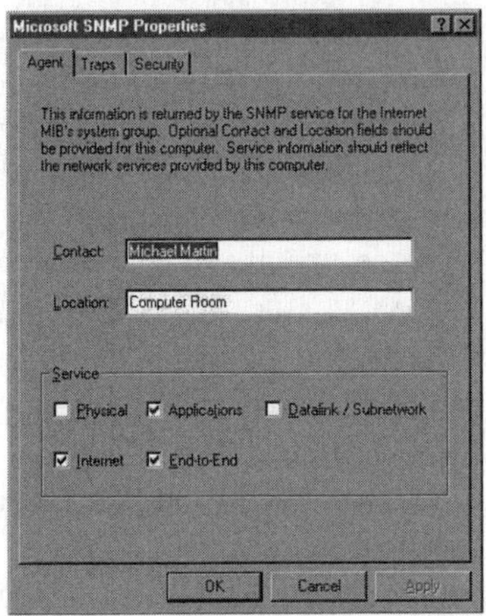

Figure 11.8 Windows NT 4.0 SNMP Agent properties window.

After the SNMP MIB-2 variables are configured, you need to configure the community name, which is set in the Security window (see Figure 11.9). Here is where you set all the SNMP communities you want to use. At a minimum, you will need at least one read-only and if you want to make management changes remotely, you will need a read-write access. Do not forget to delete the public community name. Quite often different community names are used to manage different types of devices; this, however, is a call you need to make. Using only one community name for all the devices on the network is completely acceptable. You can also specify which hosts you will accept SNMP messages from.

> **Note**
>
> To access any of the NT SNMP agent configuration variables after initial configuration, go to the Network control panel. Select the Services tab, scroll down to SNMP Service, select it, and double-click to open it.

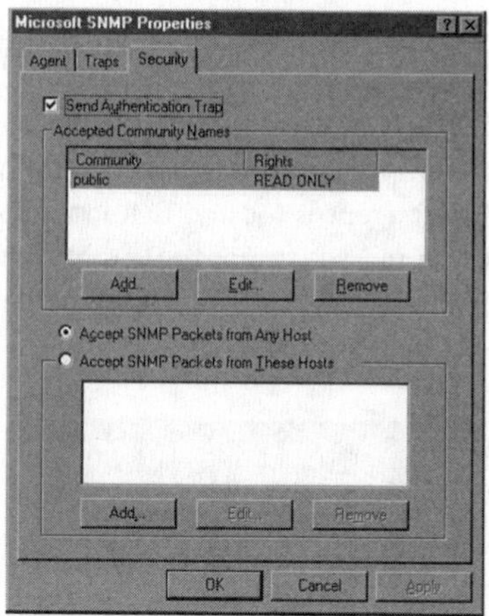

Figure 11.9 Windows NT 4.0 SNMP Security Proprieties window.

The last step is the configuration of Traps (see Figure 11.10). Here, you need to specify a Trap community name that will be used between the system and the management system receiving the Trap data. Then, you need to specify what management stations you want to send Trap messages to.

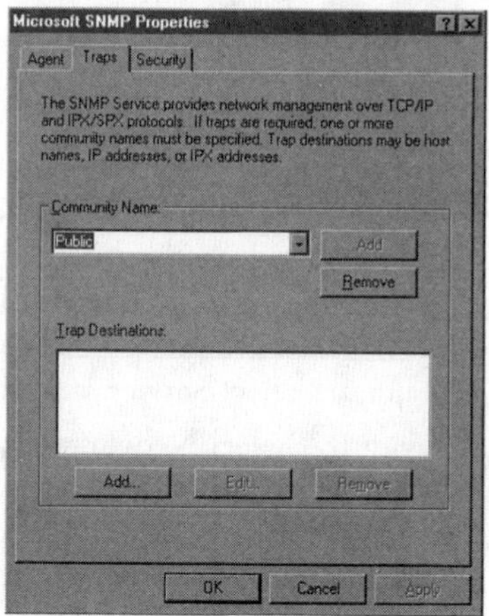

Figure 11.10 Windows NT 4.0 SNMP Trap Proprieties window.

Configuring an SNMP Agent on a Windows 95/98

With Windows 95/98 systems, two components need to be installed: the SNMP agent, and the Windows 95/98 system Policy Editor, which is used to configure the SNMP agent. Both of these components are located on the Windows installation CD in `\admin` directory. The SNMP agent is located in `\admin\nettools\snmp` directory and the Policy Editor is located in `\admin\apptools\poledit` directory.

The installation of the SNMP agent is done by adding a service through the network control panel (see Figure 11.11).

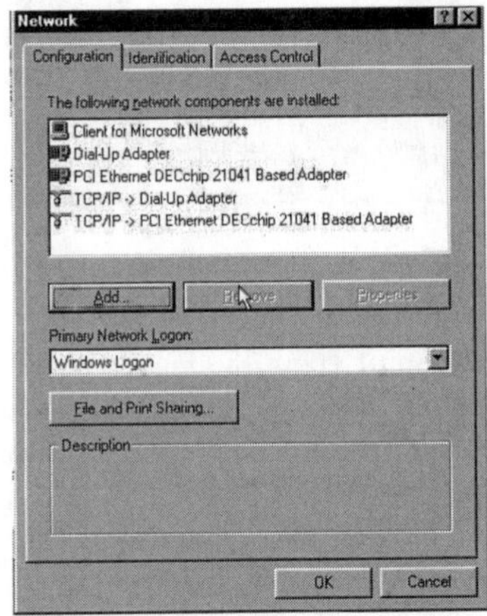

Figure 11.11 The Windows 95/98 Network control panel configuration screen.

The agent installation is accomplished by using the following steps:

1. Open the network control panel and, on the configuration screen, select the Add button. This initiates the network component installation process, which begins with the Select Component Type dialog box (see Figure 11.12). After the Component Type screen appears, select Service and click the Add button.
2. Next, you are presented with the Select Network Service dialog box (see Figure 11.13). Choose Microsoft in the Manufacture's window and then select the Microsoft SNMP Agent service in the Network Services window for installation. Then, click the OK button. At this point, you are prompted for the Windows 95/98 Install CD. Insert the disk and follow the install prompts.

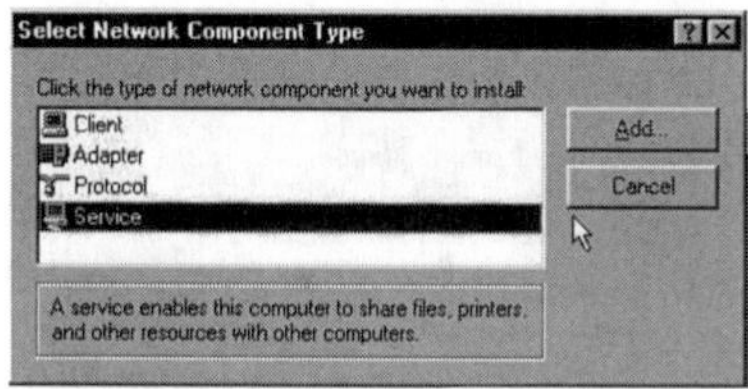

Figure 11.12 The Windows 95/98 Select Component Type dialog box.

If the Microsoft SNMP Agent service does not appear, select the Have Disk button and Browse... to the directory on the Windows 95 Installation CD, where the SNMP agent is located (`\admin\nettools\snmp`). When you reach the directory, the SNMP.inf file appears in the File Name: window; select OK and follow the install prompts.

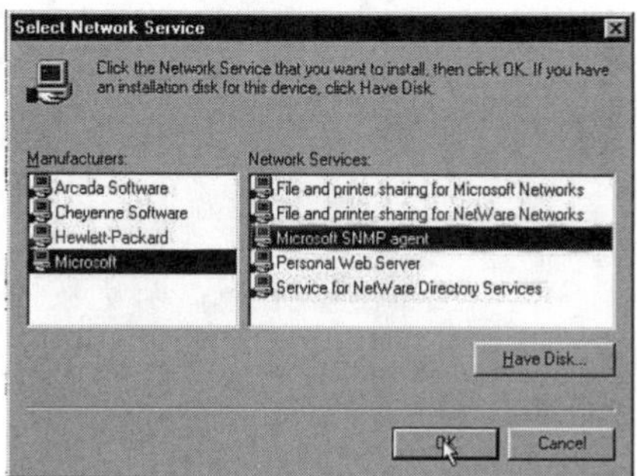

Figure 11.13 The Windows 95/98 Select Network Service dialog box.

After the SNMP agent is loaded, you need to load the Policy Editor application. Unlike Windows NT, system Policy Editor is not installed by default. It is located on the Windows 95 CD-ROM in `\admin\apptools\poledit`. The Windows 95/98 system Policy Editor is installed as a Windows 95/98 component, using the Add/Remove Programs control panel.

1. To install the Windows 95/98 system Policy Editor, first launch the Add/Remove Programs control panel and select the Windows Setup configuration tab (see Figure 11.14). After the control panel is open to the Windows Setup configuration tab, choose the Have Disk option.
2. When asked for the location in which to copy the files, choose Browse.... Now, switch to the `\admin\apptools\poledit` directory on the Windows 95/98 CD-ROM. When you reach the directory, two .inf files appear in the File Name: window, *grouppol.inf* and *poledit.inf*; click OK (see Figure 11.15). The Install From Disk dialog box should now display the `\admin\apptools\poledit` directory path, so click OK.

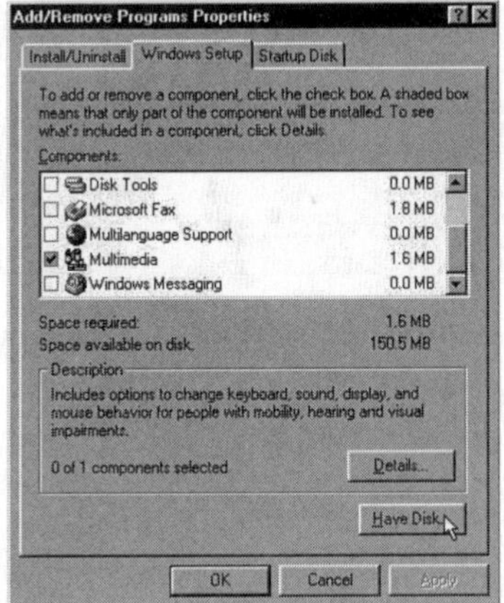

Figure 11.14 The Add/Remove Programs, Windows Setup screen.

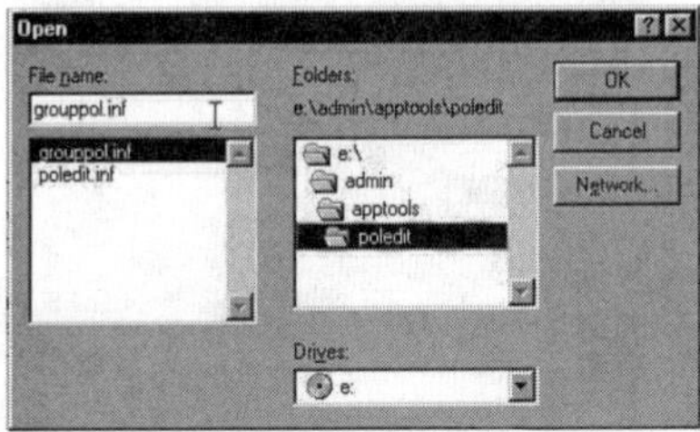

Figure 11.15 Locating the Windows 95/98 System Policy Editor.

3. The last step of the installation involves selecting the components you want to install (see Figure 11.16).

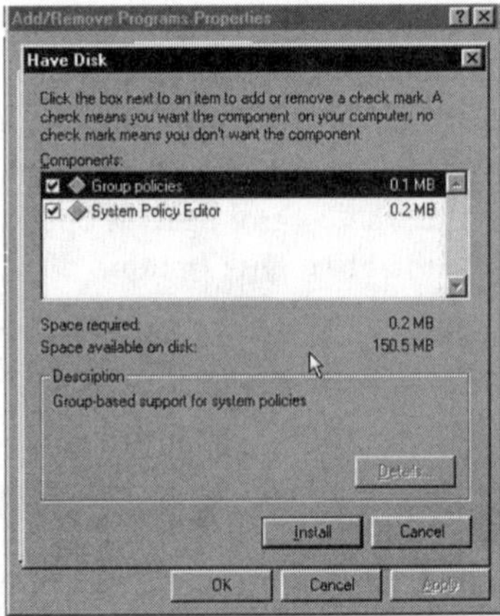

Figure 11.16 Selecting the components for installation.

Check off both the Group Policies and the System Policy Editor, and click Install. When the installation is complete, close the control panel. To launch the system Policy Editor, use the Start menu and go to \Programs\Accessories\System_Tools\ System_Policy_Editor. When you start up the Windows 95/98 system Policy Editor, go to the File menu and select Open Registry. Then, click the Local Computer icon, which opens the Local Computer Properties window (see Figure 11.17). Expand the Network properties and scroll down to SNMP policy.

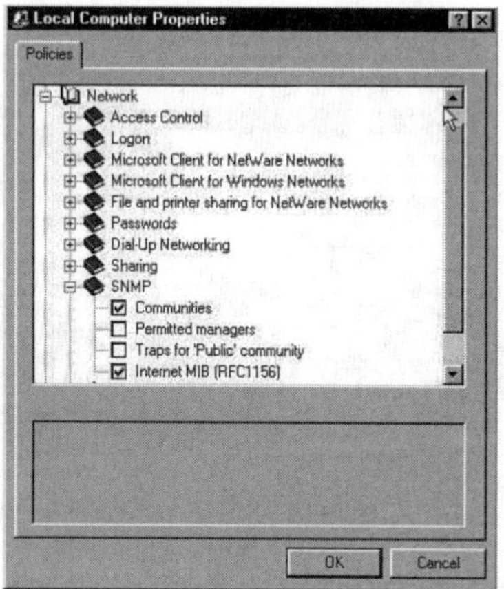

Figure 11.17 The SNMP configuration variables in the Windows 95/98 system Policy Editor.

There are four SNMP configuration variables. The Internet MIB subpolicy is where the MIB-2 system-group configurable values, `sysContact` and `sysLocation`, are set. The Communities subpolicy is where you define the SNMP community names you want the system to use. (Again, delete the public community name.) The Permitted Managers subpolicy is where you set the addresses of the management systems that are permitted to request management data from the system. The default is to accept requests from any management system. The Traps for Public community subpolicy is used to specify which hosts the Windows 95/98 system should send Trap messages to. One final note about SNMP on Windows 95: there are sight variations in the SNMP agent and system Policy Editor between the different versions of Windows 95. For best results, upgrade your Windows 95 to "B" version or later.

Configuring SNMP on the Macintosh.

SNMP support for the Macintosh has been lacking until the release of Mac OS 8.5, where it is available as custom installation option. The Macintosh SNMP client is not installed by default. The implementation itself consists of three files: an OS extension called *Open Transport SNMP*, a dynamic library called *OpenTptSNMPLib*, and a client configuration application called *SNMP Administrator*.

To install SNMP, mount the Mac OS installation disk and start the Install process by double-clicking the Install application. Proceed through the installation process as if you are performing a normal OS install. When you reach the Start Installation screen, choose the Customize radio button, which invokes the Custom Installation Options screen and perform the following:

1. Unselect all the software components except Mac OS 8.5, and choose the Customized Installation installation mode. This opens the Feature Installation window (see Figure 11.18).

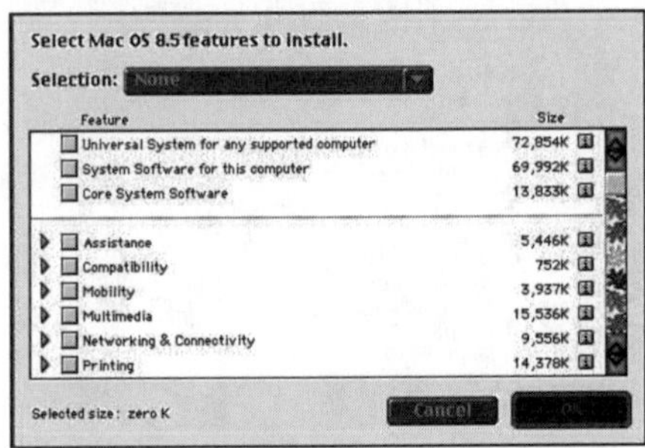

Figure 11.18 The Mac OS 8.5 custom Feature Installation window.

2. Scroll to Network Connectivity options, expand to see all the selections, scroll to SNMP, and check it off for installation (see Figure 11.19). Then, click OK, and start the installation. When the installation is complete, restart, and then configure the SNMP client.

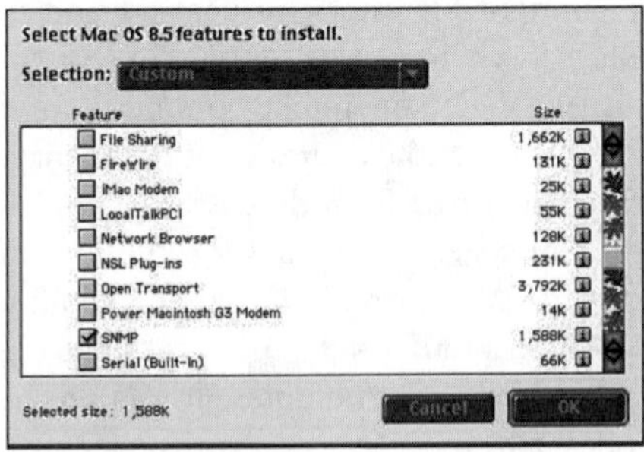

Figure 11.19 Selecting SNMP for installation on Mac OS 8.5.

To configure the Macintosh SNMP client, use the SNMP Administrator application, which, by default, is installed in the Apple Extras folder (located on the root level of the system hard disk). When the application is launched, it opens with a Community List dialog box, and the Public community is enabled by default (see Figure 11.20).

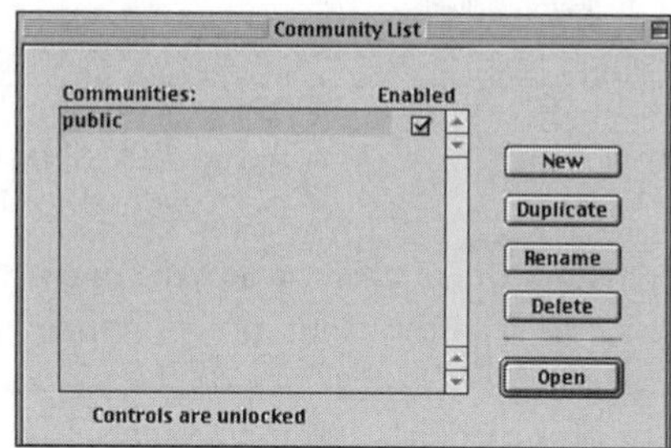

Figure 11.20 The Macintosh SNMP Agent Community List.

Disable the Public community by using the check box—do not delete it. To create a new community, click new. That prompts you to enter a new name for the community (see Figure 11.21).

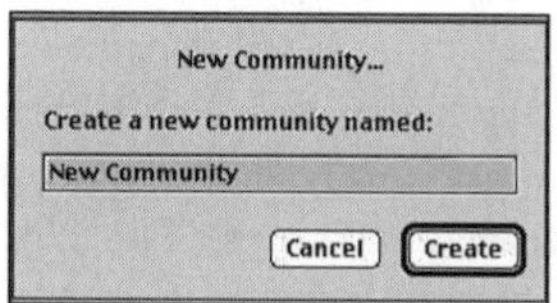

Figure 11.21 Creating a SNMP new community name.

After you enter a name for the new community, it appears already enabled in the Community List. The standard MIB II variable, `sysContact`, is derived from the network identity information entered in the File Sharing control panel. To set the `sysLocation` information, go to the Control menu and select Set Location.... Here, enter the system location information and a telephone number for the technical contact.

The Macintosh SNMP client provides several SNMP agents to collect management information. To see the SNMP agents available to a given community name, select the community you want to examine (in the Community List window), and select the Open button, or if you like, just double-click on the community in which you are interested. When the community first opens, you see a list of the available agents (see Figure 11.22).

Figure 11.22 A top level looks at the available SNMP agents.

To set the read/write permissions on an agent, select the agent, and set its access permission using the radio button's in the community's window header. To view a summary of all of the managed object data values, double click any of the agents. This displays a summary list of the objects and their values. If you want to look at each object individually, use the expansion arrows next to the agent name to see the available object groups, then double click the object you want to see.

SNMP Agents for UNIX

A number of SNMP agent implementations for UNIX exist. The most common one is the ISODE version that ships with BSD UNIX, the UCD-SNMP version from the University of California at Davis (available at `ftp://ucd-snmp.ucdavis.edu/`). The UCD-SNMP agent is a heavily modified version of the CMU-SNMP agent. There are a number of SNMP agent variations available for UNIX and each version uses a slightly different configuration. They all, however, come with documentation (some documentation being better then others). The UNIX SNMP implementations also come with a variety of SNMP tools for viewing MIBs and polling SNMP devices. Therefore, even if your UNIX implementation comes with an SNMP version, either the UCD or CMU, SNMP implementation is worth getting.

SNMP on Cisco IOS

To configure basic SNMP capability on a Cisco router, the global configuration EXEC commands are as follows:

```
<snmp-server community [community name] [ro|rw] [acl]>
<snmp-server contact [text]>
<snmp-server location [text]>
<snmp-server host [ip address] [community name] [trap type]>
<snmp-server trap-source [interface type slot/port]>
```

The first step to configuring basic SNMP capability on a Cisco router is configuring the SNMP community name. In addition to the standard read-only and read-write, the IOS also permits you to define a standard `acl` to filter which SNMP request will be processed. Here is an example that uses `acl` for the read-only and read-write community names "caller" and "listener":

```
opus(config)#access-list 20 permit 172.16.34.2
opus(config)#access-list 20 permit 172.16.34.45
opus(config)#access-list 30 permit 172.16.34.2
opus(config)#snmp-server community caller ro 20
opus(config)#snmp-server community listener rw 30
```

Now that the community names are set up, let's configure the MIB-2 system-group variables:

```
opus(config)#snmp-server contact M. Jones, AA-Office 800-555-1212
opus(config)#snmp-server location AA computer room, 5th Floor
```

The last step is to configure the Traps. In addition to the basic SNMP Traps, IOS supports Traps for Frame Relay, ISDN, config, and environmental monitoring, just to name a few (type `<snap-server enable traps ?>` for a full listing). Here is an example that just implements the basic SNMP Traps and specifies the interface that Trap messages are sent out of:

```
opus(config)#snmp-server host 172.16.34.2 traps caller
opus(config)#snmp-server trap-source e0/0
```

To display SNMP activity, use the user EXEC command `<show snmp>`.

RMON

Remote monitoring (RMON) is an SNMP MIB module that provides network fault and performance monitoring. It is designed to operate on a standalone RMON collection device or probe. After it is enabled, the probe monitors all the data packets that are sent and received on the network segment to which the probe is attached. It then generates a summary of the network segment's traffic. The probe also saves this summary information to provide a histograph of the network segment's behavior over a period of time.

The RMON probe can then be queried using an SNMP management station or a standalone RMON console. Aside from real-time data collection on the LAN, RMON also provides the ability to perform real-time network monitoring in enterprise environments. On LANs, management data exchanges between managed devices and management stations do not (normally) place any burden on network performance. Collecting management data over a WAN link, however, can generate a noticeable load, particularity if real-time monitoring is being attempted. By installing RMON probes on remote LAN segments, real-time monitoring can be performed locally on the probe. This data then can be monitored remotely over the WAN link using a RMON or SNMP management console.

Two variations of RMON exist: RMON I and RMON II. RMON I provides Layer 2 LAN monitoring. It was first defined in RFC 1271, which defined a network monitoring framework and management objects for 802.3 Ethernet. RFC 1513 added monitoring capability and management objects for 802.5 Token Ring. RFC 1757 is the current standard for RMON I; it specifies nine object-groups for monitoring Ethernet and Token Ring LAN segments. A function summary of each of the RMON I MIB module groups follows:

- (Group 1) Statistics group—Collects the 802.3 data on the following: collisions, broadcasts, CRC errors, multicast packets, runts, jabbers, and packet size statistics ranging from 64-128, 128-256, 256-512, 512-1,024, 1,024-1,518 bytes in size. It also collects 802.5 data on beacons, purge, management packets, MAC packets, soft ring errors, broadcasts, multicasts, data packets, and packet size statistics.
- (Group 2) History group—Provides historical "snapshot" views of summary data collected by statistics-group managed objects. Views and sampling intervals are user-definable.
- (Group 3) Alarm group—Provides the ability to set operational thresholds based on any collected data counter. Alarm thresholds can be set for a functional range or absolute value. It is also possible to set different alarms for the same value based on different thresholds. Alarm value determinations are made based on the comparison statistical samples, which are made at regular intervals.
- (Group 4) Host group—Collects traffic statistics on each host connected to the segment. This includes packets sent, received, errors, broadcasts, multicasts, runts, jabbers, collisions, and the order that each host was discovered by the probe.
- (Group 5) HostTopN group—Provides the ability to create a specialized list of hosts based on one or more statistics over a period of time. This provides the ability to identify hosts that generate the highest traffic or most errors, and so on.
- (Group 6) Matrix group—Provides the ability to monitor activity between node pairs. Each pair is stored and all consequent conversations are logged. This feature collects traffic and error statistics for each pair, which can then be sorted by address. This provides the ability to see which node is accessing the server the most, and so on.
- (Group 7) Filter group—Enables packet-filtering capabilities for the packet-capture group. Filters can be defined based on packet type, network event, and so on.
- (Group 8) Packet-capture group—Allows you to perform captures of all packets on the segment. Multiple buffers can be used to store different packet types using filtering capabilities defined by the filter group.
- (Group 9) Events group—Provides the ability to have local stations send standard SNMP MIB-2 Trap information to the RMON probe, which can then relay them to the management station. The events group also provides notification processing of the probe's threshold alarms to the management station and the packet-capture triggering based on traffic filter matches, and so on.

- (Group 10) Token Ring extensions—Provides Token Ring specific monitoring. The feature is essentially ring station monitoring, which includes data collection on ring state, number of stations, Active Monitor, source routing traffic, ring topology, fault domains, station order, claim tokens, and so on.

RMON II provides monitoring and collection capabilities similar to those provided in RMON I. RMON II, however, provides these collection services to OSI Layer 3 and above. This allows you to monitor network traffic by network protocol, transport protocol, service, and application type. RMON II extends the reach of RMON probes beyond the Layer 2 segment and into the enterprise. RMON II is defined in RFC 2021. A function summary of each of the RMON 2 MIB group enhancements follows:

- (Group 11) Protocol directory group—Defines the list of OSI-RM (2-7) protocols that the RMON probe can interpret.
- (Group 12) Protocol distribution group—Provides byte-level network layer traffic tracking based on network protocol type, providing a breakdown of the segment or network traffic percentages by protocol.
- (Group 13) Address-mapping group—Provides traffic-generating node discovery capabilities by Layer 2 and Layer 3 addressing. Also discovers the switch or hub port to which the node is attached.
- (Group 14) Network-layer host table group—Tracks packets, total throughput, and errors of hosts by Layer 3 addressing.
- (Group 15) Network-layer matrix table group—Tracks data throughput exchanges between Layer 3 host pairs.
- (Group 16) Application-layer host table group—Provides tracking of total throughput, packets, and errors by host and application type.
- (Group 17) Application-layer matrix table group—Tracks data exchanges between Layer 3 host pairs by application.
- (Group 18) User history group—Provides user-definable filtering capabilities and the storage of user-defined collected statistical data based on RMON 2 data collection variables.
- (Group 19) Probe configuration group—Provides standard configuration parameters for remotely configuring RMON probes through serial devices.

The development of RMON and RMON II has made network performance and fault monitoring a great deal easier, particularly in large enterprise environments. Although the use of RMON probes is the traditional method of RMON implementation, switch manufactures have been adding RMON capability to their workgroup and enterprise class switches for some time. The enabling of all the RMON I and II groups will degrade a switch's performance. For this reason, switch vendors typically only enable RMON groups I through III by default. The remaining RMON groups are available, but they need to be enabled by an administrator. This, mind you, should be done with some caution because switch performance will be impacted. So unless the switch is functioning as a probe, the higher RMON groups should only be enabled when troubleshooting. The addition of RMON support to switches is quite ironic, in a certain sense. In switched environments, traditional RMON I probes are not very effective since they can only monitor traffic they can see, which is limited to the traffic generated on the switch port they are connected to. RMON II probes overcome this problem somewhat, since they can collect traffic based on the network layer and above. Switch-based RMON implementations collect data statistics on each of the switch's ports and then provide the ability to view utilization statistics on a port by port basis. RMON probes can be valuable tool for network monitoring data collection, but they only collect data. The proper interpretation of the data for network problem resolution and network growth planning is still needed.

Summary

In terms of actual time, you will spend more time managing and troubleshooting your network than you will actually spend designing and building it. In this chapter, we looked at the different areas that make up network management. By the end of this chapter, you should have an understanding of the following:

- SNMP and its components, SMI, ASN.1, and BER
- Basic network management models
- Network baselining
- What a BERT, a TDR, and a protocol analyzer are
- What a MIB is

The topic of network management, troubleshooting, and analysis is extensive. The goal of this chapter (and the others in this book, for that matter) has been to provide you with a basic understanding of the essential concepts. There are numerous information resources available in print and on the Internet dealing with this topic, some of which have been listed below.

The essential thing you need to determine before implementing any network management solution is what you expect the solution to accomplish. Without a set of defined operational goals and expectations, you will find it difficult to determine the correct management solution for your network.

Again, the importance of documentation cannot be emphasized enough. If you and others cannot understand the relationship of the various network components in terms of the location, configuration, various hardware and software components, and their roles and interdependencies in the operation in the network, you will quickly find yourself in an untenable situation. The operation of any network, large or small, is dependent on solid planning, testing, and documentation. By following these simple concepts, you will find understanding your network a relatively easy task.

Related RFCs

RFC 1028	Simple Gateway Monitoring Protocol
RFC 1052	IAB recommendations for the development of Internet network management standards
RFC 1089	SNMP over Ethernet
RFC 1147	FYI on a network management tool catalog: Tools for monitoring and debugging TCP/IP internets and interconnected devices
RFC 1155	Structure and identification of management information for TCP/IP-based internets
RFC 1157	Simple Network Management Protocol (SNMP)
RFC 1173	Responsibilities of host and network managers: A summary of the "oral tradition" of the Internet
RFC 1189	The Common Management Information Services and Protocols for the Internet
RFC 1213	Management Information Base for network management of TCP/IP-based internets: MIB-II
RFC 1351	SNMP Administrative Model
RFC 1352	SNMP Security Protocols
RFC 1418	SNMP over OSI
RFC 1419	SNMP over AppleTalk
RFC 1420	SNMP over IPX
RFC 1445	Administrative Model for version 2 of the Simple Network Management Protocol (SNMP v2)
RFC 1446	Security Protocols for version 2 of the Simple Network Management Protocol (SNMP v2)
RFC 1447	Party MIB for version 2 of the Simple Network Management Protocol (SNMP v2)
RFC 1451	Manager to Manager Management Information Base
RFC 1503	Algorithms for Automating Administration in SNMP v2 Managers
RFC 1757	Remote Network Monitoring Management Information Base
RFC 1901	Introduction to Community-based SNMP v2
RFC 1902	Structure of Management Information for version 2 of the Simple Network Management Protocol (SNMP v2)

RFC 1907	Management Information Base for version 2 of the Simple Network Management Protocol (SNMP v2)
RFC 1910	User-based Security Model for SNMP v2
RFC 1944	Benchmarking Methodology for Network Interconnect Devices
RFC 2011	SNMP v2 Management Information Base for the Internet Protocol using SMI v2
RFC 2012	SNMP v2 Management Information Base for the Transmission Control Protocol
RFC 2013	SNMP v2 Management Information Base for the User Datagram Protocol using SMIv2
RFC 2271	An Architecture for Describing SNMP Management Frameworks
RFC 2272	Message Processing and Dispatching for the Simple Network Management Protocol (SNMP)
RFC 2273	SNMP v3 Applications
RFC 2274	User-based Security Model (USM) for version 3 of the Simple Network Management Protocol (SNMP v3)
RFC 2275	View-based Access Control Model (VACM) for the Simple Network Management Protocol (SNMP)

Additional Resources

Leinwand, Allan, and Karen Conroy. *Network Management, A Practical Perspective*. Addison-Wesley, 1996.

Miller, Mark. *Troubleshooting Internetworks*. M&T Books, 1991.

Nemzow, Martin. *Enterprise Network Performance Optimization*. McGraw-Hill, 1994.

Perkins, David, and Evan McGinnis. *Understanding SNMP MIBs*. Prentice Hall, 1997.

Rose, Marshall. *The Simple Book*, Second Edition. Prentice Hall, 1996

Rose, Marshall, and Keith McCloghrie. *How to Manage Your Network Using SNMP*. Prentice Hall, 1995.

Stallings, William. *SNMP, SNMP v2 and RMON Practical Network Management*. Addison-Wesley, 1996.

Binary Conversion Table

Table A.1 consists of a list of resources available at the time of printing.

Table A.1 **Binary Conversion Table**

A = Arabic   H = Hexadecimal   B = Binary

A	H	B	A	H	B	A	H	B	A	H	B
0	0x00	00000000	64	0x40	01000000	128	0x80	10000000	192	0xc0	11000000
1	0x01	00000001	65	0x41	01000001	129	0x81	10000001	193	0xc1	11000001
2	0x02	00000010	66	0x42	01000010	130	0x82	10000010	194	0xc2	11000010
3	0x03	00000011	67	0x43	01000011	131	0x83	10000011	195	0xc3	11000011
4	0x04	00000100	68	0x44	01000100	132	0x84	10000100	196	0xc4	11000100
5	0x05	00000101	69	0x45	01000101	133	0x85	10000101	197	0xc5	11000101
6	0x06	00000110	70	0x46	01000110	134	0x86	10000110	198	0xc6	11000110
7	0x07	00000111	71	0x47	01000111	135	0x87	10000111	199	0xc7	11000111
8	0x08	00001000	72	0x48	01001000	136	0x88	10001000	200	0xc8	11001000
9	0x09	00001001	73	0x49	01001001	137	0x89	10001001	201	0xc9	11001001
10	0x0a	00001010	74	0x4a	01001010	138	0x8a	10001010	202	0xca	11001010
11	0x0b	00001011	75	0x4b	01001011	139	0x8b	10001011	203	0xcb	11001011
12	0x0c	00001100	76	0x4c	01001100	140	0x8c	10001100	204	0xcc	11001100
13	0x0d	00001101	77	0x4d	01001101	141	0x8d	10001101	205	0xcd	11001101
14	0x0e	00001110	78	0x4e	01001110	142	0x8e	10001110	206	0xce	11001110
15	0x0f	00001111	79	0x4f	01001111	143	0x8f	10001111	207	0xcf	11001111
16	0x10	00010000	80	0x50	01010000	144	0x90	10010000	208	0xd0	11010000
17	0x11	00010001	81	0x51	01010001	145	0x91	10010001	209	0xd1	11010001
18	0x12	00010010	82	0x52	01010010	146	0x92	10010010	210	0xd2	11010010
19	0x13	00010011	83	0x53	01010011	147	0x93	10010011	211	0xd3	11010011
20	0x14	00010100	84	0x54	01010100	148	0x94	10010100	212	0xd4	11010100
21	0x15	00010101	85	0x55	01010101	149	0x95	10010101	213	0xd5	11010101
22	0x16	00010110	86	0x56	01010110	150	0x96	10010110	214	0xd6	11010110
23	0x17	00010111	87	0x57	01010111	151	0x97	10010111	215	0xd7	11010111
24	0x18	00011000	88	0x58	01011000	152	0x98	10011000	216	0xd8	11011000
25	0x19	00011001	89	0x59	01011001	153	0x99	10011001	217	0xd9	11011001
26	0x1a	00011010	90	0x5a	01011010	154	0x9a	10011010	218	0xda	11011010
27	0x1b	00011011	91	0x5b	01011011	155	0x9b	10011011	219	0xdb	11011011
28	0x1c	00011100	92	0x5c	01011100	156	0x9c	10011100	220	0xdc	11011100
29	0x1d	00011101	93	0x5d	01011101	157	0x9d	10011101	221	0xdd	11011101
30	0x1e	00011110	94	0x5e	01011110	158	0x9e	10011110	222	0xde	11011110
31	0x1f	00011111	95	0x5f	01011111	159	0x9f	10011111	223	0xdf	11011111
32	0x20	00100000	96	0x60	01100000	160	0xa0	10100000	224	0xe0	11100000

Table A.1 **Continued**

A = Arabic H = Hexadecimal B = Binary

A	H	B	A	H	B	A	H	B	A	H	B
33	0x21	00100001	97	0x61	01100001	161	0xa1	10100001	225	0xe1	11100001
34	0x22	00100010	98	0x62	01100010	162	0xa2	10100010	226	0xe2	11100010
35	0x23	00100011	99	0x63	01100011	163	0xa3	10100011	227	0xe3	11100011
36	0x24	00100100	100	0x64	01100100	164	0xa4	10100100	228	0xe4	11100100
37	0x25	00100101	101	0x65	01100101	165	0xa5	10100101	229	0xe5	11100101
38	0x26	00100110	102	0x66	01100110	166	0xa6	10100110	230	0xe6	11100110
39	0x27	00100111	103	0x67	01100111	167	0xa7	10100111	231	0xe7	11100111
40	0x28	00101000	104	0x68	01101000	168	0xa8	10101000	232	0xe8	11101000
41	0x29	00101001	105	0x69	01101001	169	0xa9	10101001	233	0xe9	11101001
42	0x2a	00101010	106	0x6a	01101010	170	0xaa	10101010	234	0xea	11101010
43	0x2b	00101011	107	0x6b	01101011	171	0xab	10101011	235	0xeb	11101011
44	0x2c	00101100	108	0x6c	01101100	172	0xac	10101100	236	0xec	11101100
45	0x2d	00101101	109	0x6d	01101101	173	0xad	10101101	237	0xed	11101101
46	0x2e	00101110	110	0x6e	01101110	174	0xae	10101110	238	0xee	11101110
47	0x2f	00101111	111	0x6f	01101111	175	0xaf	10101111	239	0xef	11101111
48	0x30	00110000	112	0x70	01110000	176	0xb0	10110000	240	0xf0	11110000
49	0x31	00110001	113	0x71	01110001	177	0xb1	10110001	241	0xf1	11110001
50	0x32	00110010	114	0x72	01110010	178	0xb2	10110010	242	0xf2	11110010
51	0x33	00110011	115	0x73	01110011	179	0xb3	10110011	243	0xf3	11110011
52	0x34	00110100	116	0x74	01110100	180	0xb4	10110100	244	0xf4	11110100
53	0x35	00110101	117	0x75	01110101	181	0xb5	10110101	245	0xf5	11110101
54	0x36	00110110	118	0x76	01110110	182	0xb6	10110110	246	0xf6	11110110
55	0x37	00110111	119	0x77	01110111	183	0xb7	10110111	247	0xf7	11110111
56	0x38	00111000	120	0x78	01111000	184	0xb8	10111000	248	0xf8	11111000
57	0x39	00111001	121	0x79	01111001	185	0xb9	10111001	249	0xf9	11111001
58	0x3a	00111010	122	0x7a	01111010	186	0xba	10111010	250	0xfa	11111010
59	0x3b	00111011	123	0x7b	01111011	187	0xbb	10111011	251	0xfb	11111011
60	0x3c	00111100	124	0x7c	01111100	188	0xbc	10111100	252	0xfc	11111100
61	0x3d	00111101	125	0x7d	01111101	189	0xbd	10111101	253	0xfd	11111101
62	0x3e	00111110	126	0x7e	01111110	190	0xbe	10111111	254	0xfe	11111110
63	0x3f	00111111	127	0x7f	01111111	191	0xbf	10111111	255	0xff	11111111

Index

Symbols

1000Base Ethernet. ***See*** **Gigabit Ethernet**

1000Base-T, Gigabit Ethernet, 166

1000Base-X CSMA/CD, Gigabit Ethernet, 167-168

100Base-FX PHY, 160-161

100Base-T, 157-159
- repeaters, 161-162

100Base-T4 PHY, Fast Ethernet, 160

100Base-TX, 160-161

100VG-AnyLAN, Fast Ethernet, 162

10Base-2, 151
- Ethernet, 138

10Base-5, 150-151, 154

10Base-FL, 157

10Base-T, 152-156

10Broad-36, 152

10Mbps, 802.3, 148-149
- 10Base-2, 151
- 10Base-5, 150-151, 154
- 10Base-FL, 157
- 10Base-T, 152-156
- 10Broad-36, 152
- AUI, 148
- MAU, 148-149

3-4-5 rule, 17

80/20 rule, 254

802 frame, 147

802.1q Layer 2 traffic prioritization, Layer 2 LAN switching services for Layer 3, 290-292

802.1q tags, 287

802.2 (IEEE subcommittee)
- LLC, 125-128
- SNAP, 146

802.2 standard, 93

802.3
- 10Mbps, 148-149
  - *10Base-2, 151*
  - *10Base-5, 150-151, 154*
  - *10Base-FL, 157*
  - *10Base-T, 152-156*
  - *10Broad-36, 152*
  - *AUI, 148*
  - *MAU, 148-149*
- Ethernet, 141-143
- Ethernet frame types, 144-146
- Ethernet MAC, operating parameters, 143-144

802.3 frame, 147

802.3 frame types, 144

802.3 PDU, LLC, 147-148

802.3 standard, Ethernet, 271

803.3 100Mbps. ***See*** **Fast Ethernet**

? command, 324

@zone, 102

A

AAA (authentication, accounting, and authorization), 366
- Kerberos (version 5), 366
- new-mode IOS authentication, 366, 370
  - *login authentication, 370-371*
- old-mode IOS authentication, 368
- RADIUS, 366

aaa accounting [service] [notice] [RADIUS/TACACS+] command, 375

aaa authorization command [privilege level 1-15] [primary authen] [secondary authen] command, 374

aaa authorization exec [primary authen] [secondary authen] command, 374

aaa authorization network primary authen] [secondary authen] command, 374

aaa net-mode command, 370

AAL (ATM Adaptation Layer), 241

AAL1, 241

AAL2, 242

AAL3/4, 242

AAL5, 242

AARP (AppleTalk Address Resolution Protocol), 88-91

AARP packets, 90

ABATS (Automated Bit Access Test System), 206

ABM (Asynchronous Balanced Mode), 245

abort delimiter transmitted error, Token Ring soft errors, 584

ABR (Available Bit Rate), 241

Abramson, Dr. Norman, 137

ABRs (area border routers), 515

Abstract Syntax Notation One. *See* ASN.1

AC (alternating current), 6-7

access control lists. *See* ACLs

access control mechanisms, LAN transmission protocols, 13

access-list [1-99/1300-1999] [permit/deny] [source address] [wildcard mask] command, 426

access-list [1-99] permit [source network address/host address] [wildcard mask] command, 449

access-list [100-199|2000-2699] [permit|deny] [protocol] {[any|host] [source|wildcard mask]} {[any|host] [destination|wildcard mask]} [op] [port-value] [options...] command, 428

access-list [600-699] [permit/deny] [match operator] [match data value] command, 434

access-list [800-899] [permit/deny] [source] [destination] command, 437

accounting, 375-376

accounting management, OSI network management model, 595

ACK (acknowledgement), 432
- TCP connections, 76

Acknowlegment Number, TCP headers, 76

ACLs (access control lists)
- action statements, 422
- AppleTalk, 434-435
  - *creating AppleTalk ACL filters, 436-437*
- applying, 424
- creating, 426-428
- deleting, 427
- denial of service attacks, 432-434
- distribute-lists, 425
- extended ACLs, traffic filters, 429
- information, displaying, 439
- IP, 422, 425-426
- IPX, 437- 439
- number ranges, 423
- policy routing, 439-441
- protocol-specific packet filtering, 424
- reflexive lists, 424
- security, 424-425
- wildcard masks, 426

action statements, ACLs, 422

active hubs, 18

Active Monitor Present (AMP), 172
- MAC frame types, 178

active monitor. *See* AM

address acquisition process, AARP, 91

Address Mapping Table (AMT), 89

address masquerading, 433

Address Resolution Protocol. ***See*** **ARP**

address signals, 215

address space
- routing tables, 390-391
- subnetting, 334
- VLSM, 389-390

address spaces, subnetting, 334

address summarization, OSPF, 517, 524

address-mapping group (Group 13), RMON II, 625

addresses. ***See also*** **IP addresses**
- binary addresses, 60
- broadcast addresses, 56
- destination network addresses, 60
- duplicate Layer 3 end-station addresses, 580
- IP datagrams, mapping, 62
- ISDN, 228-229
- loopback addresses, 56
- MAC addresses, AARP, 89
- multicast addresses, 56
- network addresses, IPX, 111
- network broadcast addresses, 56
- network protocol addresses, routers, 21
- node addresses, IPX, 111
- PRI, 228
- socket addresses, 95

addressing
- ATM, 244-245
- IPX, hexidecimal numbers, 110
- nodes, AppleTalk, 88-89

adjacencies, OSPF, 409

adjusting
- administrative distances
  - *EIGRP, 512*
  - *IGRP, 512*
  - *OSPF, 524-525*
  - *RIP, 504*
- metrics
  - *OSPF, 524-525*
  - *RIP, 504*
- OSPF, 527-528
- timers, RIP, 504-505

administrative distances
- adjusting
  - *EIGRP, 512*
  - *IGRP, 512*
  - *OSPF, 524-525*
  - *RIP, 504*
- dynamic routes, 486
- intra-autonomous BGP, 544
- IOS, 486
- static routes, 486

ADSL (Asymmetric Digital Subscriber Line), pricing, 221

ADSP (AppleTalk Data-Stream Protocol), 105-106

advantages
- of OSPF, 403
- of PCMCIA flash cards, 329
- of PSNs, 230
- of static routing (TCP/IP), 394

AEP (AppleTalk Echo Protocol), 101

AFI (Authority and Format Identifier), 244

AFP (AppleTalk Filing Protocol), 106

agents, SNMP
- configuring
  - *on Windows 95/98 systems, 616-619*
  - *on Windows NT systems, 613-615*
- UNIX, 622

AGS (Advanced Gateway Server), 306

AGS+ (Advanced Gateway Server Plus), 306

AHDLC framing, PPP, 249

alarm group (Group 3), RMON I, 624

alerting signals, 216

algorithms
- Bellman-Ford algoirthm, DVPs, 386
- Dijkstra's algorithm, LSPs, 387
- DUAL, 402
- DVPs, 386-387

ALO (At-Least-Once), ATP transactions, 103

ALOHA system, 137

alternating current (AC), 6-7

AM (active monitor), 172
- Token Ring management, 172-173

American Wire Gauge (AWG), 4

AMI coding, 207

AMP (Active Monitor Present), 172
MAC frame types, 178

AMT (Address Mapping Table), AARP, 89

analog-to-digital conversion, PCM, 198-199

analyzing networks. *See* network analysis

ansynchronous transmission, WAN transmission protocols, 12

any keyword, 427

Apple QuickDraw, 106

AppleTalk, 86
802.2 standard, 93
AARP, Layer 2 (data link), 89-91
ACLs, 434-437
ADSP, 105-106
ASP, 106
AURP, tunnel interfaces, 534
client/server model, 86
configuring, 337-338
contextual structure, 86
EtherTalk, Layer 2 (data link), 92-93
FDDITalk, 94
Frame Relay topologies, 468
Internetworks, routers, 98
Layer 3 (network), DDP, 94-97
Layer 4 (transport), 98
AEP, 101
ATP, 103
AURP, 101
NBP, 102-103
RTMP, 98-100
Layer 5 (session), 103-106
Layer 6 (presentation), 103
AFP, 106
LocalTalk, Layer 2 (data link), 91
nodes, addressing, 88-89
operational structure, 86
PAP, 106
Phase 1, 87
DDP datagram headers, 95
Phase 2, 87
cable range, 87
DDP datagram headers, 95
TokenTalk, 93
zones, 104
provisional addresses, 89
routing seeds, 99
routing tables, seed routers, 99
sockets, 95
ZIP. *See* ZIP
zones, 104-105

appletalk access-class command, 435

appletalk access-class [acl] command, 435

AppleTalk ACL filters, 436-437

AppleTalk Address Resolution Protocol. *See* AARP

AppleTalk Data-Stream Protocol (ADSP), 105-106

AppleTalk domains, AURP, 101

AppleTalk Echo Protocol (AEP), 101

AppleTalk Filing Protocol (AFP), 106

AppleTalk Phase 1, 87
DDP datagram headers, 95

AppleTalk Phase 2, 87
cable range, 87
DDP datagram headers, 95
TokenTalk, 93

appletalk routing eigrp [process id] command, 511

AppleTalk Session Protocol (ASP), 106

AppleTalk Transaction Protocol (ATP), 103

AppleTalk Update Based Routing Protocol. *See* AURP

application (Layer 4), IPX, 108

application (Layer 7), OSI-RM, 27

application services, ISDN, 221-222

application-layer host table group (Group 16), RMON II, 625

application-layer host table group (Group 17), RMON II, 625

applying ACLs, 424

architecture
ATM, 236
ATM switches, 294-295
NMS, 593-594

area authentication command, 528

area authentication [message-digest] command, 528

area border routers. ***See*** **ABRs**

area hierarchy, OSPF, 404-405

area [area id] nssa command, 527

area [area-id] stub no summary command, OSPF, 526

area [transit area id] virtual-link [router id] command, 527

aread id, 520

areas, 515
OSPF, 405

ARM (Asynchronous Response Mode), 245

ARP (Address Resolution Protocol), 38-39
commands, show arp, 489
IP, interactions with Layer 2, 38-39

ARP entries, 489

arp-server time-out [min] command, 478

ARPAnet, 37

ASBRs (Autonomous System Boundary Routers), 515

ASN.1, 601
data type name labels, 601
data types, 603-605
keywords, 602
SEQUENCE data types, 605
SMI MIB, 602

ASN.1 (Abstract Syntax Notation), 600

ASP (AppleTalk Session Protocol), 106

ASs (Autonomous Systems), 66
ERPs, 414-415
internetwork routing, 381
multihomed nontransit, 415
multihomed transit, 415
stub ASs, 415

assembling DDP datagrams, 97

Asymmetric Digital Subscriber Line (ADSL), 221

async dial-in, 343

async lines
configuring, 340-345
dial-in user accounts, 345

Async Serial line ports, 310

async serial lines, configuring, 340

Asynchronous Balanced Mode (ABM), 245

Asynchronous Response Mode (ARM), 245

Asynchronous Transfer Mode. ***See*** **ATM**

AT&T
PSTN, 193-195
T-carrier digital, 198

AT&T/Bell Labs, Transport-1, 454

ATM (Asynchronous Transfer Mode), 235
AAL, 241
AAL1, 241
AAL2, 242
AAL3/4, 242
AAL5, 242
addressing, 244-245
architecture, 236
ATM layer, 238-241
bidirectional point-to-point connections, establishing, 243
configuring, 474-477
information, displaying, 478
physical layer, 237-238
SALL, 243
serial interfaces, configuring, 475
versus IP, 478

ATM Adaptation Layer. ***See*** **AAL**

atm arp-server nasp [nasp address] command, 478

ATM cell forwarding, 295

ATM layer, 236-238
cell formatting, 239-240
QoS contracts, 240
traffic, 240-241
VPI/VCI translation, 240

ATM private-class switches, 294

ATM switches, 292-293
architecture, 294-295
cell forwarding, 295
Classical IP, limitations of, 302-303
Classical IP over ATM, 301-302
label swapping, 295-296
LANE, 296-297
BUS, 299
components, 298
functions of, 297-298
LEC, 299-300
LECS, 299
LES, 299
limitations of, 302-303

ATM-over-Serial configuration, 475

ATP (AppleTalk Transaction Protocol, 103

attentuation, 7

AUI (attachment unit interface), 148

AURP (AppleTalk Update Based Routing Protocol), 101
AppleTalk domains, 101
configuration commands, 534
display commands, 534
tunnel interfaces, 534-536

AURP tunnels, 101

auth, syslog facility definitions, 362

authentication, 374
CHAP, 456-457
OSPF, 528-529
PAP, 456
PPP, 248, 456
RIP, configuring, 505

authentication, accounting, and authorization. *See* AAA

Authority and Format Identifier (AFI), 244

authorization, new-mode IOS authentication, 374

authorization commands, new-mode IOS authentication, 374

Auto-Negotiation (AutoNeg), 158

autocommand connect command, 342

AutoInstall feature, IOS, 309

Automated Bit Access Test System (ABATS), 206

AutoNeg (Auto-Negotiation), 158

autonomous system boundary routers routers, OSPF, 406

Autonomous System Boundary Routers (ASBRs), 515

Autonomous Systems. *See* ASs

AUX ports
Cisco routers, 310
configuring, 345

Available Bit Rate (ABR), 241

average network utilization measurement values, 576

AWG (American Wire Guage standard), 4

B

B channels, ISDN, 222

B-ISDN (broadband ISDN), 222

B8ZS, T1, 207

backbone areas, 515

backbone cabling, 131

backbone IGP domains, 70

backbone networks, OSPF, 408

backbone paths, intranets, 68-69

backbone routers, 515-516
AppleTalk, internetworks, 98
OSPF, router designations, 406

backbone to backbone points, 68

backbones, OSPF, 405

backpressure, Ethernet, 270

backup designated router election, OSPF, 529

backup designated routers, OSPF, 409

backward explicit congestion notification (BECN), 234

BACP (Bandwidth Allocation Control Protocol), 247

bandwidth, 7
composite bandwidth, utilization, 575-577
composite bandwidth utilization distribution, 578
dedicated bandwidth, Frame Relay, 467
EIGRP, 511
OSPF, 526
routing protocols, 395

bandwidth [kilobytes] command, 499

baseband line signaling, 129

baseband signaling, 128

baseband transmission, 7

baselining
network analysis, 573-575
network errors, 579-580
Ethernet, 581-582
retransmission, 588
T-carriers, 586-588
Token Ring, 582-586
network throughput, 578-579

Basic Encoding Rules, 600-601

BAT (bridge address table), 20-21, 258

BCD (Binary Coded Decimal), 198

BCN (Beacon), MAC frame types, 178

beacon process, Token Ring hard errors, 584-585

bearer channels (B channels), 222

bearer services, ISDN, 221

BECN (backward explicit congestion notification), 234

Bell Labs, UNIX, 37

Bellcore, SDH, 212

Bellman-Ford algorithm, DVPs, 386

benefits of VLANs, Layer 2 LAN switching services for Layer 3, 288-289

BER (Basic Encoding Rules), 600-601

BER testing, 208

BERTs (Bit Error Rate Testers), 565, 587

BGP
configuration commands, 537
control commands, 538
debugging, 547
display commands, 537
EBGP. *See* EBGP
IBGP. *See* IBGP
inter-autonomous BGP, 538-541
inter-autonomous system routing, 538
Internet, 538
intra-autonomous BGP, 542-544
intra-autonomous system routing, 538
monitoring, 546-547
pass-through autonomous system routing, 538
peer-groups, 544
process ID, 540
reflectors, 545-546
routing types, 538

BGP configuration subcommands, distance [external] [internal] [local], 544

BGP configuration subprocess commands, 541

BGP-4 (Border Gateway Protocol version 4), 416-418

BHDLC (bit-synchronous HDLC) framing, PPP, 249

binary addresses, 60

Binary Coded Decimal (BCD), 198

binary conversion table, 629-630

binary numbers, address space, 391

binary routing tables, 60

binary stream data, synchronous data transmission, 12

bipolar violations (BPV), WAN errors, 587

Bit Error Rate (BER) testing, 208

Bit Error Rate Testers (BERTs), 565

bit streaming Token Ring hard errors, 585

bit-periods, 142

bit-synchronous HDLC (BHDLC), 249

block encoding, 130

BNCs (British Naval Connectors), coax cable, 3

Boolean logic matching, ACLs, 424

boot command, 352

boot system [source] [filename] command, 351

bootflash, Flash memory, 308

BOOTP (bootstrapping protocol), 81

bootstrap behavior, 346-348

Border Gateway Protocol, 416-418

Bosack, Len, founder of Cisco Systems, 306

Bosack, Sandy, founder of Cisco Systems, 306

BPDUs (Bridge Protocol Data Units), 261

BPV (bipolar violations), WAN errors, 587

BRI, 220
- data paths, 225

BRI interfaces
- ISDN, 457
- ISDN BRI DDR configurations, 458-461
- ISDN BRI interfaces, backup configurations, 462-465

bridge address table, 20-21, 258

Bridge Protocol Data Units (BPDUs), 261

bridges, 19, 257
- BAT, 21
- learning bridges, 20
- root bridge election process, 261-262
- source route bridges. *See* source route bridges
- switches, 21
- translation bridges, 22, 256
- translation bridges. *See* translation bridges
- transparent bridges. *See* transparent bridges

bridging
- changing to switching, 266-268
- Token Ring, 183
- versus routing, 264-265

British Naval Connectors (BNCs), 3

broadband ISDN (B-ISDN), 222

broadband signaling, 128

broadband transmission, 7

broadcast addresses, 56

Broadcast and Unknown Server (BUS), 299

broadcast keyword, 470

broadcast networks, OSPF, 406

broadcast storms, 22, 288

broadcasts, DDP, 97

buffered logging, 360-361

building routing tables, OSPF, 412

burst error, Token Ring soft errors, 584

BUS (Broadcast and Unknown Server), 299

bus topologies, 9
- non-contention-based access, LAN transmission protocols, 15
- repeateres, 17

C

C-Kermit, terminal emulation software, 313

cable range, AppleTalk Phase 2, 87

Cable Signal Fault Signature (CSFS), 564

cables
- Cisco console cable pinouts, 310
- Cisco routers, 310-314
- coaxial cable, 3
- configurations, 312
- optical fiber cable, 5
- RJ-45-M to DB-9-F pin translation, 311
- RJ-45-M to RS-232C-M pin translation, 311
- RJ45-M to DB25 pin translation, 311
- twisted-pair cables, 4-5

Cabletron, Spectrum, 612

cabling, 130
backbone cabling, 131
components of, 130
fiber optic cable, 133-136
horizontal cabling, 131
IBM Type 1 cabling, 182
IBM Type 2 cabling, 182
IBM Type 3 cabling, 182

cabling systems, Token Ring, 182

call progress tones, 215

capacity, switch backplane, 279-280

carrier class channel banks, multiplexers, 201

carrier extensions, 1000Base-X CSMA/CD, 167

carrier interfaces, multiplexers, 200

carrier sense multiple access with collision detect. *See* CSMA/CD

Castle Rock, SNMPc, 612

CAT 5 Ethernet cables, changing to Cisco cables, 310

CBR (Constant Bit Rate), 241

CCEP (Collision Consensus Enforcement Procedure), 13

CCITT, 23

CCO (Cisco Connection Online), 315

CCSL (Common Channel Signaling Links), SS7, 217

CDDI (Copper Distributed Data Interface), 184

cell formatting, ATM layer, 239-240

Cell Loss Priority (CLP), ATM cell formatting, 240

CFD (Character Framed Data), 12

CFR (Committed Frame Rate), 233

Challenge-Response Authentication Protocol (CHAP), 455-457

change logs, documentation, 563

changing conf-reg settings, 350

channel banks, multiplexers, 201

Channel Service Unit (CSU), 208-209

CHAP (Challenge-Response Authentication Protocol), 455-457

Checksum, IPX message format, 111

choosing protocols, 485

CIDR (Classless Interdomain Routing), 51, 388
Class C space, subnetting, 54
classless addressing, IP addresses, 51-52
IP addresses, classless method, 388
subnet masks, versus classful masks, 53
summarization, 392

circuit-control messages, SS7, 216

Cisco, ConfigMaker, 309

Cisco 1600 series, 306

Cisco 1600/1600-R, 306

Cisco 2500 series, 306
conf-reg, 348
IOS, 308

Cisco 2600 series, 306

Cisco 3600 series, 307

Cisco 4000 series, 307

Cisco 7x00 series, 307-309

Cisco 7x00 series routers, NTP, 358

Cisco Connection Online (CCO), 315

Cisco console cable pinouts, 310

Cisco IOS, configuring SNMP, 622-623

Cisco routers, 306-307
1600 series, 306
1600/1600-R, 306
2500 series, 306
2600 series, 306
3600 series, 307
4000 series, 307
Async Serial line ports, 310
AUX ports, 310
cables, 310-314
clocks, configuring, 356-359
configuration EXEC mode, 319
configuring, 331-332
from scratch, 332-333
IOS commands, 332

saving changes, 323
with copy command, 326-327
console ports, 310
disaster recovery. *See* disaster recovery
DRAM, 308
Flash memory, 308
initial configuration dialog, configuring, 316
interfaces, configuring, 334-335
IOS. *See* IOS
Layer 2 implementation, 307
memory, 308-309
NVRAM, 308
privileged EXEC mode, 318-319
ROM, 308
router configuration files, 316
SNMP, 309
startup configuration, 316
Telnet, 309
user EXEC mode, 317-318

Cisco Systems, 306

Cisco tunneling. *See* tunneling

cladding, 135

Claim Token (CTK), MAC frame types, 178

Class 1, PSTN hierarcy, 195

Class 2, PSTN hierarcy, 195

Class 3, PSTN hierarcy, 194

Class 4, PSTN hierarcy, 194

Class 5, PSTN hierarcy, 194

Class A addresses
classful subnetting, 49
IP addresses, 44

Class B addresses
classful subnetting, 49
IP addresses, 45

Class C addresses
classful addressing, IP addresses, 51
classful subnetting, 50
IP addresses, 45
supernetting, 51

Class C space, subnetting CIDR, 54

Class D addresses, IP addresses, 44

Class E addresses, IP addresses, 44

class types, IP addresses, 44

classes, IP addresses, 44-45

classful addressing
Class C addresses, IP addressing, 51
IP addresses, 44-45
guidelines for, 50
subnetting examples, 48-50
VLSM, 54-55

classful IP addressing, guidelines for, 56

classful method, IP addresses, 388

classful subnetting, 389
IP addresses, 48-50

Classical IP
over ATM, 301-302
limitations of, 302-303

classless addressing, IP addresses, 54
CIDR, 51-52

Classless Interdomain Routing. *See* CIDR

classless method, 388-389

classless routing, configuring, 498

clear arp command, 491

clear arp [mac address] command, 491

clear ip bgp [neighbor ip address] command, 541

clear IP net translations * command, 449

clear ip route * command, 491

clear ip route [address] [mask] command, 491

clear ip route [network] [mask] command, 491

CLEC (Competitive Local Exchange Carrier), 195

CLI (Command-Line Interface), 309

client/server model
AppleTalk, 86
TCP/IP, 36

clock set command, 357

clock timezone command, 357

closed IP networks, 492

CLP (Cell Loss Priority), ATM cell formatting, 240

CMIP (Common Management Information Protocol), 597

CMIP over TCP (CMOT), 597

CMOT (CMIP over TCP), 597

CO-to-CO transmission systems, 198

coaxial cable, 3

codecs (coder/decoder), 196

Collision Consensus Enforcement Procedure (CCEP), 13

collision domains, CSMA/CD, 13

collisions, Ethernet, 143

command lines, IOS
- more, 326
- navigating, 323-326
- size option, 325

command references, IOS (versions 9.x and higher), 323

command-line interface (CLI), 309

commands
- aaa accounting [service] [notice] [RADIUS/TACACS+], 375
- aaa authorization command [privilege level 1-15] [primary authen] [secondary authen], 374
- aaa authorization exec [primary authen] [secondary authen], 374
- aaa authorization network primary authen] [secondary authen], 374
- aaa net-mode, 370
- access-list [1-99/1300-1999] [permit/deny] [source address] [wildcard mask], 426
- access-list [1-99] permit [source network address/host address] [wildcard mask], 449
- access-list [100-199/2000-2699] [permit/deny] [protocol] {[any/host] [source/wildcard mask]} {[any/host] [destination/wildcard mask]} [op] [port-value] [options,], 428
- access-list [600-699] [permit/deny] [match operator] [match data value], 434
- access-list [800-899] [permit/deny] [source] [destination], 437
- access-list [id number], ACLs, 423
- AppleTalk, show appletalk interface [interface slot/port], 338
- appletalk access-class, 435
- appletalk access-class [acl], 435
- appletalk routing eigrp [process id], 511
- area authentication, 528
- area authentication [message-digest], 528
- area [area id] nssa, 527
- area [area-id] stub no summary, OSPF, 526
- area [transit area id] virtual-link [router id], 527
- arp-server time-out [min], 478
- atm arp-server nasp [nasp address], 478
- AURP, 534
- autocommand connect, 342
- bandwidth [kilobytes], 499
- BGP, 537-538
- boot, 352
- boot system [source] [filename], 351
- clear arp, 491
- clear arp [mac address], 491
- clear ip bgp [neighbor ip address], 541
- clear IP net translations *, 449
- clear ip route *, 491
- clear ip route [address] [mask], 491
- clear ip route [network] [mask], 491
- clock set, 357
- clock timezone, 357
- config-register, 346
- configure memory, 320
- configure network, 330
- configure terminal, 319, 321
- copy, 328-330
- copy tftp running-config, 427, 496
- copy/erase, 327
- crypto key-exchange passive [tcp port number], 453
- crypto key-exchange [passives interface ip address] [tcp port number], 453
- debug dialer, 465
- debug ip bgp events, 547
- debug ip bgp updates, 547
- debug ip ospf events, 533
- debug ppp authentication, 457

debug ppp error, 457
debug ppp negotiation, 457
debug ppp packet, 457
debug q921, 465
default-information originate always, 550
dialer load-threshold [1-255] [inbound/outbound], 460
distance eigrp [internal distance] [summary distance] [external distance], 512
distance ospf [external/inter-area] [10-255], 525
distance [1-255], 512
distance [10-255], 504
distance [external] [internal] [local], 544
distribute-list, 552
distribute-list [acl number] [in/out] [protocol/interface], 552
distribute-sap-list [IPX access-list number], 511
encapsulation PPP, 460
format, 355
frame-relay interface dcli, 470
frame-relay map, 470
GetBulkRequest, 599
GetNextRequest, 599
GetRequest, 599
GetResponse, 599
ifconfig -a, 568
InformRequest, 600
interface posi [slot/port], 479
interface serial [slot/port], 454
interface slot/port [sub-interface number], 469
interface tunnel [interface number], 452
IOS
  ?, 324
  configure, 320
  copy, 320
  IP unnumbered, 334
  show hardware, 332
  show interfaces, 332
  show interfaces accounting, 332
  show memory, 332
  show memory failures alloc, 332
  show processes cpu, 332
  show protocols, 332
  show version, 332
  terminal, 325
  terminal monitor, 332
  version 12.x, 320
  write, 320
IOS command lines, line editing commands, 325
IOS version 12.x, configure network, 321
ip classless, 498
ip default-network [ip address], 551
ip forward-protocol udp [port number], 499
ip helper address [address], 499
ip local pool [poolname] [starting address ending address], 344
ip mtu [bytes], 499
ip nat inside source list [acl] pool [pool name], 448
ip nat inside source list [acl] pool pool name] [overload], 448
ip nat inside source list [alc] pool [pool name], 449
ip nat inside source static [inside ip] [outside ip], 448
ip nat pool [name] [starting outside address range] [ending outside address range] [prefix-length], 448
ip ospf cost, 526
ip ospf demand-circuit, 531
ip ospf network [broadcast or non-broadcast], 530
ip ospf priority [0-255], 529
ip route, 492-493
ip summary-address eigrp, 510
ip tcp intercept list [ACL], 432
ip tcp intercept mode [intercept/watch], 432
IP [service] source-interface, 371
IPX, show ipx interface [interface slot/port], 338
ipx router eigrp [process id], 511
isdn switch-type [switch type], 458
locate, UNIX, 372
logging facility, 363
logging [method] level, 361
login, 368
man ping, 567
neighbor, 505

neighbor [ip address], 505
neighbor [ip address] ebgp-multihop, 541
neighbor [ip address] filter-list [acl number] weight [0 _ 65535], 544
neighbor [ip address] remote-as [as number], 543
neighbor [ip address] soft-reconfiguration inbound, 541
neighbor [ip address] version 4, 541
neighbor [ip address] [priority (0-255)], 530
neighbor [ip-address] remote-as [as number], 541
neighbor [peer-group name] filter-list [acl] [in/out] weight, 544
neighbor [peer-group name] peer-group, 544
network [ip address/ip network number], 509
no auto-summary, 503, 509
  OSPF, 516
no debug all, 514
no exec, 342
no ip source-route, 499
no ip unreachables, 492
no shutdown, 459
notify, 341
ntp update calendar, 358
offset-list [acl] [in/out] [metric] [interface], 504
overwrite-network, 320
passive-interface, 509
passive-interface [interface], 500
passive-interface), 528
peer default ip dhcp, DHCP, 344
ping [hostname/IP address, 568
ping?, 567
ppp multilink, 460
printer [name] [line#] [option], 342
privilege level, 341
redistribute [source] [process id] [metrics], 549
redistribution, 549
router eigrp [process id], 508
router igrp [process id], 508
router ospf [local process id], 519
router [protocol] [process id], 500
routing tables, 339-340
service password-encryption, 369
service timestamps debug, 359
service timestamps log, 359
session-limit, 341
session-timeout, 341
SetRequest, 599
show accounting, 376
show arp, 489
show atm interface atm [slot/port], 479
show atm map, 478
show atm traffic, 479
show atm vc, 479
show bootflash, 355
show calendar, 358
show clock, 358
show controllers bri [slot/port], 465
show crypto algorithms, 454
show crypto engine brief, 454
show crypto engine connections active, 454
show crypto map, 454
show crypto mypubkey, 454
show crypto pubkey [name/serial number], 454
show dxi map, 478
show dxi pvc, 478
show frame-relay lmi, 474
show frame-relay map, 474
show frame-relay pvc, 474
show frame-relay traffic, 474
show interface, 463
show interface tunnel [number], 453
show interface [slot/port], 568
show interface [type/port/slot], 463
show interface [type/slot/port], 499, 549
show ip accounting, 439
show ip accounting access-violations, 439
show ip bgp, 546
show ip bgp neighbors, 547
show ip bgp peer-group, 547
show ip bgp summary, 546
show ip eigrp interfaces, 513
show ip eigrp neighbors [options], 513
show ip eigrp topology, 514
show ip eigrp traffic, 513
show ip masks [address], 489
show ip nat statistics, 449

show ip nat translations, 449
show ip ospf, 523, 527
show ip ospf interface [type/slot/port], 532
show ip ospf neighbor, 532-533
show ip ospf virtual-links, 533
show ip ospf [local process id], 532
show ip protocol, 510
show ip protocols, 490, 512
show ip protocols summary, 490
show ip route, 339, 488
OSPF, 526
show ip route connected, 488-489
show isdn active, 465
show isdn history, 465
show isdn status, 459, 465
show logging, 360
show ntp associations, 358
show ntp status, 358
show ppp multilink, 457
show running-config, 340, 498
show [protocol] access-list [acl number], 439
show [protocol] route, 340
SNMP, 599-600
standby priority [0-255], 444
standby [group identifier] authentication [password], 445
standby [group identifier] track [interface type:slot:port], 444
standby [group id] prempt, 444
subnets, 550
telnet transparent, 342
terminal monitor, 359
timers basic [update] [invalid] [holddown] [flush], 504, 508
traffic-share [balanced/min], 512
transport [direction] [protocol/all], 341
user EXEC commands, 317
variance [multiplier], 512

Committed Frame Rate (CFR), 233

Common Channel Signaling Links (CCSL), SS7, 217

Common Management Information Protocol (CMIP), 597

common part convergence sublayer (CPCS), 241

comparing routing protocols, 485

compatibility, AppleTalk Phase 1 versus AppleTalk Phase 2, 87

Competitive Local Exchange Carrier (CLEC), 195

components
of cabling, 130
of fiber optic transmission, 134
of LANE, ATM switches, 298
of SNMP, 600
ASN.1. See ASN.1
BER, 601
MIB structure. See MIB
of SS7, 216-217

composite bandwidth, utilization, 575-577

composite bandwidth utilization distribution, 578

conf-reg, 346-347
Cisco 2500 series, 348
rommon, 348
settings, 349-350

conf-reg (configuration register), 346

config-register command, 346

ConfigMaker, Cisco, 309

configuration EXEC mode, 319

configuration management, OSI network management model, 595

configuration register (conf-reg), 346

Configuration Report Server (CRS), 174

configuration subprocess commands, 515
passive-interface, OSPF, 528

configurations, cables, 312

configure memory, configuration EXEC mode, 320

configure network
configuration EXEC mode, 321
IOS version 12.x commands, 321
privileged EXEC command, 330

configure terminal, configuration EXEC mode, 319

configure terminal command, 321

configuring

accounting, new-mode IOS authentication, 375-376
AppleTalk, 337-338
async serial lines, 340-345
ATM, 474-477
authentication, RIP, 505
AUX ports, 345
Cisco routers, 331-332
copy command, 326-327
from scratch, 332-333
initial configuration dialog, 316
IOS commands, 332
router configuration files, 316
saving changes, 323
setup application, 316
classless routing, 498
clocks, Cisco routers, 356-359
crypto maps, 453
default routes, 493-494
DHCP/BOOTP forwarding, 499
dynamic protocols, 500-501
EIGRP, 508-511
Frame Relay, 465
Frame Relay PVC, 465-466
IBGP reflectors, 545
IBGP/EBGP, 542-543
interfaces
AppleTalk, 337-338
Cisco routers, 334-335
DNS resolution, 336
IP unnumbered, 334
IPX, 337-338
subnetting address space, 334
viewing routing tables, 339-340
intra-autonomous BGP, 542-544
IOS commands, 320
IP, control services, 499
IP dial-in, 343-345
IPX, 337-338
ISDN, 457-458
ISDN BRI DDR, 458-461
ISDN BRI interfaces, backup configurations, 462-465
LPR service, 342-343
NAT, 447-449
OSPF, 516-517, 519-520
route announcements, 520-523
SNMP
on Cisco IOS, 622-623
on Macintoshes, 619-622
SNMP agents
on Windows 95/98 systems, 616-619
on Windows NT systems, 613-615
static routing, 494-498
terminal lines, 340-342
tunnel interfaces, AURP, 535-536
VLANs, Layer 2 LAN switching services for Layer 3, 285-286

congestion control, Frame Relay, 234

connected media, 3-5

connecting modems to console ports, 311

connection-oriented transport, OSI-RM, 29

connection-oriented transport processes, VCs, 72

connection-oriented transport protocols, OSI-RM, 29

connectionless media, 6

connectionless transport processes, datagrams, 72

connectionless transport protocols, Layer 4 (OSI-RM), 29

connections, TCP, 74-75

connectors, fiber optic cable, 136

console ports, 310-311

Constant Bit Rate (CBR), 241

constellation topologies, 8-9

contention-based access, LAN transmission protocols, 13-15

contextual structure, AppleTalk, 86

control (Layer 3), IPX, 108

control commands, 487

BGP, 538
RIP, 502
static routing, 492

control packets, ADSP, 105

control planes, ATM, 236

control plane protocols, ISDN, 223

control services, configuring IP, 499

control signals, 216

controllers, 454

controlling route redistribution, 551

convergence, 386
RIP, 399

convergence time, 386

Copper Distributed Data Interface (CDDI), 184

copy command
Cisco routers, configuring, 326-327
Flash memory, 328-329
IOS commands, 320
privileged EXEC command, installing new IOS, 352
TFTP, 329-330

copy tftp running-config command, 427, 496

copy/erase command, file systems, 327

cost, OSPF, 525

cost metric, OSPF, 413

CPCS (common part convergence sublayer), 241

CPE equipment, ISDN, 223-224

CPE T-carrier network equipment, 208-209

CRC (Cyclical Redundancy Checking), Ethernet, 581

CRC errors, WAN errors, 587

cron, syslog facility definitions, 362

CRS (Configuration Report Server), Token Ring management, 174

crypto key-exchange passive [tcp port number] command, 453

crypto key-exchange [passives interface ip address] [tcp port number] command, 453

crypto maps, configuring, 453

CSFS (Cable Signal Fault Signature), 564

CSMA/CA, LocalTalk, 91

CSMA/CD (carrier sense multiple access with collision detect), 13-15, 137
1000Base-X CSMA/CD, 167-168
contention-based access, LAN transmission protocols, 13
Ethernet, IFG, 142

CSU (Channel Service Unit), 208-209

CTK (Claim Token), MAC frame types, 178

current (I), 6

cut-through method, switch forwarding methodologies, 280

cycle, 6

Cyclical Redundancy Checking (CRC), 581

D

D channel Layer 3 protocol, ISDN, 227

D4 frame format, T1 framing, 204

D4 superframe format, T1, 204

D4-ESF/B8ZS, Transport-1, 455

D4-SF/AMI, Transport-1, 455

DAC (Dual-Attachment Concentrator), 190

daemon, syslog facility definitions, 362

DAS (Dynamically Assigned Sockets), socket addresses, 95

DAT (Duplicate Address Test), MAC frame types, 178

data, RTMP messages, 100

Data Circuit-terminating Equipment (DCE), 230

data communication reference models, 23
OSI-RM. *See* OSI-RM

Data Exchange Interface (DXI), ATM, 238

data framing, 12

data grade cable (DGC), 5

data link layer (Layer 2)
AARP, 89-91
EtherTalk, 92-93
LocalTalk, 91
OSI-RM, 31-32
data messages, RTMP, 100
Data Offset, TCP headers, 76
data paths, BRI, 225
Data Service Unit (DSU), 209
Data Terminal Equipment, 230
data type name labels, ASN.1, 601
data types, 603-605
Data-Link Connection Identifiers (DCLIs), 233
data-link framing, 245
HDLC, 245-247
PPP, 247-250
data-link protocol, 245
database-access messages, SS7, 216
Datagram Delivery Protocol. *See* DDP
datagram fragments, 42
datagram headers, DDP, 95
datagram routing, Layer 3 (network), 30
datagram service, NetBIOS, 120
datagram transports, 29
datagrams, 38. *See also* IP datagrams
connectionless transport processes, 72
DDP, assembly and delivery, 97
IPX datagrams, delivering, 113
Layer 3 (network), OSI-RM, 30
network communication protocols, 16
OSI-RM, 27
SPX, 117
daylight savings time, 357
DC (direct current), 6
DCE (Data Circuit-terminating Equipment), 230
DCI (Discrete Channel Inputs), multiplexers, 200
DCLIs (Data-Link Connection Identifiers), 233
DDP (Datagram Delivery Protocol), 94-95
broadcasts, 97
datagram headers, 95
function of, 96-97
DDP datagrams, assembly and delivery, 97
DDR networks, OSPF, 531
debug dialer command, 465
debug ip bgp events command, 547
debug ip bgp updates command, 547
debug ip ospf events command, 533
debug messages, buffered logging, 361
debug ppp authentication command, 457
debug ppp error command, 457
debug ppp negotiation command, 457
debug ppp packet command, 457
debug q921 command, 465
debugging, 514
BGP, 547
IOS, 514
OSPF, 533
dedicated bandwidth, Frame Relay, 467
default gateway, routing tables, 58
default gateways, 59
EIGRP, 510
default route distribution, OSPF/BGP redistribution, 551
default routes, configuring, 493-494
default routing, EIGRP, 510
default-information originate always command, 550
deferred releases (DR), IOS releases, 315
deleting ACLs, 427
delivering
DDP datagrams, 97
IP datagrams, 62-65
routers, 60-61

IPX datagrams, 113
messages, NetBIOS, 120

delivery of IP datagrams, 38-40

delivery models, IP datagrams, 57-58

demarcation reference points, ISDN, 223-225

denial of service attacks, ACLs, 432-434

DES, 451

designated router election, OSPF, 529

designated routers, 409-410

Destination Address, IP headers, 42

Destination Network, IPX message format, 112

destination network addresses, 60

Destination Node, IPX message format, 112

Destination Port
IPX message format, 112
TCP headers, 76

Destination Service Access Point (DSAP), 127

device information, documentation, 562-563

DGC (data grade cable), 5

DHCP (Dynamic Host Configuration Protocol), 81
commands, peer default ip address pool dhcp, 344
leases, 81

DHCP/BOOTP forwarding, configuring, 499

dial-in pools, 343

dial-in user accounts, 345

dialer load-threshold [1-255] [inbound/outbound] command, 460

diffusing update algorithm, 402

digital carrier systems, 198
CPE T-carrier network equipment, 208-209
DSH, 202
PCM, 198-199
signaling systems, 215-216
SONET, SDH, 212-214
SS7, 216-619
STDM, 199-201
T-carrier, 210-211
T1, 203-204
framing, 204-206
line coding, 206-207
TDM, 199-201

digital signal hierarchy (DSH), 202

Dijkstra's algorithm, 387

direct current (DC), 6

disadvantages of IGRP, 402

disaster recovery
bootstrap behavior, 346-347
Cisco routers, 346-347
conf-reg settings, 349
changing, 350

discontinuous subnets, 392

Discrete Channel Inputs (DCI), 200

display commands, 486
AURP, 534
BGP, 537
EIGRP, 506
IGRP, 506
OSPF, 514
static routing, 492

displaying
information
ACLs, 439
ATM, 478
Frame Relay, 474
ISDN, 465
PPP, 457
tunneling, 453-454
IP network information, 488-490
routing tables, 488

distance eigrp [internal distance] [summary distance] [external distance] command, 512

distance ospf [external/inter-area] [10-255] command, 525

distance vector protocols. *See* DVPs

distance [1-255] command, 512

distance [10-255] command, 504

distance [external] [internal] [local] command, 544

distances. *See* administrative distances

distribute-list command, 552

distribute-list [acl number] [in | out] [protocol | interface] command, 552

distribute-lists, ACLs, 425

distribute-sap-list [IPX access-list number] command, 511

Distributed Sniffer, protocol analyzers, 564

distribution lists, filtering, 551-553
OSPF, 554-555

DIX specification, 138

DMUX (DS-X multiplexers), 201

DNS (Domain Name Service), 79-80

DNS resolution, configuring interfaces, 336

documentation, network analysis, 561-563

Domain-Specific Part (DSP), 244

DRAM (dynamic random access memory), 308-309

DS-X multiplexers (DMUX), 201

DS3 service, over T3, 210-211

DSAP (Destination Service Access Point), 127

DSH (digital signal hierachy), 202

DSP (Domain-Specific Part), 244

DSU (Data Service Unit), 209

DSU/CSU, 454-455

DTE (Data Terminal Equipment), 230

DUAL (diffusing update algorithm), 402

Dual-Attachment Concentrator (DAC), 190

Duplicate Address Test (DAT), MAC frame types, 178

duplicate Layer 3 end-station addresses, 580

DVP algorithm, 386-387

DVPs (distance vector protocols), 386
IGRP. *See* IGRP
RIP, 397
versus LSPs, 387

DXI (Data Exchange Interface), ATM, 238

Dynamic Host Configuration Protocol. *See* DHCP

dynamic protocols
configuring, 500-501
routing tables, flushing, 491

dynamic random access memory. *See* DRAM

dynamic routes
administrative distances, 486
redistribution, 547

dynamic routing, 70

dynamic routing protocols, 65, 380, 383
DVPs, 386
Internet, 548
LSPs, 387
RIP. *See* RIP
route metrics, 384
routing metrics, 385

dynamic routing tables, 60

dynamic source list, NAT, 449

Dynamically Assigned Sockets (DAS), 95

E

early deployment release (ED), IOS releases, 315

EBGP (external BGP), 543

echo reply messages, 565

echo request messages, 565

EDS (Electronic Data Systems), 592

EGPs, (Exterior Gateway Protocol), 66, 69, 381

EIA (Electronics Industries Association), 130

EIGRP (Enhanced Interior Gateway Routing Protocol), 110, 402, 507-508. *See also* IGRP
- administrative distances, adjusting, 512
- bandwidth, 511
- configuring, 508-511
- default gateways, 510
- default routing, 510
- display commands, 506
- global configuration commands, 506
- global interface subprocess commands, 506
- global router subprocess commands, 506
- IP addresses, 509
- IP summarizations, 510
- IP summary-address announcements, 510
- IPX, 110, 511
- IPX routing, 511
- load balancing, 512
- monitoring, 512-514
- PID, 508
- route redistribution, 549
- subnets, 509
- updates, 509
- distance eigrp [internal distance] [summary distance] [external distance], 512

ELAP (EtherTalk Link-Access Protocol), 92

elastic buffers, Token Ring, 584

electromagnetic interference (EMI), 129

Electronic Data Systems (EDS), 592

Electronics Industries Association (EIA), 130

EMI (electromagnetic interference), 129

encapsulation, PPP, 455

encapsulation ppp command, 460

encapsulation schemes, IPX, 109

encoding, 128
- analog-to-digital conversion, 199
- block encoding, 130
- Manchester encoding, 129

encrypted GRE, tunneling, 452-453

End-Station Density. *See* ESD

end-to-end data transport, Layer 4, 28-30

end-to-end networked communication, OSI model, 26

Enhanced Interior Gateway Routing Protocol. *See* EIGRP

entering route announcements, OSPF, 520-523

enterprise switches, 268

EO (Exactly-Once), ATP transactions, 103

erase command, privileged EXEC command, 327

erasing file systems, copy | erase command, 327

ERIP (Extended RIP), 396

ERPs (exterior routing protocols), 414
- ASs, 414-415
- BGP-4, 416, 418
- versus IGPs, 415-416

errors, WAN, 586-587

errors, 579. *See also* network errors

ESA crypto engines, 452

ESD (end-station density), 274
- switch ports, Layer 2 LAN switches, 274-276

ESF (Extended Superframe Format), T1, 205

establishing bidirectional point-to-point connections, ATM, 243

ET (Exchange Termination), CPE equipment, 224

Ethernet. *See also* 802.3
- 1000Base Ethernet. *See* Gigabit Ethernet
- 100Base-FX PHY, 160-161
- 10Base-2, 138
- 10Base-5, 154
- 10Base-FL, 157

- 10Base-T, 152-156
- 802.3, 141-143
- collisions, 143
- CRC, 581
- CSMA/CD, IFG, 142
- Fast Ethernet. *See* Fast Ethernet
- frames, 582
- Gigabit Ethernet. *See* Gigabit Ethernet
- history of, 137-139, 141
- network errors, 581-582
- switch ports, 270-273
- switches, 139
- transparent bridges, 257
- UTP, 139

Ethernet collision domain rules, 392

Ethernet DIX consortium, 138

Ethernet MAC
- bit-periods, 142
- operating parameters, 143-144

Ethernet SNAP frame, 147

Ethernet/802.3 frame types, 144-146

EtherTalk, Layer 2 (data link), 92-93

EtherTalk Link-Access Protocol, 92

European Union, OSI-RM, 24

event-driven communication, packet switching, 10

events group (Group 9), RMON I, 624

examining RIP process, 506

Exchange Termination (ET), CPE equipment, 224

exec accounting record, RADIUS, 375

EXEC modes, help, 324

EXEC shell, user EXEC mode, 317-318

Expert Sniffer, protocol analyzers, 564

explicit tokens, 15

extended ACLs, 428
- filtering, 431-432
- service ports, 430-431
- traffic filters, 429

extended IP ACLs, operators, 429

Extended RIP, 396

Extended Superframe Format (ESF), T1, 205

Extended TACACS, 366

Exterior Gateway Protocol, 66, 69, 381

exterior routers, AURP, 101

exterior routing protocols (ERPs), 414

External BGP (EBGP), 543

F

Failures, gateway redundancy, 442

Farallon Corporation, PhoneNet, 91

Fast Ethernet, 139, 157-159
- 100Base-T repeaters, 161-162
- 100Base-T4 PHY, 160
- 100Base-TX, 160-161
- 100VG-AnyLAN, 162
- history of, 139-140
- MII, 158-160

fat pipe technology, 266, 271

fault management, OSI network management model, 595

FCC (Federal Communications Commission), 192

FCS (Frame Check Sequence), 235

FDDI (Fibre Distributed Data Interface)
- history of, 183-185
- specification, 185
  - *MAC, 186-187*
  - *PHY, 188*
  - *PMD, 189-190*
  - *SMT, 188*

FDDI-MIC type connectors, fiber optic cable, 136

FDDITalk, 94

FDDITalk Link-Access Protocol (FLAP), 94

FDM (Frequency Division Multiplexing), PSTN, 197

FDR (full-duplex repeaters), 169

FECN (forward explicit congestion notification), 234

Federal Communications Commission (FCC), 192

Fiber Distributed Data Interface. *See* FDDI

fiber optic cable, 133
- components of, 134
- connectors, 136
- light, 134

Fibre Channel, Gigabit Ethernet, 163-164

file systems, 329. *See also* memory
- commands, copy/erase, 327

File Transfer Protocol, 78

filter group (Group 7), RMON I, 624

filtering
- distribution lists, 551-553
  - *OSPF, 552, 554-555*
  - *RIP, 552*
- extended ACLs, 431-432
- IBGP, 544
- route-maps, 555-556
- zone list announcements, 437

filters
- AppleTalk ACL filters, 436-437
- inbound filters, ACLs, 424
- outbound filters, ACLs, 424
- output filters, 553

FIN, TCP connections, 76

finite state machine, BGP-4, 417

fixed patterns, BERTs, 565

Flags, IP headers, 41

FLAP (FDDITalk Link-Access Protocol), 94

flash cards
- PCMCIA, advantages of, 329
- [slot:], 328

Flash memory
- Cisco routers, 308
- copy command, 328-329

flash partitions, IOS, 352-354

flooding OSPF messages, 411

flow control, switch ports
- Ethernet, 270-271
- Layer 2 LAN switches, 270-271

flushing routing tables, dynamic protocols, 491

footers, SAR, 242

forward explicit congestion notification (FECN), 234

forwarding
- datagrams, Layer 3 (network) OSI-RM, 30
- IP datagrams, 57, 61-62

FPS (frames per second), 277, 574
- switch backplane, Layer 2 LAN switches, 277-279

Fragment Offset, IP headers, 41

Fragmentation, 42

Frame Check Sequence (FCS), 235

frame copy error, Token Ring soft errors, 583

frame errors, WAN errors, 587

Frame Relay, 235
- configuring, 465
- congestion control, 234
- dedicated bandwidth, 467
- frames, 234-235
- information, displaying, 474
- multipoint grouping, 470
- PSNs, 232-233
- PVC topologies, 233
- topologies, 467-468, 470
  - *AppleTalk, 468*
  - *multipoint designations, 470*
  - *multipoint VC topology, 473-474*
  - *point-to-point, 470*
  - *point-to-point VC topology, 471-473*

Frame Relay Consortium (FRC), 232

Frame Relay PVC, configuring, 465-466

frame streaming, Token Ring hard errors, 585

frame types, IPX, 109-110

Frame User Network Interface (FUNI), ATM, 238

frame-relay interface dcli command, 470

frame-relay map command, 470

framers, multiplexers, 200

frames
802 frames, 147
802.3, 147
Ethernet, 582
Ethernet SNAP frame, 147
Ethernet/802.3 frame types, 144-146
FDDI specification, MAC frames, 186-187
Frame Relay, 234-235
packet switching, 10-11
switch backplane, Layer 2 LAN switches, 277-279
Token Ring, 176-177
MAC frame types, 177-179

Frames Per Second. *See* FPS

framing. *See also* data-link framing
PPP, 249-250
T1, 204-206

FRC (Frame Relay Consortium), 232

free tokens, 170

frequency (Hz), 6

Frequency Division Multiplexing (FDM), 197

frequency error, Token Ring, 584-585

FTP (File Transfer Protocol), 78

full-duplex repeaters (FDR), 169

full-duplex transmission, 267

functions
of DDP datagram headers, 96-97
of LANE, ATM switches, 297-298
of NetBIOS, 119-120
of network management, 592-593
of routers, 21-22
of subnet masks, 45

FUNI (Frame User Network Interface), ATM, 238

G

GARP (Generic Attribute Registration Protocol), 286

gateway failures, gateway redundancy, 442

gateway redundancy, 441-442
gateway failures, 442
HSRP, 442-445

gateway servers, 306

general deployment (GD), IOS releases, 315

Generic Attribute Registration Protocol (GARP), 286

Generic Flow Control (GFC), ATM cell formatting, 239

Generic Route Encapsulation, 450-453

GetBulkRequest command, 599

GetNextRequest command, 599

GetRequest command, 599

GetResponse command, 599

Gigabit Ethernet, 163
1000Base-T, 166
1000Base-X CSMA/CD, 167-168
Fibre Channel, 163-164
Gigabit PHY, 165-166
GMII, 164
repeaters, 168-169

Gigabit PHY, Gigabit Ethernet, 165-166

global configuration commands, 426-428, 449

global configuration EXEC mode, ip classless command, 498

global configuration mode
config-register comand, 346
flash partitions, IOS, 353
loopback interfaces, 483
subcommand modes, 322

global configuration model, configuration EXEC commands, 321

global ID, MIB types, 607

global interface subprocess commands
EIGRP, 506
IGRP, 506
RIP, 502
global OID, MIB types, 607
global router subprocess commands
EIGRP, 506
IGRP, 506
RIP, 501
GMII, Gigabit Ethernet, 164
GMT (Greenwich Mean Time), 357
GOSIP (Government OSI Profile) compliant, 24
Government OSI Profile compliant), 24
Grand Junction Networks, Fast Ethernet, 139
GRE (Generic Route Encapsulation), 450-453
group-async interface, 344
guidelines
for classful IP addressing, 50, 56
for subnetting, large address spaces, 47
GxSNMP, SNMP-based management tools, 613

H

half routers, AppleTalk, 98
half-duplex repeaters (HDR), 169
hard errors, Token Ring, 584-586
hardware
bridges, 19-21
cables. See cables, 17
hubs, 18
repeaters, 17
routers, 21-22
HDLC
data-link framing, 245-247
IOS, 336
HDR (half-duplex repeaters), 169
Header Checksum, IP headers, 42
headers
SAR, 242
TCP, 76
UDP, 77
HEC (Header Error Check), ATM cell formatting, 240
help, EXEC modes, 324
Hewlett-Packared, OpenView (SNMP-based management tools), 611
hexidecimal numbers, addressing IPX, 110
hierarchies, PSTN, 194-196
history
of Ethernet, 137-139, 141
of Fast Ethernet, 139-140
of FDDI, 183-185
of ISDN, 219-221
of PSTN, 192-193
of Token Ring, 169-170
history group (Group 2), RMON I, 624
[hold down] timer, 504
Hop Count, SAP message format, 115
hop counts, 99
hop-to-hop forwarding, IP datagrams, 57
hops, RIP, 397
horizontal cabling, 131
host group (Group 4), RMON I, 624
host-to-host paradigm, 382
host/end-station addresses, 11
HostTopN group (Group 5), RMON I, 624
HSRP, gateway redundancy, 442-445
hubs, 17-18
hybrid switches, switch forwarding methodologies, 281-282
HyperTerminal, 312
Hz (frequency), 6

I

I (current), 6

IANA (Internet Assigned Numbers Authority), IP address space, 56

IBGP (Internal BGP), 542-545

IBGP/EBGP, configuring, 542-543

***IBM PC Network Technical Reference Manual*, 118**

IBM Type 1 cabling, 182

IBM Type 2 cabling, 182

IBM Type 3 cabling, 182

ICANN (Internet Corporation for Assigned Names and Numbers), IP address space, 56

ICMP (Internet Control Message Protocol), 400
- ping, 71

ICMP messages
- ping, 565-567
- traceroute, 572

Identification, IP headers, 41

identifying problems, 589-590

IDI (Initial Domain Identifier), 244

IDP (Initial Domain Part), 244

IEEE (Institute of Electrical and Electronic Engineers), 124
- Fast Ethernet, 141

IEEE (Institute of Electrical and Electronics Engineers) model, 23

IEEE 802, 124-125

IEEE 802.1q standards-based VLANs, Layer 2 LAN switching services for Layer 3, 286-288

IEEE 802.5, Token Ring
- frame formats, 176-177
- free tokens, 170
- MAC frame types, 177-179
- ring management, 172-174
- traffic priority mechanisms, 171
- TRT, 170

IETF, 37

ifconfig _a command, 568

IFG, 142

IGPs (Interior Gateway Protocols), 66, 381, 396
- EIGRP. *See* EIGRP
- IGRP. *See* IGRP
- intranetworks, 67
- OSPF. *See* OSPF
- process ID, 500
- RIP. *See* RIP
- versus ERPs, 415-416

IGRP (Interior Gateway Routing Protocol), 402. *See also* EIGRP
- administrative distances, adjusting, 512
- commands, 506
- monitoring, 512-514
- PID, 508
- route redistribution, 549
- timers, 508
- updates, 509
- versus RIP, 507

IGS (Integrated Gateway Server), 306

IHL (Internet Header Length), IP headers, 41

impedance (Z), 6

implementing VLANs, Layer 2 LAN switching services for Layer 3, 285-286

implicit tokens, 15

IMUX (inverse multiplexers), 201

inbound filters, ACLs, 424

information, displaying
- ACLs, 439
- ATM, 478
- Frame Relay, 474
- IP networks, 488-490
- IP routing, managing, 490-491
- ISDN, 465
- PPP, 457
- tunneling, 453-454

InformRequest command, 600

initial configuration dialog, configuring Cisco routers, 316

Initial Domain Identifier (IDI), 244

Initial Domain Part (IDP), 244

Initial Sequence Number (ISN), 432

inside address translation, NAT, 448

installing IOS (new), 352

Institute of Electrical and Electronic Engineers. ***See*** **IEEE**

Institute of Electrical and Electronics Engineers model (IEEE model), 23

Integrated Gateway Server (IGS), 306

inter-area paths, OSPF, 413

inter-autonomous BGP, 538, 540-541
- SIN routing, 538-539

inter-autonomous system routing, BGP, 538

inter-repeater links, 17

interCo trunks, 198

interdomain routing, 65, 69-70, 381

Interexchange Carriers (IXCs), 195

interexchange points, 67-68

interface bit values, OSPF, 525

interface configuration commands, OSPF, 514

interface configuration EXEC commands, ip ospf cost, 526

interface configuration subcommands
- ip ospf network [broadcast or non-broadcast], 530
- ip summary-address eigrp, 510
- static routing, 492

interface posi [slot/port] command, 479

interface serial [slot/port] command, 454

interface slot/port [sub-interface number] command, 469

interface specification, NAT, 449

interface tunnel [interface number] command, 452

interfaces
- AutoNeg, 158
- configuring, 334-335
  - *AppleTalk, 337-338*
  - *ATM, 475-477*
  - *DNS resolution, 336*
  - *IP unnumbered, 334*
  - *IPX, 337-338*
  - *subnetting address space, 334*
  - *viewing routing tables, 339-340*
- GMII, 164
- group-async, 344
- MII, 158-160
- PCL, 165
- PMA, 166
- PMD, 166
- Token Ring, 179

Interior Gateway Protocol. ***See*** **IGPs**

Interior Gateway Routing Protocol. ***See*** **IGRP**

interior routing, 381

Intermediate Systems (ISs), 57

Internal BGP, 542-545

internal error, Token Ring hard errors, 585

internal routers, router designations (OSPF), 406

internal routers (IR), 515

international gateways, PSTN hierarcy, 195

International Standards Organization. ***See*** **ISO**

International Telecommunication Union Telecommunications Standardization Sector, 23

Internet
- BGP, 538
- dynamic routing protocols, 548
- routing, 65

Internet (Layer 1), IPX, 108

Internet Assigned Numbers Authority (IANA), 56

Internet Control Message Protocol, 400

Internet Corporation for Assigned Names and Numbers (ICANN), 56

Internet Header Length (IHL), 41

Internet Protocol. ***See*** **IP**

Internet Society, IETF, 37

Internet-RM, 27

Internetwork Operating System. ***See*** **IOS**

internetwork routing, 381

Internetworking Exchange Protocol. *See* IPX

internetworks, 21, 65
AppleTalk, routers, 98
ASs, 66
interexchange points, 67

interruptions in VTY connectivity, 491

intra-area paths, OSPF, 413

intra-autonomous BGP, 542-544

intra-autonomous system routing, BGP, 538

intradomain routing, intranetworks, 65

intradomain routing, 381

intraLATA traffic exchanges, 195

intranet exchange points, 69

intranets, backbone paths, 68-69

intranetwork routing, 381

intranetwork/backbone router point, 67-68

intranetworks, 382
IGP, 67
intradomains, 65

[invalid] timer, 504

inverse multiplexers (IMUX), 201

IOS (Internetwork Operating System), 308
accounting, configuring, 375
administrative distances, 486
authentication
new-mode. See new-mode IOS authentication
old-mode. See old-mode
AutoInstall features, 309
command lines, 325-326
command lines navigating, 323-326
command references for IOS versions 9.x and higher, 323
commands
?, 324
configure, 320
copy, 320
IP unnumbered, 334
show hardware, 332
show interfaces, 332
show interfaces accounting, 332
show ip route, 488
show memory, 332
show memory failures alloc, 332
show processes cpu, 332
show protocols, 332
show version, 332
terminal, 325
terminal monitor, 332
write, 320
configuration EXEC mode, 319
DRAM, 309
EXEC shell, user EXEC mode, 317-318
flash partitions, 352-354
HDLC, 336
IGRP/EIGRP, debugging, 514
installing new IOS, 352
Kerberos, 367
NTP, 358
PCMCIA flash cards, partitions, 354-355
privileged EXEC mode, 318-319
protocols it supports, 314
releases, 315-316
route table display, 487
setup application, 316
TOPS-20 command shell interface, 314
troubleshooting, 355-356
upgrading, 350-351
user EXEC mode, 317-318
version 12.x, 320-321
versions, 315

IOS CLI, 317-318

IOS commands, configuring Cisco routers, 332

IOS crypto engines, 452

IOS logging messages. *See* logging messages

IOS NAT implementation, 447

IP (Internet Protocol), 11, 38
ACLs, 422, 425-426
control services, configuring, 499
discontinuous subnets, 392
Frame Relay topologies, 468
information, displaying, 488-490
interactions with Layer 2, 38-40

source routing, 499
versus ATM, 478

IP ACL protocol filtering keywords, 428

IP address spaces, 55-56

IP addresses, 43-44
Class C addresses, classful addressing, 51
classes, 44-45
classful addressing, 44-45
subnetting examples, 48-50
classful method, 388
classless addressing, 54
CIDR, 51-52
classless method, 388-389
EIGRP, 509
IPv4, 43
NAT, 445
octets, 43
routing tables, 45
subnet masks, 44
subnetting, 46-47

ip classless command, 498

IP datagrams
addressing Layer 2 and Layer 3, 62
delivering, 62-65
delivery models, 57-58
delivery of, 38-40
forwarding, 57, 61-62
Layer 2 and Layer 3, 62-64
hop-to-hop forwarding, 57
IP addresses, 43
subnetting, 46
IP fragmentation, 42-43
IP headers, 40-42
local delivery, 57
mapping addresses, 62
remote delivery, 57
routers, delivering, 60-61
routing, 57
routing tables, 45, 60
ULP, 38

ip default-network [ip address] command, 551

IP dial-in, configuring, 343-345

ip forward-protocol udp [port number] command, 499

IP fragmentation, IP datagrams, 42-43

IP headers, 40-42

ip helper address [address] command, 499

ip local pool [poolname] [starting address ending address] command, 344

ip mtu [bytes] command, 499

ip nat inside source list [acl] pool [pool name] command, 448

ip nat inside source list [acl] pool [pool name] [overload] command, 448

ip nat inside source list [alc] pool [pool name] command, 449

ip nat inside source static [inside ip] [outside ip] command, 448

ip nat pool [name] [starting outside address range] [ending outside address range] [prefix-length] command, 448

IP networks
closed, 492
statistic routing, 59

ip ospf cost command, 526

ip ospf demand-circuit command, 531

ip ospf network [broadcast or non-broadcast] command, 530

ip ospf priority [0-255] command, 529

ip route command, 492-493

IP route table, RIP, 503

IP routing, 490-491

IP routing protocols, 388
comparing, 485

IP routing tables, 58

IP summarizations, EIGRP, 510

IP summary-address announcements, EIGRP, 510

ip summary-address eigrp command, 510

ip tcp intercept list [ACL] command, 432

ip tcp intercept mode [intercept/watch] command, 432

IP unnumbered, 334

IP [service] source-interface command, 371

IPv4
- address spaces, 55
- IP addresses, 43

IPX (Internetworking Exchange Protocol), 107
- ACLs, 437-439
- addressing, hexidecimal numbers, 110
- configuring, 337-338
- EIGRP, 110
- encapsulation schemes, 109
- frame types, 109-110
- Layer 0 (transmission), 108
- Layer 1 (Internet), 108
- Layer 2 (transport), 108
- Layer 3 (control), 108
- Layer 4 (application), 108
- message format, 111-112
- NCP, 118
- NetWare architecture model, 108
- NetWare RPC, 118
- NetWare shell, 117
- network addresses, 111
- NLSP, 110
- node addresses, 111
- NWLink, 107
- port addresses, 111
- RIP, 110, 116-117
- SAP, 113
- SPX, 117
- upper layer protocols, 117

IPX datagrams
- delivering, 113
- ports, 110

IPX EIGRP, 511

IPX EIGRP subcommands, distribute-sap-list [IPX access-list number], 511

IPX end-node number, 108

IPX RIP routes, startup and shutdown procedures, 116-117

ipx router eigrp [process id] command, 511

IPX routing, EIGRP, 511

IR (internal routers), 515

ISDN, 457
- addresses, 228-229
- appliation services, 221-222
- B channels, 222
- BRI, 220
  - *data paths, 225*
- BRI interfaces, 457
  - *backup configurations, 462-465*
  - *configuring, 458-461*
- configuration information, displaying, 465
- control plane protocols, 223
- CPE equipment, 223-224
- D channel Layer 3 protocol, 227
- demarcation reference points, 223-225
- general configurations, 457-458
- history of, 219-221
- LAPD, 226-227
- Layer 1, 225-226
- Layer 2, 226-227
- Layer 3, 227
- PRI, 226
- PRI interfaces, 457
- pricing, 220
- problems with, 220
- protocol reference model, 223
- TDM digital transmission systems, 222
- user plane protocols, 223

ISDN BRI interfaces, 457
- backup configurations, 462-465
- configuring, 458-461

ISDN PRI interfaces, 457

isdn switch-type [switch type] command, 458

ISDN TelCo carrier switch types, 458

ISDN User Part (ISDN-UP), SS7, 219

ISE (isolating soft errors), 582

ISN (Initial Sequence Number), 432

ISO (International Standards Organization), 595
- security model, 343

ISO protocol suite, OSI-RM, 23

isolating soft errors (ISE), 582
ISs (Intermediate Systems), 57
ITU-TS (International Telecommunication Union Telecommunications Standarization Sector), OSI-RM, 23
IXC-POP, PSTN hierachy, 196
IXCs (Interexchange Carriers), 195

J

jabbering NIC, 582
jabs, 582
jitter, Token Ring, 584

K

KDC (Key Distribution Center), Kerberos, 366
KEEPALIVE messages, BGP-4, 418
Kerberos, 366-368
Kerberos realm, 366
kern, syslog facility definitions, 362
key chains, 505
keywords
- any, 427
- ASN.1, 602
- broadcast, 470
- IP ACL protocol filtering, 428

L

label swapping, ATM switches, 295-296
LAN
- network throughput, 578
- protocol analyzers, 564
- protocols, utilization baseline measurement guidelines, 576

LAN Bridge Server (LBS), 174
LAN Emulation Client (LEC), 299
LAN Emulation Configuration Server (LECS), 299
LAN Emulation Server (LES), 299
LAN Emulation User to Network Interface (LUNI), 298
LAN transmission protocols, 13-15, 296
LANE
- ATM switches, 296-297
  - *BUS, 299*
  - *components, 298*
  - *functions of LANE, 297-298*
  - *LEC, 299-300*
  - *LECS, 299*
  - *LES, 299*
- limitations of, 302-303

LANs (local area networks), 2
LAPD (Link Access Procedure on the D channel), ISDN, 226-227
laser diodes (LDs), 5
LATAs (local area transport areas), 195
latency, 264, 279
Layer 0 (transmission), IPX, 108
Layer 1 (Internet), IPX, 108
Layer 1 (physical), OSI-RM, 32-33
Layer 2
- addressing IP datagrams, 62
- forwarding IP datagrams, 62-64
- IP interactions, 38-40

Layer 2 (data link)
- AARP, 89, 91
- EtherTalk, 92-93
- LocalTalk, 91
- OSI-RM, 31-32

Layer 2 (transport), IPX, 108
Layer 2 implementation, Cisco routers, 307
Layer 2 LAN protocols, switching, 267-268
Layer 2 LAN switches, 268-269
- enterprise switches, 268
- switch backplane, 276
  - *capacity, 279-280*
  - *FPS, 277-279*
  - *PPS, 277*

switch forwarding methodologies, 280-282
switch ports, 269-270
ESD, 274, 276
flow control, 270-271
port configurations, 270
port mirroring, 273-274
port trunking, 271-273
workgroup switches, 268

Layer 2 LAN switching services for Layer 3, 282
802.1q Layer 2 traffic prioritization, 290-292
Layer 3 switching, 292
VLANs. *See* VLANs

Layer 2 media interface, Layer 2 LAN switches, 268

Layer 3, IP datagrams, 62-64

Layer 3 (control), IPX, 108

Layer 3 (network)
AppleTalk, DDP, 94-97
OSI-RM, 30

Layer 3 switching, Layer 2 LAN switching services for Layer 3, 292

Layer 4 (application), IPX, 108

Layer 4 (transport), 72
AppleTalk, 98
AEP, 101
ATP, 103
AURP, 101
NBP, 102-103
RTMP, 98-100
OSI-RM, 28-30
TCP. *See* TCP
UDP. *See* UDP
ULP data, 73

Layer 5 (session)
AppleTalk, 103
ADSP, 105-106
ASP, 106
ZIP, 104
OSI-RM, 28

Layer 6 (presentation)
AppleTalk, 103
AFP, 106
OSI-RM, 28

Layer 7 (application), OSI-RM, 27

layers of OSI-RM, 27
Layer 1 (physical), 32-33
Layer 2 (data link), 31-32
Layer 3 (network), 30
Layer 4 (transport), 28-30
Layer 5 (session), 28
Layer 6 (presentation), 28
Layer 7 (application), 27

LBS (LAN Bridge Server), 174

LDs (laser diodes), 5

learning bridges, 20

leases, DHCP, 81

LEC, LANE, 299-300

LEC (LAN Emulation Client), 299

LEC-CO, PSTN hierachy, 196

LEC-Tandem, PSTN hierachy, 196

LECS (LAN Emulation Configuration Server), 299

LEDs (light emitting diodes), 5

LES (LAN Emulation Server), 299

light, fiber optic transmissions, 134

lightwave transmission, connectionless media, 6

limitations
of Classical IP, 302-303
of LANe, 302-303

limited deployment (LD), IOS releases, 315

line coding, T1, 206-207

line editing commands, IOS command lines, 325

line error, Token Ring soft errors, 583

line printer daemon (LPD), 342

Line Termination (LT), CPE equipment, 224

linear PCM, 198

link aggregation, 266, 271. *See also* port trunking

link establishment, PPP, 248

link state protocols, 387

LLAP (LocalTalk Link Access Protocol), 92

LLC (Logical Link Control), 125
- 802.2, 125-126
  - *LLC PDU, 126-127*
- 802.2 (IEE subcommittee), service types, 127-128
- 802.3 PDU, 147-148

LLC PDU (LLC data unit), 126-127

LLC station service class, 128

LLC Type 1, 127

LLC Type 2, 127

LLC Type 3, 128

load balancing
- EIGRP, 512
- route metrics, 384

loading static tables, 496

local area networks. *See* LANs

local area transport areas (LATAs), 195

local delivery, IP datagrams, 57

local routers, AppleTalk, 98

local0-7, syslog facility definitions, 362

LocalTalk, 87
- Layer 2 (data link), 91

LocalTalk Link Access Protocol (LLAP), 92

locate command (UNIX), 372

logging facility command, 363

logging messages
- IOS, 359
  - *buffered logging, 360-361*
  - *trap logging, 361-364*
- mangement of UNIX shell scripts, 364-365
- syslog level classification, 360-361

logging [method] level command, 361

Logical Link Control. *See* LLC

login authentication, 370-371

login command, 368

logs, documenting change logs, 563

lookups, 60-61

loopback addresses, 56

loopback interfaces, global configuration mode, 483

loops, 259-261

lost frame error, Token Ring soft errors, 584

lost token error, Token Ring soft errors, 584

low utilization usage rate, 575

LPD (line printer daemon), 342

lpr, syslog facility definitions, 362

LPR service, configuring, 342-343

LSAs
- OSPF, 412
- OSPF messages, 408

LSPs (link state protocols), 387
- OSPF. *See* OSPF

LT (Line Termination), CPE equipment, 224

Lucent Technologies, 195

LUNI (LAN Emulation User to Network Interface), 298

M

M-Class multiplexers, 201

MAC (Media Access Control), 185
- AARP, 89
- addresses, 11
- FDDI specification, 186-187

MAC address-based VLANs, Layer 2 LAN switching services for Layer 3, 284

MAC frame types, Token Ring, 177-179

MAC-Kermit, 313

Macintoshes
- MAC-Kermit, terminal emulation software, 313
- SNMP, configuring, 619-622

mail, syslog facility definitions, 362

major release (MR), IOS releases, 315

man ping command, 567

Management Information Base. *See* MIB

management plane, ATM, 236

management tools, SNMP, 611-613

managing
IP routing information, 490-491
log messages, UNIX shell scripts, 364-365
static routing, 492

Manchester encoding, 129

MANs (Metropolitan Area networks), 2

mapping IP datagrams, addresses, 62

match lookups, 60-61

matrix, switching matrix, 254

matrix group (Group 6), RMON I, 624

MAUs (Multistation Access Units), 18, 137, 148-149, 169
Token Ring, 181

Max-repeaters, SNMP messages, 611

Maximum Transmission Unit, 42, 499

MCB (message control block), 118

MDI (medium-dependent interface), 148

mean average utilization rate, 575

measuring network throughput, 579

Media Access Control addresses. *See* MAC

media access unit. *See* MAUs

Media Independent Interface, 158-160

Medium Access Control. *See* MAC

medium dependent interface, 148

memory, 326, 329
Cisco routers, 308-309

message control block (MCB), 118

message format
IPX, 111-112
SAP, 114-115

Message framed Data, synchronous transmission, 12

message packets, ADSP, 105

Message Transfer Part (MTP), SS7, 218

messages
BGP-4, 418
NetBIOS, delivering, 120
OSPF, 410-411
RIP, 116, 398-400
RTMP, 100
SNMP, 609-611
Trap, 599

Metcalfe, Dr. Robert (Ethernet), 137

metrics
OSPF, 524-525
RIP, 504
route metrics, 384
routing metrics, 385

metropolitan area networks (MANs), 2

MGS (Mid-Range Gateway Server), 306

MIB (Management Information Base), 597, 600
structure of, 606-608

MIB modules
RMON I, 624-625
RMON II, 625-626

MIB types, global ID, 607

MII (Media Independent Interface), 158-160

MLT-3 (Multilevel 3), 129

MM-PMD (Multimode PMD), 189

modems, connecting to console ports, 311

monitoring
BGP, 546-547
EIGRP, 512-514
IGRP, 512-514
OSPF, commands, 532-533

more, IOS command lines, 326

MPLS (Multiprotocol Label Switching), 302-303
MPTs (multiprotocol tunnels), 452
MRTG (Multi-Router Traffic Grapher), SNMP-based management tools, 612
MTP (Message Transfer Part), SS7, 218
MTU (Maximum Transmission Unit), 42, 499
multi-area OSPF implementation, 407
Multi-Station Access Unit. *See* MAUs
multicast addresses, 56
multigateway networks, redistribution, 548
multihomed nontransit, ASs, 415
multihomed transit, ASs, 415
Multilevel 3, 129
Multimode PMD (MM-PMD), 189
multiple monitors, Token Ring soft errors, 583
multiplexers, 200-201
multipoint designations, Frame Relay topologies, 470
multipoint Frame Relay connections, nonbroadcast networks (OSPF), 530
multipoint frame-relay interfaces, OSPF, 530
multipoint grouping, Frame Relay, 470
multipoint tunnels, AURP tunnels, 101
multipoint VC topology, Frame Relay, 473-474
Multiprotocol Label Switching (MPLS), 302-303
multiprotocol tunnels, 452
Multistation Access Units. *See* MAUs

N

N1 (OSPF NSSA Type 1), 526
N2 (OSPF NSSA Type 2), 526
Name Binding Protocol, 95, 102-103
name confirmation, NBP, 102
name deletion, NBP, 102
name lookup, NBP, 102
name registration, NBP, 102
name service, NetBIOS, 119
naming structure, NetBIOS, 119
NANP (North American Numbering Plan), 194
NAT (Network Address Translation)
  configuring, 447
  dynamic source list, 449
  inside address translation, 448
  interface specifications, 449
  IP addresses, 445
  problems with, 449
National Internet service providers (NISPs), 65
National Terminal Number (NTN), 232
navigating IOS command lines, 323-326
NBP (Name Binding Protocol), 95, 102-103
NCP (NetWare Core Protocol), 107
  IPX, 118
NCP negotiation, PPP, 248
neighbor command, 505
neighbor notification, AM (Token Ring management), 172
neighbor [ip address] command, 505
neighbor [ip address] ebgp-multihop command, 541
neighbor [ip address] filter-list [acl number] weight [0-65535] command, 544
neighbor [ip address] remote-as [as number] command, 543
neighbor [ip address] soft-reconfiguration inbound command, 541
neighbor [ip address] version 4 command, 541
neighbor [ip address] [priority (0-255)] command, 530

neighbor [ip-address] remote-as [as number] command, 541

neighbor [peer-group name] filter-list [acl] [in|out] weight command, 544

neighbor [peer-group name] peer-group command, 544

NetBIOS (Network Basic Input/Output System), 107, 118-120

NetWare, 110
- IPX. *See* IPX

NetWare architecture model, 108

NetWare Core Protocol. *See* NCP

NetWare Link State Protocol (NLSP), 110

NetWare Remote Procedure Call (NRPC), 107

NetWare RPC, IPX, 118

NetWare Shell (NWS), 107
- IPX, 117

network (Layer 3)
- AppleTalk, DDP, 94-97
- OSI-RM, 30

Network Address, SAP message format, 115

Network Address Translation. *See* NAT

network addresses, 56
- IPX, 111

network addressing, 16

network analysis, 560
- baselining, 573-574
  - *network errors. See network errors*
  - *network throughput, 578-579*
  - *utilization, 574-575*
- BERTs, 565
- documentation, 561-563
- ping, 565-567
  - *performance monitoring tools, 569-572*
  - *ping command, 568*
- protocol analyzers, 563-564
- TDRs, 564-565
- tools, 561
- traceroute, 572

network baselining. *See* baselining

Network Basic Input/Output System. *See* NetBIOS

network broadcast addresses, 56

network broadcasts, composite bandwidth, 577

network cabling topologies, 8

network communication protocols, datagrams, 16

network errors, 579-580
- Ethernet, 581-582
- retransmission errors, 588
- T-carrier errors, 586-588
- Token Ring, 582
  - *hard errors, 584-586*
  - *soft errors, 582-584*

network loopbacks, 56

network management, 591
- ATM layer, 240-241
- functions of, 592-593
- NMS, architecture, 593-594
- OSI network management model, 595-597
- SNMP. *See* SNMP

Network Management Consoles (NMC), 594

Network Management Protocol (NMP), 594

Network Management System, 592-594

network masks, OSPF, 516

network operating systems, 86, 107

network protocol addresses, 11
- routers, 21

network segmentation, routers, 22

network switches. *See* switches

Network Termination type 1 (NT1), CPE equipment, 224

Network Termination type 2 (NT2), CPE equipment, 224

network throughput, 578-579

Network Time Protocol. *See* NTP

network topologies, 8-10

network topology maps, 562

network virtual terminal service, 78

network visible entities, 86, 102

network [ip address | ip network number] subcommand, 509

network-layer host table group (Group 14), RMON II, 625

network-layer matrix table group (Group 15), RMON II, 625

network-specific broadcasts, DDP, 97

network-to-network paradigm, 382

network-wide broadcasts, DDP, 97

networks, 2
- internetworks, 21
- LANs. *See* LANs
- MANs. *See* MANs
- OSPF, 406-408
- WANs. *See* WANs

new-mode IOS authentication
- AAA, 366, 371
- accounting, configuring, 375-376
- authorization, 374
- RADIUS, 371-372

news, syslog facility definitions, 362

NIC, jabbering, 582

NISPs (National Internet Service Providers), 65

NLSP (NetWare Link State Protocol), IPX, 110

NMC (Network Management Consoles), 594

NMP (Network Management Protocol), 594

NMS (Network Management System), 592-594

no auto-summary command, 503, 509
- OSPF, 516

no debug all command, 514

no exec command, 342

no ip redirects command, 492

no ip source-route command, 499

no ip unreachables command, 492

no shutdown command, 459

Node Address, SAP message format, 115

node addresses, 11
- IPX, 111

nodes, addressing AppleTalk, 88-89

noise, 7

non-contention-based access, LAN transmission protocols, 15

non-isolating soft errors (NSE), 582

non-linear PCM algorithms, 198

non-repeaters, SNMP messages, 611

Non-Return-to-Zero (NRZ), 129

non-volatile random access memory, 308

nonbroadcast multi-access networks, OSPF, 406

nonbroadcast networks, OSPF, 530

Normal Response Mode (NRM), 245

North American Numbering Plan (NANP), 194

NOS (network operating systems), 86, 107

notice options, accounting, 376

NOTIFICATION messages, BGP-4, 418

notify command, 341

Novell, 107
- NetBIOS. *See* NetBIOS

Novell NOS, IPX, 107

NRM (Normal Response Mode), 245

NRPC (NetWare Remote Procedure Call), 107

NRZ (Non-Return-to-Zero), 129

NSE (non-isolating soft errors), 582

NSSA (not so stubby area), 526

NT
- printers, 342
- syslog, 361

NT1 (Network Termination type 1), CPE equipment, 224-225

NT2 (Network Termination type 2), CPE equipment, 224

NTN (National Terminal Number), 232

nTOP, SNMP-based management tools, 613

NTP (Network Time Protocol), 356
- Cisco 7x00 series routers, 358
- IOS, 358

NTP server, relationships between routers, 358

ntp update calendar command, 358

NULL realm, RADIUS, 372

number ranges, ACLs, 423

NVE (network visible entities), 86, 102

NVRAM (non-volatile random access memory), Cisco routers, 308

NVT (network virtual service), Telnet, 78

NWLink, IPX, 107

NWS (NetWare Shell), 107

O

object, 102

octet-synchrous HDLC, PPP, 249

octets, IP address space, 43

ODI (Open Datalink Interface), 107

offset-list [acl] [in|out] [metric] [interface] command, 504

old-mode
- IOS authentication, 368-369
- TACACS, IOS authentication, 365

OMAP (Operations, Maintenance, Administration, and Provisioning), SS7, 219

OPEN messages, BGP-4, 418

Open Shortest Path First. *See* OSPF

Open System Interconnection. *See* OSI

Open Systems Interconnect Reference Model), 23

Open Transport SNMP, 619

OpenView (Hewlett-Packard), SNMP-based management tools, 611

operating parameters, Ethernet MAC, 143-144

Operation, SAP message format, 114

operational structure, AppleTalk, 86

Operations, Maintenance, Administration, and Provisioning (OMAP), SS7, 219

operator/port-value pair, 430

operators, extended IP ACLs, 429

optical fiber cable, 5

options
- EIGRP, configuring, 510-511
- IP headers, 42

oscillation, AC, 7

OSI (Open System Interconnection), 595
- interdomain routing, 381
- intradomain routing, 381

OSI model, end-to-end networked communication, 26

OSI network management model, 595-597

OSI standards, 24

OSI-based protocol suites, 24

OSI-RM (Open Systems Interconnect Reference Model), 23, 26
- datagrams, 27
- European Union, 24
- ISO protocol suite, 23
- Layer 4 (transport), connectionless transports, 29
- layers, 27
  - *Layer 1 (physical), 32-33*
  - *Layer 2 (data link), 31-32*
  - *Layer 3 (network), 30*

Layer 4 (transport), 28-30
Layer 5 (session), 28
Layer 6 (presentation), 28
Layer 7 (application), 27
origins of, 23-25
U.S. government, 24

OSPF (Open Shortest Path First)
address summarization, 517, 524
adjacencies, 409
adjusting, 527-528
administrative distances, adjusting, 524-525
advantages of, 403
area hierarchy, 404-405
authentication, 528-529
backup designated router election, 529
backup designated routers, 409
bandwidth, 526
configuration subprocess commands, 515
configuring, 516-517, 519-520
route announcements, 520-523
cost, 525
cost metric, 413
DDR networks, 531
debugging, 533
designated router election, 529
designated routers, 409-410
Dijkstra's algorithm, 387
display commands, 514
distribution lists, filtering, 552-555
interface bit values, 525
interface configuration commands, 514
LSAs, 412
messages, 410-411
metrics, 524-525
monitoring commands, 532-533
multiple processes, 516
multipoint frame-relay interfaces, 530
network masks, 516
networks, 406-408
nonbroadcast networks, 530
path types, 413
point-to-multipoint frame relay interfaces, 530
point-to-point interfaces, 530
route distribution, 549
route redistribution, 550
route types, 525-526
router designations, 406
routing tables, building, 412
stub networks, 526-527
summarization, 516
value of today, 414

OSPF NSSA Type 1 (N1), 526

OSPF NSSA Type 2 (N2), 526

OSPF routers, 515-516

outbound filters, ACLs, 424

output filters, 553

overwrite-network, configuration EXEC mode, 320

P

P (power), 6

packet assembler/disassembler (PAD), 230

packet filtering, ACLs, 424

packet forwarding loops, transparent bridges, 259-261

Packet Internet Groper Application. *See* PING

Packet Internet Groper. *See* ping

Packet Layer Protocol (PLP), X.25, 231

Packet Length, IPX message format, 111

Packet over SONET (POS), 479

packet sniffer applications, promiscuous mode, 258

packet sniffers, port mirroring, 273-274

packet switching, 10-11

packet tagging, 286

Packet Type, IPX message format, 112

packet-capture group (Group 8), RMON I, 624

Packet-Switched Networks. *See* PSNs

packets
AARP, 90
control packets, ADSP, 105
TCP, 76

tickle packets, PAP, 106
transparent bridges, 259

packets per second (PPS), 277, 574

PAD (Packet Assembler/Disassembler), 230

Padding, IP headers, 42

Palo Alto Research Center, 137

PAM5 (Pulse Amplitude Modulation 5), 167

PAP
authentication, 456
tickle packets, 106

PAP (Password Authentication Protocol), 455

PAP (Printer-Access Protocol), 106

parameters, Ethernet MAC, 143-144

partitions, IOS, 352-355

PAS (Port Address Support), 274

pass-through autonomous system routing, BGP, 538

passenger protocol, tunnel interface, 450

passive hubs, 18

passive-interface command, 509, 528

passive-interface [interface] command, 500

Password Authentication Protocol. *See* PAP

PAT (Port Address Translation), 446

path discovery, source route bridges, 263

path types, OSPF, 413

Payload Type (PT), ATM cell formatting, 239

PC serial COM ports, terminal emulation software, 312

PCL, Gigabit Ethernet, 165

PCM (Pulse Code Modulation), analog-to-digital conversion, 198-199

PCMCIA flash cards, 328
advantages of, 329
IOS, partitions, 354-355

peak utilization rate, 575

peer default ip address pool dhcp command, 344

peer-groups, BGP, 544

peer-to-peer communication, Layer 1 (physical), 33

peers, IBGP, 545

performance management, OSI network management model, 596-597

performance monitoring tools, ping, 569-572

performance tuning, 560. *See also* network analysis

Phase 1, AppleTalk, 87
DDP datagram headers, 95

Phase 2, AppleTalk, 87
DDP datagram headers, 95
zones, 104

PhoneNet, 91-92

PHY (Physical Layer Protocol), 185
100Base-FX PHY, 160-161
100Base-T4 PHY, 160
FDDI specification, 188
Token Ring, 179

PHY implementations, 128
Token Ring, 180-183

physical (Layer 1), OSI-RM, 32-33

physical layer, ATM, 236-238

Physical Layer Medium, 237

Physical Layer Protocol. *See* PHY

physical layer signaling (PLS), 148

physical medium attachment. *See* PMA

PID (process ID), 508

pinFET (pin field effect transistor), 5

PING (Packet Internet Groper Application), 71

ping, 565-567
commands, 568
ICMP messages, 565-567
performance monitoring tools, 569-572

ping (Packet Internet Groper), 565

ping ? command, 567

ping [hostname/IP address command, 568

placing calls, PSTN, 193-194

plain old telephone service (POTS), 192

plesiochronous multiplexing, 210

PLP (Packet Layer Protocol), X.25, 231

PLS (physical layer signaling), 148

PMA (physical medium attachment), 148
- Gigabit Ethernet, 166

PMD (Physical Layer Medium), 185
- FDDI specification, 189-190
- Gigabit Ethernet, 166

PMD sublayer, physical layer (ATM), 237

point-to-multipoint Frame Relay interfaces, OSPF, 530

point-to-multipoint topologies, Frame Relay, 233

point-to-point Frame Relay topologies, 470

point-to-point interfaces, OSPF, 530

point-to-point networks, OSPF, 406

point-to-point topologies, Frame Relay, 233

point-to-point tunneling, 450

point-to-point VC topology, Frame Relay, 471-473

poison reverse, route poisoning (RIP), 400-401

policy routing, 439-441

Port Address Support (PAS), 274

Port Address Translation (PAT), 446

port addresses, IPX, 111

port configurations, switch ports (Layer 2 LAN switches), 270

port flooding, transparent bridges, 259

port mirroring
- packet sniffers, 273-274
- switch ports, Layer 2 LAN switches, 273-274

port trunking, 271
- switch ports, Layer 2 LAN switches, 271-273

port-based VLANs, Layer 2 LAN switching services for Layer 3, 283-284

ports
- IPX datagrams, 110
- TCP, 74-75
- UDP, 77

POS (Packet over SONET), 479

PostScript, 106

POTS (plain old telephone service), 192

power (P), 6

PPP
- authentication, 456
- data-link framing, 247-250
- encapsulation, 455
- information, displaying, 457

PPP link negotiation, 248

ppp multilink command, 460

PPS (Packets per Second), 277, 574

PRAM ZIP, 104

presentation (Layer 6)
- AppleTalk, 103
  - *AFP, 106*
- OSI-RM, 28

PRG (Ring Purge), MAC frame types, 178

PRI (Primary Rate Interface), 221, 226
- addresses, 228

PRI interfaces, ISDN, 457

pricing
- ADSL, 221
- ISDN, 220
- T1, 204

printer [name] [line #] [option] command, 342

Printer-Access Protocol. ***See*** **PAP**

printers, 342

privilege level command, 341

probe configuration group (Group 19), RMON II, 625

probe messages, AARP, 90

problems
- identifying, 589-590
- with ISDN, 220
- with NAT, 449
- with transparent bridges, 259-261

process ID
- BGP, 540
- IGPs, 500

process ID (PID), 508

promiscuous mode, transparent bridges, 258

Protocol, IP headers, 41

protocol analyzers, network analysis, 563-564

protocol directory group (Group 11), RMON II, 625

protocol distribution group (Group 12), RMON II, 625

protocol match operators, IPX extended ACL, 438

protocol reference models, 23
- OSI-RM. *See* OSI-RM

protocol-specific packet filtering, ACLs, 424

protocols
- ADSP, 105-106
- AEP, 101
- AFP, 106
- ASP, 106
- ATP, 103
- AURP, 101
- BACP, 247
- BOOTP. *See* BOOTP
- CHAP, 455
- choosing the right one, 485
- connection-oriented transport protocols, Layer 4 (OSI-RM), 29
- connectionless transport protocols, Layer 4 (OSI-RM), 29
- control plane protocols, 223
- D channel Layer 3 protocol, 227
- data-link protocol, 245
- DDP, 94
- DHCP. *See* DHCP
- dynamic routing protocols, 65, 383
  - *DVPs, 386*
  - *LSPs, 387*
- dynamic routing protocols. *See* dynamic routing protocols
- EIGRP, 110
- ELAP, 92
- ERPs. *See* ERPs
- Ethernet. *See* Ethernet
- FLAP, 94
- FTP, 78
- IOS supports, 314
- IP routing protocols. *See* IP routing protocols
- IP. *See* IP
- IPX. *See* IPX
- LAN transmission protocols, 296
- LLAP, 92
- NBP, 95, 102-103
- NetWare, upper layer protocols, 117
- NLSP, 110
- NMP, 594
- Novell, 107
- NTP, 356
- PAP, 106, 455
- PLP, 231
- RADIUS. *See* RADIUS
- RARP. *See* RARP
- routing protocols
  - *deciding to use routing protocols, 382-383*
  - *summarization, 392*
- RTMP, 98-100
- SMTP. *See* SMTP
- SNMP. *See* SNMP
- SS7, 218-219
- SSCOP, 243
- TCP. *See* TCP
- TCP/IP, 396
  - *IGPs. See IGPs*
- TFTP. *See* TFTP
- TLAP, 93
- transmission protocols, 10-12
  - *LAN transmission protocols. See LAN transmission protocols*
  - *WAN transmission protocols, 12-13*

transport control protocols, Layer 2 OSI-RM, 31
UDP. *See* UDP
ULP, 124
user plane protocols, ISDN, 223
WAN transmission protocols, 297
ZIP. *See* ZIP

protocols suites, TCP/IP, 24-25

provisional addresses, AppleTalk, 89

Proxy ARP, IP interactions with Layer 2, 40

PSEs, X.25, 231

pseudorandom patterns, BERTs, 565

PSH, TCP connections, 76

PSNs (Packet-Switched Networks), 229-230
advantages of, 230
ATM, 235
AAL, 241-242
architecture, 236
ATM layer, 238-241
physical layer, 237-238
SALL, 243
Frame Relay, 232-233, 235
congestion control, 234
frames, 234-235
PVC topologies, 233
X.25, 230-232

PSTN (Public Switched Telephone Network), 191-192, 229
AT&T, 195
FDM, 197
hierarchy, 194-196
history of, 192-193
placing calls, 193-194
transport, 196-197

PSTN switches, 254

PSTN/ISUP trunks, SS7, 216

PT (Payload Type), ATM cell formatting, 239

public IP address space, 56

pulse amplitude, 5

Pulse Amplitude Modulation 5 (PAM5), 167

Pulse Code Modulation (PCM), 198

pulse frequency, 5

PVC topologies, Frame Relay, 233

Q

QoS contracts, ATM layer, 240

quantizing analog-to-digital conversion, 199

QuickDraw, 106

R

R (resistance), 6

R interface, ISDN demarcation reference points, 224

radio frequency interference (RFI), 129

radio transmission, connectionless media, 6

RADIUS (Remote Authentication Dial-In User Service), 366-367
AAA, 366
accounting, 375
dial-in users, 373
exec accounting record, 375
login authentication, 371
new-mode IOS authentication, 371-372
realms, 372-373
session users, 373
users, 372

RARP (Reverse Address Resolution Protocol), 38, 81
IP, interactions with Layer 2, 38-39

reachability information, BGP, 418

read-only memory (ROM), 308

realms, RADIUS, 372-373

receiver congestion error, Token Ring soft errors, 583

reception window size, ADSP, 106

redistribute [source] [process id] [metrics] command, 549

redistribution, 547-548

redistribution command, 549

reference models, TCP/IP, 25

reflectors, 545-546

reflexive lists, ACLs, 424

Regional Internet Service Providers (RISPs), 65

releases of IOS, 315-316

reliable transport, TCP, 73

REM (Ring Error Monitor), 173, 582

Remote Authentication Dial-In User Service. *See* RADIUS

remote delivery, IP datagrams, 57

Remote Monitoring. *See* RMON

repeaters, 17-18
- 100Base-T, 161-162
- Gigabit Ethernet, 168-169
- inter-repeater links, 17

Report Neighbor Notification Incomplete (RNNI), MAC frame types, 179

Report Soft Error (RSE), MAC frame types, 179

request messages, AARP, 90

requests
- RIP, 398
- RTMP messages, 100

resistance, 6

Resource Reservation Protocol, 293

response messages, AARP, 90

responses
- RIP, 398
- RTMP messages, 100

retransmission, network errors, 588

Reverse Address Resolution Protocol. *See* RARP

RFF routers, upgrading (IOS), 350

RFI (radio frequency interference), 129

RFR routers, upgrading (IOS), 351

Ring Error Monitor, 173, 582

ring insertion process (Token Ring), 174-175

ring management (Token Ring), 172-174

Ring Purge (PRG), MAC frame types, 178

ring recovery, AM (Token Ring management), 172

Ring Station Initialization (RSI), MAC frame types, 179

ring timing maintenance, AM (Token Ring management), 173

ring topologies, 9-10
- non-contention-based access, LAN transmission protocols, 15

RIP (Routing Information Protocol)
- administrative distances, adjusting, 504
- authentication, configuring, 505
- control commands, 502
- convergence, 399
- distribution lists, filtering, 552
- DVPs, 397
- global configuration commands, 501
- global interface subprocess commands, 502
- global router subprocess commands, 501
- hops, 397
- IP route table, 503
- IPX, 110, 116-117
- messages, 116
- metrics, adjusting, 504
- process, examining, 506
- process timers, adjusting, 504-505
- route distribution, 549
- route poisoning, 400-401
- router subprocess commands, neighbor [ip address], 505
- routing messages, 398-400
- routing subprocess commands, timers basic [update] [invalid] [holddown] [flush], 504
- subprocess commands, no auto-summary, 503
- triggered updates, 399
- value of today, 401
- version 1, 396, 502-503

version 2, 396, 502-503
versus IGRP, 507

RISPs (Regional Internet service providers), 65

RJ-45-M to DB-9-F pin translation, 311

RJ-45-M to RS-232C-M pin translation, 311

RJ45-M Cisco serial cables, 310

RJ45-M to DB25 pin translation, 311

RMON (Remote Monitoring), SNMP, 623

RMON I, SNMP, 624-625

RMON II, SNMP, 625-626

RMS (root-mean-square), 6

RNNI (Report Neighbor Notification Incomplete), MAC frame types, 179

ROM (read-only memory), Cisco routers, 308

rommon (ROM monitor), 346
conf-reg, 348

root bridge election process, 261-262

root-mean-square (RMS), 6

roots, OSPF, 405

route announcements, configuring OSPF, 520-523

route data request, RTMP messages, 100

route distribution, 549

route diversity, route metrics, 384

route metrics, 384-385

route poisoning, RIP, 400-401

route redistribution, 393, 547-549
controlling, 551
EIGRP, 549
IGRP, 549
OSPF, 550
OSPF/BGP, default route distribution, 551

route redundancy, route metrics, 384

Route Switch Processors (RSPs), 307

route table display, IOS, 487

route types, OSPF, 525-526

route-map operators, policy routing, 441

route-maps, filtering, 555-556

router configuration files, Cisco routers, 316

router configuration subcommands
distance [1-255], 512
network [ip address | ip network number], 509
no auto-summary, 509
timers basic [update] [invalid] [holddown] [flush], 508
traffic-share [balanced | min], 512
variance [multiplier], 512

router designations, OSPF, 406

router eigrp [process id] command, 508

router igrp [process id] command, 508

router ospf [local process id] command, 519

router subcommands, distance [10-255] (RIP), 504

router subprocess commands, neighbor [ip address], 505

router [protocol] [process id] command, 500

routers, 21-22
AppleTalk, ZIP, 104
AppleTalk routing tables, seed routers, 99
Cisco routers. *See* Cisco routers
designated routers, OSPF, 409-410
exterior routers, AURP, 101
functions of, 21-22
IBGP, 544
internetworks, AppleTalk, 98
IP datagrams, delivering, 60-61
network protocol addresses, 21
network segmentation, 22
NTP servers, relationships between, 358
OSPF, 515
designated routers, 409-410

routing, 70
classless routing, configuring, 498
datagrams, Layer 3 OSI-RM, 30
default routes, configuring, 493-494
dynamic routing, 70
dynamic routing protocols, 65
inter-autonomous system routing, BGP, 538
interdomain routing, 65, 69-70
Internet, 65
internetwork routing, 381
intra-autonomous system routing, BGP, 538
intradomain, 65
intranetwork routing, 381
IP datagrams, 57
OSI, 381
pass-through autonomous system routing, BGP, 538
SIN routing, 538
static routing, 393
configuring, 494-498
strengths of, 394
weaknesses of, 395-396
static routing. *See* static routing
statistic routing, 59
strategies, 70-71
versus bridging, 264-265

routing configuration subcommands, distribute-list [acl number] [in|out] [protocol|interface], 552

routing configuration subprocess commands, passive-interface [interface], 500

routing domains, 65

Routing Information Protocol. *See* RIP

routing loops
discontinuous subnets, 392
redistribution, 548

routing messages, RIP, 398-400

routing metrics, 99
dynamic routing protocols, 385

routing protocol configuration subcommands, redistribute [source] [process id] [metrics], 549

routing protocols
bandwidth, 395
deciding to use routing protocols, 382-383
dynamic routing protocols, 383
IP. See IP routing protocols, 388
summarization, 392

routing seeds, AppleTalk, 99

routing subprocess commands, timers basic [update] [invalid] [holddown] [flush], 504

routing table lookups, 60-61

Routing Table Maintenance Protocol (RTMP), 98-100

routing tables, 58
address space, 390-391
AppleTalk, seed routers, 99
binary routing tables, 60
commands, 339-340
default gateways, 58
displaying, 488
dynamic routing tables, 60
flushing, 491
IP addresses, 45
IP datagrams, 60
match lookups, 60-61
OSPF, building, 412
viewing, 339-340

routing types, BGP, 538

RPS, Token Ring management, 173

RSE (Report Soft Error), MAC frame types, 179

RSI (Ring Station Initialization), MAC frame types, 179

RSPs (Route Switch Processors), 307

RSVP (Resource Reservation Protocol), 293

RTMP (Routing Table Maintenance Protocol), 98-100

RTMP messages, 100

runts, 582

S

S interface, ISDN demarcation reference points, 224

SAC (Single-Attachment Concentrator), 190

SALL (Signaling AAL), ATM, 243

sampling analog-to-digital conversion, 199

SANs (Storage Area Networks), 164

SAP (Service Advertising Protocol), 113
- IPX, 113
- message format, 114-115
- updates, 116

SAPs (Service Access Points), 802.2 standard, 93

SAR (segmentation and reassemble), 241

SAR footers, 242

SAR headers, 242

SAS (Static Assigned Sockets), socket addresses, 95

SAT (Source Address Table), 258
- transparent bridges, 259

SC-type connectors, fiber optic cable, 136

SCCP (Signaling Connection Control Part), SS7, 219

Scotty, SNMP-based management tools, 613

SCP (Service Control Point), SS7, 217

SDH, 212-214
- transmission rates, 212

SDH (Synchronous Digital Hierarchy), 212

SDLC (Synchronous Data-Link Control), 245

security
- ACLs, filters, 424-425
- dial-in pools, 343
- IOS, 343
- SNMP, 598
- static routing, 394

security management, OSI network management model, 596

seed routers, AppleTalk routing tables, 99

segmentation and reassemble (SAR), 241

SEQUENCE data types, ASN.1, 605

Sequence Number, TCP headers, 76

Sequenced Packet Exchange, 117

serial interfaces, configuring (ATM), 475-477

Server Message Block (SMB), 119

Server Name, SAP message format, 115

Service Access Points, 93

Service Advertising Protocol. *See* SAP

Service Control Point (SCP), 217

service password-encryption command, 369

service ports, extended ACLs, 430-431

Service Switching Point (SSP), SS7, 217

service timestamps debug command, 359

service timestamps log command, 359

Service Type, SAP message format, 115

service types, LLC, 127-128

[service]-server key [word] command, 371

[service]-server host command, 371

service-modules, 454

Service-Specific Connection-Oriented Protocol (SSCOP), 243

Service-Specific Coordination Function (SSCF), 243

servtab, 367

session (Layer 5)
AppleTalk, 103
ADSP, 105-106
ASP, 106
ZIP, 104
OSI-RM, 28
session service, NetBIOS, 119-120
session-limit command, 341
session-timeout command, 341
SetRequest command, 599
setup application, IOS, 316
shifts, utilization, 575
ships in the night routing, 538
shortest path tree (SPT), 412
show, privileged EXEC command, 319
show accounting command, 376
show appletalk interface [interface slot|port] command, AppleTalk, 338
show appletalk route command, 339
show appletalk zone command, 339
show arp command, 489
show atm interface atm [slot|port] command, 479
show atm map command, 478
show atm traffic command, 479
show atm vc command, 479
show bootflash command, 355
show calendar, 358
show clock, 358
show controllers bri [slot|port] command, 465
show crypto algorithms command, 454
show crypto engine brief command, 454
show crypto engine connections active command, 454
show crypto map command, 454
show crypto mypubkey command, 454
show crypto pubkey [name|serial number] command, 454
show dxi map command, 478
show dxi pvc command, 478
show frame-relay lmi command, 474
show frame-relay map command, 474
show frame-relay pvc command, 474
show frame-relay traffic command, 474
show hardware, privileged EXEC command, 347
show hardware command, 332
show interface command, 463
show interface tunnel [number] command, 453
show interface [slot|port] command, 568
show interface [type|port|slot] command, 463
show interface [type|slot|port] command, 499, 549
show interfaces accounting command, 332
show interfaces command, 332
show ip accounting access-violations command, 439
show ip accounting command, 439
show ip bgp command, 546
show ip bgp neighbors command, 547
show ip bgp peer-group command, 547
show ip bgp summary command, 546
show ip eigrp interfaces command, 513
show ip eigrp neighbors [options] command, 513
show ip eigrp topology command, 514
show ip eigrp traffic command, 513
show ip masks [address] command, 489
show ip nat statistics command, 449
show ip nat translations command, 449
show ip ospf command, 523, 527

show ip ospf interfaces [type | slot | port] command, 532

show ip ospf neighbor command, 532-533

show ip ospf virtual-links command, 533

show ip ospf [local process id] command, 532

show ip protocol command, 490, 510-512

show ip protocols summary command, 490

show ip route, privileged EXEC command, 339

show ip route command, 488
  OSPF, 526

show ip route connected command, 488-489

show ipx interface [interface slot | port] command, IPX, 338

show ipx route command, 340

show isdn active command, 465

show isdn history command, 465

show isdn status command, 459, 465

show logging command, 360

show memory command, 332

show memory failures alloc command, 332

show ntp associations, privileged EXEC command, 358

show ntp status, privileged EXEC command, 358

show ppp multilink command, 457

show processes cpu command, 332

show protocols command, 332

show running-config command, 340, 498

show version, privileged EXEC command, 347

show version command, 332

show [filesystem]:[partition]:, privileged EXEC command, 328

show [protocol] access-list [acl number] command, 439

show [protocol] route command, 340

signal errors, WAN errors, 588

signal loss error, Token Ring hard errors, 585

Signal Quality Error (SQE), 141

signaling, 128-129

Signaling AAL (SALL), ATM, 243

Signaling Connection Control Part (SCCP), SS7, 219

Signaling System 7. *See* SS7

signaling systems, 215-216
  SS7, 216
    components of, 216-217
    protocols, 218-219

Signaling Transfer Point (STP), SS7, 217

Simple Mail Transfer Protocol, 79

Simple Monitoring Gateway Protocol (SMGP), 597

Simple Network Management Protocol. *See* SNMP

SIN (ships in the night) routing, 538

SIN routing, inter-autonomous BGP, 538-539

sine waves, AC, 7

Single-Attachment Concentrator (SAC), 190

single-mode PMD (SM-PMD), 189

size option, IOS command lines, 325

skills, troubleshooting, 588-590

[slot:], flash cards, 328

slot-time, 142

SM-PMD (single-mode PMD), 189

SMAP, SNMP-based management tools, 613

SMB (Server Message Block), 119

SMGP (Simple Monitoring Gateway Protocol), 597

SMI (Structure of Management Information), 600

SMI MIB, ASN.1, 602

SMI-defined application-wide data types, 604

SMP (Standby Monitor Present), MAC frame types, 178

SMs (standby monitors), Token Ring, 173

SMT (Station Management), 185
FDDI specification, 188

SMTP (Simple Mail Transfer Protocol), 79

SMURF attacks, 434

SNAP (Subnetwork Access Protocol), 146

SNMP (Simple Network Management Protocol)
agents, UNIX, 622
Cisco routers, 309
commands, 599-600
components, 600
ASN.1, 601
ASN.1. See ASN.1
BER, 601
MIB structure. See MIB
SMI, 600
configuring
on Cisco IOS, 622-623
on Macintoshes, 619-622
management tools, 611-613
messages, 609-611
MIB, 597, 600
RMON, 623
RMON I, 624-625
RMON II, 625-626
security, 598
Trap messages, 599
version 2, 598
version 3, 598-599

SNMP administrator, 619

SNMP Watcher, SNMP-based management tools, 613

SNMPc (Castle Rock), SNMP-based management tools, 612

Socket Address, SAP message format, 115

socket addresses, 95

socket listeners, 95

socket match operators, IPX extended ACL, 438

socket services, AppleTalk, 95

sockets, AppleTalk, 95

soft errors, Token Ring, 582-584

SONET, SDH, 212-214
transmission rates, 212

Source Address, IP headers, 42

Source Address Table (SAT), 258

Source Network, IPX message format, 112

Source Node, IPX message format, 112

Source Port
IPX message format, 112
TCP headers, 76

source route bridges, 257, 263
path discovery, 263

source route bridging (SRB), 183

source routing, IP, 499

Source Service Access Point (SSAP), 127

spanning tree algorithms, transparent bridges, 261-262, 266

Spectrum (Cabletron), SNMP-based management tools, 612

split horizon, 99
route poisoning, RIP, 400

SPT (shortest path tree), 412
routing tables, creating, 412-413

SPX (Sequenced Packet Exchange), IPX, 117

SPX datagrams, 117

SQE (Signal Quality Error), 141

SRB (source route bridging), IBM Token Ring, 183

SS7 (Signaling System 7), 216
components of, 216-217
protocols, 218-219
reference model, 218

SSAP (Source Service Access Point), 127

SSCF (Service-Specific Coordination Function, 243

SSCOP (Service-Specific Connection-Oriented Protocol), 243

SSP (Service switching Point), SS7, 217

ST-type connectors, fiber optic cable, 136

Standby Monitor Present (SMP), MAC frame types, 178

standby monitors (SMs), 173

standby priority [0-255] command, 444

standby [group identifier] authentication [password] command, 445

standby [group identifier] track [interface type:slot:port] command, 444

standby [group id] prempt command, 444

star topologies, 8-9

start-stop notice option, accounting, 376

startup and shutdown procedures, IPX RIP routes, 116-117

startup configuration, Cisco routers, 316

Static Assigned Sockets (SAS), 95

static routes
administrative distances, 486
redistribution, 548

static routing, 393, 484
configuring, 494-498
control commands, 492
display commands, 492
global configuration commands, 492
interface configuration subcommands, 492
managing, 492
security, 394
strengths of, 394
weaknesses of, 395-396

static tables, loading, 496

Station Management. *See* SMT

statistic routing, 59

statistics group (Group 1), RMON I, 624

status and control, NetBIOS, 119

STDM, 199
multiplexers, 200-201

stop-only notice option, accounting, 376

Storage Area Networks (SANs), 164

store-and-forward method, switch forwarding methodologies (Layer 2 LAN switches), 281

STP (Signaling Transfer Point), SS7, 217

strategies for routing, 70-71

structure of MIB, 606-608

Structure of Management Information, 600

structured ASN.1 data types, 604

stub ASs, ERPs, 415

stub networks, OSPF, 526-527

subcommand modes, global mode, 322

subdomains, 80

subnet masks
CIDR, versus classful masks, 53
function of, 45
IP addresses, 44

subnets
discontinuous subnets, 392
EIGRP, 509

subnets command, 550

subnetting
address space, 334
Class C space with CIDR, 54

classful addressing, VLSM, 54-55
classful subnetting, 389
guidelines for large address spaces, 47
IP addresses, 46-47
classful addressing, 48-50
VLSM, 389

Subnetwork Access Protocol, 146

summarization
OSPF, 516
routing protocols, 392

supernetting Class C addresses, 51

supervisory signals, 216

supplemental services, ISDN, 221

SVC topology, configuring (ATM), 477

switch backplanes, 252-254
ATM switches, 295
Layer 2 LAN switches, 268, 276
capacity, 279-280
FPS, 277-279
PPS, 277

switch forwarding methodologies, Layer 2 LAN switches, 280-282

switch ports, Layer 2 LAN switches, 269-270
ESD, 274, 276
flow control, 270-271
port configurations, 270
port mirroring, 273-274
port trunking, 271-273

switch types, ISDN TelCo carrier switch types, 458

switches
ATM switches. *See* ATM switches
bridges, 21
Ethernet, 139
fundamentals of, 254
hybrid switches, 281-282
Layer 2 LAN switches. *See* Layer 2 LAN switches
need for, 253-254
PSTN, 254

switching
evolution from bridging, 266-268
Layer 2 LAN protocols, 267-268

switching logic, Layer 2 LAN switches, 268

switching matrix, 254

Synchronous Data-Link Control (SDLC), 245

Synchronous Digital Hierarchy (SDH), 212

synchronous transmission, WAN transmission protocols, 12

SynOptics, Ethernet, 139

SYNTRAN (Synchronous Transmission), 211

syslog
NT, 361
syslog facility definitions, 362
trap logging, 362
UNIX, 361

syslog daemon, trap logging, 362

syslog facility definitions, 362

syslog files, 363

syslog level classification, logging messages, 360-361

syslog logging system, UNIX, 359

syslog-defined actions, 363

Sytec, NetBIOS, 118

T

T interface, ISDN demarcation reference points, 224

T-carriers
DS3 service over T3, 210-211
network errors, 586-588
SYNTRAN, 211

T-carrier digital transmission system, 198

T-carrier system, T1, 203-204
framing, 204-206
line coding, 206-207

T1, 203-204, 226
framing, 204-206
line coding, 206-207

TA (Terminal Adapter), CPE equipment, 223

TACACS (Terminal Access Controller Access Control System), 365-368
old-mode IOS authentication, 365, 369

TACACS+, 366-368
accounting, 375

TACP (Transaction Capabilities Applications Part), SS7, 219

tagged ASN.1 data types, 603

TCP (Transmission Control Protocol), 72-73
connections, 74-75
denial of service attacks, 432-434
packets, 76
ports, 74-75
services, 74

TCP headers, 76

TCP service ports, 430-431

TCP/IP (Transmission Control Protocol/Internet Protocol)
and UNIX, 37
client/server model, 36
ERPs. *See* ERPs
IETF, 37
IGPs. *See* IGPs
origins of, 36-37
protocol suites, 24-25
reference models, 25
static routing, 393
strengths of, 394
weaknesses of, 395-396

TCP/IP protocol suite
Layer 3 (network), OSI-RM, 30
UDP, 77

TDM, 199
multiplexers, 200-201

TDM digital transmission systems, ISDN, 222

TDM systems, 219

TDRs (Time Domain Reflectors), 564-565

TE1 (Terminal Equipment type 1), CPE equipment, 224

TE2 (Terminal Equipment type 2), CPE equipment, 223

Telcordia, SDH, 212

telecommunication services, ISDN, 221

Telecommunications Act of 1996, 195

Telecommunications Industry Association (TIA), 130

Telnet, 78
Cisco routers, 309
NVT, 78

telnet transparent command, 342

Terminal Access Controller Access Control System. *See* TACACS

Terminal Adapter (TA), CPE equipment, 223

terminal command, IOS, 325

terminal emulation software
C-Kermit, 313
PC serial COM ports, 312

Terminal Equipment type 1 (TE1), CPE equipment, 224

Terminal Equipment type 2 (TE2), CPE equipment, 223

terminal lines, configuring, 340-342

terminal monitor command, 332
privileged EXEC command, 359

TFEs (Time Frequency Errors), WAN errors, 588

TFSB (total frame size in bytes), 277

TFTP
copy command, 329-330
IOS, upgrading, 350-351

TFTP (Trivial File Transfer Protocol), 78

Thicknet, 150-152

TIA (Telecommunications Industry Association), 130

tickle packets, PAP, 106

Time Domain Reflectors, 564-565

Time Frequency Errors (TFEs), WAN errors, 588

Time To Live, 41

timers
IGRP, 508
RIP, adjusting, 504-505
[hold down], 504
[invalid], 504
[update], 504

timers basic [updaste] [invalid] [holddown] [flush] command, 504

timers basic [update] [invalid] [holddown] [flush] command, 508

timestamping, 359

Tivoli Systems, SNMP based management tools, 612

TLAP (TokenTalk Link-Access Protocol), 93

token passing bus topologies, non-contention-based access, 15

Token Ring, 267
bridging, 183
cabling system, 182
frame formats, 176-177
free tokens, 170
history of, 169-170
MAC frame types, 177-179
MAU, 181
network errors, 582
hard errors, 584-586
soft errors, 582-584
PHY implementations, 180-183
PHY interfaces, 179
ring insertion process, 174-175
ring management, 172-174
SMs, 173
traffic priority mechanisms, 171
TRT, 170

Token Ring extensions (Group 10), RMON I, 625

Token Ring Network Adapter Card (TR-NAC), 181

Token Ring technology, 18

Token Rotation Time (TRT), 170

TokenTalk, AppleTalk Phase 2 networks, 93

TokenTalk Link-Access Protocol (TLAP), 93

tools, network analysis, 561

topologies, 8. *See also* network topologies
Frame Relay, 467-468, 470
multipoint designations, 470
multipoint VC topology, 473-474
point-to-point, 470
point-to-point VC topology, 471-473

topology maps, 562

TOPS-20 command shell interface, IOS, 314

ToS (Type of Service), 41

total frame size in bytes (TFSB), 277

Total Length, IP headers, 41

TP-PMD (twisted-pair PMD), 189

TR-NAC (Token Ring Network Adapter Card), 181

traceroute, 572

traffic
ATM layer, 240-241
Token Ring, 171

traffic filters, ACLs, 429

traffic prioritization, 266
802.1q Layer 2 traffic prioritization, Layer 2 LAN switching services for Layer 3, 290-292

traffic-share [balanced|min] command, 512

Transaction Capabilities Applications Part (TCAP), SS7, 219

transaction time, 578

translation bridges, 22, 256-257

transmission (Layer 0), IPX, 108

Transmission Control Protocol. *See* TCP

transmission errors, WAN, 586

transmission media, 2, 132
broadband signaling, 128
connected media, 3-5
connectionless media, 6
fiber optic cable, 133-134
MLT-3, 129
telephone cable medium, 3
transmission signals, 6-7
voltage encoding techniques, 7

transmission protocol addresses, 11

transmission protocols, 10-12
LAN transmission protocols, 13
contention-based access, 13-15
non-contention-based access, 15
WAN transmission protocols, 12-13

transmission rates, SONET/SDH, 212

transmission signals, 6-7

Transmit Immediate Protocol (TXI), 183

transparent bridges, 255, 258
Ethernet, 257
packet forwarding loops, 259-261
packets, 259
port flooding, 259
problems with, 259-261
promiscuous mode, packet sniffer applications, 258
SAT, 259
spanning tree algorithms, 261-262, 266

transport, PSTN, 196-197

transport (Layer 2), IPX, 108

transport (Layer 4)
AppleTalk, 98
AEP, 101
ATP, 103
AURP, 101
NBP, 102-103
RTMP, 98-100
OSI-RM, 28-30

Transport Control, IPX message format, 111

transport control protocols, Layer 2 (data link), 31

transport layer (Layer 4), 72
TCP. *See* TCP
UDP. *See* UDP
ULP data, 73

transport protocol, tunnel interface, 450

transport [direction] [protocol | all] command, 341

Transport-1, 454-455

trap logging, 361-364

Trap messages, 599

triggered updates, RIP, 399

Trivial File Transfer Protocol (TFTP), 78

troubleshooting. 579. *See also* network analysis
developing troubleshooting skills, 588-590
IOS, 355-356
IP routing, 490
VLSM, 391

TRT (Token Rotation Time), 170

TTL (Time To Live), 41

tunnel interfaces, 450
AURP, 534
configuring, 535-536

tunneling, 450-451
encrypted GRE tunnels, 452-453
information, displaying, 453-454
point-to-point tunneling, 450
unencrypted GRE tunnels, 452
VPNs, 451

tunnels, 450

tuples
EIGRP, 507
NBP, 102
RTMP, 100
ZITs, 104

twisted-pair cable, 4-5

twisted-pair PMD (TP-PMD), 189

type, 102

Type 1, metrics (OSPF), 525

Type 1 external paths, OSPF, 413

Type 2 external paths, OSPF, 413

Type of Service (ToS), 41

U

U interface, ISDN demarcation reference points, 224

U.S. government, OSI-RM, 24

UART (Universal Asynchronous Receiver/Transmitter), 245

UBR (Unspecified Bit Rate), 241

UDP (User Datagram Protocol), 72, 77
- ports, 77

UDP service ports, 430-431

UHF (ultra-high frequency), 6

ULP (upper layer protocol), 124
- Data, Layer 4 (transport), 73
- datagrams, 38

unencrypted GRE, tunneling, 452

universal ASN.1 data types, 603

Universal Asynchronous Receiver/Transmitter (UART), 245

Universal Synchronous Asynchronous Receiver/Transmitter (USART), 245

Universal Time Coordinated (UTC), 357

UNIX
- and TCP/IP, 37
- commands, 372
- printers, 342
- SNMP agents, 622
- syslog, 361
- syslog logging system, 359

UNIX pager display, IOS command lines, 326

UNIX shell scripts, log management, 364-365

unregistered IP address space, 56

unshielded twisted-pair (UTP), 139

Unspecified Bit Rate (UBR), 241

UPDATE messages, BGP-4, 418

[update] timer, 504

updates
- EIGRP, 509
- IGRP, 509
- RTMP, 100
- SAP, 116

upgrading IOS, 350-352

UPI (user premise interfaces), 223

upper layer protocol. *See* ULP

upper-layer protocol-based VLANs, Layer 2 LAN switching services for Layer 3, 284-285

USART (Universal Synchronous Asynchronous Receiver/Transmitter), 245

usefulness of VLANs, Layer 2 LAN switching services for Layer 3, 289-290

user, syslog facility definitions, 362

User Datagram Protocol (UDP), 72

user EXEC mode, 317-318

user history group (Group 18), RMON II, 625

user plane, ATM, 236

user plane protocols, ISDN, 223

user premise interfaces (UPI), 223

users, RADIUS, 372-373

UTC (Universal Time Coordinated), 357

utilization, 574-575
- composite bandwidth, 575-577
  - *network broadcasts, 577*
- composite bandwidth utilization distribution, 578
- LAN protocols utilization baseline measurement guidelines, 576
- shifts, 575

UTP (unshielded twisted-pair), 139

uucp, syslog facility definitions, 362

V

V (voltage), 6

Variable Bit Rate (VBR), 241

Variable-Length Subnet Masks. *See* VLSM

variables, route metrics, 385

variance [multiplier] command, 512

VBR (Variable Bit Rate), 241

VCs (virtual circuits), 72

vector distance, DVPs, 386

verifying OSPF messages, 411

Versatile Interface Processors (VIPs), 307

versions
- of IOS, 315
- IP headers, 41

VGC (voice grade cable), 5

viewing routing tables, 339-340

VIPs (Versatile Interface Processors), 307

virtual circuit transport model, ATM, 235

virtual circuits (VCs), 72

virtual LANs. *See* VLANs

Virtual Path Identifier/Virtual Channel Identifier (VPI/VCI), ATM cell formatting, 239

Virtual Terminal Lines (VTYs), 317

VLANs (virtual LANs), 266
- Layer 2 LAN switching services for Layer 3, 282-283
  - *benefits of, 288-289*
  - *configuring, 285-286*
  - *IEEE 802.1q standards-based VLANs, 286-288*
  - *MAC address-based VLANs, 284*
  - *port-based VLANs, 283-284*
  - *upper-layer protocol-based VLANs, 284-285*
  - *usefulness of, 289-290*

VLSM (Variable-Length Subnet Masks), 54, 498
- address space, 389-390
- classful addressing, 54-55
- classless method, 389
- subnetting, 389
- summarization, 392
- troubleshooting, 391

voice grade cable (VGC), 5

voltage (V), 6

voltage encoding techniques, 7

VPI/VCI (Virtual Path Identifier/Virtual Channel Identifier), ATM cell formatting, 239

VPI/VCI translation, ATM layer, 240

VPNs, tunneling, 451

VT ports, old-mode IOS authentication, 369

VTY connectivity, interruptions, 491

VTYs (Virtual Terminal Lines), 317

W

wait-start notice option, accounting, 376

WAN transmission protocols, 12-13, 297

WAN transports, Transport-1, 454-455

WANs (wide area networks), 2
- error types, 587
- frame type transmission errors, 586
- protocol analyzers, 564
- telephone cable medium, 3

weaknesses of static routing (TCP/IP), 395-396

WECO (Western Electric), 193

weights, intra-autonomous BGP, 544

Western Electric (WECO), 193

wildcard masks, ACLs, 426

Window, TCP headers, 76

Windows 95/98, configuring SNMP agents, 616-619

Windows Networking, NetBIOS, 119

Windows NT, configuring SNMP agents, 613-615

workgroup switches, 268

write IOS commands, 320

X

X.200 standard. *See* OSI-RM

X.25, 230-232
PSEs, 231

X.25 network standards, 24

xmodem, 355-356

XNS (Xerox Network Systems), 107

Y

ymodem, 355-356

Z

Z (impedance), 6

Z-Term, 312

ZIP (Zone Information Protocol)
AppleTalk, 89
PRAM ZIP, 104
zone tables, 104

ZIP extended reply messages, 104

ZIP GetNetInfo, 104

ZIP GetNetInfoReply, 104

ZIP query messages, 104

ZIP response messages, 104

ZITs (Zone Information Tables), 104, 435

zone information distribution filtering, 435

zone list announcements, filtering, 437

zone-specific broadcasts, DDP, 97

zones, 86
AppleTalk, 104-105
tables, ZIP, 104

Networking Answers

Selected Networking Titles from New Riders Publishing/MTP

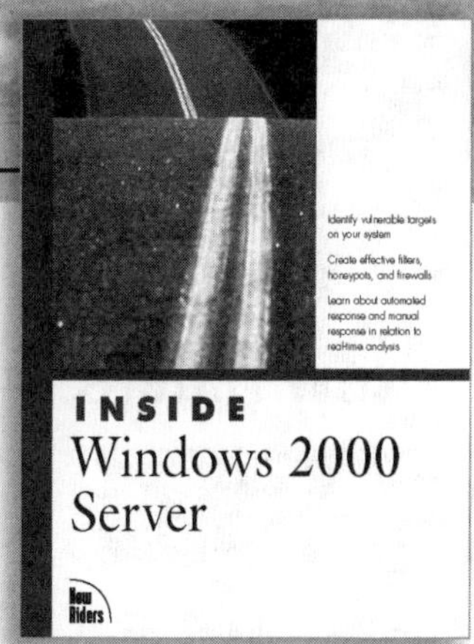

This is the updated edition of New Riders' best-selling *Inside Windows NT 4 Server.* Taking the author-driven, no-nonsense approach that we pioneered with our Windows NT *Landmark* books, New Riders proudly offers something unique for Windows 2000 administrators—an interesting and discriminating book on Windows 2000 Server, written by someone in the trenches who can anticipate your situation and provide answers you can trust.

ISBN: 1-56205-929-7

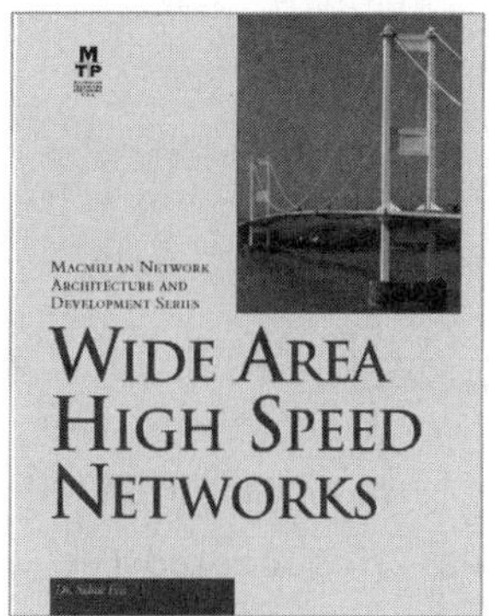

ISBN: 1-57870-114-7

Networking is in a transitional phase between long-standing, conventional wide area services and new technologies and services. This book presents current and emerging wide area technologies and services, makes them understandable, and puts them into perspective so that their merits and disadvantages are clear.

Windows 2000 and Mainframe Integration provides mainframe computing professionals with the practical know-how to build and integrate Windows 2000 technologies into their current environment.

ISBN: 1-57870-200-3

Advanced Information on Networking Technologies

New Riders Books Offer Advice and Experience

LANDMARK

Rethinking Computer Books

We know how important it is to have access to detailed, solution-oriented information on core technologies. *Landmark* books contain the essential information you need to solve technical problems. Written by experts and subjected to rigorous peer and technical reviews, our *Landmark* books are hard-core resources for practitioners like you.

ESSENTIAL REFERENCE

Smart, Like You

The *Essential Reference* series from New Riders provides answers when you know what you want to do but need to know how to do it. Each title skips extraneous material and assumes a strong base of knowledge. These are indispensable books for the practitioner who wants to find specific features of a technology quickly and efficiently. Avoiding fluff and basic material, these books present solutions in an innovative, clean format—and at a great value.

MCSE CERTIFICATION

Engineered for Test Success

New Riders offers a complete line of test preparation materials to help you achieve your certification. With books like the *MCSE Training Guide*, *TestPrep*, and *Fast Track*, and software like the acclaimed *MCSE Complete* and the revolutionary *ExamGear*, New Riders offers comprehensive products built by experienced professionals who have passed the exams and instructed hundreds of candidates.

Books for Networking Professionals

Windows NT Titles

Windows NT TCP/IP

By Karanjit S. Siyan, Ph.D.
1st Edition
480 pages, $29.99
ISBN: 1-56205-887-8

If you're still looking for good documentation on Microsoft TCP/IP, look no further—this is your book. *Windows NT TCP/IP* cuts through the complexities and provides the most informative and complete reference book on Windows-based TCP/IP. Concepts essential to TCP/IP administration are explained thoroughly and then are related to the practical use of Microsoft TCP/IP in a real-world networking environment. The book begins by covering TCP/IP architecture and advanced installation and configuration issues and then moves on to routing with TCP/IP, DHCP Management, and WINS/DNS Name Resolution.

Windows NT DNS

By Michael Masterson, Herman Knief, Scott Vinick, and Eric Roul
1st Edition
340 pages, $29.99
ISBN: 1-56205-943-2

Have you ever opened a Windows NT book looking for detailed information about DNS only to discover that it doesn't even begin to scratch the surface? DNS is probably one of the most complicated subjects for NT administrators, and there are few books on the market that address it in detail. This book answers your most complex DNS questions, focuses on the implementation of the Domain Name Service within Windows NT, and treats it thoroughly from the viewpoint of an experienced Windows NT professional. Many detailed, real-world examples illustrate the understanding of the material throughout. The book covers the details of how DNS functions within NT and then explores specific interactions with critical network components. Finally, proven procedures to design and set up DNS are demonstrated. You'll also find coverage of related topics, such as maintenance, security, and troubleshooting.

Windows NT Registry: A Settings Reference

By Sandra Osborne
1st Edition
550 pages, $29.99
ISBN: 1-56205-941-6

The NT Registry can be a very powerful tool for those capable of using it wisely. Unfortunately, there is little information regarding the NT Registry due to Microsoft's insistence that their source code be kept secret. If you're looking to optimize your use of the Registry, you're usually forced to search the Web for bits of information. This book is your resource. It covers critical issues and settings used for configuring network protocols, including NWLink, PTP, TCP/IP, and DHCP. This book approaches the material from a unique point of view. It discusses the problems related to a particular component and then discusses settings, which are the actual changes necessary for implementing robust solutions.

Windows NT Performance: Monitoring, Benchmarking, and Tuning

By Mark T. Edmead and Paul Hinsberg
1st Edition
288 pages, $29.99
ISBN: 1-56205-942-4

Performance monitoring is a little like preventive medicine for the administrator: No one enjoys a checkup, but it's a good thing to do on a regular basis. This book helps you focus on the critical aspects of improving the performance of your NT system by showing you how to monitor the system, implement benchmarking, and tune your network. The book is organized by resource components, which makes it easy to use as a reference tool.

Windows NT Terminal Server and Citrix MetaFrame

By Ted Harwood
1st Edition
416 pages, $29.99
ISBN: 1-56205-944-0

It's no surprise that most administration headaches revolve around integration with other networks and clients. This book addresses these types of real-world issues on a case-by-case basis, giving tools and advice on solving each problem. The author also offers the real nuts and bolts of thin client administration on multiple systems, covering relevant issues such as installation, configuration, network connection, management, and application distribution.

Windows NT Power Toolkit

By Stu Sjouwerman and Ed Tittel
1st Edition
900 pages, $49.99
ISBN: 0-7357-0922-X

This book covers the analysis, tuning, optimization, automation, enhancement, maintenance, and troubleshooting of Windows NT Server 4.0 and Windows NT Workstation 4.0. In most cases, the two operating systems overlap completely and will be discussed together; in other cases, where the two systems diverge, each platform will be covered separately. This advanced title comprises a task-oriented treatment of the Windows NT 4 environment, including both Windows NT Server 4.0 and Windows NT Workstation 4.0. Thus, this book is aimed squarely at power users to guide them to painless, effective use of Windows NT both inside and outside the workplace. By concentrating on the use of operating system tools and utilities, Resource Kit elements, and selected third-party tuning, analysis, optimization, and productivity tools, this book will show its readers how to carry out everyday and advanced tasks.

Windows NT Network Management: Reducing Total Cost of Ownership

By Anil Desai
1st Edition
400 pages, $34.99
ISBN: 1-56205-946-7

Administering a Windows NT network is kind of like trying to herd cats—an impossible task characterized by constant motion, exhausting labor, and lots of hairballs. Author Anil Desai knows all about it; he's a consulting engineer for Sprint Paranet who specializes in

Windows NT implementation, integration, and management. So, we asked him to put together a concise manual of the best practices—a book of tools and ideas that other administrators can turn to again and again in managing their own NT networks.

Planning for Windows 2000

By Eric K. Cone, Jon Boggs, and Sergio Perez
1st Edition
400 pages, $29.99
ISBN: 0-7357-0048-6

Windows 2000 is poised to be one of the largest and most important software releases of the next decade, and you are charged with planning, testing, and deploying it in your enterprise. Are you ready? With this book, you will be. *Planning for Windows 2000* lets you know what the upgrade hurdles will be, informs you how to clear them, guides you through effective Active Directory design, and presents you with detailed rollout procedures. Eric K. Cone, Jon Boggs, and Sergio Perez give you the benefit of their extensive experiences as Windows 2000 Rapid Deployment Program members by sharing problems and solutions they've encountered on the job.

Inside Windows 2000 Server

By William Boswell
1st Edition
1533 pages, $49.99
ISBN: -56205-929-7

Finally, a totally new edition of New Riders' best-selling *Inside Windows NT Server 4*. Taking the author-driven, no-nonsense approach we pioneered with our Windows NT *Landmark* books, New Riders proudly offers something unique for Windows 2000 administrators—an interesting, discriminating book on Windows 2000 Server written by someone who can anticipate your situation and give you workarounds that won't leave a system unstable or sluggish.

BackOffice Titles

Implementing Exchange Server

By Doug Hauger, Marywynne Leon, and William C. Wade III
1st Edition
400 pages, $29.99
ISBN: 1-56205-931-9

If you're interested in connectivity and maintenance issues for Exchange Server, this book is for you. Exchange's power lies in its capability to be connected to multiple email subsystems to create a "universal email backbone." It's not unusual to have several different and complex systems all connected via email gateways, including Lotus Notes or cc:Mail, Microsoft Mail, legacy mainframe systems, and Internet mail. This book covers all of the problems and issues associated with getting an integrated system running smoothly, and it addresses troubleshooting and diagnosis of email problems with an eye toward prevention and best practices.

Exchange System Administration

By Janice Rice Howd
1st Edition
400 pages, $34.99
ISBN: 0-7357-0081-8

Okay, you've got your Exchange Server installed and connected; now what? Email administration is one of the most critical networking jobs, and Exchange can be particularly troublesome in large, heterogeneous environments. Janice Howd, a noted consultant and teacher with over a decade of email administration experience, has put together this advanced, concise handbook for daily, periodic, and emergency administration. With in-depth coverage of topics like managing disk resources, replication, and disaster recovery, this is the one reference book every Exchange administrator needs.

SQL Server System Administration

By Sean Baird, Chris Miller, et al.
1st Edition
352 pages, $29.99
ISBN: 1-56205-955-6

How often does your SQL Server go down during the day when everyone wants to access the data? Do you spend most of your time being a "report monkey" for your coworkers and bosses? *SQL Server System Administration* helps you keep data consistently available to your users. This book omits introductory information. The authors don't spend time explaining queries and how they work. Instead, they focus on the information you can't get anywhere else, like how to choose the correct replication topology and achieve high availability of information.

Internet Information Services Administration

By Kelli Adam
1st Edition,
300 pages, $29.99
ISBN: 0-7357-0022-2

Are the new Internet technologies in Internet Information Server giving you headaches? Does protecting security on the Web take up all of your time? Then this is the book for you. With hands-on configuration training, advanced study of the new protocols in IIS, and detailed instructions on authenticating users with the new Certificate Server and implementing and managing the new e-commerce features, *Internet Information Services Administration* gives you the real-life solutions you need. This definitive resource also prepares you for the release of Windows 2000 by giving you detailed advice on working with Microsoft Management Console, which was first used by IIS.

SMS 2 Administration

By Michael Lubanski and Darshan Doshi
1st Edition, Winter 2000
350 pages, $39.99
ISBN: 0-7357-0082-6

Microsoft's new version of its Systems Management Server (SMS) is starting to turn heads. Although complex, it allows administrators to lower their total cost of ownership and more efficiently manage clients, applications, and support operations. So if your organization is using or implementing SMS, you'll need some expert advice. Darshan Doshi and Michael Lubanski can help you get the most bang for your buck, with insight, expert tips, and real-world examples. Darshan and

Michael are consultants specializing in SMS and have worked with Microsoft on one of the most complex SMS rollouts in the world, involving 32 countries, 15 languages, and thousands of clients.

UNIX/Linux Titles

Solaris Essential Reference

By John P. Mulligan
1st Edition
350 pages, $24.95
ISBN: 0-7357-0023-0

Looking for the fastest, easiest way to find the Solaris command you need? Need a few pointers on shell scripting? How about advanced administration tips and sound, practical expertise on security issues? Are you looking for trustworthy information about available third-party software packages that will enhance your operating system? Author John Mulligan—creator of the popular Unofficial Guide to Solaris Web site (`sun.icsnet.com`)—delivers all that and more in one attractive, easy-to-use reference book. With clear and concise instructions on how to perform important administration and management tasks and key information on powerful commands and advanced topics, *Solaris Essential Reference* is the book you need when you know what you want to do and only need to know how.

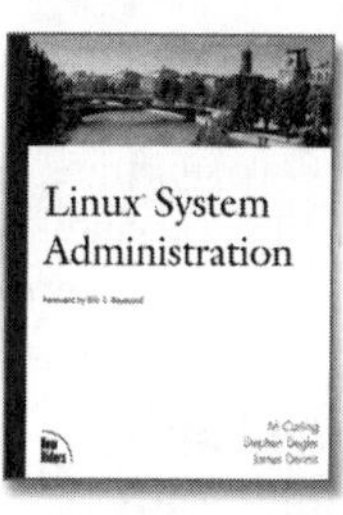

Linux System Administration

By M Carling, et al.
1st Edition
450 pages, $29.99
ISBN: 1-56205-934-3

As an administrator, you probably feel that most of your time and energy is spent in endless firefighting. If your network has become a fragile quilt of temporary patches and work-arounds, this book is for you. For example, have you had trouble sending or receiving email lately? Are you looking for a way to keep your network running smoothly with enhanced performance? Are your users always hankering for more storage, services, and speed? *Linux System Administration* advises you on the many intricacies of maintaining a secure, stable system. In this definitive work, the author addresses all the issues related to system administration from adding users and managing file permissions, to Internet services and Web hosting, to recovery planning and security. This book fulfills the need for expert advice that will ensure a trouble-free Linux environment.

GTK+/Gnome Application Development

By Havoc Pennington
1st Edition
492 pages, $39.99
ISBN: 0-7357-0078-8

This title is for the reader who is conversant with the C programming language and UNIX/Linux development. It provides detailed and solution-oriented information designed to meet the needs of programmers and application developers using the GTK+/Gnome libraries. Coverage complements existing GTK+/Gnome documentation, going into more depth on pivotal issues such as uncovering the GTK+ object system, working with the event loop, managing the Gdk substrate, writing custom widgets, and mastering GnomeCanvas.

Developing Linux Applications with GTK+ and GDK

By Eric Harlow
1st Edition
400 pages, $34.99
ISBN: 0-7357-0021-4

We all know that Linux is one of the most powerful and solid operating systems in existence. And as the success of Linux grows, there is an increasing interest in developing applications with graphical user interfaces that take advantage of the power of Linux. In this book, software developer Eric Harlow gives you an indispensable development handbook focusing on the GTK+ toolkit. More than an overview of the elements of application or GUI design, this is a hands-on book that delves deeply into the technology. With in-depth material on the various GUI programming tools and loads of examples, this book's unique focus will give you the information you need to design and launch professional-quality applications.

Linux Essential Reference

By Ed Petron
1st Edition
400 pages, $24.95
ISBN: 0-7357-0852-5

This book is all about getting things done as quickly and efficiently as possible by providing a structured organization to the plethora of available Linux information. We can sum it up in one word—value. This book has it all: concise instructions on how to perform key administration tasks, advanced information on configuration, shell scripting, hardware management, systems management, data tasks, automation, and tons of other useful information. All of this coupled with an unique navigational structure and a great price. This book truly provides groundbreaking information for the growing community of advanced Linux professionals.

Lotus Notes and Domino Titles

Domino System Administration

By Rob Kirkland, CLP, CLI
1st Edition
850 pages, $49.99
ISBN: 1-56205-948-3

Your boss has just announced that you will be upgrading to the newest version of Notes and Domino when it ships. As a Premium Lotus Business Partner, Lotus has offered a substantial price break to keep your company away from Microsoft's Exchange Server. How are you supposed to get this new system installed, configured, and rolled out to all your end users? You understand how Lotus Notes works—you've been administering it for years. What you need is a concise, practical explanation of the new features and how to make some of the advanced stuff work smoothly. You need answers and solutions from someone like you, who has worked with the product for years and understands what you need to know. *Domino System Administration* is the answer—the first book on Domino that attacks the technology at the professional level with practical, hands-on assistance to get Domino running in your organization.

Lotus Notes and Domino Essential Reference

By Tim Bankes and Dave Hatter
1st Edition
500 pages, $45.00
ISBN: 0-7357-0007-9

You're in a bind because you've been asked to design and program a new database in Notes for an important client that will keep track of and itemize a myriad of inventory and shipping data. The client wants a user-friendly interface without sacrificing speed or functionality. You are experienced (and could develop this application in your sleep) but feel that you need to take your talents to the next level. You need something to facilitate your creative and technical abilities, something to perfect your programming skills. The answer is waiting for you: *Lotus Notes and Domino Essential Reference*. It's compact and simply designed. It's loaded with information. All of the objects, classes, functions, and methods are listed. It shows you the object hierarchy and the relationship between each one. It's perfect for you. Problem solved.

Networking Titles

Cisco Router Configuration & Troubleshooting

By Mark Tripod
1st Edition
300 pages, $34.99
ISBN: 0-7357-0024-9

Want the real story on making your Cisco routers run like a dream? Why not pick up a copy of *Cisco Router Configuration & Troubleshooting* and see what Mark Tripod has to say? They're the folks responsible for making some of the largest sites on the Net scream, like Amazon.com, Hotmail, USAToday, Geocities, and Sony. In this book, they provide advanced configuration issues, sprinkled with advice and preferred practices. You won't see a general overview on TCP/IP. They talk about more meaty issues, like security, monitoring, traffic management, and more. In the troubleshooting section, the authors provide a unique methodology and lots of sample problems to illustrate. By providing real-world insight and examples instead of rehashing Cisco's documentation, Mark gives network administrators information they can start using today.

Network Intrusion Detection: An Analyst's Handbook

By Stephen Northcutt
1st Edition
267 pages, $39.99
ISBN: 0-7357-0868-1

Get answers and solutions from someone who has been in the trenches. Author Stephen Northcutt, original developer of the Shadow intrusion detection system and former Director of the United States Navy's Information System Security Office at the Naval Security Warfare Center, gives his expertise to intrusion detection specialists, security analysts, and consultants responsible for setting up and maintaining an effective defense against network security attacks.

Other Books By New Riders

Windows Technologies

Planning for Windows 2000
0-7357-0048-6
Windows NT Network Management: Reducing Total Cost of Ownership
1-56205-946-7
Windows NT DNS
1-56205-943-2
Windows NT Performance Monitoring, Benchmarking, and Tuning
1-56205-942-4
Windows NT Power Toolkit
0-7357-0922-X
Windows NT Registry: A Settings Reference
1-56205-941-6
Windows NT TCP/IP
1-56205-887-8
Windows NT Terminal Server and Citrix MetaFrame
1-56205-944-0
Implementing Exchange Server
1-56205-931-9
Inside Window 2000 Server
1-56205-929-7
Exchange Server Admninistration
0-7357-0081-8
SQL Server System Administration
1-56205-955-6

Networking

Cisco Router Configuration and Troubleshooting
0-7357-0024-9
Understanding Data Communications, Sixth Edition
0-7357-0036-2
Network Intrusion Detection An Analyst's Handbook
0-7357-0968-1

Certification

A+ Certification TestPrep
1-56205-892-4
A+ Certification Top Score Software
0-7357-0017-6
A+ Certification Training Guide, 2E
0-7357-0907-6
A+ Complete v1.1
0-7357-0045-1
A+ Fast Track
0-7357-0028-1
MCSD Fast Track: Visual Basic 6, Exam 70-176
0-7357-0019-2
MCSE Fast Track: Internet Information Server 4
1-56205-936-X
MCSE Fast Track: Networking Essentials
1-56205-939-4
MCSE Fast Track: TCP/IP
1-56205-937-8
MCSD Fast Track: Visual Basic 6, Exam 70-175
0-7357-0018-4
MCSE Fast Track: Windows 98
0-7357-0016-8
MCSE Fast Track: Windows NT Server 4
1-56205-935-1
MCSE Fast Track: Windows NT Server 4 Enterprise
1-56205-940-8
MCSE Fast Track: Windows NT Workstation 4
1-56205-938-6
MCSE Simulation Guide: Windows NT Server 4 Enterprise
1-56205-914-9
MCSE Simulation Guide: Windows NT Workstation 4
1-56205-925-4
MCSE TestPrep: Core Exam Bundle, Second Edition
0-7357-0030-3
MCSE TestPrep: Networking Essentials, Second Edition
0-7357-0010-9
MCSE TestPrep: TCP/IP, Second Edition
0-7357-0025-7
MCSE TestPrep: Windows 95, Second Edition
0-7357-0011-7
MCSE TestPrep: Windows 98
1-56205-922-X
MCSE TestPrep: Windows NT Server 4 Enterprise, Second Edition
0-7357-0009-5
MCSE TestPrep: Windows NT Server 4, Second Edition
0-7357-0012-5
MCSE TestPrep: Windows NT Workstation 4, Second Edition
0-7357-0008-7
MCSD TestPrep: Visual Basic 6 Exams
0-7357-0032-X
MCSE Training Guide: Core Exams Bundle, Second Edition
1-56205-926-2
MCSE Training Guide: Networking Essentials, Second Edition
1-56205-919-X
MCSE Training Guide: TCP/IP, Second Edition
1-56205-920-3
MCSE Training Guide: Windows 98
1-56205-890-8
MCSE Training Guide: Windows NT Server 4, Second Edition
1-56205-916-5
MCSE Training Guide: Windows NT Server Enterprise, Second Edition
1-56205-917-3
MCSE Training Guide: Windows NT Workstation 4, Second Edition
1-56205-918-1
MCSD Training Guide: Visual Basic 6 Exams
0-7357-0002-8
MCSE Top Score Software: Core Exams
0-7357-0033-8
MCSE + Complete, v1.1
0-7897-1564-3
MCSE + Internet Complete, v1.2
0-7357-0072-9

Graphics

Inside 3D Studio MAX 2, Volume I
1-56205-857-6
Inside 3D Studio MAX 2, Volume II: Modeling and Materials
1-56205-864-9
Inside 3D Studio MAX 2, Volume III: Animation
1-56205-865-7
Inside 3D Studio MAX 2 Resource Kit
1-56205-953-X
Inside AutoCAD 14, Limited Edition
1-56205-898-3
Inside Softimage 3D
1-56205-885-1
HTML Web Magic, Second Edition
1-56830-475-7
Dynamic HTML Web Magic
1-56830-421-8
Designing Web Graphics.3
1-56205-949-1
Illustrator 8 Magic
1-56205-952-1
Inside trueSpace 4
1-56205-957-2
Inside Adobe Photoshop 5
1-56205-884-3
Inside Adobe Photoshop 5, Limited Edition
1-56205-951-3
Photoshop 5 Artistry
1-56205-895-9
Photoshop 5 Type Magic
1-56830-465-X
Photoshop 5 Web Magic
1-56205-913-0

We Want to Know What You Think

To better serve you, we would like your opinion on the content and quality of this book. Please complete this card, and mail it to us or fax it to 317-581-4663.

Name ______________________________

Address ______________________________

City__________ State__________ Zip __________

Phone ______________________________

Email Address ______________________________

Occupation ______________________________

Operating system(s) that you use ______________________________

What influenced your purchase of this book?

- ❑ Recommendation ❑ Cover Design
- ❑ Table of Contents ❑ Index
- ❑ Magazine Review ❑ Advertisement
- ❑ New Riders' Reputation ❑ Author Name

How would you rate the contents of this book?

- ❑ Excellent ❑ Very Good
- ❑ Good ❑ Fair
- ❑ Below Average ❑ Poor

How do you plan to use this book?

- ❑ Quick Reference ❑ Self-Training
- ❑ Classroom ❑ Other

What do you like most about this book? Check all that apply.

- ❑ Content ❑ Writing Style
- ❑ Accuracy ❑ Examples
- ❑ Listings ❑ Design
- ❑ Index ❑ Page Count
- ❑ Price ❑ Illustrations

What do you like least about this book? Check all that apply.

- ❑ Content ❑ Writing Style
- ❑ Accuracy ❑ Examples
- ❑ Listings ❑ Design
- ❑ Index ❑ Page Count
- ❑ Price ❑ Illustrations

What would be a useful follow-up book for you? ______________________________

Where did you purchase this book? ______________________________

Can you name a similar book that you like better than this one, or one that is as good? Why?

____

How many New Riders books do you own? ______________________________

What are your favorite computer books? ______________________________

What other titles would you like to see us develop? ______________________________

Any comments for us? ______________________________

Understanding the Network: 0-7357-977-7

Fold here and tape to mail

Place
Stamp
Here

New Riders Publishing
201 W. 103rd St.
Indianapolis, IN 46290

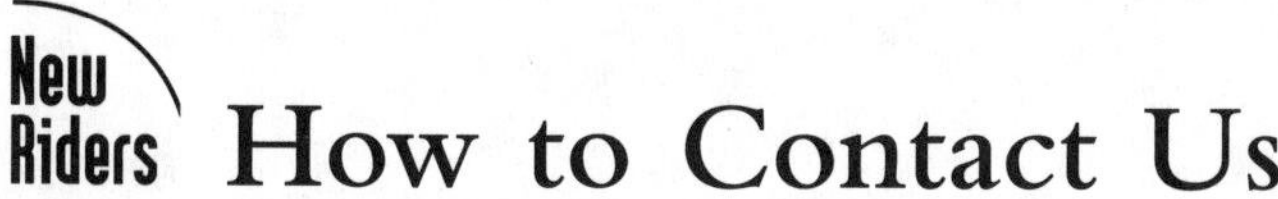

How to Contact Us

Visit Our Web Site

www.newriders.com

On our Web site you'll find information about our other books, authors, tables of contents, indexes, and book errata.

Email Us

Contact us at this address:

nrfeedback@newriders.com

- If you have comments or questions about this book
- To report errors that you have found in this book
- If you have a book proposal to submit or are interested in writing for New Riders
- If you would like to have an author kit sent to you
- If you are an expert in a computer topic or technology and are interested in being a technical editor who reviews manuscripts for technical accuracy

nrfeedback@newriders.com

- To find a distributor in your area, please contact our international department at this address.

nrmedia@newriders.com

- For instructors from educational institutions who want to preview New Riders books for classroom use. Email should include your name, title, school, department, address, phone number, office days/hours, text in use, and enrollment in the body of your text, along with your request for desk/examination copies and/or additional information.
- For members of the media who are interested in reviewing copies of New Riders books send your name, mailing address, and email address, along with the name of the publication or Web site you work for.

Write to Us

New Riders Publishing

201 W. 103rd St.

Indianapolis, IN 46290-1097

Call Us

Toll-free (800) 571-5840 + 9 +4511

If outside U.S. (317) 581-3500. Ask for New Riders.

Fax Us

(317) 581-4663